D0204324

APPLIED HYDROLOGY

McGraw-Hill Series in Water Resources and Environmental Engineering

Rolf Eliassen, Paul H. King, and Ray K. Linsley —*Consulting Editors*

Bailey and Ollis: *Biochemical Engineering Fundamentals*
Bishop: *Marine Pollution and its Control*
Biswas: *Models for Water Quality Management*
Bouwer: *Groundwater Hydrology*
Canter: *Environmental Impact Assessment*
Chanlett: *Environmental Protection*
Chow, Maidment, and Mays: *Applied Hydrology*
Linsley and Franzini: *Water Resources Engineering*
Linsley, Kohler, and Paulhus: *Hydrology for Engineers*
Metcalf and Eddy, Inc.: *Wastewater Engineering Collection and Pumping of Wastewater*
Metcalf and Eddy, Inc.: *Wastewater Engineering: Treatment, Disposal, Rease*
Peavy, Rowe, and Tchobanoglous: *Environmental Engineering*
Rich: *Low-Maintenance, Mechanically-Simple Wastewater Treatment Systems*
Sawyer and McCatthy: *Chemistry for Environmental Engineering*
Steel and McGhee: *Water Supply and Sewerage*
Tchobanoglous, Theisen, and Eliassen: *Solid Wastes: Engineering Principles and Management Issues*

APPLIED HYDROLOGY

Ven Te Chow
Late Professor of Hydrosystems Engineering
University of Illinois, Urbana-Champaign

David R. Maidment
Associate Professor of Civil Engineering
The University of Texas at Austin

Larry W. Mays
Professor of Civil Engineering
The University of Texas at Austin

McGraw-Hill Publishing Company
New York St. Louis San Francisco Auckland Bogotá Caracas
Hamburg Lisbon London Madrid Mexico Milan
Montreal New Delhi Oklahoma City Paris San Juan
São Paulo Singapore Sydney Tokyo Toronto

This book was set in Times Roman by Publication Services.
The editors were B. J. Clark and John Morriss;
the cover was designed by Amy Becker;
the production supervisor was Leroy A. Young.
Project supervision was done by Publication Services.
Arcata Graphics/Halliday was printer and binder.

APPLIED HYDROLOGY

4567890 HDHD 99876543210

ISBN 0-07-010810-2

Library of Congress Cataloging-in-Publication Data

Chow, Ven Te
Applied hydrology.
(McGraw-Hill series in water resources and environmental engineering)
Includes index.
1. Hydrology. I. Maidment, David R. II. Mays, Larry W. III. Title.
IV. Series.
GB661.2.C43 1988 627 87-16860
ISBN 0-07-010810-2

ABOUT THE AUTHORS

The late Ven Te Chow was a professor in the Civil Engineering Department of the University of Illinois at Urbana-Champaign from 1951 to 1981. He gained international prominence as a scholar, educator, and diplomat in hydrology, hydraulics, and hydrosystems engineering. He received his B.S. degree in 1940 from Chaio Tung University in Shangai, spent several years in China as an instructor and professor, then went to Pennsylvania State University from which he received his M.S. degree in 1948 and the University of Illinois where he received his Ph.D. degree in 1950. He also received four honorary doctoral degrees and many other awards and honors including membership in the National Academy of Engineering. He was a prolific author, writing his first book at the age of 27 on the theory of structures (in Chinese). He authored *Open-Channel Hydraulics* in 1959 and was editor-in-chief of the *Handbook of Applied Hydrology* in 1964; both books are still considered standard reference works. He was active in professional societies, especially the International Water Resources Association of which he was a principal founder and the first President.

David R. Maidment is Associate Professor of Civil Engineering at the University of Texas at Austin where he has been on the faculty since 1981. Prior to that time he taught at Texas A & M University and carried out hydrology research at the International Institute for Applied Systems Analysis in Vienna, Austria, and at the Ministry of Works and Development in New Zealand. He obtained his bachelor's degree from the University of Canterbury, Christchurch New Zealand, and his M.S. and Ph.D degrees from the University of Illinois at Urbana-Champaign. Dr. Maidment serves as a consultant in hydrology to government and industry and is an associate editor of the *Hydrological Sciences Journal*.

Larry W. Mays is a Professor of Civil Engineering and holder of an Engineering Foundation Endowed Professorship at the University of Texas at Austin where he has been on the faculty since 1976. Prior to that he was a graduate research assistant and then a Visiting Research Assistant Professor at the University of Illinois at Urbana-Champaign where he received his Ph.D. He received his

B.S. (1970) and M.S. (1971) degrees from the University of Missouri at Rolla, after which he served in the U.S. Army stationed at the Lawrence Livermore Laboratory in California. Dr. Mays has been very active in research and teaching at the University of Texas in the areas of hydrology, hydraulics, and water resource systems analysis. In addition he has served as a consultant in these areas to various government agencies and industries including the U.S. Army Corps of Engineers, the Attorney General's Office of Texas, the United Nations, NATO, the World Bank, and the Government of Taiwan. He is a registered engineer in seven states and has been active in committees with the American Society of Civil Engineers and other professional organizations.

CONTENTS

PREFACE

Applied Hydrology is a textbook for upper level undergraduate and graduate courses in hydrology and is a reference for practicing hydrologists. Surface water hydrology is the focus of the book which is presented in three sections: Hydrologic Processes, Hydrologic Analysis, and Hydrologic Design.

Hydrologic processes are covered in Chapters 1 to 6, which describe the scientific principles governing hydrologic phenomena. The hydrologic system is visualized as a generalized control volume, and the Reynold's Transport Theorem (or general control volume equation) from fluid mechanics is used to apply the physical laws governing mass, momentum, and energy to the flow of atmospheric water, subsurface water, and surface water. This section is completed by a chapter on hydrologic measurement.

Hydrologic analysis is treated in the next six chapters (7 to 12), which emphasize computational methods in hydrology for specific tasks such as rainfall-runoff modeling, flow routing, and analysis of extreme events. These chapters are organized in a sequence according to the way the analysis treats the space and time variability and the randomness of the hydrologic system behavior. Special attention is given in Chapters 9 and 10 to the subject of flow routing by the dynamic wave method where the recent availability of standardized computer programs has made possible the general application of this method.

Hydrologic design is presented in the final three chapters (13 to 15), which focus on the risks inherent in hydrologic design, the selection of design storms including probable maximum precipitation, and the calculation of design flows for various problems including the design of storm sewers, flood control works, and water supply reservoirs.

How is *Applied Hydrology* different from other available books in this field? First, this is a book with a general coverage of surface water hydrology. There are a number of recently published books in special fields such as evaporation, statistical hydrology, hydrologic modelling, and stormwater hydrology. Although this book covers these subjects, it emphasizes a sound foundation for the subject of hydrology as a whole. Second, *Applied Hydrology* is organized around a central theme of using the hydrologic system or control volume as a framework for analysis in order to unify the subject of hydrology so that its various analytical methods are seen as different views of hydrologic system operation rather than as separate and unrelated topics. Third, we believe that the reader learns by doing, so 90 example problems are solved in the text and 400 additional problems are presented at the end of chapters for homework or self-study. In some cases, theoretical developments too extensive for inclusion in the text are presented as problems at the end of the chapter so that by solving these problems the reader can play a part in the development of the subject. Some of the problems are intended for solution by using a spreadsheet program, by developing a computer code, or by use of standard hydrologic simulation programs.

This book is used for three courses at the University of Texas at Austin: an undergraduate and a graduate course in surface water hydrology, and an undergraduate course in hydrologic design. At the undergraduate level a selection of topics is presented from throughout the book, with the hydrologic design course focusing on the analysis and design chapters. At the graduate level, the chapters on hydrologic processes and analysis are emphasized. There are conceivably many different courses that could be taught from the book at the undergraduate or graduate levels, with titles such as surface water hydrology, hydrologic design, urban hydrology, physical hydrology, computational hydrology, etc.

Any hydrology book reflects a personal perception of the subject evolved by its authors over many years of teaching, research, and professional experience. And *Applied Hydrology* is our view of the subject. We have aimed at making it rigorous, unified, numerical, and practical. We believe that the analytical approach adopted will be sufficiently sound so that as new knowledge of the field becomes available it can be built upon the basis established here. Hydrologic events such as floods and droughts have a significant impact on public welfare, and a corresponding responsibility rests upon the hydrologist to provide the best information that current knowledge and available data will permit. This book is intended to be a contribution toward the eventual goal of better hydrologic practice.

A special word is appropriate concerning the development of this book. The work was initiated many years ago by Professor Ven Te Chow of the University of Illinois Urbana-Champaign, who developed a considerable volume of manuscript for some of the chapters. Following his death in 1981, his wife, Lora, asked us to carry this work to completion. We both obtained our graduate degrees at the University of Illinois Urbana-Champaign and shared the hydrologic system perspective which Ven Te Chow was so instrumental in fostering during his lifetime. During the years required for us to write this book, it occurred, perhaps inevitably, that we had to start almost from the beginning again so that the resulting work would be consistent and complete. As we used the text in teaching our hydrology courses at the University of Texas at Austin, we gradually evolved the concepts to the point they are presented here. We believe we have retained the concept which animated Ven Te Chow's original work on the subject.

We express our thanks to Becky Brudniak, Jan Hausman, Suzi Jimenez, Amy Phillips, Carol Sellers, Fidel Saenz de Ormijana, and Ellen Wadsworth, who helped us prepare the manuscript. We also want to acknowledge the assistance provided to us by reviewers of the manuscript including Gonzalo Cortes-Rivera of Bogotá, Colombia, L. Douglas James of Utah State University, Jerome C. Westphal, University of Missouri-Rolla, Ben Chie Yen of the University of Illinois Urbana Champaign, and our colleagues and students at the University of Texas at Austin.

A book is a companion along the pathway of learning. We wish you a good journey.

David R. Maidment
Larry W. Mays

Austin, Texas
December, 1987

APPLIED HYDROLOGY

CHAPTER 1

INTRODUCTION

Water is the most abundant substance on earth, the principal constituent of all living things, and a major force constantly shaping the surface of the earth. It is also a key factor in air-conditioning the earth for human existence and in influencing the progress of civilization. Hydrology, which treats all phases of the earth's water, is a subject of great importance for people and their environment. Practical applications of hydrology are found in such tasks as the design and operation of hydraulic structures, water supply, wastewater treatment and disposal, irrigation, drainage, hydropower generation, flood control, navigation, erosion and sediment control, salinity control, pollution abatement, recreational use of water, and fish and wildlife protection. The role of applied hydrology is to help analyze the problems involved in these tasks and to provide guidance for the planning and management of water resources.

The hydrosciences deal with the waters of the earth: their distribution and circulation, their physical and chemical properties, and their interaction with the environment, including interaction with living things and, in particular, human beings. Hydrology may be considered to encompass all the hydrosciences, or defined more strictly as the study of the hydrologic cycle, that is, the endless circulation of water between the earth and its atmosphere. Hydrologic knowledge is applied to the use and control of water resources on the land areas of the earth; ocean waters are the domain of ocean engineering and the marine sciences.

Changes in the distribution, circulation, or temperature of the earth's waters can have far-reaching effects; the ice ages, for instance, were a manifestation of such effects. Changes may be caused by human activities. People till the soil, irrigate crops, fertilize land, clear forests, pump groundwater, build dams, dump wastes into rivers and lakes, and do many other constructive or destructive things that affect the circulation and quality of water in nature.

1.1 HYDROLOGIC CYCLE

Water on earth exists in a space called the hydrosphere which extends about 15 km up into the atmosphere and about 1 km down into the lithosphere, the crust of the earth. Water circulates in the hydrosphere through the maze of paths constituting the hydrologic cycle.

The hydrologic cycle is the central focus of hydrology. The cycle has no beginning or end, and its many processes occur continuously. As shown schematically in Fig. 1.1.1, water *evaporates* from the oceans and the land surface to become part of the atmosphere; water vapor is transported and lifted in the atmosphere until it condenses and *precipitates* on the land or the oceans; precipitated water may be *intercepted* by vegetation, become *overland flow* over the ground surface, *infiltrate* into the ground, flow through the soil as *subsurface flow*, and discharge into streams as *surface runoff*. Much of the intercepted water and surface runoff returns to the atmosphere through evaporation. The infiltrated water may percolate deeper to *recharge* groundwater, later emerging in springs or seeping into streams to form surface runoff, and finally flowing out to the sea or evaporating into the atmosphere as the hydrologic cycle continues.

Estimating the total amount of water on the earth and in the various processes of the hydrologic cycle has been a topic of scientific exploration since the second half of the nineteenth century. However, quantitative data are scarce, particularly over the oceans, and so the amounts of water in the various components of the global hydrologic cycle are still not known precisely.

Table 1.1.1 lists estimated quantities of water in various forms on the earth. About 96.5 percent of all the earth's water is in the oceans. If the earth were a uniform sphere, this quantity would be sufficient to cover it to a depth of about 2.6 km (1.6 mi). Of the remainder, 1.7 percent is in the polar ice, 1.7 percent in groundwater and only 0.1 percent in the surface and atmospheric water systems. The atmospheric water system, the driving force of surface water hydrology, contains only 12,900 km^3 of water, or less than one part in 100,000 of all the earth's water.

Of the earth's *fresh water*, about two-thirds is polar ice and most of the remainder is groundwater going down to a depth of 200 to 600 m. Most groundwater is saline below this depth. Only 0.006 percent of fresh water is contained in rivers. Biological water, fixed in the tissues of plants and animals, makes up about 0.003 percent of all fresh water, equivalent to half the volume contained in rivers.

Although the water content of the surface and atmospheric water systems is relatively small at any given moment, immense quantities of water annually pass through them. The global annual water balance is shown in Table 1.1.2; Fig. 1.1.1 shows the major components in units relative to an annual land precipitation volume of 100. It can be seen that evaporation from the land surface consumes 61 percent of this precipitation, the remaining 39 percent forming runoff to the oceans, mostly as surface water. Evaporation from the oceans contributes nearly 90 percent of atmospheric moisture. Analysis of the flow and storage of water in the global water balance provides some insight into the dynamics of the hydrologic cycle.

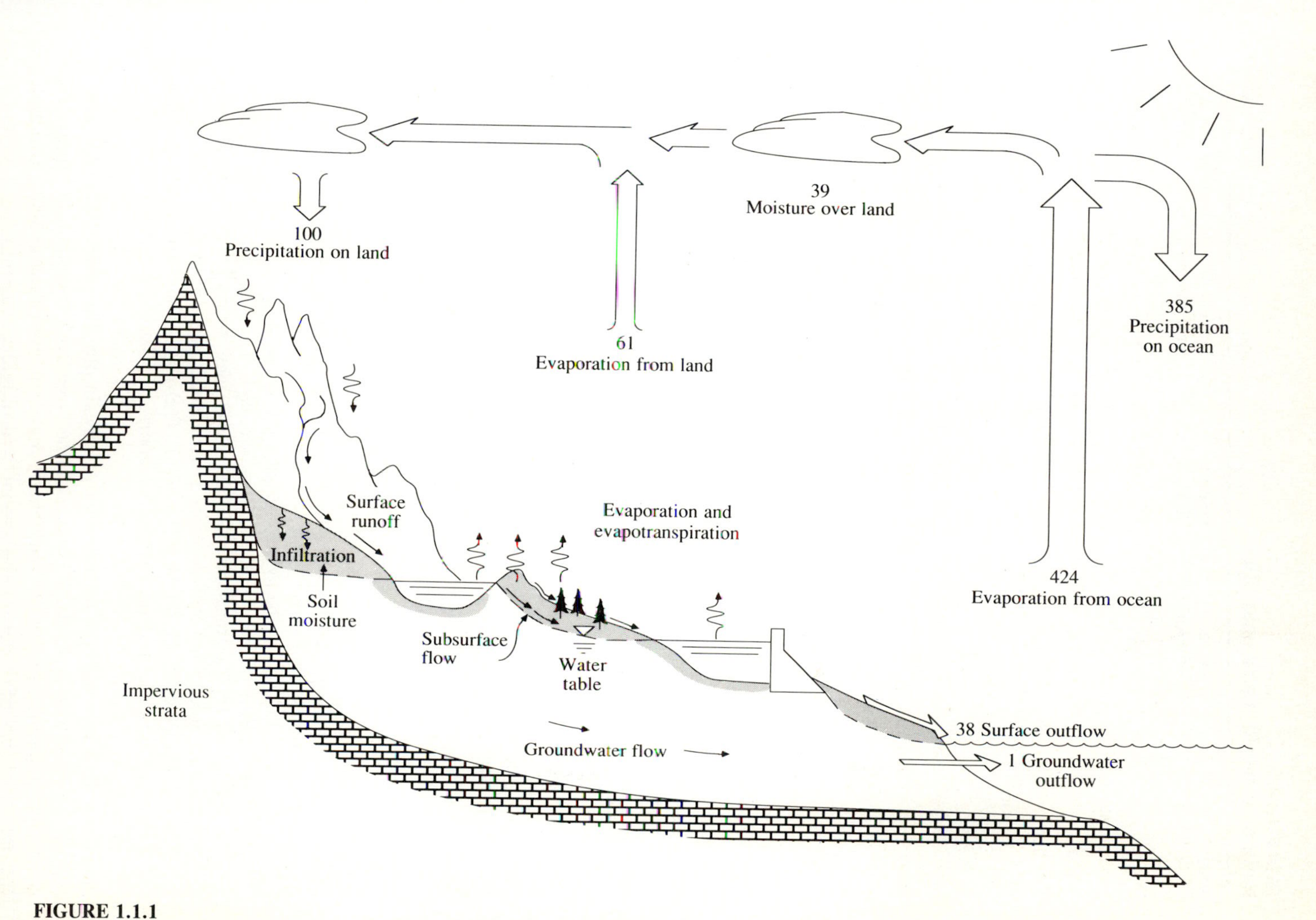

FIGURE 1.1.1
Hydrologic cycle with global annual average water balance given in units relative to a value of 100 for the rate of precipitation on land.

TABLE 1.1.1
Estimated world water quantities

Item	Area (10^6 km^2)	Volume (km^3)	Percent of total water	Percent of fresh water
Oceans	361.3	1,338,000,000	96.5	
Groundwater				
Fresh	134.8	10,530,000	0.76	30.1
Saline	134.8	12,870,000	0.93	
Soil Moisture	82.0	16,500	0.0012	0.05
Polar ice	16.0	24,023,500	1.7	68.6
Other ice and snow	0.3	340,600	0.025	1.0
Lakes				
Fresh	1.2	91,000	0.007	0.26
Saline	0.8	85,400	0.006	
Marshes	2.7	11,470	0.0008	0.03
Rivers	148.8	2,120	0.0002	0.006
Biological water	510.0	1,120	0.0001	0.003
Atmospheric water	510.0	12,900	0.001	0.04
Total water	510.0	1,385,984,610	100	
Fresh water	148.8	35,029,210	2.5	100

Table from World Water Balance and Water Resources of the Earth, Copyright, UNESCO, 1978.

Example 1.1.1. Estimate the *residence time* of global atmospheric moisture.

Solution. The residence time T_r is the *average* duration for a water molecule to pass through a subsystem of the hydrologic cycle. It is calculated by dividing the volume of water S in storage by the flow rate Q.

$$T_r = \frac{S}{Q} \tag{1.1.1}$$

The volume of atmospheric moisture (Table 1.1.1) is 12,900 km^3. The flow rate of moisture from the atmosphere as precipitation (Table 1.1.2) is 458,000 + 119,000 = 577,000 km^3/yr, so the average residence time for moisture in the atmosphere is T_r = 12,900/577,000 = 0.022 yr = 8.2 days. The very short residence time for moisture in the atmosphere is one reason why weather cannot be forecast accurately more than a few days ahead. Residence times for other components of the hydrologic cycle are similarly computed. These values are averages of quantities that may exhibit considerable spatial variation.

Although the concept of the hydrologic cycle is simple, the phenomenon is enormously complex and intricate. It is not just one large cycle but rather is composed of many interrelated cycles of continental, regional, and local extent. Although the total volume of water in the global hydrologic cycle remains essentially

TABLE 1.1.2
Global annual water balance

		Ocean	Land
Area (km^2)		361,300,000	148,800,000
Precipitation	(km^3/yr)	458,000	119,000
	(mm/yr)	1270	800
	(in/yr)	50	31
Evaporation	(km^3/yr)	505,000	72,000
	(mm/yr)	1400	484
	(in/yr)	55	19
Runoff to ocean			
Rivers	(km^3/yr)	–	44,700
Groundwater	(km^3/yr)	–	2200
Total runoff	(km^3/yr)	–	47,000
	(mm/yr)	–	316
	(in/yr)	–	12

Table from World Water Balance and Water Resources of the Earth, Copyright, UNESCO, 1978

constant, the distribution of this water is continually changing on continents, in regions, and within local drainage basins.

The hydrology of a region is determined by its weather patterns and by physical factors such as topography, geology and vegetation. Also, as civilization progresses, human activities gradually encroach on the natural water environment, altering the dynamic equilibrium of the hydrologic cycle and initiating new processes and events. For example, it has been theorized that because of the burning of fossil fuels, the amount of carbon dioxide in the atmosphere is increasing. This could result in a warming of the earth and have far-reaching effects on global hydrology.

1.2 SYSTEMS CONCEPT

Hydrologic phenomena are extremely complex, and may never be fully understood. However, in the absence of perfect knowledge, they may be represented in a simplified way by means of the *systems* concept. A system is a set of connected parts that form a whole. The hydrologic cycle may be treated as a system whose components are precipitation, evaporation, runoff, and other phases of the hydrologic cycle. These components can be grouped into subsystems of the overall cycle; to analyze the total system, the simpler subsystems can be treated separately and the results combined according to the interactions between the subsystems.

In Fig. 1.2.1, the global hydrologic cycle is represented as a system. The dashed lines divide it into three subsytems: the *atmospheric water system* containing the processes of precipitation, evaporation, interception, and transpiration; the *surface water system* containing the processes of overland flow, surface runoff, subsurface and groundwater outflow, and runoff to streams and the ocean; and the *subsurface water system* containing the processes of infiltration, groundwater recharge, subsurface flow and groundwater flow. Subsurface flow takes place in the soil near the land surface; groundwater flow occurs deeper in the soil or rock strata.

For most practical problems, only a few processes of the hydrologic cycle are considered at a time, and then only considering a small portion of the earth's surface. A more restricted system definition than the global hydrologic system is appropriate for such treatment, and is developed from a concept of the control volume. In fluid mechanics, the application of the basic principles of mass, momentum, and energy to a fluid flow system is accomplished by using a control volume, a reference frame drawn in three dimensions through which the fluid

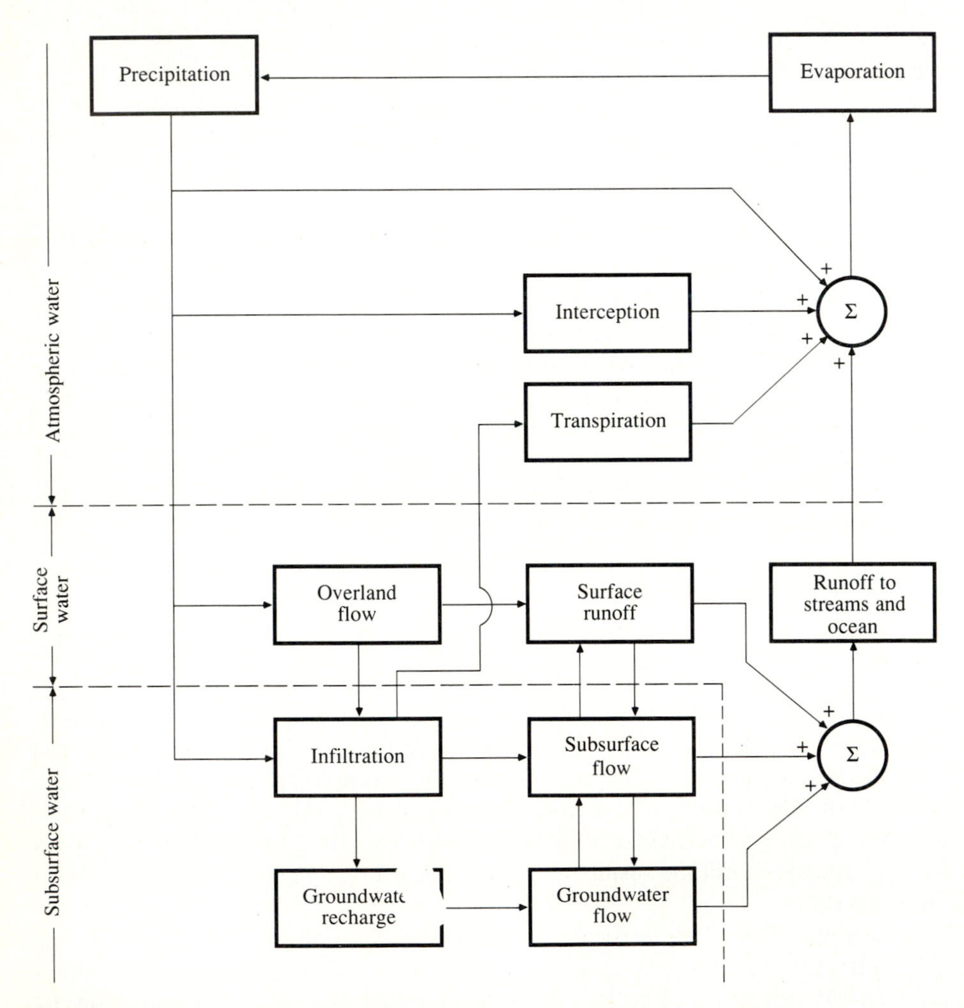

FIGURE 1.2.1
Block-diagram representation of the global hydrologic system.

flows. The control volume provides the framework for applying the laws of conservation of mass and energy and Newton's second law to obtain practical equations of motion. In developing these equations, it is not necessary to know the precise flow pattern inside the control volume. What must be known are the properties of the fluid flow at the *control surface*, the boundary of the control volume. The fluid inside the control volume is treated as a mass, which may be represented as being concentrated at one point in space when considering the action of external forces such as gravity.

By analogy, a *hydrologic system* is defined as *a structure or volume in space, surrounded by a boundary, that accepts water and other inputs, operates on them internally, and produces them as outputs* (Fig. 1.2.2). The structure (for surface or subsurface flow) or volume in space (for atmospheric moisture flow) is the totality of the flow paths through which the water may pass as *throughput* from the point it enters the system to the point it leaves. The boundary is a continuous surface defined in three dimensions enclosing the volume or structure. A *working medium* enters the system as input, interacts with the structure and other media, and leaves as output. Physical, chemical, and biological processes operate on the working media within the system; the most common working media involved in hydrologic analysis are water, air, and heat energy.

The procedure of developing working equations and models of hydrologic phenomena is similar to that in fluid mechanics. In hydrology, however, there is generally a greater degree of approximation in applying physical laws because the systems are larger and more complex, and may involve several working media. Also, most hydrologic systems are inherently random because their major input is precipitation, a highly variable and unpredictable phenomenon. Consequently, statistical analysis plays a large role in hydrologic analysis.

Example 1.2.1. Represent the storm rainfall-runoff process on a watershed as a hydrologic system.

Solution. A *watershed* is the area of land draining into a stream at a given location. The *watershed divide* is a line dividing land whose drainage flows toward the given stream from land whose drainage flows away from that stream. The system boundary is drawn around the watershed by projecting the watershed divide vertically upwards and downwards to horizontal planes at the top and bottom (Fig. 1.2.3). Rainfall is the input, distributed in space over the upper plane; streamflow is the output, concentrated in space at the watershed outlet. Evaporation and subsurface flow could also be considered as outputs, but they are small compared with streamflow during a storm. The structure of the system is the set of flow paths over or through the soil and includes the tributary streams which eventually merge to become streamflow at the watershed outlet.

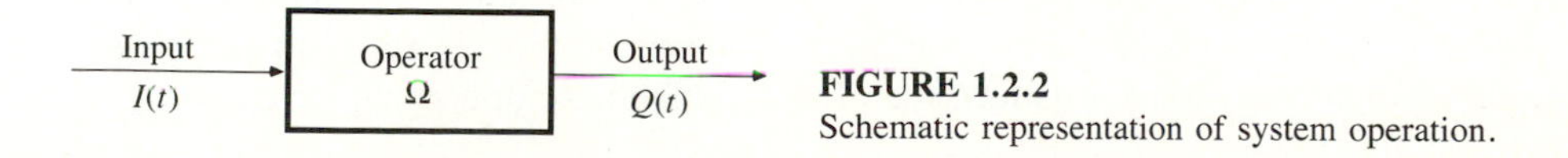

FIGURE 1.2.2
Schematic representation of system operation.

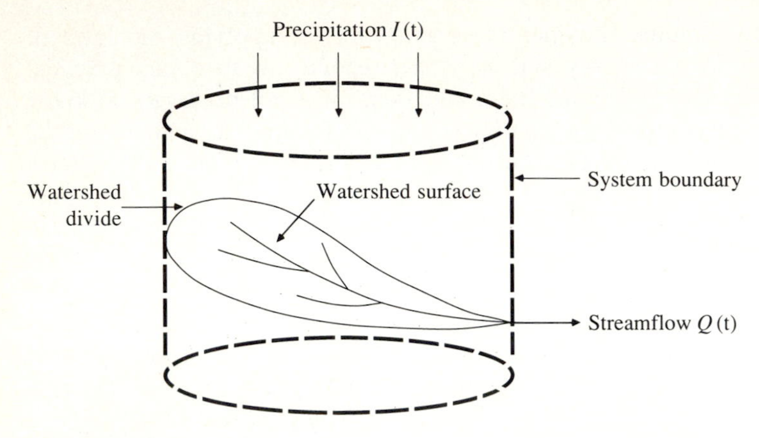

FIGURE 1.2.3
The watershed as a hydrologic system.

If the surface and soil of a watershed are examined in great detail, the number of possible flow paths becomes enormous. Along any path, the shape, slope, and boundary roughness may be changing continuously from place to place and these factors may also vary in time as the soil becomes wet. Also, precipitation varies randomly in space and time. Because of these great complications, it is not possible to describe some hydrologic processes with exact physical laws. By using the system concept, effort is directed to the construction of a model relating inputs and outputs rather than to the extremely difficult task of exact representation of the system details, which may not be significant from a practical point of view or may not be known. Nevertheless, knowledge of the physical system helps in developing a good model and verifying its accuracy.

1.3 HYDROLOGIC SYSTEM MODEL

The objective of hydrologic system analysis is to study the system operation and predict its output. A hydrologic system model is an approximation of the actual system; its inputs and outputs are measurable hydrologic variables and its structure is a set of equations linking the inputs and outputs. Central to the model structure is the concept of a *system transformation*.

Let the input and output be expressed as functions of time, $I(t)$ and $Q(t)$ respectively, for t belonging to the time range T under consideration. The system performs a transformation of the input into the output represented by

$$Q(t) = \Omega I(t) \tag{1.3.1}$$

which is called the *transformation equation* of the system. The symbol Ω is a *transfer function* between the input and the output. If this relationship can be expressed by an algebraic equation, then Ω is an algebraic operator. For example, if

$$Q(t) = CI(t) \tag{1.3.2}$$

where C is a constant, then the transfer function is the operator

$$\Omega = \frac{Q(t)}{I(t)} = C \tag{1.3.3}$$

If the transformation is described by a differential equation, then the transfer function serves as a *differential operator*. For example, a *linear reservoir* has its storage S related to its outflow Q by

$$S = kQ \tag{1.3.4}$$

where k is a constant having the dimensions of time. By continuity, the time rate of change of storage dS/dt is equal to the difference between the input and the output

$$\frac{dS}{dt} = I(t) - Q(t) \tag{1.3.5}$$

Eliminating S between the two equations and rearranging,

$$k\frac{dQ}{dt} + Q(t) = I(t) \tag{1.3.6}$$

so

$$\Omega = \frac{Q(t)}{I(t)} = \frac{1}{1 + kD} \tag{1.3.7}$$

where D is the differential operator d/dt. If the transformation equation has been determined and can be solved, it yields the output as a function of the input. Equation (1.3.7) describes a *linear* system if k is a constant. If k is a function of the input I or the output Q then (1.3.7) describes a *nonlinear system* which is much more difficult to solve.

1.4 HYDROLOGIC MODEL CLASSIFICATION

Hydrologic models may be divided into two categories: *physical* models and *abstract* models. Physical models include *scale* models which represent the system on a reduced scale, such as a hydraulic model of a dam spillway; and *analog* models, which use another physical system having properties similar to those of the prototype. For example, the Hele-Shaw model is an analog model that uses the movement of a viscous fluid between two closely spaced parallel plates to model seepage in an aquifer or embankment.

Abstract models represent the system in mathematical form. The system operation is described by a set of equations linking the input and the output variables. These variables may be functions of space and time, and they may also be *probabilistic* or *random* variables which do not have a fixed value at a particular point in space and time but instead are described by probability distributions. For example, tomorrow's rainfall at a particular location cannot be forecast exactly but the probability that there will be some rain can be estimated. The most general representation of such variables is a *random field*, a region of space and time within which the value of a variable at each point is defined by a probability distribution (Vanmarcke, 1983). For example, the precipitation intensity in a thunderstorm varies rapidly in time, and from one location to another, and cannot be predicted accurately, so it is reasonable to represent it by a random field.

Trying to develop a model with random variables that depend on all three space dimensions and time is a formidable task, and for most practical purposes it is necessary to simplify the model by neglecting some sources of variation. Hydrologic models may be classified by the ways in which this simplification is accomplished. Three basic decisions to be made for a model are: Will the model variables be random or not? Will they vary or be uniform in space? Will they vary or be constant in time? The model may be located in a "tree" according to these choices, as shown in Fig. 1.4.1.

A *deterministic* model does not consider randomness; a given input always produces the same output. A *stochastic* model has outputs that are at least partially random. One might say that deterministic models make *forecasts* while stochastic models make *predictions*. Although all hydrologic phenomena involve some randomness, the resulting variability in the output may be quite small when compared to the variability resulting from known factors. In such cases, a deterministic model is appropriate. If the random variation is large, a stochastic model is more suitable, because the actual output could be quite different from the single value a deterministic model would produce. For example, reasonably good deterministic models of daily evaporation at a given location can be developed using energy supply and vapor transport data, but such data cannot be used to make reliable models of daily precipitation at that location because precipitation is largely random. Consequently, most daily precipitation models are stochastic.

At the middle level of the tree in Fig. 1.4.1, the treatment of spatial variation is decided. Hydrologic phenomena vary in all three space dimensions, but explicitly accounting for all of this variation may make the model too cumbersome for practical application. In a deterministic *lumped* model, the system is spatially averaged, or regarded as a single point in space without dimensions. For example, many models of the rainfall-runoff process shown in Fig. 1.2.3 treat the precipitation input as uniform over the watershed and ignore the internal spatial variation of watershed flow. In contrast, a deterministic *distributed* model considers the hydrologic processes taking place at various points in space and defines the model variables as functions of the space dimensions. Stochastic models are classified as space-independent or space-correlated according to whether or not random variables at different points in space influence each other.

At the third level of the tree, time variability is considered. Deterministic models are classified as *steady-flow* (the flow rate not changing with time) or *unsteady-flow* models. Stochastic models always have outputs that are variable in time. They may be classified as *time-independent* or *time-correlated*; a time-

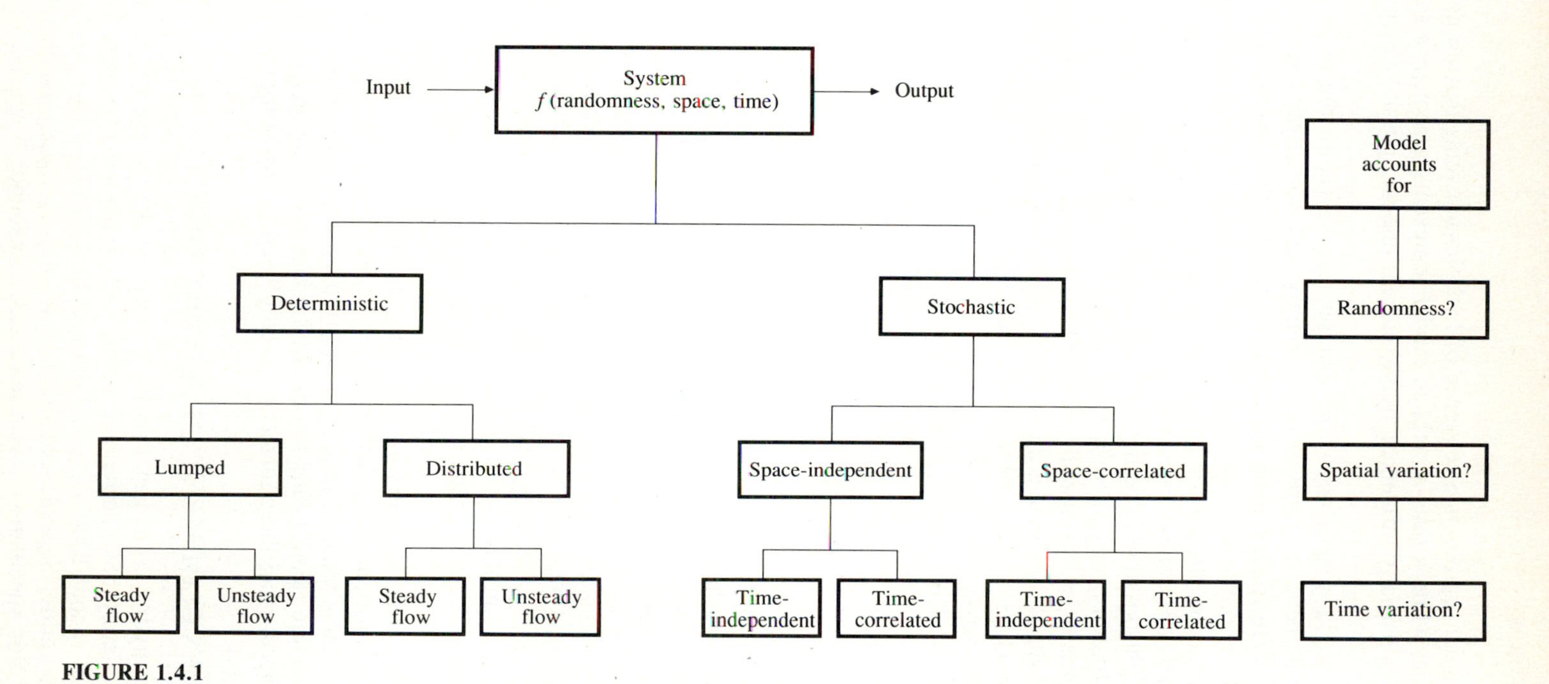

FIGURE 1.4.1
Classification of hydrologic models according to the way they treat the randomness and space and time variability of hydrologic phenomena.

independent model represents a sequence of hydrologic events that do not influence each other, while a time-correlated model represents a sequence in which the next event is partially influenced by the current one and possibly by others in the sequence.

All hydrologic models are approximations of reality, so the output of the actual system can never be forecast with certainty; likewise, hydrologic phenomena vary in all three space dimensions, and in time, but the simultaneous consideration of all five sources of variation (randomness, three space dimensions, and time) has been accomplished for only a few idealized cases. A practical model usually considers only one or two sources of variation.

Of the eight possible hydrologic model types shown along the bottom line of Fig. 1.4.1, four are considered in detail in this book. In Fig. 1.4.2, a section of a river channel is used to illustrate these four cases and the differences between them. On the right of the figure is a *space-time domain* in which space, or distance along the channel, is shown on the horizontal axis and time on the vertical axis for each of the four cases.

The simplest case, (*a*), is a deterministic lumped steady-flow model. The inflow and outflow are equal and constant in time, as shown by the equally sized dots on the lines at $x = 0$ and $x = L$. Many of the equations in the first six chapters of this book are of this type (see Ex. 1.1.1, for example). The next case, (*b*), is a deterministic lumped unsteady-flow model. The inflow $I(t)$ and outflow $Q(t)$ are now allowed to vary in time, as shown by the varying sized dots at $x=0$ and $x=L$. A lumped model does not illuminate the variation in space between the ends of the channel section so no dots are shown there. The lumped model representation is used in Chaps. 7 and 8 to describe the conversion of storm rainfall into runoff and the passage of the resulting flow through reservoirs and river channels. The third case, (*c*), is a deterministic distributed unsteady-flow model; here, variation along the space axis is also shown and the flow rate calculated for a mesh of points in space and time. Chapters 9 and 10 use this method to obtain a more accurate model of channel flow than is possible with a lumped model. Finally, in case (*d*), randomness is introduced. The system output is shown not as a single-valued dot, but as a distribution assigning a probability of occurrence to each possible value of the variable. This is a stochastic space-independent time-independent model where the probability distribution is the same at every point in the space-time plane and values at one point do not influence values elsewhere. This type of model is used in Chaps. 11 and 12 to describe extreme hydrologic events such as annual maximum rainfalls and floods. In the last three chapters, 13 to 15, the models developed using these methods are employed for hydrologic design.

1.5 THE DEVELOPMENT OF HYDROLOGY

The science of hydrology began with the concept of the hydrologic cycle. From ancient times, many have speculated about the circulation of water, including the poet Homer (about 1000 B.C.), and philosophers Thales, Plato, and Aristotle in Greece; Lucretius, Seneca, and Pliny in Rome; and many medieval scholars. Much of this speculation was scientifically unsound; however, the Greek

Space-time domain

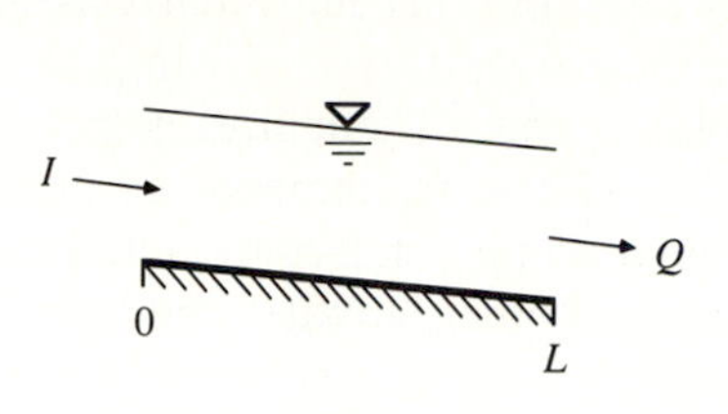

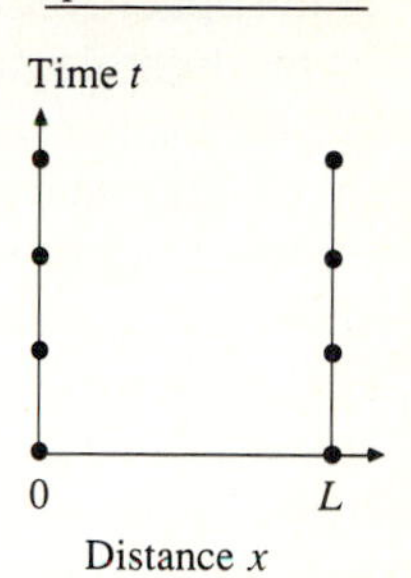

(a) Deterministic lumped steady-flow model, $I = Q$.

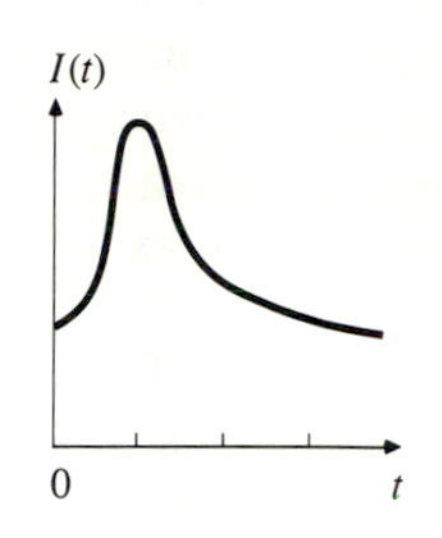

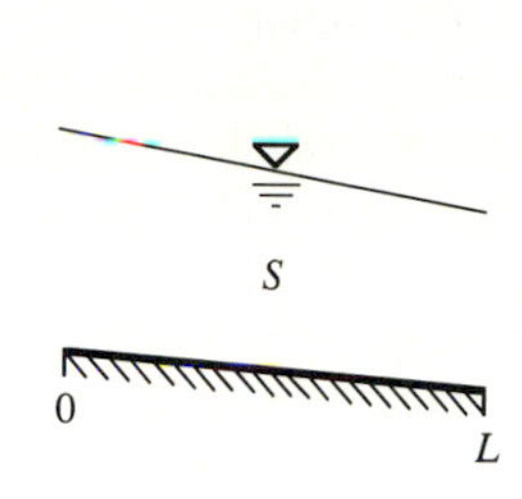

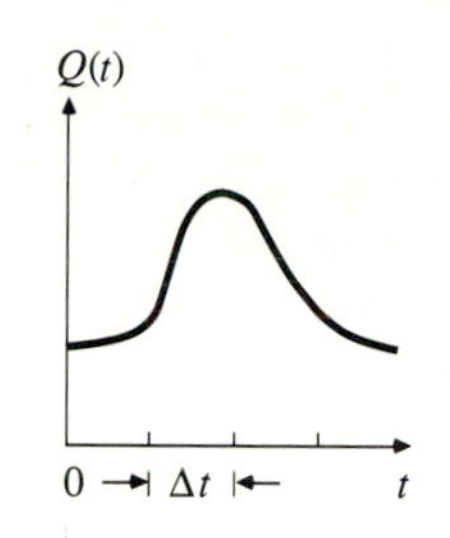

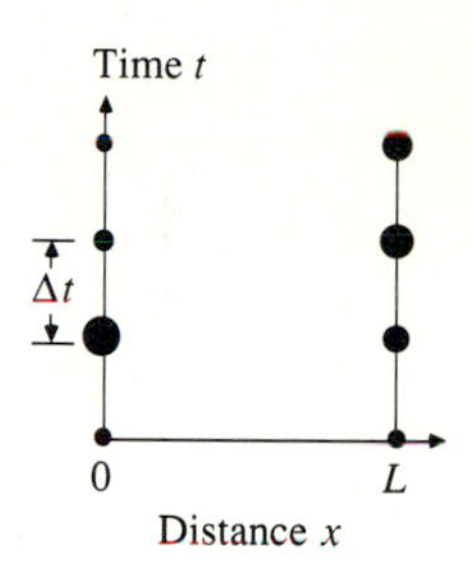

(b) Deterministic lumped unsteady-flow model, $dS/dt = I(t) - Q(t)$.

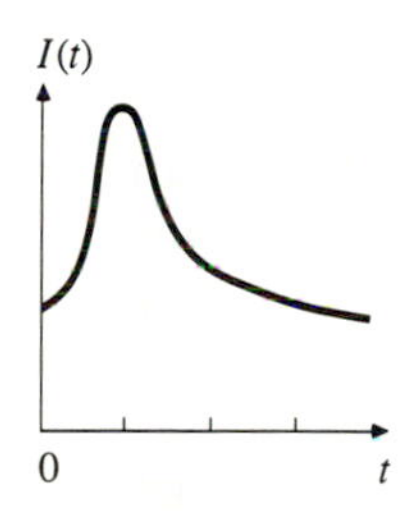

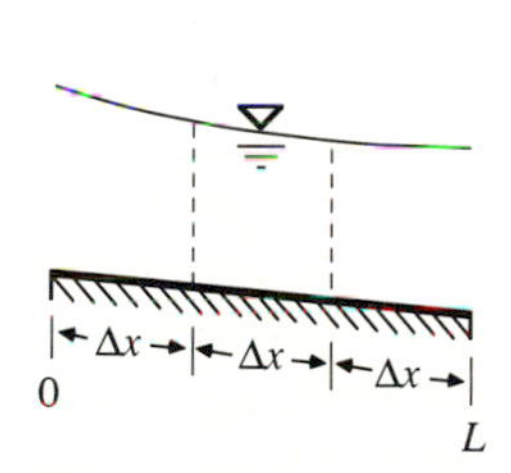

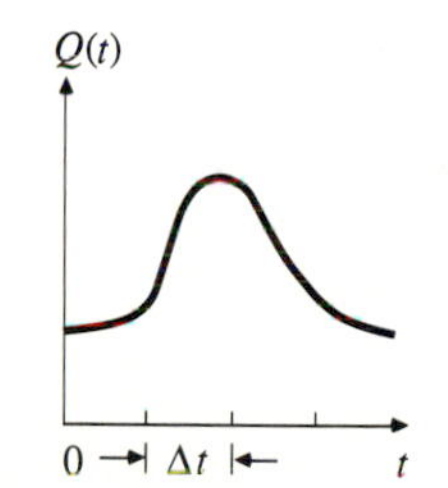

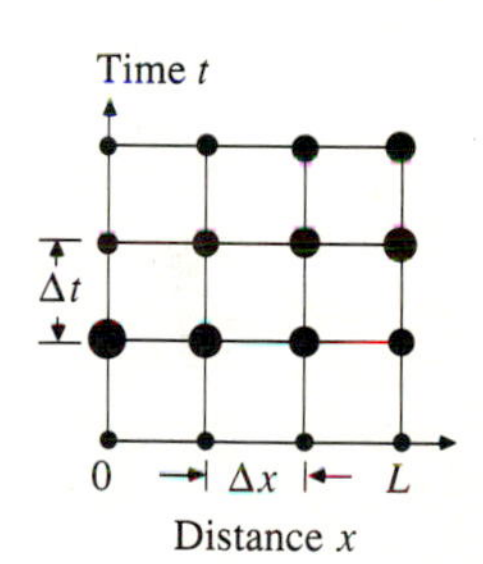

(c) Deterministic distributed unsteady-flow model.

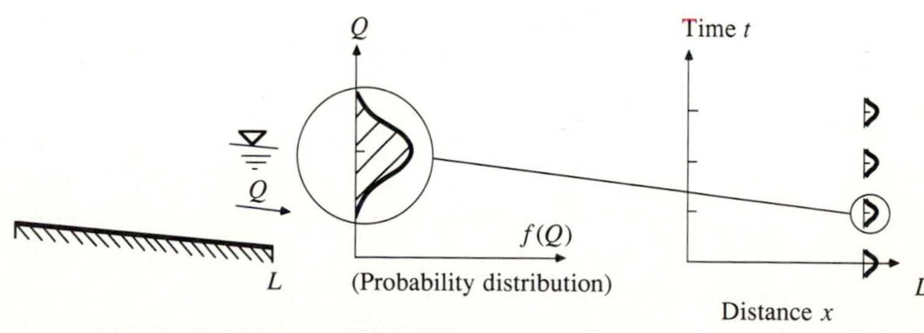

(d) Stochastic space-independent time-independent model.

FIGURE 1.4.2

The four types of hydrologic models used in this book are illustrated here by flow in a channel. For the three deterministic models (a) to (c), the size of the dots indicates the magnitude of the flow, the change of inflow and outflow with time being shown on the vertical lines at $x = 0$ and $x = L$, respectively. For the stochastic system (d), the flow is represented by a probability distribution that is shown only at $x = L$ because the model is independent in space.

philosopher Anaxagoras of Clazomenae (500–428 B.C.) formed a primitive version of the hydrologic cycle. He believed that the sun lifted water from the sea into the atmosphere, from which it fell as rain, and that rainwater was then collected in underground reservoirs, which fed the river flows. An improvement of this theory was made by another Greek philosopher, Theophrastus (c. 372–287 B.C.), who correctly described the hydrologic cycle in the atmosphere; he gave a sound explanation of the formation of precipitation by condensation and freezing. After studying the works of Theophrastus, the Roman architect and engineer Marcus Vitruvius, who lived about the time of Christ, conceived the theory that is now generally accepted: he extended Theophrastus' explanation, claiming that groundwater was largely derived from rain and snow through infiltration from the ground surface. This may be considered a forerunner of the modern version of the hydrologic cycle.

Independent thinking occurred in ancient Asian civilizations (UNESCO, 1974). The Chinese recorded observations of rain, sleet, snow, and wind on Anyang oracle bones as early as 1200 B.C. They probably used rain gages around 1000 B.C., and established systematic rain gaging about 200 B.C. In India, the first quantitative measurements of rainfall date back to the latter part of the fourth century B.C. The concept of a dynamic hydrologic cycle may have arisen in China by 900 B.C.,[1] in India by 400 B.C.,[2] and in Persia by the tenth century,[3] but these ideas had little impact on Western thought.

During the Renaissance, a gradual change occurred from purely philosophical concepts of hydrology toward observational science. Leonardo da Vinci (1452–1519) made the first systematic studies of velocity distribution in streams, using a weighted rod held afloat by an inflated animal bladder. The rod would be released at a point in the stream, and Leonardo would walk along the bank marking its progress with an odometer (Fig. 1.5.1) and judging the difference between the surface and bottom velocities by the angle of the rod. By releasing the rod at different points in the stream's cross section, Leonardo traced the

[1] In the volume "Minor Folksongs" of the "Book of Odes" (anonymous, 900–500 B.C.) is written: "Rain and snow are interchangeable and becoming sleet through first (fast) condensation." Also, Fan Li (400 B.C., Chi Ni tzu or "The Book of Master Chi Ni") said: "...the wind (containing moisture) is ch'i (moving force or energy) in the sky, and the rain is ch'i of the ground. Wind blows according to the time of the year and rain falls due to the wind (by condensation). We can say that the ch'i in the sky moves downwards (by precipitation) while the ch'i of the ground moves upwards (through evaporation)."

[2] Upanisads, dating from as early as 400 B.C. (Micropaedia, Vol. X, The New Encyclopaedia Britannica, p. 283, 1974), translated from Sanskrit to English by Swami Prabhavananda and Frederick Manchester, Mentor Books, No. MQ921, p. 69. In this work is written: "The rivers in the east flow eastwards, the rivers in the west flow westward, and all enter into the sea. From sea to sea they pass, the clouds lifting them to the sky as vapor and sending them down as rain."

[3] Karaji, M., "Extraction of Hidden Water", ca. 1016 A.D., translated from Arabic to Persian by H. Khadiv-Djam, Iranian Culture Foundation, Tehran, Iran. In this work is written: "Springs come from waters hidden inside the earth while waters on the ground surface from rains and snows ... and rain and snowmelt percolate the earth while only excess waters run off into the sea...."

FIGURE 1.5.1
Leonardo da Vinci measured the velocity distribution across a stream section by repeated experiments of the type shown in (*a*). He would release a weighted rod (*b*) held afloat by an inflated bladder and follow its progress downstream, measuring distance with the odometer and time by rhythmic chanting. (*Source:* Frazier, 1974, Figs. 6 and 7. Used with permission.)

velocity distribution across the channel. According to Frazier (1974), the 8000 existing pages of Leonardo's notes contain more entries concerning hydraulics than about any other subject. Concerning the velocity distribution in streams, he wrote, "Of water of uniform weight, depth, breadth and declivity [slope], that portion is swifter which is nearest to the surface; and this occurs because the water that is uppermost is contiguous to the air, which offers but little resistance through its being lighter than water; the water that is below is contiguous to the earth, which offers great resistance through being immovable and heavier than water" (MacCurdy, 1939). Prior to Leonardo, it was thought that water flowed more rapidly at the bottom of a stream, because if two holes were pierced in a wall holding back a body of water, the flow from the lower hole was more rapid than the flow from the upper one.

The French Huguenot scientist Bernard Palissy (1510–1589) showed that rivers and springs originate from rainfall, thus refuting an age-old theory that streams were supplied directly by the sea. The French naturalist Pierre Perrault (1608–1680) measured runoff and found it to be only a fraction of rainfall.

He recognized that rainfall is a source for runoff and correctly concluded that the remainder of the precipitation was lost by transpiration, evaporation, and diversion.

Hydraulic measurements and experiments flourished during the eighteenth century. New hydraulic principles were discovered such as the Bernoulli equation and Chezy's formula, and better instruments were developed, including the tipping bucket rain gage and the current meter. Hydrology advanced more rapidly during the nineteenth century. Dalton established a principle for evaporation (1802), the theory of capillary flow was described by the Hagen-Poiseuille equation (1839), and the rational method for determining peak flood flows was proposed by Mulvaney (1850). Darcy developed his law of porous media flow (1856), Rippl presented his diagram for determining storage requirements (1883), and Manning proposed his open-channel flow formula (1891).

However, quantitative hydrology was still immature at the beginning of the twentieth century. Empirical approaches were employed to solve practical hydrological problems. Gradually hydrologists replaced empiricism with rational analysis of observed data. Green and Ampt (1911) developed a physically based model for infiltration, Hazen (1914) introduced frequency analysis of flood peaks and water storage requirements, Richards (1931) derived the governing equation for unsaturated flow, Sherman devised the unit hydrograph method to transform effective rainfall to direct runoff (1932), Horton developed infiltration theory (1933) and a description of drainage basin form (1945), Gumbel proposed the extreme value law for hydrologic studies (1941), and Hurst (1951) demonstrated that hydrologic observations may exhibit sequences of low or high values that persist over many years.

Like many sciences, hydrology was recognized only recently as a separate discipline. About 1965, the United States Civil Service Commission recognized hydrologist as a job classification. The "hydrology series" of positions in the Commission list of occupations was described as follows:

> This series includes professional scientific positions that have as their objective the study of the interrelationship and reaction between water and its environment in the hydrologic cycle. These positions have the functions of investigation, analysis, and interpretation of the phenomena of occurrence, circulation, distribution, and quality of water in the Earth's atmosphere, on the Earth's surface, and in the soil and rock strata. Such work requires the application of basic principles drawn from and supplemented by fields such as meteorology, geology, soil science, plant physiology, hydraulics, and higher mathematics.

The advent of the computer revolutionized hydrology and made hydrologic analysis possible on a larger scale. Complex theories describing hydrologic processes are now applied using computer simulations, and vast quantities of observed data are reduced to summary statistics for better understanding of hydrologic phenomena and for establishing hydrologic design levels. More recently, developments in electronics and data transmission have made possible instantaneous data retrieval from remote recorders and the development of "real-time" programs for forecasting floods and other water operations. Microcomputers and spreadsheet programs now provide many hydrologists with new computational convenience and power. The evolution of hydrologic knowledge and methods brings about continual improvement in the scope and accuracy of solutions to hydrologic problems.

Hydrologic problems directly affect the life and activities of large numbers of people. An element of risk is always present — a more extreme event than any historically known can occur at any time. A corresponding responsibility rests upon the hydrologist to provide the best analysis that knowledge and data will permit.

REFERENCES

Dalton, J., Experimental essays on the constitution of mixed gases; on the force of steam or vapor from waters and other liquids, both in a Torricellian vacuum and in air; on evaporation; and on the expansion of gases by heat, *Mem. Proc. Manch. Lit. Phil. Soc.*, vol. 5, pp. 535–602, 1802.

Darcy, H., *Les fontaines publiques de la ville de Dijon,* V. Dalmont, Paris, 1856.

Frazier, A. H., Water current meters, *Smithsonian Studies in History and Technology* no. 28, Smithsonian Institution Press, Washington, D.C., 1974.

Green, W. H., and G. A. Ampt, Studies on soil physics, *J. Agric. Sci.*, vol. 4, part 1, pp. 1–24, 1911.

Gumbel, E. J., The return period of flood flows, *Ann. Math. Stat.*, vol. 12, no. 2, pp. 163–190, 1941.

Hagen, G. H. L., Über die Bewegung des Wassers in engen cylindrischen Rohren, *Poggendorfs Annalen der Physik und Chemie,* vol. 16, 1839.

Hazen, A., Storage to be provided in impounding reservoirs for municipal water supply, *Trans. Am. Soc. Civ. Eng.*, vol. 77, pp. 1539–1640, 1914.

Horton, R. E., The role of infiltration in the hydrologic cycle, *Trans. Am. Geophys. Union,* vol. 14, pp. 446–460, 1933.

Horton, R. E., Erosional development of streams and their drainage basins; Hydrophysical approach to quantitative morphology, *Bull. Geol. Soc. Am.*, vol. 56, pp. 275–370, 1945.

Hurst, H. E., Long-term storage capacity of reservoirs, *Trans. Am. Soc. Civ. Eng.*, vol. 116, paper no. 2447, pp. 770–799, 1951.

MacCurdy, E., *The Notebooks of Leonardo da Vinci,* vol. 1, Reynal and Hitchcock, New York, 1939.

Manning, R., On the flow of water in open channels and pipes, *Trans. Inst. Civ. Eng. Ireland,* vol. 20, pp. 161–207, 1891; supplement vol. 24, pp. 179–207, 1895.

Mulvaney, T. J., On the use of self-registering rain and flood gauges in making observations of the relations of rainfall and of flood discharges in a given catchment, *Proc. Inst. Civ. Eng. Ireland,* vol. 4, pp. 18–31, 1850.

Richards, L. A., Capillary conduction of liquids through porous mediums, *Physics, A Journal of General and Applied Physics,* American Physical Society, Minneapolis, Minn., vol. 1, pp. 318–333, July–Dec. 1931.

Rippl, W., Capacity of storage reservoirs for water supply, *Minutes of Proceedings, Institution of Civil Engineers,* vol. 71, pp. 270–278, 1883.

Sherman, L. K., Streamflow from rainfall by the unit-graph method, *Eng. News Rec.*, vol. 108, pp. 501–505, 1932.

UNESCO, Contributions to the development of the concept of the hydrological cycle, Sc.74/Conf.804/Col.1, Paris, August 1974.

U.S.S.R. National Committee for the International Hydrological Decade, World water balance and water resources of the earth, English translation, *Studies and Reports in Hydrology*, 25, UNESCO, Paris, 1978.

Vanmarcke, E., *Random Fields: Analysis and Synthesis*, MIT Press, Cambridge, Mass., 1983.

BIBLIOGRAPHY

General

Chow, V. T. (ed.), *Handbook of Applied Hydrology*, McGraw-Hill, New York, 1964.

Eagleson, P. S., *Dynamic Hydrology*, McGraw-Hill, New York, 1970.

Gray, D. M. (ed.), *Principles of Hydrology*, Water Information Center, Syosset, N.Y., 1970.

Hjelmfelt, A. T., Jr., and J. J. Cassidy, *Hydrology for Engineers and Planners*, Iowa State University Press, Ames, Iowa, 1975.

Linsley, R. K., M. A. Kohler, and J. L. H. Paulhus, *Hydrology for Engineers*, McGraw-Hill, New York, 1982.

Meinzer, O. E. (ed.), *Hydrology*, Physics of the Earth Series, vol. IX, McGraw-Hill, New York, 1942; reprinted by Dover, New York, 1949.

Raudkivi, A. J., *Hydrology*, Pergamon Press, Oxford, 1979.

Shaw, E. M., *Hydrology in Practice*, Van Nostrand Reinhold (U.K.), Wokingham, England, 1983.

Viessman, W., Jr., J. W. Knapp, G. L. Lewis, and T. E. Harbaugh, *Introduction to Hydrology*, Harper and Row, New York, 1977.

Hydrologic cycle

Baumgartner, A., and E. Reichel, *The World Water Balance*, trans. by R. Lee, Elsevier Scientific Publishing Company, 1975.

Chow, V. T., Hydrologic cycle, in *Encyclopaedia Britannica*, 15th ed., Macropaedia vol. 9, Chicago, 1974, pp. 116–125.

L'vovich, M. I., *World Water Resources and Their Future*, Mysl' P. H. Moscow, 1974; trans. ed. by R. L. Nace, American Geophysical Union, Washington, D.C., 1979.

Hydrologic history

Biswas, A. K., *History of Hydrology*, North-Holland Publishing Company, Amsterdam, 1970.

Chow, V. T., Hydrology and its development, pp. 1–22 in *Handbook of Applied Hydrology*, ed. by V. T. Chow, McGraw-Hill, New York, 1964.

Rouse, H., and S. Ince, *History of Hydraulics*, Iowa Institute of Hydraulic Research, University of Iowa, Iowa City, Iowa, 1957.

Hydrologic systems

Cadzow, J. A., *Discrete Time Systems: An Introduction with Interdisciplinary Applications*, Prentice-Hall, Englewood Cliffs, N.J., 1973.

Chow, V. T., Hydrologic modeling, *J. Boston Soc. Civ. Eng.*, vol. 60, no. 5, pp. 1–27, 1972.

Dooge, J. C. I., The hydrologic cycle as a closed system, *IASH Bull.*, vol. 13, no. 1, pp. 58–68, 1968.

Eykhoff, P., *System Identification*, Wiley, New York, 1974.

Rich, L. G., *Environmental Systems Engineering*, McGraw-Hill, New York, 1973.

Rodriguez-Iturbe, I., and R. Bras, *Random Functions in Hydrology*, Addison-Wesley, Reading, Mass., 1985.

Salas, J. D., et al., *Applied Modeling of Hydrologic Time Series,* Water Resources Publications, Littleton, Colo., 1980.

PROBLEMS

1.1.1. Assuming that all the water in the oceans is involved in the hydrologic cycle, calculate the average residence time of ocean water.

1.1.2. Assuming that all surface runoff to the oceans comes from rivers, calculate the average residence time of water in rivers.

1.1.3. Assuming that all groundwater runoff to the oceans comes from fresh groundwater, calculate the average residence time of this water.

1.1.4. The world population in 1980 has been estimated at about 4.5 billion. The annual population increase during the preceding decade was about 2 percent. At this rate of population growth, predict the year when there will be a shortage of fresh-water resources if everyone in the world enjoyed the present highest living standard, for which fresh-water use is about 6.8 m^3/day (1800 gal/day) per capita including public water supplies and water withdrawn for irrigation and industry. Assume that 47,000 km^3 of surface and subsurface runoff is available for use annually.

1.1.5. Calculate the global average precipitation and evaporation (cm/yr).

1.1.6. Calculate the global average precipitation and evaporation (in/yr).

1.2.1. Take three hydrologic systems with which you are familiar. For each, draw the system boundary and identify the inputs, outputs, and working media.

1.3.1. The equation $k(dQ/dt) + Q(t) = I(t)$ has been used to describe the gradual depletion of flow in a river during a rainless period. In this case, $I(t) = 0$ and $Q(t) = Q_0$ for $t = 0$. Solve the differential equation for $Q(t)$ for $t > 0$ and plot the result over a 20-day period if $k = 10$ days and $Q_0 = 100$ cfs.

1.3.2. The equation $k(dQ/dt) + Q(t) = I(t)$ has been used to describe the response of streamflow to a constant rate of precipitation continuing indefinitely on a watershed. In this case, let $I(t) = 1$ for $t > 0$, and $Q(t) = 0$ for $t = 0$. Solve the differential equation and plot the values of $I(t)$ and $Q(t)$ over a 10-hour period if $k = 2$ h.

1.4.1. Classify the following hydrologic phenomena according to the structure given in Fig. 1.4.1: (*a*) steady, uniform flow in an open channel; (*b*) a sequence of daily average flows at a stream-gaging site; (*c*) the annual maximum values of daily flow at a site; (*d*) the longitudinal profile of water surface elevation for steady flow in a stream channel upstream of a bridge; (*e*) the same as (*d*) but with a flood passing down the channel; (*f*) a sequence of annual precipitation values at a site; (*g*) a sequence of annual precipitation values at a group of nearby locations.

1.5.1. Select a major water resources project in your area. Explain the purposes of the project and describe its main features.

1.5.2. Select a water resources project of national or international significance. Explain the purposes of the project and describe its main features.

1.5.3. Select three major agencies in your area that have hydrologic responsibilities and explain what those responsibilities are.

1.5.4. Select a major hydrologic event such as a flood or drought that occurred in your area and describe its effects.

CHAPTER 2

HYDROLOGIC PROCESSES

Hydrologic processes transform the space and time distribution of water throughout the hydrologic cycle. The motion of water in a hydrologic system is influenced by the physical properties of the system, such as the size and shape of its flow paths, and by the interaction of the water with other working media, including air and heat energy. Phase changes of water between liquid, solid, and vapor are important in some cases. Many physical laws govern the operation of hydrologic systems.

A consistent mechanism needed for developing hydrologic models is provided by the *Reynolds transport theorem*, also called the *general control volume equation*. The Reynolds transport theorem is used to develop the continuity, momentum, and energy equations for various hydrologic processes.

2.1 REYNOLDS TRANSPORT THEOREM

The Reynolds transport theorem takes physical laws that are normally applied to a discrete mass of a substance and applies them instead to a fluid flowing continuously through a control volume. For this purpose, two types of fluid properties can be distinguished: *extensive* properties, whose values depend on the amount of mass present, and *intensive* properties, which are independent of mass. For any extensive property B, a corresponding intensive property β can be defined as the quantity of B per unit mass of fluid, that is $\beta = dB/dm$. B and β can be scalar or vector quantities depending on the property being considered.

The Reynolds transport theorem relates the time rate of change of an extensive property in the fluid, dB/dt, to the external causes producing this change. Consider fluid momentum; in this case, $\mathbf{B} = m\mathbf{V}$ and $\boldsymbol{\beta} = d(m\mathbf{V})/dm = \mathbf{V}$, the fluid velocity, where bold face type indicates a vector quantity. By Newton's second law, the time rate of change of momentum is equal to the net applied force on the fluid: $d\mathbf{B}/dt = d(m\mathbf{V})/dt = \Sigma\mathbf{F}$. The extensive properties discussed in this book are the mass, momentum, and energy of liquid water, and the mass of water vapor.

When Newton's second law or other physical laws are applied to a solid body, the focus is on the motion of the body and the analysis follows the body wherever it moves. This is the *Lagrangian* view of motion. Although this concept can be applied to fluids, it is more common to consider that fluids form a continuum wherein the motion of individual particles is not traced. The focus is then on a control volume, a fixed frame in space through which the fluid passes, called the *Eulerian* view of motion. The theorem separates the action of external influences on the fluid, expressed by dB/dt, into two components: the time rate of change of the extensive property stored within the control volume, and the net outflow of the extensive property across the control surface. The Reynolds transport theorem is commonly used in fluid mechanics (White, 1979; Shames, 1982; Fox and MacDonald, 1985; and Roberson and Crowe, 1985). Although it has not been widely used in hydrology up to this time it provides a consistent means for applying physical laws to hydrologic systems.

To derive the governing equation of the theorem, consider the control volume shown in Fig. 2.1.1, whose boundary is defined by the dashed control surface. Within the control volume there is a shaded element of volume $d\forall$. If the density of the fluid is ρ, the mass of fluid in the element is $dm = \rho d\forall$, the amount of extensive property B contained in the fluid element is $dB = \beta dm = \beta\rho d\forall$, and the total amount of extensive property within any volume is the integral of these elemental amounts over that volume:

$$B = \iiint \beta\rho \, d\forall \tag{2.1.1}$$

where $\iiint$ indicates integration over a volume.

Fluid flows from left to right through the control volume in Fig. 2.1.1, but no fluid passes through the upper or lower boundary. After a small interval of time Δt, the fluid mass inside the control volume at time t has moved to the right and occupies the space delineated by dotted lines. Three regions of space can then be identified: region I, to the left, which the fluid mass occupies at time t but not at $t + \Delta t$; region II, in the center, filled by the fluid mass at both points in time; and region III, on the right, outside the control volume, which the fluid mass occupies at $t + \Delta t$ but not at t. For the cross-hatched fluid mass initially within the control volume, the time rate of change of the extensive property can be defined by

$$\frac{dB}{dt} = \lim_{\Delta t \to 0} \frac{1}{\Delta t}[(B_{\mathrm{II}} + B_{\mathrm{III}})_{t+\Delta t} - (B_{\mathrm{I}} + B_{\mathrm{II}})_t] \tag{2.1.2}$$

where the subscripts t and $t + \Delta t$ are used to denote the values of the subscripted

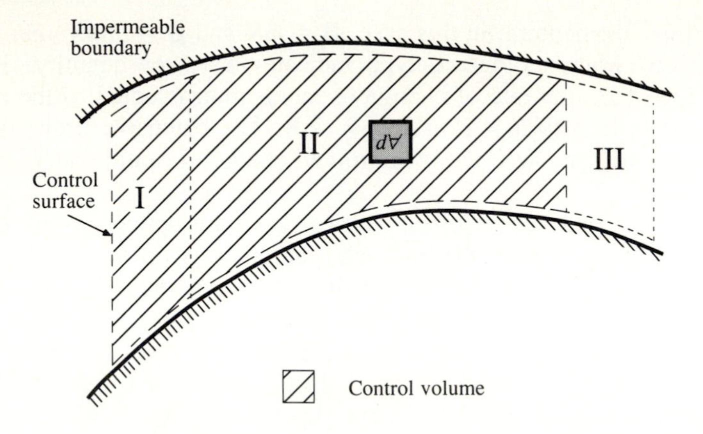

(a) Fluid in regions I and II (the control volume) at time t occupies regions II and III at time $t + \Delta t$.

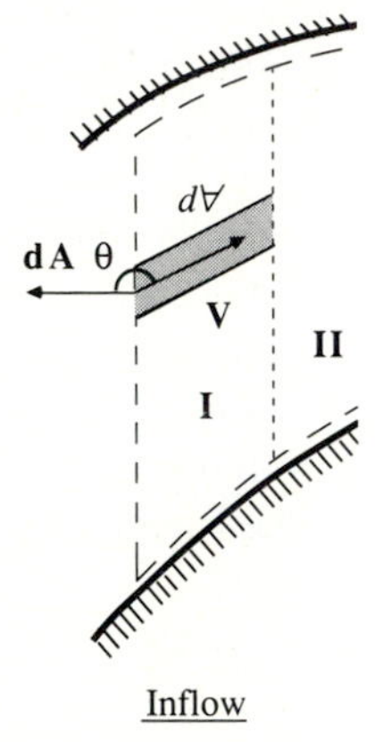

Inflow

$\mathbf{V} \cdot \mathbf{dA} = V \cos\theta \; dA \; (\cos\theta < 0)$

(b) Expanded view of inflow region.

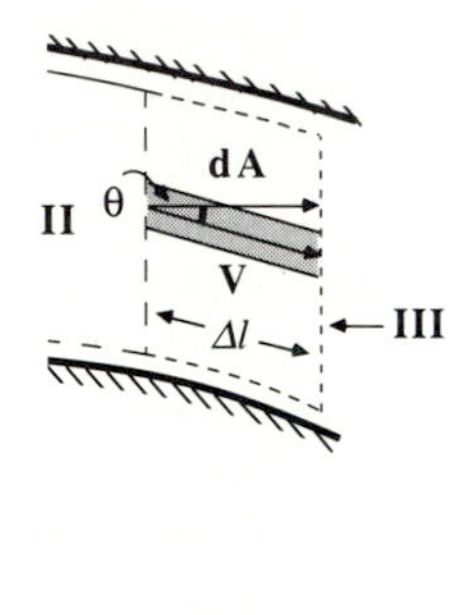

Outflow

$d\forall = \Delta l \; \cos\theta \, dA$

$\mathbf{V} \cdot \mathbf{dA} = V \cos\theta \; dA \; (\cos\theta > 0)$

(c) Expanded view of outflow region.

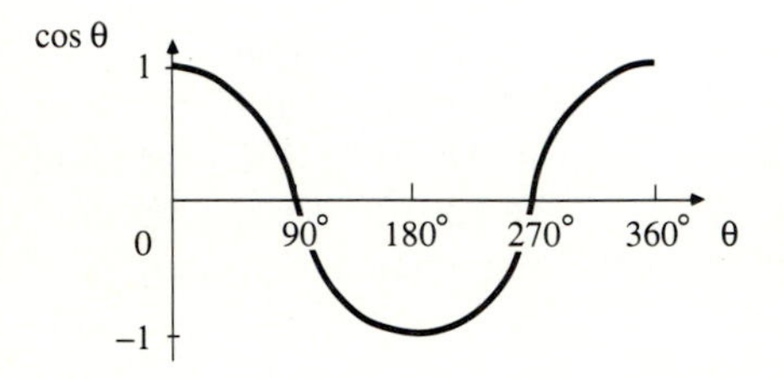

$\cos\theta > 0$ for outflow
$\cos\theta = 0$ at impermeable boundaries
$\cos\theta < 0$ for inflow

(d) $\cos\theta$ vs. θ for inflow and outflow.

FIGURE 2.1.1
Fluid control volume for derivation of the Reynolds transport theorem.

quantities at these two time points. Rearranging (2.1.2) to separate the extensive property remaining within the control volume (B_{II}) from that passing across the control surface (B_{I}) and (B_{III}) yields

$$\frac{dB}{dt} = \lim_{\Delta t \to 0} \left\{ \frac{1}{\Delta t}\left[(B_{\text{II}})_{t+\Delta t} - (B_{\text{II}})_t\right] + \frac{1}{\Delta t}\left[(B_{\text{III}})_{t+\Delta t} - (B_{\text{I}})_t\right] \right\} \tag{2.1.3}$$

As Δt approaches 0, region II becomes coincident with the control volume, and the first term in (2.1.3) becomes the time derivative (d/dt) of the amount of B stored within the control volume:

$$\lim_{\Delta t \to 0} \frac{1}{\Delta t}\left[(B_{\text{II}})_{t+\Delta t} - (B_{\text{II}})_t\right] = \frac{d}{dt} \iiint_{\text{c.v.}} \beta \rho d\forall \tag{2.1.4}$$

In this equation, the total derivative d/dt is used to account for the case when the control volume is *deformable* (i.e., changes in size and shape as time passes). If the control volume is *fixed* in space and time, the total derivative can be replaced by the partial derivative $\partial/\partial t$ because the focus is on the time rate of change of the extensive property stored in the control volume without regard for its internal spatial distribution.

The second term in Eq. (2.1.3), involving B_{I} and B_{III}, represents the flow of the extensive property across the control surface. Figure 2.1.1(c) shows a close-up view of region III at the outlet from the control volume. An element of area in the outlet control surface is labeled dA, and the element of volume $d\forall$ is the volume of the tube containing all the fluid passing through dA in time Δt. The length of the tube is $\Delta l = V\Delta t$, the length of the flow path in time Δt. The volume of the tube is $d\forall = \Delta l \cos\theta dA$ where θ is the angle between the velocity vector $\mathbf{V}$ and the direction normal to the area element dA. The amount of extensive property B in the tube is $\beta\rho d\forall = \beta\rho\Delta l \cos\theta dA$. The total amount of fluid in region III is found by integrating these elemental amounts over the entire outlet control surface. Thus the term in B_{III} in (2.1.3) can be written as

$$\lim_{\Delta t \to 0} \left\{ \frac{1}{\Delta t}(B_{\text{III}})_{t+\Delta t} \right\} = \lim_{\Delta t \to 0} \frac{\iint \beta\rho\Delta l \cos\theta dA}{\Delta t} \tag{2.1.5}$$

where the double integral $\iint$ indicates that the integral is over a surface.

As Δt approaches 0, the limit of the ratio $\Delta l/\Delta t$ is the magnitude of the fluid velocity V. Let the *normal area vector* $\mathbf{dA}$ be defined as a vector of magnitude dA with direction normal to the area dA pointing *outward* from the control surface; then the term $V\cos\theta dA$ can be expressed as the vector dot product $\mathbf{V}\cdot\mathbf{dA}$. So Eq. (2.1.5) can be rewritten to give the flow rate of the extensive property leaving the control surface as

$$\lim_{\Delta t \to 0} \left\{ \frac{1}{\Delta t}(B_{\text{III}})_{t+\Delta t} \right\} = \iint_{\text{III}} \beta\rho\mathbf{V}\cdot\mathbf{dA} \tag{2.1.6}$$

A similar analysis may be made for fluid entering the control volume in region I [see Fig. 2.1.1(*b*)]. In this case, $\cos\theta$ is negative and $d\forall = \Delta l \cos(180° - \theta)\, dA = -\Delta l \cos\theta\, dA$, so that

$$\lim_{\Delta t \to 0}\left\{\frac{1}{\Delta t}(B_{\mathrm{I}})_t\right\} = \lim_{\Delta t \to 0}\frac{\iint \beta\rho(-\Delta l \cos\theta\, dA)}{\Delta t}$$

$$= -\iint_{\mathrm{I}} \beta\rho \mathbf{V}\cdot\mathbf{dA} \tag{2.1.7}$$

Substituting (2.1.4), (2.1.6), and (2.1.7) into (2.1.3) gives

$$\frac{dB}{dt} = \frac{d}{dt}\iiint_{\text{c.v.}} \beta\rho\, d\forall + \iint_{\mathrm{III}} \beta\rho\mathbf{V}\cdot\mathbf{dA} + \iint_{\mathrm{I}} \beta\rho\mathbf{V}\cdot\mathbf{dA} \tag{2.1.8}$$

For fluid entering the control volume, the angle between the velocity vector **V**, pointing into the control volume, and the area vector **dA**, pointing out, is in the range $90° < \theta < 270°$ for which $\cos\theta$ is negative [see Fig 2.1.1 (*d*)]. Consequently, $\mathbf{V}\cdot\mathbf{dA}$ is always negative for inflow. For fluid leaving the control volume $\cos\theta$ is positive, so $\mathbf{V}\cdot\mathbf{dA}$ is always positive for outflow. At the impermeable boundaries, **V** and **dA** are perpendicular and therefore $\mathbf{V}\cdot\mathbf{dA} = 0$. Thus, the integrals in (2.1.8) over inlet I and outlet III can be replaced by a single integral over the entire control surface representing the outflow minus inflow, or *net outflow*, of extensive property B:

$$\frac{dB}{dt} = \frac{d}{dt}\iiint_{\text{c.v.}} \beta\rho\, d\forall + \iint_{\text{c.s.}} \beta\rho\mathbf{V}\cdot\mathbf{dA} \tag{2.1.9}$$

Equation (2.1.9) is the governing equation of the Reynolds transport theorem. It is used a number of times in this book, and it is worthwhile to review the meaning of each term. As stated previously, the equation will be used to provide a mechanism for taking physical laws normally applied to a discrete mass and applying them instead to continuously flowing fluid. In words, the Reynolds transport theorem states that the *total rate of change of an extensive property* of a fluid is equal to the *rate of change of extensive property stored in the control volume*, $d/dt \iiint \beta\rho\, d\forall$, plus the net *outflow of extensive property through the control surface*, $\iint \beta\rho\, \mathbf{V}\cdot\mathbf{dA}$. When using the theorem, inflows are always considered negative and outflows positive.

In the following sections, the Reynolds transport theorem is applied to develop continuity, momentum, and energy equations for hydrologic processes.

2.2 CONTINUITY EQUATIONS

The conservation of mass is the most useful physical principle in hydrologic analysis and is required in almost all applied problems. *Continuity equations* expressing this principle can be developed for a fluid volume, for a flow cross-section, and for a point within a flow. In this chapter, only the *integral equation of continuity* for a flow volume is developed. The equation for continuity at a point will be derived in Chap. 4 to describe flow in a porous medium, and the continuity equation at a cross section will be derived in Chap. 9 to describe flow at a river section. The integral equation of continuity is the basis for the other two forms.

Integral Equation of Continuity

The integral equation of continuity applies to a volume of fluid. If mass is the extensive property being considered in the Reynolds transport theorem, then $B = m$, and $\beta = dB/dm = 1$. By the law of conservation of mass, $dB/dt = dm/dt = 0$ because mass cannot be created or destroyed. Substituting these values into the Reynolds transport theorem (2.1.9) gives

$$0 = \frac{d}{dt}\iiint_{\text{c.v.}} \rho \, d\forall + \iint_{\text{c.s.}} \rho \mathbf{V}\cdot\mathbf{dA} \tag{2.2.1}$$

which is the integral equation of continuity for an unsteady, variable-density flow.

If the flow has constant density, ρ can be divided out of both terms of (2.2.1), leaving

$$\frac{d}{dt}\iiint_{\text{c.v.}} d\forall + \iint_{\text{c.s.}} \mathbf{V}\cdot\mathbf{dA} = 0 \tag{2.2.2}$$

The integral $\iiint d\forall$ is the volume of fluid stored in the control volume, denoted by S, so the first term in (2.2.2) is the time rate of change of storage dS/dt. The second term, the net outflow, can be split into inflow $I(t)$ and outflow $Q(t)$:

$$\iint_{\text{c.s.}} \mathbf{V}\cdot\mathbf{dA} = \iint_{\text{outlet}} \mathbf{V}\cdot\mathbf{dA} + \iint_{\text{inlet}} \mathbf{V}\cdot\mathbf{dA} = Q(t) - I(t) \tag{2.2.3}$$

and the integral equation of continuity can be rewritten

$$\frac{dS}{dt} + Q(t) - I(t) = 0$$

or

$$\frac{dS}{dt} = I(t) - Q(t) \tag{2.2.4}$$

which is the integral equation of continuity for an unsteady, constant density flow, used extensively in this book. When the flow is steady, $dS/dt = 0$, and (2.2.2) reduces to

$$\iint_{\text{c.s.}} \mathbf{V} \cdot \mathbf{dA} = 0 \tag{2.2.5}$$

which states that the volumetric inflow rate and outflow rate are equal; that is, $I(t) = Q(t)$. A steady flow is one in which the velocity at every point in the flow is constant in time. A simple way of thinking about this is to imagine taking a "snapshot" of the flow now, and again five minutes later; if the the flow is steady, the two snapshots will be identical.

If the total amounts of inflow and outflow are equal, the system is said to be *closed* so that

$$\int_{-\infty}^{\infty} I(t)dt = \int_{-\infty}^{\infty} Q(t)dt \tag{2.2.6}$$

When this condition does not hold, the system is *open*. The hydrologic cycle is a closed system for water, but the rainfall-runoff process on a watershed is an open system, because not all the rainfall becomes runoff; some is returned to the atmosphere through evaporation.

The continuity equations above are derived for *single phase* flow, that is, a liquid or a gas, but not both together. In multiphase situations, such as when water is evaporating, the liquid and gaseous phases of water must be carefully distinguished. A continuity equation should be written separately for each phase of the flow; for each phase dB/dt is the rate at which mass is being added to, or taken from, that phase.

2.3 DISCRETE TIME CONTINUITY

Because most hydrologic data are available only at discrete time intervals, it is necessary to reformulate the continuity equation (2.2.4) on a discrete time basis. Suppose that the time horizon is divided into intervals of length Δt, indexed by j. Equation (2.2.4) can be rewritten as $dS = I(t)\,dt - Q(t)\,dt$ and integrated over the jth time interval to give

$$\int_{S_{j-1}}^{S_j} dS = \int_{(j-1)\Delta t}^{j\Delta t} I(t)dt - \int_{(j-1)\Delta t}^{j\Delta t} Q(t)dt \tag{2.3.1}$$

or

$$S_j - S_{j-1} = I_j - Q_j \qquad j = 1, 2, \ldots \tag{2.3.2}$$

where I_j and Q_j are the volumes of inflow and outflow in the jth time interval. Note that in Eq. (2.2.4), $I(t)$ and $Q(t)$ are flow rates, having dimensions $[L^3/T]$, while S is a volume, having dimensions $[L^3]$. In (2.3.2), all the variables have dimensions $[L^3]$. If the incremental change in storage is denoted by ΔS_j, then one writes $\Delta S_j = I_j - Q_j$, and

$$S_j = S_{j-1} + \Delta S_j \tag{2.3.3}$$

If the initial storage at time 0 is S_0, then $S_1 = S_0 + I_1 - Q_1$, $S_2 = S_1 + I_2 - Q_2$, and so on. By substituting for intermediate storage values, one obtains

$$S_j = S_0 + \sum_{i=1}^{j} (I_i - Q_i) \tag{2.3.4}$$

which is the discrete-time continuity equation.

Data Representation

The functions $Q(t)$ and $I(t)$ are defined on a continuous time domain; that is, a value of the function is defined at every instant of the time domain, and these values can change from one instant to the next [Fig. 2.3.1(*a*)]. Figure 2.3.1 shows two methods by which a continuous time function can be represented on a discrete time domain. The first method [Fig. 2.3.1(*b*)] uses a *sample data function* in which the value of a function $Q(t)$ in the jth time interval, Q_j, is given simply by the instantaneous value of $Q(t)$ at time $j\Delta t$:

$$Q_j = Q(t_j) = Q(j\Delta t) \tag{2.3.5}$$

The dimensions of $Q(t)$ and Q_j are the same, either [L^3/T] or [L/T].

The second method uses a *pulse data function* [Fig. 2.3.1(*c*)], in which the value of the discrete time function Q_j is given by the area under the continuous time function:

$$Q_j = \int_{(j-1)\Delta t}^{j\Delta t} Q(t)\, dt \tag{2.3.6}$$

Here Q_j has dimensions of [L^3] or [L] for $Q(t)$ in dimensions of [L^3/T] or [L/T], respectively. Alternatively, the dimensions of Q_j and $Q(t)$ can be kept the same if Q_j is calculated as the *average rate* over the interval:

$$Q_j = \frac{1}{\Delta t} \int_{(j-1)\Delta t}^{j\Delta t} Q(t)\, dt \tag{2.3.7}$$

The two principle variables of interest in hydrology, streamflow and precipitation, are measured as sample data and pulse data respectively. When the values of streamflow and precipitation are recorded by gages at a given instant, the streamflow gage value is the flow rate at that instant, while the precipitation gage value is the accumulated depth of precipitation which has occurred up to that instant. The successive differences of the measurements of accumulated precipitation form a pulse data series (in inches or centimeters). When divided by the time interval Δt, as in Eq. (2.3.7), the resulting data give the *precipitation intensity* (in inches per hour or centimeters per hour). The continuity equation must be applied carefully when using such discrete time data.

Example 2.3.1 Calculate the storage of water on a watershed as a function of time given the data in columns 3 and 4 of Table 2.3.1 for incremental precipitation

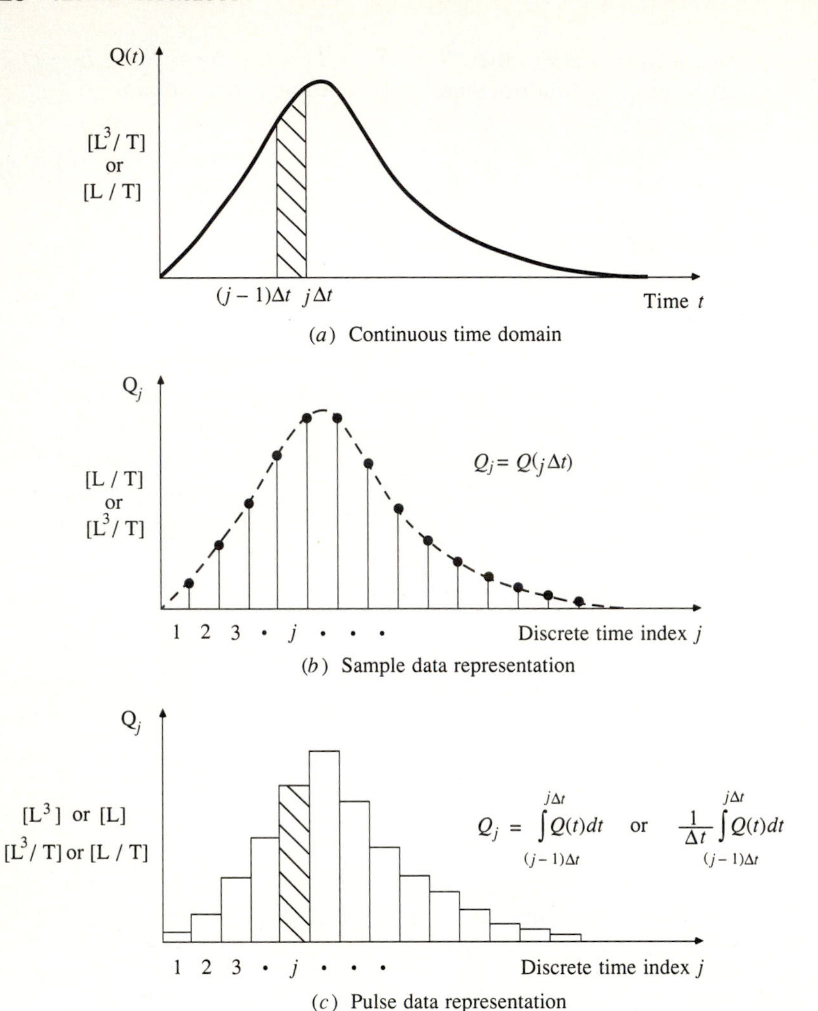

FIGURE 2.3.1
A continuous time function $Q(t)$, (a), can be defined on a discrete time domain either by a sampled data system (b), in which instantaneous values of the continuous time function are used, or by a pulse data system (c), in which the integral or average value of the function over the interval is used.

over the watershed and streamflow measured at its outlet. These data are adapted from a flood that occurred on Shoal Creek at Northwest Park in Austin, Texas on May 24–25, 1981. The watershed area is 7.03 mi^2. Assume that the initial storage is zero.

Solution. The precipitation input is recorded as a pulse data sequence in column 3; the value shown is the incremental depth for the preceding time interval (e.g., the value shown at $t = 0.5$ h, 0.15 in, is the precipitation depth occurring during the first 0.5 h and the value shown at $t = 1$ h, 0.26 in, is the incremental precipitation between $t = 0.5$ h and $t = 1$ h, and so on). The streamflow output is recorded as a sample data sequence; the value shown is the instantaneous flow rate (e.g., the streamflow rate is 246 cfs at $t = 0.5$ h, 283 cfs at $t = 1$ h, and so on). To apply the discrete time continuity equation (2.3.4), the streamflow must be converted to a pulse data sequence. The time interval is $\Delta t = 0.5 \text{ h} = 0.5 \times 3600 \text{ s} = 1800$ s. For each 0.5 h interval, the volume of streamflow is calculated by averaging the streamflow rates at the ends of the interval and multiplying by Δt. The equivalent depth over the watershed of incremental streamflow is then calculated by dividing the streamflow volume by the watershed area, which is $7.03 \text{ mi}^2 = 7.03 \times 5280^2 \text{ ft}^2 = 1.96 \times 10^8 \text{ ft}^2$.

For example, during the first time interval, between 0 and 0.5 h, the streamflows [Col.(4)] are $Q(0) = 203$ cfs and $Q(0.5) = 246$ cfs, so the incremental volume in this interval is $[(203+246)/2] \times \Delta t = 224.5 \times 1800 = 4.04 \times 10^5 \text{ ft}^3$. The equivalent depth over the watershed is $Q_1 = 4.04 \times 10^5/1.96 \times 10^8 = 2.06 \times 10^{-3}$ ft $= 2.06 \times 10^{-3} \times 12$ in $= 0.02$ in, as shown in column 5.

The incremental precipitation I_1 for the same time interval is 0.15 in, so the incremental change in storage is found from Eq. (2.3.2) with $j = 1$:

$$\begin{aligned} \Delta S_1 &= I_1 - Q_1 \\ &= 0.15 - 0.02 \\ &= 0.13 \text{ in} \end{aligned}$$

as shown in column 6. The cumulative storage on the watershed is found from (2.3.3) with $j = 1$ and initial storage $S_0 = 0$:

$$\begin{aligned} S_1 &= S_0 + \Delta S_1 \\ &= 0 + 0.13 \\ &= 0.13 \text{ in} \end{aligned}$$

as shown in column 7. The calculations for succeeding time intervals are similar. Table 2.3.1 shows that of the 6.31 in total precipitation, 5.45 in, or 86 percent, appeared as streamflow at the watershed outlet in the eight hours after precipitation began. The remaining 0.86 in was retained in storage on the watershed. In columns 5 and 6 it can be seen that after precipitation ceased, all streamflow was drawn directly from storage.

The values of incremental precipitation and streamflow, change in storage, and cumulative storage are plotted in Fig. 2.3.2. The critical time is $t = 2.5$ h, when the maximum storage occurs. Before 2.5 h, precipitation exceeds streamflow and there is a gain in storage; after 2.5 h, the reverse occurs and there is a loss in storage.

2.4 MOMENTUM EQUATIONS

When the Reynolds transport theorem is applied to fluid momentum, the extensive property is $\mathbf{B} = m\mathbf{V}$, and $\boldsymbol{\beta} = d\mathbf{B}/dm = \mathbf{V}$. By Newton's second law, the time rate of change of momentum is equal to the net force applied in a given direction, so $d\mathbf{B}/dt = d(m\mathbf{V})/dt = \Sigma\, \mathbf{F}$. Substituting into the Reynolds transport theorem (2.1.9), results in

TABLE 2.3.1
The time distribution of storage on a watershed calculated using the discrete-time continuity equation (Example 2.3.1)

1 Time interval j	2 Time t (h)	3 Incremental precipitation I_j (in)	4 Instantaneous streamflow $Q(t)$ (cfs)	5 Incremental streamflow Q_j (in)	6 Incremental storage ΔS_j (in)	7 Cumulative storage S_j (in)
	0.0		203			0.00
1	0.5	0.15	246	0.02	0.13	0.13
2	1.0	0.26	283	0.03	0.23	0.36
3	1.5	1.33	828	0.06	1.27	1.62
4	2.0	2.20	2323	0.17	2.03	3.65
5	2.5	2.08	5697	0.44	1.64	5.29
6	3.0	0.20	9531	0.84	−0.64	4.65
7	3.5	0.09	11025	1.13	−1.04	3.61
8	4.0		8234	1.06	−1.06	2.55
9	4.5		4321	0.69	−0.69	1.85
10	5.0		2246	0.36	−0.36	1.49
11	5.5		1802	0.22	−0.22	1.27
12	6.0		1230	0.17	−0.17	1.10
13	6.5		713	0.11	−0.11	1.00
14	7.0		394	0.06	−0.06	0.93
15	7.5		354	0.04	−0.04	0.89
16	8.0		303	0.04	−0.04	0.86
Total		6.31		5.45		

$$\sum \mathbf{F} = \frac{d}{dt}\iiint_{\text{c.v.}} \mathbf{V}\rho\, d\forall + \iint_{\text{c.s.}} \mathbf{V}\rho\mathbf{V}\cdot d\mathbf{A} \tag{2.4.1}$$

the integral momentum equation for an unsteady, nonuniform flow. A *nonuniform flow* is one in which the velocity does vary in space; in a *uniform flow* there is no spatial variation.

If a nonuniform flow is steady (in time), the time derivative in Eq. (2.4.1) drops out, leaving

$$\sum \mathbf{F} = \iint_{\text{c.s.}} \mathbf{V}\rho\mathbf{V}\cdot d\mathbf{A} \tag{2.4.2}$$

For a steady uniform flow the velocity is the same at all points on the control surface, and therefore the integral over the control surface is zero and the forces applied to the system are in equilibrium:

$$\sum \mathbf{F} = 0 \tag{2.4.3}$$

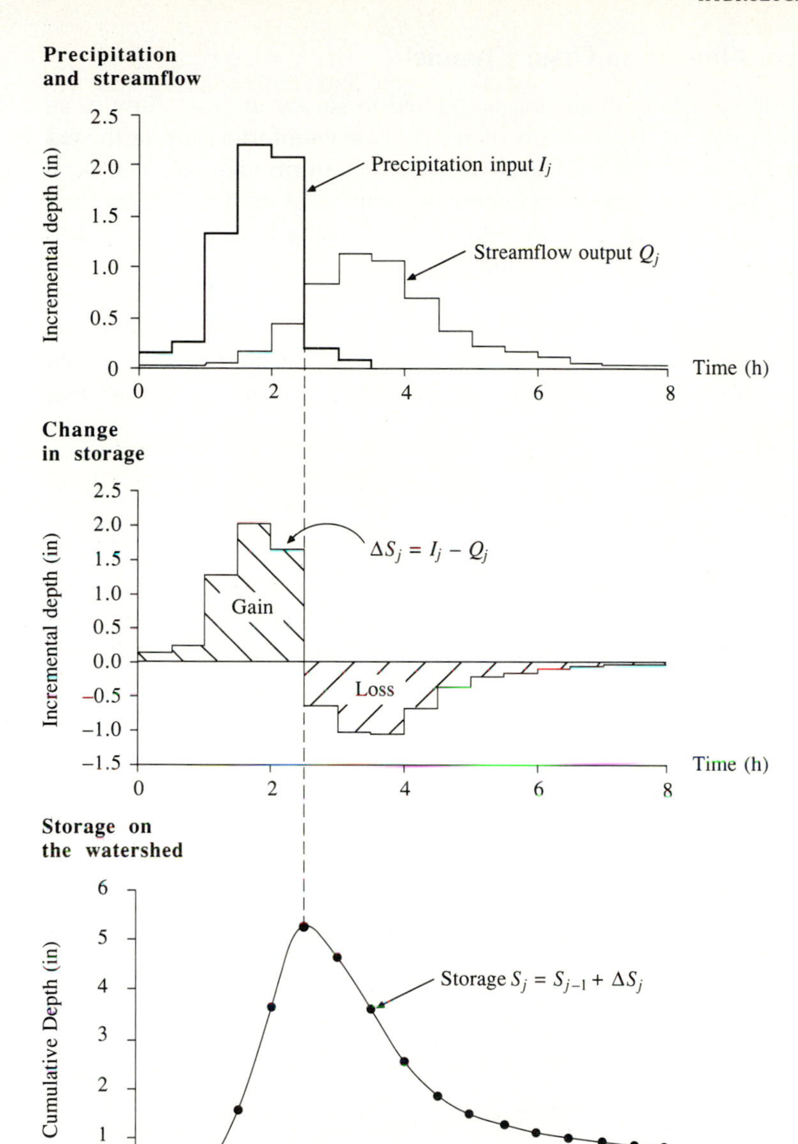

FIGURE 2.3.2
The time distribution of storage on a watershed calculated using the discrete-time continuity equation (Example 2.3.1).

Steady Uniform Flow in an Open Channel

In this section, the momentum equation is applied to steady uniform flow in an open channel. The more complex case of unsteady nonuniform flow is treated in Sec. 9.1. Figure 2.4.1 shows a steady flow in a uniform channel, that is, a channel whose cross section, slope, and boundary roughness do not change along its length. The continuity, momentum, and energy equations can be applied to the control volume between sections 1 and 2.

Continuity. For steady flow, Eq. (2.2.5) holds and $Q_1 = Q_2$; for uniform flow, the velocity is the same everywhere in the flow, so $V_1 = V_2$. Hence, cross-sectional area $A_1 = Q_1/V_1 = Q_2/V_2 = A_2$, and since the channel is uniform, it follows that the depths are also equal, $y_1 = y_2$.

Energy. The energy equation from fluid mechanics (Roberson and Crowe, 1985) is written for sections 1 and 2 as

$$z_1 + y_1 + V_1^2/2g = z_2 + y_2 + V_2^2/2g + h_f \tag{2.4.4}$$

where z is the bed elevation, g is the acceleration due to gravity, and h_f is the head loss between the two sections. Head loss is the energy lost due to friction effects per unit weight of fluid. With $V_1 = V_2$ and $y_1 = y_2$, (2.4.4) reduces to

$$h_f = z_1 - z_2 \tag{2.4.5}$$

Dividing both sides by L, the length of the channel, the following is obtained.

$$\frac{h_f}{L} = \frac{z_1 - z_2}{L} \tag{2.4.6}$$

The bed slope $S_0 = \tan \theta$ where θ is the angle of inclination of the channel bed. If θ is small ($< 10°$), then $\tan \theta \approx \sin \theta = (z_1 - z_2)/L$. In this case, the

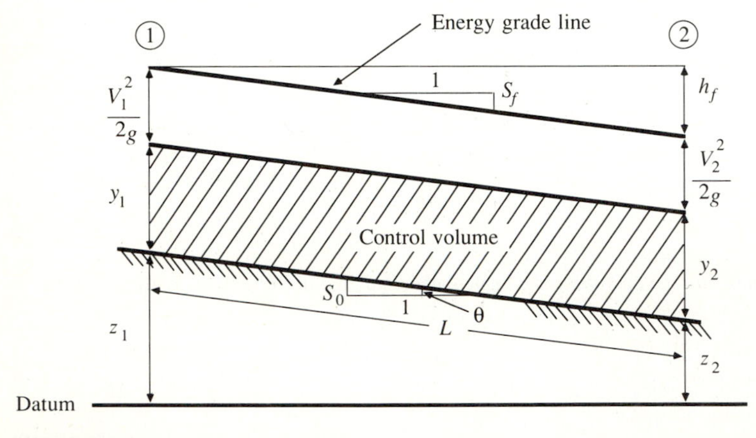

FIGURE 2.4.1
Steady uniform flow in an open channel.

friction slope, $S_f = h_f/L$, is equal to the bed slope S_0. It is assumed in this analysis that the only source of energy loss is friction between the flow and the channel wall. In general, energy can also be lost because of such factors as wind shear on the surface and eddy motion arising from abrupt changes in the channel geometry, but these effects will not be discussed until Chap. 9. When wall friction is the only source of energy loss, the slope of the energy grade line is equal to the friction slope S_f, as shown in Fig. 2.4.1.

Momentum. There are three forces acting on the fluid control volume: friction, gravity, and pressure. Of these, the pressure forces at the two ends of the section are equal and cancel each other for uniform flow (because $y_1 = y_2$). So the friction and gravity forces must be balanced, because, with the flow steady and uniform, Eq. (2.4.3) applies ($\Sigma \mathbf{F} = 0$). The friction force $\mathbf{F}_f$ is equal to the product of the *wall shear stress* τ_0 and the area over which it acts, PL, where P is the *wetted perimeter* of the cross section; that is, $\mathbf{F}_f = -\tau_0 PL$, where the negative sign indicates that the friction force acts opposite to the direction of flow. The weight of fluid in the control volume is γAL, where γ is the *specific weight* of the fluid (weight per unit volume); the gravity force on the fluid, $\mathbf{F}_g$, is the component of the weight acting in the direction of flow, that is, $\mathbf{F}_g = \gamma AL \sin\theta$. Hence

$$\sum \mathbf{F} = 0 = -\tau_0 PL + \gamma AL \sin\theta \tag{2.4.7}$$

When θ is small, $\sin\theta \approx S_0$, so the approximation is made that

$$\begin{aligned} \tau_0 &= \frac{\gamma ALS_0}{PL} \\ &= \gamma RS_0 \end{aligned} \tag{2.4.8}$$

where $R = A/P$ is the *hydraulic radius*. For a steady uniform flow, $S_0 = S_f$, so

$$\tau_0 = \gamma RS_f \tag{2.4.9}$$

By a similar analysis, Henderson (1966) showed that (2.4.9) is also valid for nonuniform flow, although the bed slope S_0 and friction slope S_f are no longer equal. Equation (2.4.9) expresses a linkage between the momentum and energy principles in that the effects of friction are represented from the momentum viewpoint as the wall shear stress τ_0 and from the energy viewpoint as a rate of energy dissipation S_f.

2.5 OPEN CHANNEL FLOW

Open channel flow is channel flow with a free surface, such as flow in a river or in a partially full pipe. In this section the *Manning equation* to determine the velocity of open channel flow is derived, on the basis of the *Darcy-Weisbach* equation for head losses due to wall friction.

In fluid mechanics, the head loss h_f over a length L of pipe of diameter D for a flow with velocity V is given by the Darcy-Weisbach equation

$$h_f = f\frac{L}{D}\frac{V^2}{2g} \tag{2.5.1}$$

where f is the Darcy-Weisbach friction factor and g is the acceleration due to gravity (Roberson and Crowe, 1985). Using the definition of friction slope, $S_f = h_f/L$, (2.5.1) can be solved for V:

$$V = \sqrt{\frac{2g}{f}DS_f} \tag{2.5.2}$$

The hydraulic radius R of a circular pipe is $R = A/P = (\pi D^2/4)/\pi D = D/4$, so the pipe diameter D can be replaced in (2.5.2) by

$$D = 4R \tag{2.5.3}$$

to give the Darcy-Weisbach equation:

$$V = \sqrt{\frac{8g}{f}RS_f} \tag{2.5.4}$$

The *Chezy C* is defined as $C = \sqrt{8g/f}$; using this symbol, (2.5.4) is rewritten

$$V = C\sqrt{RS_f} \tag{2.5.5}$$

which is *Chezy's equation* for open channel flow. Manning's equation is produced from Chezy's equation by setting $C = R^{1/6}/n$, where n is the *Manning roughness coefficient*:

$$V = \frac{R^{2/3}S_f^{1/2}}{n} \tag{2.5.6}$$

Manning's equation (2.5.6) is valid for SI units, with R in meters and V in meters per second (S_f is dimensionless). Values of Manning's n for various surfaces are listed in Table 2.5.1. For V in feet per second and R in feet, Manning's equation is rewritten

$$V = \frac{1.49}{n}R^{2/3}S_f^{1/2} \tag{2.5.7}$$

[$1.49 = (3.281)^{1/3}$ and 3.281 ft = 1 m]. By comparing Eqs. (2.5.4) and (2.5.6), Manning's n can be expressed in terms of the Darcy-Weisbach friction factor f, as follows:

$$n = \sqrt{\frac{f}{8g}}R^{1/6} \tag{2.5.8}$$

with all values in SI units.

Manning's equation is valid for *fully turbulent flow*, in which the Darcy-Weisbach friction factor f is independent of the *Reynolds number Re*. Henderson (1966) gives the following criterion for fully turbulent flow:

TABLE 2.5.1
Manning roughness coefficients for various open channel surfaces

Material	Typical Manning roughness coefficient
Concrete	0.012
Gravel bottom with sides — concrete	0.020
— mortared stone	0.023
— riprap	0.033
Natural stream channels	
Clean, straight stream	0.030
Clean, winding stream	0.040
Winding with weeds and pools	0.050
With heavy brush and timber	0.100
Flood Plains	
Pasture	0.035
Field crops	0.040
Light brush and weeds	0.050
Dense brush	0.070
Dense trees	0.100

Source: Chow, 1959.

$$n^6 \sqrt{RS_f} \geq 1.9 \times 10^{-13} \text{ with R in feet} \qquad (2.5.9a)$$

or

$$n^6 \sqrt{RS_f} \geq 1.1 \times 10^{-13} \text{ with R in meters} \qquad (2.5.9b)$$

Example 2.5.1 There is uniform flow in a 200-ft wide rectangular channel with bed slope 0.03 percent and Manning's n is 0.015. If the depth is 5 ft, calculate the velocity and flow rate, and verify that the flow is fully turbulent so that Manning's equation applies.

Solution. The wetted perimeter in the channel is $P = 200 + 2 \times 5 = 210$ ft. The hydraulic radius is $R = A/P = 200 \times 5/210 = 4.76$ ft. The flow velocity is given by Manning's equation with $n = 0.015$ and $S_f = S_0$ (for uniform flow) $= 0.03\% = 0.0003$.

$$V = \frac{1.49}{n} R^{2/3} S_f^{1/2}$$

$$= \frac{1.49}{0.015} (4.76)^{2/3} (0.0003)^{1/2}$$

$$= 4.87 \text{ ft/s}$$

The flow rate is $Q = VA = 4.87 \times 200 \times 5 = 4870$ cfs. The criterion for fully turbulent flow is calculated from (2.5.9a):

$$n^6 \sqrt{RS_f} = (0.015)^6(4.76 \times 0.0003)^{1/2}$$
$$= 4.3 \times 10^{-13}$$

which is greater than 1.9×10^{-13} so the criterion is satisfied and Manning's equation is applicable.

In the event that the flow is not fully turbulent, the flow velocity may be computed with the Darcy-Weisbach equation (2.5.4), calculating the friction factor f as a function of the Reynolds number Re and the boundary roughness. Figure (2.5.1) shows a modified form of the *Moody diagram* for pipe flow; the pipe diameter D is replaced by $4R$. The Reynolds number is given by

$$Re = \frac{4VR}{\nu} \tag{2.5.10}$$

where ν is the *kinematic viscosity* of water, given in Table 2.5.2 as a function of temperature. The *relative roughness* ϵ is defined by

$$\epsilon = \frac{k_s}{4R} \tag{2.5.11}$$

where k_s is the size of sand grains resulting in a surface resistance equivalent to that observed in the channel.

Figure 2.5.1 for open channel flow was constructed from equations presented by Chow (1959) and Henderson (1966). For Reynolds number less than 2000, the flow is *laminar*, and

$$f = \frac{C_L}{Re} \tag{2.5.12}$$

where $C_L = 96$ for a smooth-surfaced channel of infinite width and larger if the surface is rough (Chow, 1959; Emmett, 1978). As the Reynolds number increases past 2000, the flow enters a region where both laminar and turbulent effects govern friction losses and the friction factor is given by a modified form of the Colebrook-White equation (Henderson, 1966):

$$\begin{aligned} \frac{1}{\sqrt{f}} &= -2\log_{10}\left[\frac{k_s}{12R} + \frac{2.5}{Re\sqrt{f}}\right] \\ &= -2\log_{10}\left[\frac{\epsilon}{3} + \frac{2.5}{Re\sqrt{f}}\right] \end{aligned} \tag{2.5.13}$$

For large Reynolds numbers, that is, in the upper right region of the Moody diagram, the flow is fully turbulent, and the friction factor is a function of the relative roughness alone. Eq. (2.5.13) reduces to

$$\frac{1}{\sqrt{f}} = -2\log_{10}\left(\frac{\epsilon}{3}\right) \tag{2.5.14}$$

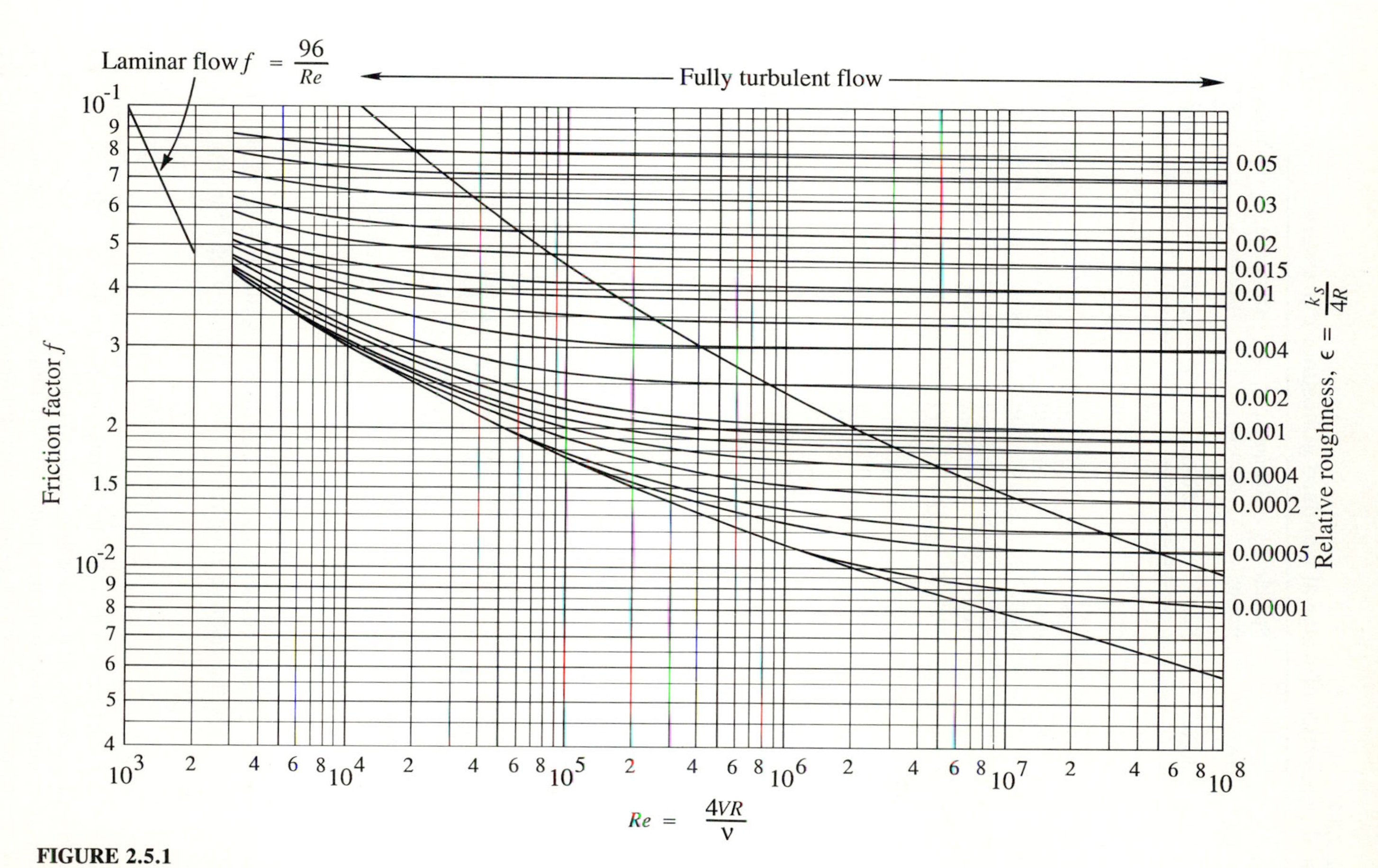

FIGURE 2.5.1
Moody diagram for open channel flow.

TABLE 2.5.2
Physical properties of water at standard atmospheric pressure

Temperature	Density	Specific weight	Dynamic viscosity	Kinematic viscosity	Vapor pressure
	kg/m^3	N/m^3	$N \cdot s/m^2$	m^2/s	N/m^2 abs.
0°C	1000	9810	1.79×10^{-3}	1.79×10^{-6}	611
5°C	1000	9810	1.51×10^{-3}	1.51×10^{-6}	872
10°C	1000	9810	1.31×10^{-3}	1.31×10^{-6}	1230
15°C	999	9800	1.14×10^{-3}	1.14×10^{-6}	1700
20°C	998	9790	1.00×10^{-3}	1.00×10^{-6}	2340
25°C	997	9781	8.91×10^{-4}	8.94×10^{-7}	3170
30°C	996	9771	7.96×10^{-4}	7.99×10^{-7}	4250
35°C	994	9751	7.20×10^{-4}	7.24×10^{-7}	5630
40°C	992	9732	6.53×10^{-4}	6.58×10^{-7}	7380
50°C	988	9693	5.47×10^{-4}	5.54×10^{-7}	12,300
60°C	983	9643	4.66×10^{-4}	4.74×10^{-7}	20,000
70°C	978	9594	4.04×10^{-4}	4.13×10^{-7}	31,200
80°C	972	9535	3.54×10^{-4}	3.64×10^{-7}	47,400
90°C	965	9467	3.15×10^{-4}	3.26×10^{-7}	70,100
100°C	958	9398	2.82×10^{-4}	2.94×10^{-7}	101,300
	$slugs/ft^3$	lb/ft^3	$lb\text{-}sec/ft^2$	ft^2/sec	psia
40° F	1.94	62.43	3.23×10^{-5}	1.66×10^{-5}	0.122
50° F	1.94	62.40	2.73×10^{-5}	1.41×10^{-5}	0.178
60° F	1.94	62.37	2.36×10^{-5}	1.22×10^{-5}	0.256
70° F	1.94	62.30	2.05×10^{-5}	1.06×10^{-5}	0.363
80° F	1.93	62.22	1.80×10^{-5}	0.930×10^{-5}	0.506
100° F	1.93	62.00	1.42×10^{-5}	0.739×10^{-5}	0.949
120° F	1.92	61.72	1.17×10^{-5}	0.609×10^{-5}	1.69
140° F	1.91	61.38	0.981×10^{-5}	0.514×10^{-5}	2.89
160° F	1.90	61.00	0.838×10^{-5}	0.442×10^{-5}	4.74
180° F	1.88	60.58	0.726×10^{-5}	0.385×10^{-5}	7.51
200° F	1.87	60.12	0.637×10^{-5}	0.341×10^{-5}	11.53
212° F	1.86	59.83	0.593×10^{-5}	0.319×10^{-5}	14.70

Source: Roberson, J. A., and C. T. Crowe, Engineering Fluid Mechanics, 2nd ed., Houghton Mifflin, Boston, 1980, Table A-5, p. 642. Used with permission.

For this case, the friction factor f can be eliminated between Eqs. (2.5.14) and (2.5.8) to solve for the relative roughness ϵ as a function of Manning's n and hydraulic radius R:

$$\epsilon = 3 \times 10^{-\phi R^{1/6}/(4n\sqrt{2g})} \tag{2.5.15}$$

where $\phi = 1$ for SI units and 1.49 for English units. To use the Moody diagram given R and V, ϵ is calculated using (2.5.15) with the given value of n, then the Reynolds number is computed using (2.5.10) and the corresponding value of f read from Fig. 2.5.1. An estimate of V is obtained from Eq. (2.5.4), and the process is repeated iteratively until the values for V converge.

The Moody diagram given here for open channel flow has some limitations. First, it accounts for resistance due to friction elements randomly distributed on the channel wall, but it does not account for form drag associated with nonuniformities in the channel. Emmett (1978) found that the friction factor for thin sheet flows on soil or grass surfaces could be as much as a factor of 10 greater than the value for friction drag alone. Also, the Moody diagram is valid only for fixed bed channels, not for erodable ones. The shape of the cross section (rectangular, triangular, circular, etc.) has some influence on the friction factor but the effect is not large. Because of these limitations, the Moody diagram shown should be applied only to lined channels with uniform cross section.

2.6 POROUS MEDIUM FLOW

A porous medium is an interconnected structure of tiny conduits of various shapes and sizes. For steady uniform flow in a circular pipe of diameter D, (2.4.9) remains valid:

$$\tau_0 = \gamma R S_f \tag{2.6.1}$$

with the hydraulic radius $R = D/4$. For laminar flow in a circular conduit, the wall shear stress is given by

$$\tau_0 = \frac{8\mu V}{D} \tag{2.6.2}$$

where μ is the dynamic viscosity of the fluid. Combining (2.6.1) and (2.6.2) gives

$$V = \left(\frac{\gamma D^2}{32\mu}\right) S_f \tag{2.6.3}$$

which is the *Hagen-Poiseulle equation* for laminar flow in a circular conduit.

For flow in a porous medium, part of the cross-sectional area A is occupied by soil or rock strata, so the ratio Q/A does not equal the actual fluid velocity, but defines a volumetric flux q called the *Darcy flux*. *Darcy's law* for flow in a porous medium is written from (2.6.3) as

$$\frac{Q}{A} = q = K S_f \tag{2.6.4}$$

where K is the *hydraulic conductivity* of the medium, $K = \gamma D^2/32\mu$. Values of the hydraulic conductivity for various porous media are shown in Table 2.6.1 along with values of the *porosity* η, the ratio of the volume of voids to the total volume of the medium. The actual average fluid velocity in the medium is

$$V_a = \frac{q}{\eta} \tag{2.6.5}$$

Darcy's law is valid so long as flow is laminar. Flow in a circular conduit is laminar when its Reynolds number

TABLE 2.6.1
Hydraulic conductivity and porosity of unconsolidated porous media

Material	Hydraulic conductivity K (cm/s)	Porosity η (%)
Gravel	10^{-1}–10^{2}	25–40
Sand	10^{-5}–1	25–50
Silt	10^{-7}–10^{-3}	35–50
Clay	10^{-9}–10^{-5}	40–70

Source: Freeze and Cherry, 1979.

$$Re = \frac{VD}{\nu} \tag{2.6.6}$$

is less than 2000, a condition satisfied by almost all naturally occurring flows in porous media.

Example 2.6.1 Water is percolating through a fine sand aquifer with hydraulic conductivity 10^{-2} cm/s and porosity 0.4 toward a stream 100 m away. If the slope of the water table is 1 percent, calculate the travel time of water to the stream.

Solution. The Darcy flux q is calculated by (2.6.4) with $K = 0.01$ cm/sec $=$ 8.64 m/day and $S_f = 1\% = 0.01$; hence $q = KS_f = 8.64 \times 0.01 = 0.086$ m/day. The water velocity V_a is given by (2.6.5): $V_a = q/\eta = 0.086/0.4 = 0.216$ m/day. The travel time to the stream 100 m away is $100/V_a = 100/0.216 = 463$ days $=$ 1.3 years.

2.7 ENERGY BALANCE

The energy balance of a hydrologic system is an accounting of all inputs and outputs of energy to and from a system, taking the difference between the rates of input and output as the rate of change of storage, as was done for the continuity or mass balance equation in Sec. 2.2. In the basic Reynolds transport theorem, Eq. (2.1.9), the extensive property is now taken as $B = E$, the amount of energy in the fluid system, which is the sum of *internal energy* E_u, *kinetic energy* $\frac{1}{2}mV^2$, and *potential energy* mgz (z represents elevation):

$$B = E = E_u + \frac{1}{2}mV^2 + mgz \tag{2.7.1}$$

Hence,

$$\beta = \frac{dB}{dm} = e_u + \frac{1}{2}V^2 + gz \tag{2.7.2}$$

where e_u is the internal energy per unit mass. By the *first law of thermodynamics*,

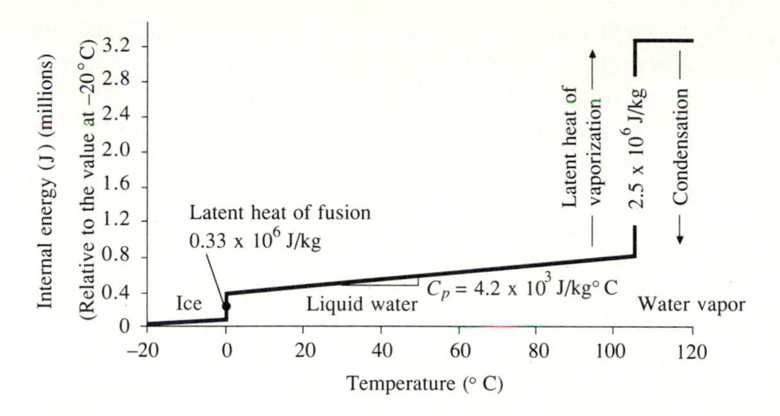

FIGURE 2.7.1
Specific and latent heats for water. Latent heat is absorbed or given up when water changes its state of being solid, liquid, or gas.

the net rate of energy transfer into the fluid, dE/dt, is equal to the rate at which heat is transferred into the fluid, dH/dt, less the rate at which the fluid does work on its surroundings, dW/dt:

$$\frac{dB}{dt} = \frac{dE}{dt} = \frac{dH}{dt} - \frac{dW}{dt} \tag{2.7.3}$$

Substituting for dB/dt and β in the Reynolds transport theorem

$$\frac{dH}{dt} - \frac{dW}{dt} = \frac{d}{dt}\iiint_{\text{c.v.}} (e_u + \frac{1}{2}V^2 + gz)\rho d\forall + \iint_{\text{c.s.}} (e_u + \frac{1}{2}V^2 + gz)\rho \mathbf{V}\cdot\mathbf{dA} \tag{2.7.4}$$

This is the energy balance equation for an unsteady variable-density flow.

Internal Energy

Sensible heat. *Sensible heat* is that part of the internal energy of a substance that is proportional to the substance's temperature. Temperature changes produce proportional changes in internal energy, the coefficient of proportionality being the *specific heat* C_p

$$de_u = C_p dT \tag{2.7.5}$$

The subscript p denotes that the specific heat is measured at constant pressure.

Latent heat. When a substance changes phase (solid, liquid, or gaseous state) it gives up or absorbs *latent heat*. The three latent heats of interest are those for *fusion*, or melting, of ice to water; for *vaporization* of liquid water to water vapor; and for *sublimation*, or direct conversion, of ice to water vapor. Figure 2.7.1 shows how the internal energy of water varies as the result of sensible and latent heat transfer. Latent heat transfers at phase changes are indicated by the vertical jumps in internal energy at melting and vaporization. Internal energy changes due to sensible heat transfer are shown by the sloping lines.

Phase changes can occur at temperatures other than the normal ones of 0°C for melting and 100°C for boiling. Evaporation, for example, can occur at any temperature below the boiling point. At any given temperature, the latent heat of sublimation (solid to gas) equals the sum of the latent heats of fusion (solid to liquid) and vaporization (liquid to gas).

Latent heat transfers are the dominant cause of internal energy changes for water in most hydrologic applications; the amount of latent heat involved is much larger than the sensible heat transfer for a change in temperature of a few degrees, which is the usual case in hydrologic processes. The latent heat of vaporization l_v varies slightly with temperature according to

$$l_v = 2.501 \times 10^6 - 2370T \text{ (J/kg)} \tag{2.7.6}$$

where T is temperature in °C and l_v is given in joules (J) per kilogram (Raudkivi, 1979). A joule is an SI unit representing the amount of energy required to exert a force of 1 newton through a distance of 1 meter.

2.8 TRANSPORT PROCESSES

Heat energy transport takes place in three ways: *conduction, convection*, and *radiation*. Conduction results from random molecular motion in substances; heat is transferred as molecules in higher temperature zones collide with and transfer energy to molecules in lower temperature zones, as in the gradual warming along an iron bar when one end is placed in a fire. Convection is the transport of heat energy associated with mass motion of a fluid, such as eddy motion in a fluid stream. Convection transports heat on a much larger scale than conduction in fluids, but its extent depends on fluid turbulence so it cannot be characterized as precisely. Radiation is the direct transfer of energy by means of electromagnetic waves, and can take place in a vacuum.

The conduction and convection processes that transfer heat energy also transport mass and momentum (Bird, Stewart, and Lightfoot, 1960; Fahien, 1983). For each of the extensive properties mass, momentum, and energy, the rate of flow of extensive property per unit area of surface through which it passes is called the *flux*. For example, in Darcy's law, volumetric flow rate is Q across area A, so the volumetric flux is $q = Q/A$; the corresponding mass flow rate is $\dot{m} = \rho Q$, so the mass flux is $\rho Q/A$. By analogy the momentum flow rate is $\dot{m}V = \rho QV$ and the momentum flux is $\dot{m}V/A = \rho QV/A = \rho V^2$. The corresponding energy flow rate is dE/dt and the energy flux is $(dE/dt)/A$, measured in watts per meter squared in the SI system; a watt (W) is one joule per second. In general, a flux is given by

$$\text{Flux} = \frac{\text{flow rate}}{\text{area}} \tag{2.8.1}$$

Conduction

In conduction the flux is directly proportional to the gradient of a *potential* (Fahien, 1983). For example, the lateral transfer of momentum in a laminar flow is described by *Newton's law of viscosity,* in which the potential is the flow velocity:

$$\tau = \mu \frac{du}{dz} \tag{2.8.2}$$

Here τ is momentum flux, μ is a proportionality coefficient called the dynamic viscosity (measured in lb·s/ft^2 or N·s/m^2), and du/dz is the gradient of the velocity u as a function of distance z from the boundary. The symbol τ is usually used to represent a shear stress, but it can be shown that the dimensions of shear stress and momentum flux are the same, and τ can be thought of as the lateral momentum flux in a fluid flow occurring through the action of shear stress between elements of fluid having different velocities, as shown in Fig. 2.8.1.

Analogous to Newton's law of viscosity for momentum, the laws of conduction for mass and energy are *Fick's law of diffusion*, and *Fourier's law of heat conduction*, respectively (Carslaw and Jaeger, 1959). Their governing equations have the same form as (2.8.2), as shown in Table 2.8.1. The measure of potential for mass conduction is the mass concentration C of the substance being transported. In Chap. 4, for example, when the transport of water vapor in air is described, C is the mass of water vapor per unit mass of moist air. The proportionality constant for mass conduction is the *diffusion coefficient D*. The measure of potential for heat energy transport is the temperature T and the proportionality constant is the heat conductivity k of the substance.

The proportionality constant can also be written in a *kinematic* form. For example, the dynamic viscosity μ and the kinematic viscosity ν are related by

$$\mu = \rho\nu \tag{2.8.3}$$

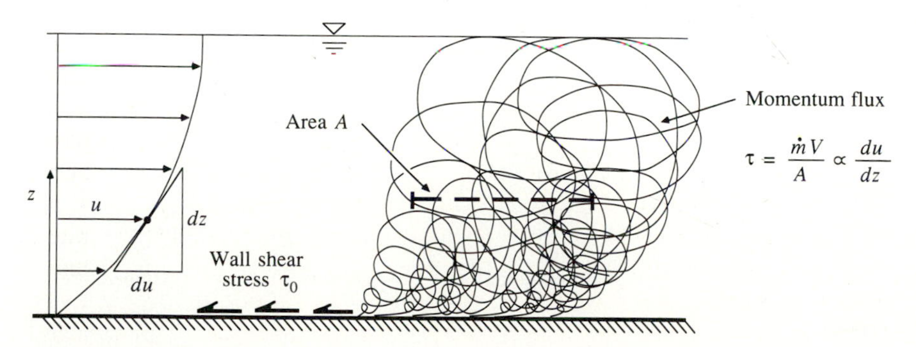

FIGURE 2.8.1
The relationship between the momentum flux and the velocity gradient in a free surface flow. Momentum is transferred between the wall and the interior of the flow through molecular and turbulent eddy motion. The shear stress in the interior of the flow is the same as the momentum flux through a unit area (dashed line) parallel to the boundary.

so Eq. (2.8.2) can be rewritten

$$\tau = \rho \nu \frac{du}{dz} \tag{2.8.4}$$

The dimensions of ν are $[L^2/T]$.

Convection

For convection, transport occurs through the action of turbulent eddies, or the mass movement of elements of fluid with different velocities, rather than through the movement of individual molecules as in conduction. Convection requires a flowing fluid, while conduction does not. The momentum flux in a turbulent flow is not governed by Newton's law of viscosity but is related to the instantaneous departures of the turbulent velocity from its time-averaged value. It is convenient, however, to write equations describing convection in the same form as those for conduction. For momentum transfer, the flux in a turbulent flow is written as

$$\tau_{turb} = \rho K_m \frac{du}{dz} \tag{2.8.5}$$

TABLE 2.8.1
Laws of conduction and corresponding equations for convection of mass, momentum, and heat energy in a fluid

	Extensive property transported		
	Mass	**Momentum**	**Heat energy**
Conduction:			
Name of law	Fick's	Newton's	Fourier's
Equation	$f_m = -D\frac{dC}{dz}$	$\tau = \mu\frac{du}{dz}$	$f_h = -k\frac{dT}{dz}$
Flux	f_m	τ	f_h
Constant of proportionality	D (diffusion coeff.)	μ (viscosity)	k (heat conductivity)
Potential gradient	$\frac{dC}{dz}$ (concentration)	$\frac{du}{dz}$ (velocity)	$\frac{dT}{dz}$ (temperature)
Convection:			
Equation	$f_m = -\rho K_w\frac{dC}{dz}$	$\tau = \rho K_m\frac{du}{dz}$	$f_h = -\rho C_p K_h\frac{dT}{dz}$
Diffusivity $[L^2/T]$	K_w	K_m	K_h

where K_m is the *momentum diffusivity*, or *eddy viscosity*, with dimensions [L^2/T]. K_m is four to six orders of magnitude greater than ν (Priestley, 1959), and turbulent momentum flux is the dominant form of momentum transfer in surface water flow and in air flow over the land surface. Equations analogous to (2.8.5) can be written for mass and energy transport as shown in Table 2.8.1.

It should be noted that the direction of transport of extensive properties described by the equations in Table 2.8.1 is transverse to the direction of flow. For example, in Fig. 2.8.1, the flow is horizontal while the transport process is vertical through the dashed area shown. Extensive property transport in the direction of motion is called *advection* and is described by the term $\int\int \beta\rho \mathbf{V}\cdot\mathbf{dA}$ in the Reynolds transport theorem, Eq. (2.1.9).

Velocity Profile

Determination of the rates of conduction and convection of momentum requires knowledge of the velocity profile in the boundary layer. For flow of air over land or water, the *logarithmic velocity profile* is applicable (Priestley, 1959). The wind velocity u is given as a function of the elevation z by

$$\frac{u}{u^*} = \frac{1}{k}\ln\left(\frac{z}{z_0}\right) \tag{2.8.6}$$

where the shear velocity $u^* = \sqrt{\tau_0/\rho}$ (τ_0 is the boundary shear stress and ρ is the fluid density), k is von Karman's constant (≈ 0.4), and z_0 is the *roughness height* of the surface. Table 2.8.2 gives values of the roughness height for some surfaces. By differentiating (2.8.6), the velocity gradient is found to be

$$\frac{du}{dz} = \frac{u^*}{kz} \tag{2.8.7}$$

This equation can be used to determine the laminar and turbulent momentum fluxes at various elevations.

TABLE 2.8.2
Approximate values of the roughness height of natural surfaces

Surface	Roughness height z_0 (cm)
Ice, mud flats	0.001
Water	0.01 – 0.06
Grass (up to 10 cm high)	0.1 – 2.0
Grass (10 – 50 cm high)	2 – 5
Vegetation (1 – 2 m high)	20
Trees (10 – 15 m high)	40 – 70

Source: Brutsaert, W., Evaporation into the atmosphere, D. Reidel, Dordrecht, Holland, 1982, Table 5.1, p. 114 (adapted).

Example 2.8.1 The wind speed has been measured at 3 m/s at a height of 2 m above a short grass field ($z_0 = 1$ cm). Plot the velocity profile and calculate the rates of laminar and turbulent momentum flux at 20 cm, and the turbulent momentum flux at 2 m elevation. For air, $\rho = 1.20$ kg/m^3, $\nu = 1.51 \times 10^{-5}$ m^2/s, and $K_m = 1.5$ m^2/s.

Solution. The shear velocity is calculated from Eq. (2.8.6) using the known velocity u = 3 m/s at z = 2 m:

$$\frac{u}{u^*} = \frac{1}{k} \ln\left(\frac{z}{z_0}\right)$$

$$\frac{3}{u^*} = \frac{1}{0.4} \ln\left(\frac{2}{0.01}\right)$$

Solving, $u^* = 0.226$ m/s.

The velocity profile is found by substituting values for z in (2.8.6); for example, for z = 20 cm = 0.2 m, then

$$\frac{u}{0.226} = \frac{1}{0.4} \ln\left(\frac{0.2}{0.01}\right)$$

Solving, $u = 1.7$ m/s at z = 0.2 m. Similarly computed values for other values of z are plotted in Fig. 2.8.2. The velocity gradient at z = 0.2 m is given by Eq. (2.8.7):

$$\frac{du}{dz} = \frac{u^*}{kz} = \frac{0.226}{0.4 \times 0.2} = 2.83 \text{ s}^{-1}$$

and the laminar momentum flux τ is given by Newton's law of viscosity (2.8.4) with air density ρ = 1.20 kg/m^3 and the kinematic viscosity $\nu = 1.51 \times 10^{-5}$ m^2/s.

$$\begin{aligned} \tau &= \rho\nu\frac{du}{dz} \\ &= 1.20 \times 1.51 \times 10^{-5} \times 2.83 \\ &= 5.1 \times 10^{-5} \text{ N/m}^2 \end{aligned}$$

at z = 0.2 m. The turbulent momentum flux is given by Eq. (2.8.5):

$$\begin{aligned} \tau_{turb} &= \rho K_m \frac{du}{dz} \\ &= 1.20 \times 1.5 \times 2.83 \\ &= 5.1 \text{ N/m}^2 \end{aligned}$$

at z = 0.2 m. The ratio $\tau_{turb}/\tau = K_m/\nu = 5.1/(5.1 \times 10^{-5}) = 10^5$; hence, turbulent momentum flux (convection by eddy diffusion) is the dominant transport mechanism in this air stream. At z = 2 m, $du/dz = u^*/kz = 0.226/(0.4 \times 2) = 0.28 \text{ s}^{-1}$ and $\tau_{turb} = \rho K_m(du/dz) = 1.20 \times 1.50 \times 0.28 = 0.51$ N/m^2. Note that the ratio of the convective momentum fluxes at 0.2 and 2.0 m is 5.1/0.51 = 10; the momentum flux (or shear stress) is inversely proportional to elevation in a logarithmic velocity profile. The momentum flux, therefore, is largest near the ground surface and diminishes as elevation increases.

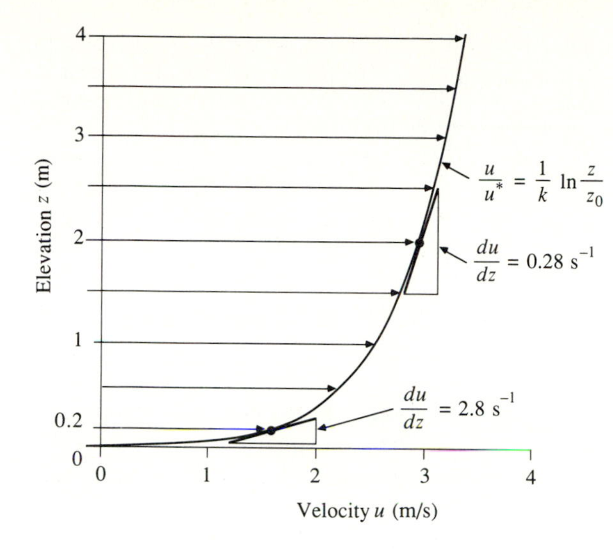

FIGURE 2.8.2
Logarithmic velocity profile for roughness height $z_0 = 1$ cm and measured velocity 3 m/s at height 2 m in a flow of air (Example 2.8.1). The resulting velocity gradient and shear stress are inversely proportional to elevation.

Radiation

When radiation strikes a surface (see Fig. 2.8.3), it is either reflected or absorbed. The fraction reflected is called the *albedo* α $(0 \leq \alpha \leq 1)$. For example, deep water bodies absorb most of the radiation they receive, having $\alpha \approx 0.06$, while fresh snow reflects most of the incoming radiation, with α as high as 0.9 (Brutsaert, 1982). Radiation is also continuously emitted from all bodies at rates depending on their surface temperatures. The *net radiation* R_n is the net input of radiation at the surface at any instant; that is, the difference between the radiation absorbed, $R_i(1-\alpha)$ (where R_i is the incident radiation), and that emitted, R_e:

$$R_n = R_i(1-\alpha) - R_e \tag{2.8.8}$$

Net radiation at the earth's surface is the major energy input for evaporation of water.

Emission. Radiation emission is governed by the Stefan-Boltzmann law

$$R_e = e\sigma T^4 \tag{2.8.9}$$

where e is the *emissivity* of the surface, σ is the Stefan-Boltzmann constant (5.67×10^{-8} W/m$^2 \cdot$K^4) and T is the absolute temperature of the surface in degrees Kelvin (Giancoli, 1984); the Kelvin temperature equals the Celsius temperature plus 273. For a perfect radiator, or *black body*, the emissivity is $e = 1$; for water surfaces $e \approx 0.97$. The *wavelength* λ of emitted radiation is inversely proportional to the surface temperature, as given by *Wien's law*:

$$\lambda = \frac{2.90 \times 10^{-3}}{T} \tag{2.8.10}$$

where T is in degrees Kelvin and λ is in meters (Giancoli, 1984). As a

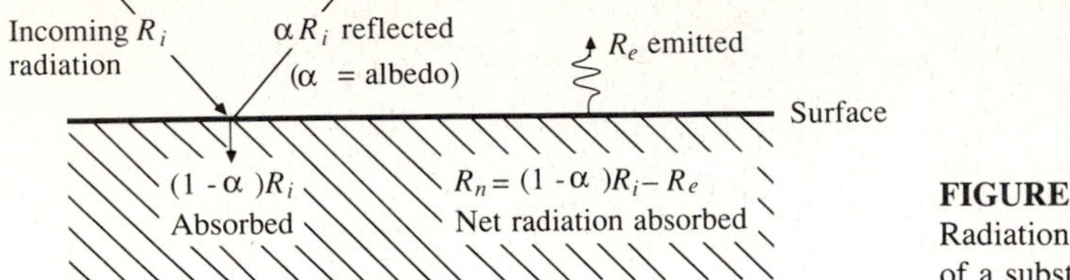

FIGURE 2.8.3
Radiation balance at the surface of a substance.

consequence of Wien's law, the radiation emitted by the sun has a much shorter wavelength than that emitted by the cooler earth.

Reflection and scattering. The albedo α in Eq. (2.8.8) measures the proportion of incoming radiation that is reflected back into the atmosphere. The albedo varies somewhat depending on the wavelength of the radiation and its angle of incidence, but it is customary to adopt a single value typical of the type of surface.

When radiation strikes tiny particles in the atmosphere of a size on the same order of magnitude as the radiation wavelength, the radiation is scattered randomly in all directions. Small groups of molecules called *aerosols* scatter light in this way. The addition of aerosols and dust particles to the atmosphere from human activity in modern times has given rise to concern about the *greenhouse effect*, in which some of the radiation emitted by the earth is scattered back by the atmosphere; increased scattering results in a general warming of the earth's surface. However, the precise magnitude of the earth's warming by this mechanism is not yet known.

Net radiation at Earth's surface. The intensity of solar radiation arriving at the top of the atmosphere is decreased by three effects before reaching a unit area of the earth's surface: scattering in the atmosphere, absorption by clouds, and the obliqueness of the earth's surface to the incoming radiation (a function of latitude, season, and time of day). The intensity of solar radiation received per unit area of the earth's surface is denoted by R_s. The atmosphere also acts as a radiator, especially on cloudy days, emitting longer wave radiation than the sun because its temperature is lower; the intensity of this radiation is denoted R_l. The incoming radiation at the earth's surface is thus $R_i = R_s + R_l$. The earth emits radiation R_e (of a wavelength close to that of the atmospheric radiation), and the net radiation received at the earth's surface is

$$R_n = (R_s + R_l)(1 - \alpha) - R_e \qquad (2.8.11)$$

The interaction of radiation processes between the atmosphere and the earth's surface is complex. Figure 2.8.4 presents a summary of relative values for the various components of the annual average atmospheric and surface heat balance. It can be seen that for 100 units of incoming solar radiation at the top of the atmosphere, about half (51 units) reaches the earth's surface and is absorbed there; of these 51 units, 21 are emitted as longwave radiation, leaving a net

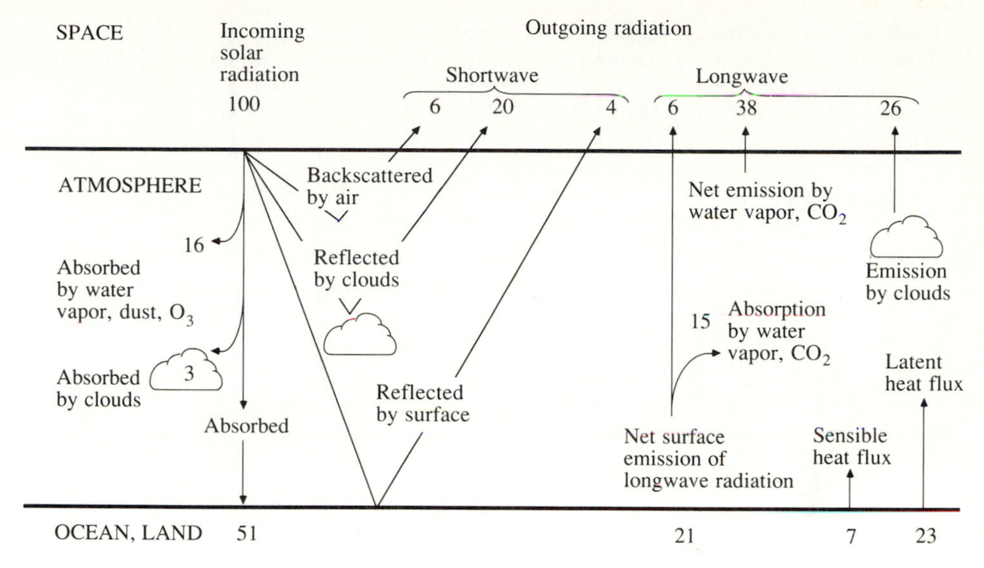

FIGURE 2.8.4
Radiation and heat balance in the atmosphere and at the earth's surface. (*Source:* "Understanding Climatic Change," p. 18, National Academy of Sciences, Washington, D.C., 1975. Used with permission.)

radiation of 30 units at the earth's surface; 23 units of this energy input are used to evaporate water, and thus returned to the atmosphere as latent heat flux; the remaining 7 units go to heat the air overlying the earth's surface, as sensible heat flux.

REFERENCES

Bird, R. B., W. E. Stewart, and E. N. Lightfoot, *Transport Phenomena*, Wiley, New York, 1960.

Brutsaert, W., *Evaporation into the Atmosphere*, D. Reidel, Dordrecht, Holland, 1982.

Carslaw, H. R., and J. C. Jaeger, *Conduction of Heat in Solids*, 2nd ed., Oxford University Press, Oxford, 1959.

Chow, V. T., *Open-channel Hydraulics*, McGraw-Hill, New York, 1959.

Emmett, W. W., Overland flow, in *Hillslope Hydrology*, ed. by M. J. Kirkby, Wiley, Chichester, England, pp. 145–176, 1978.

Fahien, R. W., *Fundamentals of Transport Phenomena*, McGraw-Hill, New York, 1983.

Fox, R. W., and A. T. MacDonald, *Introduction to Fluid Mechanics*, 3rd ed., Wiley, New York, 1985.

Freeze, R. A., and J. A. Cherry, *Groundwater*, Prentice-Hall, Englewood Cliffs, N.J., 1979.

French, R. H., *Open-channel Hydraulics*, McGraw-Hill, New York, 1985.

Giancoli, D. C., *General Physics*, Prentice-Hall, Englewood Cliffs, N.J., 1984.

Henderson, F. M., *Open Channel Flow*, Macmillan, New York, 1966.

Oke, T. R., *Boundary Layer Climates*, Methuen, London, 1978.

Priestley, C. H. B., *Turbulent Transfer in the Lower Atmosphere*, University of Chicago Press, 1959.

Raudkivi, A. J., *Hydrology*, Pergamon Press, Oxford, 1979.
Roberson, J. A., and C. T. Crowe, *Engineering Fluid Mechanics*, 3rd ed., Houghton-Mifflin, Boston, 1985.
Shames, I. V., *Mechanics of Fluids*, 2nd ed., McGraw-Hill, New York, 1982.
White, F. M., *Fluid Mechanics*, McGraw-Hill, New York, 1979.

PROBLEMS

2.2.1 A reservoir has the following inflows and outflows (in cubic meters) for the first three months of the year. If the storage at the beginning of January is 60 m^3, determine the storage at the end of March.

Month	Jan	Feb	Mar
Inflow	4	6	9
Outflow	8	11	5

2.2.2 Compute the constant draft from a 500–hectare reservoir for a 30–day period during which the reservoir level dropped half a meter despite an average upstream inflow of 200,000 m^3/day. During the period, the total seepage loss was 2 cm, the total precipitation was 10.5 cm, and the total evaporation was 8.5 cm. (1 hectare = 10^4 m^2).

2.2.3 Solve Prob. 2.2.2 if the reservoir area is 1200 acres, the drop in level 2 ft, the inflow 25 ft^3/s, the seepage loss 1 in, the precipitation 4 in and the evaporation 3 in (1 acre = 43,560 ft^2).

2.2.4 From the hydrologic records of over 50 years on a drainage basin of area 500 km^2, the average annual rainfall is estimated as 90 cm and the average annual runoff as 33 cm. A reservoir in the basin, having an average surface area of 1700 hectares, is planned at the basin outlet to collect available runoff for supplying water to a nearby community. The annual evaporation over the reservoir surface is estimated as 130 cm. There is no groundwater leakage or inflow to the basin. Determine the available average annual withdrawal from the reservoir for water supply.

2.2.5 Solve Prob. 2.2.4 if the drainage basin area is 200 mi^2, annual rainfall 35 in^2, runoff 13 in, reservoir area 4200 acres, and evaporation 50 in.

2.2.6 The consecutive monthly flows into and out of a reservoir in a given year are the following, in relative units:

Month	J	F	M	A	M	J	J	A	S	O	N	D
Inflow	3	5	4	3	4	10	30	15	6	4	2	1
Outflow	6	8	7	10	6	8	20	13	4	5	7	8

The reservoir contains 60 units at the beginning of the year. How many units of water are in the reservoir at the middle of August? At the end of the year?

2.3.1 Specify whether the following variables are usually recorded as sample data or pulse data: (*a*) daily maximum air temperature, (*b*) daily precipitation, (*c*) daily wind speed, (*d*) annual precipitation, (*e*) annual maximum discharge.

2.3.2 The precipitation and streamflow for the storm of May 12, 1980, on Shoal Creek at Northwest Park in Austin, Texas, are shown below. Calculate the time distribution of storage on the watershed assuming that the initial storage is 0. Compute the total depth of precipitation and the equivalent depth of streamflow which occurred during the 8-hour period. How much storage remained in the watershed at the end of the period? What percent of the precipitation appeared as streamflow during this period? What was the maximum storage? Plot the time distribution of incremental precipitation, streamflow, change in storage, and cumulative storage. The watershed area is 7.03 mi^2.

Time (h)	0	0.5	1.0	1.5	2.0	2.5	3.0	3.5	
Incremental Precipitation (in)		0.18	0.42	0.21	0.16				
Instantaneous Streamflow (cfs)	25	27	38	109	310	655	949	1060	
Time	4.0	4.5	5.0	5.5	6.0	6.5	7.0	7.5	8.0
Instantaneous Streamflow (cfs)	968	1030	826	655	466	321	227	175	160

2.5.1 Calculate the velocity and flow rate of a uniform flow 3 ft deep in a 100-ft-wide stream with approximately rectangular cross section, bed slope 1 percent, and Manning's n of 0.035. Check that the criterion for fully turbulent flow is satisfied.

2.5.2 Solve Prob. 2.5.1 for a channel 30 m wide with flow 1 m deep.

2.5.3 Solve Prob. 2.5.1 for a stream channel with approximately trapezoidal cross section with 100-ft bottom width and sides of slope 3 horizontal : 1 vertical.

2.5.4 Solve Prob. 2.5.3 if the bottom width is 30 m and the depth 1 m.

2.5.5 Water is flowing over an asphalt parking lot with slope 0.5 percent and Manning's n is 0.015. Calculate the velocity and flow rate if the flow is 1 in deep. Check that the criterion for fully turbulent flow is satisfied.

2.5.6 Solve Prob. 2.5.5 if the flow depth is 1 cm. Assume kinematic viscosity is 1×10^{-6} m^2/s.

2.5.7 Solve Prob. 2.5.5 for a flow depth of 0.5 in. Show that the criterion for fully turbulent flow is not satisfied, and compute the velocity and flow rate using the Darcy-Weisbach equation and the Moody diagram. By what percent is the velocity obtained from this procedure different from what would have been obtained if Manning's equation had been used? Assume kinematic viscosity is 1×10^{-5} ft^2/s.

2.5.8 Solve Prob. 2.5.7 if the flow depth is 1 mm.

2.5.9 For a steady uniform flow in a circular conduit of diameter D, show that the wall shear stress τ_0 is given by

$$\tau_0 = \frac{\gamma D h_f}{4L}$$

where γ is the specific weight of the fluid and h_f is the head loss over a length L of the conduit.

2.5.10 For laminar flow in a circular conduit use Newton's law of viscosity, $\tau_0 = \mu\, du/dy$, where u is the fluid velocity at distance y from the wall, to establish that the velocity distribution in the conduit is given by $u = u_{max}(1 - r^2/R^2)$, where r is the distance from the center of a pipe of radius R and u_{max} is the velocity at the pipe center.

2.5.11 Use the parabolic velocity distribution formula for laminar flow in a circular conduit, given in Prob. 2.5.10, to establish that the wall shear stress is $\tau_0 = 8\mu V/D$, in which V is the average pipe velocity.

2.5.12 A rectangular open channel 12 m wide and 1 m deep has a slope of 0.001 and is lined with cemented rubble (n =0.025). Determine (*a*) its maximum discharge capacity, and (*b*) the maximum discharge obtainable by changing the cross-sectional dimensions without changing the rectangular form of the section, the slope, and the volume of excavation. Hint: the best hydraulic rectangular section has a minimum wetted perimeter or a width-depth ratio of 2.

2.5.13 Solve Prob. 2.5.12 if the channel is 30 ft wide and 4 ft deep.

2.6.1 Water is flowing with a friction slope S_f = 0.01. Determine (*a*) the velocity of flow in a thin capillary tube of diameter 1 mm ($\nu = 1.00 \times 10^{-6}$ m^2/s), (*b*) the Darcy flux Q/A and the actual velocity of flow through a fine sand and (*c*) gravel.

2.6.2 Compute the rate of flow of water at 20°C through a 10-m-long conduit filled with fine sand of effective diameter 0.01 mm under a pressure head difference of 0.5 m between the ends of the conduit. The cross-sectional area of the conduit is 2.0 m^2.

2.6.3 Solve Example 2.6.1 in the text if the water is flowing through: (*a*) gravel with a hydraulic conductivity of 10 cm/s and a porosity of 30 percent, (*b*) silt with a hydraulic conductivity of 10^{-4} cm/s and a porosity of 45 percent, and (*c*) clay with a hydraulic conductivity of 10^{-7} cm/s and a porosity of 50 percent. Compare your answers with that obtained in the example.

2.8.1 Air is flowing over a short grass surface, and the velocity measured at 2 m elevation is 1 m/s. Calculate the shear velocity and plot the velocity profile from the surface to height 4 m. Assume z_0 = 1 cm. Calculate the turbulent momentum flux at heights 20 cm and 2 m and compare the values. Assume K_m = 0.07 m^2/s and ρ = 1.20 kg/m^3 for air.

2.8.2 Solve Prob. 2.8.1 if the fluid is water. Assume K_m =0.15 m^2/s and ρ = 1000 kg/m^3. Calculate and compare the laminar and turbulent momentum fluxes at 20 cm elevation if $\nu = 1.51 \times 10^{-6}$ m^2/s for water.

2.8.3 Assuming the sun to be a black body radiator with a surface temperature of 6000 K, calculate the intensity and wavelength of its emitted radiation.

2.8.4 Solve Prob. 2.8.3 for the earth and compare the intensity and wavelength of the earth's radiation with that emitted by the sun. Assume the earth has a surface temperature of 300 K.

2.8.5 The incoming radiation intensity on a lake is 200 W/m^2. Calculate the net radiation into the lake if the albedo is α = 0.06, the surface temperature is 30°C, and the emissivity is 0.97.

2.8.6 Solve Prob. 2.8.5 for fresh snow if the albedo is α =0.8, the emissivity is 0.97, and the surface temperature is 0°C.

2.8.7 Solve Prob. 2.8.5 for a grassy field with albedo α = 0.2, emissivity 0.97, and surface temperature 30°C.

CHAPTER 3

ATMOSPHERIC WATER

Of the many meteorological processes occurring continuously within the atmosphere, the processes of precipitation and evaporation, in which the atmosphere interacts with surface water, are the most important for hydrology. Much of the water precipitated on the land surface is derived from moisture evaporated from the oceans and transported long distances by atmospheric circulation. The two basic driving forces of atmospheric circulation result from the rotation of the earth and the transfer of heat energy between the equator and the poles.

3.1 ATMOSPHERIC CIRCULATION

The earth constantly receives heat from the sun through solar radiation and emits heat through re-radiation, or *back radiation* into space. These processes are in balance at an average rate of approximately 210 W/m^2. The heating of the earth is uneven; near the equator, the incoming radiation is almost perpendicular to the land surface and averages about 270 W/m^2, while near the poles, it strikes the earth at a more oblique angle at a rate of about 90 W/m^2. Because the rate of radiation is proportional to the absolute temperature at the earth's surface, which does not vary greatly between the equator and the poles, the earth's emitted radiation is more uniform than the incoming radiation. In response to this imbalance, the atmosphere functions as a vast heat engine, transferring energy from the equator toward the poles at an average rate of about 4×10^9 MW.

If the earth were a nonrotating sphere, atmospheric circulation would appear as in Fig. 3.1.1. Air would rise near the equator and travel in the upper atmosphere toward the poles, then cool, descend into the lower atmosphere, and return toward the equator. This is called *Hadley circulation*.

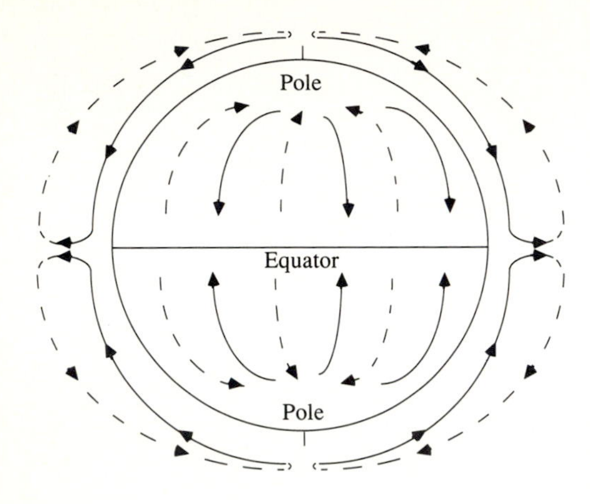

FIGURE 3.1.1
One-cell atmospheric circulation pattern for a nonrotating planet.

The rotation of the earth from west to east changes the circulation pattern. As a ring of air about the earth's axis moves toward the poles, its radius decreases. In order to maintain angular momentum, the velocity of air increases with respect to the land surface, thus producing a westerly air flow. The converse is true for a ring of air moving toward the equator—it forms an easterly air flow. The effect producing these changes in wind direction and velocity is known as the *Coriolis force*.

The actual pattern of atmospheric circulation has three cells in each hemisphere, as shown in Fig. 3.1.2. In the *tropical cell*, heated air ascends at the equator, proceeds toward the poles at upper levels, loses heat and descends toward the ground at latitude 30°. Near the ground, it branches, one branch moving toward the equator and the other toward the pole. In the *polar cell*, air rises at 60° and flows toward the poles at upper levels, then cools and flows back to 60° near the earth's surface. The *middle cell* is driven frictionally by the other two; its surface air flows toward the pole, producing prevailing westerly air flow in the mid-latitudes.

The uneven distribution of ocean and land on the earth's surface, coupled with their different thermal properties, creates additional spatial variation in atmospheric circulation. The annual shifting of the thermal equator due to the earth's revolution around the sun causes a corresponding oscillation of the three-cell circulation pattern. With a larger oscillation, exchanges of air between adjacent cells can be more frequent and complete, possibly resulting in many flood years. Also, monsoons may advance deeper into such countries as India and Australia. With a smaller oscillation, intense high pressure may build up around 30° latitude, thus creating extended dry periods. Since the atmospheric circulation is very complicated, only the general pattern can be identified.

The atmosphere is divided vertically into various zones. The atmospheric circulation described above occurs in the *troposphere*, which ranges in height from about 8 km at the poles to 16 km at the equator. The temperature in the troposphere decreases with altitude at a rate varying with the moisture content of

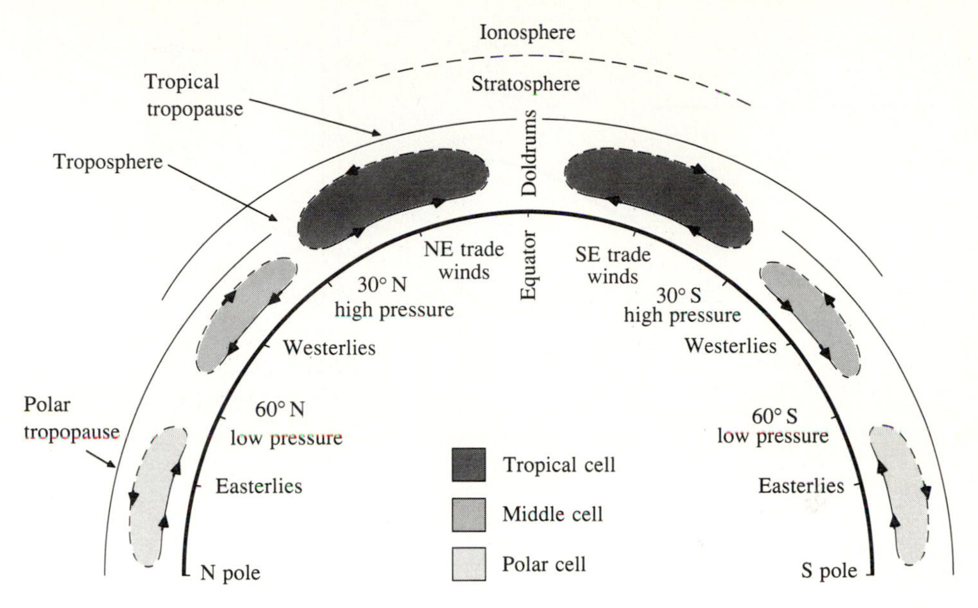

FIGURE 3.1.2
Latitudinal cross section of the general atmospheric circulation.

the atmosphere. For dry air the rate of decrease is called the *dry adiabatic lapse rate* and is approximately 9.8°C/km (Brutsaert,1982). The *saturated adiabatic lapse rate* is less, about 6.5°C/km, because some of the vapor in the air condenses as it rises and cools, releasing heat into the surrounding air. These are average figures for lapse rates that can vary considerably with altitude. The *tropopause* separates the troposphere from the *stratosphere* above. Near the tropopause, sharp changes in temperature and pressure produce strong narrow air currents known as *jet streams* with speeds ranging from 15 to 50 m/s (30 to 100 mi/h). They flow for thousands of kilometers and have an important influence on air mass movement.

An *air mass* in the general circulation is a large body of air that is fairly uniform horizontally in properties such as temperature and moisture content. When an air mass moves slowly over land or sea areas, its characteristics reflect those of the underlying surface. The region where an air mass acquires its characteristics is its source region; the tropics and the poles are two source regions. Where a warm air mass meets a cold air mass, instead of their simply mixing, a definite surface of discontinuity appears between them, called a *front*. Cold air, being heavier, underlies warm air. If the cold air is advancing toward the warm air, the leading edge of the cold air mass is a *cold front* and is nearly vertical in slope. If the warm air is advancing toward the cold air, the leading edge is a *warm front*, which has a very flat slope, the warm air flowing up and over the cold air.

A *cyclone* is a region of low pressure around which air flows in a counterclockwise direction in the northern hemisphere, clockwise in the southern hemisphere. *Tropical cyclones* form at low latitudes and may develop into

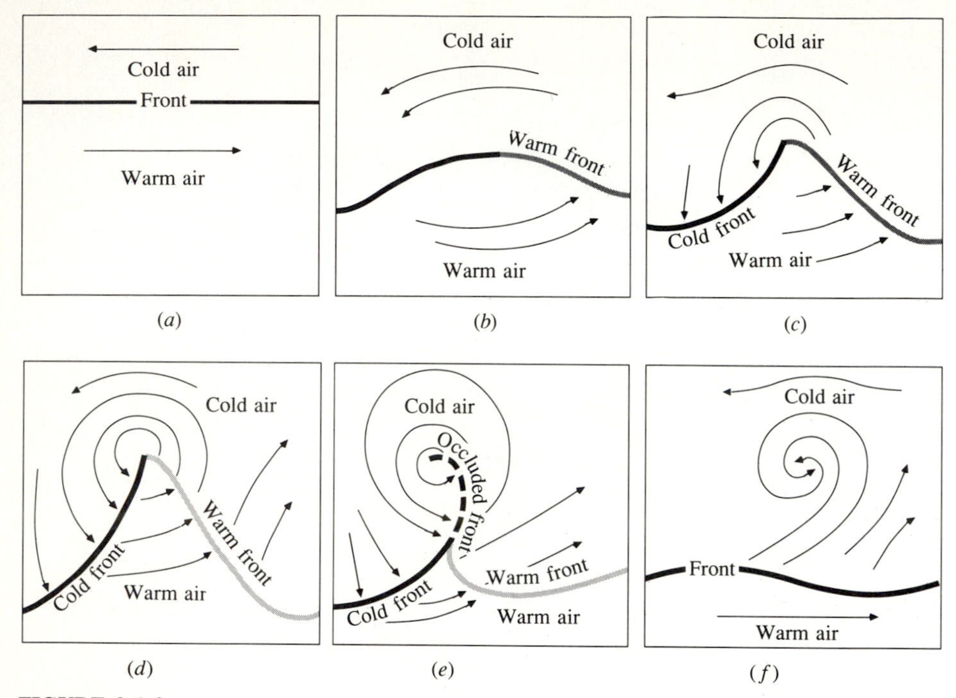

FIGURE 3.1.3
A plan view of the life cycle of a Northern Hemisphere frontal cyclone: (*a*) surface front between cold and warm air; (*b*) wave beginning to form; (*c*) cyclonic circulation and wave have developed; (*d*) faster-moving cold front is overtaking retreating warm front and reducing warm sector; (*e*) warm sector has been eliminated and (*f*) cyclone is dissipating.

hurricanes or *typhoons*. *Extratropical cyclones* are formed when warm and cold air masses, initially flowing in opposite directions adjacent to one another, begin to interact and whirl together in a circular motion, creating both a warm front and a cold front centered on a low pressure zone (Fig. 3.1.3). An *anticyclone* is a region of high pressure around which air flows clockwise in the northern hemisphere, counterclockwise in the Southern hemisphere. When air masses are lifted in atmospheric motion, their water vapor can condense and produce precipitation.

3.2 WATER VAPOR

Atmospheric water mostly exists as a gas, or vapor, but briefly and locally it becomes a liquid in rainfall and in water droplets in clouds, or it becomes a solid in snowfall, in hail, and in ice crystals in clouds. The amount of water vapor in the atmosphere is less than 1 part in 100,000 of all the waters of the earth, but it plays a vital role in the hydrologic cycle.

Vapor transport in air through a hydrologic system can be described by the Reynolds transport theorem [Eq. (2.1.9)] letting the extensive property B be the mass of water vapor. The intensive property $\beta = dB/dm$ is the mass of water vapor per unit mass of moist air; this is called the *specific humidity* q_v, and equals the ratio of the densities of water vapor (ρ_v) and moist air (ρ_a):

$$q_v = \frac{\rho_v}{\rho_a} \tag{3.2.1}$$

By the law of conservation of mass, $dB/dt = \dot{m}_v$, the rate at which water vapor is being added to the system. For evaporation from a water surface, $\dot{m}_v$ is positive and represents the mass flow rate of evaporation; conversely, for condensation, $\dot{m}_v$ is negative and represents the rate at which vapor is being removed from the system. The Reynolds transport equation for this system is the continuity equation for water vapor transport:

$$\dot{m}_v = \frac{d}{dt}\iiint_{\text{c.v.}} q_v\rho_a \, d\forall + \iint_{\text{c.s.}} q_v\rho_a \, \mathbf{V}\cdot d\mathbf{A} \tag{3.2.2}$$

Vapor Pressure

Dalton's law of partial pressures states that the pressure exerted by a gas (its vapor pressure) is independent of the presence of other gases; the vapor pressure e of the water vapor is given by the *ideal gas law* as

$$e = \rho_v R_v T \tag{3.2.3}$$

where T is the absolute temperature in K, and R_v is the gas constant for water vapor. If the total pressure exerted by the moist air is p, then $p - e$ is the partial pressure due to the dry air, and

$$p - e = \rho_d R_d T \tag{3.2.4}$$

where ρ_d is the density of dry air and R_d is the gas constant for dry air (287 J/kg·K). The density of moist air ρ_a is the sum of the densities of dry air and water vapor, that is, $\rho_a = \rho_d + \rho_v$, and the gas constant for water vapor is $R_v = R_d/0.622$, where 0.622 is the ratio of the molecular weight of water vapor to the average molecular weight of dry air. Combining (3.2.3) and (3.2.4) using the above definitions gives

$$p = \left[\rho_d + \left(\frac{\rho_v}{0.622}\right)\right] R_d T \tag{3.2.5}$$

By taking the ratio of Eqs. (3.2.3) and (3.2.5), the specific humidity q_v is approximated by

$$q_v = 0.622\frac{e}{p} \tag{3.2.6}$$

Also, (3.2.5) can be rewritten in terms of the gas constant for moist air, R_a, as

$$p = \rho_a R_a T \tag{3.2.7}$$

The relationship between the gas constants for moist air and dry air is given by

$$\begin{aligned} R_a &= R_d(1 + 0.608q_v) \\ &= 287(1 + 0.608q_v) \text{ J/kg}\cdot\text{K} \end{aligned} \tag{3.2.8}$$

The gas constant of moist air increases with specific humidity, but even for a large specific humidity (e.g., $q_v = 0.03$ kg water/kg of moist air), the difference between the gas constants for moist and dry air is only about 2 percent.

For a given air temperature, there is a maximum moisture content the air can hold, and the corresponding vapor pressure is called the *saturation vapor pressure* e_s. At this vapor pressure, the rates of evaporation and condensation are equal. Over a water surface the saturation vapor pressure is related to the air temperature as shown in Fig. 3.2.1; an approximate equation is:

$$e_s = 611 \exp\left(\frac{17.27T}{237.3 + T}\right) \tag{3.2.9}$$

where e_s is in pascals (Pa = N/m^2) and T is in degrees Celsius (Raudkivi, 1979). Some values of the saturation vapor pressure of water are listed in Table 3.2.1.

The gradient $\Delta = de_s/dT$ of the saturated vapor pressure curve is found by differentiating (3.2.9):

$$\Delta = \frac{4098e_s}{(237.3 + T)^2} \tag{3.2.10}$$

where Δ is the gradient in pascals per degree Celsius.

The *relative humidity* R_h is the ratio of the actual vapor pressure to its saturation value at a given air temperature (see Fig. 3.2.1):

$$R_h = \frac{e}{e_s} \tag{3.2.11}$$

The temperature at which air would just become saturated at a given specific humidity is its *dew-point temperature* T_d.

Example 3.2.1 At a climate station, air pressure is measured as 100 kPa, air temperature as 20°C, and the wet-bulb, or dew-point, temperature as 16°C. Calculate

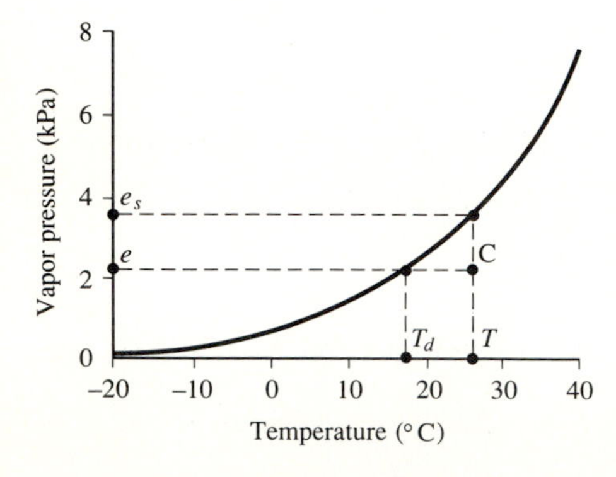

FIGURE 3.2.1
Saturated vapor pressure as a function of temperature over water. Point C has vapor pressure e and temperature T, for which the saturated vapor pressure is e_s. The relative humidity is $R_h = e/e_s$. The temperature at which the air is saturated for vapor pressure e is the dew-point temperature T_d.

the corresponding vapor pressure, relative humidity, specific humidity, and air density.

Solution. The saturated vapor pressure at T = 20°C is given by Eq. (3.2.9)

$$e_s = 611 \exp\left(\frac{17.27T}{237.3 + T}\right)$$

$$= 611 \exp\left(\frac{17.27 \times 20}{237.3 + 20}\right)$$

$$= 2339 \text{ Pa}$$

and the actual vapor pressure e is calculated by the same method substituting the dew-point temperature T_d = 16°C:

$$e = 611 \exp\left(\frac{17.27T_d}{237.3 + T_d}\right)$$

$$= 611 \exp\left(\frac{17.27 \times 16}{237.3 + 16}\right)$$

$$= 1819 \text{ Pa}$$

The relative humidity from (3.2.11) is

$$R_h = \frac{e}{e_s}$$

$$= \frac{1819}{2339}$$

$$= 0.78$$

$$= 78\%$$

TABLE 3.2.1
Saturated vapor pressure of water vapor over liquid water

Temperature °C	Saturated Vapor Pressure Pa
–20	125
–10	286
0	611
5	872
10	1227
15	1704
20	2337
25	3167
30	4243
35	5624
40	7378

Source: Brutsaert, 1982, Table 3.4, p. 41. Used with permission.

and the specific humidity is given by (3.2.6) with $p = 100$ kPa $= 100 \times 10^3$ Pa:

$$
\begin{aligned}
q_v &= 0.622\frac{e}{p} \\
&= 0.622\left(\frac{1819}{100 \times 10^3}\right) \\
&= 0.0113 \text{ kg water/kg moist air}
\end{aligned}
$$

The air density is calculated from the ideal gas law (3.2.7). The gas constant R_a is given by (3.2.8) with $q_v = 0.0113$ kg/kg as $R_a = 287(1 + 0.608q_v) = 287(1 + 0.608 \times 0.0113) = 289$ J/kg·K, and $T = 20°\text{C} = (20 + 273)$ K $= 293$ K, so that

$$
\begin{aligned}
\rho_a &= \frac{p}{R_a T} \\
&= \frac{100 \times 10^3}{289 \times 293} \\
&= 1.18 \text{ kg/m}^3
\end{aligned}
$$

Water Vapor in a Static Atmospheric Column

Two laws govern the properties of water vapor in a static column, the ideal gas law

$$p = \rho_a R_a T \tag{3.2.12}$$

and the *hydrostatic pressure law*

$$\frac{dp}{dz} = -\rho_a g \tag{3.2.13}$$

The variation of air temperature with altitude is described by

$$\frac{dT}{dz} = -\alpha \tag{3.2.14}$$

where α is the lapse rate. As shown in Fig. 3.2.2, a linear temperature variation combined with the two physical laws yields a nonlinear variation of pressure with altitude. Density and specific humidity also vary nonlinearly with altitude. From (3.2.12), $\rho_a = p/R_a T$, and substituting this into (3.2.13) yields

$$\frac{dp}{dz} = \frac{-pg}{R_a T}$$

or

$$\frac{dp}{p} = \left(\frac{-g}{R_a T}\right) dz$$

Substituting $dz = -dT/\alpha$ from (3.2.14):

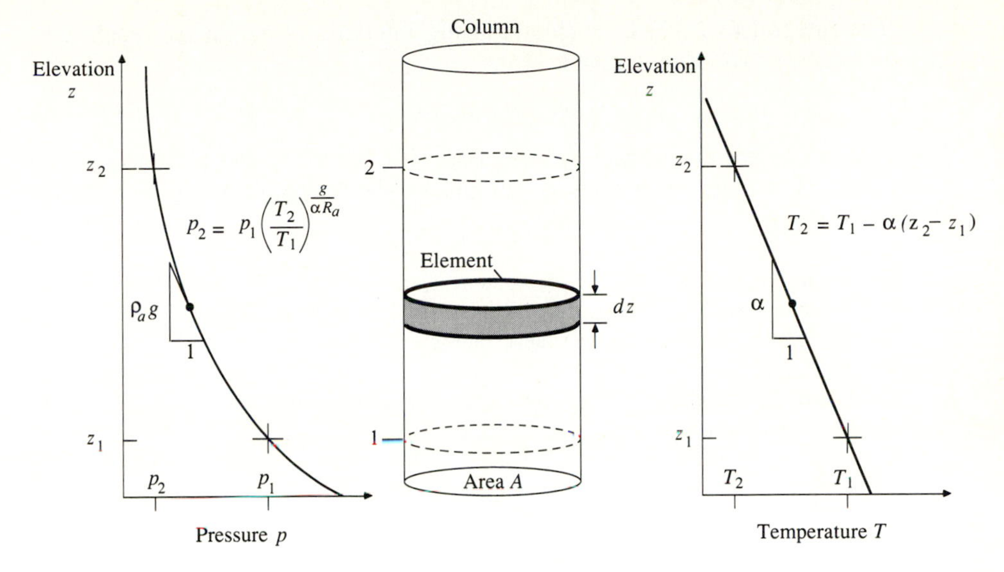

FIGURE 3.2.2
Pressure and temperature variation in an atmospheric column.

$$\frac{dp}{p} = \left(\frac{g}{\alpha R_a}\right)\frac{dT}{T}$$

and integrating both sides between two levels 1 and 2 in the atmosphere gives

$$\ln\left(\frac{p_2}{p_1}\right) = \left(\frac{g}{\alpha R_a}\right)\ln\left(\frac{T_2}{T_1}\right)$$

or

$$p_2 = p_1\left(\frac{T_2}{T_1}\right)^{g/\alpha R_a} \tag{3.2.15}$$

From (3.2.14) the temperature variation between altitudes z_1 and z_2 is

$$T_2 = T_1 - \alpha(z_2 - z_1) \tag{3.2.16}$$

Precipitable Water

The amount of moisture in an atmospheric column is called its *precipitable water*. Consider an element of height dz in a column of horizontal cross-sectional area A (Fig. 3.2.2). The mass of air in the element is $\rho_a A\,dz$ and the mass of water contained in the air is $q_v \rho_a A\,dz$. The total mass of precipitable water in the column between elevations z_1 and z_2 is

$$m_p = \int_{z_1}^{z_2} q_v \rho_a A\,dz \tag{3.2.17}$$

The integral (3.2.17) is calculated using intervals of height Δz, each with an incremental mass of precipitable water

$$\Delta m_p = \overline{q}_v \overline{\rho}_a A \Delta z \tag{3.2.18}$$

where $\overline{q}_v$ and $\overline{\rho}_a$ are the average values of specific humidity and air density over the interval. The mass increments are summed over the column to give the total precipitable water.

Example 3.2.2. Calculate the precipitable water in a saturated air column 10 km high above 1 m^2 of ground surface. The surface pressure is 101.3 kPa, the surface air temperature is 30°C, and the lapse rate is 6.5°C/km.

Solution. The results of the calculation are summarized in Table 3.2.2. The increment in elevation is taken as Δz = 2 km = 2000 m. For the first increment, at $z_1 = 0$ m, T_1 = 30°C = (30 + 273) K = 303 K; at $z_2 = 2000$ m, by Eq. (3.2.16) using α = 6.5°C/km = 0.0065°C/m,

$$\begin{aligned} T_2 &= T_1 - \alpha(z_2 - z_1) \\ &= 30 - 0.0065(2000 - 0) \\ &= 17^\circ\text{C} \end{aligned}$$

TABLE 3.2.2
Calculation of precipitable water in a saturated air column (Example 3.2.2)

Column	1 Elevation z (km)	2 Temperature (°C)	3 Temperature (°K)	4 Air pressure p (kPa)	5 Density ρ_a (kg/m³)	6 Vapor pressure e (kPa)
	0	30	303	101.3	1.16	4.24
	2	17	290	80.4	0.97	1.94
	4	4	277	63.2	0.79	0.81
	6	–9	264	49.1	0.65	0.31
	8	–22	251	37.6	0.52	0.10
	10	–35	238	28.5	0.42	0.03

Column	7 Specific humidity q_v (kg/kg)	8 Average over increment $\overline{q}_v$ (kg/kg)	9 Average over increment $\overline{\rho}_a$ (kg/m³)	10 Incremental mass Δm (kg)	11 % of total mass
	0.0261				
	0.0150	0.0205	1.07	43.7	57
	0.0080	0.0115	0.88	20.2	26
	0.0039	0.0060	0.72	8.6	11
	0.0017	0.0028	0.59	3.3	4
	0.0007	0.0012	0.47	1.1	2
				77.0	

$$= (17 + 273)\ \text{K}$$

$$= 290\ \text{K}$$

as shown in column 3 of the table. The gas constant R_a can be taken as 287 J/kg·K in this example because its variation with specific humidity is small [see Eq. (3.2.8)]. The air pressure at 2000 m is then given by (3.2.15) with $g/\alpha R_a = 9.81/(0.0065 \times 287) = 5.26$, as

$$p_2 = p_1\left(\frac{T_2}{T_1}\right)^{g/\alpha R_a}$$

$$= 101.3\left(\frac{290}{303}\right)^{5.26}$$

$$= 80.4\ \text{kPa}$$

as shown in column 4.

The air density at the ground is calculated from (3.2.12):

$$\rho_a = \frac{p}{R_a T}$$

$$= \frac{101.3 \times 10^3}{(287 \times 303)}$$

$$= 1.16\ \text{kg/m}^3$$

and a similar calculation yields the air density of 0.97 kg/m^3 at 2000 m. The average density over the 2 km increment is therefore $\overline{\rho}_a = (1.16 + 0.97)/2 = 1.07$ kg/m^3 (see columns 5 and 9).

The saturated vapor pressure at the ground is determined using (3.2.9):

$$e = 611\ \exp\left(\frac{17.27T}{237.3 + T}\right)$$

$$= 611\ \exp\left(\frac{17.27 \times 30}{237.3 + 30}\right)$$

$$= 4244\ \text{Pa}$$

$$= 4.24\ \text{kPa}$$

The corresponding value at 2000 m where $T = 17°\text{C}$, is $e = 1.94$ kPa (column 6). The specific humidity at the ground surface is calculated by Eq. (3.2.6):

$$q_v = 0.622\frac{e}{p}$$

$$= 0.622 \times \frac{4.24}{101.3}$$

$$= 0.026\ \text{kg/kg}$$

At 2000 m $q_v = 0.015$ kg/kg. The average value of specific humidity over the 2-km increment is therefore $\overline{q}_v = (0.026 + 0.015)/2 = 0.0205$ kg/kg (column 8). Substituting into (3.2.18), the mass of precipitable water in the first 2-km increment is

$$\Delta m_p = \overline{q}_v \overline{\rho}_a A \, \Delta z$$
$$= 0.0205 \times 1.07 \times 1 \times 2000$$
$$= 43.7 \text{ kg}$$

By adding the incremental masses, the total mass of precipitable water in the column is found to be m_p = 77 kg (column 10). The equivalent depth of liquid water is $m_p / \rho_w A = 77/(1000 \times 1) \doteq 0.077$ m = 77 mm.

The numbers in column 11 of Table 3.2.2 for percent of total mass in each increment show that more than half of the precipitable water is located in the first 2 km above the land surface in this example. There is only a very small amount of precipitable water above 10 km elevation. The depth of precipitable water in this column is sufficient to produce a small storm, but a large storm would require inflow of moisture from surrounding areas to sustain the precipitation.

3.3 PRECIPITATION

Precipitation includes rainfall, snowfall, and other processes by which water falls to the land surface, such as hail and sleet. The formation of precipitation requires the lifting of an air mass in the atmosphere so that it cools and some of its moisture condenses. The three main mechanisms of air mass lifting are *frontal lifting*, where warm air is lifted over cooler air by frontal passage; *orographic lifting*, in which an air mass rises to pass over a mountain range; and *convective lifting*, where air is drawn upwards by convective action, such as in the center of a thunderstorm cell. Convective cells are initiated by surface heating, which causes a vertical instability of moist air, and are sustained by the latent heat of vaporization given up as water vapor rises and condenses.

The formation of precipitation in clouds is illustrated in Fig. 3.3.1. As air rises and cools, water condenses from the vapor to the liquid state. If the temperature is below the freezing point, then ice crystals are formed instead. Condensation requires a seed called a *condensation nucleus* around which the water molecules can attach or *nucleate* themselves. Particles of dust floating in air can act as condensation nuclei; particles containing ions are effective nuclei because the ions electrostatically attract the polar-bonded water molecules. Ions in the atmosphere include particles of salt derived from evaporated sea spray, and sulphur and nitrogen compounds resulting from combustion. The diameters of these particles range from 10^{-3} to 10 μm and the particles are known as *aerosols*. For comparison, the size of an atom is about 10^{-4} μm, so the smallest aerosols may be composed of just a few hundred atoms.

The tiny droplets grow by condensation and impact with their neighbors as they are carried by turbulent air motion, until they become large enough so that the force of gravity overcomes that of friction and they begin to fall, further increasing in size as they hit other droplets in the fall path. However, as the drop falls, water evaporates from its surface and the drop size diminishes, so the drop may be reduced to the size of an aerosol again and be carried upwards in the

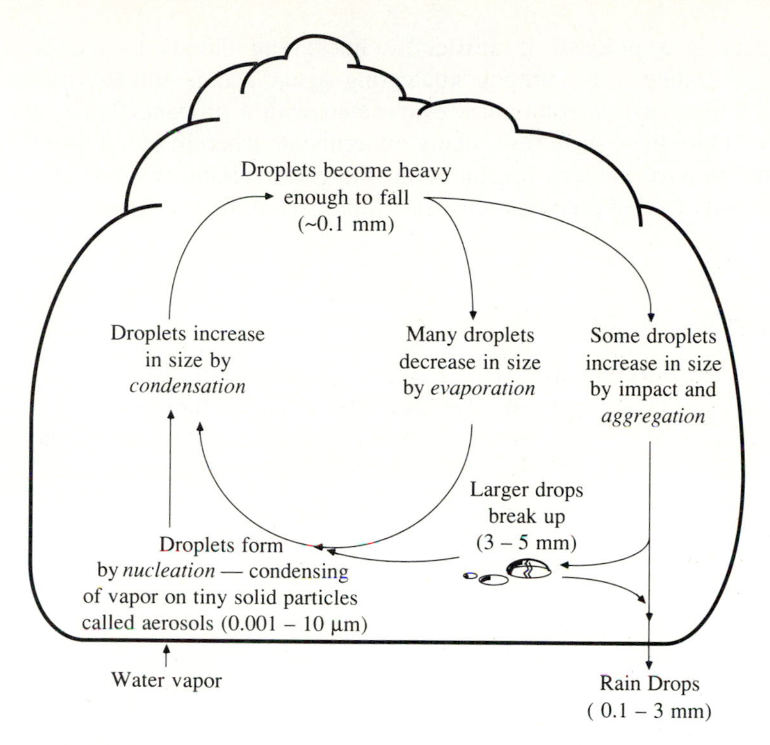

FIGURE 3.3.1
Water droplets in clouds are formed by nucleation of vapor on aerosols, then go through many condensation-evaporation cycles as they circulate in the cloud, until they aggregate into large enough drops to fall through the cloud base.

cloud through turbulent action. An upward current of only 0.5 cm/s is sufficient to carry a 10 μm droplet. Ice crystals of the same weight, because of their shape and larger size, can be supported by even lower velocities. The cycle of condensation, falling, evaporation, and rising occurs on average about ten times before the drop reaches a critical size of about 0.1 mm, which is large enough to fall through the bottom of the cloud.

Up to about 1 mm in diameter, the droplets remain spherical in shape, but beyond this size they begin to flatten out on the bottom until they are no longer stable falling through air and break up into small raindrops and droplets. Normal raindrops falling through the cloud base are 0.1 to 3 mm in diameter.

Observations indicate that water droplets may exist in clouds at subfreezing temperatures down to –35°C. At this temperature, the supercooled droplets will freeze even without freezing nuclei. The saturation vapor pressure of water vapor is lower over ice than over liquid water, so if ice particles are mixed with water droplets, the ice particles will grow by evaporation from the droplets and condensation on the ice crystals. By collision and coalescence, ice crystals typically form clusters and fall as snow flakes. However, single ice crystals may grow so large that they fall directly to the earth as hail or sleet.

Cloud seeding is a process of artificially nucleating clouds to induce precipitation. Silver iodide is a common nucleating agent and is spread from aircraft in which a silver iodide solution is evaporated with a propane flame to produce particles. While there have been many experiments wherein cloud seeding was considered to have induced precipitation, the great variability of meteorological processes involved in producing precipitation make it difficult to achieve consistent results.

Terminal Velocity

Three forces act on a falling raindrop (Fig. 3.3.2): a gravity force F_g due to its weight, a buoyancy force F_b due to the displacement of air by the drop, and a drag force F_d due to friction between the drop and the surrounding air. If the drop is a sphere of diameter D, its volume is $(\pi/6)D^3$ so the weight force is

$$F_g = \rho_w g\left(\frac{\pi}{6}\right)D^3 \tag{3.3.1}$$

and the bouyancy force is

$$F_b = \rho_a g\left(\frac{\pi}{6}\right)D^3 \tag{3.3.2}$$

where ρ_w and ρ_a and are the densities of water and air, respectively. The friction drag force is given by

$$F_d = C_d \rho_a A \frac{V^2}{2} \tag{3.3.3}$$

where C_d is a dimensionless *drag coefficient*, $A = (\pi/4)D^2$ is the cross-sectional area of the drop, and V is the fall velocity.

If the drop is released from rest, it will accelerate until it reaches its terminal velocity V_t, at which the three forces are balanced. In this condition,

$$F_d = F_g - F_b$$

Hence, letting $V = V_t$ in Eqs. (3.3.1–3),

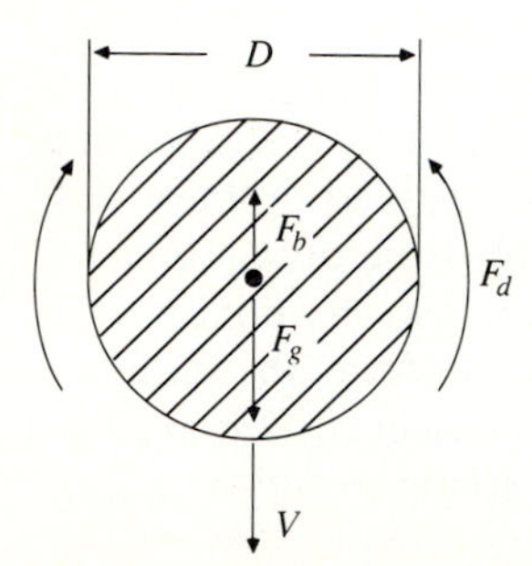

FIGURE 3.3.2
Forces on a falling raindrop: F_g = weight; F_b = buoyancy; F_d = drag force of surrounding air.

$$C_d \rho_a D^2 \left(\frac{\pi}{4}\right) \frac{V_t^2}{2} = \rho_w g \left(\frac{\pi}{6}\right) D^3 - \rho_a g \left(\frac{\pi}{6}\right) D^3$$

which, solved for V_t, is:

$$V_t = \left[\frac{4gD}{3C_d}\left(\frac{\rho_w}{\rho_a} - 1\right)\right]^{1/2} \tag{3.3.4}$$

The assumption of a spherical raindrop shape is valid for drops up to 1 mm in diameter. Beyond this size, the drops become flattened on the bottom and more oval in cross section; then they are characterized by the equivalent diameter of a spherical raindrop having the same volume as the actual drop (Pruppacher and Klett, 1978). Raindrops can range up to 6 mm in diameter, but drops larger than 3 mm are unusual, especially in low-intensity rainfall.

For tiny droplets in clouds, up to 0.1 mm diameter, the drag force is specified by *Stokes' law* for which the drag coefficient is $C_d = 24/Re$, where Re is the Reynolds number $\rho_a VD/\mu_a$ with μ_a being the air viscosity. Falling raindrops are beyond the range of Stokes' law; values of C_d developed experimentally by observation of raindrops are given in Table 3.3.1.

Example 3.3.1. Calculate the terminal velocity of a 1-mm-diameter raindrop falling in still air at standard atmospheric pressure (101.3 kPa) and temperature 20°C.

Solution. The terminal velocity is given by Eq. (3.3.4) with $C_d = 0.671$ from Table 3.3.1. At 20°C, $\rho_w = 998\ \text{kg/m}^3$, and $\rho_a = 1.20\ \text{kg/m}^3$ at pressure 101.3 kPa:

$$V_t = \left[\frac{4gD}{3C_d}\left(\frac{\rho_w}{\rho_a} - 1\right)\right]^{1/2}$$

$$= \left[\frac{4 \times 9.81 \times 0.001}{3 \times 0.671}\left(\frac{998}{1.20} - 1\right)\right]^{1/2}$$

$$= 4.02\ \text{m/s}$$

Values of V_t similarly computed for various diameters are plotted in Fig. 3.3.3. It can be seen that the terminal velocity increases with drop size up to a plateau level of about 5 mm drop size, for which the terminal velocity is approximately 9 m/s.

TABLE 3.3.1
Drag coefficients for spherical raindrops of diameter *D*, at standard atmospheric pressure (101.3 kPa) and 20°C air temperature

Drop diameter D (mm)	0.2	0.4	0.6	0.8	1.0	2.0	3.0	4.0	5.0
Drag coefficient C_d	4.2	1.66	1.07	0.815	0.671	0.517	0.503	0.559	0.660

Source: Mason, 1957, Table 8.2, p. 436.

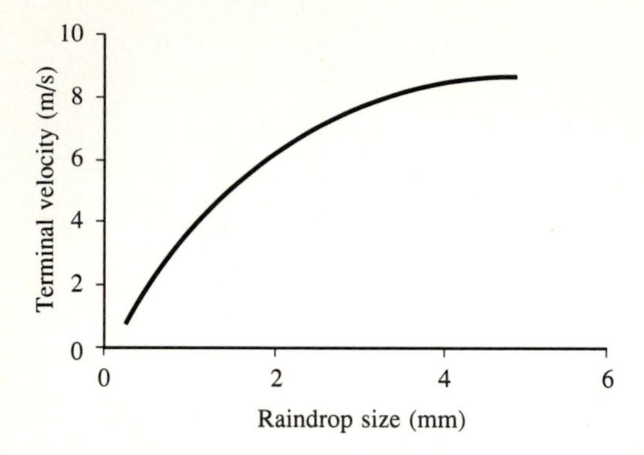

FIGURE 3.3.3
Terminal velocity of raindrops as calculated from Eq. (3.3.4) using drag coefficients in Table (3.3.1). Results are for standard atmospheric conditions at sea level.

The preceding computations are for sea level conditions. Higher in the atmosphere, the air density ρ_a decreases, and Eq. (3.3.4) shows that there will be a corresponding increase in V_t; raindrops fall faster in thinner air. At air pressure 50 kPa and temperature –10°C, the plateau velocity of large drops increases from 9 m/s to a little more than 12 m/s.

Thunderstorm Cell Model

The mechanisms underlying air mass lifting and precipitation are illustrated by considering a schematic model of a thunderstorm cell, as shown in Fig. 3.3.4. The thunderstorm is visualized as a vertical column made up of three parts, an *inflow region* near the ground where warm, moist air is drawn into the cell, an *uplift region* in the middle where moisture condenses as air rises, producing precipitation, and an *outflow region* in the upper atmosphere where outflow of cooler, dryer air occurs. Outside the cell column, the outflow air may descend over a wide area, pick up more moisture, and reenter the cell at the bottom. This entire pattern, called *convective cell circulation*, is driven by the vast amount of

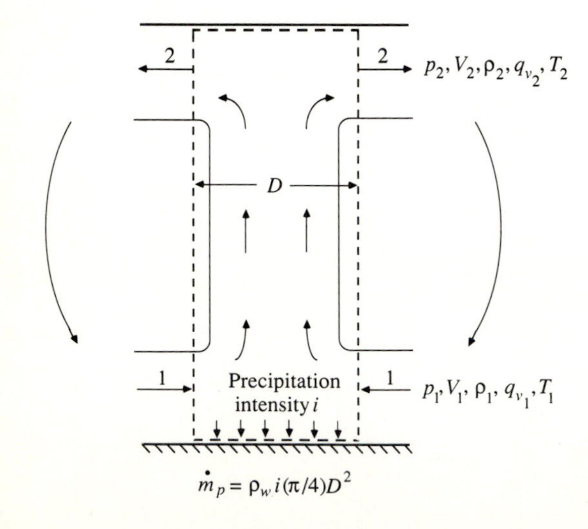

FIGURE 3.3.4
A convective thunderstorm cell visualized as a cylindrical column of diameter D having inflow, uplift, and outflow regions.

heat energy released by the condensing moisture in the uplift region. Observations of cumulonimbus clouds producing thunderstorms indicate that the elevation of the top of the convective cell ranges from 8 km to 16 km (5 to 10 mi) in the atmosphere (Wiesner, 1970), and at times the tops of these clouds may even penetrate through the tropopause into the stratosphere.

The thunderstorm is analyzed using the continuity equation for water vapor:

$$\dot{m}_v = \frac{d}{dt}\iiint_{\text{c.v.}} q_v \rho_a \, d\forall + \iint_{\text{c.s.}} q_v \rho_a \, \mathbf{V} \cdot \mathbf{dA} \tag{3.3.5}$$

If precipitation of intensity i (in/h or cm/h) is falling on an area A beneath the storm cell, the mass flow rate of water leaving the cell is $\dot{m}_v = -\rho_w i A$, where ρ_w is the density of liquid water. Under the assumption of steady flow, the time derivative term in (3.3.5) is zero, and the mass flow rate of precipitation is equal to the difference between the mass flow rates of water vapor entering the cell (1) and leaving (2) (see Fig. 3.3.4), so

$$-\rho_w i A = \iint_2 q_v \rho_a \mathbf{V} \cdot \mathbf{dA} + \iint_1 q_v \rho_a \mathbf{V} \cdot \mathbf{dA} \tag{3.3.6}$$

The cell is a cylinder of diameter D, and air enters through height increment Δz_1 and leaves through height increment Δz_2. If air density and specific humidity are assumed constant within each increment (in the manner shown in Example 3.2.2), then

$$\rho_w i A = (q_v \rho_a V)_1 \pi D \Delta z_1 - (q_v \rho_a V)_2 \pi D \Delta z_2 \tag{3.3.7}$$

A continuity equation may be written similarly for the dry air carrying the vapor:

$$0 = \iint_{\text{c.s.}} \rho_d \mathbf{V} \cdot \mathbf{dA} \tag{3.3.8}$$

where ρ_d is the density of dry air, which may be expressed using Eq. (3.2.1) as $\rho_d = \rho_a(1 - q_v)$. Substituting into (3.3.8):

$$0 = [\rho_a(1 - q_v)V\Delta z]_2 \pi D - [\rho_a(1 - q_v)V\Delta z]_1 \pi D$$

or

$$(\rho_a V \Delta z)_2 = (\rho_a V \Delta z)_1 \left(\frac{1 - q_{v_1}}{1 - q_{v_2}}\right) \tag{3.3.9}$$

Substituting (3.3.9) into (3.3.7) and noting that the area on which precipitation is falling is $A = (\pi/4)D^2$, it follows that

$$i = \frac{4\rho_{a_1} V_1 \Delta z_1}{\rho_w D}\left(\frac{q_{v_1} - q_{v_2}}{1 - q_{v_2}}\right) \tag{3.3.10}$$

Example 3.3.2 A thunderstorm cell 5 km in diameter has a cloud base of 1.5 km, and surface conditions recorded nearby indicate saturated air conditions with air temperature 30°C, pressure 101.3 kPa, and wind speed 1 m/s. Assuming a lapse rate of 7.5°C/km and an average outflow elevation of 10 km, calculate the precipitation intensity from this storm. Also determine what proportion of the incoming moisture is precipitated as air passes through the storm cell and calculate the rate of release of latent heat through moisture condensation in the column.

Solution. The precipitation intensity is given by (3.3.10) where $V_1 = 1$ m/s, $\Delta z_1 = 1500$ m, $\rho_w = 1000$ kg/m^3, and $D = 5000$ m. The quantities ρ_{a_1}, q_{v_1}, and q_{v_2} are found by the method outlined in Ex. 3.2.2 using $\alpha = 0.0075$°C/m for the lapse rate. A table may be set up for the required values at $z = 0$, 1.5, and 10 km.

Elevation (km)	Temperature (°C)	(K)	Air Pressure (kPa)	Air Density (kg/m^3)	Vapor Pressure (kPa)	Specific Humidity (kg/kg)
0	30	303	101.3	1.16	4.24	0.0261
1.5	19	292	85.6	1.02	2.20	0.0160
10	−45	228	27.7	0.42	0.01	0.0002

From the table, $q_{v_2} = 0.0002$ kg/kg; the values for ρ_{a_1} and q_{v_1} are taken as averages between 0 and 1.5 km: $\rho_{a_1} = (1.16 + 1.02)/2 = 1.09$ kg/m^3, and $q_{v_1} = (0.0261 + 0.0160)/2 = 0.021$ kg/kg. Substituting into (3.3.10):

$$
\begin{aligned}
i &= \frac{4\rho_{a_1} V_1 \Delta z_1}{\rho_w D}\left(\frac{q_{v_1} - q_{v_2}}{1 - q_{v_2}}\right) \\
&= \frac{4 \times 1.09 \times 1 \times 1500}{1000 \times 5000}\left(\frac{0.0210 - 0.0002}{1 - 0.0002}\right) \\
&= 2.72 \times 10^{-5} \text{ m/s} \\
&= 9.8 \text{ cm/h}
\end{aligned}
$$

The mass flow rate of precipitation is given by $\dot{m}_p = \rho_w i A$, where $A = (\pi/4)D^2 = (\pi/4) \times 5000^2 = 1.96 \times 10^7$ m^2 and $\rho_w = 1000$ kg/m^3; $\dot{m}_p = 1000 \times 2.72 \times 10^{-5} \times 1.96 \times 10^7 = 5.34 \times 10^5$ kg/s.

The mass flow rate of incoming moisture is given by

$$
\begin{aligned}
\dot{m}_{v_1} &= (\rho_a q_v V \Delta z)_1 \pi D \\
&= 1.09 \times 0.021 \times 1.00 \times 1500 \times \pi \times 5000 \\
&= 5.39 \times 10^5 \text{ kg/s}
\end{aligned}
$$

The proportion of the incoming moisture precipitated is $\dot{m}_p/\dot{m}_{v_1} = (5.34 \times 10^5)/(5.39 \times 10^5) = 0.99$!

The rate of release of latent heat due to moisture condensation is $l_v \dot{m}_p$, where l_v is the latent heat of vaporization of water, 2.5×10^6 J/kg:

$$
\begin{aligned}
l_v \dot{m}_p &= 2.5 \times 10^6 \times 5.34 \times 10^5 \\
&= 1.335 \times 10^{12} \text{ W}
\end{aligned}
$$

$$= 1,335,000 \text{ MW}$$

This heat energy can be compared to large thermal power plants, which may have a capacity of 3000 MW. It can be seen that the energy released in thunderstorms is immense.

Variability of Precipitation

Precipitation varies in space and time according to the general pattern of atmospheric circulation and according to local factors. The average over a number of years of observations of a weather variable is called its *normal* value. Figure 3.3.5 shows the normal monthly precipitation for a number of locations in the United States. Higher precipitation occurs near the coasts than inland because the oceans supply the bulk of the atmospheric moisture for precipitation. Areas to the east of the Cascade mountains (e.g., Boise, Idaho) have lower precipitation than those to the west (e.g., Seattle, Washington) because much of the moisture in the predominantly westerly air flow in the mid-latitudes is extracted as the air rises over the mountains.

Pronounced seasonal variation in precipitation occurs where the annual oscillation in the atmospheric circulation changes the amount of moisture inflow over those regions (e.g., San Francisco and Miami). This pattern is illustrated in Fig. 3.3.6, which shows the normal monthly precipitation for various locations in the United States. Precipitation is very variable in the mountain states in the west where orographic effects influence precipitation. Precipitation increases going east across the great plains and is spatially more uniform in the east than in the west. Precipitation variability for the world is shown in Fig. 3.3.7. The average annual precipitation on the land surface of the earth is about 800 mm (32 in), but great variability exists, from Arica, Chile, with an annual average of 0.5 mm (0.02 in) to Mt. Waialeale, Hawaii, which receives 11,680 mm (460 in) per year on average.

3.4 RAINFALL

Rainstorms vary greatly in space and time. They can be represented by *isohyetal maps*; an *isohyet* is a contour of constant rainfall. Figure 3.4.1 shows an isohyetal map of total rainfall depth measured for two storms: one a storm of May 30–June 1, 1889, which caused about 2000 deaths in Johnstown, Pennsylvania, following a dam failure, and the other a storm of May 24–25, 1981, in Austin, Texas, which caused 13 deaths and $35 million in property damage (Moore, et al., 1982). The Johnstown storm is plotted on a scale 50 times larger than the Austin storm. The maximum depth of precipitation in both storms is nearly the same ($\approx$ 10 in), but the Austin storm was briefer and more localized than the Johnstown storm. The Austin storm was caused by a convective cell thunderstorm of the type analyzed in Example 3.3.2.

Isohyetal maps are prepared by interpolating rainfall data recorded at gaged points. A rain gage record consists of a set of rainfall depths recorded for

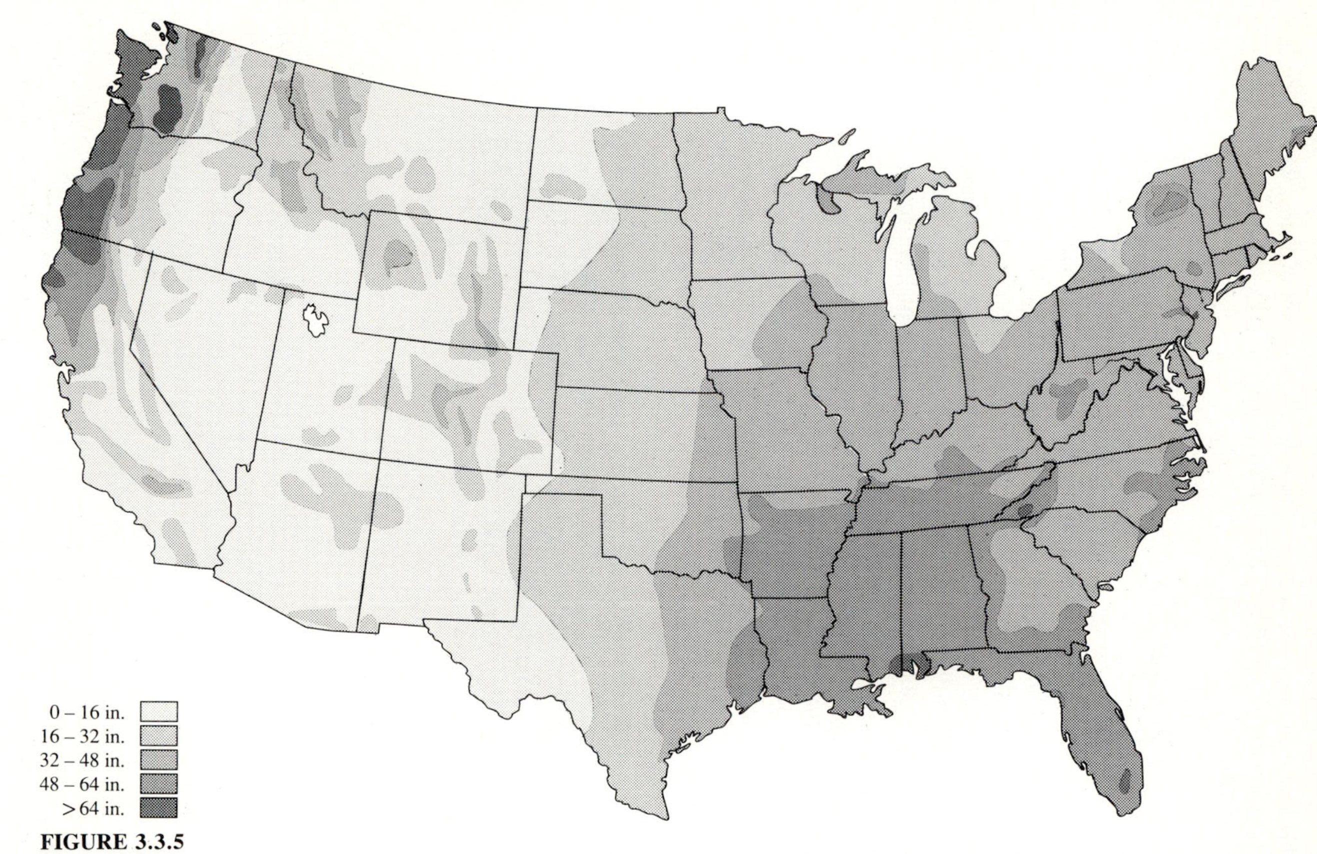

FIGURE 3.3.5
Mean annual precipitation in the U.S.A. in inches (1 in = 25.4 mm). (Adapted from *Climatic Atlas of the U.S.*, U.S. Environmental Data Service, U.S.G.P.O., pp. 43–44, June, 1968.)

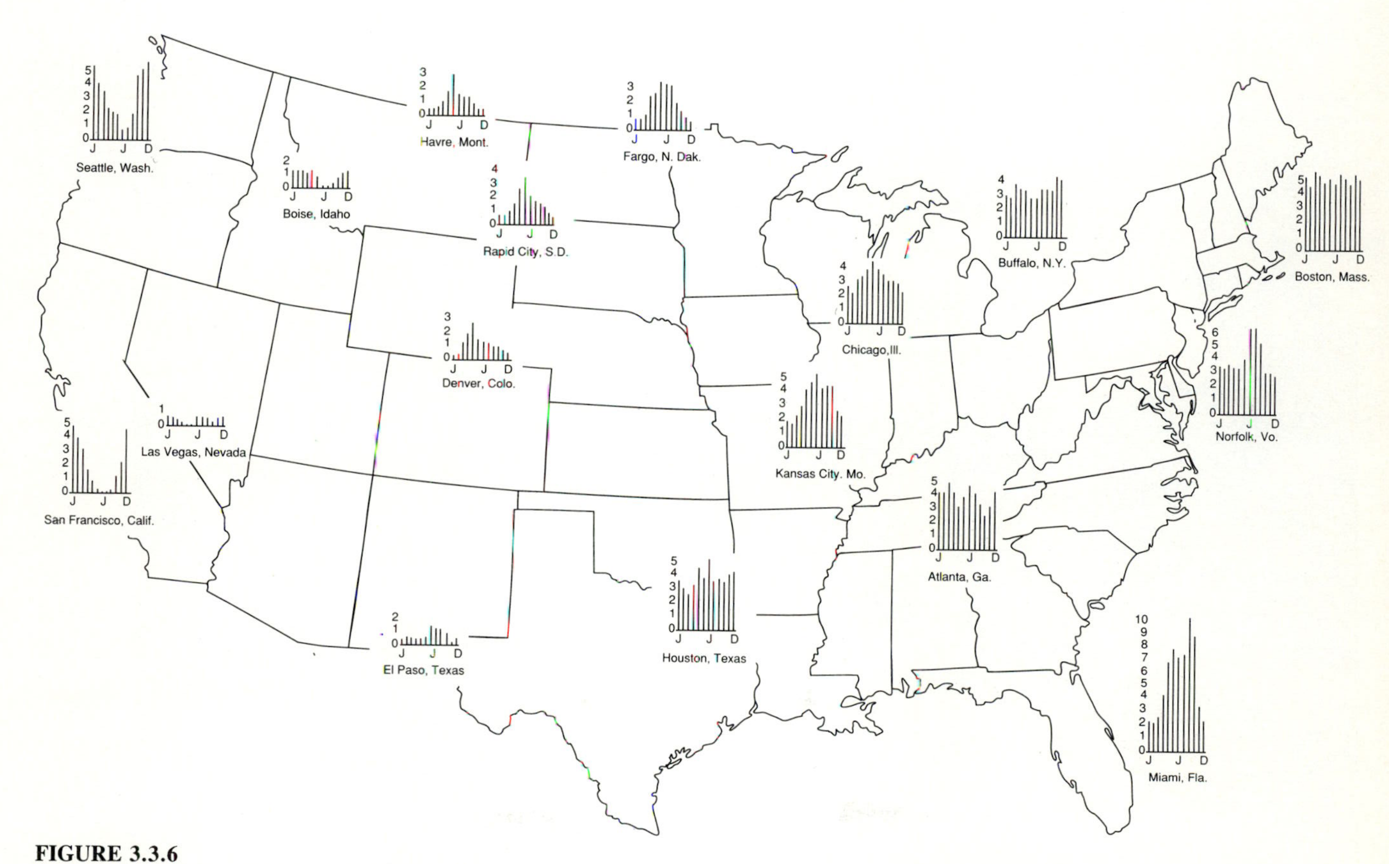

FIGURE 3.3.6
Normal monthly distribution of precipitation in the U. S. A. in inches (1 in = 25.4 mm). (Adapted from *Climatic Atlas of the U.S.*, U.S. Environmental Data Service, U.S.G.P.O., pp. 43–44, June, 1968.)

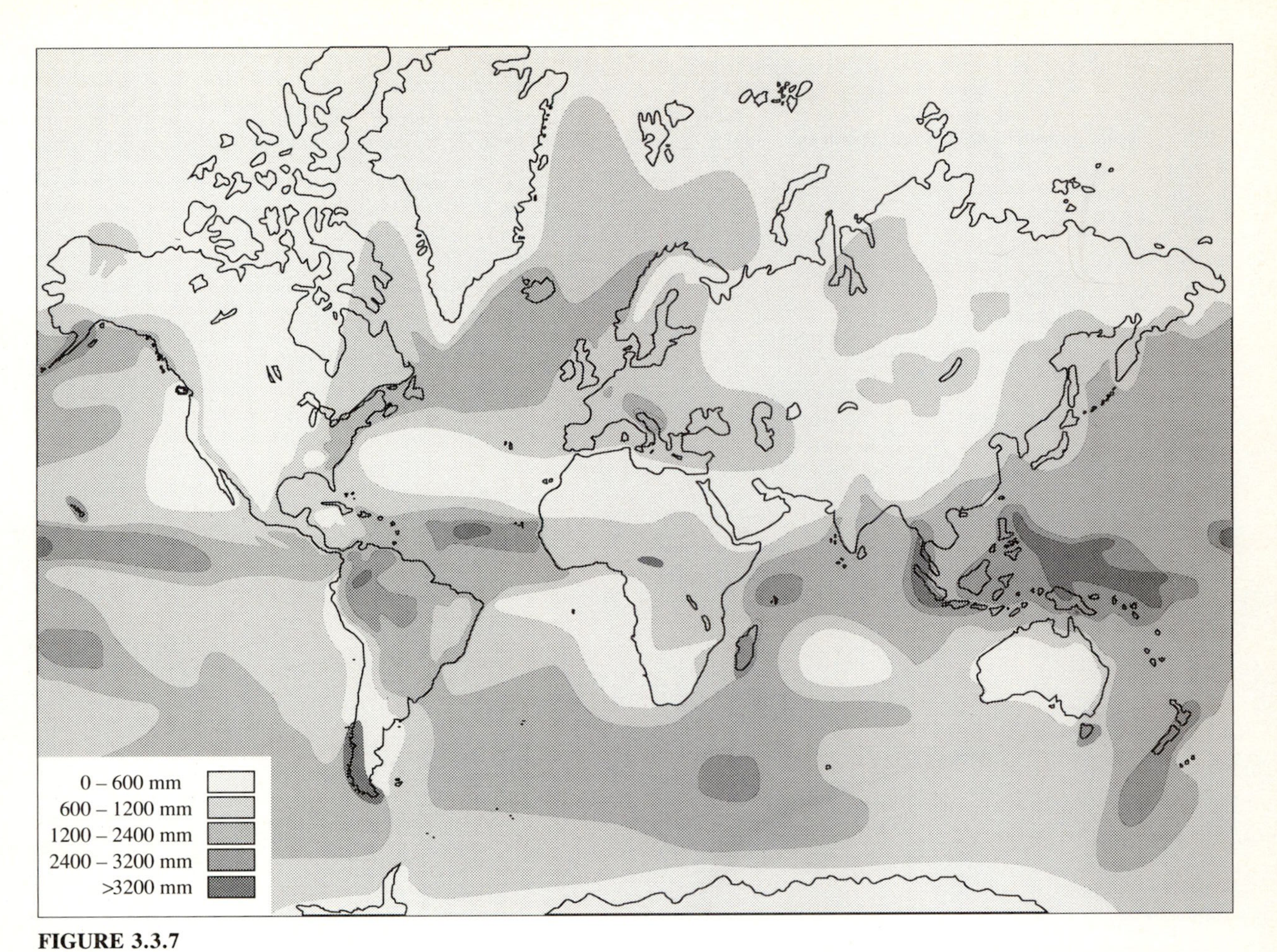

FIGURE 3.3.7
Mean annual precipitation of the world in millimeters (1mm = 0.04 in). (Sheet 1/2 from the Atlas of the World Water Balance. copyright UNESCO, 1977)

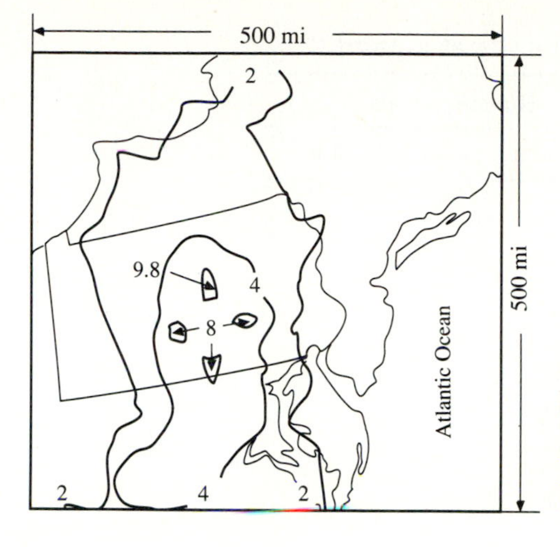

(*a*) Storm of May 30—June 1, 1889, which produced the Johnstown flood in Pennsylvania. Maximum rainfall of 9.8 in. recorded over 18 hour period at Wellsboro, Pennsylvania. Isohyets are in inches depth of total rainfall in the storm. (*Source*: U.S. Army Corps of Engineers, 1943.)

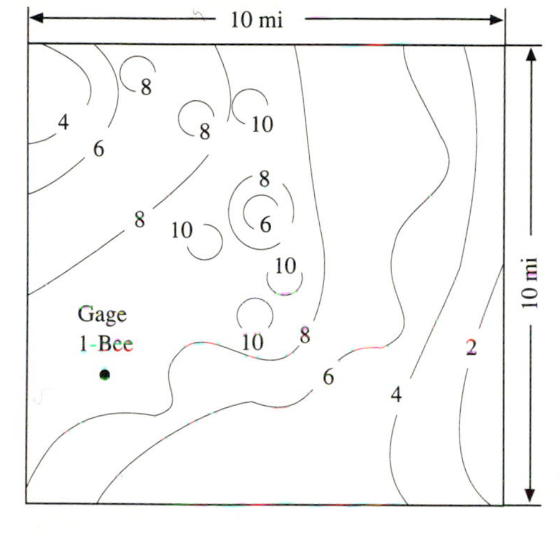

(*b*) Storm of May 24—25, 1981, in Austin, Texas. Maximum rainfall of 11 in. recorded over 3 hours. Isohyets are in inches depth of total rainfall in the storm. (*Source*: Massey, Reeves, and Lear, 1982.)

FIGURE 3.4.1
Isohyetal maps for two storms. The storms have about the same maximum depth of point rainfall, but the Johnstown storm covered a much larger area and had a longer duration than did the Austin storm.

successive increments in time, as shown in Table 3.4.1 for the data in 5-minute increments from gage 1-Bee in the Austin storm. A *rainfall hyetograph* is a plot of rainfall depth or intensity as a function of time, shown in the form of a histogram in Fig. 3.4.2(*a*) for the 1-Bee data. By summing the rainfall increments through time, a *cumulative rainfall hyetograph*, or *rainfall mass curve*, is produced, as shown in Table 3.4.1 and Fig. 3.4.2(*b*).

The maximum rainfall depth, or intensity, (depth/time) recorded in a given time interval in a storm is found by computing a series of running totals of rainfall depth for that time interval starting at various points in the storm, then selecting the maximum value of this series. For example, for a 30-minute time interval, Table 3.4.1 shows running totals beginning with 1.17 inches recorded in the first

TABLE 3.4.1
Computation of rainfall depth and intensity at a point

Time (min)	Rainfall (in)	Cumulative rainfall	Running Totals 30 min	1 h	2 h
0		0.00			
5	0.02	0.02			
10	0.34	0.36			
15	0.10	0.46			
20	0.04	0.50			
25	0.19	0.69			
30	0.48	1.17	1.17		
35	0.50	1.67	1.65		
40	0.50	2.17	1.81		
45	0.51	2.68	2.22		
50	0.16	2.84	2.34		
55	0.31	3.15	2.46		
60	0.66	3.81	2.64	3.81	
65	0.36	4.17	2.50	4.15	
70	0.39	4.56	2.39	4.20	
75	0.36	4.92	2.24	4.46	
80	0.54	5.46	2.62	4.96	
85	0.76	6.22	***3.07***	5.53	
90	0.51	6.73	2.92	***5.56***	
95	0.44	7.17	3.00	5.50	
100	0.25	7.42	2.86	5.25	
105	0.25	7.67	2.75	4.99	
110	0.22	7.89	2.43	5.05	
115	0.15	8.04	1.82	4.89	
120	0.09	8.13	1.40	4.32	8.13
125	0.09	8.22	1.05	4.05	***8.20***
130	0.12	8.34	0.92	3.78	7.98
135	0.03	8.37	0.70	3.45	7.91
140	0.01	8.38	0.49	2.92	7.88
145	0.02	8.40	0.36	2.18	7.71
150	0.01	8.41	0.28	1.68	7.24
Max. depth	0.76		3.07	5.56	8.20
Max. intensity (in/h)	9.12		6.14	5.56	4.10

30 minutes, 1.65 inches from 5 min to 35 min, 1.81 inches from 10 min to 40 min, and so on. The maximum 30 minute recorded depth is 3.07 inches recorded between 55 min and 85 min, corresponding to an average intensity of 3.07 in/0.5 h = 6.14 in/h over this interval. Table 3.4.1 shows similarly computed maximum depths and intensities for one and two-hour intervals. It can be seen that as the time period increases, the average intensity sustained by the storm decreases (5.56 in/h for one hour, 4.10 in/h for two hours), just as the average intensity over an area decreases as the area increases, as shown in Fig. 3.4.1. Computations

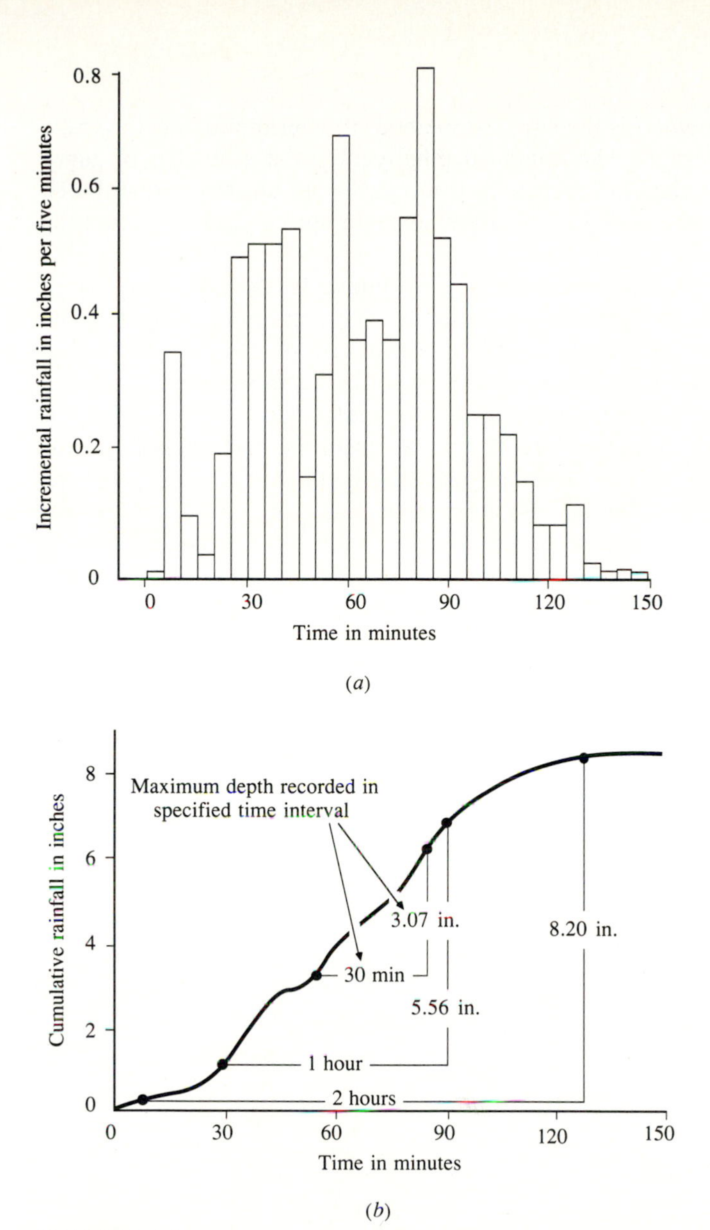

FIGURE 3.4.2
Incremental and cumulative rainfall hyetographs at gage 1-Bee for storm of May 24–25, 1981 in Austin, Texas.

of maximum rainfall depth and intensity performed in this way give an index of how severe a particular storm is, compared to other storms recorded at the same location, and they provide useful data for design of flow control structures. An important fact to be determined from historical rainfall records is the average depth of rainfall over an area such as a watershed.

Areal Rainfall

The *arithmetic-mean method* is the simplest method of determining areal average rainfall. It involves averaging the rainfall depths recorded at a number of gages [Fig. 3.4.3(*a*)]. This method is satisfactory if the gages are uniformly distributed over the area and the individual gage measurements do not vary greatly about the mean.

If some gages are considered more representative of the area in question than others, then relative weights may be assigned to the gages in computing the areal average. The *Thiessen method* assumes that at any point in the watershed the rainfall is the same as that at the nearest gage so the depth recorded at a given gage is applied out to a distance halfway to the next station in any direction. The relative weights for each gage are determined from the corresponding areas of application in a *Thiessen polygon* network, the boundaries of the polygons being formed by the perpendicular bisectors of the lines joining adjacent gages [Fig. 3.4.3(*b*)]. If there are J gages, and the area within the watershed assigned to each is A_j, and P_j is the rainfall recorded at the jth gage, the areal average precipitation for the watershed is

$$\overline{P} = \frac{1}{A}\sum_{j=1}^{J} A_j P_j \tag{3.4.1}$$

where the watershed area $A = \Sigma_{j=1}^{J} A_j$. The Thiessen method is generally more

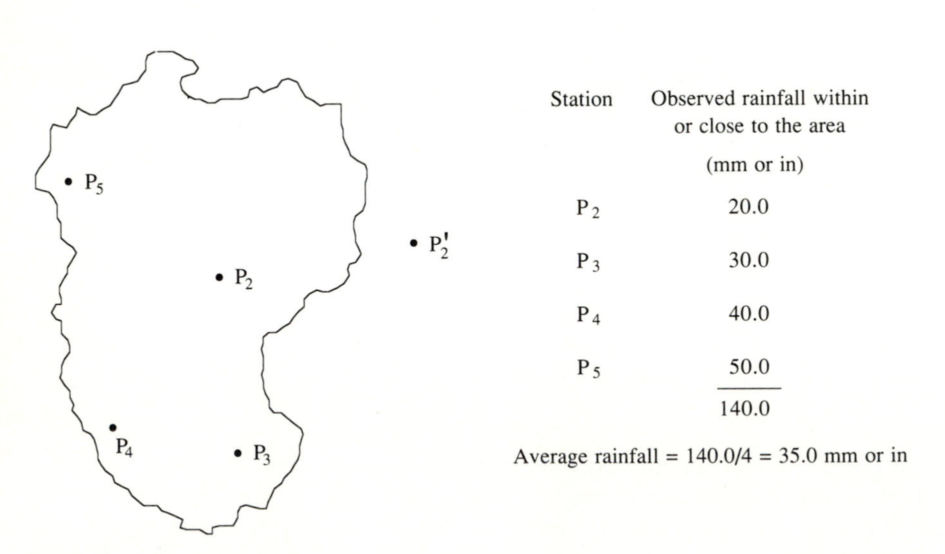

Station	Observed rainfall within or close to the area (mm or in)
P 2	20.0
P 3	30.0
P 4	40.0
P 5	50.0
	140.0

Average rainfall = 140.0/4 = 35.0 mm or in

FIGURE 3.4.3(*a*)
Computation of areal average rainfall by the arithmetic-mean method.

accurate than the arithmetic mean method, but it is inflexible, because a new Thiessen network must be constructed each time there is a change in the gage network, such as when data is missing from one of the gages. Also, the Thiessen method does not directly account for orographic influences on rainfall.

The *isohyetal method* overcomes some of these difficulties by constructing isohyets, using observed depths at rain gages and interpolation between adjacent gages [Fig. 3.4.3(c)]. Where there is a dense network of raingages, isohyetal maps can be constructed using computer programs for automated contouring. Once the isohyetal map is constructed, the area A_j between each pair of isohyets, within the watershed, is measured and multiplied by the average P_j of the rainfall depths of the two boundary isohyets to compute the areal average precipitation by Eq. (3.4.1). The isohyetal method is flexible, and knowledge of the storm pattern can influence the drawing of the isohyets, but a fairly dense network of gages is needed to correctly construct the isohyetal map from a complex storm.

Other methods of weighting rain gage records have been proposed, such as the *reciprocal-distance-squared method* in which the influence of the rainfall at a gaged point on the computation of rainfall at an ungaged point is inversely proportional to the distance between the two points (Wei and McGuinness, 1973). Singh and Chowdhury (1986) studied the various methods for calculating areal average precipitation, including the ones described here, and concluded that all the methods give comparable results, especially when the time period is long;

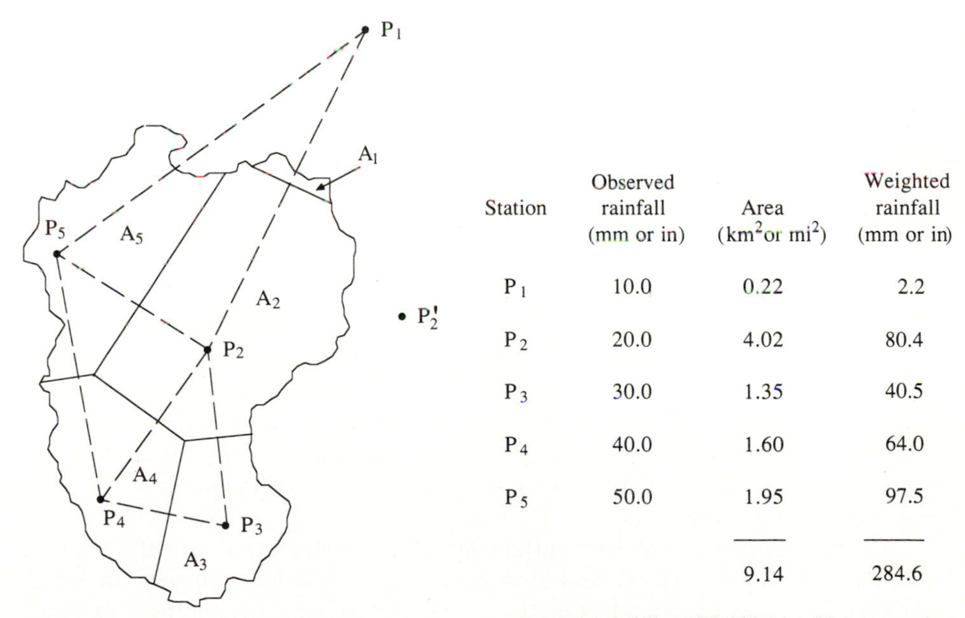

Station	Observed rainfall (mm or in)	Area (km^2 or mi^2)	Weighted rainfall (mm or in)
P_1	10.0	0.22	2.2
P_2	20.0	4.02	80.4
P_3	30.0	1.35	40.5
P_4	40.0	1.60	64.0
P_5	50.0	1.95	97.5
		9.14	284.6

Average rainfall = 284.6/9.14 = 31.1 mm or in

FIGURE 3.4.3(*b*)
Computation of areal average rainfall by the Thiessen method.

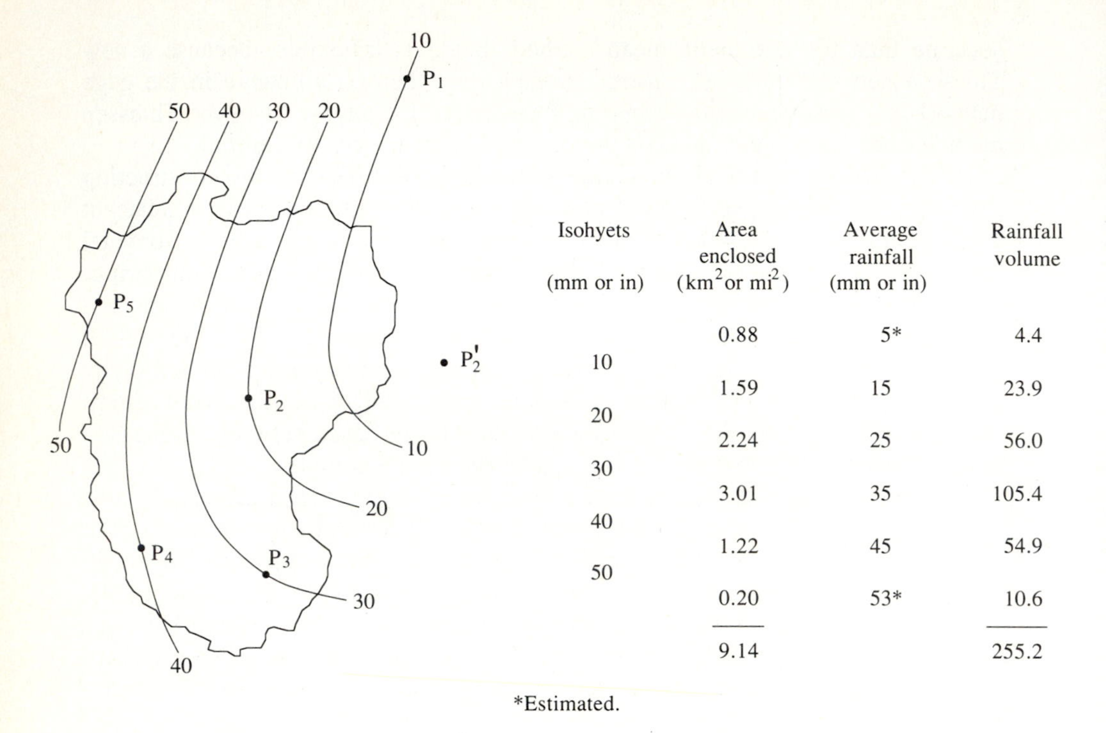

Isohyets (mm or in)	Area enclosed (km^2 or mi^2)	Average rainfall (mm or in)	Rainfall volume
	0.88	5*	4.4
10			
	1.59	15	23.9
20			
	2.24	25	56.0
30			
	3.01	35	105.4
40			
	1.22	45	54.9
50			
	0.20	53*	10.6
	9.14		255.2

*Estimated.

Average rainfall = 255.2 / 9.14 = 27.9 mm or in

FIGURE 3.4.3(*c*)
Computation of areal average rainfall by the isohyetal method.

that is, the different methods vary more from one to another when applied to daily rainfall data than when applied to annual data.

3.5 EVAPORATION

The two main factors influencing evaporation from an open water surface are the supply of energy to provide the latent heat of vaporization and the ability to transport the vapor away from the evaporative surface. Solar radiation is the main source of heat energy. The ability to transport vapor away from the evaporative surface depends on the wind velocity over the surface and the specific humidity gradient in the air above it.

Evaporation from the land surface comprises evaporation directly from the soil and vegetation surface, and *transpiration* through plant leaves, in which water is extracted by the plant's roots, transported upwards through its stem, and diffused into the atmosphere through tiny openings in the leaves called stomata. The processes of evaporation from the land surface and transpiration from vegetation are collectively termed *evapotranspiration*. Evapotranspiration is

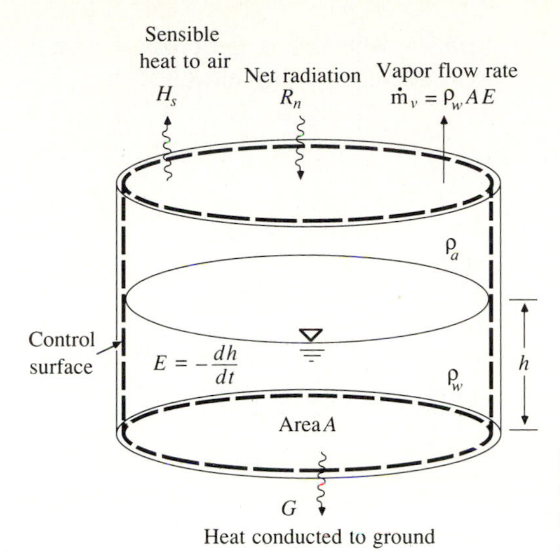

FIGURE 3.5.1
Control volume defined for continuity and energy equation development for an evaporation pan.

influenced by the two factors described previously for open water evaporation, and also by a third factor, the supply of moisture at the evaporative surface. The *potential evapotranspiration* is the evapotranspiration that would occur from a well vegetated surface when moisture supply is not limiting, and this is calculated in a way similar to that for open water evaporation. Actual evapotranspiration drops below its potential level as the soil dries out.

Energy Balance Method

To develop the continuity and energy equations applicable for evaporation, consider evaporation from an *evaporation pan* as shown in Fig. 3.5.1. An evaporation pan is a circular tank containing water, in which the rate of evaporation is measured by the rate of fall of the water surface. A control surface is drawn around the pan enclosing both the water in the pan and the air above it.

Continuity. Because the control volume contains water in both the liquid and vapor phases, the integral continuity equation must be written separately for the two phases. For the liquid phase, the extensive property is B = mass of liquid water; $\beta = 1, \rho = \rho_w$ (the density of water), and $dB/dt = -\dot{m}_v$, which is the mass flow rate of evaporation . The continuity equation for the liquid phase is

$$-\dot{m}_v = \frac{d}{dt}\iiint_{\text{c.v.}} \rho_w d\forall + \iint_{\text{c.s.}} \rho_w \mathbf{V}\cdot\mathbf{dA} \tag{3.5.1}$$

The pan has impermeable sides, so there is no flow of liquid water across the control surface and $\iint \rho_w \mathbf{V}\cdot\mathbf{dA} = 0$. The rate of change of storage within the system is given by $(d/dt)\iiint \rho_w d\forall = \rho_w A\, dh/dt$, where A is the cross-sectional area of the pan and h is the depth of water in it. Substituting into (3.5.1):

$$-\dot{m}_v = \rho_w A\left(\frac{dh}{dt}\right)$$

or

$$\dot{m}_v = \rho_w A E \tag{3.5.2}$$

where $E = -dh/dt$ is the *evaporation rate*.

For the vapor phase, B = mass of water vapor; $\beta = q_v$, the specific humidity, $\rho = \rho_a$, the air density, and $dB/dt = \dot{m}_v$, so the continuity equation for this phase is

$$\dot{m}_v = \frac{d}{dt}\iiint_{\text{c.v.}} q_v \rho_a d\forall + \iint_{\text{c.s.}} q_v \rho_a \mathbf{V}\cdot\mathbf{dA} \tag{3.5.3}$$

For a steady flow of air over the evaporation pan, the time derivative of water vapor stored within the control volume is zero. Thus, after substituting for $\dot{m}_v$ from (3.5.2), (3.5.3) becomes

$$\rho_w A E = \iint_{\text{c.s.}} q_v \rho_a \mathbf{V}\cdot\mathbf{dA} \tag{3.5.4}$$

which is the continuity equation for an evaporation pan, considering both water and water vapor. In a more general sense, (3.5.4) can be used to define the evaporation or evapotranspiration rate from any surface when written in the form

$$E = \left(\frac{1}{\rho_w A}\right)\iint_{\text{c.s.}} q_v \rho_a \mathbf{V}\cdot\mathbf{dA} \tag{3.5.5}$$

where E is the equivalent depth of water evaporated per unit time (in/day or mm/day).

Energy. The heat energy balance of a hydrologic system, as expressed by Eq. (2.7.4) can be applied to the water in the control volume:

$$\frac{dH}{dt} - \frac{dW}{dt} = \frac{d}{dt}\iiint_{\text{c.v.}}\left(e_u + \frac{1}{2}V^2 + gz\right)\rho\, d\forall + \iint_{\text{c.s.}}\left(e_u + \frac{1}{2}V^2 + gz\right)\rho\mathbf{V}\cdot\mathbf{dA} \tag{3.5.6}$$

where dH/dt is the rate of heat input to the system from external sources, dW/dt

is the rate of work done by the system (zero in this case), e_u is the specific internal heat energy of the water, and the two terms on the right hand side are, respectively, the rate of change of heat energy stored in the control volume and the net outflow of heat energy carried across the control surface with flowing water. Because $V = 0$ for the water in the evaporation pan, and the rate of change of its elevation, z, is very small, (3.5.6) can be simplified to

$$\frac{dH}{dt} = \frac{d}{dt}\iiint_{\text{c.v.}} e_u \rho_w d\forall \tag{3.5.7}$$

Considering a unit area of water surface, the source of heat energy is net radiation flux R_n, measured in watts per meter squared; the water supplies a sensible heat flux H_s to the air stream and a ground heat flux G to the ground surface, so $dH/dt = R_n - H_s - G$. If it is assumed that the temperature of the water within the control volume is constant in time, the only change in the heat stored within the control volume is the change in the internal energy of the water evaporated, which is equal to $l_v \dot{m}_v$, where l_v is the latent heat of vaporization. Hence, (3.5.7) can be rewritten as

$$R_n - H_s - G = l_v \dot{m}_v \tag{3.5.8}$$

By substituting for $\dot{m}_v$ from (3.5.2) with $A = 1\ \text{m}^2$, (3.5.8) may be solved for E:

$$E = \frac{1}{l_v \rho_w}(R_n - H_s - G) \tag{3.5.9}$$

which is the *energy balance equation* for evaporation. If the sensible heat flux H_s and the ground heat flux G are both zero, then an evaporation rate E_r can be calculated as the rate at which all the incoming net radiation is absorbed by evaporation:

$$E_r = \frac{R_n}{l_v \rho_w} \tag{3.5.10}$$

Example 3.5.1. Calculate by the energy balance method the evaporation rate from an open water surface, if the net radiation is 200 W/m^2 and the air temperature is 25°C, assuming no sensible heat or ground heat flux.

Solution. From (2.7.6) the latent heat of vaporization at 25°C is $l_v = 2500 - 2.36 \times 25 = 2441$ kJ/kg. From Table 2.5.2, water density $\rho_w = 997$ kg/m^3, and substitution into (3.5.10) gives

$$\begin{aligned} E_r &= \frac{200}{2441 \times 10^3 \times 997} \\ &= 8.22 \times 10^{-8} \text{ m/s} \\ &= 8.22 \times 10^{-8} \times 1000 \times 86400 \text{ mm/day} \\ &= 7.10 \text{ mm/day} \end{aligned}$$

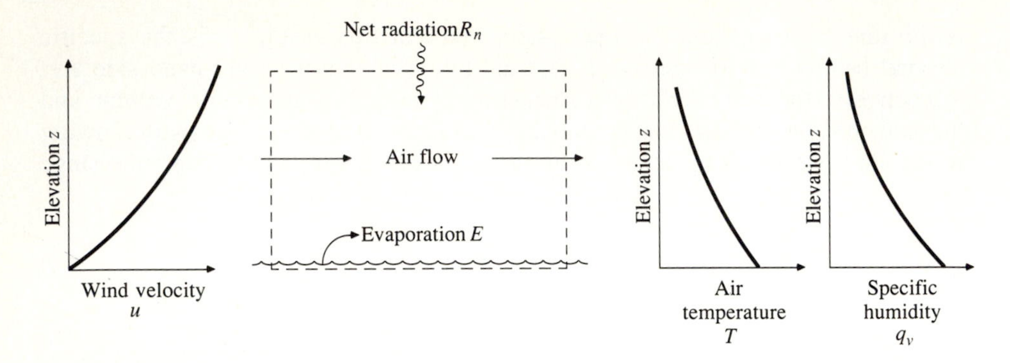

FIGURE 3.5.2
Evaporation from an open water surface.

Aerodynamic Method

Besides the supply of heat energy, the second factor controlling the evaporation rate from an open water surface is the ability to transport vapor away from the surface. The transport rate is governed by the humidity gradient in the air near the surface and the wind speed across the surface, and these two processes can be analyzed by coupling the equations for mass and momentum transport in air.

In the control volume shown in Fig. 3.5.2, consider a horizontal plane of unit area located at height z above the surface. The vapor flux $\dot{m}_v$ passing upward by convection through this plane is given by the equation (from Table 2.8.1 with $c = q_v$):

$$\dot{m}_v = -\rho_a K_w \frac{dq_v}{dz} \tag{3.5.11}$$

where K_w is the vapor eddy diffusivity. The momentum flux upward through the plane is likewise given by an equation from Table 2.8.1:

$$\tau = \rho_a K_m \frac{du}{dz} \tag{3.5.12}$$

Suppose the wind velocity u_1 and specific humidity q_{v_1} are measured at elevation z_1, and u_2 and q_{v_2} at elevation z_2, the elevations being sufficiently close that the transport rates $\dot{m}_v$ and τ are constant between them. Then the substitutions $dq_v/dz = (q_{v_2} - q_{v_1})/(z_2 - z_1)$ and $du/dz = (u_2 - u_1)/(z_2 - z_1)$ can be made in (3.5.11) and (3.5.12), respectively, and a ratio of the resulting equations taken to give

$$\frac{\dot{m}_v}{\tau} = -\frac{K_w(q_{v_2} - q_{v_1})}{K_m(u_2 - u_1)}$$

or

$$\dot{m}_v = \tau \frac{K_w(q_{v_1} - q_{v_2})}{K_m(u_2 - u_1)} \tag{3.5.13}$$

The wind velocity in the boundary layer near the earth's surface (up to about 50 m) is well described by the logarithmic profile law [Eq. (2.8.5)]

$$\frac{u}{u^*} = \frac{1}{k} \ln\left(\frac{z}{z_0}\right) \tag{3.5.14}$$

where u^* = shear velocity= $\sqrt{\tau/\rho_a}$, k is the von Karman constant, usually taken as 0.4, and z_0 is the roughness height of the surface given in Table 2.8.2. Hence,

$$u_2 - u_1 = \frac{u^*}{k}\left[\ln\left(\frac{z_2}{z_0}\right) - \ln\left(\frac{z_1}{z_0}\right)\right]$$

$$= \frac{u^*}{k} \ln\left(\frac{z_2}{z_1}\right)$$

and

$$u^* = \frac{k(u_2 - u_1)}{\ln(z_2/z_1)}$$

But u^* = $\sqrt{\tau/\rho_a}$ by definition, hence

$$\tau = \rho_a\left[\frac{k(u_2 - u_1)}{\ln(z_2/z_1)}\right]^2$$

Substituting this result into (3.5.13) and rearranging gives

$$\dot{m}_v = \frac{K_w k^2 \rho_a (q_{v_1} - q_{v_2})(u_2 - u_1)}{K_m\left[\ln(z_2/z_1)\right]^2} \tag{3.5.15}$$

which is the *Thornthwaite-Holzman equation* for vapor transport, first developed by Thornthwaite and Holzman (1939). In application it is usually assumed that the ratio $K_w/K_m = 1$ and is constant. Thornthwaite and Holzman set up measurement towers to sample q_v and u at different heights and computed the corresponding evaporation rate, and many subsequent investigators have made similar experiments.

For operational application where such apparatus is not available and measurements of q_v and u are made at only one height in a standard climate station, Eq. (3.5.15) is simplified by assuming that the wind velocity $u_1 = 0$ at the roughness height $z_1 = z_0$ and that the air is saturated with moisture there. From Eq. (3.2.6), $q_v = 0.622\, e/p$, where e is the vapor pressure and p is the ambient air pressure (the same at both heights), so measurements of vapor pressure can be substituted for those of specific humidity. At height z_2, the vapor pressure is e_a, the ambient vapor pressure in air, and the vapor pressure at the surface is taken to be e_{as}, the saturated vapor pressure corresponding to the ambient air temperature. Under these assumptions (3.5.15) is rewritten as

$$\dot{m}_v = \frac{0.622k^2\rho_a(e_{as} - e_a)u_2}{p\left[\ln (z_2/z_0)\right]^2} \qquad (3.5.16)$$

Recalling that $\dot{m}_v$ is defined here for a unit area of surface, an equivalent evaporation rate E_a, expressed in dimensions of [L/T], can be found by setting $\dot{m}_v = \rho_w E_a$ in (3.5.16) and rearranging:

$$E_a = B(e_{as} - e_a) \qquad (3.5.17)$$

where

$$B = \frac{0.622k^2\rho_a u_2}{p\rho_w\left[\ln (z_2/z_0)\right]^2} \qquad (3.5.18)$$

Eq. (3.5.17) is a common basis for many evaporation equations, with the form of the vapor transfer coefficient B varying from one place to another. This type of equation was first proposed by Dalton in 1802.

Example 3.5.2 Calculate the evaporation rate from an open water surface by the aerodynamic method with air temperature 25°C, relative humidity 40 percent, air pressure 101.3 kPa, and wind speed 3 m/s, all measured at height 2 m above the water surface. Assume a roughness height $z_0 = 0.03$ cm.

Solution. The vapor transfer coefficient B is given by (3.5.18), using $k = 0.4$, $\rho_a = 1.19$ kg/m^3 for air at 25°C, and $\rho_w = 997$ kg/m^3. Hence

$$\begin{aligned} B &= \frac{0.622k^2\rho_a u_2}{p\rho_w\left[\ln (z_2/z_0)\right]^2} \\ &= \frac{0.622 \times 0.4^2 \times 1.19 \times 3}{101.3 \times 10^3 \times 997\{\ln\left[2/(3 \times 10^{-4})\right]\}^2} \\ &= 4.54 \times 10^{-11} \text{ m/Pa·s} \end{aligned}$$

The evaporation rate is given by (3.5.17), using $e_{as} = 3167$ Pa at 25°C from Table (3.2.1) and, from (3.2.11), $e_a = R_h e_{as} = 0.4 \times 3167 = 1267$ Pa:

$$\begin{aligned} E_a &= B(e_{as} - e_a) \\ &= 4.54 \times 10^{-11}(3167 - 1267) \\ &= 8.62 \times 10^{-8} \text{ m/s} \\ &= 8.62 \times 10^{-8} \times \left(\frac{1000 \text{ mm}}{1 \text{ m}}\right) \times \left(\frac{86400 \text{ s}}{\text{day}}\right) \\ &= 7.45 \text{ mm/day} \end{aligned}$$

Combined Aerodynamic and Energy Balance Method

Evaporation may be computed by the aerodynamic method when energy supply is not limiting and by the energy balance method when vapor transport is not limiting. But, normally, both of these factors are limiting, so a combination of the two methods is needed. In the energy balance method, the sensible heat flux H_s is difficult to quantify. But since the heat is transferred by convection through the air overlying the water surface, and water vapor is similarly transferred by convection, it can be assumed that the vapor heat flux $l_v \dot{m}_v$ and the sensible heat flux H_s are proportional, the proportionality constant being called the *Bowen ratio* β (Bowen, 1926):

$$\beta = \frac{H_s}{l_v \dot{m}_v} \tag{3.5.19}$$

The energy balance equation (3.5.9) with ground heat flux $G = 0$ can then be written as

$$R_n = l_v \dot{m}_v (1 + \beta) \tag{3.5.20}$$

The Bowen ratio is calculated by coupling the transport equations for vapor and heat, this is similar to the coupling of the vapor and momentum transport equations used in developing the Thornthwaite-Holzman equation. From Table 2.8.1, the transport equations for vapor and heat are

$$\dot{m}_v = -\rho_a K_w \frac{dq_v}{dz} \tag{3.5.21}$$

$$H_s = -\rho_a C_p K_h \frac{dT}{dz} \tag{3.5.22}$$

where C_p is the specific heat at constant pressure and K_h is the heat diffusivity. Using measurements of q_v and T made at two levels z_1 and z_2 and assuming the transport rate is constant between these levels, division of (3.5.22) by (3.5.21) gives

$$\frac{H_s}{\dot{m}_v} = \frac{C_p K_h (T_2 - T_1)}{K_w (q_{v_2} - q_{v_1})} \tag{3.5.23}$$

Dividing (3.5.23) by l_v and substituting $0.622\, e/p$ for q_v provides the expression for the Bowen ratio β from (3.5.19)

$$\beta = \frac{C_p K_h p (T_2 - T_1)}{0.622 l_v K_w (e_2 - e_1)}$$

or

$$\beta = \gamma \left(\frac{T_2 - T_1}{e_2 - e_1} \right) \tag{3.5.24}$$

where γ is the *psychrometric constant*

$$\gamma = \frac{C_p K_h p}{0.622 l_v K_w} \tag{3.5.25}$$

The ratio K_h/K_w of the heat and vapor diffusivities is commonly taken to be 1 (Priestley and Taylor, 1972).

If the two levels 1 and 2 are taken at the evaporative surface and in the overlying air stream, respectively, it can be shown that the evaporation rate E_r computed from the rate of net radiation [as given by Eq. (3.5.10)] and the evaporation rate computed from aerodynamic methods [Eq. (3.5.17)] can be combined to yield a weighted estimate of evaporation E, by

$$E = \frac{\Delta}{\Delta + \gamma} E_r + \frac{\gamma}{\Delta + \gamma} E_a \tag{3.5.26}$$

where γ is the psychrometric constant and Δ is the gradient of the saturated vapor pressure curve at air temperature T_a, as given by (3.2.10); the weighting factors $\Delta/(\Delta + \gamma)$ and $\gamma/(\Delta + \gamma)$ sum to unity. Equation (3.5.26) is the basic equation for the combination method of computing evaporation, which was first developed by Penman (1948). Its derivation is lengthy (see Wiesner, 1970), and will not be presented here.

The combination method of calculating evaporation from meteorological data is the most accurate method when all the required data are available and the assumptions are satisfied. The chief assumptions of the energy balance are that steady state energy flow prevails and that changes in heat storage over time in the water body are not significant. This assumption limits the application of the method to daily time intervals or longer, and to situations not involving large heat storage capacity, such as a large lake possesses. The chief assumption of the aerodynamic method is associated with the form of the vapor transfer coefficient B in Eq. (3.5.17). Many empirical forms of B have been proposed, locally fitted to observed wind and other meteorological data.

The combination method is well suited for application to small areas with detailed climatological data. The required data include net radiation, air temperature, humidity, wind speed, and air pressure. When some of these data are unavailable, simpler evaporation equations requiring fewer variables must be used (American Society of Civil Engineers, 1973; Doorenbos and Pruitt, 1977). For evaporation over very large areas, energy balance considerations largely govern the evaporation rate. For such cases Priestley and Taylor (1972) found that the second term of the combination equation (3.5.26) is approximately 30 percent of the first, so that (3.5.26) can be rewritten as the *Priestley-Taylor evaporation equation*

$$E = \alpha \frac{\Delta}{\Delta + \gamma} E_r \tag{3.5.27}$$

where $\alpha = 1.3$. Other investigators have confirmed the validity of this approach, with the value of α varying slightly from one location to another.

Pan evaporation data provide the best indication of nearby open water evaporation where such data are available. The observed values of pan evaporation E_p are multiplied by a *pan factor* k_p $(0 \leq k_p \leq 1)$ to convert them to equivalent open water evaporation values. Usually $k_p \approx 0.7$, but this factor varies by season and location.

The formulas for the various methods of calculating evaporation are summarized in Table 3.5.1.

Example 3.5.3. Use the combination method to calculate the evaporation rate from an open water surface subject to net radiation of 200 W/m^2, air temperature 25°C, relative humidity 40 percent, and wind speed 3 m/s, all recorded at height 2 m, and atmospheric pressure 101.3 kPa.

Solution. From Example 3.5.1 the evaporation rate corresponding to a net radiation of 200 W/m^2 is E_r = 7.10 mm/day, and from Example 3.5.2, the aerodynamic method yields E_a = 7.45 mm/day for the given air temperature, humidity, and wind speed conditions. The combination method requires values for Δ and γ in Eq. (3.5.26). The psychrometric constant γ is given by (3.5.25), using C_p = 1005 J/kg·K for air, K_h/K_w = 1.00, and $l_v = 2441 \times 10^3$ J/kg at 25°C (from Example 3.5.1):

$$
\begin{aligned}
\gamma &= \frac{C_p K_h p}{0.622 l_v K_w} \\
&= \frac{1005 \times 1.00 \times 101.3 \times 10^3}{0.622 \times 2441 \times 10^3} \\
&= 67.1 \text{ Pa/°C}
\end{aligned}
$$

Δ is the gradient of the saturated vapor pressure curve at 25°C, given by (3.2.10) with $e_s = e_{as}$ = 3167 Pa for T = 25°C:

$$
\begin{aligned}
\Delta &= \frac{4098 e_s}{(237.3 + T)^2} \\
&= \frac{4098 \times 3167}{(237.3 + 25)^2} \\
&= 188.7 \text{ Pa/°C}
\end{aligned}
$$

The weights in the combination equation, then, are $\gamma/(\Delta + \gamma)$ = 67.1/(188.7 + 67.1) = 0.262 and $\Delta/(\Delta + \gamma)$ = 188.7/(188.7 + 67.1) = 0.738. The evaporation rate is then computed by (3.5.26):

$$
\begin{aligned}
E &= \frac{\Delta}{\Delta + \gamma} E_r + \frac{\gamma}{\Delta + \gamma} E_a \\
&= 0.738 \times 7.10 + 0.262 \times 7.45 \\
&= 7.2 \text{ mm/day}
\end{aligned}
$$

Example 3.5.4 Use the Priestley-Taylor method to calculate the evaporation rate for a water body with net radiation 200 W/m^2 and air temperature 25°C.

Solution. The Priestley-Taylor method uses Eq. (3.5.27) with E_r = 7.10 mm/day from Example 3.5.1, $\Delta/(\Delta + \gamma)$ = 0.738 at 25°C from Example 3.5.3, and α = 1.3. Hence,

$$
E = \alpha \frac{\Delta}{\Delta + \gamma} E_r
$$

TABLE 3.5.1
Summary of equations for calculating evaporation*

(1) Energy balance method

$$E_r = 0.0353 R_n \text{ (mm/day)}$$

where

$$R_n = \text{net radiation (W/m}^2\text{)}$$

(2) Aerodynamic method

$$E_a = B(e_{as} - e_a) \text{ (mm/day)}$$

where

$$B = \frac{0.102 u_2}{\left[\ln\left(\frac{z_2}{z_0}\right)\right]^2} \text{ (mm/day} \cdot \text{Pa)}$$

u_2 is wind velocity (m/s) measured at height z_2 (cm), and z_0 is from Table 2.8.2. Also,

$$e_{as} = 611 \exp\left(\frac{17.27T}{237.3 + T}\right) \text{ (Pa)}$$

$$T = \text{air temperature (°C)}$$

$$e_a = R_h e_{as} \text{ (Pa)}$$

in which R_h is the relative humidity ($0 \leq R_h \leq 1$).

(3) Combination method

$$E = \frac{\Delta}{\Delta + \gamma} E_r + \frac{\gamma}{\Delta + \gamma} E_a \text{ (mm/day)}$$

where

$$\Delta = \frac{4098 e_{as}}{(237.3 + T)^2} \text{ (Pa/°C)}$$

and

$$\gamma = 66.8 \text{ (Pa/°C)}$$

(4) Priestley-Taylor method

$$E = \alpha \frac{\Delta}{\Delta + \gamma} E_r$$

where $\alpha = 1.3$

* The values shown are valid for standard atmospheric pressure and air temperature 20°C.

$$= 1.3 \times 0.738 \times 7.10$$

$$= 6.8 \text{ mm/day}$$

which is close to the result from the more complicated combination method shown in the previous example.

3.6 EVAPOTRANSPIRATION

Evapotranspiration is the combination of evaporation from the soil surface and transpiration from vegetation. The same factors governing open water evaporation also govern evapotranspiration, namely energy supply and vapor transport. In addition, a third factor enters the picture: the supply of moisture at the evaporative surface. As the soil dries out, the rate of evapotranspiration drops below the level it would have maintained in a well watered soil.

Calculations of the rate of evapotranspiration are made using the same methods described previously for open water evaporation, with adjustments to account for the condition of the vegetation and soil (Van Bavel, 1966; Monteith, 1980). For given climatic conditions, the basic rate is the *reference crop evapotranspiration,* this being "the rate of evapotranspiration from an extensive surface of 8 cm to 15 cm tall green grass cover of uniform height, actively growing, completely shading the ground and not short of water" (Doorenbos and Pruitt, 1977).

Comparisons of computed and measured values of evapotranspiration have been made at many locations by the American Society of Civil Engineers (1973) and by Doorenbos and Pruitt (1977). They concluded that the combination method of Eq. (3.5.26) is the best approach, especially if the vapor transport coefficient B in Eq. (3.5.18) is calibrated for local conditions. For example, Doorenbos and Pruitt recommend

$$B = 0.0027\left(1 + \frac{u}{100}\right) \tag{3.6.1}$$

in which B is in mm/day·Pa and u is the 24-hour wind run in kilometers per day measured at height 2 m. The 24-hr wind run is the cumulative distance a particle would move in the airstream in 24 hours under the prevailing wind conditions. Note that the dimensions of u given here are not meters per second as used in the equation for B given in Table 3.5.1, but the resulting value of E_a is in millimeters per day in both cases.

The potential evapotranspiration of another crop growing under the same conditions as the reference crop is calculated by multiplying the reference crop evapotranspiration E_{tr} by a *crop coefficient* k_c, the value of which changes with the stage of growth of the crop. The actual evapotranspiration E_t is found by multiplying the potential evapotranspiration by a *soil coefficient* k_s $(0 \leq k_s \leq 1)$:

$$E_t = k_s k_c E_{tr} \tag{3.6.2}$$

The values of the crop coefficient k_c vary over a range of about $0.2 \leq k_c \leq 1.3$, as shown in Fig. 3.6.1 (Doorenbos and Pruitt, 1977). The initial value of k_c, for well-watered soil with little vegetation, is approximately 0.35. As the vegetation develops, k_c increases to a maximum value, which can be greater than 1 for crops with large vegetative cover, such as corn, which transpire at a greater rate than grass. As the crop matures or ripens, its moisture requirements diminish. The precise shape of the crop coefficient curve varies with the agricultural practices of a region, such as the times of plowing and harvest. Some vegetation, such as orchards or permanent ground cover, may not exhibit all the growth stages shown in Fig. 3.6.1.

Example 3.6.1. (From Gouevsky, Maidment, and Sikorski, 1980) The monthly values of reference crop evapotranspiration E_{tr}, calculated using the combination

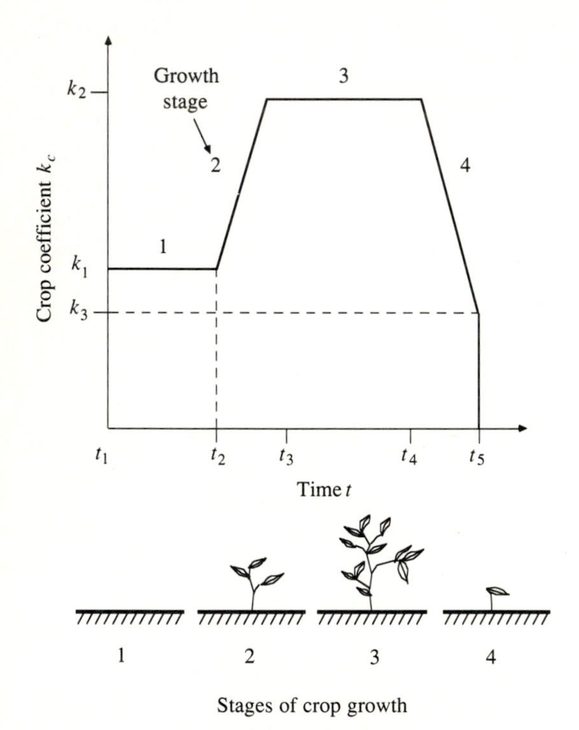

Stage	Crop condition
1	Initial stage — less than 10% ground cover.
2	Development stage — from initial stage to attainment of effective full ground cover (70 — 80%).
3	Mid-season stage — from full ground cover to maturation.
4	Late season stage — full maturity and harvest.

FIGURE 3.6.1
The relationship between the crop coefficient k_c and the stage of crop growth.

method, for average conditions in Silistra, Bulgaria, are shown in the table below. The crop coefficients for corn (see Fig. 3.6.1) are $k_1 = 0.38, k_2 = 1.00$, and $k_3 = 0.55$; t_1 = April 1, t_2 = June 1, t_3 = July 1, t_4 = September 1, and t_5 = October 1. Calculate the actual evapotranspiration from this crop assuming a well-watered soil.

Solution.

Month	Apr	May	Jun	Jul	Aug	Sep	Oct	Apr–Oct total
E_{tr} (mm/day)	4.14	5.45	5.82	6.60	5.94	4.05	2.34	34.3 mm
k_c	0.38	0.38	0.69	1.00	1.00	0.78	0.55	
E_t (mm/day)	1.57	2.07	4.02	6.60	5.94	3.16	1.29	24.7 mm

Monthly average values of k_c are specified following the curve in Fig. 3.6.1 using the given values. In June, k_c rises from 0.38 at t_2 = June 1 to 1.00 at t_3 = July 1, so k_c is taken as $(0.38 + 1.00)/2 = 0.69$. The values of E_t are computed using Eq. (3.6.2) with $k_s = 1$ for a well-watered soil; that is, $E_t = k_c E_{tr}$. The total evapotranspiration for the growing season from April to October for corn, 24.7 mm, is 72 percent of the value a grass cover would have yielded under the same conditions, 34.3 mm.

REFERENCES

American Society of Civil Engineers, Consumptive use of water and irrigation water requirements, ed. by M. E. Jensen, *Technical Committee on Irrigation Water Requirements*, New York, 1973.

Bowen, I. S., The ratio of heat losses by conduction and by evaporation from any water surface, *Phys. Rev.*, vol. 27, no. 6, pp. 779–787, 1926.

Brutsaert, W., *Evaporation into the Atmosphere*, D. Reidel, Dordrecht, Holland, 1982.

Doorenbos, J., and W. O. Pruitt, Crop water requirements, *Irrigation and Drainage Paper 24,* U. N. Food and Agriculture Organization, Rome, Italy, 1977.

Gouevsky, I. V., D. R. Maidment, and W. Sikorski, Agricultural water demands in the Silistra region, RR-80-38, Int. Inst. App. Sys. Anal., Laxenburg, Austria, 1980.

Mason, B. J., *The Physics of Clouds*, Oxford University Press, London, 1957.

Massey, B. C., W. E. Reeves, and W. A. Lear, Flood of May 24–25, 1981, in the Austin, Texas, metropolitan area, *Hydrologic Investigations Atlas, HA-656*, U. S. Geological Survey, 1982.

Monteith, J. L., The development and extension of Penman's evaporation formula, in *Applications of Soil Physics*, ed. by D. Hillel, Academic Press, Orlando, Fla., pp. 247–253, 1980.

Moore, W. L., et al., The Austin, Texas, flood of May 24–25, 1981, *Report for Committee on Natural Disasters*, National Academy of Sciences, Washington, D.C., 1982.

Penman, H. L., Natural evaporation from open water, bare soil, and grass, *Proc. R. Soc. London A*, vol. 193, pp. 120–146, 1948.

Priestley, C. H. B., and R. J. Taylor, On the assessment of surface heat flux and evaporation using large-scale parameters, *Monthly Weather Rev.*, vol. 100, pp. 81–92, 1972.

Pruppacher, H. R., and J. D. Klett, *Microphysics of Clouds and Precipitation,* D. Reidel, Dordrecht, Holland, 1978.

Raudkivi, A. J., *Hydrology*, Pergamon Press, Oxford, 1979.
Singh, V. P., and P. K. Chowdhury, Comparing some methods of estimating mean areal rainfall, *Water Resour. Bull.*, vol. 22, no. 2, pp. 275–282, 1986.
Thornthwaite, C. W., and B. Holzman, The determination of evaporation from land and water surfaces, *Monthly Weather Rev.*, vol. 67, pp. 4–11, 1939.
U. S. Army Corps of Engineers, Storm studies data sheet, Baltimore District Office, Mid-Atlantic Division, Baltimore, Md., 1943.
Van Bavel, C. H. M., Potential evaporation: the combination concept and its experimental verification, *Water Resour. Res.*, vol. 2, no. 3, pp. 455–467, 1966.
Wei, T. C., and J. L. McGuinness, Reciprocal distance squared method, a computer technique for estimating area precipitation, *ARS-NC-8*, U. S. Agricultural Research Service, North Central Region, Coshocton, Ohio, 1973.
Wiesner, C. J., *Hydrometeorology*, Chapman and Hall, London, 1970.

PROBLEMS

3.1.1 A parcel of air at the equator is at rest relative to the earth's surface. Considering the effects on air motion of the conservation of angular momentum, calculate the theoretical eastward velocity of the air relative to the earth's surface if the parcel is moved to 30°N latitude. Mean radius of earth = 6371 km.

3.1.2 A parcel of air initially at rest relative to the earth's surface, is moved to a latitude α° (either north or south). Considering only the effect of the conservation of angular momentum, show that the velocity of the parcel of air relative to the earth's surface is $2\pi r_e \sin\alpha \tan\alpha/T$, where r_e is the mean radius of the earth and T is the period of the earth's rotation about its own axis.

3.2.1 At a climate station, the following measurements are made: air pressure = 101.1 kPa, air temperature = 25°C, and dew point temperature = 20°C. Calculate the corresponding vapor pressure, relative humidity, specific humidity, and air density.

3.2.2 Calculate the vapor pressure, air pressure, specific humidity, and air density at elevation 1500 m if the surface conditions are as specified in Prob. 3.2.1 and the lapse rate is 9°C/km.

3.2.3 If the air temperature is 15°C and the relative humidity 35 percent, calculate the vapor pressure, specific humidity, and air density. Assume standard atmospheric pressure (101.3 kPa).

3.2.4 Solve Prob. 3.2.3 if the air temperature rises to 30°C. By what percentage does the specific humidity increase as a result of the temperature rise from 15 to 30°C?

3.2.5 Calculate the precipitable water (mm) in a 10-km-high saturated atmospheric column if the surface conditions are temperature = 20°C, pressure = 101.3 kPa, and the lapse rate is 6.5°C/km.

3.2.6 Solve Prob. 3.2.5 for surface temperatures of 0, 10, 20, 30 and 40°C and plot a graph showing the variation of precipitable water depth with surface temperature.

3.3.1 Calculate the terminal velocity of a 0.8-mm raindrop at standard atmospheric pressure and air temperature 20°C. Air density = 1.20 kg/m^3.

3.3.2 An air current moving vertically upward at 5 m/s carries raindrops of various sizes. Calculate the velocity of a 2-mm-diameter drop and determine whether it is rising or falling. Repeat this exercise for a 0.2-mm-diameter drop. Assume standard atmospheric pressure and air temperature 20°C. Air density = 1.20 kg/m^3.

3.3.3 If a spherical raindrop of diameter D, density ρ_w, and drag coefficient C_d, is released from rest in an atmosphere of density ρ_a, show that the distance z it falls to attain velocity V is given by

$$z = -\frac{2\rho_w D}{3\rho_a C_d} \ln\left[1 - \frac{3\rho_a C_d V^2}{4Dg(\rho_w - \rho_a)}\right]$$

Assume $V \leq$ terminal velocity.

3.3.4 Using the equation given in Prob. 3.3.3, calculate the distance a 0.8-mm raindrop would need to fall to attain 50 percent, 90 percent, and 99 percent of its terminal velocity at standard atmospheric pressure and 20°C air temperature.

3.3.5 Raindrops of diameter 1 mm are falling on an erodable soil. Estimate the impact energy of each drop. Assume standard atmospheric conditions of 20°C temperature and 101.3 kPa air pressure. Hint: the drop will lose its kinetic energy on impact.

3.3.6 Solve Prob. 3.3.5 for drop sizes of 0.1, 0.5, 1, and 5 mm and plot a graph showing the variation of impact energy with drop size.

3.3.7 For the thunderstorm cell model, show that the proportion of incoming moisture precipitated is given by $(q_{v_1} - q_{v_2})/q_{v_1}(1 - q_{v_2})$, where q_{v_1} and q_{v_2} are the specific humidities of the inflow and outflow air streams, respectively.

3.3.8 Solve Example 3.3.2 in the text to determine the precipitation intensity if the surface temperature is 20°C. By what percentage is the precipitation intensity reduced by lowering the surface temperature from 30 to 20°C? Calculate the rate of release of latent heat in the thunderstorm through condensation of water vapor to produce precipitation.

3.3.9 Solve Example 3.3.2 in the text to determine the precipitation from a thunderstorm if the moisture outflow is at elevation 5 km. What percentage of the incoming moisture is now precipitated?

3.4.1 Lay a piece of graph paper over the isohyetal map for the Johnstown storm [Fig. 3.4.1(*a*)] and trace the isohyets. Calculate the volume of precipitation in this storm and the average depth of precipitation within the area bounded by the 2-in isohyet.

3.4.2 Calculate the average depth of precipitation over the 10 mi × 10 mi area shown for the Austin storm [Fig. 3.4.1(*b*)].

3.4.3 The following rainfall data were recorded at gage 1-Bol for the storm of May 24–25, 1981, Austin, Texas:

Time (min)	0	5	10	15	20	25	30	35	40
Rainfall (in)	–	0.07	0.20	0.25	0.22	0.21	0.16	0.12	0.03

Plot the rainfall hyetograph. Compute and plot the cumulative rainfall hyetograph. Calculate the maximum depth and intensity recorded in 10, 20, and 30 minutes for this storm. Compare the 30-minute intensity with the value found in Table 3.4.1 in the text for gage 1-Bee.

3.4.4 The following incremental rainfall data were recorded at gage 1-WLN in Austin, Texas, on May 24, 1981. Plot the rainfall hyetograph. Compute and plot the cumulative rainfall hyetograph. Calculate the maximum depth and intensity of rainfall for 5, 10, 30, 60, 90, 120 minutes for this storm. Compare the results for 30, 60, and 120 minutes with the values given in Table 3.4.1 for gage 1-Bee in the same storm. Which gage experienced the more severe rainfall?

Time (min)	0	5	10	15	20	25	30	35	40	45	50
Rainfall (in)	–	0.09	0.00	0.03	0.13	0.10	0.13	0.21	0.37	0.22	0.30
Time (min)	55	60	65	70	75	80	85	90	95	100	105
Rainfall (in)	0.20	0.10	0.13	0.14	0.12	0.16	0.14	0.18	0.25	0.48	0.40
Time (min)	110	115	120	125	130	135	140	145	150		
Rainfall (in)	0.39	0.24	0.41	0.44	0.27	0.17	0.17	0.14	0.10		

3.4.5 The shape of a drainage basin can be approximated by a polygon whose vertices are located at the following coordinates: (5,5), (−5,5), (−5,−5), (0, −10), and (5,−5). The rainfall amounts of a storm were recorded by a number of rain gages situated within and nearby the basin as follows:

Gage number	Coordinates	Recorded rainfall (mm)
1	(7, 4)	62
2	(3, 4)	59
3	(−2, 5)	41
4	(−10, 1)	39
5	(−3, −3)	105
6	(−7, −7)	98
7	(2, −3)	60
8	(2,−10)	41
9	(0, 0)	81

All coordinates are expressed in kilometers. Determine the average rainfall on the basin by (*a*) the arithmetic-mean method, (*b*) the Thiessen method, and (*c*) the isohyetal method. Hints: For the Thiessen method, begin by drawing a polygon around gage 9, then draw polygons around gages 2, 3, 5, and 7; for the isohyetal method, draw the isohyets with maximum rainfall on a ridge running southwest to northeast through (−3,−3).

3.4.6 Compute the average rainfall over the drainage area in Fig. 3.4.3 if gage station P_2 is moved to P_2' using (*a*) the arithmetic-mean method, (*b*) the Thiessen method, and (*c*) the isohyetal method.

3.4.7 Four rain gages located within a rectangular area with four corners at (0,0), (0,13), (14,13), and (14,0) have the following coordinates and recorded rainfalls:

Raingage location	Rainfall (in)
(2, 9)	0.59
(7,11)	0.79
(12,10)	0.94
(6, 2)	1.69

All coordinates are expressed in miles. Compute the average rainfall in the area by the Thiessen method.

3.5.1 Compute by the Priestley-Taylor method the evaporation rate in millimeters per day from a lake on a winter day when the air temperature is 5°C and the net radiation 50 W/m^2, and on a summer day when the net radiation is 250 W/m^2 and the temperature is 30°C.

3.5.2 For Cairo, Egypt, in July, average net radiation is 185 W/m^2, air temperature 28.5°C, relative humidity 55 percent, and wind speed 2.7 m/s at height 2 m. Calculate the open water evaporation rate in millimeters per day using the energy method (E_r), the aerodynamic method (E_a), the combination method, and the Priestley-Taylor method. Assume standard atmospheric pressure (101.3 kPa) and $z_0 = 0.03$ cm.

3.5.3 For Cairo in January, the average weather conditions are: net radiation 40 W/m^2, temperature 14°C, relative humidity 65 percent, and wind speed 2.0 m/s measured at height 2 m. Calculate the open water evaporation rate by the energy method (E_r), the aerodynamic method (E_a), the combination method, and the Priestley-Taylor method. Assume standard atmospheric pressure (101.3 kPa) and $z_0 = 0.03$ cm.

3.6.1 For the meteorological data for Cairo in July given in Prob. 3.5.2, calculate the reference crop evapotranspiration using the Doorenbos and Pruitt vapor transfer coefficient $B = 0.0027[1 + (u/100)]$ where u is the wind run in kilometers per day.

3.6.2 Compute the reference crop evapotranspiration (mm/day) in January in Cairo using the meteorological data given in Prob. 3.5.3 and the Doorenbos and Pruitt vapor transfer coefficient $B = 0.0027[1 + (u/100)]$, where u is the wind run in kilometers per day.

3.6.3 The following data (from the American Society of Civil Engineers, 1973) show climatic conditions over a well-watered grass surface in May, July, and September in Davis, California, (latitude 38°N). Calculate the corresponding evapotranspiration rate (mm/day) by the energy balance method, the aerodynamic method, the combination method, and the Priestley-Taylor method. Assume standard atmospheric pressure. Use Eq. (3.6.1) for the coefficient B.

	Temperature (°C)	Vapor Pressure (kPa)	Net radiation (W/m^2)	Wind run (km/day)
May	17	1.1	169	167
July	23	1.4	189	121
September	20	1.2	114	133

3.6.4 Solve Prob. 3.6.3 for Coshocton, Ohio, where the meteorological conditions are:

	Temperature (°C)	Vapor Pressure (kPa)	Net radiation (W/m^2)	Wind run (km/day)
May	16	1.3	135	110
July	23	2.0	112	89
September	18	1.5	59	94

3.6.5 Use the aerodynamic method to calculate the evapotranspiration rate (mm/day) from a well-watered, short grass area on a day when the average air temperature is 25°C, relative humidity is 30 percent, 24-hour wind run is 100 km, and normal atmospheric pressure (101.3 kPa) prevails. Assume the Doorenbos-Pruitt wind function (3.6.1) is valid. By what percentage would the evapotranspiration rate change if the relative humidity were doubled and the temperature, wind speed, and air pressure remained constant?

CHAPTER 4

SUBSURFACE WATER

Subsurface water flows beneath the land surface. In this chapter, only subsurface flow processes important to surface water hydrology are described. The broader field of groundwater flow is covered in a number of other textbooks (Freeze and Cherry, 1979; de Marsily, 1986).

4.1 UNSATURATED FLOW

Subsurface flow processes and the zones in which they occur are shown schematically in Fig. 4.1.1. Three important processes are *infiltration* of surface water into the soil to become *soil moisture*, *subsurface flow* or unsaturated flow through the soil, and *groundwater flow* or saturated flow through soil or rock strata. Soil and rock strata which permit water flow are called *porous media*. Flow is unsaturated when the porous medium still has some of its voids occupied by air, and saturated when the voids are filled with water. The *water table* is the surface where the water in a saturated porous medium is at atmospheric pressure. Below the water table, the porous medium is saturated and at greater pressure than atmospheric. Above the water table, capillary forces can saturate the porous medium for a short distance in the *capillary fringe*, above which the porous medium is usually unsaturated except following rainfall, when infiltration from the land surface can produce saturated conditions temporarily. *Subsurface* and *groundwater outflow* occur when subsurface water emerges to become surface flow in a stream or spring. Soil moisture is extracted by evapotranspiration as the soil dries out.

Consider a cross section through an unsaturated soil as shown in Fig. 4.1.2. A portion of the cross section is occupied by solid particles and the remainder by

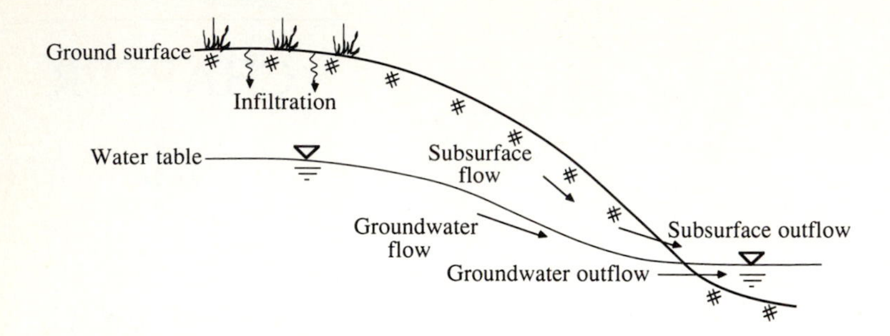

FIGURE 4.1.1
Subsurface water zones and processes.

voids. The *porosity* η is defined as

$$\eta = \frac{\text{volume of voids}}{\text{total volume}} \tag{4.1.1}$$

The range for η is approximately $0.25 < \eta < 0.75$ for soils, the value depending on the soil texture (see Table 2.6.1).

A part of the voids is occupied by water and the remainder by air, the volume occupied by water being measured by the *soil moisture content* θ defined as

$$\theta = \frac{\text{volume of water}}{\text{total volume}} \tag{4.1.2}$$

Hence $0 \leq \theta \leq \eta$; the soil moisture content is equal to the porosity when the soil is saturated.

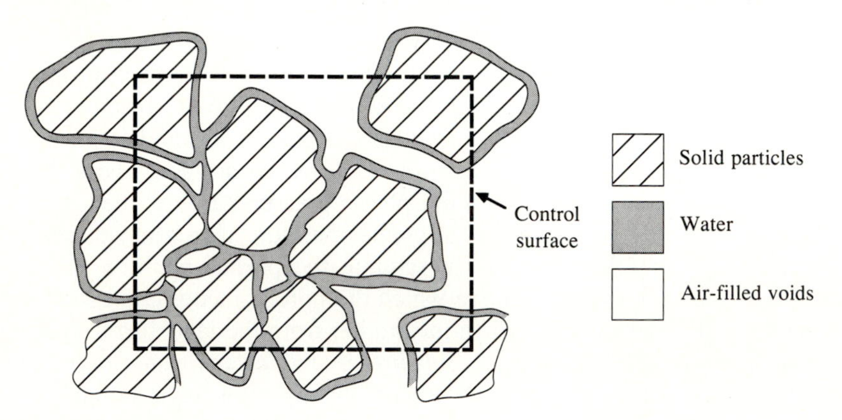

FIGURE 4.1.2
Cross section through an unsaturated porous medium.

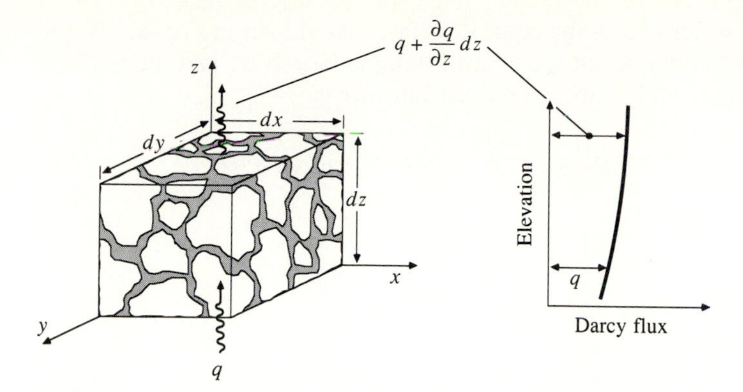

FIGURE 4.1.3
Control volume for development of the continuity equation in an unsaturated porous medium.

Continuity Equation

A control volume containing unsaturated soil is shown in Figure 4.1.3. Its sides have length dx, dy, and dz, in the coordinate directions, so its volume is $dx\,dy\,dz$, and the volume of water contained in the control volume is $\theta\,dx\,dy\,dz$. The flow of water through the soil is measured by the *Darcy flux* $q = Q/A$, the volumetric flow rate per unit area of soil. The Darcy flux is a vector, having components in each of the coordinate directions, but in this presentation the horizontal fluxes are assumed to be zero, and only the vertical or z component of the Darcy flux is considered. As the z axis is postive upward, upward flow is considered positive and downward flow negative.

In the Reynolds transport theorem, the extensive property B is the mass of soil water, hence $\beta = dB/dm = 1$, and $dB/dt = 0$ because no phase changes are occurring in the water. The Reynolds transport theorem thus takes the form of the integral equation of continuity (2.2.1):

$$0 = \frac{d}{dt}\iiint_{\text{c.v.}} \rho_w \, d\forall + \iint_{\text{c.s.}} \rho_w \mathbf{V}\cdot\mathbf{dA} \tag{4.1.3}$$

where ρ_w is the density of water. The first term in (4.1.3) is the time rate of change of the mass of water stored within the control volume, which is given by

$$\begin{aligned}\frac{d}{dt}\iiint_{\text{c.v.}} \rho_w \, d\forall &= \frac{d}{dt}(\rho_w \theta \, dx\,dy\,dz) \\ &= \rho_w \, dx\,dy\,dz \, \frac{\partial \theta}{\partial t}\end{aligned} \tag{4.1.4}$$

where the density is assumed constant and the partial derivative suffices because the spatial dimensions of the control volume are fixed. The second term in (4.1.3) is the net outflow of water across the control surface. As shown in Fig. 4.1.3, the volumetric inflow at the bottom of the control volume is $q\,dx\,dy$ and the outflow at the top is $[q + (\partial q/\partial z)dz]\,dx\,dy$, so the net outflow is

$$\iint_{c.s.} \rho_w \mathbf{V}\cdot\mathbf{dA} = \rho_w\left(q + \frac{\partial q}{\partial z}dz\right)dx\,dy - \rho_w q\,dx\,dy$$
$$= \rho_w\,dx\,dy\,dz\,\frac{\partial q}{\partial z} \tag{4.1.5}$$

Substituting (4.1.4) and (4.1.5) into (4.1.3) and dividing by $\rho_w\,dx\,dy\,dz$ gives

$$\frac{\partial \theta}{\partial t} + \frac{\partial q}{\partial z} = 0 \tag{4.1.6}$$

This is the continuity equation for one-dimensional unsteady unsaturated flow in a porous medium. This equation is applicable to flow at shallow depths below the land surface. At greater depth, such as in deep aquifers, changes in the water density and in the porosity can occur as the result of changes in fluid pressure, and these must also be accounted for in developing the continuity equation.

Momentum Equation

In Eq. (2.6.4) *Darcy's Law* was developed to relate the Darcy flux q to the rate of head loss per unit length of medium, S_f:

$$q = KS_f \tag{4.1.7}$$

Consider flow in the vertical direction and denote the total head of the flow by h; then $S_f = -\partial h/\partial z$ where the negative sign indicates that the total head is decreasing in the direction of flow because of friction. Darcy's law is then expressed as

$$q = -K\frac{\partial h}{\partial z} \tag{4.1.8}$$

Darcy's Law applies to a cross section of the porous medium found by averaging over an area that is large compared with the cross section of individual pores and grains of the medium (Philip, 1969). At this scale, Darcy's law describes a steady uniform flow of constant velocity, in which the net force on any fluid element is zero. For unconfined saturated flow the only two forces involved are gravity and friction, but for unsaturated flow the *suction force* binding water to soil particles through surface tension must also be included.

The porous medium is made up of a matrix of particles, as shown in Fig. 4.1.2. When the void spaces are only partially filled with water, the water is attracted to the particle surfaces through electrostatic forces between the water molecules' polar bonds and the particle surfaces. This surface adhesion draws the water up around the particle surfaces, leaving the air in the center of the voids. As more water is added to the porous medium, the air exits upwards and the area of free surfaces diminishes within the medium, until the medium is saturated and there are no free surfaces within the voids and, therefore, no soil suction force. The effect of soil suction can be seen if a column of dry soil is placed vertically with its bottom in a container of water—moisture will be drawn up into the dry soil to a height above the water surface at which the soil suction and gravity forces are just equal. This height ranges from a few millimeters for a coarse sand to several meters for a clay soil.

The head h of the water is measured in dimensions of height but can also be thought of as the energy per unit weight of the fluid. In an unsaturated porous medium, the part of the total energy possessed by the fluid due to the soil suction forces is referred to as the *suction head* ψ. From the preceding discussion, it is evident that the suction head will vary with the moisture content of the medium, as illustrated in Fig. 4.1.4, which shows that for this clay soil, the suction head and hydraulic conductivity can range over several orders of magnitude as the moisture content changes. The total head h is the sum of the suction and gravity heads

$$h = \psi + z \tag{4.1.9}$$

No term is included for the velocity head of the flow because the velocity is so small that its head is negligible.

Substituting for h in (4.1.8)

$$q = -K\frac{\partial(\psi + z)}{\partial z} \tag{4.1.10}$$

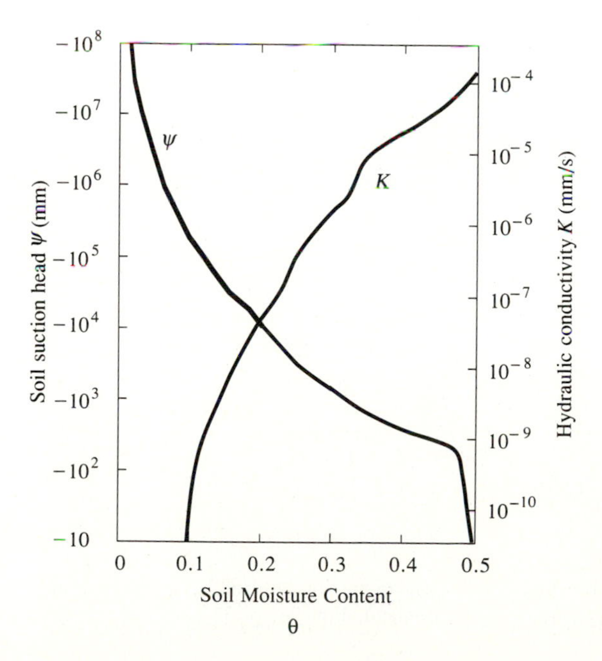

FIGURE 4.1.4
Variation of soil suction head ψ and hydraulic conductivity K with moisture content θ for Yolo light clay. (Reprinted with permission from A. J. Raudkivi, *Hydrology*, Copyright 1979, Pergamon Books Ltd.)

$$= -\left(K\frac{d\psi}{d\theta}\frac{\partial\theta}{\partial z} + K\right)$$

$$= -\left(D\frac{\partial\theta}{\partial z} + K\right) \quad (4.1.11)$$

where D is the *soil water diffusivity* $K(d\psi/d\theta)$ which has dimensions [L^2/T]. Substituting this result into the continuity equation (4.1.6) gives

$$\frac{\partial\theta}{\partial t} = \frac{\partial}{\partial z}\left(D\frac{\partial\theta}{\partial z} + K\right) \quad (4.1.12)$$

which is a one-dimensional form of *Richard's equation*, the governing equation for unsteady unsaturated flow in a porous medium, first presented by Richards (1931).

Computation of Soil Moisture Flux

The flow of moisture through the soil can be calculated by Eq. (4.1.8) given measurements of soil suction head ψ at different depths z in the soil and knowledge

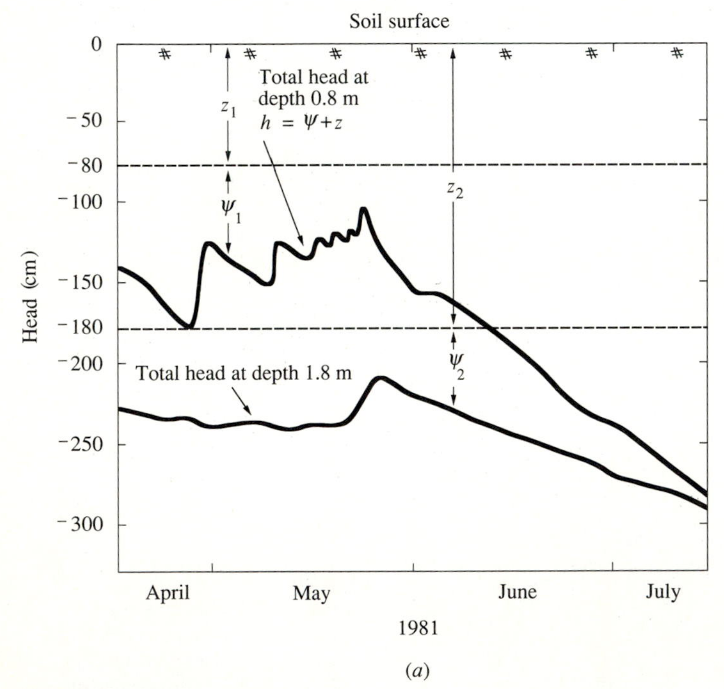

FIGURE 4.1.5(*a*)
Profiles of total soil moisture head through time at Deep Dean in Sussex, England. (*Source:* Research Report 1981–84, Institute of Hydrology, Wallingford, England, Fig. 36, p. 33, 1984. Used with permission.)

of the relationship between hydraulic conductivity K and ψ. Figure 4.1.5(a) shows profiles through time of soil moisture head measured by tensiometers located at depths 0.8 m and 1.8 m in a soil at Deep Dean, Sussex, England. The total head h is found by adding the measured suction head ψ to the depth z at which it was measured. These are both negative: z because it is taken as positive upward with 0 at the soil surface, and ψ because it is a suction force which resists flow of moisture away from the location.

Example 4.1.1. Calculate the soil moisture flux q (cm/day) between depths 0.8 m and 1.8 m in the soil at Deep Dean. The data for total head at these depths

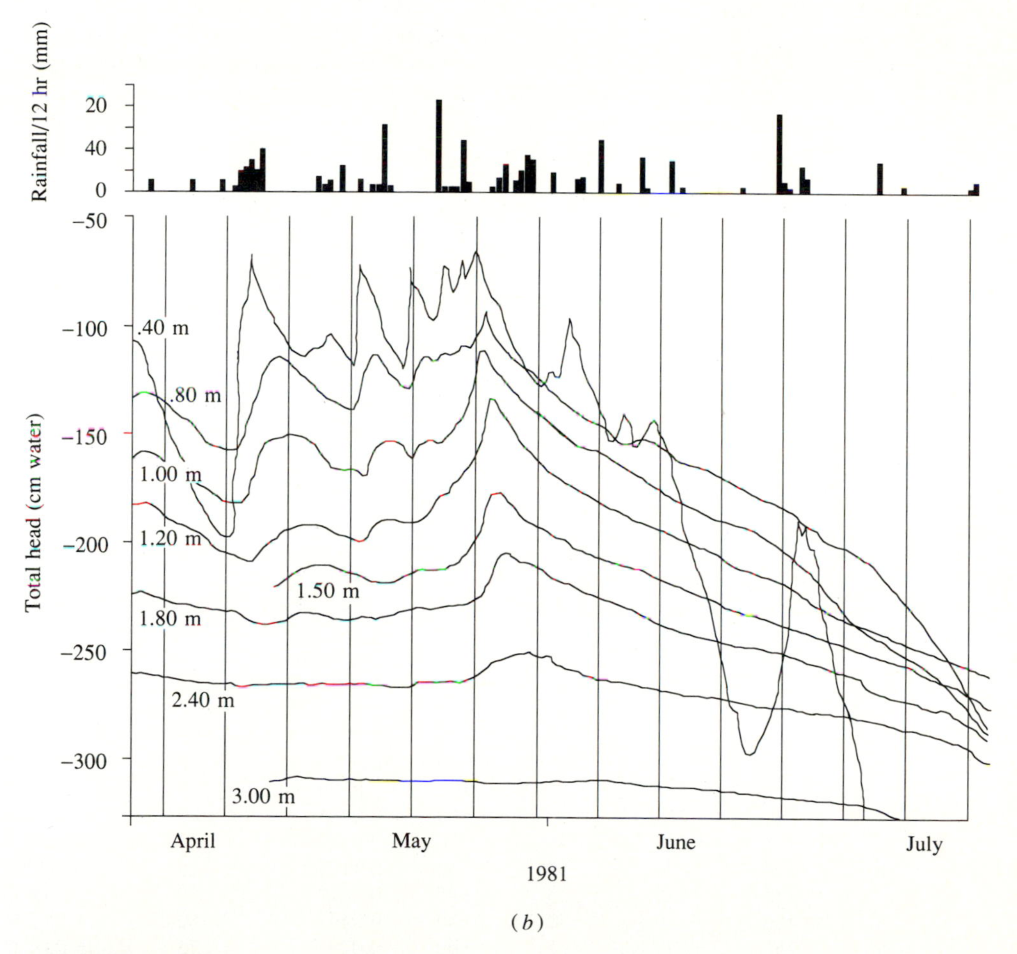

(b)

FIGURE 4.1.5(b)
Variation through time of total soil water head h at various depths in a loam soil at Deep Dean, Sussex, England. The infiltration of rainfall reduces soil suction which increases again an evapotranspiration dries out the soil. Soil suction head is the difference betwen the total head and the value for evaluation shown on each line. (*Source:* Research Report 1981–84, Institute of Hydrology, Wallingford, England, Fig. 36, p. 33, 1984. Used with permission.)

are given at weekly time intervals in columns 2 and 3 of Table 4.1.1. For this soil the relationship between hydraulic conductivity and soil suction head is $K = 250(-\psi)^{-2.11}$, where K is in centimeters per day and ψ is in centimeters.

Solution. Equation (4.1.8) is rewritten for an average flux q_{12} between measurement points 1 and 2 as

$$q_{12} = -K\frac{h_1 - h_2}{z_1 - z_2}$$

In this case, measurement point 1 is at 0.8 m and point 2 at 1.8 m, so $z_1 = -80$ cm, $z_2 = -180$ cm, and $z_1 - z_2 = -80 - (-180) = 100$ cm. The suction head at each depth is $\psi = h - z$. For example, for week 1 at 0.8 m, $h_1 = -145$, so $\psi_1 = h_1 - z_1 = -145 - (-80) = -65$ cm, and $\psi_2 = -230 - (-180) = -50$ cm, as shown in columns 4 and 5 of the table. The hydraulic conductivity K varies with ψ, so the value corresponding to the average of the ψ values at 0.8 and 1.8 m is used. For week 1, the average suction head is $\psi_{av} = [(-50) + (-65)]/2 = -57.5$ cm; and the corresponding hydraulic conductivity is $K = 250(-\psi_{av})^{-2.11} = 250(57.5)^{-2.11} = 0.0484$ cm/day, as shown in column 6. The head difference $h_1 - h_2 = (-145) - (-230) = 85$ cm. The soil moisture flux between 0.8 and 1.8 m for week 1 is

$$\begin{aligned} q &= -K\frac{h_1 - h_2}{z_1 - z_2} \\ &= -0.0484\frac{85}{100} \\ &= -0.0412 \text{ cm/day} \end{aligned}$$

TABLE 4.1.1
Computation of soil moisture flux between 0.8 m and 1.8 m depth at Deep Dean (Example 4.1.1)

Column: 1 Week	2 Total head h_1 at 0.8 m (cm)	3 Total head h_2 at 1.8 m (cm)	4 Suction head ψ_1 at 0.8 m (cm)	5 Suction ψ_2 at 1.8 m (cm)	6 Unsaturated hydraulic conductivity K (cm/day)	7 Head difference $h_1 - h_2$ (cm)	8 Moisture flux q (cm/day)
1	−145	−230	−65	−50	0.0484	85	−0.0412
2	−165	−235	−85	−55	0.0320	70	−0.0224
3	−130	−240	−50	−60	0.0532	110	−0.0585
4	−140	−240	−60	−60	0.0443	100	−0.0443
5	−125	−240	−45	−60	0.0587	115	−0.0675
6	−105	−230	−25	−50	0.1193	125	−0.1492
7	−135	−215	−55	−35	0.0812	80	−0.0650
8	−150	−230	−70	−50	0.0443	80	−0.0354
9	−165	−240	−85	−60	0.0297	75	−0.0223
10	−190	−245	−110	−65	0.0200	55	−0.0110
11	−220	−255	−140	−75	0.0129	35	−0.0045
12	−230	−265	−150	−85	0.0107	35	−0.0038
13	−255	−275	−175	−95	0.0080	20	−0.0016
14	−280	−285	−200	−105	0.0062	5	−0.0003

as shown in column 8. The flux is negative because the moisture is flowing downward.

The Darcy flux has dimensions [L/T] because it is a flow per unit area of porous medium. If the flux is passing through a horizontal plane of area $A = 1 \text{ m}^2$, then the volumetric flow rate in week 1 is

$$
\begin{aligned}
Q &= qA \\
&= -0.0412 \text{ cm/day} \times 1 \text{ m}^2 \\
&= -4.12 \times 10^{-4} \text{ m}^3\text{/day} \\
&= -0.412 \text{ liters/day } (-0.11 \text{ gal/day})
\end{aligned}
$$

Table 4.1.1 shows the flux q calculated for all time periods, and the computed values of q, K, and $h_1 - h_2$ are plotted in Fig. 4.1.6. In all cases the head at 0.8 m is greater than that at 1.8 m so moisture is always being driven downward between these two depths in this example. It can be seen that the flux reaches a maximum in week 6 and diminishes thereafter, because both the head difference and the hydraulic conductivity diminish as the soil dries out. The figure shows the importance of the variability of the unsaturated hydraulic conductivity K in affecting the moisture flux q. As the soil becomes wetter, its hydraulic conductivity increases, because there are more continuous fluid-filled pathways through which the flow can move.

The complete picture of rainfall on the soil at Deep Dean and the soil moisture head at various depths is presented in Fig. 4.1.5(b). Rainfall during April and May flows down into the soil, reducing the soil suction head, but later the soil dries out by evapotranspiration, causing the soil suction head to increase again. The head profile at the shallowest depth (0.4 m) shows the greatest variability and the fact that it falls below the profile at 0.8 m from the beginning of June onwards shows that during this period, soil moisture flows upwards between these two depths to supply moisture for evapotranspiration (Wellings, 1984).

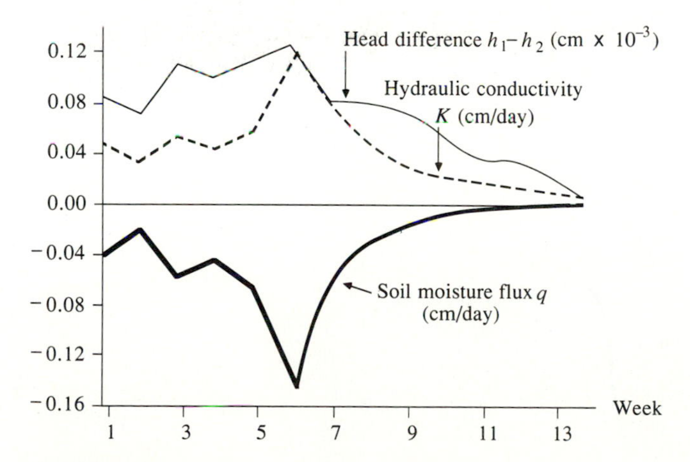

FIGURE 4.1.6
Computation of the soil moisture flux at Deep Dean (Example 4.1.1).

4.2 INFILTRATION

Infiltration is the process of water penetrating from the ground surface into the soil. Many factors influence the infiltration rate, including the condition of the soil surface and its vegetative cover, the properties of the soil, such as its porosity and hydraulic conductivity, and the current moisture content of the soil. Soil strata with different physical properties may overlay each other, forming *horizons*; for example, a silt soil with relatively high hydraulic conductivity may overlay a clay zone of low conductivity. Also, soils exhibit great spatial variability even within relatively small areas such as a field. As a result of these great spatial variations and the time variations in soil properties that occur as the soil moisture content changes, infiltration is a very complex process that can be described only approximately with mathematical equations.

The distribution of soil moisture within the soil profile during the downward movement of water is illustrated in Fig. 4.2.1. There are four moisture zones: a *saturated zone* near the surface, a *tranmission zone* of unsaturated flow and fairly uniform moisture content, a *wetting zone* in which moisture decreases with depth, and a *wetting front* where the change of moisture content with depth is so great as to give the appearance of a sharp discontinuity between the wet soil above and the dry soil below. Depending on the amount of infiltration and the physical properties of the soil, the wetting front may penetrate from a few inches to several feet into a soil (Hillel, 1980).

The *infiltration rate f*, expressed in inches per hour or centimeters per hour, is the rate at which water enters the soil at the surface. If water is ponded on the surface, the infiltration occurs at the *potential infiltration rate*. If the rate of supply of water at the surface, for example by rainfall, is less than the potential infiltration rate then the actual infiltration rate will also be less than the potential rate. Most infiltration equations describe the potential rate. The *cumulative infiltration F* is the accumulated depth of water infiltrated during a given time period and is equal to the integral of the infiltration rate over that period:

$$F(t) = \int_0^t f(\tau)\, d\tau \tag{4.2.1}$$

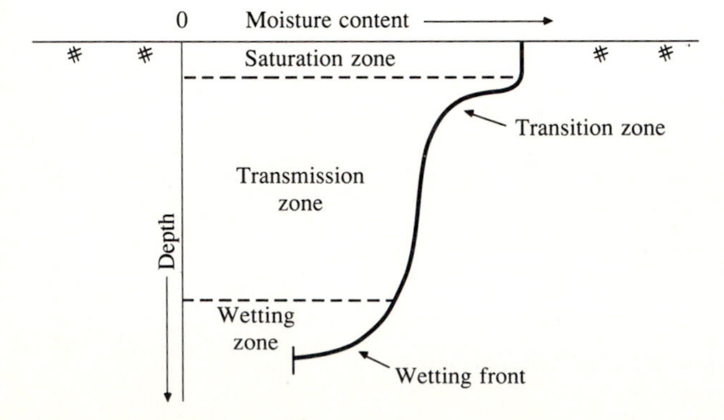

FIGURE 4.2.1
Moisture zones during infiltration.

where τ is a dummy variable of time in the integration. Conversely, the infiltration rate is the time derivative of the cumulative infiltration:

$$f(t) = \frac{dF(t)}{dt} \tag{4.2.2}$$

Horton's Equation

One of the earliest infiltration equations was developed by Horton (1933, 1939), who observed that infiltration begins at some rate f_0 and exponentially decreases until it reaches a constant rate f_c (Fig. 4.2.2):

$$f(t) = f_c + (f_0 - f_c)e^{-kt} \tag{4.2.3}$$

where k is a decay constant having dimensions $[T^{-1}]$. Eagleson (1970) and Raudkivi (1979) have shown that Horton's equation can be derived from Richard's equation (4.1.12) by assuming that K and D are constants independent of the moisture content of the soil. Under these conditions (4.1.12) reduces to

$$\frac{\partial \theta}{\partial t} = D\frac{\partial^2 \theta}{\partial z^2} \tag{4.2.4}$$

which is the standard form of a diffusion equation and may be solved to yield the moisture content θ as a function of time and depth. Horton's equation results from solving for the rate of moisture diffusion $D(\partial\theta/\partial z)$ at the soil surface.

Philip's Equation

Philip (1957, 1969) solved Richard's equation under less restrictive conditions by assuming that K and D can vary with the mositure content θ. Philip employed the Boltzmann transformation $B(\theta) = zt^{-1/2}$ to convert (4.1.12) into an ordinary differential equation in B, and solved this equation to yield an infinite series for

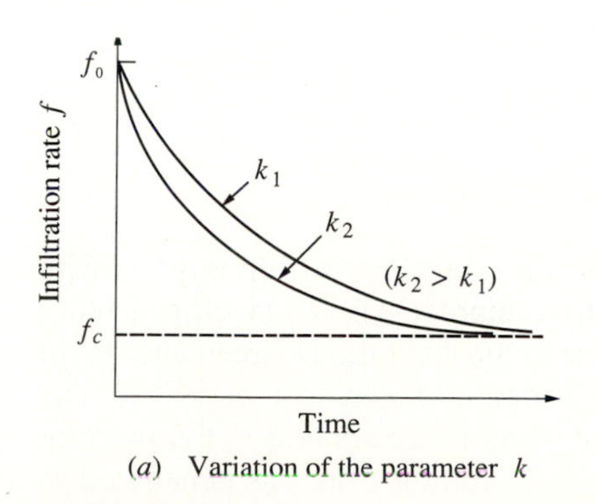

(*a*) Variation of the parameter k

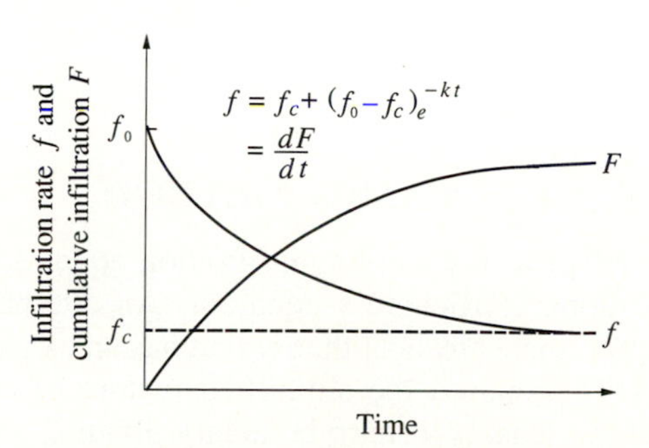

(*b*) Infiltration rate and cumulative infiltration.

FIGURE 4.2.2
Infiltration by Horton's equation.

cumulative infiltration $F(t)$, which is approximated by

$$F(t) = St^{1/2} + Kt \tag{4.2.5}$$

where S is a parameter called *sorptivity*, which is a function of the soil suction potential, and K is the hydraulic conductivity.

By differentiation

$$f(t) = \frac{1}{2}St^{-1/2} + K \tag{4.2.6}$$

As $t \rightarrow \infty$, $f(t)$ tends to K. The two terms in Philip's equation represent the effects of soil suction head and gravity head, respectively. For a horizontal column of soil, soil suction is the only force drawing water into the column, and Philip's equation reduces to $F(t) = St^{1/2}$.

Example 4.2.1. A small tube with a cross-sectional area of 40 cm^2 is filled with soil and laid horizontally. The open end of the tube is saturated, and after 15 minutes, 100 cm^3 of water have infiltrated into the tube. If the saturated hydraulic conductivity of the soil is 0.4 cm/h, determine how much infiltration would have taken place in 30 minutes if the soil column had initially been placed upright with its upper surface saturated.

Solution. The cumulative infiltration depth in the horizontal column is $F = 100$ $cm^3/40$ $cm^2 = 2.5$ cm. For horizontal infiltration, cumulative infiltration is a function of soil suction alone so that after $t = 15$ min $= 0.25$ h,

$$F(t) = St^{1/2}$$

and

$$2.5 = S(0.25)^{1/2}$$

$$S = 5 \text{ cm}\cdot\text{h}^{-1/2}$$

For infiltration down a vertical column, (4.2.5) applies with $K = 0.4$ cm/h. Hence, with $t = 30$ min $= 0.5$ h

$$\begin{aligned} F(t) &= St^{1/2} + Kt \\ &= 5(0.5)^{1/2} + 0.4(0.5) \\ &= 3.74 \text{ cm} \end{aligned}$$

4.3 GREEN-AMPT METHOD

In the previous section, infiltration equations were developed from approximate solutions of Richard's equation. An alternative approach is to develop a more approximate physical theory that has an exact analytical solution. Green and Ampt (1911) proposed the simplified picture of infiltration shown in Fig. 4.3.1. The wetting front is a sharp boundary dividing soil of moisture content θ_i below from saturated soil with moisture content η above. The wetting front has penetrated to a depth L in time t since infiltration began. Water is ponded to a small depth h_0 on the soil surface.

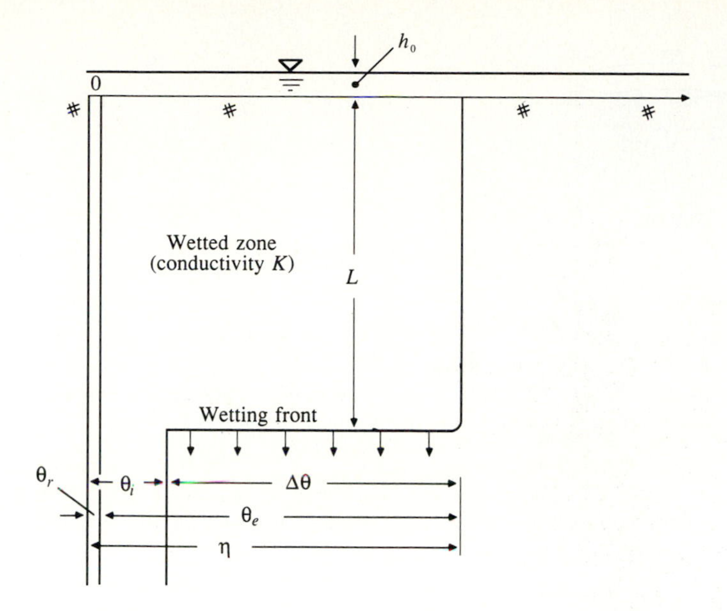

FIGURE 4.3.1
Variables in the Green-Ampt infiltration model. The vertical axis is the distance from the soil surface, the horizontal axis is the moisture content of the soil.

Continuity

Consider a vertical column of soil of unit horizontal cross-sectional area (Fig. 4.3.2) and let a control volume be defined around the wet soil between the surface and depth L. If the soil was initially of moisture content θ_i throughout its entire depth, the moisture content will increase from θ_i to η (the porosity) as the wetting front passes. The moisture content θ is the ratio of the volume of water to the total volume within the control surface, so the increase in the water stored within the control volume as a result of infiltration is $L(\eta - \theta_i)$ for a unit cross section. By definition this quantity is equal to F, the cumulative depth of water infiltrated into the soil. Hence

$$
\begin{aligned}
F(t) &= L(\eta - \theta_i) \\
&= L\Delta\theta
\end{aligned}
\tag{4.3.1}
$$

where $\Delta\theta = \eta - \theta_i$.

Momentum

Darcy's law may be expressed

$$
q = -K\frac{\partial h}{\partial z} \tag{4.3.2}
$$

In this case the Darcy flux q is constant throughout the depth and is equal to $-f$, because q is positive upward while f is positive downward. If points 1 and 2 are

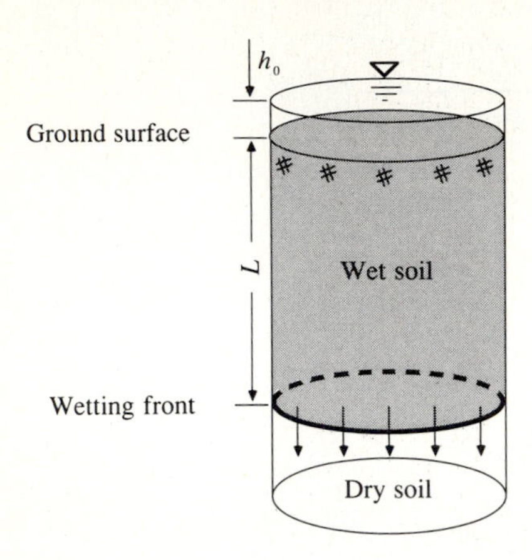

FIGURE 4.3.2
Infiltration into a column of soil of unit cross-sectional area for the Green-Ampt model.

located respectively at the ground surface and just on the dry side of the wetting front, (4.3.2) can be approximated by

$$f = K\left[\frac{h_1 - h_2}{z_1 - z_2}\right] \tag{4.3.3}$$

The head h_1 at the surface is equal to the ponded depth h_0. The head h_2, in the dry soil below the wetting front, equals $-\psi - L$. Darcy's law for this system is written

$$\begin{aligned} f &= K\left[\frac{h_0 - (-\psi - L)}{L}\right] \\ &\approx K\left[\frac{\psi + L}{L}\right] \end{aligned} \tag{4.3.4}$$

if the ponded depth h_0, is negligible compared to ψ and L. This assumption is usually appropriate for surface water hydrology problems because it is assumed that ponded water becomes surface runoff. Later, it will be shown how to account for h_0 if it is not negligible.

From (4.3.1) the wetting front depth is $L = F/\Delta\theta$, and assuming $h_0 = 0$, substitution into (4.3.4) gives

$$f = K\left[\frac{\psi\Delta\theta + F}{F}\right] \tag{4.3.5}$$

Since $f = dF/dt$, (4.3.5) can be expressed as a differential equation in the one unknown F:

$$\frac{dF}{dt} = K\left[\frac{\psi\Delta\theta + F}{F}\right]$$

To solve for F, cross-multiply to obtain

$$\left[\frac{F}{F + \psi\Delta\theta}\right]dF = K\,dt$$

then split the left-hand side into two parts

$$\left[\left(\frac{F + \psi\Delta\theta}{F + \psi\Delta\theta}\right) - \left(\frac{\psi\Delta\theta}{F + \psi\Delta\theta}\right)\right]dF = K\,dt$$

and integrate

$$\int_0^{F(t)} \left(1 - \frac{\psi\Delta\theta}{F + \psi\Delta\theta}\right) dF = \int_0^t K\,dt$$

to obtain

$$F(t) - \psi\Delta\theta\left\{\ln\left[F(t) + \psi\Delta\theta\right] - \ln\left(\psi\Delta\theta\right)\right\} = Kt$$

or

$$F(t) - \psi\Delta\theta\,\ln\left(1 + \frac{F(t)}{\psi\Delta\theta}\right) = Kt \tag{4.3.6}$$

This is the *Green-Ampt equation* for cumulative infiltration. Once F is found from Eq. (4.3.6), the infiltration rate f can be obtained from (4.3.5) or

$$f(t) = K\left(\frac{\psi\Delta\theta}{F(t)} + 1\right) \tag{4.3.7}$$

In the case when the ponded depth h_0 is not negligible, the value of $\psi - h_0$ is substituted for ψ in (4.3.6) and (4.3.7).

Equation (4.3.6) is a nonlinear equation in F. It may be solved by the *method of successive substitution* by rearranging (4.3.6) to read

$$F(t) = Kt + \psi\Delta\theta\ln\left(1 + \frac{F(t)}{\psi\Delta\theta}\right) \tag{4.3.8}$$

Given K, t, ψ and $\Delta\theta$, a trial value F is substituted on the right-hand side (a good trial value is $F = Kt$), and a new value of F calculated on the left-hand side, which is substituted as a trial value on the right-hand side, and so on, until the calculated values of F converge to a constant. The final value of cumulative infiltration F is substituted into (4.3.7) to determine the corresponding potential infiltration rate f.

Equation (4.3.6) can also be solved by *Newton's iteration method,* which is more complicated than the method of successive substitution but converges in fewer iterations. Newton's iteration method is explained in Sec. 5.6.

Green-Ampt Parameters

Application of the Green-Ampt model requires estimates of the hydraulic conductivity K, the porosity η, and the wetting front soil suction head ψ. The variation with moisture content θ of the suction head and hydraulic conductivity was studied by Brooks and Corey (1964). They concluded, after laboratory tests of many soils, that ψ can be expressed as a logarithmic function of an *effective saturation* s_e (see Fig. 4.3.3). If the residual moisture content of the soil after it has been thoroughly drained is denoted by θ_r, the effective saturation is the ratio of the available moisture $\theta - \theta_r$ to the maximum possible available moisture content $\eta - \theta_r$:

$$s_e = \frac{\theta - \theta_r}{\eta - \theta_r} \tag{4.3.9}$$

where $\eta - \theta_r$ is called the *effective porosity* θ_e.

The effective saturation has the range $0 \leq s_e \leq 1.0$, provided $\theta_r \leq \theta \leq \eta$. For the initial condition, when $\theta = \theta_i$, cross-multiplying (4.3.9) gives $\theta_i - \theta_r = s_e \theta_e$, and the change in the moisture content when the wetting front passes is $\Delta\theta = \eta - \theta_i = \eta - (s_e\theta_e + \theta_r)$; therefore

$$\Delta\theta = (1 - s_e)\theta_e \tag{4.3.10}$$

The logarithmic relationship shown in Fig. 4.3.3 can be expressed by the *Brooks-Corey equation*

$$s_e = \left[\frac{\psi_b}{\psi}\right]^\lambda \tag{4.3.11}$$

in which ψ_b and λ are constants obtained by draining a soil in stages, measuring the values of s_e and ψ at each stage, and fitting (4.3.11) to the resulting data.

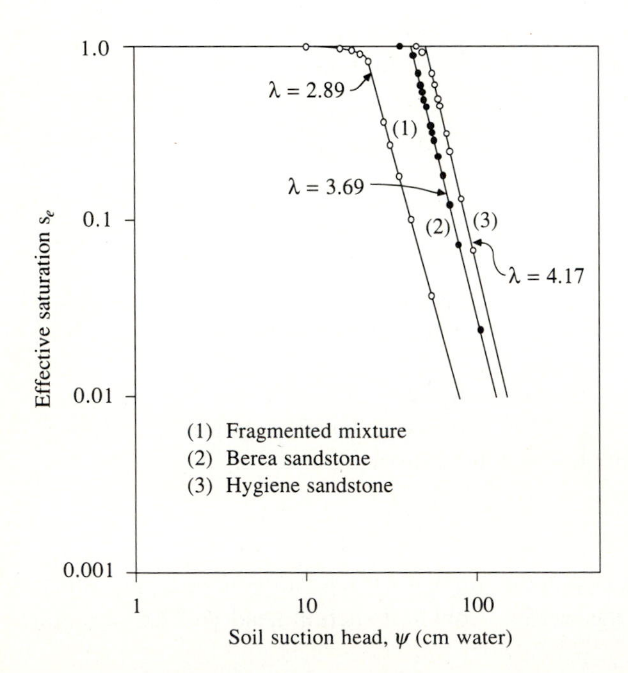

FIGURE 4.3.3
The Brooks-Corey relationship between soil suction head and effective saturation. (*Source:* Brooks and Corey, 1964, Fig. 2, p. 5. Used with permission.)

Bouwer (1966) also studied the variation of hydraulic conductivity with moisture content and concluded that the effective hydraulic conductivity for an unsaturated flow is approximately half the corresponding value for saturated flow.

Brakensiek, Engleman, and Rawls (1981) presented a method for determining the Green-Ampt parameters using the Brooks-Corey equation. Rawls, Brakensiek, and Miller (1983) used this method to analyze approximately 5000 soil horizons across the United States and determined average values of the Green-Ampt parameters η, θ_e, ψ, and K for different soil classes, as shown in Table 4.3.1. As the soil becomes finer moving from sand to clay the wetting front soil suction head increases while the hydraulic conductivity decreases. Table 4.3.1 also shows typical ranges for η, θ_e, and ψ. The ranges are not large for η and θ_e, but ψ can vary over a wide range for a given soil. As was shown in Example 4.1.1, K varies along with ψ, so the values given in Table 4.3.1 for both ψ and K should be considered typical values that may show a considerable degree of variability in application (American Society of Agricultural Engineers, 1983; Devaurs and Gifford, 1986).

TABLE 4.3.1
Green-Ampt infiltration parameters for various soil classes

Soil class	Porosity η	Effective porosity θ_e	Wetting front soil suction head ψ (cm)	Hydraulic conductivity K (cm/h)
Sand	0.437 (0.374–0.500)	0.417 (0.354–0.480)	4.95 (0.97–25.36)	11.78
Loamy sand	0.437 (0.363–0.506)	0.401 (0.329–0.473)	6.13 (1.35–27.94)	2.99
Sandy loam	0.453 (0.351–0.555)	0.412 (0.283–0.541)	11.01 (2.67–45.47)	1.09
Loam	0.463 (0.375–0.551)	0.434 (0.334–0.534)	8.89 (1.33–59.38)	0.34
Silt loam	0.501 (0.420–0.582)	0.486 (0.394–0.578)	16.68 (2.92–95.39)	0.65
Sandy clay loam	0.398 (0.332–0.464)	0.330 (0.235–0.425)	21.85 (4.42–108.0)	0.15
Clay loam	0.464 (0.409–0.519)	0.309 (0.279–0.501)	20.88 (4.79–91.10)	0.10
Silty clay loam	0.471 (0.418–0.524)	0.432 (0.347–0.517)	27.30 (5.67–131.50)	0.10
Sandy clay	0.430 (0.370–0.490)	0.321 (0.207–0.435)	23.90 (4.08–140.2)	0.06
Silty clay	0.479 (0.425–0.533)	0.423 (0.334–0.512)	29.22 (6.13–139.4)	0.05
Clay	0.475 (0.427–0.523)	0.385 (0.269–0.501)	31.63 (6.39–156.5)	0.03

The numbers in parentheses below each parameter are one standard deviation around the parameter value given. *Source:* Rawls, Brakensiek, and Miller, 1983.

Two-layer Green-Ampt Model

Consider a soil with two layers, as shown in Fig. 4.3.4. The upper layer has thickness H_1 and Green-Ampt parameters K_1, ψ_1, and $\Delta\theta_1$, and the lower layer has thickness H_2 and parameters K_2, ψ_2, and $\Delta\theta_2$. Water is ponded on the surface and the wetting front has penetrated through the upper layer and a distance L_2 into the lower layer ($L_2 < H_2$). It is required that $K_1 > K_2$ for the upper layer to remain saturated while water infiltrates into the lower layer. By a method similar to that described previously for one layer of soil, it can be shown that the infiltration rate is given by

$$f = \frac{K_1 K_2}{H_1 K_2 + L_2 K_1}(\psi_2 + H_1 + L_2) \tag{4.3.12}$$

and that the cumulative infiltration is given by

$$F = H_1 \Delta\theta_1 + L_2 \Delta\theta_2 \tag{4.3.13}$$

By combining Eqs. (4.3.12) and (4.3.13) into a differential equation for L_2 and integrating, one arrives at

$$L_2 \frac{\Delta\theta_2}{K_2} + \frac{1}{K_1 K_2}[\Delta\theta_2 H_1 K_2 - \Delta\theta_2 K_1(\psi_2 + H_1)] \ln\left[1 + \frac{L_2}{\psi_2 + H_1}\right] = t \tag{4.3.14}$$

from which the cumulative infiltration and infiltration rate can be determined. This approach can be employed when a more permeable upper soil layer overlies a less permeable lower layer. The normal Green-Ampt equations are used while the wetting front is in the upper layer; (4.3.12) to (4.3.14) are used once the wetting front enters the lower layer.

Example 4.3.1. Compute the infiltration rate f and cumulative infiltration F after one hour of infiltration into a silt loam soil that initially had an effective saturation of 30 percent. Assume water is ponded to a small but negligible depth on the surface.

Solution. From Table 4.3.1, for a silt loam soil $\theta_e = 0.486$, $\psi = 16.7$ cm, and $K = 0.65$ cm/h. The initial effective saturation is $s_e = 0.3$, so in (4.3.10)

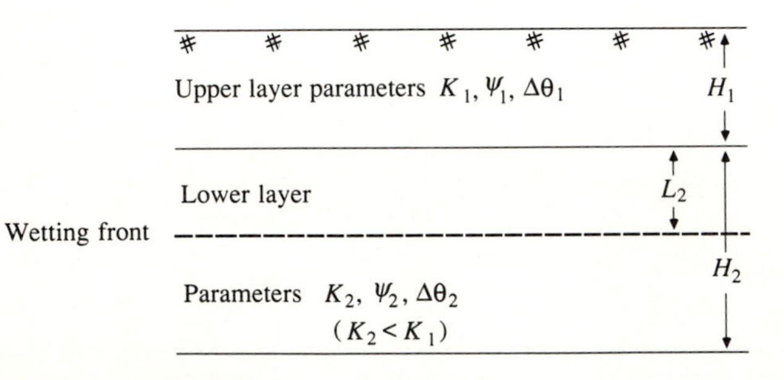

FIGURE 4.3.4
Parameters in a two-layer Green-Ampt model.

$$\Delta\theta = (1 - s_e)\theta_e$$
$$= (1 - 0.3)(0.486)$$
$$= 0.340$$

and

$$\psi\Delta\theta = 16.7 \times 0.340$$
$$= 5.68 \text{ cm}$$

The cumulative infiltration at t = 1 h is calculated employing the method of successive substitution in Eq. (4.3.8). Take a trial value of $F(t) = Kt = 0.65$ cm, then calculate

$$F(t) = Kt + \psi\Delta\theta \ln\left(1 + \frac{F(t)}{\psi\Delta\theta}\right)$$
$$= 0.65 \times 1 + 5.68 \ln\left(1 + \frac{0.65}{5.68}\right)$$
$$= 1.27 \text{ cm}$$

Substituting $F = 1.27$ into the right-hand side of (4.3.8) gives $F = 1.79$ cm, and after a number of iterations F converges to a constant value of 3.17 cm. The infiltration rate after one hour is found from Eq. (4.3.7):

$$f = K\left(\frac{\psi\Delta\theta}{F} + 1\right)$$
$$= 0.65\left(\frac{5.68}{3.17} + 1\right)$$
$$= 1.81 \text{ cm/h}$$

4.4 PONDING TIME

In the preceding sections several methods for computing the rate of infiltration into the soil were presented. All of these methods used the assumption that water is ponded to a small depth on the soil surface so all the water the soil can infiltrate is available at the surface. However, during a rainfall, water will pond on the surface only if the rainfall intensity is greater than the infiltration capacity of the soil. The *ponding time* t_p is the elapsed time between the time rainfall begins and the time water begins to pond on the soil surface.

If rainfall begins on dry soil, the vertical moisture profile in the soil may appear as in Fig. 4.4.1. Prior to the ponding time ($t < t_p$), the rainfall intensity is less than the potential infiltration rate and the soil surface is unsaturated. Ponding begins when the rainfall intensity exceeds the potential infiltration rate. At this time ($t = t_p$), the soil surface is saturated. As rainfall continues ($t > t_p$), the saturated zone extends deeper into the soil and overland flow occurs from the ponded water. How can the infiltration equations developed previously be used to describe this situation?

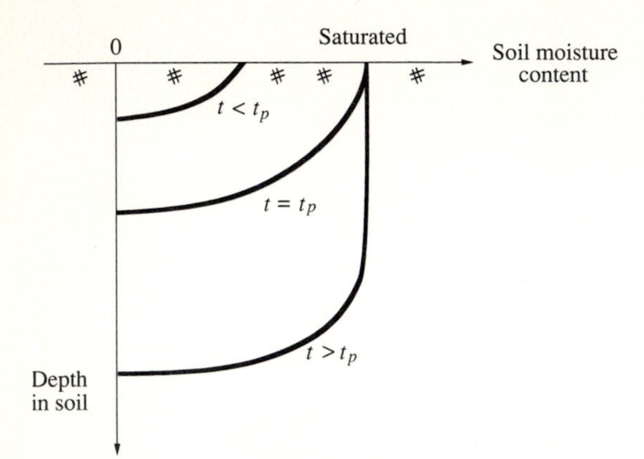

FIGURE 4.4.1
Soil moisture profiles before, during, and after ponding occurs.

Mein and Larson (1973) presented a method for determining the ponding time with infiltration into the soil described by the Green-Ampt equation for rainfall of intensity i starting instantaneously and continuing indefinitely. There are three principles involved: (1) prior to the time ponding occurs, all the rainfall is infiltrated; (2) the potential infiltration rate f is a function of the cumulative infiltration F; and (3) ponding occurs when the potential infiltration rate is less than or equal to the rainfall intensity.

In the Green-Ampt equation, the infiltration rate f and cumulative infiltration F are related by

$$f = K\left(\frac{\psi \Delta\theta}{F} + 1\right) \tag{4.4.1}$$

where K is the hydraulic conductivity of the soil, ψ is the wetting front capillary pressure head, and $\Delta\theta$ is the difference between the initial and final moisture contents of the soil. As shown in Fig. 4.4.2, the cumulative infiltration at the ponding time t_p is given by $F_p = i t_p$ and the infiltration rate by $f = i$; substituting into Eq. (4.4.1),

$$i = K\left(\frac{\psi \Delta\theta}{i t_p} + 1\right)$$

solving, (4.4.2)

$$t_p = \frac{K \psi \Delta\theta}{i(i - K)}$$

gives the ponding time under constant rainfall intensity using the Green-Ampt infiltration equation.

Example 4.4.1. Compute the ponding time and the depth of water infiltrated at ponding for a silt loam soil of 30 percent initial effective saturation, subject to rainfall intensities of (a) 1 cm/h and (b) 5 cm/h.

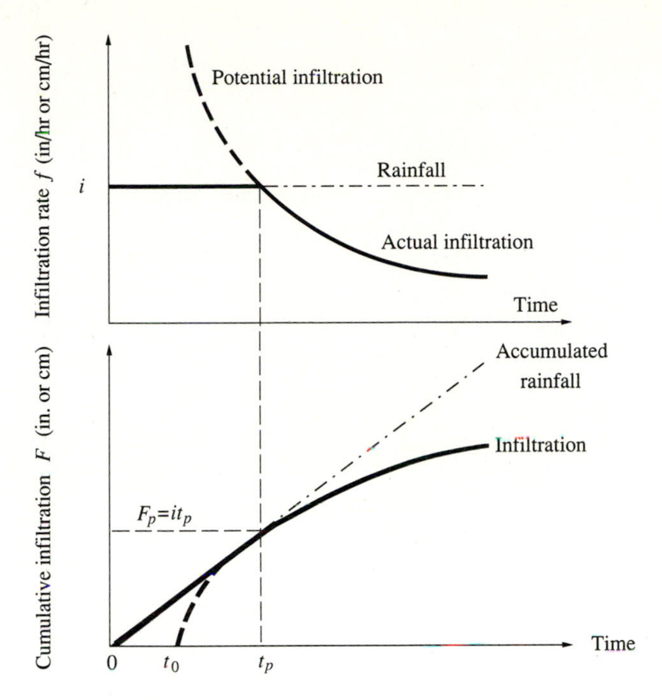

FIGURE 4.4.2
Infiltration rate and cumulative infiltration for ponding under constant intensity rainfall.

Solution. From Example 4.3.1, for a silt loam soil $\psi\,\Delta\theta = 5.68$ cm and $K = 0.65$ cm/h. The ponding time is given by (4.4.2):

$$t_p = \frac{K\,\psi\,\Delta\theta}{i(i-K)}$$

(*a*) For $i = 1$ cm/h,

$$t_p = \frac{0.65 \times 5.68}{1.0(1.0 - 0.65)}$$
$$= 10.5 \text{ h}$$

and

$$F_p = it_p$$
$$= 1.0 \times 10.5$$
$$= 10.5 \text{ cm}$$

(*b*) For $i = 5$ cm/h,

$$t_p = \frac{0.65 \times 5.68}{5(5 - 0.65)}$$
$$= 0.17 \text{ h (10 min)}$$

and

$$F_p = i t_p$$
$$= 5.0 \times 0.17$$
$$= 0.85 \text{ cm}$$

In each case the infiltration rate f equals the rainfall intensity i at ponding.

To obtain the actual infiltration rate after ponding, a curve of potential infiltration is constructed beginning at a time t_0 such that the cumulative infiltration and the infiltration rate at t_p are equal to those observed under rainfall beginning at time 0 (see dashed line in Fig. 4.4.2). Substituting $t = t_p - t_0$ and $F = F_p$ into Eq. (4.3.6) gives

$$F_p - \psi \Delta\theta \ln\left(1 + \frac{F_p}{\psi \Delta\theta}\right) = K(t_p - t_0) \tag{4.4.3}$$

For $t > t_p$,

$$F - \psi \Delta\theta \ln\left(1 + \frac{F}{\psi \Delta\theta}\right) = K(t - t_0) \tag{4.4.4}$$

and subtracting (4.4.3) from (4.4.4),

$$F - F_p - \psi \Delta\theta \left[\ln\left(\frac{\psi \Delta\theta + F}{\psi \Delta\theta}\right) - \ln\left(\frac{\psi \Delta\theta + F_p}{\psi \Delta\theta}\right)\right] = K(t - t_p)$$

or

$$F - F_p - \psi \Delta\theta \ln\left[\frac{\psi \Delta\theta + F}{\psi \Delta\theta + F_p}\right] = K(t - t_p) \tag{4.4.5}$$

Equation (4.4.5) can be used to calculate the depth of infiltration after ponding, and then (4.3.7) can be used to obtain the infiltration rate f.

Example 4.4.2. Calculate the cumulative infiltration and the infiltration rate after one hour of rainfall of intensity 5 cm/h on a silt loam soil with an initial effective saturation of 30 percent.

Solution. From Example 4.3.1, $\psi \Delta\theta = 5.68$ cm and $K = 0.65$ cm/h for this soil, and from Example 4.4.1, $t_p = 0.17$ h and $F_p = 0.85$ cm under rainfall intensity 5 cm/h. For $t = 1.0$ h, the infiltration depth is given by (4.4.5):

$$F - F_p - \psi \Delta\theta \ln\left(\frac{\psi \Delta\theta + F}{\psi \Delta\theta + F_p}\right) = K(t - t_p)$$

$$F - 0.85 - 5.68 \ln\left(\frac{5.68 + F}{5.68 + 0.85}\right) = 0.65(1.0 - 0.17)$$
$$= 0.54$$

F is obtained by the method of successive substitution in the manner used in Example 4.3.1. The solution converges to $F = 3.02$ cm. The corresponding rate is given by (4.3.7):

$$f = K\left(\frac{\psi \Delta\theta}{F} + 1\right)$$

$$= 0.65\left(\frac{5.68}{3.02} + 1\right)$$

$$= 1.87 \text{ cm/h}$$

These results may be compared with the cumulative infiltration of 3.17 cm obtained in Example 4.3.1 for infiltration under continuous ponding. Less water is infiltrated after one hour under the 5 cm/h rainfall because it took 10 minutes for ponding to occur, and the infiltration rate during this period was less than its potential value.

Table 4.4.1 summarizes the equations needed for computing various quantities for constant rainfall intensity; a set of equations is given for each of three approaches, based respectively on the Green-Ampt, Horton, and Philip infiltration equations. Equations (1) and (2) are the methods for computing infiltration under ponded conditions. Equation (3) gives the ponding time under constant rainfall intensity, and Equation (4) gives the equivalent time origin t_o, from which the same infiltration rate and cumulative infiltration as those observed at ponding time would be produced under continuously ponded conditions. After ponding has occurred, the infiltration functions can be found for Horton's and Philip's equations by substituting $t - t_0$ into Eqs. (1) and (2). For the Green-Ampt equation, the method illustrated in Ex. 4.4.2 can be used. In Eq. (4.4.2) under the Green-Ampt method, ponding time t_p is positive and finite only if $i > K$; ponding will never occur if the rainfall intensity is less than or equal to the hydraulic conductivity of the soil. Table 4.4.1 indicates that the same condition holds for Philip's equation, while Horton's equation requires $i > f_c$ to achieve ponding. If, in the Horton equation, $i > f_0$, ponding will occur immediately and $t_p = 0$.

The condition $i < K$ holds for most rainfalls on very permeable soils and for light rainfall on less permeable soils. In such cases, streamflow results from subsurface flow, especially from areas near the stream channel.

Determination of ponding times under rainfall of variable intensity can be done by an approach similar to that for constant intensity. Cumulative infiltration is calculated from rainfall as a function of time. A potential infiltration rate can be calculated from the cumulative infiltration using the Green-Ampt or other infiltration formulas. Whenever rainfall intensity is greater than the potential infiltration rate, ponding is occurring (Bouwer, 1978; Morel-Seytoux, 1981). For sites where estimates of a constant infiltration rate are available, the estimates can be used as a guide to decide whether surface or subsurface flow is the primary mechanism producing flood flows (Pearce and McKerchar, 1979). This subject is developed further in Chap. 5.

TABLE 4.4.1
Equations for calculating ponding time and infiltration after ponding occurs

Equation	Variable calculated	Green-Ampt equation	Horton's equation	Philip's equation
(1)	Potential infiltration rate as a function of time	Solve for F from (2) then use (6).	$f = f_c + (f_0 - f_c)e^{-kt}$	$f = \frac{1}{2}St^{-1/2} + K$
(2)	Potential cumulative infiltration as a function of time	$F - \psi\Delta\theta \ln\left(1 + \frac{F}{\psi\Delta\theta}\right) = Kt$	$F = f_c t + \frac{f_0 - f_c}{k}\left(1 - e^{-kt}\right)$	$F = St^{1/2} + Kt$
(3)	Ponding time under constant rainfall intensity i	$t_p = \frac{K\psi\Delta\theta}{i(i - K)}$ $(i > K)$	$t_p = \frac{1}{ik}\left[f_0 - i + f_c \ln\left(\frac{f_0 - f_c}{i - f_c}\right)\right]$ $(f_c < i < f_0)$	$t_p = \frac{S^2(i - K/2)}{2i(i - K)^2}$ $(i > K)$
(4)	Equivalent time origin for potential infiltration after ponding	$t_0 = t_p - \frac{1}{K}\left[F_p - \psi\Delta\theta \ln\left(1 + \frac{\psi\Delta\theta}{F_p}\right)\right]$	$t_0 = t_p - \frac{1}{k} \ln\left(\frac{f_0 - f_c}{i - f_c}\right)$	$t_0 = t_p - \frac{1}{4K^2}\left(\sqrt{S^2 + 4KF_p} - S\right)^2$
(5)	Cumulative infiltration after ponding	$F - F_p - \psi\Delta\theta \ln\left(\frac{\psi\Delta\theta + F}{\psi\Delta\theta + F_p}\right) = K(t - t_p)$	Substitute $(t - t_0)$ for t in (2).	Substitute $(t - t_0)$ for t in (2).
(6)	Infiltration rate after ponding	$f = K\left(\frac{\psi\Delta\theta}{F} + 1\right)$	Substitute $(t - t_0)$ for t in (1).	Substitute $(t - t_0)$ for t in (1).

REFERENCES

American Society of Agricultural Engineers, Advances in infiltration, *Proc. Nat. Conf. on Advances in Infiltration, Chicago, Ill.*, ASAE Publ. 11-83, St. Joseph, Mich., 1983.

Bouwer, H., Rapid field measurement of air entry value and hydraulic conductivity of soil as significant parameters in flow system analysis, *Water Resour. Res.*, vol. 2, 729–738, 1966.

Bouwer, H., Surface-subsurface water relations, Chap. 8 in *Groundwater Hydrology,* McGraw-Hill, New York, 1978.

Brakensiek, D. L., R. L. Engleman, and W. J. Rawls, Variation within texture classes of soil water parameters, *Trans. Am. Soc. Agric. Eng.*, vol. 24, no. 2, pp. 335–339, 1981.

Brooks, R. H., and A. T. Corey, Hydraulic properties of porous media, *Hydrology Papers,* no. 3, Colorado State Univ., Fort Collins, Colo., 1964.

de Marsily, G., *Quantitative Hydrogeology*, Academic Press, Orlando, Fla., 1986.

Devaurs, M., and G. F. Gifford, Applicability of the Green and Ampt infiltration equation to rangelands, *Water Resour. Bull.,* vol. 22, no. 1, pp. 19–27, 1986.

Eagleson, P. S., *Dynamic Hydrology,* McGraw-Hill, New York, 1970.

Freeze, R. A. and J. A. Cherry, *Groundwater,* Prentice-Hall, Englewood Cliffs, N.J., 1979.

Green, W. H., and G. A. Ampt, Studies on soil physics, part I, the flow of air and water through soils, *J. Agric. Sci.,* vol. 4, no. 1, pp. 1–24, 1911.

Hillel, D., *Applications of Soil Physics,* Academic Press, Orlando, Fla., 1980.

Horton, R. E., The role of infiltration in the hydrologic cycle, *Trans. Am. Geophys. Union,* vol. 14, pp. 446–460, 1933.

Horton, R. E., Analysis of runoff plat experiments with varying infiltration capacity, *Trans. Am. Geophys. Union,* Vol. 20, pp. 693–711, 1939.

Mein, R. G., and C. L. Larson, Modeling infiltration during a steady rain, *Water Resour. Res.,* vol. 9, no. 2, pp. 384–394, 1973.

Morel-Seytoux, H. J., Application of infiltration theory for the determination of the excess rainfall hyetograph, *Water Resour. Bull.,* vol. 17, no. 6, pp. 1012–1022, 1981.

Pearce, A. J., and A. I. McKerchar, Upstream generation of storm runoff, *Physical Hydrology, New Zealand Experience*, ed. by D. L. Murray and P. Ackroyd, New Zealand Hydrological Society, Wellington, New Zealand, pp. 165–192, 1979.

Philip, J. R., The theory of infiltration: 1. The infiltration equation and its solution, *Soil Sci.,* vol. 83, no. 5, pp. 345–357, 1957.

Philip, J. R., Theory of infiltration, in *Advances in Hydroscience,* ed. by V. T. Chow, vol. 5, pp. 215–296, 1969.

Raudkivi, A. J., *Hydrology,* Pergamon Press, Oxford, 1979.

Rawls, W. J., D. L. Brakensiek, and N. Miller, Green-Ampt infiltration parameters from soils data, *J. Hydraul. Div., Am. Soc. Civ. Eng.,* vol. 109, no. 1, pp. 62–70, 1983.

Richards, L. A., Capillary conduction of liquids through porous mediums, *Physics,* vol. 1, pp. 318–333, 1931.

Skaggs, R. W., Infiltration, Chap. 4 in *Hydrologic Modelling of Small Watersheds,* ed. by C. T. Haan, H. P. Johnson, and D. L. Brakensiek, Am. Soc. Agric. Eng. Mon. no. 5, St. Joseph, Mich., 1982.

Terstriep, M. L., and J. B. Stall, The Illinois urban drainage area simulator, *ILLUDAS, Bull. 58,* Ill. State Water Survey, Urbana, Ill., 1974.

Wellings, S. R., Recharge of the Upper Chalk Aquifer at a site in Hampshire, England: 1. Water balance and unsaturated flow, *J. Hydrol.,* vol. 69, pp. 259-273, 1984.

PROBLEMS

4.1.1 Figure 4.1.5(b) shows the profiles through time of soil moisture head h, with vertical lines at weekly intervals. Calculate the soil moisture flux q between 0.8

m and 1.0 m at weekly intervals using the relationship $K = 250(-\psi)^{-2.11}$, where K is hydraulic conductivity (cm/day) and ψ is soil suction head (cm).

4.1.2 Solve Prob. 4.1.1 for the soil moisture flux between 1.0 m and 1.2 m.

4.1.3 Take each pair of successive soil moisture head profiles in Fig. 4.1.5(*b*) (i.e., the profiles at 0.4 m and 0.8 m, 0.8 m and 1.0 m, . . . , 2.40 m and 3.0 m). Use the relationship $K = 250(-\psi)^{-2.11}$ with K in centimeters per day and ψ in centimeters to calculate the soil moisture flux between each pair of levels. Plot the soil moisture flux profiles and discuss how the moisture flows in the soil in relation to the rainfall and evapotranspiration at the surface.

4.1.4 Using the Yolo light clay data shown in Fig. 4.1.4, calculate values of the soil water diffusivity $D = K(d\psi/d\theta)$, for $\theta = 0.1, 0.2, 0.3$, and 0.4. Plot a graph of D vs θ.

4.2.1 Suppose that the parameters for Horton's equation are $f_0 = 3.0$ in/h, $f_c = 0.53$ in/h, and $k = 4.182\ \text{h}^{-1}$. Determine the infiltration rate and cumulative infiltration after 0, 0.5, 1.0, 1.5, and 2 h. Plot both as functions of time. Plot the infiltration rate as a function of cumulative infiltration. Assume continuously ponded conditions.

4.2.2 For the same conditions as in Prob. 4.2.1, determine the incremental depth of infiltration between 0.75 and 2.0 h.

4.2.3 For Horton's equation suppose $f_0 = 5$ cm/h, $f = 1$ cm/h, and $k = 2\ \text{h}^{-1}$. Determine the cumulative infiltration after 0, 0.5, 1.0, 1.5, and 2.0 h. Plot the infiltration rate and cumulative infiltration as functions of time. Plot the infiltration rate as a function of the cumulative infiltration. Assume continuously ponded conditions.

4.2.4 The infiltration rate at the beginning of a storm was $f_0 = 4.0$ in/h and it decreased to 0.5 in/h after two hours. A total of 1.7 in infiltrated during these two hours. Determine the value of k for Horton's equation. Assume continuously ponded conditions.

4.2.5 For the same conditions as in Prob. 4.2.4, determine the value of k for Horton's equation if a total of 1.2 in infiltrated during the two-hour period.

4.2.6 Suppose the parameters for Philip's equation are sorptivity $S = 5\ \text{cm}\cdot\text{h}^{-1/2}$ and $K = 0.4$ cm/h. Determine the cumulative infiltration after 0, 0.5, 1.0, 1.5, and 2.0 h. Plot the infiltration rate and the cumulative infiltration as functions of time. Plot the infiltration rate as a function of the cumulative infiltration. Assume continuously ponded conditions.

4.2.7 The infiltration rate as a function of time for an Alexis silt loam is as follows (Terstriep and Stall, 1974):

Time (h)	0	0.07	0.16	0.27	0.43	0.67	1.10	2.53
Infiltration rate (in/h)	0.26	0.21	0.17	0.13	0.09	0.05	0.03	0.01

Determine the best values for the parameters f_0, f_c, and k for Horton's equation to describe the infiltration for Alexis silt loam.

4.2.8 The infiltration into a Yolo light clay as a function of time for a steady rainfall rate of 0.5 cm/h is as follows (Skaggs, 1982):

Time (h)	0	1.07	1.53	2.30	3.04	3.89	4.85	7.06
Cumulative infiltration (cm)	0	0.54	0.75	1.0	1.2	1.4	1.6	2.0
Infiltration rate (cm/h)	0.5	0.5	0.37	0.29	0.25	0.22	0.20	0.17

Determine the parameters f_0, f_c, and k for Horton's equation. Assume that ponding occurs at $t = 1.07$ h.

4.2.9 Determine the parameters for Philip's equation for the infiltration data given in Prob. 4.2.8.

4.2.10 Parameters in Philip's equation for a clay soil are $S = 45$ cm·h$^{-1/2}$ and $K = 10$ cm/h. Determine the cumulative infiltration and the infiltration rate at 0.5-hour increments for a 3-hour period. Plot both as functions of time. Plot the infiltration rate as a function of the cumulative infiltration. Assume continuously ponded conditions.

4.2.11 Solve Prob. 4.2.10 for a sandy soil with parameters $S = 9.0$ cm·h$^{-1/2}$ and $K = 10$ cm/h. Assume continuously ponded conditions.

4.3.1 For a sandy loam soil, calculate the infiltration rate (cm/h) and depth of infiltration (cm) after one hour if the effective saturation is initially 40 percent, using the Green-Ampt method. Assume continuously ponded conditions.

4.3.2 For the same conditions as in Prob. 4.3.1, plot curves of cumulative infiltration depth F and infiltration rate f vs. time t for the first three hours of infiltration using 0.5-h increments. Plot the infiltration rate as a function of the cumulative infiltration for the same period.

4.3.3 Use the Green-Ampt method to evaluate the infiltration rate and cumulative infiltration depth for a silty clay soil at 0.1-hour increments up to 6 hours from the beginning of infiltration. Assume initial effective saturation 20 percent and continous ponding.

4.3.4 For the soil of Prob. 4.3.3, compute the cumulative infiltration after one hour for initial effective saturations of 0, 20, 40, 60, 80, and 100 percent. Draw a graph of cumulative infiltration vs. initial effective saturation.

4.3.5 Show that the depth L_2 to the wetting front in the lower layer of the two-layer Green-Ampt model satisfies

$$L_2\frac{\Delta\theta_2}{K_2} + \frac{1}{K_1K_2}[\Delta\theta_2 H_1 K_2 - \Delta\theta_2 K_1(\psi_2 + H_1)] \ln\left[1 + \frac{L_2}{\psi_2 + H_1}\right] = t$$

4.3.6 A soil comprises two layers, an upper layer six centimeters thick of silt loam overlying a very deep layer of clay. The initial effective saturation in each layer is 10 percent. As the wetting front penetrates into the soil, calculate, at 1-cm increments of wetting front depth, the values of f, F, and t. For the clay layer use the relationships given in Eqs. (4.3.12) to (4.3.14). Stop the calculations once the wetting front reaches 10 cm. Plot graphs of the infiltration rate and cumulative infiltration as functions of time.

4.3.7 Using the parameter values in Table 4.3.1, determine points on the infiltration rate curves for sand, loam, clay loam, and clay from time 0 to 4 h, at 0.5-h increments. Plot and compare these curves. Assume an initial effective saturation of 30 percent in each soil and continuous ponding.

4.3.8 Solve Prob. 4.3.7 for cumulative infiltration curves.

4.3.9 Solve Prob. 4.3.7 using an initial effective saturation of 15 percent in each soil.

4.3.10 Solve Prob. 4.3.8 using an initial effective saturation of 15 percent in each soil.

4.4.1 Solve Example 4.4.1 in the text with an initial effective saturation of 20 percent.

4.4.2 Solve Example 4.4.2 in the text with an initial effective saturation of 20 percent.

4.4.3 Compute the ponding time and cumulative infiltration at ponding for a clay loam soil with a 25 percent initial effective saturation subject to a rainfall intensity of (a) 1 cm/h (b) 3 cm/h.

4.4.4 Calculate the cumulative infiltration and the infiltration rate after one hour of rainfall at 3 cm/h on a clay loam with a 25 percent initial effective saturation.

4.4.5 Compute the ponding time and depth of water infiltrated at ponding for a silty clay soil with a 20 percent initial effective saturation subject to a rainfall intensity of (a) 1 cm/h (b) 3 cm/h.

4.4.6 Calculate the cumulative infiltration and the infiltration rate on a silty clay soil after one hour of rainfall at 1 cm/h if the initial effective saturation is 20 percent. Assume ponding depth h_0 is negligible in the calculations.

4.4.7 Solve Prob. 4.4.6 assuming that any ponded water remains stationary over the soil so that h_0 must be accounted for in the calculations.

4.4.8 Rainfall of intensity 2 cm/h falls on a clay loam soil, and ponding occurs after five minutes. Calculate the ponding time on a nearby sandy loam soil if both soils initially had the same effective saturation s_e.

4.4.9 A soil has sorptivity $S = 5$ cm·h$^{-1/2}$ and conductivity $K = 0.4$ cm/h. Calculate the ponding time and cumulative infiltration at ponding under a rainfall of 6 cm/h.

4.4.10 A soil has Horton's equation parameters $f_0 = 10$ cm/h, $f = 4$ cm/h and $k = 2$ h^{-1}. Calculate the ponding time and cumulative infiltration at ponding under a rainfall of 6 cm/h.

4.4.11 Show that the ponding time under rainfall of constant intensity i for a soil described by Philip's equation with parameters S and K is given by

$$t_p = \frac{S^2(i - K/2)}{2i(i - K)^2}$$

4.4.12 Show that the ponding time under rainfall of intensity i for a soil described by Horton's equation with parameters f_0, f_c, and k is given by

$$t_p = \frac{1}{ik}\left[f_0 - i + f_c \ln\left(\frac{f_0 - f_c}{i - f_c}\right)\right]$$

Indicate the range of values of rainfall intensity for which this equation is valid and explain what happens if i is outside this range.

CHAPTER
5

SURFACE WATER

Surface water is water stored or flowing on the earth's surface. The surface water system continually interacts with the atmospheric and subsurface water systems described in previous chapters. This chapter describes the physical laws governing surface water flow and shows how hydrologic data are analyzed to provide input information for models of surface flow.

5.1 SOURCES OF STREAMFLOW

The *watershed*, or *catchment*, is the area of land draining into a stream at a given location. To describe how the various surface water processes vary through time during a storm, suppose that precipitation of a constant rate begins and continues indefinitely on a watershed. Precipitation contributes to various storage and flow processes, as illustrated in Fig. 5.1.1. The vertical axis of this diagram represents, relative to the rate of precipitation, the rate at which water is flowing or being added to storage in each of the processes shown at any instant of time.

Initially, a large proportion of the precipitation contributes to *surface storage*; as water infiltrates into the soil, there is also *soil moisture storage*. There are two types of storage: *retention* and *detention*; retention is storage held for a long period of time and depleted by evaporation, and detention is short-term storage depleted by flow away from the storage location.

As the detention storages begin filling, flow away from them occurs: *unsaturated flow* through the unsaturated soil near the land surface, *groundwater flow* through saturated aquifers deeper down, and *overland flow* across the land surface.

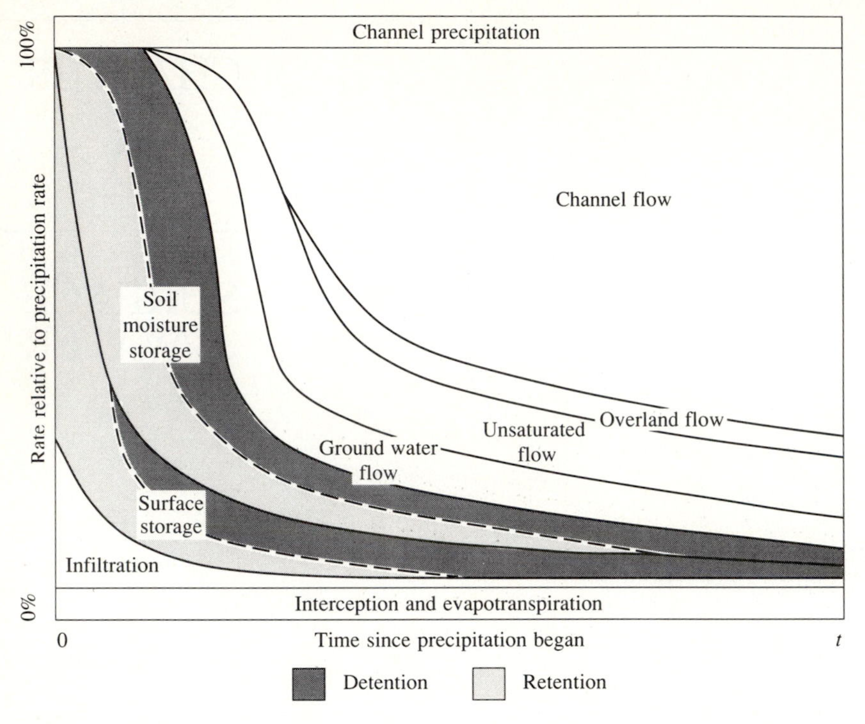

FIGURE 5.1.1
Schematic illustration of the disposal of precipitation during a storm on a watershed.

Channel flow is the main form of surface water flow, and all the other surface flow processes contribute to it. Determining flow rates in stream channels is a central task of surface water hydrology. The precipitation which becomes streamflow may reach the stream by overland flow, subsurface flow, or both.

Hortonian Overland Flow

Horton (1933) described overland flow as follows: "Neglecting interception by vegetation, surface runoff is that part of the rainfall which is not absorbed by the soil by infiltration. If the soil has an infiltration capacity f, expressed in inches depth absorbed per hour, then when the rain intensity i is less than f the rain is all absorbed and there is no surface runoff. It may be said as a first approximation that if i is greater than f, surface runoff will occur at the rate $(i - f)$." Horton termed this difference $(i - f)$ "rainfall excess." Horton considered surface runoff to take the form of a sheet flow whose depth might be measured in fractions of an inch. As flow accumulates going down a slope, its depth increases until discharge into a stream channel occurs (Fig. 5.1.2). Along with overland flow there is *depression storage* in surface hollows and *surface detention storage* proportional to the depth of the overland flow itself. The soil stores infiltrated water and

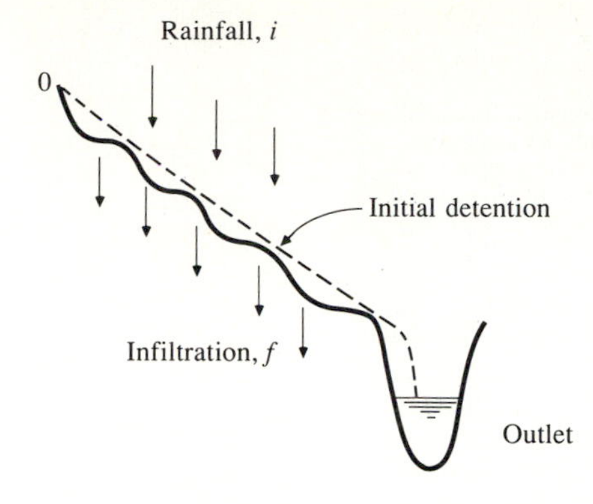

FIGURE 5.1.2
Overland flow on a slope produced by the excess of rainfall over infiltration. (After Horton, 1945, Fig. 13, p. 314.)

then slowly releases it as subsurface flow to enter the stream as baseflow during rainless periods.

Hortonian overland flow is applicable for impervious surfaces in urban areas, and for natural surfaces with thin soil layers and low infiltration capacity as in semiarid and arid lands.

Subsurface Flow

Hortonian overland flow occurs rarely on vegetated surfaces in humid regions (Freeze, 1972, 1974; Dunne, Moore, and Taylor, 1975). Under these conditions, the infiltration capacity of the soil exceeds observed rainfall intensities for all except the most extreme rainfalls. Subsurface flow then becomes a primary mechanism for transporting stormwater to streams. The process of subsurface flow is illustrated in Fig. 5.1.3, using the results of numerical simulations carried out by Freeze (1974). Part (*a*) shows an idealized cross section of a hillside draining into a stream. Prior to rainfall, the stream surface is in equilibrium with the water table and no saturated subsurface flow occurs. Parts (*b*)–(*d*) show how a seepage pattern develops from rainfall on surface DE, which serves to raise the water table (*e*) until inflow ceases (t = 277 min), after which the water table declines (*f*). All of the rainfall is infiltrated along surface DE until t = 84 min, when the soil first becomes saturated at D; as time continues, decreasing infiltration occurs along DE as progressively more of the surface becomes saturated (*g*). The total outflow (*h*) partly comprises saturated groundwater flow contributed directly to the stream and partly unsaturated subsurface flow seeping from the hillside above the water table.

Subsurface flow velocities are normally so low that subsurface flow alone cannot contribute a significant amount of storm precipitation directly to streamflow except under special circumstances where the hydraulic conductivity of the soil is very high (Pearce, Stewart, and Sklash, 1986). However, Moseley (1979) has suggested that flow through root holes in a forested soil can be much more rapid than flow through the adjacent soil mass.

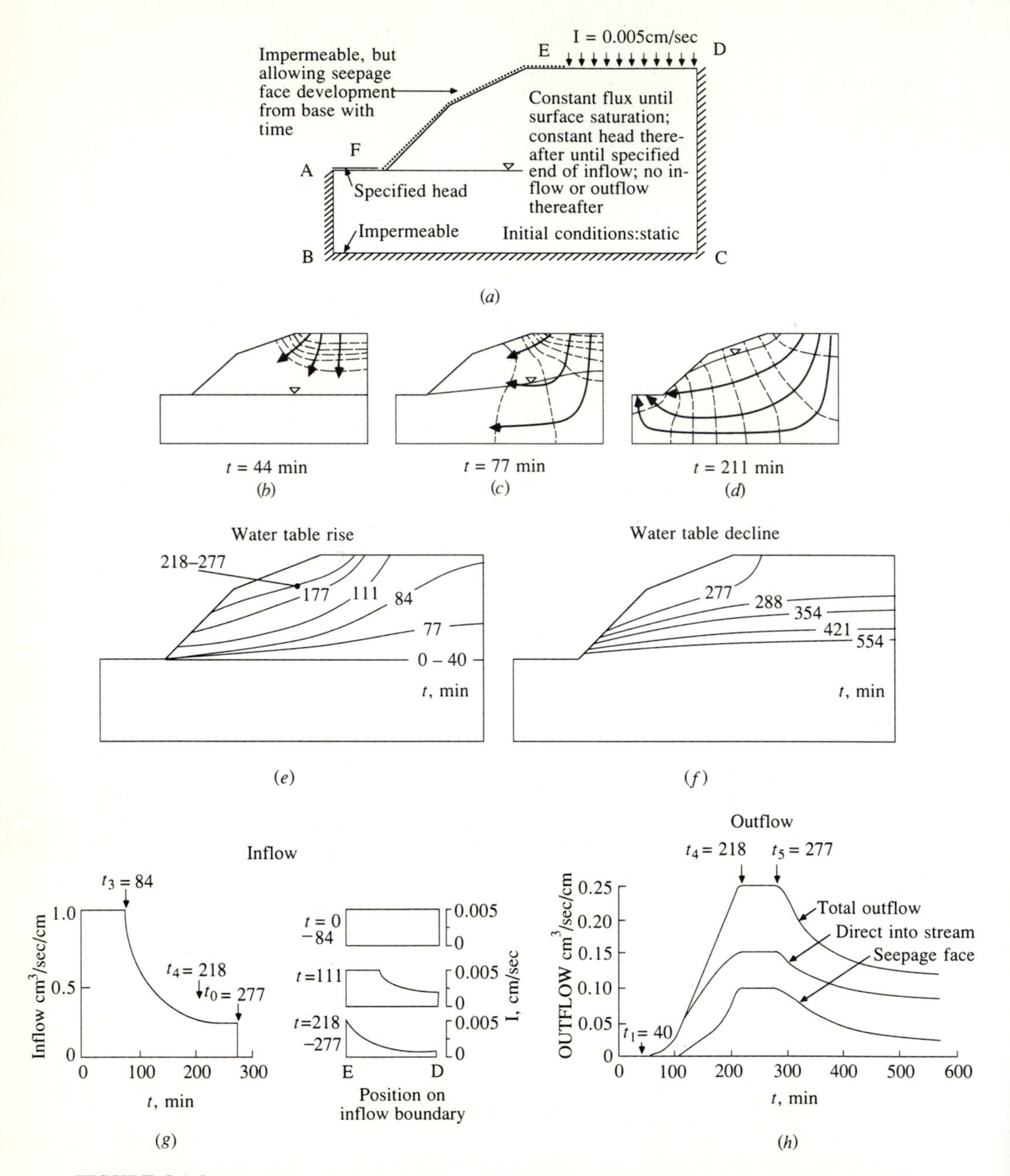

FIGURE 5.1.3
Saturated-unsaturated subsurface flow in a small idealized two-dimensional flow system (*a*) Boundary and initial conditions. (*b*)-(*d*) Transient hydraulic head contours (broken line) and stream lines (solid). (*e*) Water table rise. (*f*) Water table decline. (*g*) Inflow as a function of time and position. (*h*) Outflow hydrograph. (*Source:* Freeze, 1974, p. 644. Copyright by the American Geophysical Union.)

Saturation Overland Flow

Saturation overland flow is produced when subsurface flow saturates the soil near the bottom of a slope and overland flow then occurs as rain falls onto saturated soil. Saturation overland flow differs from Hortonian overland flow in that in Hortonian overland flow the soil is saturated from above by infiltration, while in saturation overland flow it is saturated from below by subsurface flow. Saturation overland flow occurs most often at the bottom of hill slopes and near stream banks.

The velocity of subsurface flow is so low that not all of a watershed can contribute subsurface flow or saturation overland flow to a stream during a storm. Forest hydrologists (Hewlett, 1982) have coined the terms *variable source areas*, or *partial areas*, to denote the area of the watershed actually contributing flow to the stream at any time (Betson, 1964; Ragan, 1968; Harr, 1977; Pearce and McKerchar, 1979; Hewlett, 1982). As shown in Fig. 5.1.4, the variable source area expands during rainfall and contracts thereafter. The source area for streamflow may constitute only 10 percent of the watershed during a storm in a humid, well vegetated region.

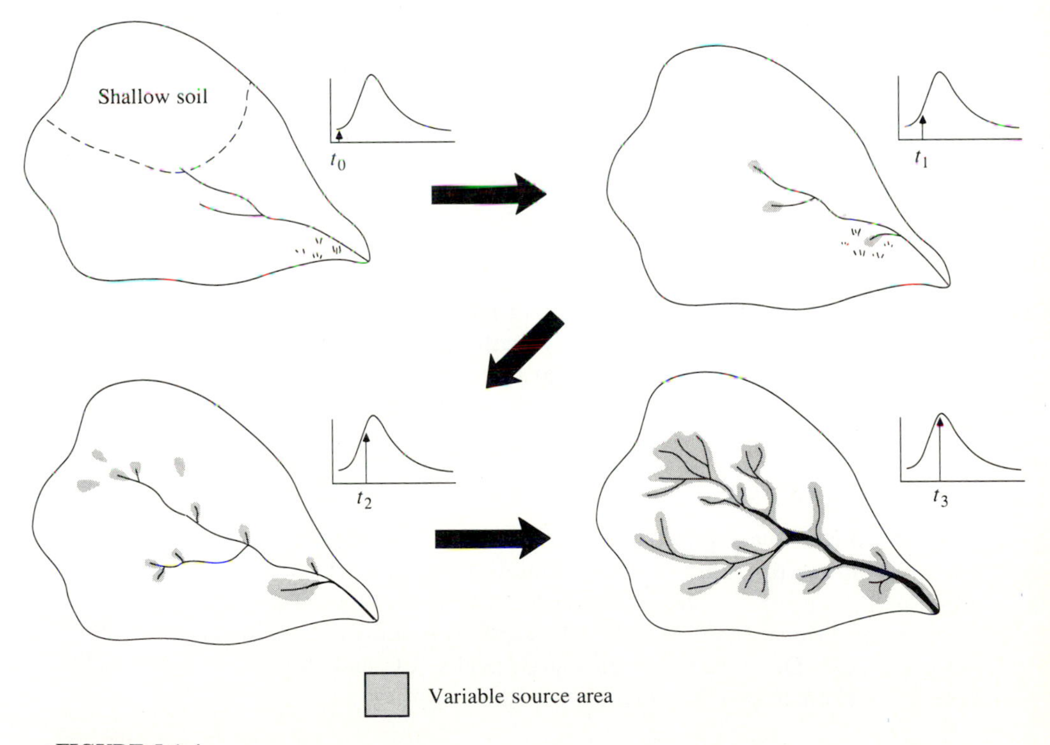

FIGURE 5.1.4
The small arrows in the hydrographs show how streamflow increases as the variable source extends into swamps, shallow soils and ephemeral channels. The process reverses as streamflow declines. (Reprinted from *Principles of Forest Hydrology* by J. D. Hewlett, Copyright 1982 the University of Georgia Press. Reprinted by permission of the University of Georgia Press.)

5.2 STREAMFLOW HYDROGRAPH

A streamflow or discharge *hydrograph* is a graph or table showing the flow rate as a function of time at a given location on the stream. In effect, the hydrograph is "an integral expression of the physiographic and climatic characteristics that govern the relations between rainfall and runoff of a particular drainage basin" (Chow, 1959). Two types of hydrographs are particularly important: the annual hydrograph and the storm hydrograph.

Annual Hydrograph

The annual hydrograph, a plot of streamflow vs. time over a year, shows the long-term balance of precipitation, evaporation, and streamflow in a watershed. Examples typical of three main types of annual hydrographs are shown in Fig. 5.2.1.

The first hydrograph, from Mill Creek near Belleville, Texas, has a *perennial* or continuous flow regime typical of a humid climate. The spikes, caused by rain storms, are called *direct runoff* or *quickflow*, while the slowly varying flow in rainless periods is called *baseflow*. The total volume of flow under the annual hydrograph is the *basin yield*. For a river with perennial flow most of the basin yield usually comes from baseflow, indicating that a large proportion of the rainfall is infiltrated into the basin and reaches the stream as subsurface flow.

The second hydrograph, from the Frio River near Uvalde, Texas, is an example of an *ephemeral* river in an arid climate. There are long periods when the river is dry. Most storm rainfall becomes direct runoff and little infiltration occurs. Basin yield from this watershed is the result of direct runoff from large storms.

The third hydrograph, from the East River near Almont, Colorado, is produced by a snow-fed river. The bulk of the basin yield occurs in the spring and early summer from snowmelt. The large volume of water stored in the snowpack, and its steady release, create an annual hydrograph which varies more smoothly over the year than for the perennial or ephemeral streams illustrated.

Storm Hydrograph

Study of annual hydrographs shows that peak streamflows are produced infrequently, and are the result of storm rainfall alone or storm rainfall and snowmelt combined. Figure 5.2.2 shows four components of a streamflow hydrograph during a storm. Prior to the time of intense rainfall, baseflow is gradually diminishing (segment AB). Direct runoff begins at B, peaks at C and ends at D. Segment DE follows as normal baseflow recession begins again.

Baseflow Separation

A variety of techniques have been suggested for separating baseflow and direct runoff. One of the oldest is the *normal depletion curve* described by Horton

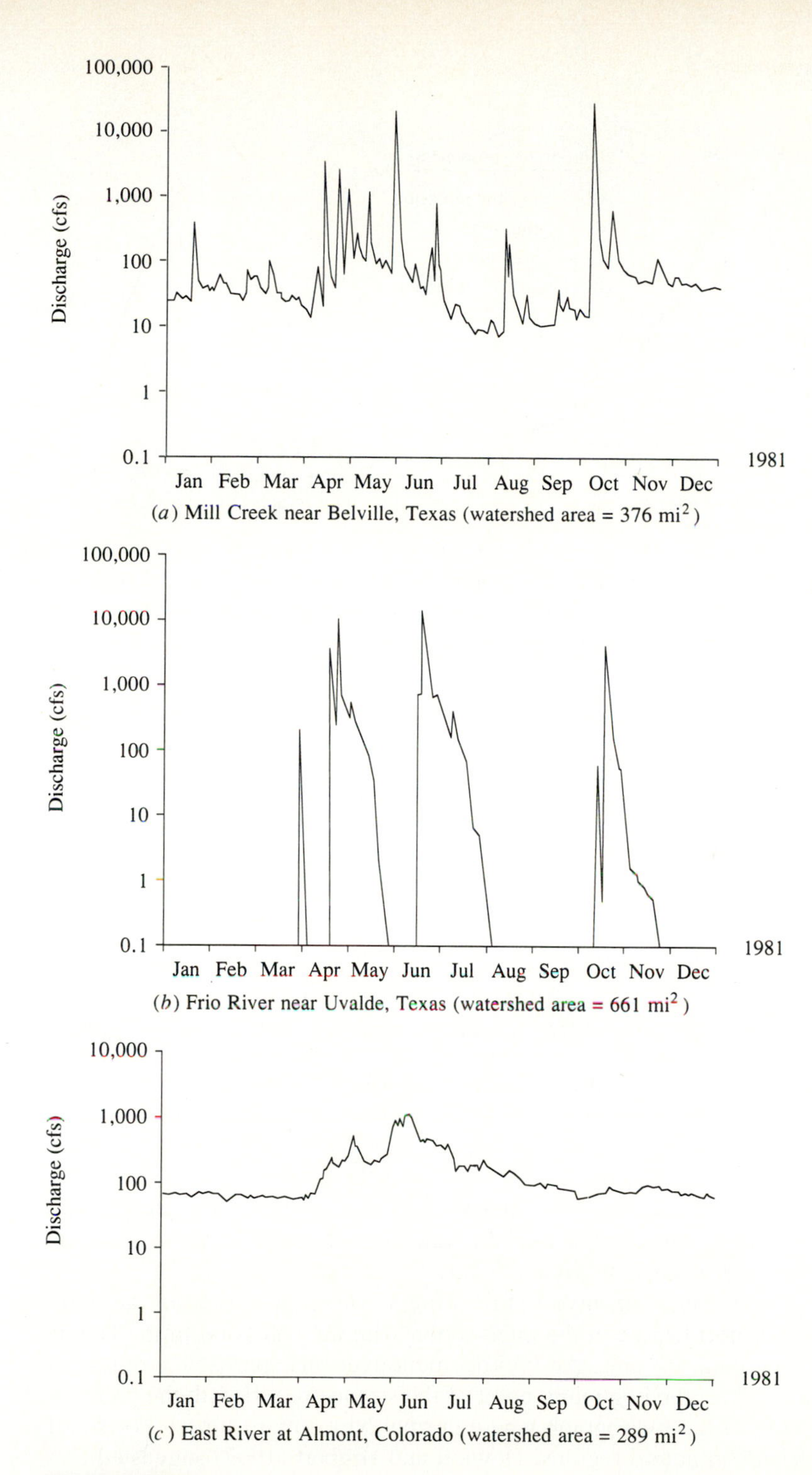

FIGURE 5.2.1
The annual streamflow hydrographs for 1981 from three different gaging stations illustrating the main types of hydrologic regimes: (*a*) perennial river, (*b*) ephemeral river, (*c*) snow-fed river. (Data provided by the U. S. Geological Survey).

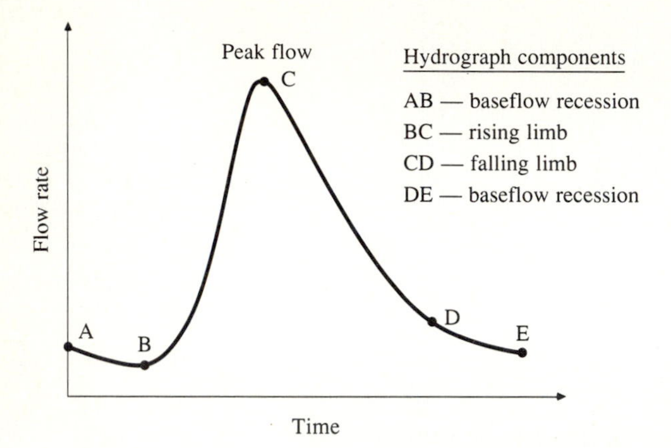

FIGURE 5.2.2
Components of the streamflow hydrograph during a storm.

(1933). The normal depletion curve, or *master baseflow recession curve*, is a characteristic graph of flow recessions compiled by superimposing many of the recession curves observed on a given stream. Recession curves often take the form of exponential decay:

$$Q(t) = Q_0 e^{-(t-t_0)/k} \tag{5.2.1}$$

where Q_0 is the flow at time t_0 and k is an exponential decay constant having the dimensions of time (Singh and Stall, 1971). Equation (5.2.1) is linearized by plotting the logarithm of $Q(t)$ against time on a linear scale. In Northland, New Zealand, a typical value for k is 6×10^{-3} days, which corresponds to a "half-life" of 116 days (Martin, 1973). The half-life is the time for baseflow to recede to the point where $Q(t)/Q_0 = 0.5$. The concept underlying Eq. (5.2.1) is that of a *linear reservoir*, whose outflow rate is proportional to the current storage (see Sec. 8.5):

$$S(t) = kQ(t) \tag{5.2.2}$$

By noting the periods of time when the streamflow hydrograph is coincident with the normal baseflow recession curve, the points where direct runoff begins and ceases can be identified (B and D on Fig. 5.2.2). Between these points direct runoff and baseflow can be separated by various methods.

Some alternative methods of baseflow separation are: (*a*) the straight line method, (*b*) the fixed base length method, and (*c*) the variable slope method. These methods are illustrated in Figure 5.2.3.

The *straight line method*, involves drawing a horizontal line from the point at which surface runoff begins to the intersection with the recession limb. This is applicable to ephemeral streams. An improvement over this approach is to use an inclined line to connect the beginning point of the surface runoff with the point on the recession limb of the hydrograph where normal baseflow resumes. For small forested watersheds in humid regions, Hewlett and Hibbert (1967) suggested that baseflow during a storm can be assumed to be increasing at a rate of 0.0055 l/s·ha·h (0.05 cfs/mi^{2}h).

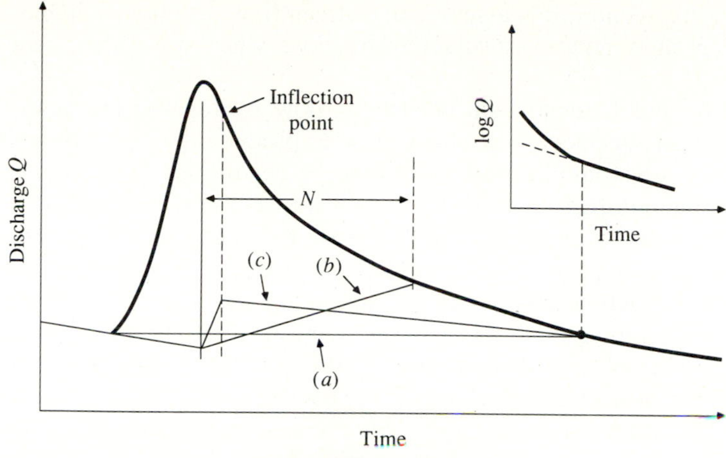

(a) Straight line method.
(b) Fixed base method.
(c) Variable slope method.

FIGURE 5.2.3
Baseflow separation techniques.

In the *fixed base method*, the surface runoff is assumed to end a fixed time N after the hydrograph peak. The baseflow before the surface runoff began is projected ahead to the time of the peak. A straight line is used to connect this projection at the peak to the point on the recession limb at time N after the peak.

In the *variable slope method*, the baseflow curve before the surface runoff began is extrapolated forward to the time of peak discharge, and the baseflow curve after surface runoff ceases is extrapolated backward to the time of the point of inflection on the recession limb. A straight line is used to connect the endpoints of the extrapolated curves.

5.3 EXCESS RAINFALL AND DIRECT RUNOFF

Excess rainfall, or effective rainfall, is that rainfall which is neither retained on the land surface nor infiltrated into the soil. After flowing across the watershed surface, excess rainfall becomes *direct runoff* at the watershed outlet under the assumption of Hortonian overland flow. The graph of excess rainfall vs. time, or *excess rainfall hyetograph* (ERH), is a key component of the study of rainfall-runoff relationships. The difference between the observed total rainfall hyetograph and the excess rainfall hyetograph is termed *abstractions*, or *losses*. Losses are primarily water absorbed by infiltration with some allowance for interception and surface storage.

The excess rainfall hytograph may be determined from the rainfall hyetograph in one of two ways, depending on whether streamflow data are available for the storm or not. In this section, it is assumed that streamflow data are available. Sections 5.4 and 5.5 show how to calculate abstractions when streamflow data are not available.

Suppose that a rainfall hyetograph and streamflow hyetograph are available, baseflow has been separated from streamflow to produce the direct runoff hydrograph, and the excess rainfall hyetograph is to be determined. The parameters of infiltration equations can be determined by optimization techniques such as nonlinear programming (Unver and Mays, 1984), but these techniques are complicated. There is a simpler alternative, called a ϕ-index. The ϕ-index is that constant rate of abstractions (in/h or cm/h) that will yield an excess rainfall hyetograph (ERH) with a total depth equal to the depth of direct runoff r_d over the watershed. The value of ϕ is determined by picking a time interval length Δt, judging the number of intervals M of rainfall that actually contribute to direct runoff, subtracting $\phi\Delta t$ from the observed rainfall in each interval, and adjusting the values of ϕ and M as necessary so that the depths of direct runoff and excess rainfall are equal:

$$r_d = \sum_{m=1}^{M} (R_m - \phi\Delta t) \tag{5.3.1}$$

where R_m is the observed rainfall (in) in time interval m.

Example 5.3.1. Determine the direct runoff hydrograph, the ϕ-index, and the excess rainfall hyetograph from the observed rainfall and streamflow data given in Table 5.3.1. The watershed area is 7.03 mi^2.

Solution. The basin-average rainfall data given in column 2 of Table 5.3.1 were obtained by taking Thiessen-weighted averages of the rainfall data from two rainfall gages in the watershed. (Ideally, data from several more gages would be used.) The pulse data representation is used for rainfall with a time interval of Δt = 1/2 h, so each value shown in column 2 is the incremental precipitation that occurred during the half-hour up to the time shown. The streamflow data shown were recorded as sample data; the value shown in column 3 is the streamflow recorded at that instant of time. The observed rainfall and streamflow data are plotted in Fig. 5.3.1, from which it is apparent that rainfall prior to 9:30 P.M. produced a small flow in the stream (approximately 400 cfs) and that the direct runoff occurred following intense rainfall between 9:30 and 11:30 P.M.

The computation of the effective rainfall hyetograph and the direct runoff hydrograph uses the following procedure:

Step 1. Estimate the baseflow. A constant baseflow rate of 400 cfs is selected.

Step 2. Calculate the direct runoff hydrograph (DRH). The DRH, in column 6 of Table 5.3.1, is found by the straight line method, by subtracting the 400 cfs baseflow from the observed streamflow (column 3). Eleven half-hour time intervals in column 4 are labeled from the first period of non-zero direct runoff, beginning at 9:30 P.M.

Step 3. Compute the volume V_d and depth r_d of direct runoff.

TABLE 5.3.1

Rainfall and streamflow data dapted from the storm of May 24–25, 1981, on Shoal Creek at Northwest Park, Austin, Texas

	Time	Observed Rainfall (in)	Observed Streamflow (cfs)	Time $\left(\frac{1}{2}\text{ h}\right)$	Excess rainfall hyetograph (ERH) (in)	Direct runoff hydrograph (DRH) (cfs)
Column:	1	2	3	4	5	6
24 May	8:30 P.M.		203			
	9:00	0.15	246			
	9:30	0.26	283			
	10:00	1.33	828	1	1.06	428
	10:30	2.20	2323	2	1.93	1923
	11:00	2.08	5697	3	1.81	5297
	11:30	0.20	9531	4		9131
25 May	12:00 A.M.	0.09	11025	5		10625
	12:30		8234	6		7834
	1:00		4321	7		3921
	1:30		2246	8		1846
	2:00		1802	9		1402
	2:30		1230	10		830
	3:00		713	11		313
	3:30		394			
	4:00		354	Total	4.80	43550
	4:30		303			

Excess rainfall = observed rainfall − abstractions (0.27 in per half-hour)

Direct runoff = observed streamflow − baseflow (400 cfs)

$$V_d = \sum_{n=1}^{11} Q_n \Delta t$$

$$= 43,550 \text{ cfs } \times 1/2 \text{ h}$$

$$= 43,550 \frac{\text{ft}^3}{\text{s}} \times \frac{3600}{1}\frac{\text{s}}{\text{h}} \times \frac{1}{2}\text{ h}$$

$$= 7.839 \times 10^7 \text{ ft}^3$$

$$r_d = \frac{V_d}{\text{watershed area}}$$

$$= \frac{7.839 \times 10^7 \text{ ft}^3}{7.03 \text{ mi}^2 \times 5280^2 \text{ ft}^2/\text{mi}^2}$$

$$= 0.400 \text{ ft}$$

$$= 4.80 \text{ in}$$

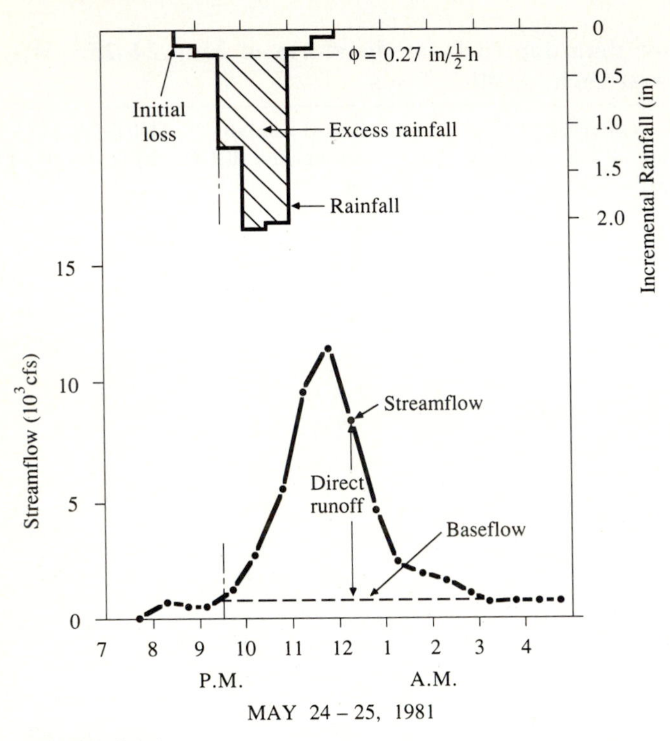

FIGURE 5.3.1
Rainfall and streamflow for the storm of May 24–25, 1981, on Shoal Creek at Northwest Park, Austin, Texas.

Step 4. Estimate the rate of rainfall abstractions by infiltration and surface storage in the watershed. Any rainfall prior to the beginning of direct runoff is taken as *initial abstraction* (i.e., that rainfall prior to 9:30 P.M. in Table 5.3.1). The abstraction rate ϕ, and M, the number of nonzero pulses of excess rainfall, are found by trial and error.

1. If $M = 1$, the largest rainfall pulse, $R_m = 2.20$ in, is selected, substituted into Eq. (5.3.1) using $r_d = 4.80$ in and $\Delta t = 0.5$ h, and solved for a trial value of ϕ:

$$r_d = \sum_{m=1}^{M} (R_m - \phi\Delta t)$$

$$4.80 = (2.20 - \phi \times 0.5)$$

$$\phi = -5.20 \text{ in/h}$$

which is not physically possible.

2. If $M = 2$, the one-hour period having the highest rainfall is selected (between 10:00 P.M. and 11:00 P.M.) and substituted into (5.3.1) to solve for a new trial

value of ϕ:

$$r_d = \sum_{m=1}^{M} (R_m - \phi \Delta t)$$

$$4.80 = (2.20 + 2.08 - \phi \times 2 \times 0.5)$$

$$\phi = -0.52 \text{ in/h}$$

again impossible.

3. If $M = 3$, the $1\frac{1}{2}$ hour period having pulses 1.33, 2.20, and 2.08 in is selected, and the data is substituted into (5.3.1):

$$r_d = \sum_{m=1}^{M} (R_m - \phi \Delta t)$$

$$4.80 = (1.33 + 2.20 + 2.08 - \phi \times 3 \times 0.5)$$

$$\phi = 0.54 \text{ in/h}$$

This value of ϕ is satisfactory because it gives $\phi \Delta t = 0.27$ in, which is greater than all of the rainfall pulses in column 2 outside of the three assumed to contribute to direct runoff.

Step 5. Calculate the excess rainfall hyetograph. The ordinates (column 5) are found by subtracting $\phi \Delta t = 0.27$ in from the ordinates of the observed rainfall hyetograph (column 2), neglecting all intervals in which the observed rainfall depth is less than $\phi \Delta t$. The duration of excess rainfall is 1.5 h in this example (9:30 to 11:00 P.M.). The depth of excess rainfall is checked to ensure that it equals r_d (total of column 5 = 4.80 in). The excess portion of the observed rainfall hyetograph is cross-hatched in Fig. 5.3.1.

Runoff Coefficients

Abstractions may also be accounted for by means of *runoff coefficients*. The most common definition of a runoff coefficient is that it is the ratio of the peak rate of direct runoff to the average intensity of rainfall in a storm. Because of highly variable rainfall intensity, this value is difficult to determine from observed data. A runoff coefficient can also be defined to be the ratio of runoff to rainfall over a given time period. These coefficients are most commonly applied to storm rainfall and runoff, but can also be used for monthly or annual rainfall and streamflow data. If $\Sigma_{m=1}^{M} R_m$ is the total rainfall and r_d the corresponding depth of runoff, then a runoff coefficient can be defined as

$$C = \frac{r_d}{\sum_{m=1}^{M} R_m} \tag{5.3.2}$$

Example 5.3.2. Determine the runoff coefficient for the storm in Example 5.3.1.

Solution. Considering only the rainfall that occurred after the beginning of direct runoff (9:30 P.M.):

$$\sum_{m=1}^{M} R_m = 1.33 + 2.20 + 2.08 + 0.20 + 0.09$$

$$= 5.90 \text{ in}$$

$$C = \frac{r_d}{\sum_{m=1}^{M} R_m}$$

$$= \frac{4.80}{5.90}$$

$$= 0.81$$

5.4 ABSTRACTIONS USING INFILTRATION EQUATIONS

Abstractions include *interception* of precipitation on vegetation above the ground, *depression storage* on the ground surface as water accumulates in hollows over the surface, and *infiltration* of water into the soil. Interception and depression storage abstractions are estimated based on the nature of the vegetation and ground surface or are assumed to be negligible in a large storm.

In the previous section, the rate of abstractions from rainfall was determined by using a known streamflow hydrograph. In most hydrologic problems, the streamflow hydrograph is not available and the abstractions must be determined by calculating infiltration and accounting separately for other forms of abstraction, such as interception, and detention or depression storage. In this section, it is assumed that all abstractions arise from infiltration, and a method for determining the ponding time and infiltration under a variable intensity rainfall is developed based on the Green-Ampt infiltration equation. Equivalent relationships for use with the Horton and Philip equations are presented in Table 5.4.1. The problem considered is: given a rainfall hyetograph defined using the pulse data representation, and the parameters of an infiltration equation, determine the ponding time, the infiltration after ponding occurs, and the excess rainfall hyetograph.

The basic principles used for determining ponding time under constant rainfall intensity in Sec. 4.4 are also employed here: in the absence of ponding, cumulative infiltration is calculated from cumulative rainfall; the potential infiltration rate at a given time is calculated from the cumulative infiltration at that time; and ponding has occurred when the potential infiltration rate is less than or equal to the rainfall intensity.

TABLE 5.4.1
Equations for calculating infiltration at and following ponding

Equation	Green-Ampt equation	Horton's equation	Philip's equation
(1) Cumulative infiltration $F_{t+\Delta t}$			
	$F_{t+\Delta t} = F_t + K\Delta t + \psi\Delta\theta \ln\left[\frac{F_{t+\Delta t} + \psi\Delta\theta}{F_t + \psi\Delta\theta}\right]$	$F_{t+\Delta t} = F_t + f_c\Delta t + (f_t - f_c)\frac{(1 - e^{-k\Delta t})}{k}$	$F_{t+\Delta t} = F_t + K\Delta t - \frac{S^2}{2(f_t - K)} + S\left[\Delta t + \frac{S^2}{4(f_t - K)^2}\right]^{1/2}$
(2) Infiltration rate $f_{t+\Delta t}$			
	$f_{t+\Delta t} = K\left(\frac{\psi\Delta\theta}{F_{t+\Delta t}} + 1\right)$	$f_{t+\Delta t} = f_t - k(F_{t+\Delta t} - F_t - f_c\Delta t)$	$f_{t+\Delta t} = K + S\left(\frac{S + \sqrt{S^2 + 4KF_{t+\Delta t}}}{4F_{t+\Delta t}}\right)$
(3) Cumulative infiltration at ponding F_p			
	$F_p = \frac{K\psi\Delta\theta}{i_t - K}$ $(i_t > K)$	$F_p = \frac{1}{k}\left[f_o - i_t + f_c \ln\left(\frac{f_o - f_c}{i_t - f_c}\right)\right]$ $(f_c < i_t < f_o)$	$F_p = \frac{S^2(i_t - K/2)}{2(i_t - K)^2}$ $(i_t > K)$

Consider a time interval from t to $t + \Delta t$. The rainfall intensity during this interval is denoted i_t and is constant throughout the interval. The potential infiltration rate and cumulative infiltration at the beginning of the interval are f_t and F_t, respectively, and the corresponding values at the end of the interval are $f_{t+\Delta t}$, and $F_{t+\Delta t}$. It is assumed that F_t is known from given initial conditions or previous computation.

A flow chart for determining ponding time is presented in Fig 5.4.1. There are three cases to be considered: (1) ponding occurs throughout the interval; (2) there is no ponding throughout the interval; and (3) ponding begins part-way through the interval. The infiltration rate is always either decreasing or constant with time, so once ponding is established under a given rainfall intensity, it will continue. Hence, ponding cannot cease in the middle of an interval, but only at its end point, when the value of the rainfall intensity changes.

Following the flow chart, the first step is to calculate the current potential infiltration rate f_t from the known value of cumulative infiltration F_t. For the Green-Ampt method, one uses

$$f_t = K\left(\frac{\psi \Delta \theta}{F_t} + 1\right) \tag{5.4.1}$$

The result f_t is compared to the rainfall intensity i_t. If f_t is less than or equal to i_t, case (1) arises and there is ponding throughout the interval. In this case, for the Green-Ampt equation, the cumulative infiltration at the end of the interval, $F_{t+\Delta t}$, is calculated from

$$F_{t+\Delta t} - F_t - \psi \Delta \theta \ln \left[\frac{F_{t+\Delta t} + \psi \Delta \theta}{F_t + \psi \Delta \theta}\right] = K \Delta t \tag{5.4.2}$$

This equation is derived in a manner similar to that shown in Sec. 4.4 for Eq. (4.4.5).

Both cases (2) and (3) have $f_t > i_t$ and no ponding at the beginning of the interval. Assume that this remains so throughout the interval; then, the infiltration rate is i_t and a tentative value for cumulative infiltration at the end of the time interval is

$$F'_{t+\Delta t} = F_t + i_t \Delta t \tag{5.4.3}$$

Next, a corresponding infiltration rate $f'_{t+\Delta t}$ is calculated from $F'_{t+\Delta t}$. If $f'_{t+\Delta t}$ is greater than i_t, case (2) occurs and there is no ponding throughout the interval. Thus $F_{t+\Delta t} = F'_{t+\Delta t}$ and the problem is solved for this interval.

If $f'_{t+\Delta t}$ is less than or equal to i_t, ponding occurs during the interval (case (3)). The cumulative infiltration F_p at ponding time is found by setting $f_t = i_t$ and $F_t = F_p$ in (5.4.1) and solving for F_p to give, for the Green-Ampt equation,

$$F_p = \frac{K\psi \Delta \theta}{i_t - K} \tag{5.4.4}$$

The ponding time is then $t + \Delta t'$, where

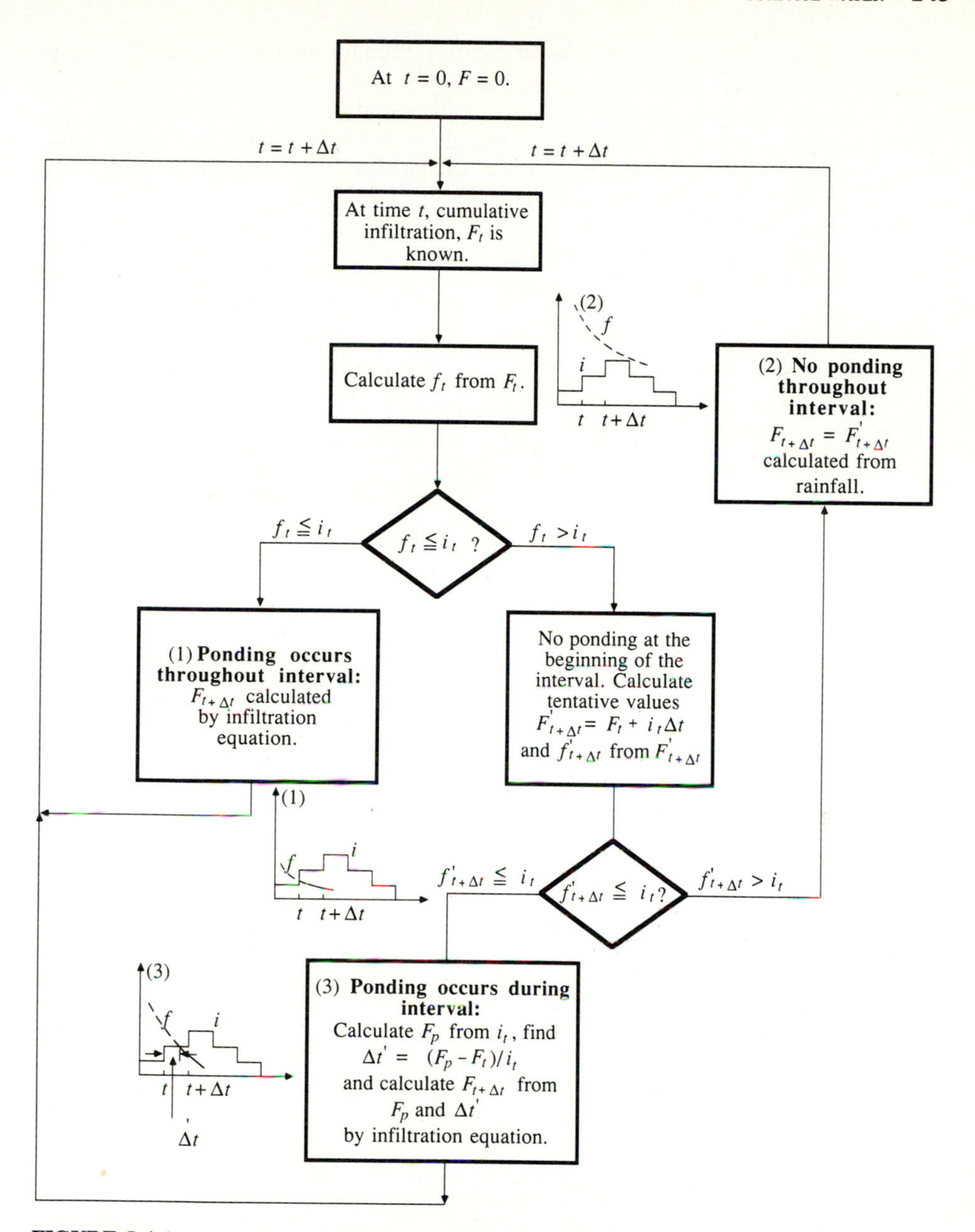

FIGURE 5.4.1
Flow chart for determining infiltration and ponding time under variable rainfall intensity.

$$\Delta t' = \frac{F_p - F_t}{i_t} \tag{5.4.5}$$

and the cumulative infiltration $F_{t+\Delta t}$ is found by substituting $F_t = F_p$ and $\Delta t = \Delta t - \Delta t'$ in (5.4.2). The excess rainfall values are calculated by subtracting cumulative infiltration from cumulative rainfall, then taking successive differences of the resulting values.

Example 5.4.1 A rainfall hyetograph is given in columns 1 and 2 of Table 5.4.2. If this rain falls on a sandy loam soil of initial effective saturation 40 percent, determine the excess rainfall hyetograph.

Solution. From Table 4.3.1, for a sandy loam soil, $K = 1.09$ cm/h, $\psi = 11.01$ cm and $\theta_e = 0.412$. From Eq. (4.3.10)

$$\begin{aligned}\Delta\theta &= (1 - s_e)\theta_e \\ &= (1 - 0.4)(0.412) \\ &= 0.247\end{aligned}$$

and

$$\begin{aligned}\psi\,\Delta\theta &= 0.247 \times 11.01 \\ &= 2.72 \text{ cm}\end{aligned}$$

The time interval in Table 5.4.2 is $\Delta t = 10$ min $= 0.167$ h. Column 3 of the table shows the cumulative rainfall depths found by summing the incremental values in column 2. The rainfall hyetograph and the cumulative rainfall hyetograph are shown in Fig. 5.4.2. The rainfall intensity in column 4 is found from column 2 by dividing by Δt. For example, during the first time interval, 0.18 cm of rainfall occurs, so $i_t = 0.18/0.167 = 1.08$ cm/h as shown. Initially, $F = 0$, so $f = \infty$ from (5.4.1) and ponding does not occur at time 0. Hence F at time 10 min is calculated by (5.4.3), thus: $F'_{t+\Delta t} = F_t + i_t\,\Delta t = 0 + 0.18 = 0.18$ cm. The corresponding value of $f'_{t+\Delta t}$ is, from (5.4.1),

$$\begin{aligned}f'_{t+\Delta t} &= K\left(\frac{\psi\Delta\theta}{F'_{t+\Delta t}} + 1\right) \\ &= 1.09\left(\frac{2.72}{0.18} + 1\right) \\ &= 17.57 \text{ cm/h}\end{aligned}$$

as shown in column 5 of the table. This value is greater than i_t; therefore, no ponding occurs during this interval and cumulative infiltration equals cumulative rainfall as shown in column 6. It is found that ponding does not occur up to 60 minutes of rainfall, but at 60 min,

$$\begin{aligned}f_t &= K\left(\frac{\psi\,\Delta\theta}{F_t} + 1\right) \\ &= 1.09\left(\frac{2.72}{1.77} + 1\right) \\ &= 2.77 \text{ cm/h}\end{aligned}$$

which is less than $i_t = 3.84$ cm/h for the interval from 60 to 70 minutes, so ponding begins at 60 min [see Fig. 5.4.2(*a*)].

TABLE 5.4.2
Calculation of excess rainfall hyetograph using the Green-Ampt infiltration equation (Example 5.4.1)

Column:	1	2	3	4	5	6	7	8
		Rainfall			Infiltration		Excess Rainfall	
	Time (min)	Incremental (cm)	Cumulative (cm)	Intensity (cm/h)	Rate (cm/h)	Cumulative (cm)	Cumulative (cm)	Incremental (cm)
	0		0.00	1.08		0.00		
	10	0.18	0.18	1.26	17.57	0.18		
	20	0.21	0.39	1.56	8.70	0.39		
	30	0.26	0.65	1.92	5.65	0.65		
	40	0.32	0.97	2.22	4.15	0.97		
	50	0.37	1.34	2.58	3.30	1.34		
	60	0.43	1.77	3.84	2.77	1.77	0.00	
Ponding	70	0.64	2.41	6.84	2.43	2.21	0.20	0.20
Ponding	80	1.14	3.55	19.08	2.23	2.59	0.96	0.76
Ponding	90	3.18	6.73	9.90	2.09	2.95	3.78	2.82
Ponding	100	1.65	8.38	4.86	1.99	3.29	5.09	1.31
Ponding	110	0.81	9.19	3.12	1.91	3.62	5.57	0.48
Ponding	120	0.52	9.71	2.52	1.84	3.93	5.78	0.21
Ponding	130	0.42	10.13	2.16	1.79	4.24	5.90	0.12
Ponding	140	0.36	10.49	1.68	1.74	4.53	5.96	0.06
	150	0.28	10.77	1.44	1.71	4.81		
	160	0.24	11.01	1.14	1.68	5.05		
	170	0.19	11.20	1.02	1.66	5.24		
	180	0.17	11.37		1.64	5.41		

During the ponded period, (5.4.2) is used to calculate infiltration. The value of $F_{t+\Delta t}$ at 70 mins is given by

$$F_{t+\Delta t} - F_t - \psi\,\Delta\theta \ln\left[\frac{F_{t+\Delta t} + \psi\,\Delta\theta}{F_t + \psi\,\Delta\theta}\right] = K\Delta t$$

$$F_{t+\Delta t} - 1.77 - 2.72 \ln\left[\frac{F_{t+\Delta t} + 2.72}{1.77 + 2.72}\right] = 1.09 \times 0.167$$

or

$$F_{t+\Delta t} = 1.95 + 2.72 \ln\left(\frac{F_{t+\Delta t} + 2.72}{4.49}\right)$$

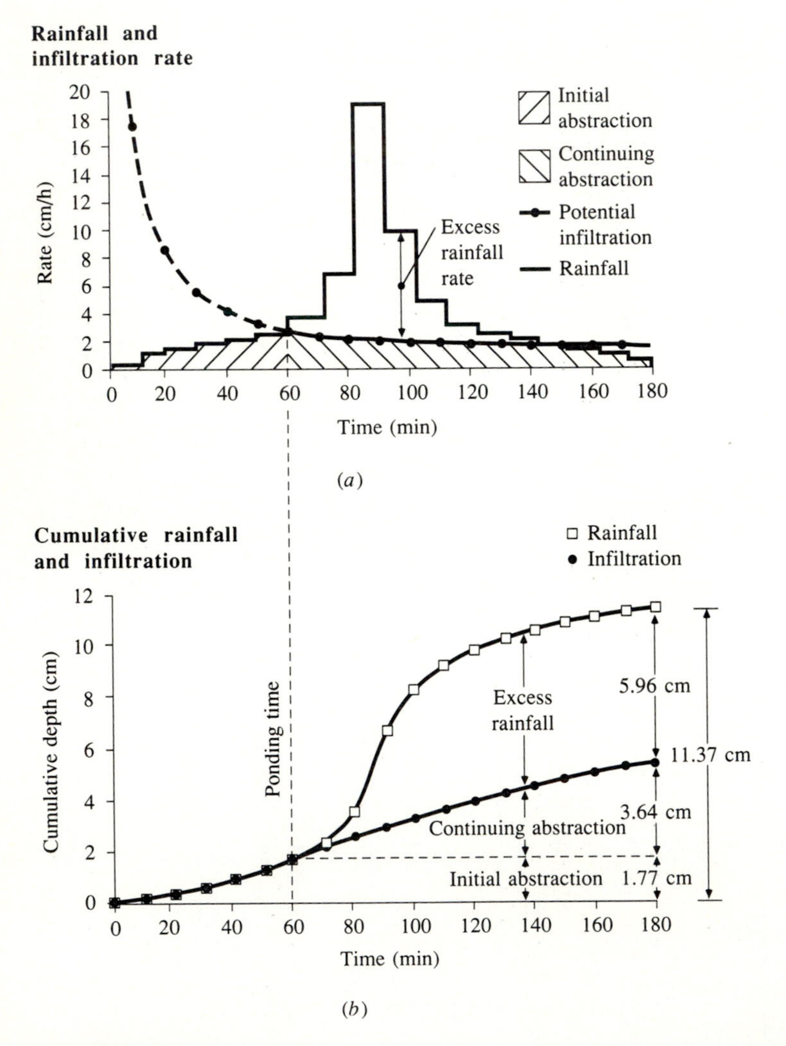

FIGURE 5.4.2
Infiltration and excess rainfall under variable rainfall intensity (Example 5.4.1).

which is solved by the method of successive approximation to give $F_{t+\Delta t} = 2.21$ cm as shown in column 6 of Table 5.4.2. The cumulative excess rainfall (column 7) is found by subtracting cumulative infiltration (column 6) from cumulative rainfall (column 3). And the excess rainfall values in column 8 are found by taking differences of successive cumulative rainfall values. Ponding ceases at 140 min when the rainfall intensity falls below the potential infiltration rate. After 140 min, cumulative infiltration is computed from rainfall by (5.4.3). For example, at 150 min $F_{t+\Delta t} = F_t + i_t\Delta t = 4.53 + 0.28 = 4.81$ in as shown in column 6.

As shown in Fig. 5.4.2, the total rainfall of 11.37 cm is disposed of as an initial abstraction of 1.77 cm (cumulative infiltration at ponding time), a continuing abstraction of 3.64 cm (5.41 cm total infiltration − 1.77 cm initial abstraction), and an excess rainfall of 5.96 cm.

5.5 SCS METHOD FOR ABSTRACTIONS

The Soil Conservation Service (1972) developed a method for computing abstractions from storm rainfall. For the storm as a whole, the depth of excess precipitation or direct runoff P_e is always less than or equal to the depth of precipitation P; likewise, after runoff begins, the additional depth of water retained in the watershed, F_a, is less than or equal to some potential maximum retention S (see Fig. 5.5.1). There is some amount of rainfall I_a (initial abstraction before ponding) for which no runoff will occur, so the potential runoff is $P - I_a$. The hypothesis of the SCS method is that the ratios of the two actual to the two potential quantities are equal, that is,

$$\frac{F_a}{S} = \frac{P_e}{P - I_a} \tag{5.5.1}$$

From the continuity principle

$$P = P_e + I_a + F_a \tag{5.5.2}$$

Combining (5.5.1) and (5.5.2) to solve for P_e gives

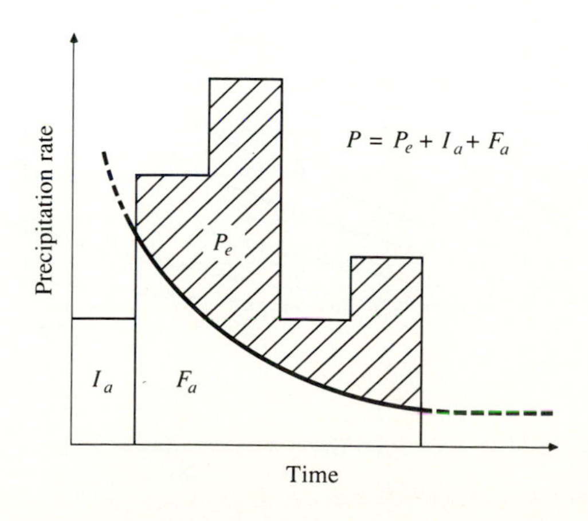

FIGURE 5.5.1
Variables in the SCS method of rainfall abstractions: I_a = initial abstraction, P_e = rainfall excess, F_a = continuing abstraction, P = total rainfall.

$$P_e = \frac{(P - I_a)^2}{P - I_a + S} \tag{5.5.3}$$

which is the basic equation for computing the depth of excess rainfall or direct runoff from a storm by the SCS method.

By study of results from many small experimental watersheds, an empirical relation was developed.

$$I_a = 0.2S \tag{5.5.4}$$

On this basis

$$P_e = \frac{(P - 0.2S)^2}{P + 0.8S} \tag{5.5.5}$$

Plotting the data for P and P_e from many watersheds, the SCS found curves of the type shown in Fig. 5.5.2. To standardize these curves, a dimensionless curve number CN is defined such that $0 \leq \text{CN} \leq 100$. For impervious and water surfaces CN = 100; for natural surfaces CN < 100. As an illustration, the rainfall event of Example 5.3.2 has $P_e = 4.80$ in. and $P = 5.80$ in. From Fig. 5.5.2, it can be seen that CN = 91 for this event.

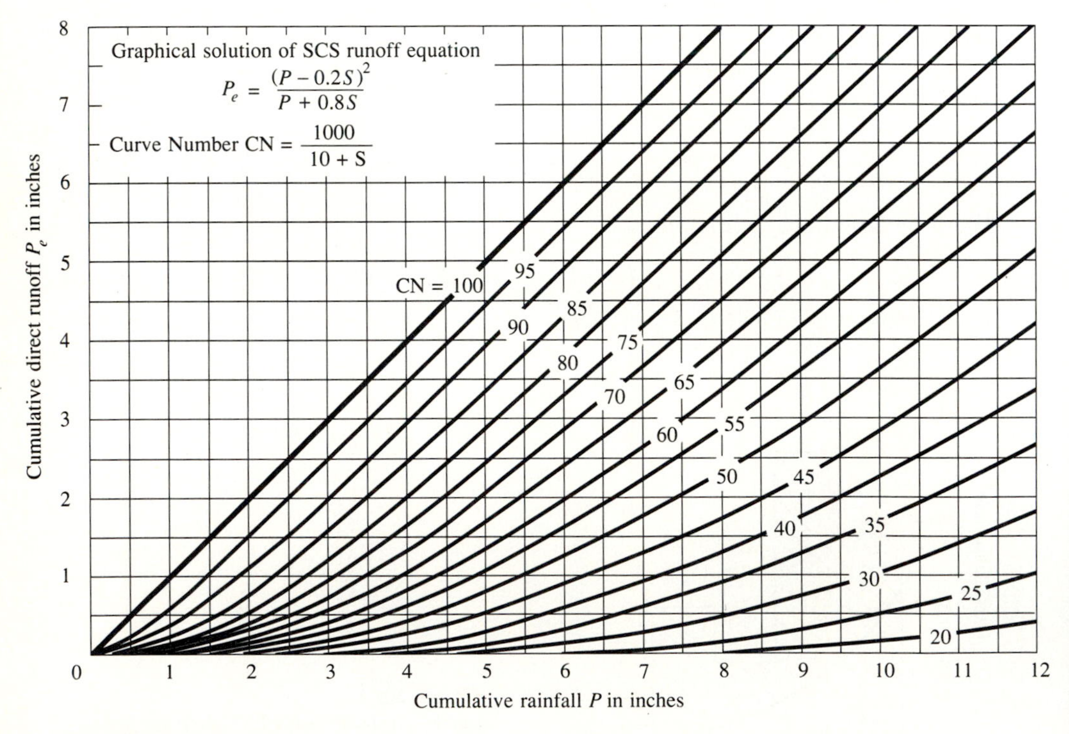

FIGURE 5.5.2
Solution of the SCS runoff equations. (*Source:* Soil Conservation Service, 1972, Fig. 10.1, p. 10.21)

The curve number and S are related by

$$S = \frac{1000}{\text{CN}} - 10 \tag{5.5.6}$$

where S is in inches. The curve numbers shown in Fig. 5.5.2 apply for normal *antecedent moisture conditions* (AMC II). For dry conditions (AMC I) or wet conditions (AMC III), equivalent curve numbers can be computed by

$$\text{CN(I)} = \frac{4.2\text{CN(II)}}{10 - 0.058\text{CN(II)}} \tag{5.5.7}$$

and

$$\text{CN(III)} = \frac{23\text{CN(II)}}{10 + 0.13\text{CN(II)}} \tag{5.5.8}$$

The range of antecedent moisture conditions for each class is shown in Table 5.5.1.

Curve numbers have been tabulated by the Soil Conservation Service on the basis of soil type and land use. Four soil groups are defined:

Group A: Deep sand, deep loess, aggregated silts
Group B: Shallow loess, sandy loam
Group C: Clay loams, shallow sandy loam, soils low in organic content, and soils usually high in clay
Group D: Soils that swell significantly when wet, heavy plastic clays, and certain saline soils

The values of CN for various land uses on these soil types are given in Table 5.5.2. For a watershed made up of several soil types and land uses, a composite CN can be calculated.

Example 5.5.1 (After Soil Conservation Service, 1975). Compute the runoff from 5 inches of rainfall on a 1000-acre watershed. The hydrologic soil group is 50 percent Group B and 50 percent Group C interspersed throughout the watershed. Antecedent moisture condition II is assumed. The land use is:

40 percent residential area that is 30 percent impervious
12 percent residential area that is 65 percent impervious

TABLE 5.5.1
Classification of antecedent moisture classes (AMC) for the SCS method of rainfall abstractions

	Total 5-day antecedent rainfall (in)	
AMC group	**Dormant season**	**Growing season**
I	Less than 0.5	Less than 1.4
II	0.5 to 1.1	1.4 to 2.1
III	Over 1.1	Over 2.1

(*Source:* Soil Conservation Service, 1972, Table 4.2, p. 4.12.)

TABLE 5.5.2
Runoff curve numbers for selected agricultural, suburban, and urban land uses (antecedent moisture condition II, $I_a = 0.2S$)

Land Use Description		Hydrologic Soil Group			
		A	B	C	D
Cultivated land[1]: without conservation treatment		72	81	88	91
with conservation treatment		62	71	78	81
Pasture or range land: poor condition		68	79	86	89
good condition		39	61	74	80
Meadow: good condition		30	58	71	78
Wood or forest land: thin stand, poor cover, no mulch		45	66	77	83
good cover[2]		25	55	70	77
Open Spaces, lawns, parks, golf courses, cemeteries, etc.					
good condition: grass cover on 75% or more of the area		39	61	74	80
fair condition: grass cover on 50% to 75% of the area		49	69	79	84
Commercial and business areas (85% impervious)		89	92	94	95
Industrial districts (72% impervious)		81	88	91	93
Residential[3]:					
Average lot size	Average % impervious[4]				
1/8 acre or less	65	77	85	90	92
1/4 acre	38	61	75	83	87
1/3 acre	30	57	72	81	86
1/2 acre	25	54	70	80	85
1 acre	20	51	68	79	84
Paved parking lots, roofs, driveways, etc.[5]		98	98	98	98
Streets and roads:					
paved with curbs and storm sewers[5]		98	98	98	98
gravel		76	85	89	91
dirt		72	82	87	89

[1]For a more detailed description of agricultural land use curve numbers, refer to Soil Conservation Service, 1972, Chap. 9

[2]Good cover is protected from grazing and litter and brush cover soil.

[3]Curve numbers are computed assuming the runoff from the house and driveway is directed towards the street with a minimum of roof water directed to lawns where additional infiltration could occur.

[4]The remaining pervious areas (lawn) are considered to be in good pasture condition for these curve numbers.

[5]In some warmer climates of the country a curve number of 95 may be used.

18 percent paved roads with curbs and storm sewers

16 percent open land with 50 percent fair grass cover and 50 percent good grass cover

14 percent parking lots, plazas, schools, and so on (all impervious)

Solution. Compute the weighted curve number using Table 5.5.2.

	Hydrologic soil group					
	B			C		
Land Use	**%**	**CN**	**Product**	**%**	**CN**	**Product**
Residential (30% impervious)	20	72	1440	20	81	1620
Residential (65% impervious)	6	85	510	6	90	540
Roads	9	98	882	9	98	882
Open land: Good cover	4	61	244	4	74	296
Fair cover	4	69	276	4	79	316
Parking lots, etc	7	98	686	7	98	686
	50		4038	50		4340

Thus,

$$\text{Weighted CN} = \frac{4038 + 4340}{100} = 83.8$$

$$S = \frac{1000}{\text{CN}} - 10$$

$$= \frac{1000}{83.8} - 10$$

$$= 1.93 \text{ in}$$

$$P_e = \frac{(P - 0.2S)^2}{(P + 0.8S)}$$

$$= \frac{(5 - 0.2 \times 1.93)^2}{5 + 0.8 \times 1.93}$$

$$= 3.25 \text{ in}$$

Example 5.5.2. Recompute the runoff from this watershed if the wet conditions of antecedent moisture condition III are applicable.

Solution. Find a curve number for AMC III equivalent to CN = 83.8 under AMC II using Eq. (5.5.8):

$$\text{CN(III)} = \frac{23\text{CN(II)}}{10 + 0.13\text{CN(II)}}$$

$$= \frac{23 \times 83.8}{10 + 0.13 \times 83.8}$$

$$= 92.3$$

Then,

$$S = \frac{1000}{\text{CN}} - 10$$

$$= \frac{1000}{92.3} - 10$$

$$= 0.83 \text{ in}$$

$$P_e = \frac{(P - 0.2S)^2}{P + 0.8S}$$

$$= \frac{(5 - 0.2 \times 0.83)^2}{5 + 0.8 \times 0.83}$$

$$= 4.13 \text{ in}$$

The change in runoff caused by the change in antecedent moisture condition is $4.13 - 3.25 = 0.88$ in, a 27 percent increase.

Urbanization Effects

During the past 15 to 20 years, hydrologists have paid considerable attention to the effects of urbanization. Early works in urban hydrology were concerned with the effects of urbanization on the flood potential of small urban watersheds. The effects of urbanization on the flood hydrograph include increased total runoff volumes and peak flow rates, as depicted in Fig. 5.5.3. In general, the major changes in flow rates in urban watersheds are due to the following:

1. The volume of water available for runoff increases because of the increased impervious cover provided by parking lots, streets, and roofs, which reduce the amount of infiltration.

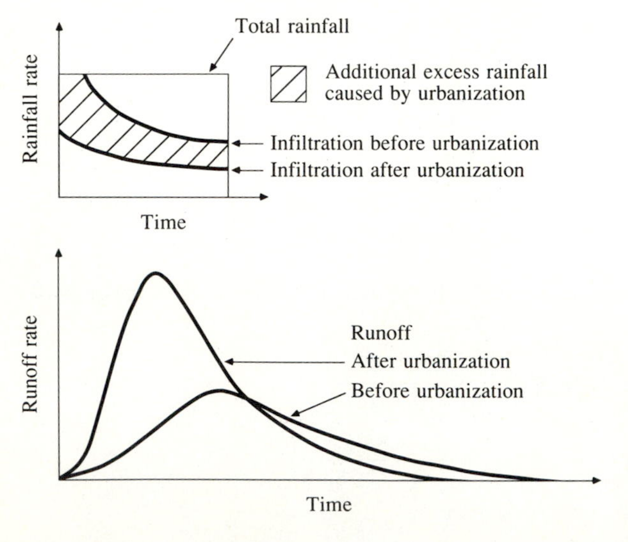

FIGURE 5.5.3
The effect of urbanization on storm runoff.

2. Changes in hydraulic efficiency associated with artificial channels, curbing, gutters, and storm drainage collection systems increase the velocity of flow and the magnitude of flood peaks.

The SCS method for rainfall-runoff analysis can be applied to determine the increase in the amount of runoff caused by urbanization.

Example 5.5.3 Calculate the runoff from 5 inches of rainfall on a 1000-acre watershed. The soil is 50 percent Group B and 50 percent Group C. Assume antecedent moisture condition II. The land use is open land with fair grass cover before urbanization; after urbanization it is as specified in Example 5.5.1. How much additional runoff is caused by urbanization?

Solution. The curve numbers for open land with fair grass cover are CN = 69 for Group B and 79 for Group C, so the average curve number for the watershed is CN = (69 + 79)/2 = 74. From (5.5.6), S = (1000/74) − 10 = 3.51 in. The excess rainfall or direct runoff P_e is calculated from (5.5.5) with $P = 5.0$ in:

$$P_e = \frac{(P - 0.2S)^2}{P + 0.8S}$$

$$= \frac{(5.0 - 0.2 \times 3.51)^2}{5.0 + 0.8 \times 3.51}$$

$$= 2.37 \text{ in (before urbanization)}$$

After urbanization, Example 5.5.1 shows $P_e = 3.25$ in, so the impact of urbanization is to cause 3.25 − 2.37 = 0.88 in of additional runoff from this storm, a 27 percent increase.

Time Distribution of SCS Abstractions

To this point, only the depth of excess rainfall or direct runoff during a storm has been computed. By extension of the previous method, the time distribution of abstractions F_a within a storm can be found. Solving for F_a from Eqs. (5.5.1) and (5.5.2),

$$F_a = \frac{S(P - I_a)}{P - I_a + S} \qquad P \geq I_a \tag{5.5.9}$$

Differentiating, and noting that I_a and S are constants,

$$\frac{dF_a}{dt} = \frac{S^2 dP/dt}{(P - I_a + S)^2} \tag{5.5.10}$$

As $P \to \infty$, $(dF_a/dt) \to 0$ as required, but the presence of dP/dt (rainfall intensity) in the numerator means that as the rainfall intensity increases, the rate of retention of water within the watershed tends to increase. This property of the SCS method may not have a strong physical basis (Morel-Seytoux and Verdin, 1981).

In application, cumulative abstractions and rainfall excess may be determined either from (5.5.9) or from (5.5.5).

Example 5.5.4. Storm rainfall occurred on a watershed as shown in column 2 of Table 5.5.3. The value of CN is 80 and Antecedent Moisture Condition II applies. Calculate the cumulative abstractions and the excess rainfall hyetograph.

Solution. For CN = 80, S = (1000/80) − 10 = 2.50 in; $I_a = 0.2S = 0.5$ in. The initial abstraction absorbs all of the rainfall up to $P = 0.5$ in. This includes the 0.2 in of rainfall occurring during the first hour and 0.3 in of the rain falling during the second hour. For $P > 0.5$ in, the continuing abstraction F_a is computed from (5.5.9):

$$F_a = \frac{S(P - I_a)}{P - I_a + S}$$

$$= \frac{2.50(P - 0.5)}{P - 0.5 + 2.50}$$

$$= \frac{2.50(P - 0.5)}{P + 2.0}$$

For example, after two hours, the cumulative rainfall is $P = 0.90$ in, so

$$F_a = \frac{2.50(0.9 - 0.5)}{0.9 + 2.0}$$

$$= 0.34 \text{ in}$$

as shown in column 4 of the table. The excess rainfall is that remaining after initial and continuing abstractions. From (5.5.2)

TABLE 5.5.3

Computation of abstractions and excess rainfall hyetograph by the SCS method (Example 5.5.4)

Column:	1	2	3	4	5	6
			Cumulative abstractions (in)		Cumulative excess rainfall P_e (in)	Excess rainfall hyetograph (in)
	Time (h)	Cumulative rainfall P (in)	I_a	F_a		
	0	0	0	–	0	
						0
	1	0.20	0.20	–	0	
						0.06
	2	0.90	0.50	0.34	0.06	
						0.12
	3	1.27	0.50	0.59	0.18	
						0.58
	4	2.31	0.50	1.05	0.76	
						1.83
	5	4.65	0.50	1.56	2.59	
						0.56
	6	5.29	0.50	1.64	3.15	
						0.06
	7	5.36	0.50	1.65	3.21	

$$P_e = P - I_a - F_a$$
$$= 0.90 - 0.50 - 0.34$$
$$= 0.06 \text{ in}$$

as shown in column 5. The excess rainfall hyetograph is determined by taking the difference of successive values of P_e (column 6).

5.6 FLOW DEPTH AND VELOCITY

The flow of water over a watershed surface is a complicated process varying in all three space dimensions and time. It begins when water becomes ponded on the surface at sufficient depth to overcome surface retention forces and begins to flow. Two basic flow types may be distinguished: overland flow and channel flow. Overland flow has a thin layer of water flowing over a wide surface. Channel flow has a much narrower stream of water flowing in a confined path. Chapter 2 gave the physical laws applicable to these two types of flow. On a natural watershed, overland flow is the first mechanism of surface flow but it may persist for only a short distance (say up to 100 ft) before nonuniformities in the watershed surface concentrate the flow into tortuous channels. Gradually, the outflows from these small channels combine to produce recognizable stream channel flows which accumulate going downstream to form streamflow at the watershed outlet.

Surface water flow is governed by the principles of continuity and momentum. The application of these principles to three-dimensional unsteady flow on a watershed surface is possible only in very simplified situations, so one- or two-dimensional flow is usually assumed.

Overland Flow

Overland flow is a very thin sheet flow which occurs at the upper end of slopes before the flow concentrates into recognizable channels. Figure 5.6.1 shows flow down a uniform plane on which rain is falling at intensity i and infiltration occurring at rate f. Sufficient time has passed since rainfall began that all flows are steady. The plane is of unit width and length L_0, and is inclined at angle θ to the horizontal with slope $S_0 = \tan \theta$.

Continuity. The continuity equation (2.2.5) for steady, constant density flow is

$$\iint_{c.s.} \mathbf{V} \cdot \mathbf{dA} = 0 \tag{5.6.1}$$

The inflow to the control volume from rainfall is $iL_0 \cos \theta$, and the outflow is $fL_0 \cos \theta$ from infiltration plus Vy from overland flow. The depth y is measured perpendicular to the bed and the velocity V parallel to the bed. Thus the continuity equation is written

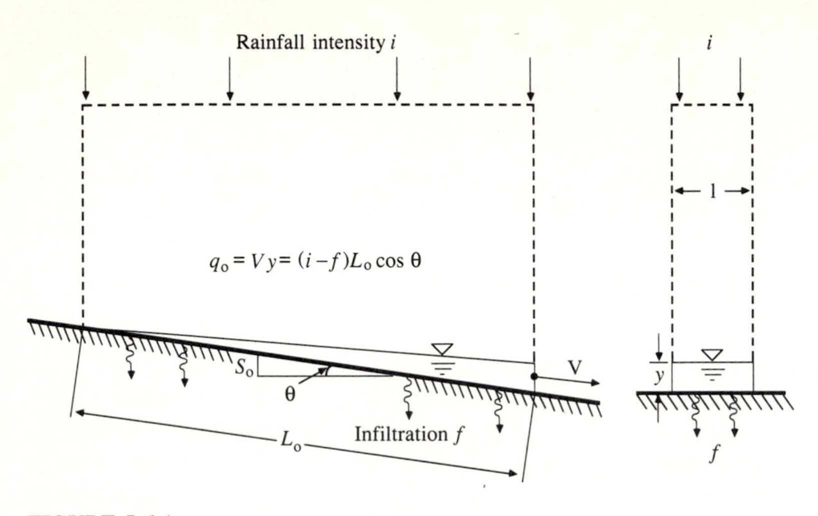

FIGURE 5.6.1
Steady flow on a uniform plane under rainfall.

$$\iint_{\text{c.s.}} \mathbf{V \cdot dA} = fL_0 \cos\theta + Vy - iL_0 \cos\theta = 0$$

The discharge per unit width, q_0, is given by

$$q_0 = Vy = (i - f)L_0 \cos\theta \tag{5.6.2}$$

Momentum. For uniform laminar flow on an inclined plane, it can be shown (Roberson and Crowe, 1985), that the average velocity V is given by

$$V = \frac{gS_0 y^2}{3\nu} \tag{5.6.3}$$

where g is acceleration due to gravity and ν is the kinematic viscosity of the fluid. For uniform flow, $S_0 = S_f = h_f/L$, and (5.6.3) can be rearranged to yield

$$h_f = \frac{24\nu}{Vy}\frac{L}{4y}\frac{V^2}{2g} \tag{5.6.4}$$

which is in the form of the Darcy-Weisbach equation (2.5.1) for flow resistance

$$h_f = f\frac{L}{4R}\frac{V^2}{2g} \tag{5.6.5}$$

with the friction factor $f = 96/Re$ in which the Reynolds number $Re = 4VR/\nu$, and the hydraulic radius $R = y$. For a unit width sheet flow, $R =$ area/(wetted perimeter) $= y \times 1/1 = y$, as required. The flow remains laminar provided $Re \leq 2000$.

For laminar sheet flow under rainfall, the friction factor increases with the rainfall intensity. If it is assumed that f has the form C_L/Re, where C_L is a resistance coefficient, experimentation carried out at the University of Illinois (Chow and Yen, 1976) gave

$$C_L = 96 + 108i^{0.4} \tag{5.6.6}$$

where i is the rainfall intensity in inches per hour.

Solving for y from (5.6.5) and using the fact that $h_f/L = S_0$ for uniform flow, one finds

$$y = \frac{fV^2}{8gS_0} \tag{5.6.7}$$

then $q_0 = Vy$ from (5.6.2) is used to substitute for V, yielding

$$y = \left(\frac{fq_0^2}{8gS_0}\right)^{1/3} \tag{5.6.8}$$

which specifies the depth of sheet flow on a uniform plane.

Example 5.6.1. A rainfall of intensity 1 in/h falls on a uniform, smooth, impervious plane 100 feet long at 5 percent slope. Calculate the discharge per unit width, the depth, and the velocity at the lower end of the plane. Take $\nu = 1.2 \times 10^{-5}$ ft²/s.

Solution. The discharge per unit width is given by (5.6.2) with $i = 1$ in/h $= 2.32 \times 10^{-5}$ ft/s, and $f = 0$. The angle $\theta = \tan^{-1}(S_0) = \tan^{-1}(0.05) = 2.86°$, so $\cos\theta = 0.999$.

$$\begin{aligned} q_0 &= (i - f)L_0 \cos\theta \\ &= (2.32 \times 10^{-5} - 0) \times 100 \times 0.999 \\ &= 2.31 \times 10^{-3} \text{ ft}^2/\text{s} \end{aligned}$$

The Reynolds number is

$$\begin{aligned} Re &= \frac{4Vy}{\nu} \\ &= \frac{4q_0}{\nu} \\ &= \frac{4 \times 2.31 \times 10^{-3}}{1.2 \times 10^{-5}} \\ &= 770 \end{aligned}$$

and the flow is laminar. The resistance coefficient C_L is given by (5.6.6):

$$\begin{aligned} C_L &= 96 + 108i^{0.4} \\ &= 96 + 108(1)^{0.4} \\ &= 204 \end{aligned}$$

The friction factor is $f = C_L/Re = 204/770 = 0.265$, and the depth is calculated from Eq. (5.6.8),

$$y = \left(\frac{fq_0^2}{8gS_0}\right)^{1/3}$$

$$= \left[\frac{0.265 \times (2.31 \times 10^{-3})^2}{8 \times 32.2 \times 0.05}\right]^{1/3}$$

$$= 0.0048 \text{ ft } (0.06 \text{ in})$$

The velocity V is given by

$$V = \frac{q_0}{y}$$

$$= 2.31 \times 10^{-3}/0.0048$$

$$= 0.48 \text{ ft/s}$$

Field studies of overland flow (Emmett, 1978) indicate that the flow is laminar but that the flow resistance is about ten times larger than for laboratory studies on uniform planes. The increase in flow resistance results primarily from the unevenness in the topography and surface vegetation. Equation (5.6.8) can be rewritten in the more general form

$$y = \alpha q_0^m \tag{5.6.9}$$

For laminar flow $m = 2/3$ and $\alpha = (f/8gS_0)^{1/3}$. Emmett's studies indicate that the Darcy-Weisbach friction factor f is in the range 20–200 for overland flow at field sites.

When the flow becomes turbulent, the friction factor becomes independent of the Reynolds number and dependent only on the roughness of the surface. In this case, Manning's equation (2.5.7) is applicable to describe the flow:

$$V = \frac{1.49}{n} R^{2/3} S_f^{1/2} \tag{5.6.10}$$

with $R = y$, $S_f = S_0$ for uniform flow, and $q_0 = Vy$. This can be solved for y to yield

$$y = \left(\frac{nq_0}{1.49 S_0^{1/2}}\right)^{3/5} \tag{5.6.11}$$

which is in the general form of (5.6.9) with $\alpha = (n/1.49S_0^{1/2})^{3/5}$ and $m = 3/5$. For SI units, $\alpha = n^{0.6}/S_0^{0.3}$.

Example 5.6.2. Calculate the depth and velocity of a discharge of 2.31×10^{-3} cfs/ft (width) on turf having Darcy-Weisbach $f = 75$ and a slope of 5 percent. Take $\nu = 1.2 \times 10^{-5}$ ft²/s.

Solution. The Reynolds number is $Re = 4q_0/\nu = 4 \times 2.31 \times 10^{-3}/1.2 \times 10^{-5} = 770$ (laminar flow), and $\alpha = (f/8gS_0)^{1/3} = (75/(8 \times 32.2 \times 0.05))^{1/3} = 1.80$. From (5.6.9) with $m = 2/3$

$$y = \alpha q_0^m$$
$$= 1.80(2.31 \times 10^{-3})^{2/3}$$
$$= 0.031 \text{ ft (0.4 in)}$$

Velocity $V = q_0/y = 2.31 \times 10^{-3}/0.031 = 0.075$ ft/s. It can be seen that this flow is much deeper and slower flowing than flow on the smooth plane of Example 5.6.1.

Example 5.6.3. Calculate the discharge per unit width, depth, and velocity at the end of a 200-ft strip of asphalt, of slope 0.02, subject to rainfall of 10 in/h, with Manning's $n = 0.015$ and kinematic viscosity $\nu = 1.2 \times 10^{-5}$ ft²/s.

Solution. The discharge per unit width is given by Eq. (5.6.2) with $i = 10$ in/h $= 2.32 \times 10^{-4}$ ft/s, $f = 0$, and $\theta = \tan^{-1}(0.02) = 1.15°$, for which $\cos\theta = 1.00$:

$$q_0 = (i - f)L_0 \cos\theta$$
$$= 2.32 \times 10^{-4} \times 200 \times 1.00$$
$$= 0.0464 \text{ cfs/ft}$$

The Reynolds number is $Re = 4q_0/\nu = 4 \times 0.046/(1.2 \times 10^{-5}) = 15333$, so the flow is turbulent. The depth of flow is given by Eq. (5.6.9) with $\alpha = (n/1.49S_0^{1/2})^{3/5} = [0.015/(1.49 \times 0.02^{1/2})]^{3/5} = 0.205$ and $m = 0.6$:

$$y = \alpha q_0^m$$
$$= 0.205 \times (0.0464)^{0.6}$$
$$= 0.032 \text{ ft (0.4 in)}$$

Also,

$$V = \frac{q_0}{y}$$
$$= \frac{0.0464}{0.032}$$
$$= 1.43 \text{ ft/s.}$$

Channel Flow

The passage of overland flow into a channel can be viewed as a lateral flow in the same way that the previous examples have considered rainfall as a lateral flow onto the watershed surface.

Consider a channel of length L_c that is fed by overland flow from a plane as shown in Fig. 5.6.2. The overland flow has discharge q_0 per unit width, so the discharge in the channel is $Q = q_0 L_c$. To find the depth and velocity at various points along the channel, an iterative solution of Manning's equation is necessary. Manning's equation is

$$Q = \frac{1.49}{n} S_0^{1/2} A R^{2/3} \tag{5.6.12}$$

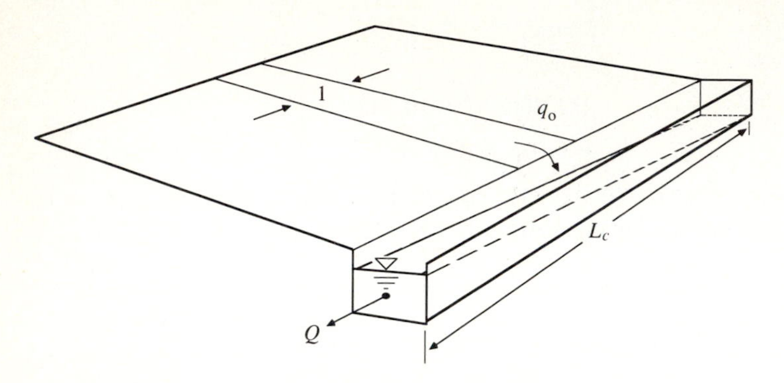

FIGURE 5.6.2
Overland flow from a plane into a channel.

Solution of Manning's Equation by Newton's Method

There is no general analytical solution to Manning's equation for determining the flow depth given the flow rate because the area A and hydraulic radius R may be complicated functions of the depth. Newton's method can be applied iteratively to give a numerical solution. Suppose that at iteration j the depth y_j is selected and the flow rate Q_j is computed from (5.6.12), using the area and hydraulic radius corresponding to y_j. This Q_j is compared with the actual flow Q; the object is to select y so that the error

$$f(y_j) = Q_j - Q \tag{5.6.13}$$

is acceptably small. The gradient of f with respect to y is

$$\frac{df}{dy_j} = \frac{dQ_j}{dy_j} \tag{5.6.14}$$

because Q is a constant. Hence, assuming Manning's n is constant,

$$\begin{aligned}\left(\frac{df}{dy}\right)_j &= \left(\frac{1.49}{n} S_0^{1/2} A_j R_j^{2/3}\right) \\ &= \frac{1.49}{n} S_0^{1/2}\left(\frac{2AR^{-1/3}}{3}\frac{dR}{dy} + R^{2/3}\frac{dA}{dy}\right)_j \\ &= \frac{1.49}{n} S_0^{1/2} A_j R_j^{\ 2/3}\left(\frac{2}{3R}\frac{dR}{dy} + \frac{1}{A}\frac{dA}{dy}\right)_j \\ &= Q_j\left(\frac{2}{3R}\frac{dR}{dy} + \frac{1}{A}\frac{dA}{dy}\right)_j\end{aligned} \tag{5.6.15}$$

where the subscript j outside the parentheses indicates that the contents are evaluated for $y = y_j$.

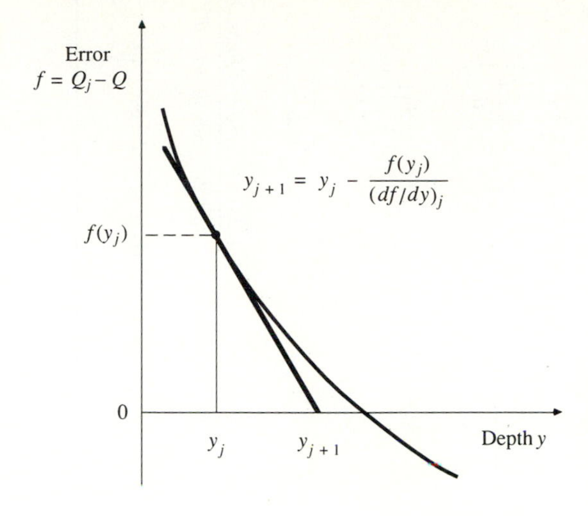

FIGURE 5.6.3
Newton's method extrapolates the tangent of the error function at the current depth y_j to obtain the depth y_{j+1} for the next iteration.

This expression for the gradient is useful for Newton's method, where, given a choice of y_j, y_{j+1} is chosen to satisfy

$$\left(\frac{df}{dy}\right)_j = \frac{0 - f(y)_j}{y_{j+1} - y_j} \tag{5.6.16}$$

This y_{j+1} is the value of y, in a plot of f vs. y, where the tangent to the curve at $y = y_j$ intersects the horizontal axis, as illustrated in Fig. 5.6.3.

Solving (5.6.16) for y_{j+1},

$$y_{j+1} = y_j - \frac{f(y_j)}{(df/dy)_j} \tag{5.6.17}$$

which is the fundamental equation of the Newton's method. Iterations are continued until there is no significant change in y; this will happen when the error $f(y)$ is very close to zero.

Substituting into (5.6.17) from Eqs. (5.6.13) and (5.6.15) gives the Newton's-method equation for solving Manning's equation:

$$y_{j+1} = y_j - \frac{1 - Q/Q_j}{\left(\frac{2}{3R}\frac{dR}{dy} + \frac{1}{A}\frac{dA}{dy}\right)_j} \tag{5.6.18}$$

For a rectangular channel $A = B_w y$ and $R = B_w y/(B_w + 2y)$ where B_w is the channel width; after some manipulation, (5.6.18) becomes

$$y_{j+1} = y_j - \frac{1 - Q/Q_j}{\left(\frac{5B_w + 6y_j}{3y_j(B_w + 2y_j)}\right)}$$

Values for the *channel shape function* $[(2/3R)(dR/dy) + (1/A)(dA/dy)]$ for other cross sections are given in Table 5.6.1.

TABLE 5.6.1
Geometric functions for channel elements

Section:	Rectangle	Trapezoid	Triangle	Circle
Area A	$B_w y$	$(B_w + zy)y$	zy^2	$\frac{1}{8}(\theta - \sin\theta)d_o^2$
Wetted perimeter P	$B_w + 2y$	$B_w + 2y\sqrt{1+z^2}$	$2y\sqrt{1+z^2}$	$\frac{1}{2}\theta d_o$
Hydraulic radius R	$\frac{B_w y}{B_w + 2y}$	$\frac{(B_w + zy)y}{B_w + 2y\sqrt{1+z^2}}$	$\frac{zy}{2\sqrt{1+z^2}}$	$\frac{1}{4}\left(1 - \frac{\sin\theta}{\theta}\right)d_o$
Top width B	B_w	$B_w + 2zy$	$2zy$	$\left[\sin\left(\frac{\theta}{2}\right)\right]d_o$ or $2\sqrt{y(d_o - y)}$
$\frac{2}{3R}\frac{dR}{dy} + \frac{1}{A}\frac{dA}{dy}$	$\frac{5B_w + 6y}{3y(B_w + 2y)}$	$\frac{(B_w + 2zy)(5B_w + 6y\sqrt{1+z^2}) + 4zy^2\sqrt{1+z^2}}{3y(B_w + zy)(B_w + 2y\sqrt{1+z^2})}$	$\frac{8}{3y}$	$\frac{4(2\sin\theta + 3\theta - 5\theta\cos\theta)}{3d_o\theta(\theta - \sin\theta)\sin(\theta/2)}$ where $\theta = 2\cos^{-1}\left(1 - \frac{2y}{d_o}\right)$

Source: Chow, V. T., *Open-Channel Hydraulics*, McGraw-Hill, New York, 1959, Table 2.1, p. 21 (with additions).

Example 5.6.4. Calculate the flow depth in a two-foot-wide rectangular channel having $n = 0.015$, $S_0 = 0.025$, and $Q = 9.26$ cfs.

Solution.

$$Q_j = \frac{1.49}{n} S_0^{1/2} \frac{(B_w y_j)^{5/3}}{(B_w + 2y_j)^{2/3}}$$

$$= \frac{1.49}{0.015}(0.025)^{1/2} \frac{(2y_j)^{5/3}}{(2 + 2y_j)^{2/3}}$$

$$= \frac{31.41 y_j^{5/3}}{(1 + y_j)^{2/3}} \tag{5.6.19}$$

Also,

$$\frac{2}{3R}\frac{dR}{dy} + \frac{1}{A}\frac{dA}{dy} = \frac{5B_w + 6y_j}{3y_j(B_w + 2y_j)} = \frac{10 + 6y_j}{3y_j(2 + 2y_j)}$$

$$= \frac{1.667 + y_j}{y_j(1 + y_j)}$$

From Eq. (5.6.18)

$$y_{j+1} = y_j - \frac{(1 - 9.26/Q_j)\, y_j (1 + y_j)}{1.667 + y_j} \tag{5.6.20}$$

From an arbitrarily chosen starting guess of $y_1 = 1.00$ ft, the solution to three significant figures is achieved after three iterations by successively solving (5.6.19) and (5.6.20) for Q_j and y_{j+1}. The result is $y = 0.58$ ft.

Iteration j	**1**	**2**	**3**	**4**
y_j (ft)	1.00	0.601	0.577	0.577
Q_j (cfs)	19.79	9.82	9.26	9.26

Example 5.6.5. Compute the velocity and depth of flow at 200-foot increments along a 1000-foot-long rectangular channel having width 2 ft, roughness $n = 0.015$, and slope $S = 0.025$, supplied by a lateral flow of 0.00926 cfs/ft.

Solution. The method of Example 5.6.4 is applied repetitively to compute y for $Q = 0.00926L$. The velocity is $V = Q/B_w y = Q/2y$.

Distance along channel, L (ft)	**0**	**200**	**400**	**600**	**800**	**1000**
Flow rate (cfs)	0	1.85	3.70	5.56	7.41	9.26
Depth y (ft)	0	0.20	0.31	0.41	0.49	0.58
Velocity V (ft/s)	0	4.63	5.97	6.86	7.56	8.02

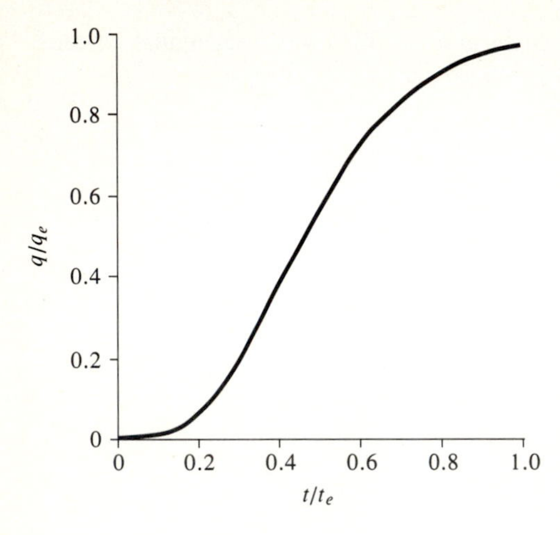

FIGURE 5.6.4
Dimensionless hydrograph of overland flow. The steady flow q_c is attained at time of equilibrium t_e. (After Izzard, 1946.)

The examples in this section have assumed steady flow on the watershed. In reality, under a constant intensity rainfall, the steady flow at equilibrium is approached asymptotically in the manner illustrated by Fig. 5.6.4. Thus, the flow is varying both in space and time on the watershed surface and in the stream channel.

5.7 TRAVEL TIME

The travel time of flow from one point on a watershed to another can be deduced from the flow distance and velocity. If two points on a stream are a distance L apart and the velocity along the path connecting them is $v(l)$, where l is distance along the path, then the travel time t is given by

$$dl = v(l)\, dt$$

$$\int_0^t dt = \int_0^L \frac{dl}{v(l)} \tag{5.7.1}$$

or

$$t = \int_0^L \frac{dl}{v(l)} \tag{5.7.2}$$

If the velocity can be assumed constant at v_i in an increment of length Δl_i, $i = 1, 2, \ldots, I$, then

$$t = \sum_{i=1}^{I} \frac{\Delta l_i}{v_i} \tag{5.7.3}$$

Velocities for use in Eq. (5.7.3) may be computed using the methods described in Sec. 5.6 or by reference to Table 5.7.1.

TABLE 5.7.1
Approximate average velocities in ft/s of runoff flow for calculating time of concentration

Description of water course	Slope in percent 0–3	4–7	8–11	12–
Unconcentrated*				
Woodlands	0–1.5	1.5– 2.5	2.5– 3.25	3.25–
Pastures	0–2.5	2.5– 3.5	3.5– 4.25	4.25–
Cultivated	0–3.0	3.0– 4.5	4.5– 5.5	5.5–
Pavements	0–8.5	8.5–13.5	13.5–17	17–
Concentrated**				
Outlet channel—determine velocity by Manning's formula				
Natural channel not well defined	0–2	2–4	4–7	7–

*This condition usually occurs in the upper extremities of a watershed prior to the overland flows accumulating in a channel.

**These values vary with the channel size and other conditions. Where possible, more accurate determinations should be made for particular conditions by the Manning channel formula for velocity.

(*Source*: Drainage Manual, Texas Highway Department, Table VII, p. II-28, 1970.)

Because of the travel time to the watershed outlet, only part of the watershed may be contributing to surface water flow at any time t after precipitation begins. The growth of the contributing area may be visualized as in Fig. 5.7.1. If rainfall of constant intensity begins and continues indefinitely, then the area bounded by the dashed line labeled t_1 will contribute to streamflow at the watershed outlet after time t_1; likewise, the area bounded by the line labeled t_2 will contribute to

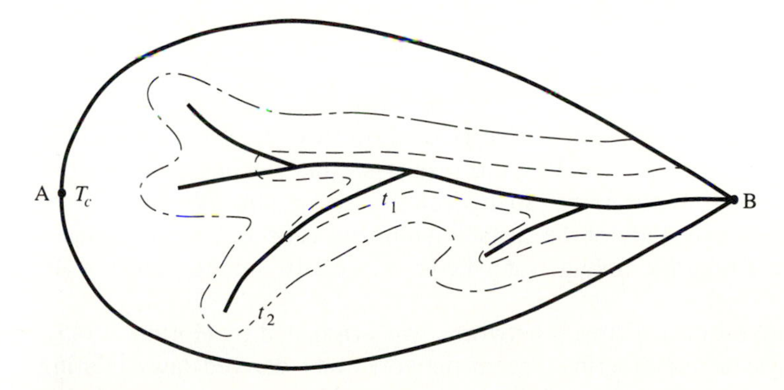

FIGURE 5.7.1
Isochrones at t_1 and t_2 define the area contributing to flow at the outlet for rainfall of durations t_1 and t_2. Time of concentration t_c is the time of flow from the farthest point in the watershed (A) to the outlet (B).

TABLE 5.7.2
Travel time in a channel (Example 5.7.1)

Distance along channel, l (ft)	**0**		**200**		**400**		**600**		**800**		**1000**
Δl		**200**		**200**		**200**		**200**		**200**	
Calculated velocity V (ft/s)	0		4.63		5.97		6.86		7.56		8.02
Average velocity $\bar{V}$ (ft/s)		2.32		5.30		6.42		7.21		7.79	
Travel time $\Delta t = \Delta l/\bar{V}$ (s)		86.2		37.7		31.2		27.7		25.7	

($\Sigma\ \Delta t = 208.5$ s)

streamflow after time t_2. The boundaries of these contributing areas are lines of equal time of flow to the outlet and are called *isochrones*. The time at which all of the watershed begins to contribute is the *time of concentration* T_c; this is the time of flow from the farthest point on the watershed to the outlet.

Example 5.7.1. Calculate the time of concentration of a watershed in which the longest flow path covers 100 feet of pasture at a 5 percent slope, then enters a 1000-foot-long rectangular channel having width 2 ft, roughness $n = 0.015$, and slope 2.5 percent, and receiving a lateral flow of 0.00926 cfs/ft.

Solution. From Table 5.7.1, pasture at 5 percent slope has a velocity of flow in the range 2.5–3.5 ft/s; use a velocity of 3.0 ft/s. The travel time over the 100 feet of pasture is $\Delta t = \Delta l/v = 100/3.0 = 33$ s. For the rectangular channel, the velocity at 200-foot intervals was calculated in Example 5.6.5. The travel time over each interval is found from the average velocity in that interval. For example, for the first 200 ft, $\Delta t = \Delta l/\bar{v} = 200/2.32 = 86.2$ s. This yields a total travel time for the channel of 208.5 s, as shown in Table 5.7.2. The time of concentration t_c is the sum of the travel times over pasture and in the channel, or $33 + 209 = 242s = 4$ min.

5.8 STREAM NETWORKS

In fluid mechanics, the study of the similarity of fluid flow in systems of different sizes is an important tool in relating the results of small-scale model studies to large-scale prototype applications. In hydrology, the *geomorphology* of the watershed, or quantitative study of the surface landform, is used to arrive at measures of geometric similarity among watersheds, especially among their stream networks.

The quantitative study of stream networks was originated by Horton (1945). He developed a system for ordering stream networks and derived laws relating the number and length of streams of different order. Horton's stream ordering system, as slightly modified by Strahler (1964), is as follows:

The smallest recognizable channels are designated order 1; these channels normally flow only during wet weather.

Where two channels of order 1 join, a channel of order 2 results downstream; in general, where two channels of order i join, a channel of order $i + 1$ results.

Where a channel of lower order joins a channel of higher order, the channel downstream retains the higher of the two orders.

The order of the drainage basin is designated as the order of the stream draining its outlet, the highest stream order in the basin, I.

An example of this classification system for a small watershed in Texas is shown in Fig. 5.8.1.

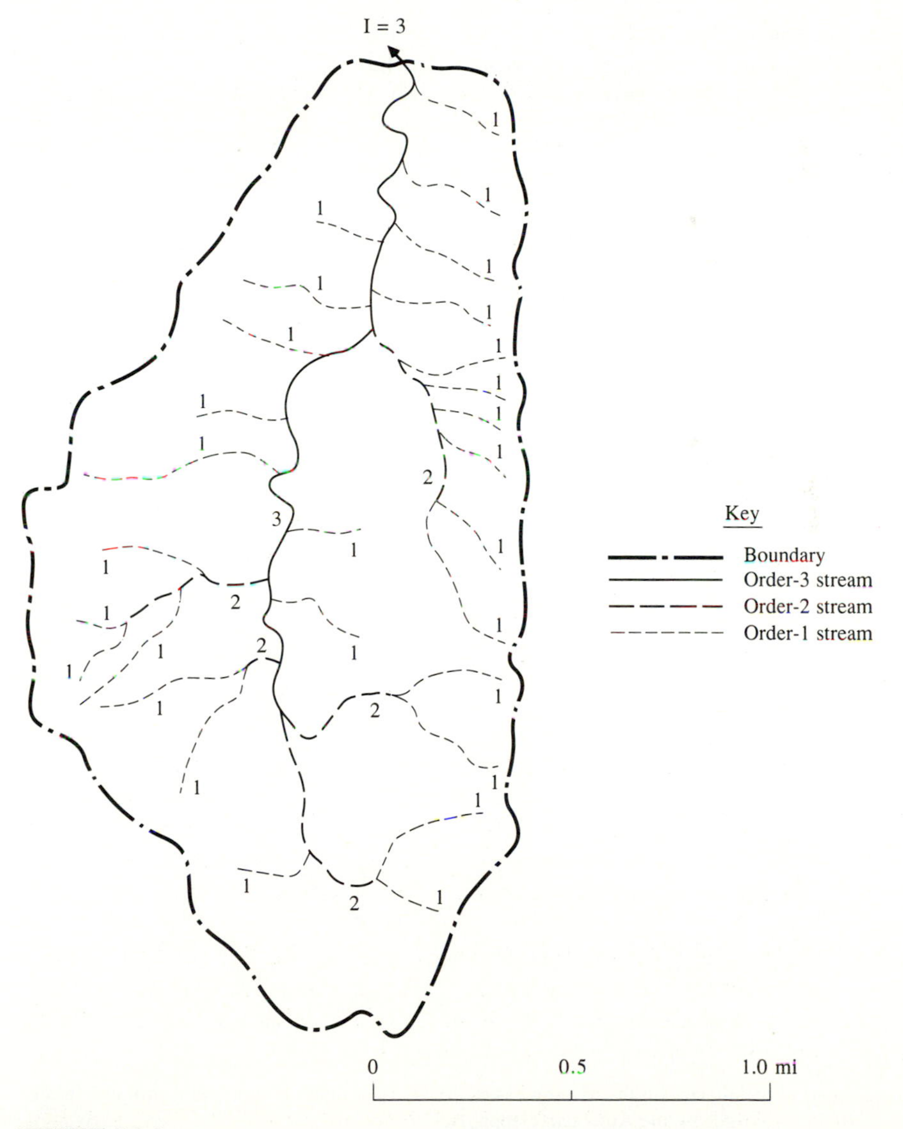

FIGURE 5.8.1
Watershed of Miller Creek, Blanco County, Texas, showing the delineation of stream orders.

Horton (1945) found empirically that the *bifurcation ratio* R_B, or ratio of the number N_i, of channels of order i to the number N_{i+1} of channels of order $i + 1$ is relatively constant from one order to another. This is *Horton's Law of Stream Numbers:*

$$\frac{N_i}{N_{i+1}} = R_B \qquad i = 1, 2, \ldots, I - 1 \tag{5.8.1}$$

As an example, in Fig. 5.8.1, $N_1 = 28$, $N_2 = 5$, and $N_3 = 1$; so $N_1/N_2 = 5.6$ and $N_2/N_3 = 5.0$. The theoretical minimum value of the bifurcation ratio is 2, and values typically lie in the range 3–5 (Strahler, 1964).

By measuring the length of each stream, the average length of streams of each order, L_i, can be found. Horton proposed a *Law of Stream Lengths* in which the average lengths of streams of successive orders are related by a *length ratio* R_L:

$$\frac{L_{i+1}}{L_i} = R_L \tag{5.8.2}$$

By a similar reasoning, Schumm (1956) proposed a *Law of Stream Areas* to relate the average areas A_i drained by streams of successive order

$$\frac{A_{i+1}}{A_i} = R_A \tag{5.8.3}$$

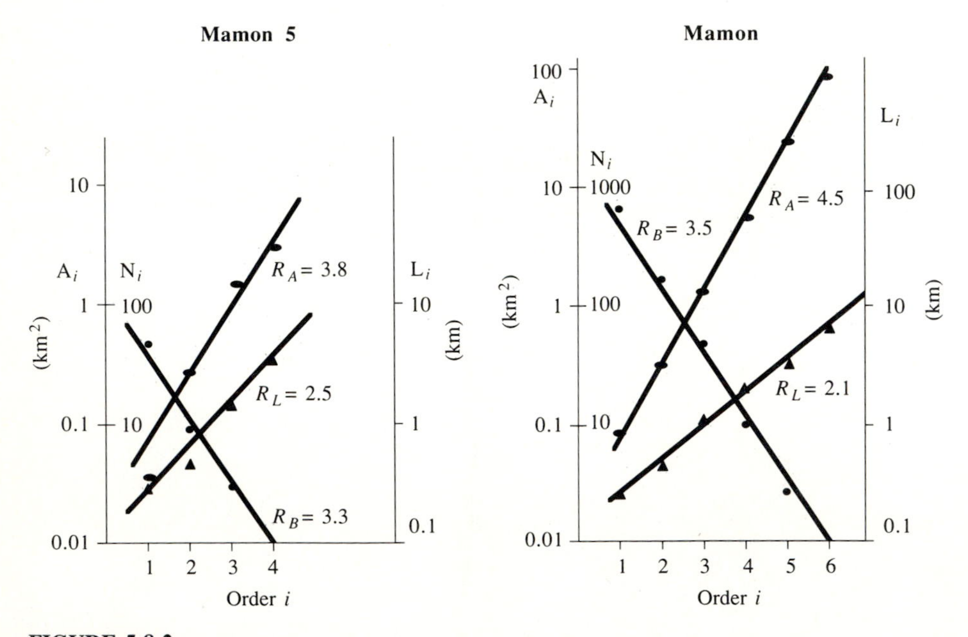

FIGURE 5.8.2
Geomorphological parameters for the Mamon basins. (*Source:* Valdes, Fiallo, and Rodriguez-Iturbe, p. 1123, 1979. Copyright by the American Geophysical Union.)

These ratios are computed by plotting the values for N_i, L_i, and A_i on a logarithmic scale against stream order on a linear scale, as shown for two Venezuelan watersheds in Fig. 5.8.2. The ratios R_B, R_L, and R_A are computed from the slopes of the lines on these graphs. The Mamon 5 watershed is a subbasin of the Mamon watershed (Fig. 5.8.3). The consistency of R_B, R_L, and R_A between the two watersheds demonstrates their geometric similarity. Studies have been made to relate the characteristics of flood hydrographs to stream network parameters (Rodriguez-Iturbe and Valdes, 1979; Gupta, Waymire, and Wang, 1980; Gupta, Rodriguez-Iturbe and Wood, 1986).

Other parameters useful for hydrologic analysis are the drainage density and the length of overland flow (Smart, 1972). The *drainage density D* is the ratio of the total length of stream channels in a watershed to its area

$$D = \frac{\sum_{i=1}^{I} \sum_{j=1}^{N_i} L_{ij}}{A_I} \tag{5.8.4}$$

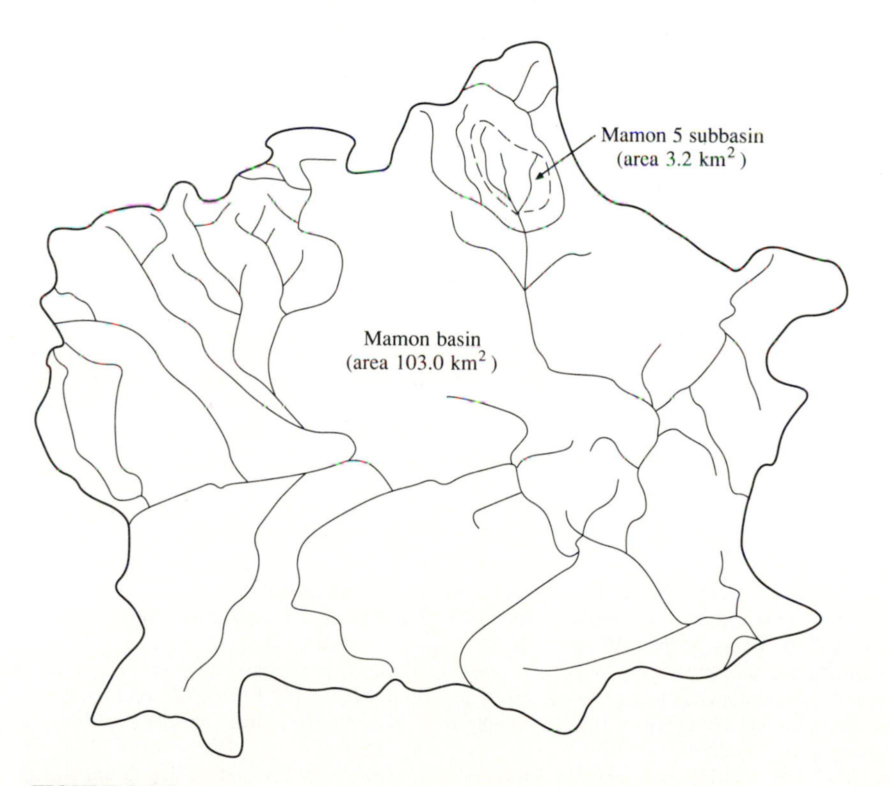

FIGURE 5.8.3
Drainage basin of the Mamon watershed in Venezuala. (*Source:* Valdes, Fiallo, and Rodriguez-Iturbe, p. 1123, 1979. Copyright by the American Geophysical Union.)

where L_{ij} is the length of the jth stream of order i. If the streams are fed by Hortonian overland flow from all of their contributing area, then the average length of overland flow, L_o, is given approximately by

$$L_o = \frac{1}{2D} \tag{5.8.5}$$

Shreve (1966) showed that Horton's stream laws result from the most likely combinations of channels into a network if random selection is made among all possible combinations.

REFERENCES

Betson, R. P., What is watershed runoff? *J. Geophys. Res.*, vol. 69, no. 8, pp. 1541–1552, 1964.

Chow, V. T., *Open-channel Hydraulics,* McGraw-Hill, New York, 1959.

Chow, V. T., and B. C. Yen, Urban stormwater runoff: determination of volumes and flowrates, report EPA-600/2-76-116, Municipal Environmental Research Laboratory, Office of Research and Development, U. S. Environmental Protection Agency, Cincinnati, Ohio, May 1976.

Dunne, T., T. R. Moore, and C. H. Taylor, Recognition and prediction of runoff-producing zones in humid regions, *Hydrol. Sci. Bull.*, vol. 20, no. 3, pp. 305–327, 1975.

Emmett, W. W., Overland flow, in *Hillslope Hydrology,* ed. by M. J. Kirkby, Wiley, New York, pp. 145–176, 1978.

Freeze, R. A., Role of subsurface flow in generating surface runoff 2. Upstream source areas, *Water Resour. Res.* vol. 8, no. 5, pp. 1272–1283, 1972.

Freeze, R. A., Streamflow generation, *Rev. Geophys. Space Phys.*, vol. 12, no. 4, pp. 627–647, 1974.

Gupta, V. K., E. Waymire, C. T. Wang, A representation of an instantaneous unit hydrograph from geomorphology, *Water Resour. Res.*, vol. 16, no. 5, pp. 855–862, 1980.

Gupta, V. K., I. Rodriguez-Iturbe, and E. F. Wood (eds.), *Scale Problems in Hydrology,* D. Reidel, Dordrecht, Holland, 1986.

Harr, R. D., Water flux in soil and subsoil on a steep forested slope, *J. Hydrol.*, vol. 33, 37–58, 1977.

Hewlett, J. D., and A. R. Hibbert, Factors affecting the response of small watersheds to precipitation in humid areas, in *Int. Symp. on Forest Hydrology,* ed. by W. E. Sopper and H. W. Lull, Pergamon Press, Oxford, pp. 275–290, 1967.

Hewlett, J. D., *Principles of Forest Hydrology,* Univ. of Georgia Press, Athens, Ga., 1982.

Horton, R. E., The role of infiltration in the hydrologic cycle, *Trans. Am. Geophys. Union,* vol. 14, pp. 446–460, 1933.

Horton, R. E., Erosional development of streams and their drainage basins; hydrophysical approach to quantitative morphology, *Bull. Geol. Soc. Am.*, vol. 56, pp. 275-370, 1945.

Izzard, C. F., Hydraulics of runoff from developed surfaces, *Proceedings,* 26th Annual Meeting of the Highway Research Board, vol. 26, pp. 129–146, December 1946.

Martin, G. N., Characterization of simple exponential baseflow recession, *J. Hydrol.* (New Zealand), vol. 12, no. 1, pp. 57–62, 1973.

Morel-Seytoux, H. J., and J. P. Verdin, Extension of the Soil Conservation Service rainfall runoff methodology for ungaged watersheds, report no. FHWA/RD-81/060, Federal Highway Administration, Washington, D.C., available from National Technical Information Service, Springfield, Va. 22616, 1981.

Moseley, M. P., Streamflow generation in a forested watershed, New Zealand, *Water Resour. Res.* vol. 15, no. 4, pp. 795–806, 1979.

Pearce, A. J., and A. I. McKerchar, Upstream generation of storm runoff, in *Physical Hydrology, New Zealand Experience,* ed. by D. L. Murray and P. Ackroyd, New Zealand Hydrological Society, Wellington, New Zealand, pp. 165–192, 1979.

Pearce, A. J., M. K. Stewart, and M. G. Sklash, Storm runoff generation in humid headwater catchments, 1, Where does the water come from? *Water Resour. Res.*, vol. 22, no. 8, pp. 1263–1272, 1986.

Ragan, R. M., An experimental investigation of partial area contributions, *Int. Ass. Sci. Hydrol. Publ. 76*, pp. 241–249, 1968.

Roberson, J. A., and C. T. Crowe, *Engineering fluid mechanics*, 3rd ed., Houghton-Mifflin, Boston, 1985.

Rodriguez-Iturbe, I., and J. B. Valdes, The geomorphologic structure of hydrologic response, *Water Resour. Res.* vol. 15, no. 6, pp. 1409–1420, 1979.

Schumm, S. A., Evolution of drainage systems and slopes in badlands at Perth Amboy, New Jersey, *Bull. Geol. Soc. Am.*, vol. 67, pp. 597–646, 1956.

Shreve, R. L., Statistical law of stream numbers, *J. of Geol.*, vol. 74, pp. 17–37, 1966.

Singh, K. P., and J. B. Stall, Derivation of baseflow recession curves and parameters, *Water Resour. Res.*, vol. 7, no. 2, pp. 292–303, 1971.

Smart, J. S., Channel networks, in *Advances in Hydroscience*, ed. by V. T. Chow, Academic Press, Orlando, Fla., vol. 8, pp. 305–346, 1972.

Soil Conservation Service, Urban hydrology for small watersheds, *tech. rel. no. 55*, U. S. Dept. of Agriculture, Washington, D.C., 1975.

Soil Conservation Service, National Engineering Handbook, section 4, Hydrology, U. S. Dept. of Agriculture, available from U. S. Government Printing Office, Washington, D.C., 1972.

Strahler, A. N., Quantitative geomorphology of drainage basins and channel networks, section 4-II, in *Handbook of Applied Hydrology*, ed. by V. T. Chow, pp. 4–39, 4–76, McGraw-Hill, New York, 1964.

Terstriep, M. L., and J. B. Stall, The Illinois urban drainage area simulator, *ILLUDAS, Bull. 58*, Illinois State Water Survey, Urbana, Ill., 1974.

Unver, O., and L. W. Mays, Optimal determination of loss rate functions and unit hydrographs, *Water Resour. Res.*, vol. 20, no. 2, pp. 203–214, 1984.

Valdes, J. B., Y. Fiallo, and I. Rodriguez-Iturbe, A rainfall-runoff analysis of the geomorphologic IUH, *Water Resour. Res.*, vol. 15, no. 6, pp. 1421–1434, 1979.

PROBLEMS

5.2.1 If $Q(t) = Q_0 e^{-(t-t_0)/k}$ describes baseflow recession in a stream, prove that the storage $S(t)$ supplying baseflow is given by $S(t) = kQ(t)$.

5.2.2 Baseflow on a river is 100 cfs on July 1 and 80 cfs on July 10. Previous study of baseflow recession on this river has shown that it follows the linear reservoir model. If there is no rain during July, estimate the flow rate on July 31 and the volume of water in subsurface storage on July 1 and July 31.

5.2.3 The streamflow hydrograph at the outlet of a 300-acre drainage area is as shown:

Time (h)	0	1	2	3	4	5	6	7	8	9	10	11	12
Discharge (cfs)	102	100	98	220	512	630	460	330	210	150	105	75	60

Determine the base flow using the straight line method, the fixed base method, and the variable slope method. Assume $N = 5$ hours for the fixed base method.

5.3.1 For the following rainfall-runoff data, determine the ϕ-index and the cumulative infiltration curve based upon the ϕ-index. Also, determine the cumulative excess rainfall as a function of time. Plot these curves. The watershed area is 0.2 mi^2.

Time (h)	1	2	3	4	5	6	7
Rainfall rate (in/h)	1.05	1.28	0.80	0.75	0.70	0.60	0
Direct runoff (cfs)	0	30	60	45	30	15	0

5.3.2 Determine the direct runoff hydrograph, the ϕ-index, and the excess rainfall hyetograph for the storm of May 12, 1980, on Shoal Creek in Austin, Texas, for which the rainfall and streamflow data are given in Prob. 2.3.2. The watershed area is 7.03 mi^2.

5.4.1 Determine the excess rainfall hyetograph for the data given in Example 5.4.1 in the text if the initial effective saturation of the soil is 60 percent.

5.4.2 Determine the excess rainfall hyetograph for the data given in the text in Example 5.4.1 if the rain falls on a clay soil of initial effective saturation 40 percent.

5.4.3 Solve Example 5.4.1 in the text if the soil is described by Philip's equation with $S = 5$ cm·h$^{-1/2}$ and $K = 2$ cm/h.

5.4.4 Solve Example 5.4.1 in the text if the soil is described by Horton's equation with $f_0 = 5$ cm/h, $f_c = 1$ cm/h, and $k = 2$ h^{-1}.

5.4.5 Using the cumulative rainfall hyetograph given below for a 150-km^2 watershed, determine the abstractions and the excess rainfall hyetograph using Horton's equation with $f_0 = 40$ mm/h, $f_c = 10$ mm/h, and $k = 2$ h^{-1}. Assume that an interception storage of 10 mm is satisfied before infiltration begins. Also, determine the depth and volume of excess rainfall and its duration.

Time (h)	1	2	3	4	5	6
Cumulative rainfall (mm)	25	70	115	140	160	180

5.4.6 Solve Prob. 5.4.5 if the soil is described by Philip's equation with $S = 50$ mm·h$^{-1/2}$ and $K = 20$ mm/h.

5.4.7 Determine the excess rainfall hyetograph for the following storm hyetograph.

Time (h)	0–0.5	0.5–1.0	1.0–1.5	1.5–2.0
Rainfall intensity (in/h)	3.0	1.5	1.0	0.5

Horton's equation is applicable, with $f_0 = 3.0$ in/h, $f_c = 0.53$ in/h, and $k = 4.182$ h^{-1}. Determine the cumulative infiltration and rainfall curves and plot them. Also plot the infiltration rate and excess rainfall hyetograph. What is the total depth of excess rainfall?

5.4.8 Terstriep and Stall (1974) developed standard infiltration curves for bluegrass turf for each of the U.S. Soil Conservation Service hydrologic soil groups. These standard infiltration curves, which are used in the ILLUDAS model (Chap. 15), are based on Horton's equation with the following parameters:

SCS hydrological soil group	**A**	**B**	**C**	**D**
f_c (in/h)	1.00	0.50	0.25	0.10
f_o (in/h)	10.00	8.00	5.00	3.00
k (h^{-1})	2.00	2.00	2.00	2.00
Depression storage (in)	0.2	0.2	0.2	0.2

For the following storm hyetograph, determine the excess rainfall hyetograph, the cumulative infiltration, and the depth of excess rainfall for hydrologic soil group A.

Time (h)	0–0.5	0.5–1.0	1.0–1.5	1.5–2.0	2.0–5.0
Rainfall rate (in/h)	10	5	3	2	0.5

5.4.9 Solve Prob. 5.4.8 for each of the hydrologic soil groups and compare the depths of excess rainfall determined.

5.4.10 Solve Prob. 5.4.8 for the following rainfall hyetograph.

Time (h)	0–2.0	2.0–4.0	4.0–5.0
Rainfall rate (in/h)	2	1.5	0.5

5.4.11 Show that for infiltration under ponded conditions described by the Green-Ampt equation, the cumulative infiltration at the end of a time interval, $F_{t+\Delta t}$, is given by

$$F_{t+\Delta t} - F_t - \psi \Delta\theta \ln \left[\frac{F_{t+\Delta t} + \psi \Delta\theta}{F_t + \psi \Delta\theta} \right] = K \Delta t$$

5.4.12 Derive Eqs. (1) and (2) in Table 5.4.1, for $F_{t+\Delta t}$ and $f_{t+\Delta t}$ respectively, for the Horton infiltration equation.

5.4.13 Derive Eqs. (1) and (2) in Table 5.4.1, for $F_{t+\Delta t}$ and $f_{t+\Delta t}$ respectively, for the Philip infiltration equation.

5.5.1 Determine the cumulative abstractions for the Austin, Texas, 25-year design storm given below, for SCS curve numbers of 75 and 90. Use the SCS method and plot these two curves of cumulative abstractions on one graph. Also compute the cumulative excess rainfall vs. time and the excess rainfall hyetograph for each curve number. Plot the excess rainfall hyetographs for the two curve numbers on one graph.

Design storm rainfall depths (in)

Minutes	10	20	30	40	50	60	70	80	90	100
10-year	0.070	0.083	0.104	0.126	0.146	0.170	0.250	0.450	1.250	0.650
25-year	0.105	0.122	0.140	0.167	0.173	0.225	0.306	0.510	1.417	0.783
100-year	0.138	0.155	0.168	0.203	0.250	0.332	0.429	0.665	1.700	0.935

Minutes	110	120	130	140	150	160	170	180	Totals
10-year	0.317	0.203	0.164	0.142	0.112	0.093	0.073	0.067	4.470
25-year	0.417	0.297	0.192	0.170	0.143	0.126	0.119	0.099	5.511
100-year	0.513	0.373	0.293	0.243	0.182	0.159	0.147	0.135	7.020

5.5.2 Solve Prob. 5.5.1 for the 10-year design storm.

5.5.3 Solve Prob. 5.5.1 for the 100-year design storm.

5.5.4 (*a*) Compute the runoff from a 7-in rainfall on a 1500-acre watershed that has hydrologic soil groups that are 40 percent group A, 40 percent group B, and 20 percent group C interspersed throughout the watershed. The land use is 90 percent residential area that is 30 percent impervious, and 10 percent paved roads with curbs. Assume AMC II conditions.

(*b*) What was the runoff for the same watershed and same rainfall before development occurred? The land use prior to development was pasture and range land in poor condition.

5.5.5 A 200-acre watershed is 40 percent agricultural and 60 percent urban land. The agricultural area is 40 percent cultivated land with conservation treatment, 35 percent meadow in good condition, and 25 percent forest land with good cover. The urban area is residential: 60 percent is $\frac{1}{3}$-acre lots, 25 percent is $\frac{1}{4}$-acre lots, and 15 percent is streets and roads with curbs and storm sewers. The entire watershed is in hydrologic soil group B. Compute the runoff from the watershed for 5 in of rainfall. Assume AMC II conditions.

5.5.6 Solve Prob. 5.5.5 if the moisture condition is (a) AMC I, and (b) AMC III.

5.5.7 For the rainfall-runoff data given in Prob. 5.3.1, use the SCS method for abstractions to determine the representative SCS curve number for this watershed, assuming AMC II.

5.5.8 Considering the rainfall-runoff data in Prob. 5.3.1 and using the curve number determined in Prob. 5.5.7, determine the cumulative infiltration and the cumulative rainfall excess as functions of time. Plot these curves.

5.6.1 Compute the uniform flow depth in a trapezoidal channel having $n = 0.025$, $S_0 = 0.0005$, and $Q = 30$ cfs. The base width is 4 ft, and the side slopes are $1{:}z = 1{:}3$.

5.6.2 Compute the uniform flow depth in a triangular channel having $n = 0.025$, $S_0 = 0.0004$, $Q = 10$ cfs, and side slopes $1{:}z = 1{:}4$.

5.6.3 A rainfall of 3 in/h falls on a uniform, smooth, impervious plane that is 50 feet long and has a slope of 1 percent. Calculate discharge per unit width, depth, and velocity at the bottom end of the plane. Take $\nu = 1.2 \times 10^{-5}$ ft/s and $n = 0.015$.

5.6.4 Solve Prob. 5.6.3 if the rainfall has intensity 10 in/h.

5.6.5 Solve Prob. 5.6.3 if the rain falls on grass with an infiltration rate of 0.5 in/h and a Darcy-Weisbach roughness $f = 100$.

5.7.1 Solve Example 5.7.1 in the text if the flow length over pasture is 50 ft, and the channel is 500 feet long.

5.8.1 Determine the length ratio R_L for the Miller Creek watershed in Fig. 5.8.1.

5.8.2 Determine the drainage density and average overland flow length for the Miller Creek watershed in Fig. 5.8.1.

CHAPTER 6

HYDROLOGIC MEASUREMENT

Hydrologic measurements are made to obtain data on hydrologic processes. These data are used to better understand these processes and as a direct input into hydrologic simulation models for design, analysis, and decision making. A rapid expansion of hydrologic data collection worldwide was fostered by the International Hydrologic Decade (1965–1974), and it has become a routine practice to store hydrologic data on computer files and to make the data available in a machine-readable form, such as on magnetic tapes or disks. These two developments, the expansion and computerization of hydrologic data, have made available to hydrologists a vast array of information, which permits studies of greater detail and precision than was formerly possible. Recent advances in electronics allow data to be measured and analyzed as the events occur, for purposes such as flood forecasting and flood warning. The purpose of this chapter is to review the sequence of steps involved in hydrologic measurement, from the observation of the process to the receipt of the data by the user.

Hydrologic processes vary in space and time, and are random, or probabilistic, in character. Precipitation is the driving force of the land phase of the hydrologic cycle, and the random nature of precipitation means that prediction of the resulting hydrologic processes (e.g., surface flow, evaporation, and streamflow) at some future time is always subject to a degree of uncertainty that is large in comparison to prediction of the future behavior of soils or building structures, for example. These uncertainties create a requirement for hydrologic measurement to provide observed data at or near the location of interest so that conclusions can be drawn directly from on-site observations.

6.1 HYDROLOGIC MEASUREMENT SEQUENCE

Although hydrologic processes vary continuously in time and space, they are usually measured as *point samples,* measurements made through time at a fixed location in space. For example, rainfall varies continuously in space over a watershed, but a rain gage measures the rainfall at a specific point in the watershed. The resulting data form a *time series,* which may be subjected to statistical analysis.

In recent years, some progress has been made in measuring *distributed samples* over a line or area in space at a specific point in time. For example, estimates of winter snow cover are made by flying an aircraft over the snow field and measuring the radiation reflected from the snow. The resulting data form a *space series*. Distributed samples are most often measured at some distance from the phenomenon being observed; this is termed *remote sensing*. Whether the data are measured as a time series or as a space series, a similar sequence of steps is followed.

The sequence of steps commonly followed for hydrologic measurement is shown in Fig. 6.1.1, beginning with the physical device which senses or reacts to the physical phenomenon and ending with the delivery of data to a user. These steps are now described.

1. *Sensing*. A *sensor* is an instrument that translates the level or intensity of the phenomenon into an observable signal. For example, a mercury thermometer senses temperature through the expansion or contraction of the volume of mercury within a thin tube; a *storage rain gage* collects the incoming rainfall in a can or tube. Sensors may be *direct* or *indirect*.

A direct sensor measures the phenomenon itself, as with the storage rain gage; an indirect sensor measures a variable related to the phenomenon, as with the mercury thermometer. Many hydrologic variables are measured indirectly, including streamflow, temperature, humidity, and radiation. Sensors for the major hydrologic variables are discussed in the subsequent sections of this chapter.

2. *Recording*. A recorder is a device or procedure for preserving the signal produced by the sensor. *Manual recording* simply involves an observer taking readings off the sensor and tabulating them for future reference. Most of the available rainfall data are produced by observers who read the level in a storage rain gage each day at a fixed time (e.g., 9 A.M.). *Automatic recording* requires a device which accepts the signal from the sensor and stores it on a paper chart or punched tape, or an electronic memory including magnetic disks or tapes. Paper records require a mechanical system of pulleys or levers to translate the motion of the sensor to the motion of a pen on a chart or a punching mechanism for a paper tape. Fig. 6.1.2 shows hydrologic paper chart and tape recorders in common use. Historically, charts were the first recorders widely used in hydrology; they are still used when there is a need to have a direct visual image of the record, but charts have a great disadvantage in that translation of the chart record into a computerized form is a tedious procedure, involving manually tracking the line on the chart and recording the points where it changes direction. By contrast, paper tape recorders can be directly read by a computer. Sixteen-track paper

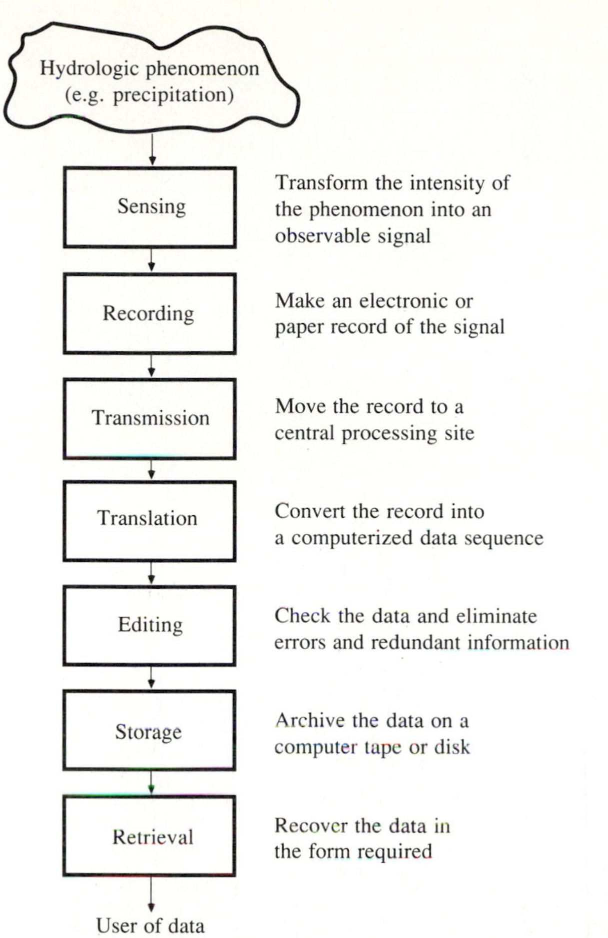

FIGURE 6.1.1
The hydrologic measurement sequence.

tapes are currently the most widespread form of automatic hydrologic recorders, but electronic storages are beginning to be adopted; their use can be expected to spread because of their convenience and because they need no mechanical system to translate the signal from sensor to recorder.

3. *Transmission*. Transmission is the transfer of a record from a remote recording site to a central location. Transmission may be done routinely, such as by manually changing the chart or tape on a recorder at regular intervals (from one week to several months in duration) and carrying the records to the central location. A rapidly developing area of hydrology is *real-time transmission* of data through microwave networks, satellites, or telephone lines. The recorder site is "polled" by the central location when data are needed; the recorder has the data already electronically stored and sends them back to the central location immediately. Microwave transmitters operate with relatively short-wavelength electromagnetic waves (10^{-1} to 10^{-3} m) traveling directly over the land surface with the aid of repeater stations; satellite data transmission uses radio waves (1 to

FIGURE 6.1.2(*a*)
Digital recorder for hydrologic data. The 16-track paper tape moves vertically behind the metal plate shown in the center of the picture. At predetermined intervals (usually every 15 minutes) the plate pushes the tape back against a set of needles, one for each track on the tape. The needles, in turn, are pushed back against the two rimmed wheels—if the needle hits one of the raised rims, it punches through the tape, if not, the tape is left blank. In this way, a pattern of holes emerges across the 16 tracks, 8 of which are used for recording the level of the phenomenon, and 8 for recording the time at which the punch was made. The rimmed disks, one for time and one for measurement level, rotate as time passes and the level of the phenomenon changes. (*Source:* T. J. Buchanan, U.S. Geological Survey. First published as Fig. 3 in "Techniques of water-resources investigations of the United States Geological Survey," Book 3, Chapter A6, U.S. Geological Survey, 1968.)

10^4 m in wavelength) reflected off a satellite whose position is fixed relative to the earth's surface. Microwave and satellite transmission of data are valuable for producing flood forecasts and for providing continuous access to remote recording sites which are difficult to reach by land travel.

4. *Translation*. Translation is the conversion of a record from a field instrument form into a computerized record for permanent electronic storage. For example, translators are available which read 16-track hydrologic paper tape records and produce an electronic signal in a form readable by computers. Cassette readers and chart followers are other devices of this type.

5. *Editing*. Editing is the procedure of checking the records translated into the computer to correct any obvious errors which have occurred during any of

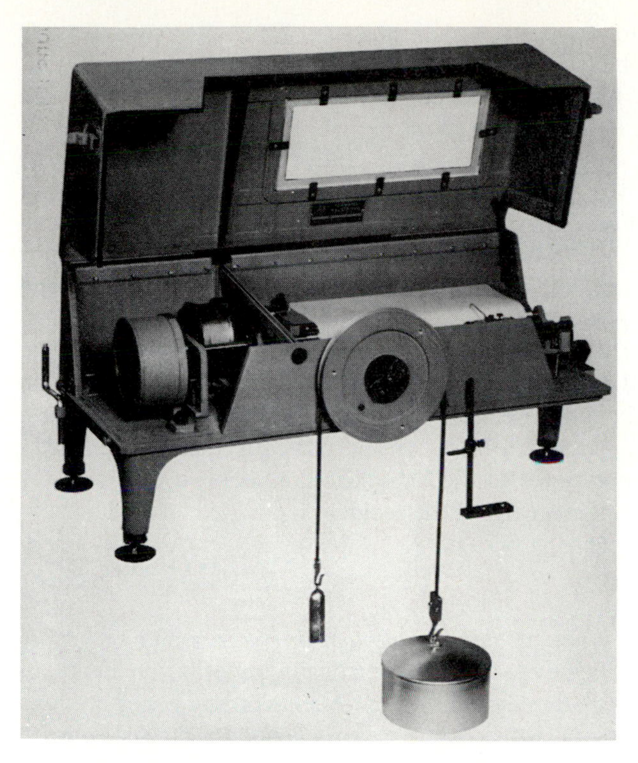

FIGURE 6.1.2(*b*)
Paper chart recorder attached to a float for recording water level variation. Rises or falls in the float level move the pen horizontally, parallel to the front of the recorder case. The paper is driven toward the back of the case continuously at a slow rate, thereby allowing the pen to trace out a record of water level against time on the chart. (*Source:* T. J. Buchanan, U.S. Geological Survey.)

the previous steps. Common errors include mistakes in the automatic timing of recorded measurements and information lost in transmission and translation, which is filled in by directly analyzing the record made at the recorder site.

6. *Storage*. Edited data are stored in a computerized data archive such as WATSTORE, operated by the U.S. Geological Survey, or TNRIS (Texas Natural Resource Information System). Such archives contain many millions of hydrologic data systematically compiled into files indexed by location and sequenced by the time of measurement.

7. *Retrieval*. Data are retrieved for users either in a machine-readable form, such as magnetic tape or diskette, or as a paper printout.

6.2 MEASUREMENT OF ATMOSPHERIC WATER

Atmospheric Moisture

The measurement of moisture high in the atmosphere is made by means of a *radiosonde*, which is a balloon filled with helium that is attached to a measuring device recording temperature, humidity, and air pressure. The balloon is released, and as it rises in the atmosphere, it sends the data back to a field tracking station. At the tracking station, the balloon is tracked by radar as it rises, and the wind speed at various elevations is thereby observed.

The measurement of atmospheric moisture and climate parameters near the ground is accomplished at a *climate station*. A climate station commonly contains, within a screened box, thermometers for measuring the maximum and minimum air temperatures each day, and a wet- and dry-bulb thermometer or *hygrometer,* to measure humidity; nearby are located precipitation gages, and sometimes an *evaporation pan* and an *anemometer*. For detailed measurements of climate variables, special weather stations are installed at the testing site, and the data can be accumulated and sent by microwave to a central recording station as described previously. The measurement of radiation is accomplished with a device known as a *radiometer*, which relies on the principle that a black body will have a temperature proportional to the amount of radiation it receives. By measuring this temperature, the intensity of the incident radiation can be deduced.

Rainfall

Rainfall is recorded by two types of gages: *nonrecording gages* and *recording gages*. A recording gage is a device that automatically records the depth of rainfall in intervals down to one minute in duration. Nonrecording gages are read manually at longer time intervals. Nonrecording gages generally consist of open receptacles with vertical sides, in which the depth of precipitation is measured by a graduated measuring cylinder or dipstick. The two types of nonrecording gages are *standard gages* and *storage gages*. Standard gages are ordinarily used for daily rainfall readings and consist of a collector above a funnel leading into a receiver. Rain gages for locations where only weekly or monthly readings are used are similar in design to the daily type but have a larger capacity receiver. Storage gages are used to measure rainfall over an entire season, usually in remote, sparsely inhabited areas. These rain gages consist of a collector above a funnel that leads into a storage area large enough for the season rainfall volume. Standard gages are the most widespread rainfall data measurement devices used in hydrology. Many thousands of these gages are read by voluntary observers, and their data are recorded by weather services.

There are three types of recording rain gages in general use: the *weighing type,* the *float type,* and the *tipping bucket type*. A weighing type rain gage continuously records the weight of the receiving can plus the accumulated rainfall by means of a spring mechanism or a system of balance weights (Fig. 6.2.1). These gages are designed to prevent excessive evaporation losses by the addition of oil or other evaporation-suppressing material to form a film over the surface. Weighing rain gages are useful in recording snow, hail, and mixtures of snow and rain.

A float type rain gage has a chamber containing a float that rises vertically as the water level in the chamber rises. Vertical movement of the float is translated

FIGURE 6.2.1
Recording rain gage with the top removed. The gage records the weight of precipitation received through the circular opening, which is 4 inches in diameter. (*Source:* L. A. Reed, U.S. Geological Survey. First published as Fig. 2 in USGS Water Supply Paper 1798-M, 1976.)

into movement of a pen on a chart. A device for siphoning the water out of the gage is used so that the total amount of rainfall falling can be collected.

A tipping bucket type rain gage operates by means of a pair of buckets (Fig. 6.2.2). The rainfall first fills one bucket, which overbalances, directing the flow of water into the second bucket. The flip-flop motion of the tipping buckets is transmitted to the recording device and provides a measure of the rainfall intensity.

Whether a rain gage operates by the vertical rise of a float, the accumulation of weight, or the tipping of a bucket, the movement can be recorded. A drum or strip chart is rotated by a spring or electrically driven clock past a pen whose motion is linked to that of the float, weighing device, or tipping bucket system. The motion of the mechanism can also be converted into an electrical signal and

FIGURE 6.2.2
Tipping bucket rain gage. (*Source:* Ministry of Works and Development, New Zealand.)

transmitted to a distant receiver. Rain gages commonly have a windbreak device constructed around them in order to minimize the amount of distortion in the measurement of rainfall caused by the wind flow pattern around the gage.

Radar can be used to observe the location and movement of areas of precipitation. Certain types of radar equipment can provide estimates of rainfall rates over areas within the range of the radar (World Meteorological Organization, 1981). Radar is sometimes used to get a visual image of the pattern of rainfall-producing thunderstorms and is particularly useful for tracking the movement of tornadoes. The introduction of color digital radar has made it possible to measure rainfall in distant thunderstorms with more precision than was formerly possible. The phenomenon upon which weather radar depends is the reflection of microwaves emitted by the radar transmitter by the droplets of water in the storm. The degree of reflection is related to the density of the droplets and therefore to the rainfall intensity.

Snowfall

Snowfall is recorded as part of precipitation in rain gages. In regions where there is a continuous snow cover, the measurement of the depth and density of this snow cover is important in predicting the runoff which will result when the snow cover melts. This is accomplished by means of surveyed *snow courses,* which are sections of the snow cover whose depth is determined by means of gages that have been installed prior to the snowfall. The density of snow in the snowpack may be determined by boring a hole through the pack or into the pack and measuring the amount of liquid water obtained from the sample. Automated devices for measuring the weight of the snow above a certain point in the ground have been developed—these include *snow pillows,* which measure the pressure of snow on a plastic pillow filled with a nonfreezing fluid.

Interception

The amount of precipitation captured by vegetation and trees is determined by comparing the precipitation in gages beneath the vegetation with that recorded nearby under the open sky. The precipitation detained by interception is dissipated as stem flow down the trunks of the trees and evaporation from the leaf surface. Stem flow may be measured by catch devices around tree trunks.

Evaporation

The most common method of measuring evaporation is by means of an evaporation pan. There are various types of evaporation pans; however, the most widely used are the U.S. Class A pan, the U.S.S.R. GGI-3000 pan, and the 20-m^2 tank (World Meteorological Organization, 1981). The Class A pan measures 25.4 cm (10 in) deep and 120.67 cm (4 ft) in diameter and is constructed of Monel metal or unpainted galvanized iron. The pan is placed on timber supports so that air

TABLE 6.2.1
Summary of pan coefficients (after Linsley, Kohler, and Paulhus, 1982)

Location:	Class A Pan Coefficient
Felt Lake, California	0.77
Ft. Collins, Colorado	0.70
Lake Colorado City, Texas	0.72
Lake Elsinora, California	0.77
Lake Hefner, Oklahoma	0.69
Lake Okeechobee, Florida	0.81
Red Bluff Res., Texas	0.68

circulates beneath it. The U.S.S.R. GGI-3000 pan is a 61.8-cm diameter tank with a conical base fabricated of galvanized sheet iron. The surface area is 0.3 m^2; the tank is 60 cm deep at the wall and 68.5 cm deep at the center. The tank is sunk in the ground with the rim projecting approximately 7.5 cm above ground level.

In addition to the pan, several other instruments are used at evaporation stations: (1) an anemometer located 1 to 2 meters above the pan, for determining wind movement; (2) a nonrecording precipitation gage; (3) a thermometer to measure water temperature in the pan; and (4) a thermometer for air temperature, or a *psychrometer* where temperature and humidity of the air are desired.

By measuring the water level in the pan each day, the amount of evaporation which has occurred can be deduced after accounting for the precipitation during that day. The depth of the water in the pan is measured to the nearest hundredth of an inch by means of a hook gage or by adding the amount of water necessary to raise its level to a fixed point. The evaporation recorded in a pan is greater than that which would be recorded from the same area of water surface in a very large lake. Adjustment factors or pan coefficients have been determined to convert the data recorded in evaporation pans so that they correspond to the evaporation from large open water surfaces. Table 6.2.1 lists pan evaporation coefficients for various locations.

Evapotranspiration

Evaporation from the land surface plus transpiration through the plant leaves, or evapotranspiration, may be measured by means of *lysimeters*. A lysimeter is a tank of soil in which vegetation is planted that resembles the surrounding ground cover. The amount of evapotranspiration from the lysimeter is measured by means of a water balance of all moisture inputs and outputs. The precipitation on the lysimeter, the drainage through its bottom, and the changes in the soil moisture within the lysimeter are all measured. The amount of evapotranspiration is the amount necessary to complete the water balance.

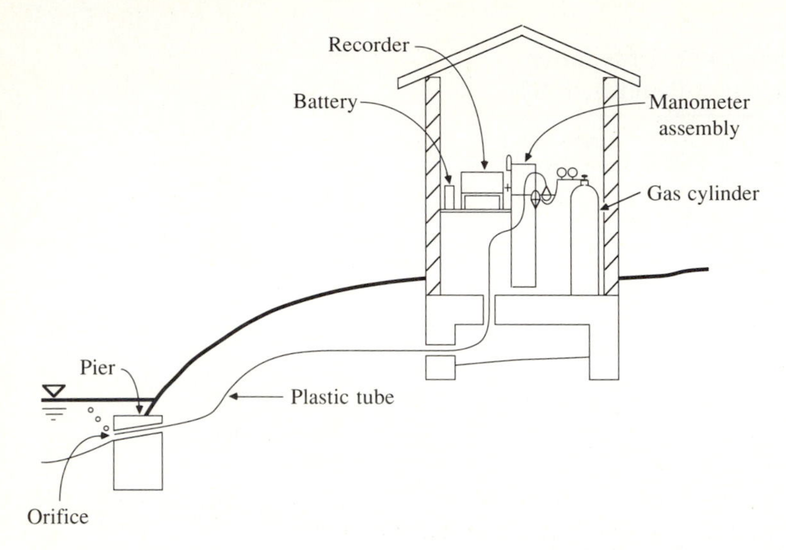

FIGURE 6.3.1(*a*)
Water level measurement using a bubble gage recorder. The water level is measured as the back pressure on the bubbling stream of gas by using a mercury manometer. (*Source:* Rantz, et al., vol. 1, Fig. 31, p. 52, 1982.)

6.3 MEASUREMENT OF SURFACE WATER

Water Surface Elevation

Water surface elevation measurements include both peak levels (flood crest elevations) and the stage as a function of time. These measurements can be made manually or automatically. *Crest stage gages* are used to obtain a record of flood crests at sites where recording gages are not installed. A crest stage gage consists of a wooden *staff gage* or scale, situated inside a pipe that has small holes for the entry of water. A small amount of cork is placed in the pipe, floats as the water rises, and adheres to the staff or scale at the highest water level.

Manual observations of water level are made using staff gages, which are graduated boards set in the water surface, or by means of sounding devices that signal the level at which they reach the water surface, such as a weight on a wire suspended from a bridge over the surface of a river.

Automatic records of water levels are made at about 10,000 locations in the United States; the *bubble gage* is the sensor most widely used [Fig. 6.3.1(*a*)]. A bubble gage senses the water level by bubbling a continuous stream of gas (usually carbon dioxide) into the water. The pressure required to continuously push the gas stream out beneath the water surface is a measure of the depth of the water over the nozzle of the bubble stream. This pressure is measured by a manometer in the recorder house [Fig. 6.3.1(*b*)]. Continuous records of water levels are maintained for the calculation of stream flow rates. The level of water in a stream at any time is referred to as the *gage height*.

FIGURE 6.3.1(*b*)
The mercury manometer used to measure the gas pressure in a water level recorder. As the water level and gas pressure change, an electric motor drives a pair of sensor wires up or down to follow the motion of the mercury surface. (*Source:* G. N. Mesnier, US Geological Survey, USGS Water Supp. Pap. 1669-Z, Fig. 7, 1963.)

Flow Velocity

The velocity of flow in a stream can be measured with a *current meter*. Current meters are propeller devices placed in the flow, the speed with which the propeller rotates being proportional to the flow velocity (Fig. 6.3.2). The current meter can be hand-held in the flow in a small stream, suspended from a bridge or cable-way across a larger stream, or lowered from the bow of a boat (Fig. 6.3.3). The flow velocity varies with depth in a stream as shown in Fig. 6.3.4. Figure 6.3.5 shows *isovels* (lines of equal velocity) for sections of the Kaskaskia River in Illinois. The velocity rises from 0 at the bed to a maximum near the surface, with an average value occurring at about 0.6 of the depth. It is a standard practice of the U.S. Geological Survey to measure velocity at 0.2 and 0.8 of the depth when the depth is more than 2 ft and to average the two velocities to determine the average velocity for the vertical section. For shallow rivers and near the banks on deeper rivers where the depths are less than 2 ft, velocity measurements are made at 0.6 depth. On some occasions, it is desired to know the travel time of flow from one location to another some distance away, perhaps several days flow

FIGURE 6.3.2
Current meters for measuring water velocity. The smaller one mounted on the base in the foreground is attached to a vertical rod and used when wading across a shallow stream. The larger one in the background is suspended on a wire and used for gaging a deeper river from a bridge or boat. Both meters work on the principle that the speed of rotation of the cups is proportional to the flow velocity. The operator attaches electrical wires to the two screws on the vertical shaft holding the cups. Each time the cups complete a rotation, a contact is closed inside the shaft and the operator hears a click in headphones to which the wires are attached. By counting the number of clicks in a fixed time interval (say 40 seconds), the velocity is determined. (*Source:* T. J. Buchanan, U. S. Geological Survey. First published as Fig. 4 in "Techniques of water-resources investigations of the United States Geological Survey," Book 3, Chapter A8, 1969.)

FIGURE 6.3.3
Current meter suspended from bow of a boat. (*Source:* Ministry of Works and Development, New Zealand).

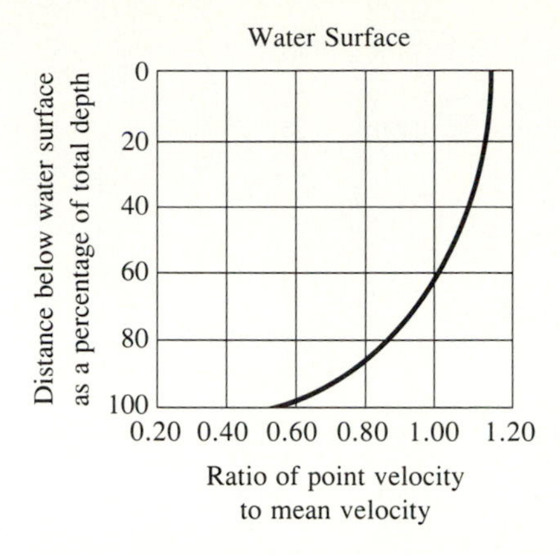

FIGURE 6.3.4
Typical vertical variation of the flow velocity in a stream. (*Source:* Rantz, et al., vol. 1, Fig. 88, p. 133, 1982.)

time. For these purposes a float is used which is carried along with the water at approximately its average velocity.

Velocity measurements can also be made based upon electromagnetic sensing. The Velocity Modified Flow Measurement (VMFM) meter is a velocity-sensing instrument based upon such principles (Marsh-McBirney, 1979). The portable meter shown in Fig. 6.3.6 has a solid state electronics system housed in a small box, an electromagnetic sensor, and connecting cable. The sensor is placed on the same rod used for propeller-type current meters and the rod is hand held for making velocity measurements. When the sensor is immersed in flowing water, a magnetic field within the sensor is altered by the water flow, creating a voltage variation which is measured by electrodes imbedded in the sensor. The amplitude of the voltage variation is proportional to the water velocity. The voltage variation is transmitted through the cable to the electronic processor system, which automatically averages point velocity measurements made at different locations in a stream cross section. The sensor also monitors water depth using a bubble gage, and the processor integrates velocity and depth measurements to produce discharge data. This meter can also be used to measure flows in sewer pipes and in other types of open channels.

Stream Flow Rate

Stream flow is not directly recorded, even though this variable is perhaps the most important in hydrologic studies. Instead, water level is recorded and stream flow is deduced by means of a *rating curve* (Riggs, 1985). The rating curve is developed using a set of measurements of discharge and gage height in the stream, these measurements being made over a period of months or years so as to obtain an accurate relationship between the stream flow rate, or discharge, and the gage height at the gaging site.

DISCHARGE COMPUTATION. The discharge of a stream is calculated from measurements of velocity and depth. A marked line is stretched across the stream.

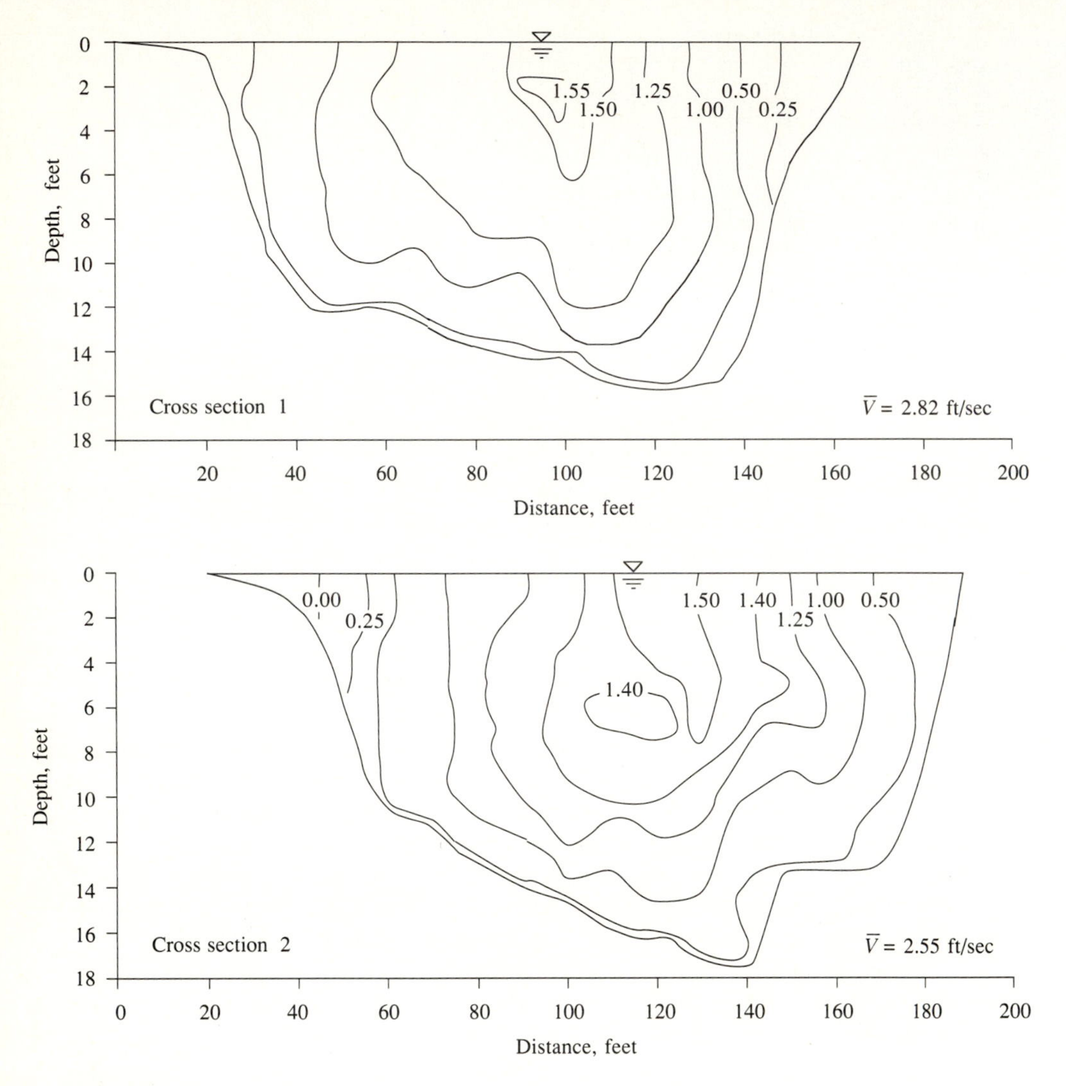

FIGURE 6.3.5
Velocity profiles for sections of the Kaskaskia River, Illinois. These profiles are based upon point velocity data which were converted to nondimensional velocities by dividing the point velocities by the average velocity of the section. The nondimensional velocities were used to draw the isovels (lines of equal velocity). The discharge for the isovels shown was 4000 cfs. (*Source:* Bhowmik, 1979. Used with permission.)

At regular intervals along the line, the depth of the water is measured with a graduated rod or by lowering a weighted line from the surface to the stream bed, and the velocity is measured using a current meter. The discharge at a cross section of area A is found by

$$Q = \iint_A \mathbf{V} \cdot d\mathbf{A} \tag{6.3.1}$$

FIGURE 6.3.6
VMFM (Velocity Modified Flow Measurement) meter. (Courtesy of Marsh-McBirney, Inc., 1987. Used with permission.)

in which the integral is approximated by summing the incremental discharges calculated from each measurement i, $i = 1, 2, \ldots, n$, of velocity V_i and depth d_i (Fig. 6.3.7). The measurements represent average values over width Δw_i of the stream, so the discharge is computed as

$$Q = \sum_{i=1}^{n} V_i d_i \Delta w_i \tag{6.3.2}$$

Example 6.3.1 At known distances from an initial point on the stream bank, the measured depth and velocity of a stream are shown in Table 6.3.1. Calculate the corresponding discharge at this location.

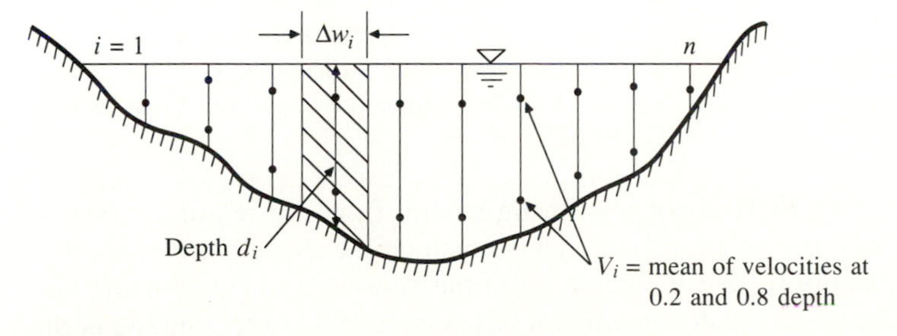

FIGURE 6.3.7
Computation of discharge from stream gaging data.

TABLE 6.3.1
Computation of discharge from stream gaging.

Measurement number i	Distance from initial point (ft)	Width Δw (ft)	Depth d (ft)	Mean velocity V (ft/s)	Area $d\Delta w$ (ft^2)	Discharge $Vd\Delta w$ (cfs)
1	0	6.0	0.0	0.00	4.7	0.0
2	12	16.0	3.1	0.37	49.6	18.4
3	32	20.0	4.4	0.87	88.0	76.6
4	52	20.0	4.6	1.09	92.0	100.3
5	72	20.0	5.7	1.34	114.0	152.8
6	92	20.0	4.5	0.71	90.0	63.9
7	112	20.0	4.4	0.87	88.0	76.6
8	132	20.0	5.4	1.42	108.0	153.4
9	152	17.5	6.1	2.03	106.8	216.7
10	167	15.0	5.8	2.22	87.0	193.1
11	182	15.0	5.7	2.51	85.5	214.6
12	197	15.0	5.1	3.06	76.5	234.1
13	212	15.0	6.0	3.12	90.0	280.8
14	227	15.0	6.5	2.96	97.5	288.6
15	242	15.0	7.2	2.62	108.0	283.0
16	257	15.0	7.2	2.04	108.0	220.3
17	272	15.0	8.2	1.56	123.0	191.9
18	287	15.0	5.5	2.04	82.5	168.3
19	302	15.0	3.6	1.57	54.0	84.8
20	317	11.5	3.2	1.18	36.8	43.4
21	325	4.0	0.0	0.00	3.2	0.0
Total		325.0			1693.0	3061.4

Data were provided by the U. S. Geological Survey from a gaging made on the Colorado River at Austin, October 5, 1983.

Solution. Each measurement represents the conditions up to halfway between this measurement and the adjacent measurements on either side. For example, the first three measurements were made 0, 12, and 32 feet from the initial point, and so $\Delta w_2 = [(32 - 12)/2] + [(12 - 0)/2] = 16.0$ ft. The corresponding area increment is $d_2\Delta w_2 = 3.1 \times 16.0 = 49.6$ ft^2, and the resulting discharge increment is $V_2 d_2 \Delta w_2 = 0.37 \times 49.6 = 18.4$ ft^3/s. The other incremental areas and discharges are similarly computed as shown in Table 6.3.1 and summed to yield discharge $Q = 3061$ ft^3/s, and total cross-sectional area $A = 1693$ ft^2. The average velocity at this cross section is $V = Q/A = 3061/1693 = 1.81$ ft/s.

There are indirect methods of measuring stream flow not requiring the use of current meters or water level records. These include the *dye gaging method* in which a known quantity of dye is injected into the flow at an upstream site and measured some distance downstream when it has become completely mixed in the water. By comparing the concentrations at the downstream site with the mass of

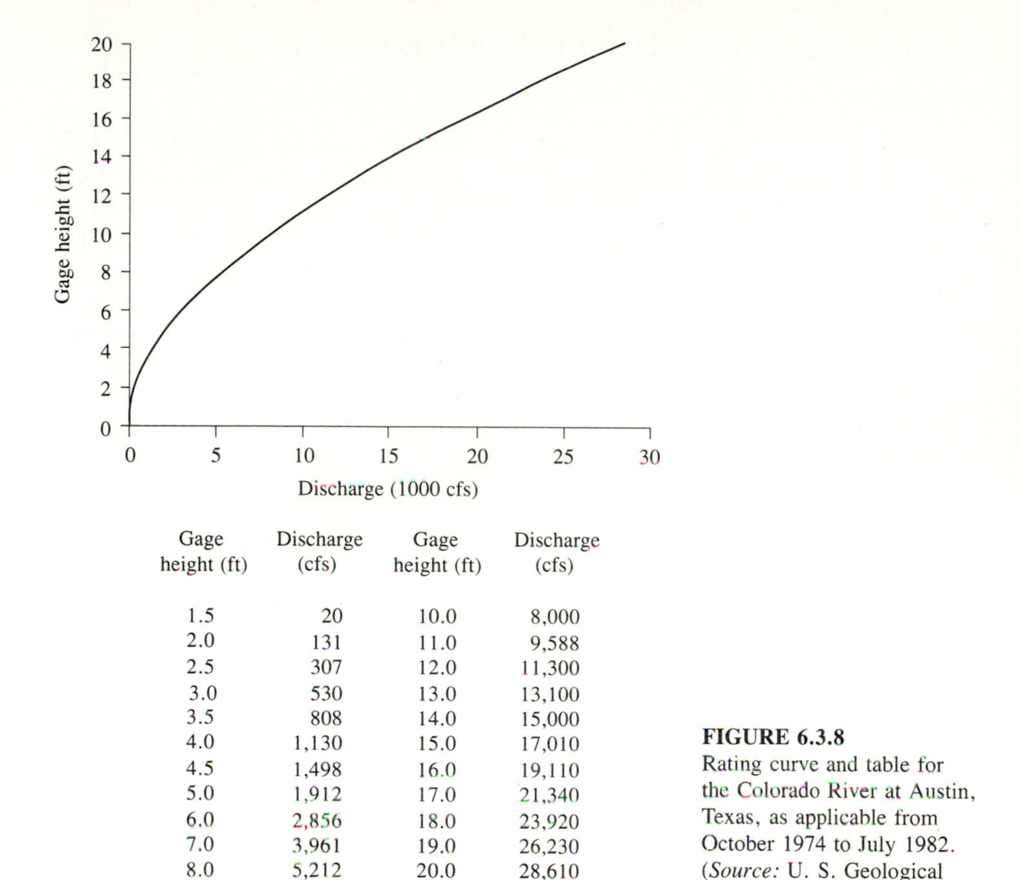

Gage height (ft)	Discharge (cfs)	Gage height (ft)	Discharge (cfs)
1.5	20	10.0	8,000
2.0	131	11.0	9,588
2.5	307	12.0	11,300
3.0	530	13.0	13,100
3.5	808	14.0	15,000
4.0	1,130	15.0	17,010
4.5	1,498	16.0	19,110
5.0	1,912	17.0	21,340
6.0	2,856	18.0	23,920
7.0	3,961	19.0	26,230
8.0	5,212	20.0	28,610
9.0	6,561		

FIGURE 6.3.8
Rating curve and table for the Colorado River at Austin, Texas, as applicable from October 1974 to July 1982. (*Source:* U. S. Geological Survey, Austin, Texas.)

the dye injected at the upstream site, the flow rate can be deduced. This method is particularly suitable for stony mountain streams, where the dye is mixed quickly and measurements by other methods are difficult.

RATING CURVE. The rating curve is constructed by plotting successive measurements of the discharge and gage height on a graph such as that shown in Fig. 6.3.8. The rating curve is then used to convert records of water level into flow rates. The rating curve must be checked periodically to ensure that the relationship between the discharge and gage height has remained constant; scouring of the stream bed or deposition of sediment in the stream can cause the rating curve to change so that the same recorded gage height produces a different discharge.

The relationship between water level and the flow rate at a given site can be maintained consistently by constructing a special flow control device in the stream, such as a sharp crested weir or a flume.

6.4 MEASUREMENT OF SUBSURFACE WATER

Soil Moisture

The amount of moisture in the soil can be found by taking a sample of soil and oven drying it. By comparing the weight of the sample before and after the drying and measuring the volume of the sample, the moisture content of the soil can be determined. Some recording devices which record soil moisture directly in the field have been developed, particularly for irrigation studies. These include *gypsum blocks* and *neutron probes*. Neutron probes rely on the reflection of neutrons emitted from a probe device inserted in a hole in the ground by the moisture in the surrounding soil, the degree of reflection of the neutrons being proportional to the moisture content (Shaw, 1983).

Infiltration

Measurements of infiltration are made using a *ring infiltrometer*, which is a metal ring approximately two feet in diameter that is driven into the soil; water is placed inside the ring and the level of the water is recorded at regular time intervals as it recedes. This permits the construction of the cumulative infiltration curve, and from this the infiltration rate as a function of time may be calculated. Sometimes a second ring is added outside the first, filled with water and maintained at a constant level so that the infiltration from the inner ring goes vertically down into the soil. In some cases, measurements of infiltration can be made by using tracers introduced at the surface of the soil and extracted from probes placed below the surface.

Ground Water

The level of water in the saturated flow or ground water zone is determined by means of observation wells. An observation well has a float device so that the vertical movement of water in the well is transmitted by means of a pulley system to the recorder house at the surface. Devices that drop a probe down the well on a wire to sense the water level can be used to obtain instantaneous measurements. The velocity of ground water flow can also be determined by tracers, including common salt. A quantity of the tracer is introduced at an upstream well, and the time for the pulse of tracer to reach a well somewhat downstream of the first is recorded. This is the *actual velocity* and not the apparent or Darcy velocity. Such measurements also assist in determining the amount of dispersion of contaminants introduced into ground water.

6.5 HYDROLOGIC MEASUREMENT SYSTEMS

Urban Hydrology Monitoring Systems

Urban stormwater investigations require well-designed data collection systems and instrumentation, both for water quantity and water quality. Besides conventional stream gaging and precipitation measurement, elaborate instruments employing microprocessor technology are used to collect and record information at remote locations such as in underground storm drains.

An instrumentation package called an *urban hydrology monitoring system,* as used by the U. S. Geological Survey for urban stormwater investigations, is shown in Fig. 6.5.1 (Jennings, 1982). This system is designed to collect storm rainfall and runoff quantity and quality data. It was specifically designed for flow gaging in underground storm sewers using a flow constriction as the discharge control. The system is composed of five components: the system control unit, rain gages, atmospheric sampling, stage sensing, and water quality sampling.

The system control unit is a microprocessor that records data at a central site, controls an automatic water-sampling device, records rain gage readings via telephone lines, and continuously monitors the stage in the storm sewers. The

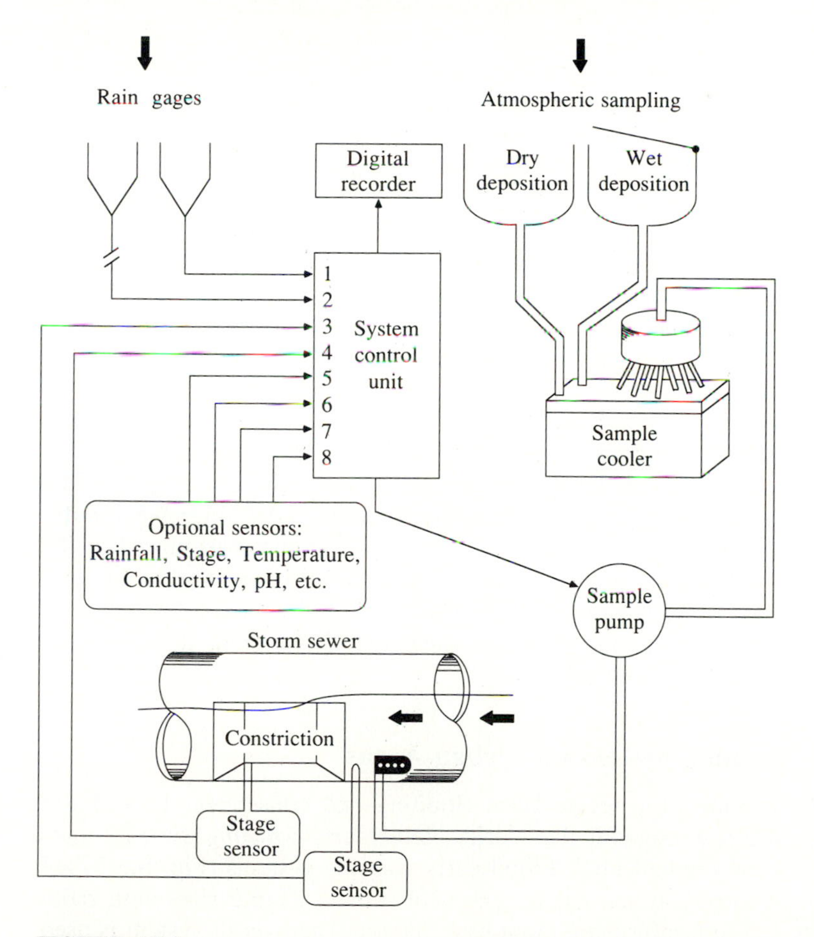

FIGURE 6.5.1
Typical installation of a U. S. Geological Survey urban hydrology monitoring system (*Source:* Jennings, 1982. Used with permission.)

control unit operates in a standby mode between storms, so that data are collected only if there is rainfall. The rain gage has an 8-in diameter orifice and a tipping bucket mechanism coupled to a mercury switch. The buckets are calibrated to tip after each 0.01 in of rainfall.

The atmospheric sampler is used to collect samples of atmospheric constituents affecting water quality, as for acid rain studies. Two rectangular collectors are used, one for sampling rainfall and the other for dry deposition of dust and other constituents between rainfalls. Water quality samples are also taken from storm sewers using automatic pump samplers. The samples are stored in a freezer which maintains the water temperature at approximately 5°C.

Real-time Data Collection Systems for River-Lake Systems

Real-time data collection and transmission can be used for flood forecasting on large river-lake systems covering thousands of square miles, as shown in Fig. 6.5.2 for the lower Colorado River in central Texas. The data collection system used there is called a Hydrometeorological Data Acquisition System (Hydromet, EG&G Washington Analytical Services Center, Inc., 1981) and is used to provide information for a flood forecasting model (Section 15.5). This information is of two types: (a) the water surface elevations at various locations throughout the river-lake system, and (b) rainfall from a rain gage network for the ungaged drainage areas around the lakes. The Hydromet system consists of (a) remote terminal unit (RTU) hydrometeorological data acquisition stations installed at U.S. Geological Survey river gage sites, (b) microwave terminal unit (MTU) microwave-to-UHF radio interface units located at microwave repeater sites, which convert radio signals to microwave signals, and (c) a central control station located at the operations control center in Austin, Texas, which receives its information from the microwave repeating stations. The system is designed to automatically acquire river level and meteorological data from each RTU; telemeter this data on request to the central station via the UHF/microwave radio system; determine the flow rate at each site by using rating tables stored in the central system memory; format and output the data for each site; and maintain a historical file of data for each site which may be accessed by the local operator, a computer, or a remote dial-up telephone line terminal. The system also functions as a self-reporting flood alarm network.

Flood Early Warning System for Urban Areas

Because of the potential for severe flash flooding and consequent loss of life in many urban areas throughout the world, *flood early warning systems* have been constructed and implemented. Flood early warning systems (Fig. 6.5.3) are real-time event reporting systems that consist of remote gaging sites with radio repeater sites to transmit information to a base station. The overall system is used to collect, transport, and analyze data, and make a flood forecast in order to maximize the warning time to occupants in the flood plain. Such systems have

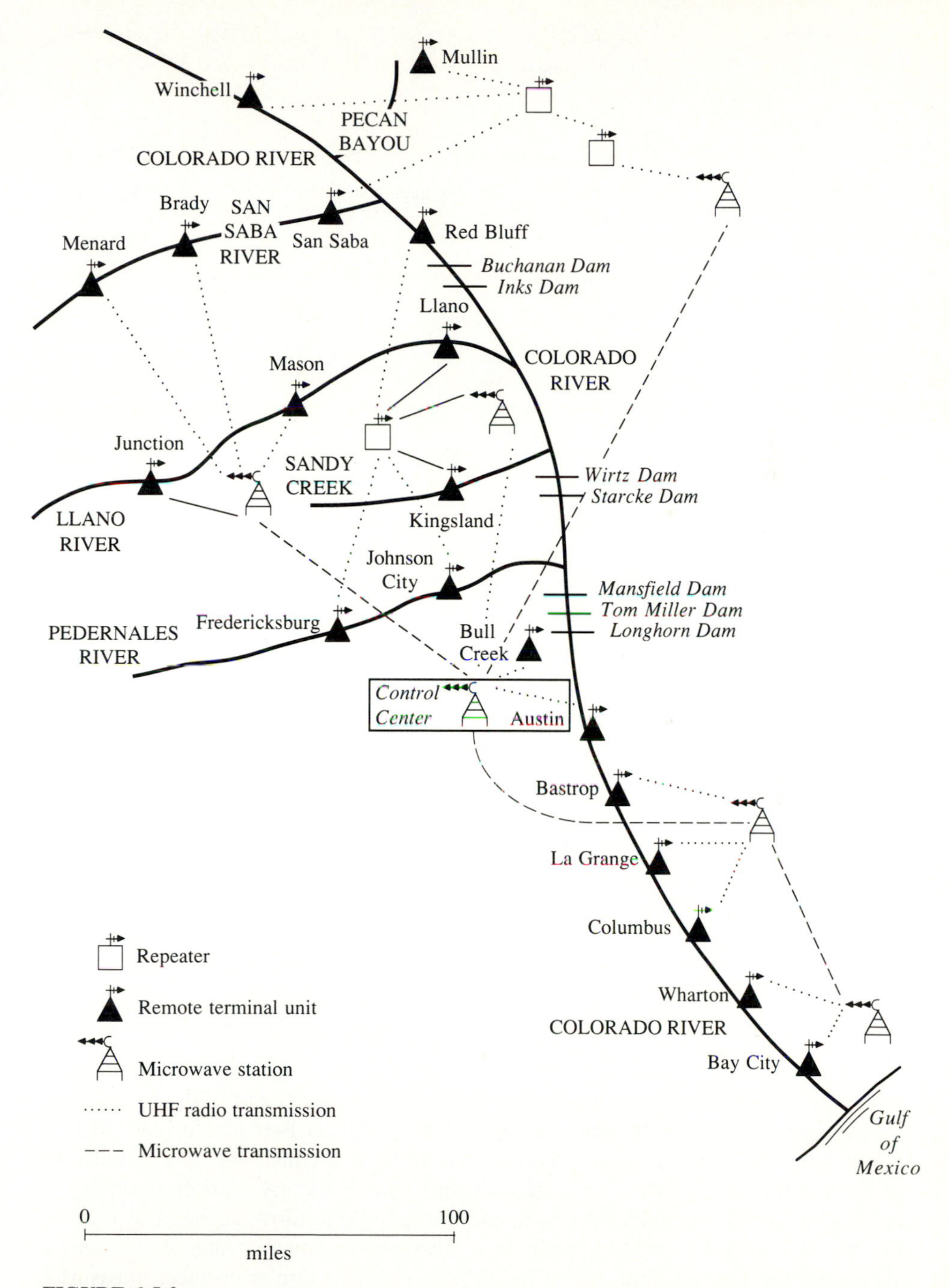

FIGURE 6.5.2
Real-time data transmission network on the lower Colorado River, Texas. Water level and rainfall data are automatically transmitted to the control center in Austin every 3 hours to guide releases from the dams. During floods data are updated every 15 minutes.

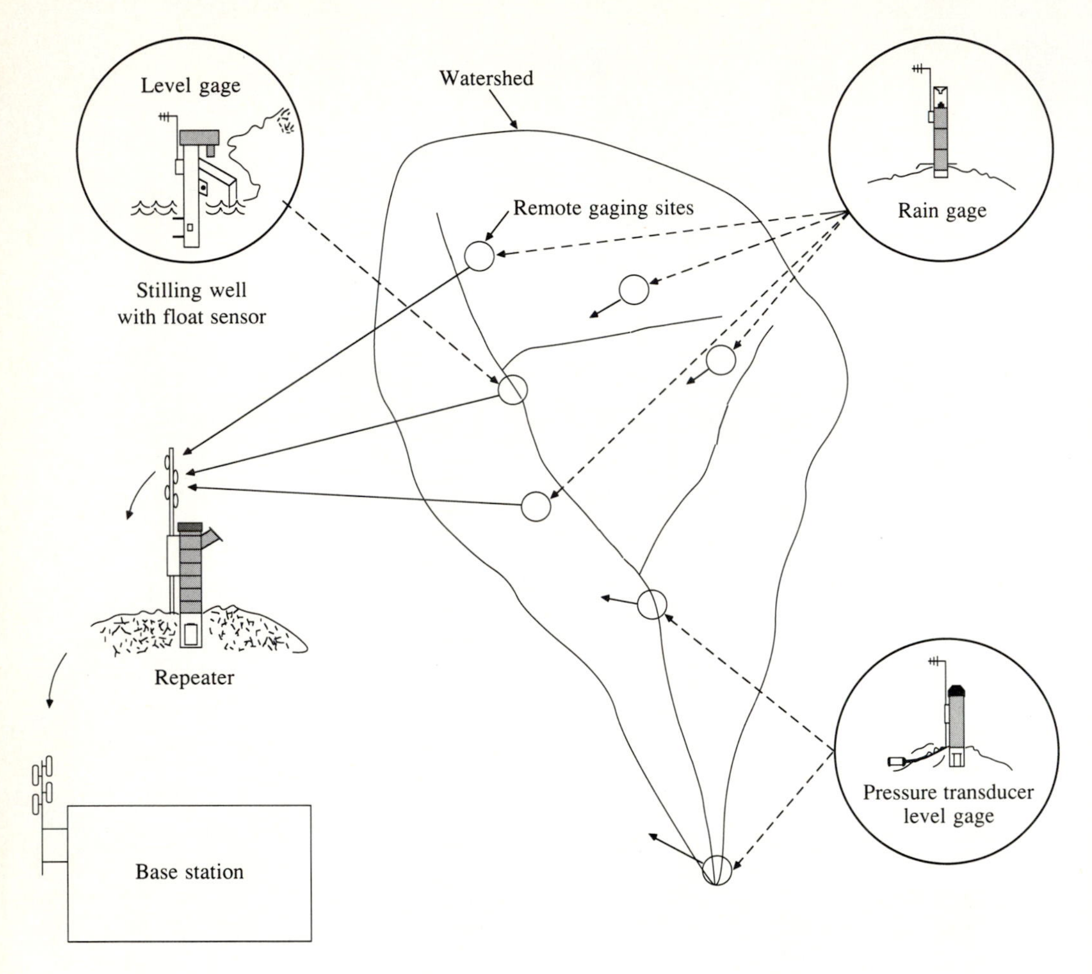

FIGURE 6.5.3
Example of a flood early warning system for urban areas.

been installed in Austin and Houston, Texas, and elsewhere (Sierra/Misco, Inc., 1986).

The remote stations (Fig. 6.5.4) each have a tipping bucket rain gage, which generates a digital input to a transmitter whenever 1 mm of rainfall drains through the funnel assembly. A transmission to the base station is made for each tip of the bucket. The rain gage is completely self-contained, consisting of a cylindrical stand pipe housing for the rain gage, antenna mount, battery, and electronics.

Some remote stations have both rainfall and streamflow gages. The remote stations can include a stilling well or a pressure transducer water level sensor similar to the one illustrated in Fig. 6.5.5. The pressure transducer measures changes of the water level above the pressure sensor's orifice. The electronic differential pressure transducer automatically compensates for temperature and barometric pressure changes with a one percent accuracy over the measured range.

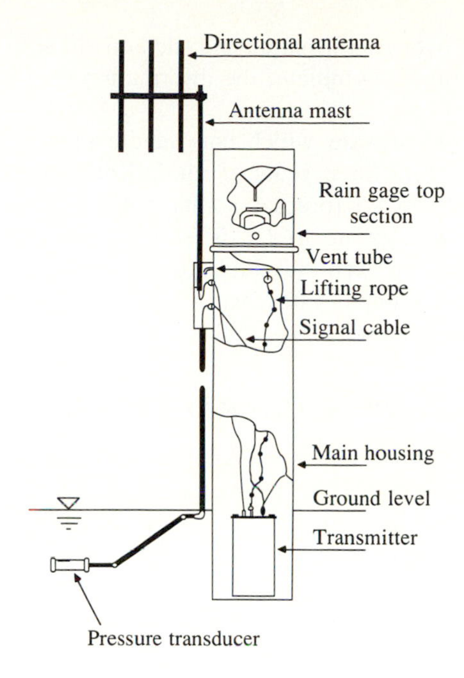

FIGURE 6.5.4
Remote station combining precipitation and stream gages. (Courtesy of Sierra/Misco, Inc., 1986. Used with permission.)

Automatic repeater stations, located between the remote stations and the base station, receive data from the remote stations, check the data for validity, and transmit the data to the base station.

Incoming radio signals are transformed from radio analog form to digital format and are forwarded to the base station computer through a communications

FIGURE 6.5.5
Self-reporting rain and water level gage on the Navidad River, Texas. (Courtesy of Sierra/Misco, Inc., 1986. Used with permission.)

port. After data quality checks are made, the data are formatted and filed on either hard or floppy disk media. Once the data filing is complete, the information can be displayed or saved for analysis.

The base station has data management software which can handle up to 700 sensors with enough on-line storage to store three years of rainfall data. It can cover 12 separate river systems with up to 25 forecast points possible in each; each forecast point can receive inflow from up to 10 different sources. Different future rainfall scenarios can be input for each individual forecast point, and optional features can be added to control pumps, gates, wall maps, remote alarms, and voice synthesized warnings (Sierra/Misco, Inc., 1986).

6.6 MEASUREMENT OF PHYSIOGRAPHIC CHARACTERISTICS

In hydrologic studies in which gaged data are sometimes not available, for example in rainfall-runoff analysis, runoff characteristics are estimated from physiographic characteristics. Watershed physiographic information can be obtained from maps describing land use, soils maps, geologic maps, topographic maps, and aerial photography. A typical inventory of physiographic characteristics is the following list of 22 factors compiled for the USGS-EPA National Urban Studies Program (Jennings, 1982):

1. Total drainage area in square miles (excluding noncontributing areas).
2. Impervious area as a percentage of drainage area.
3. Effective impervious area as a percentage of drainage area. Only impervious surfaces connected directly to a sewer pipe or other stormwater conveyance are included.
4. Average basin slope, in feet per mile, determined from an average of terrain slopes at 50 or more equispaced points using the best available topographic map.
5. Main conveyance slope, in feet per mile, measured at points 10 and 85 percent of the distance from the gaging station to the drainage divide along the main conveyance channel.
6. Permeability of the A horizon of the soil profile, in inches per hour.
7. Soil moisture capacity average over the A, B, and C soil horizons, in inches of water per inch of soil.
8. Soil water pH in the A horizon.
9. Hydrologic soil group (A, B, C, or D) according to the U.S. Soil Conservation Service methodology.
10. Population density in persons per square mile.
11. Street density, in lane miles per square mile (approximately 12-ft lanes).
12. Land use of the basins as a percentage of drainage area including: (a) rural and pasture, (b) agricultural, (c) low-density residential ($\frac{1}{2}$ to 2 acres per dwelling), (d) medium-density residential (3 to 8 dwellings per acre),

(e) high-density residential (9 or more dwellings per acre), (f) commercial, (g) industrial, (h) under construction (bare surface), (i) vacant land, (j) wetland, and (k) parkland.

13. Detention storage, in acre-feet of storage.
14. Percent of watershed upstream from detention storage.
15. Percent of area drained by a storm sewer system.
16. Percent of streets with ditch and gutter drainage.
17. Percent of streets with ditch and swale drainage.
18. Mean annual precipitation, in inches (long term).
19. Ten-year frequency, one-hour duration, rainfall intensity, in inches per hour (long term).
20. Mean annual loads of water quality constituents in runoff, in pounds per acre.
21. Mean annual loads of constituents in precipitation, in pounds per acre.
22. Mean annual loads of constituents in dry deposition, in pounds per acre.

These data are employed in modeling the water quantity and quality characteristics of urban watersheds so that the conclusions drawn from field studies can be extended to other locations.

REFERENCES

Bhowmik, N., Hydraulics of flow in the Kaskaskia River, Illinois, *Report of Investigation 91*, Illinois State Water Survey, Urbana, Ill., 1979.

EG&G Washington Analytical Services Center, Inc., *Lower Colorado River Authority Software User's Manual,* Albuquerque, N. Mex., December 1981.

Jennings, M. E., Data collection and instrumentation, in *Urban Stormwater Hydrology,* ed. by D. F. Kibler, Water Resources Monograph 7, American Geophysical Union, pp. 189–217, Washington, D.C., 1982.

Linsley, R. K., M. A. Kohler, and J. L. H. Paulhus, *Hydrology for Engineers,* McGraw-Hill, New York, 1982.

Marsh-McBirney, Inc., The UMFM flowmeter, Product brochure, Gaithersburg, Md., 1979.

Rantz, S. E., et al., Measurement and computation of streamflow, vol. 1, Measurement of stage and discharge, *Water Supply Paper 2175*, U. S. Geological Survey, 1982.

Riggs, H. C., *Streamflow Characteristics,* Elsevier, Amsterdam, Holland, 1985.

Shaw, E. M., *Hydrology in Practice,* Van Nostrand Reinhold (UK), Wokingham, England, 1983.

Sierra/Misco, Inc., Flood early warning system for city of Austin, Texas; Berkeley, Calif., 1986.

U. S. Geological Survey, *National Handbook of Recommended Methods for Water-data Acquisition,* Office of Water Data Coordination, U. S. Geological Survey, Reston, Va., 1977.

World Meteorological Organization, *Guide to Hydrological Practices,* vol. 1: *Data Acquisition and Processing,* Report no. 168, Geneva, Switzerland, 4th ed., 1981.

PROBLEMS

6.3.1 A discharge measurement made on the Colorado River at Austin, Texas, on June 11, 1981, yielded the following results. Calculate the discharge in ft^3/s.

Distance from bank (ft)	0	30	60	80	100	120	140	160
Depth (ft)	0	18.5	21.5	22.5	23.0	22.5	22.5	22.0
Velocity (ft/s)	0	0.55	1.70	3.00	3.06	2.91	3.20	3.36

Distance	180	200	220	240	260	280	300	320	340
Depth	22.0	23.0	22.0	22.5	23.0	22.8	21.5	19.2	18.0
Velocity	3.44	2.70	2.61	2.15	1.94	1.67	1.44	1.54	0.81

Distance	360	380	410	450	470	520	570	615
Depth	14.7	12.0	11.4	9.0	5.0	2.6	1.3	0
Velocity	1.10	1.52	1.02	0.60	0.40	0.33	0.29	0

6.3.2 Plot a graph of velocity vs. distance from the bank for the data given in Prob. 6.3.1. Plot a graph of velocity vs. depth of flow.

6.3.3 The observed gage height during a discharge measurement of the Colorado River at Austin is 11.25 ft. If the measured discharge was 9730 ft^3/s, calculate the percent difference between the discharge given by the rating curve (Fig. 6.3.8) and that obtained in this discharge measurement.

6.3.4 The bed slope of the Colorado River at Austin is 0.03 percent. Determine, for the data given in Example 6.3.1, what value of Manning's n would yield the observed discharge for the data shown.

6.3.5 A discharge measurement on the Colorado River at Austin, Texas, on June 16, 1981, yielded the following results. Calculate the discharge in ft^3/s.

Distance from bank (ft)	0	35	55	75	95	115	135	155
Depth (ft)	0	18.0	19.0	21.0	20.5	18.5	18.2	19.5
Velocity (ft/s)	0	0.60	2.00	3.22	3.64	3.74	4.42	3.49

Distance	175	195	215	235	255	275	295
Depth	20.0	21.5	21.5	21.5	22.0	21.5	20.5
Velocity	5.02	4.75	4.92	4.44	3.94	2.93	2.80

Distance	325	355	385	425	465	525	575
Depth	17.0	13.5	10.6	9.0	6.1	2.0	0
Velocity	2.80	1.52	1.72	0.95	0.50	0.39	0

6.3.6 If the bed slope is 0.0003, determine the value of Manning's n that would yield the same discharge as the value you found in Problem 6.3.5.

6.3.7 The observed gage height for the discharge measurement in Prob. 6.3.5 was 19.70 ft above datum. The rating curve at this site is shown in Fig. 6.3.8. Calculate the percent difference between the discharge found from the rating curve for this gage height and the value found in Prob. 6.3.5.

CHAPTER 7

UNIT HYDROGRAPH

In the previous chapters of this book, the physical laws governing the operation of hydrologic systems have been described and working equations developed to determine the flow in atmospheric, subsurface, and surface water systems. The Reynolds transport theorem applied to a control volume provided the mathematical means for consistently expressing the various applicable physical laws. It may be remembered that the control volume principle does not call for a description of the internal dynamics of flow within the control volume; all that is required is knowledge of the inputs and outputs to the control volume and the physical laws regulating their interaction.

In Chap. 1, a tree classification was presented (Fig. 1.4.1), distinguishing the various types of models of hydrologic systems according to the way each deals with the randomness and the space and time variability of the hydrologic processes involved. Up to this point in the book, most of the working equations developed have been for the simplest type of model shown in this diagram, namely a deterministic (no randomness) lumped (one point in space) steady-flow model (flow does not change with time). This chapter takes up the subject of deterministic lumped unsteady flow models; subsequent chapters (8–12) cover a range of models in the classification tree from left to right. Where possible, use is made of knowledge of the governing physical laws of the system. In addition to this, methods drawn from other fields of study such as linear systems analysis, optimization, and applied statistics are employed to analyze the input and output variables of hydrologic systems.

In the development of these models, the concept of control volume remains as it was introduced in Chap. 1: "A volume or structure in space, surrounded by a boundary, which accepts water and other inputs, operates on them internally and produces them as outputs." In this chapter, the interaction between rainfall and runoff on a watershed is analyzed by viewing the watershed as a lumped linear system.

7.1 GENERAL HYDROLOGIC SYSTEM MODEL

The amount of water stored in a hydrologic system, S may be related to the rates of inflow I and outflow Q by the integral equation of continuity (2.2.4):

$$\frac{dS}{dt} = I - Q \tag{7.1.1}$$

Imagine that the water is stored in a hydrologic system, such as a reservoir (Fig. 7.1.1), in which the amount of storage rises and falls with time in response to I and Q and their rates of change with respect to time: dI/dt, $d^2I/dt^2, \ldots,$ $dQ/dt, d^2Q/dt^2, \ldots$. Thus, the amount of storage at any time can be expressed by a *storage function* as:

$$S = f\left(I, \frac{dI}{dt}, \frac{d^2I}{dt^2}, \ldots, Q, \frac{dQ}{dt}, \frac{d^2Q}{dt^2}, \ldots\right) \tag{7.1.2}$$

The function f is determined by the nature of the hydrologic system being examined. For example, the linear reservoir introduced in Chap. 5 as a model for baseflow in streams relates storage and outflow by $S = kQ$, where k is a constant.

The continuity equation (7.1.1) and the storage function equation (7.1.2) must be solved simultaneously so that the output Q can be calculated given the input I, where I and Q are both functions of time. This can be done in two ways: by differentiating the storage function and substituting the result for dS/dt in (7.1.1), then solving the resulting differential equation in I and Q by integration; or by applying the finite difference method directly to Eqs. (7.1.1) and (7.1.2) to solve them recursively at discrete points in time. In this chapter, the first, or integral, approach is taken, and in Chap. 8, the second, or differential, approach is adopted.

Linear System in Continuous Time

For the storage function to describe a *linear system*, it must be expressed as a linear equation with constant coefficients. Equation (7.1.2) can be written

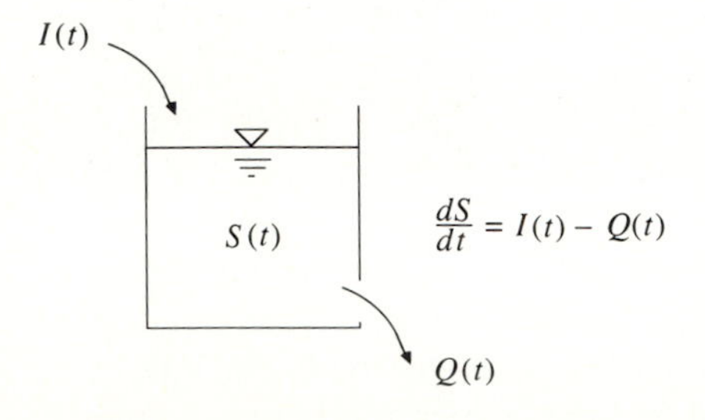

FIGURE 7.1.1
Continuity of water stored in a hydrologic system.

$$S = a_1 Q + a_2 \frac{dQ}{dt} + a_3 \frac{d^2Q}{dt^2} + \ldots + a_n \frac{d^{n-1}Q}{dt^{n-1}}$$
$$+ b_1 I + b_2 \frac{dI}{dt} + b_3 \frac{d^2I}{dt^2} + \ldots + b_m \frac{d^{m-1}I}{dt^{m-1}} \tag{7.1.3}$$

in which $a_1, a_2, \ldots, a_n, b_1, b_2, \ldots, b_m$ are constants and derivatives of higher order than those shown are neglected. The constant coefficients also make the system *time-invariant* so that the way the system processes input into output does not change with time.

Differentiating (7.1.3), substituting the result for dS/dt in (7.1.1), and rearranging yields

$$a_n \frac{d^nQ}{dt^n} + a_{n-1} \frac{d^{n-1}Q}{dt^{n-1}} + \ldots + a_2 \frac{d^2Q}{dt^2} + a_1 \frac{dQ}{dt} + Q =$$
$$I - b_1 \frac{dI}{dt} - b_2 \frac{d^2I}{dt^2} - \ldots - b_{m-1} \frac{d^{m-1}I}{dt^{m-1}} - b_m \frac{d^mI}{dt^m} \tag{7.1.4}$$

which may be rewritten in the more compact form

$$N(D)Q = M(D)I \tag{7.1.5}$$

where $D = d/dt$ and $N(D)$ and $M(D)$ are the differential operators

$$N(D) = a_n \frac{d^n}{dt^n} + a_{n-1} \frac{d^{n-1}}{dt^{n-1}} + \ldots + a_1 \frac{d}{dt} + 1$$

and

$$M(D) = -b_m \frac{d^m}{dt^m} - b_{m-1} \frac{d^{m-1}}{dt^{m-1}} - \ldots - b_1 \frac{d}{dt} + 1$$

Solving (7.1.5) for Q yields

$$Q(t) = \frac{M(D)}{N(D)} I(t) \tag{7.1.6}$$

The function $M(D)/N(D)$ is called the *transfer function* of the system; it describes the response of the output to a given input sequence.

Equation (7.1.4) was presented by Chow and Kulandaiswamy (1971) as a general hydrologic system model. It describes a lumped system because it contains derivatives with respect to time alone and not spatial dimensions. Chow and Kulandaiswamy showed that many of the previously proposed models of lumped hydrologic systems were special cases of this general model. For example, for a linear reservoir, the storage function (7.1.3) has $a_1 = k$ and all other coefficients zero, so (7.1.4) becomes

$$k \frac{dQ}{dt} + Q = I \tag{7.1.7}$$

7.2 RESPONSE FUNCTIONS OF LINEAR SYSTEMS

The solution of (7.1.6) for the transfer function of hydrologic systems follows two basic principles for linear system operations which are derived from methods for solving linear differential equations with constant coefficients (Kreyszig, 1968):

1. If a solution $f(Q)$ is multiplied by a constant c, the resulting function $cf(Q)$ is also a solution (*principle of proportionality*).
2. If two solutions $f_1(Q)$ and $f_2(Q)$ of the equation are added, the resulting function $f_1(Q) + f_2(Q)$ is also a solution of the equation (*principle of additivity* or *superposition*).

The particular solution adopted depends on the *input function* $N(D)I$, and on the specified *initial conditions* or values of the output variables at $t = 0$.

Impulse Response Function

The response of a linear system is uniquely characterized by its *impulse response function*. If a system receives an input of unit amount applied instantaneously (a unit impulse) at time τ, the response of the system at a later time t is described by the unit impulse response function $u(t-\tau)$; $t-\tau$ is the time lag since the impulse was applied [Fig. 7.2.1(*a*)]. The response of a guitar string when it is plucked is one example of a response to an impulse; another is the response of the shock absorber in a car after the wheel passes over a pothole. If the storage reservoir in Fig. 7.1.1 is initially empty, and then the reservoir is instantaneously filled with a unit amount of water, the resulting outflow function $Q(t)$ is the impulse response function.

Following the two principles of linear system operation cited above, if two impulses are applied, one of 3 units at time τ_1 and the other of 2 units at time τ_2, the response of the system will be $3u(t-\tau_1) + 2u(t-\tau_2)$, as shown in Fig. 7.2.1(*b*). Analogously, continuous input can be treated as a sum of infinitesimal impulses. The amount of input entering the system between times τ and $\tau + d\tau$ is $I(\tau)\,d\tau$. For example, if $I(\tau)$ is the precipitation intensity in inches per hour and $d\tau$ is an infinitesimal time interval measured in hours, then $I(\tau)\,d\tau$ is the depth in inches of precipitation input to the system during this interval. The direct runoff $t-\tau$ time units later resulting from this input is $I(\tau)u(t-\tau)d\tau$. The response to the complete input time function $I(\tau)$ can then be found by integrating the response to its constituent impulses:

$$Q(t) = \int_0^t I(\tau)u(t-\tau)\,d\tau \qquad (7.2.1)$$

This expression, called the *convolution integral*, is the fundamental equation for solution of a linear system on a continuous time scale. Figure 7.2.2 illustrates the response summation process for the convolution integral.

For most hydrologic applications, solutions are needed at discrete intervals of time, because the input is specified as a discrete time function, such as an

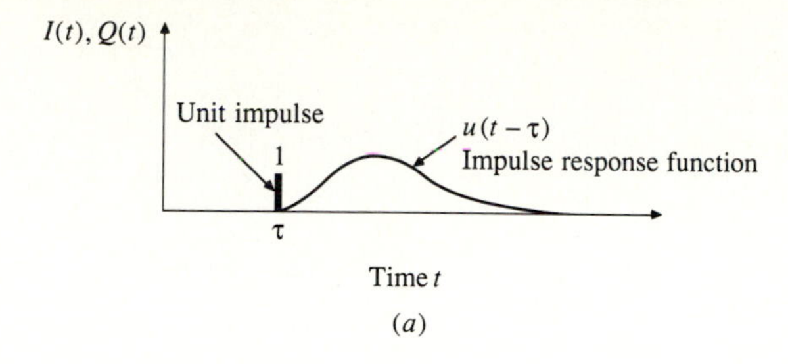

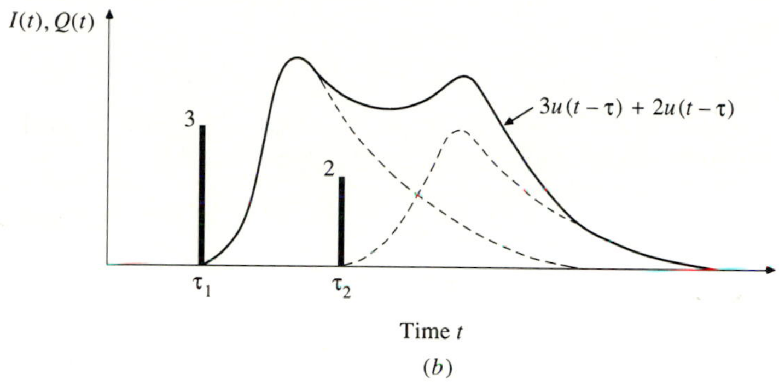

FIGURE 7.2.1
Responses of a linear system to impulse inputs. (*a*) Unit impulse response function. (*b*) The response to two impulses is found by summing the individual response functions.

excess rainfall hyetograph. To handle such input, two further functions are needed, the unit step response function and the unit pulse response function, as shown in Fig. 7.2.3.

Step Response Function

A *unit step input* is an input that goes from a rate of 0 to 1 at time 0 and continues indefinitely at that rate thereafter [Fig. 7.2.3(*b*)]. The output of the system, its *unit step response function* $g(t)$ is found from (7.2.1) with $I(\tau) = 1$ for $\tau \geq 0$, as

$$Q(t) = g(t) = \int_0^t u(t - \tau)\, d\tau \tag{7.2.2}$$

If the substitution $l = t - \tau$ is made in (7.2.2) then $d\tau = -dl$, the limit $\tau = t$ becomes $l = t - t = 0$, and the limit $\tau = 0$ becomes $l = t - 0 = t$. Hence,

$$g(t) = -\int_t^0 u(l)\, dl$$

or

$$g(t) = \int_0^t u(l)\, dl \tag{7.2.3}$$

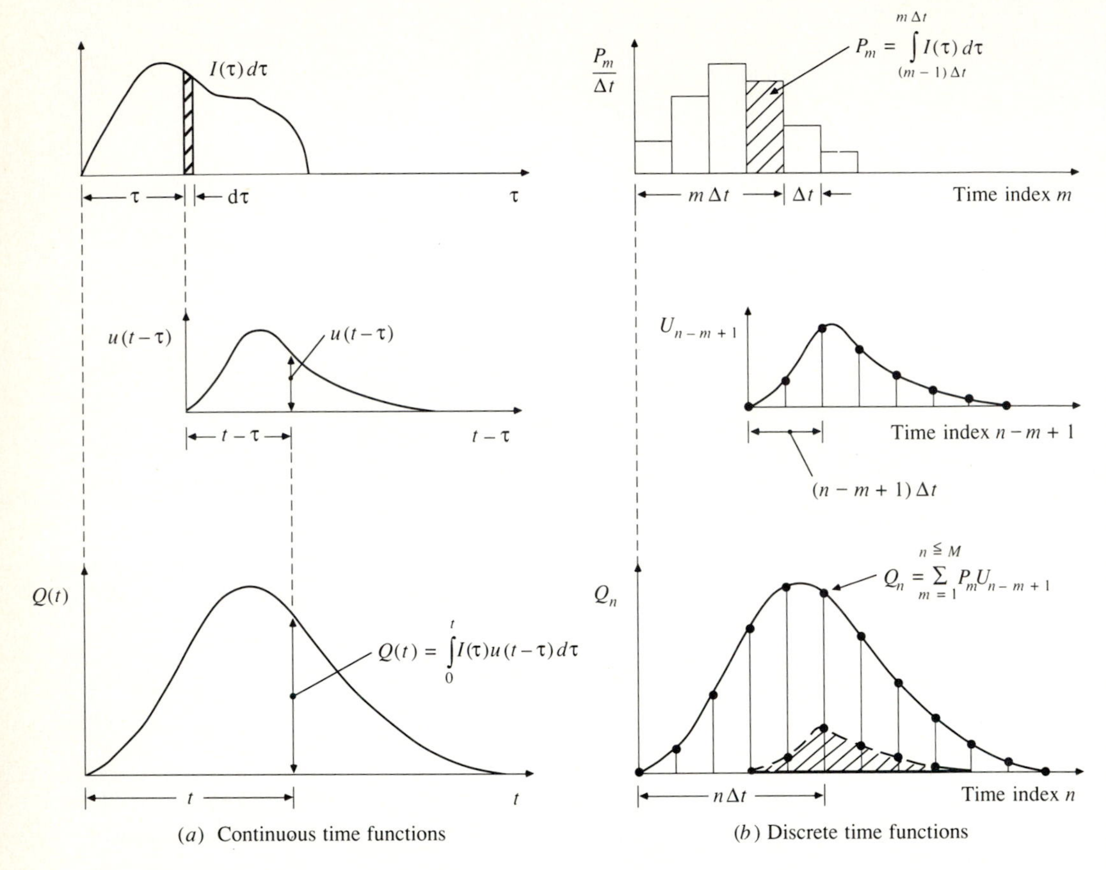

(*a*) Continuous time functions (*b*) Discrete time functions

FIGURE 7.2.2
The relationship between continuous and discrete convolution.

In words, the value of the unit step response function $g(t)$ at time t equals the integral of the impulse response function up to that time, as shown in Fig. 7.2.3(*a*) and (*b*).

Pulse Response Function

A *unit pulse input* is an input of unit amount occurring in duration Δt. The rate is $I(\tau) = 1/\Delta t$, $0 \leq \tau \leq \Delta t$, and zero elsewhere. The *unit pulse response function* produced by this input can be found by the two linear system principles cited earlier. First, by the principle of proportionality, the response to a unit step input of rate $1/\Delta t$ beginning at time 0 is $(1/\Delta t)g(t)$. If a similar unit step input began at time Δt instead of at 0, its response function would be lagged by time interval Δt, and would have a value at time t equal to $(1/\Delta t)g(t - \Delta t)$. Then, using the principle of superposition, the response to a unit pulse input duration Δt is found by subtracting the response to a step input of rate $1/\Delta t$ beginning at time Δt from

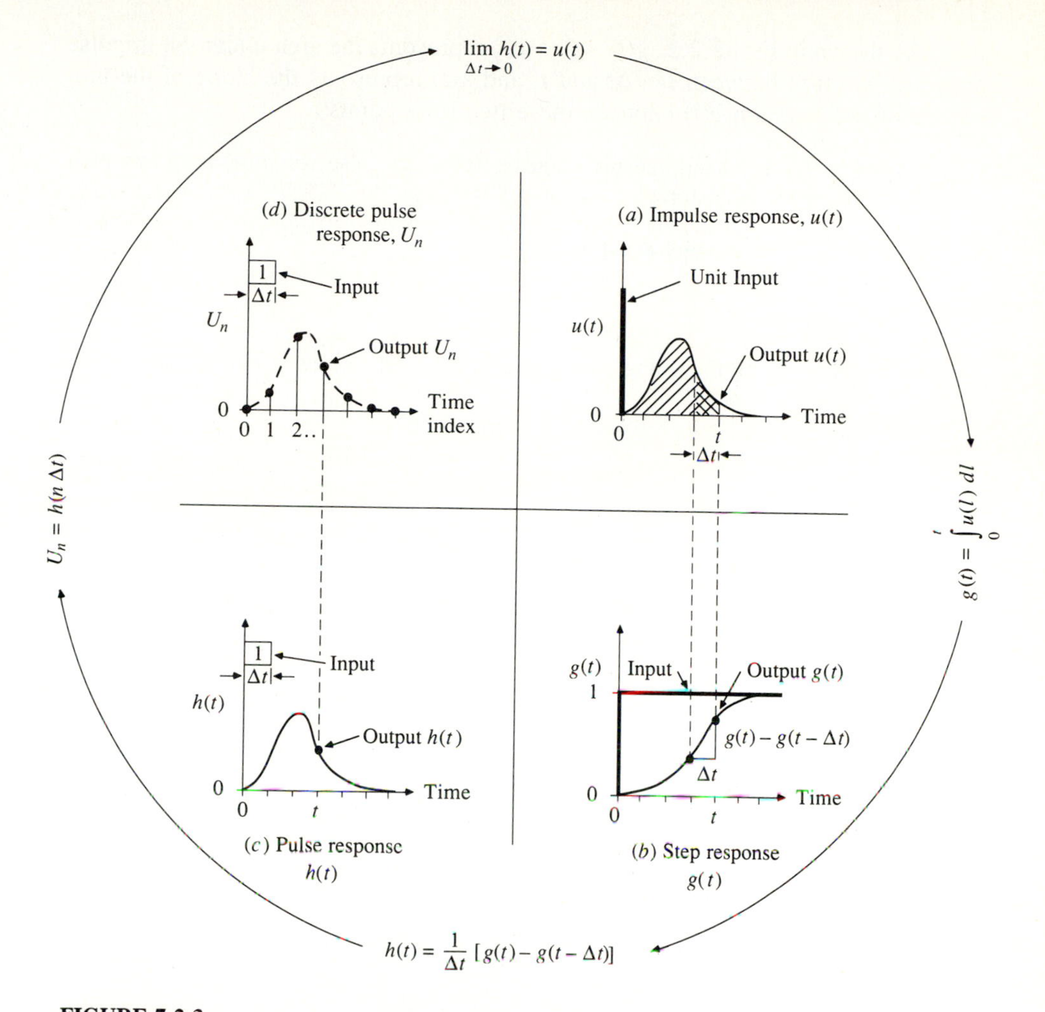

FIGURE 7.2.3
Response functions of a linear system. The response functions in (a), (b), and (c) are on a continuous time domain and that in (d) on a discrete time domain.

the response to a step input of the same rate beginning at time 0, so that the *unit pulse response function* $h(t)$ is

$$h(t) = \frac{1}{\Delta t}[g(t) - g(t - \Delta t)] \tag{7.2.4}$$

$$= \frac{1}{\Delta t}\left[\int_0^t u(l)\,dl - \int_0^{t-\Delta t} u(l)\,dl\right]$$

$$= \frac{1}{\Delta t}\int_{t-\Delta t}^{t} u(l)\,dl \tag{7.2.5}$$

As shown in Fig. 7.2.3, $g(t)-g(t-\Delta t)$ represents the area under the impulse response function between $t - \Delta t$ and t, and $h(t)$ represents the slope of the unit step response function $g(t)$ between these two time points.

Example 7.2.1. Determine the impulse, step and pulse response functions of a linear reservoir with storage constant $k\,(S = kQ)$.

Solution. The continuity equation (7.1.1) is

$$\frac{dS}{dt} = I(t) - Q(t)$$

and differentiating the storage function $S = kQ$ yields $dS/dt = k\,dQ/dt$, so

$$k\frac{dQ}{dt} = I(t) - Q(t)$$

or

$$\frac{dQ}{dt} + \frac{1}{k}Q(t) = \frac{1}{k}I(t)$$

This is a first-order linear differential equation, and can be solved by multiplying both sides of the equation by the *integrating factor* $e^{t/k}$:

$$e^{t/k}\frac{dQ}{dt} + \frac{1}{k}e^{t/k}Q(t) = \frac{1}{k}e^{t/k}I(t)$$

so that the two terms on the left-hand side of the equation can be combined as

$$\frac{d}{dt}(Qe^{t/k}) = \frac{1}{k}e^{t/k}I(t)$$

Integrating from the initial conditions $Q = Q_o$ at $t = 0$

$$\int_{Q_o,0}^{Q(t),t} d(Qe^{t/k}) = \int_0^t \frac{1}{k}e^{\tau/k}I(\tau)\,d\tau$$

where τ is a dummy variable of time in the integration. Solving,

$$Q(t)e^{t/k} - Q_o = \int_0^t \frac{1}{k}e^{\tau/k}I(\tau)\,d\tau$$

and rearranging,

$$Q(t) = Q_o e^{-t/k} + \int_0^t \frac{1}{k}e^{-(t-\tau)/k}I(\tau)\,d\tau$$

Comparing this equation with the convolution integral (7.2.1), it can be seen that the two equations are the same provided $Q_o = 0$ and

$$u(t-\tau) = \frac{1}{k}e^{-(t-\tau)/k}$$

So if l is defined as the lag time $t - \tau$, the impulse response function of a linear reservoir is

$$u(l) = \frac{1}{k}e^{-l/k}$$

The requirement that $Q_o = 0$ implies that the system starts from rest when the convolution integral is applied.

The unit step response is given by (7.2.3):

$$
\begin{aligned}
g(t) &= \int_o^t u(l)\,dl \\
&= \int_o^t \frac{1}{k} e^{-l/k} dl \\
&= [-e^{-l/k}]_o^t \\
&= 1 - e^{-t/k}
\end{aligned}
$$

The unit pulse response is given by (7.2.4):

$$h(t) = \frac{1}{\Delta t}[g(t) - g(t - \Delta t)]$$

1. For $0 \leq t \leq \Delta t$, $g(t - \Delta t) = 0$, so

$$h(t) = \frac{1}{\Delta t} g(t) = \frac{1}{\Delta t}(1 - e^{-t/k})$$

2. For $t > \Delta t$,

$$
\begin{aligned}
h(t) &= \frac{1}{\Delta t}[1 - e^{-t/k} - (1 - e^{-(t-\Delta t)/k})] \\
&= \frac{e^{-t/k}}{\Delta t}(e^{\Delta t/k} - 1)
\end{aligned}
$$

The impulse and step response functions of a linear reservoir with $k = 3$ h are plotted in Fig. 7.2.4, along with the pulse response function for $\Delta t = 2$ h.

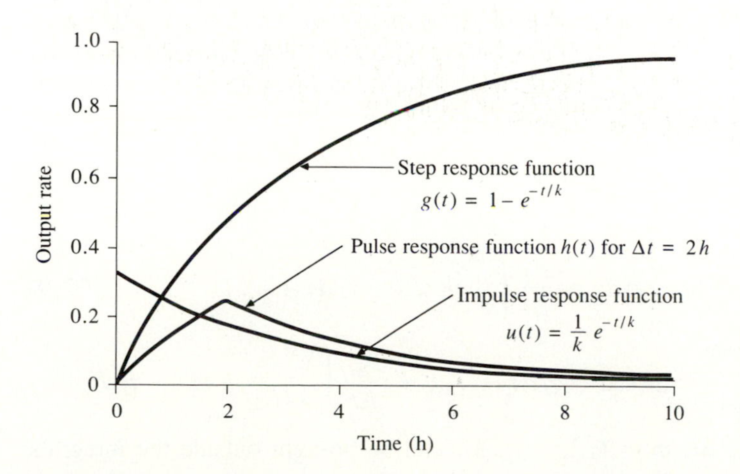

FIGURE 7.2.4
Response function of a linear reservoir with $k = 3$ h. Pulse response function is for a pulse input of two hours duration. (from Example 7.2.1.)

Linear System in Discrete Time

The impulse, step, and pulse response functions have all been defined on a continuous time domain. Now let the time domain be broken into discrete intervals of duration Δt. As shown in Sec. 2.3, there are two ways to represent a continuous time function on a discrete time domain, as a *pulse data system* or as a *sample data system*. The pulse data system is used for precipitation and the value of its discrete input function for the mth time interval is

$$P_m = \int_{(m-1)\Delta t}^{m\Delta t} I(\tau)\,dt \qquad m = 1, 2, 3, \ldots \tag{7.2.6}$$

P_m is the depth of precipitation falling during the time interval (in inches or centimeters). The sample data system is used for streamflow and direct runoff, so that the value of the system output in the nth time interval ($t = n\,\Delta t$) is

$$Q_n = Q(n\Delta t) \qquad n = 1, 2, 3, \ldots \tag{7.2.7}$$

Q_n is the instantaneous value of the flow rate at the end of the nth time interval (in cfs or m^3/s). Thus the input and output variables to a watershed system are recorded with different dimensions and using different discrete data representations. The effect of an input pulse of duration Δt beginning at time $(m-1)\Delta t$ on the output at time $t = n\Delta t$ is measured by the value of the unit pulse response function $h[t-(m-1)\Delta t] = h[n\Delta t - (m-1)\Delta t] = h[(n-m+1)\Delta t]$, given, following Eq. (7.2.5), as

$$h[(n-m+1)\Delta t] = \frac{1}{\Delta t}\int_{(n-m)\Delta t}^{(n-m+1)\Delta t} u(l)\,dl \tag{7.2.8}$$

On a discrete time domain, the input function is a series of M pulses of constant rate: for pulse m, $I(\tau) = P_m/\Delta t$ for $(m-1)\,\Delta t \le \tau \le m\,\Delta t$. $I(\tau) = 0$ for $\tau > M\Delta t$. Consider the case where the output is being calculated after all the input has ceased, that is, at $t = n\Delta t > M\,\Delta t$ [see Fig. 7.2.2(*b*)]. The contribution to the output of each of the M input pulses can be found by breaking the convolution integral (7.2.1) at $t = n\Delta t$ into M parts:

$$\begin{aligned} Q_n &= \int_0^{n\Delta t} I(\tau)u(n\Delta t - \tau)d\tau \\ &= \frac{P_1}{\Delta t}\int_0^{\Delta t} u(n\Delta t - \tau)\,d\tau + \frac{P_2}{\Delta t}\int_{\Delta t}^{2\Delta t} u(n\Delta t - \tau)\,d\tau + \ldots \\ &\quad + \frac{P_m}{\Delta t}\int_{(m-1)\Delta t}^{m\Delta t} u(n\Delta t - \tau)\,d\tau + \ldots + \frac{P_M}{\Delta t}\int_{(M-1)\Delta t}^{M\Delta t} u(n\Delta t - \tau)\,d\tau \end{aligned} \tag{7.2.9}$$

where the terms $P_m/\Delta t$, $m = 1, 2, \ldots, M$, can be brought outside the integrals because they are constants.

In each of these integrals, the substitution $l = n\Delta t - \tau$ is made, so $d\tau = -dl$, the limit $\tau = (m-1)\,\Delta t$ becomes $l = n\Delta t - (m-1)\,\Delta t = (n-m+1)\Delta t$, and the limit $\tau = m\Delta t$ becomes $l = (n-m)\Delta t$. The mth integral in (7.2.9) is now written

$$\frac{P_m}{\Delta t}\int_{(m-1)\Delta t}^{m\Delta t} u(n\Delta t - \tau)\, d\tau = \frac{P_m}{\Delta t}\int_{(n-m+1)\Delta t}^{(n-m)\Delta t} -u(l)\, dl$$

$$= \frac{P_m}{\Delta t}\int_{(n-m)\Delta t}^{(n-m+1)\Delta t} u(l)\, dl \qquad (7.2.10)$$

$$= P_m h[(n-m+1)\Delta t]$$

by substitution from (7.2.7). After making these substitutions for each term in (7.2.9),

$$\begin{aligned} Q_n &= P_1 h[(n\Delta t)] + P_2 h[(n-1)\Delta t] + \dots \\ &\quad + P_m h[(n-m+1)\Delta t] + \dots \\ &\quad + P_M h[(n-M+1)\Delta t] \end{aligned} \qquad (7.2.11)$$

which is a convolution equation with input P_m in pulses and output Q_n as a sample data function of time.

Discrete Pulse Response Function

As shown in Fig. 7.2.3(*d*), the continuous pulse response function $h(t)$ may be represented on a discrete time domain as a sample data function U where

$$U_{n-m+1} = h[(n-m+1)\Delta t] \qquad (7.2.12)$$

It follows that $U_n = h[n\Delta t]$, $U_{n-1} = h[(n-1)\Delta t], \dots,$ and $U_{n-M+1} = h[(n-M+1)\Delta t]$. Substituting into (7.2.11), the discrete-time version of the convolution integral is

$$\begin{aligned} Q_n &= P_1 U_n + P_2 U_{n-1} + \dots + P_m U_{n-m+1} + \dots + P_M U_{n-M+1} \\ &= \sum_{m=1}^{M} P_m U_{n-m+1} \end{aligned} \qquad (7.2.13)$$

Equation (7.2.13) is valid provided $n \geq M$; if $n < M$, then, in (7.2.9), one would only need to account for the first n pulses of input, since these are the only pulses that can influence the output up to time $n\,\Delta t$. In this case, (7.2.13) is rewritten

$$Q_n = \sum_{m=1}^{n} P_m U_{n-m+1} \qquad (7.2.14)$$

Combining (7.2.13) and (7.2.14) gives the final result

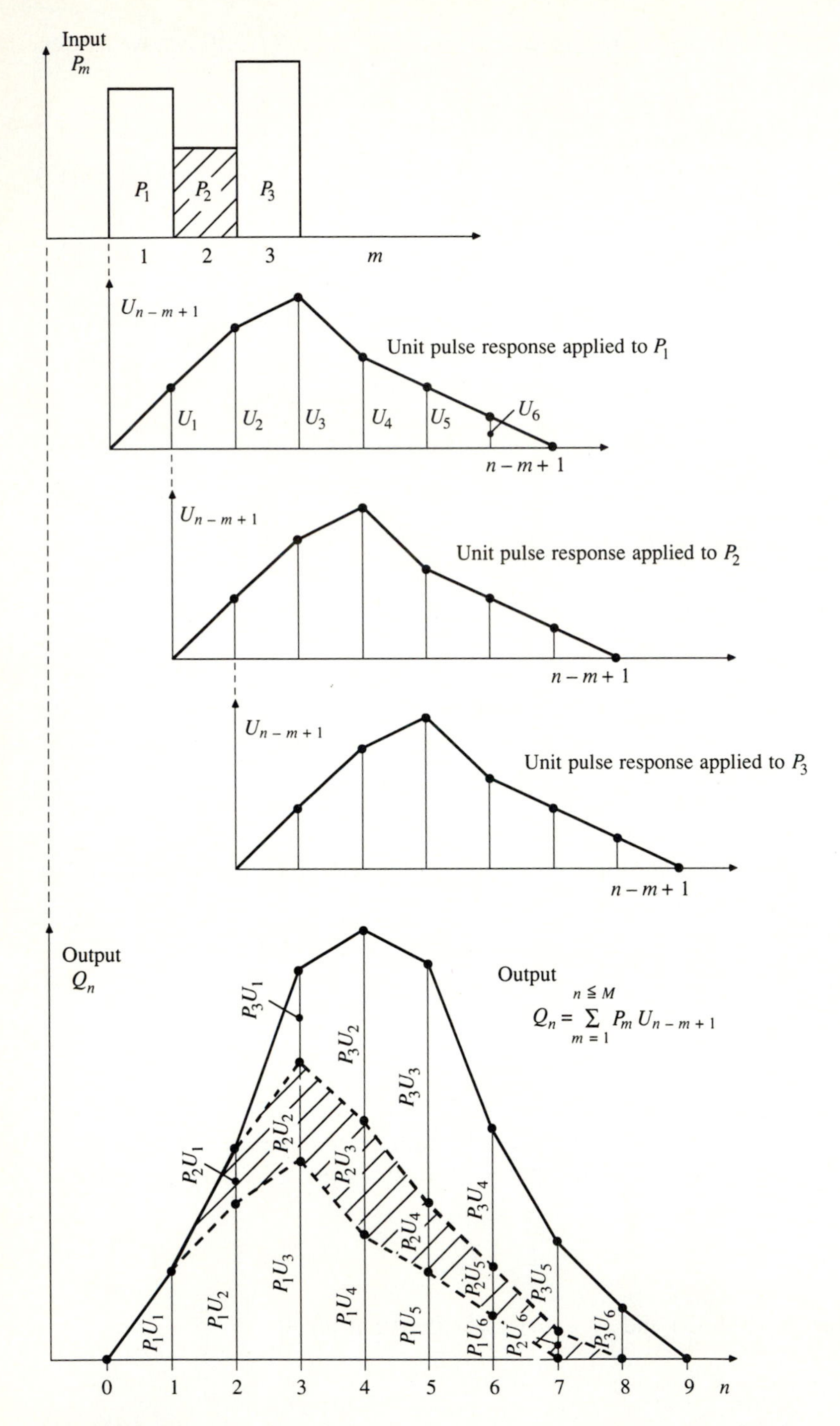

FIGURE 7.2.5
Application of the discrete convolution equation to the output from a linear system.

$$Q_n = \sum_{m=1}^{n \le M} P_m U_{n-m+1} \tag{7.2.15}$$

which is the *discrete convolution equation* for a linear system. The notation $n \le M$ as the upper limit of the summation shows that the terms are summed for $m = 1, 2, \ldots, n$ for $n \le M$, but for $n > M$, the summation is limited to $m = 1, 2, \ldots, M$.

As an example, suppose there are $M = 3$ pulses of input: P_1, P_2, and P_3. For the first time interval ($n = 1$), there is only one term in the convolution, that for $m = 1$;

$$Q_1 = P_1 U_{1-1+1} = P_1 U_1$$

For $n = 2$, there are two terms, corresponding to $m = 1, 2$:

$$Q_2 = P_1 U_{2-1+1} + P_2 U_{2-2+1} = P_1 U_2 + P_2 U_1$$

For $n = 3$, there are three terms:

$$Q_3 = P_1 U_{3-1+1} + P_2 U_{3-2+1} + P_3 U_{3-3+1} = P_1 U_3 + P_2 U_2 + P_3 U_1$$

And for $n = 4, 5, \ldots$ there continue to be just three terms:

$$Q_n = P_1 U_n + P_2 U_{n-1} + P_3 U_{n-2}$$

The results of the calculation are shown diagramatically in Fig. 7.2.5. The sum of the subscripts in each term on the right-hand side of the summation is always one greater than the subscript of Q.

In the example shown in the diagram, there are 3 input pulses and 6 non-zero terms in the pulse response function U, so there are $3 + 6 - 1 = 8$ non-zero terms in the output function Q. The values of the output for the final three periods are:

$$Q_6 = P_1 U_6 + P_2 U_5 + P_3 U_4$$

$$Q_7 = P_2 U_6 + P_3 U_5$$

$$Q_8 = P_3 U_6$$

Q_n and P_m are expressed in different dimensions, and U has dimensions that are the ratio of the dimensions of Q_n and P_m to make (7.2.15) dimensionally consistent. For example, if P_m is measured in inches and Q_n in cfs, then the dimensions of U are cfs/in, which may be interpreted as cfs of output per inch of input.

7.3 THE UNIT HYDROGRAPH

The unit hydrograph is the unit pulse response function of a linear hydrologic system. First proposed by Sherman (1932), the unit hydrograph (originally named *unit-graph*) of a watershed is defined as a direct runoff hydrograph (DRH) resulting from 1 in (usually taken as 1 cm in SI units) of excess rainfall generated uniformly over the drainage area at a constant rate for an effective duration. Sherman originally used the word "unit" to denote a unit of time, but since that time it has often been interpreted as a unit depth of excess rainfall. Sherman classified runoff into surface runoff and groundwater runoff and defined the unit hydrograph for use only with surface runoff. Methods of calculating excess rainfall and direct runoff from observed rainfall and streamflow data are presented in Chap. 5.

The unit hydrograph is a simple linear model that can be used to derive the hydrograph resulting from any amount of excess rainfall. The following basic assumptions are inherent in this model:

1. The excess rainfall has a constant intensity within the effective duration.
2. The excess rainfall is uniformly distributed throughout the whole drainage area.
3. The base time of the DRH (the duration of direct runoff) resulting from an excess rainfall of given duration is constant.
4. The ordinates of all DRH's of a common base time are directly proportional to the total amount of direct runoff represented by each hydrograph.
5. For a given watershed, the hydrograph resulting from a given excess rainfall reflects the unchanging characteristics of the watershed.

Under natural conditions, the above assumptions cannot be perfectly satisfied. However, when the hydrologic data to be used are carefully selected so that they come close to meeting the above assumptions, the results obtained by the unit hydrograph model are generally acceptable for practical purposes (Heerdegen, 1974). Although the model was originally devised for large watersheds, it has been found applicable to small watersheds from less than 0.5 hectares to 25 km^2 (about 1 acre to 10 mi^2). Some cases do not support the use of the model because one or more of the assumptions are not well satisfied. For such reasons, the model is considered inapplicable to runoff originating from snow or ice.

Concerning assumption (1), the storms selected for analysis should be of short duration, since these will most likely produce an intense and nearly constant excess rainfall rate, yielding a well-defined single-peaked hydrograph of short time base.

Concerning assumption (2), the unit hydrograph may become inapplicable when the drainage area is too large to be covered by a nearly uniform distribution of rainfall. In such cases, the area has to be divided and each subarea analyzed for storms covering the whole subarea.

Concerning assumption (3), the base time of the direct runoff hydrograph (DRH) is generally uncertain but depends on the method of baseflow separation (see Sec. 5.2). The base time is usually short if the direct runoff is considered to include the surface runoff only; it is long if the direct runoff also includes subsurface runoff.

Concerning assumption (4), the principles of superposition and proportionality are assumed so that the ordinates Q_n of the DRH may be computed by Eq.

TABLE 7.3.1
Comparison of linear system and unit hydrograph concepts

Linear system	Unit hydrograph
1. 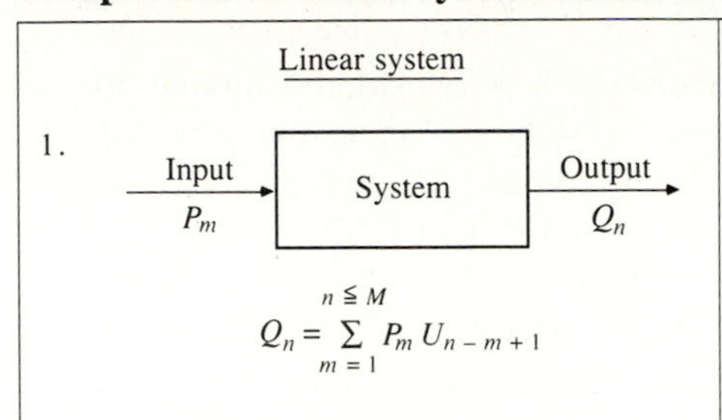 $Q_n = \sum_{m=1}^{n \leq M} P_m U_{n-m+1}$	1. 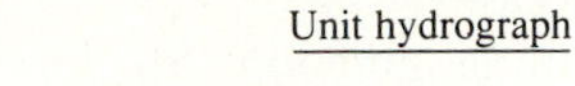 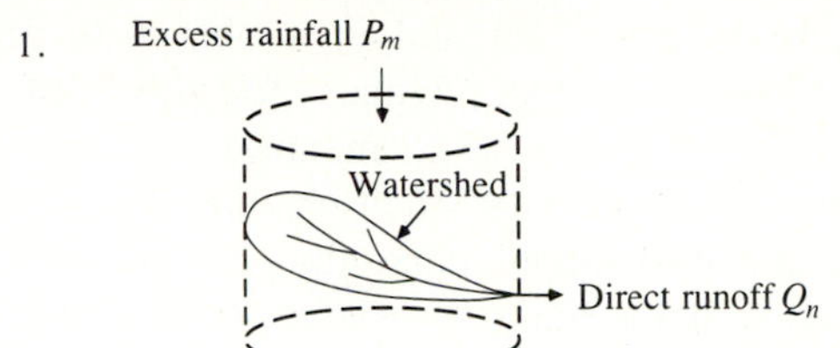
2. 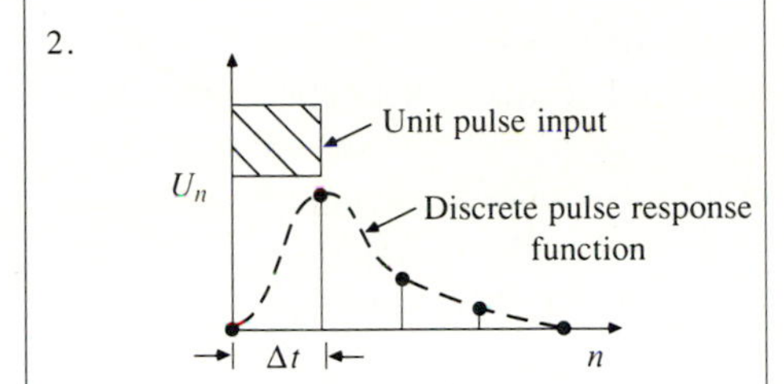	2. 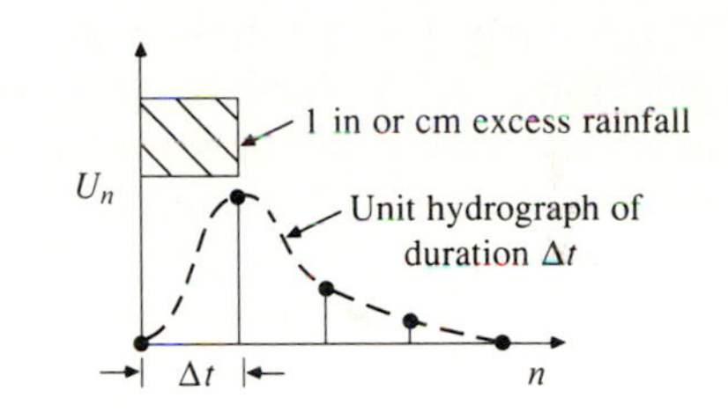
3. 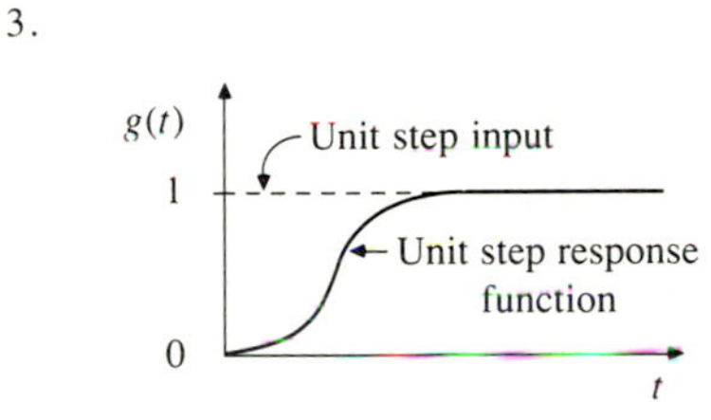	3. 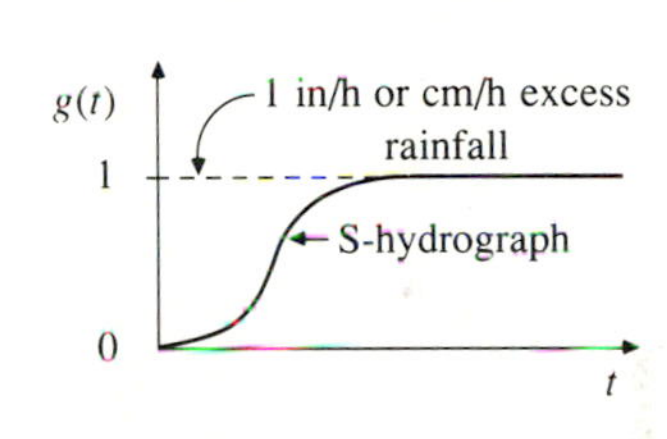
4. 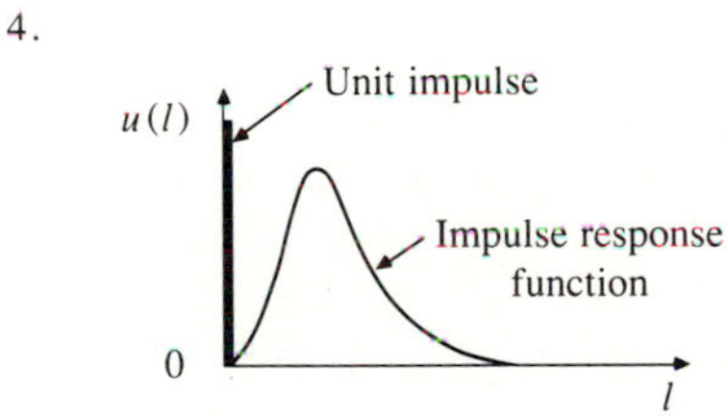	4. 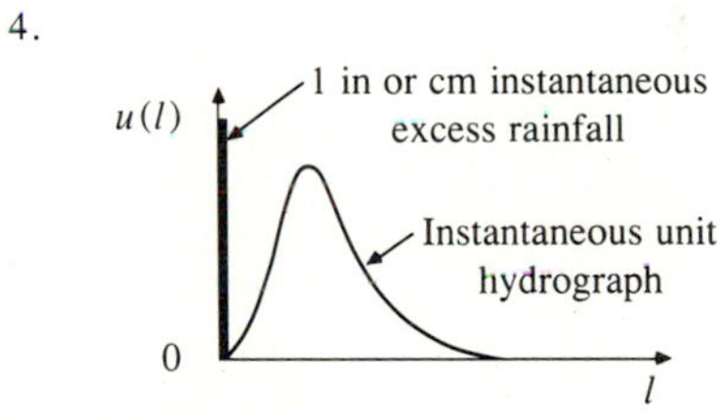
5. System starts from rest.	5. Direct runoff hydrograph starts from zero. All previous rainfall is absorbed by watershed (initial abstraction or loss).
6. System is linear.	6. Direct runoff hydrograph is calculated using principles of proportionality and superposition.
7. Transfer function has constant coefficients.	7. Watershed response is time invariant, not changing from one storm to another.
8. System obeys continuity. $\frac{dS}{dt} = I(t) - Q(t)$	8. Total depths of excess rainfall and direct runoff are equal. $\sum_n Q_n = \sum_m P_m$

(7.2.15). Actual hydrologic data are not truly linear; when applying (7.2.15) to them, the resulting hydrograph is only an approximation, which is satisfactory in many practical cases.

Concerning assumption (5), the unit hydrograph is considered unique for a given watershed and invariable with respect to time. This is the *principle of time invariance*, which, together with the principles of superposition and proportionality, is fundamental to the unit hydrograph model. Unit hydrographs are applicable only when channel conditions remain unchanged and watersheds do not have appreciable storage. This condition is violated when the drainage area contains many reservoirs, or when the flood overflows into the flood plain, thereby producing considerable storage.

The principles of linear system analysis form the basis of the unit hydrograph method. Table 7.3.1 shows a comparison of linear system concepts with the corresponding unit hydrograph concepts. In hydrology, the step response function is commonly called the *S-hydrograph*, and the impulse response function is called the *instantaneous unit hydrograph* which is the hypothetical response to a unit depth of excess rainfall deposited instantaneously on the watershed surface.

7.4 UNIT HYDROGRAPH DERIVATION

The discrete convolution equation (7.2.15) allows the computation of direct runoff Q_n given excess rainfall P_m and the unit hydrograph U_{n-m+1}

$$Q_n = \sum_{m=1}^{n \le M} P_m U_{n-m+1} \tag{7.4.1}$$

The reverse process, called *deconvolution*, is needed to derive a unit hydrograph given data on P_m and Q_n. Suppose that there are M pulses of excess rainfall and N pulses of direct runoff in the storm considered; then N equations can be written for $Q_n, n = 1, 2, \ldots, N$, in terms of $N - M + 1$ unknown values of the unit hydrograph, as shown in Table 7.4.1.

If Q_n and P_m are given and U_{n-m+1} is required, the set of equations in Table 7.4.1 is *overdetermined*, because there are more equations (N) than unknowns ($N - M + 1$).

Example 7.4.1. Find the half-hour unit hydrograph using the excess rainfall hyetograph and direct runoff hydrograph given in Table 7.4.2. (these were derived in Example 5.3.1.)

Solution. The ERH and DRH in Table 7.4.2 have $M = 3$ and $N = 11$ pulses respectively. Hence, the number of pulses in the unit hydrograph is $N - M + 1 = 11 - 3 + 1 = 9$. Substituting the ordinates of the ERH and DRH into the equations in Table 7.4.1 yields a set of 11 simultaneous equations. These equations may be solved by *Gauss elimination* to give the unit hydrograph ordinates. Gauss elimination involves isolating the unknown variables one by one and successively solving for them. In this case, the equations can be solved from top to bottom, working with just the equations involving the first pulse P_1, starting with

TABLE 7.4.1

The set of equations for discrete time convolution $Q_n = \sum_{m=1}^{n \le M} P_m U_{n-m+1}$; $n = 1, 2, \ldots, N$

$$
\begin{array}{lllllllll}
Q_1 & = P_1U_1 & & & & & & & \\
Q_2 & = P_2U_1 & + P_1U_2 & & & & & & \\
Q_3 & = P_3U_1 & + P_2U_2 & + P_1U_3 & & & & & \\
\cdots & & & & & & & & \\
Q_M & = P_MU_1 & + P_{M-1}U_2 & + & + P_1U_M & & & & \\
Q_{M+1} & = \; 0 & + P_MU_2 & + \ldots & + P_2U_M & + P_1U_{M+1} & & & \\
\cdots & & & & & & & & \\
Q_{N-1} & = \; 0 & + \; 0 & + \ldots + & 0 & + \; 0 & + \ldots & + P_MU_{N-M} & + P_{M-1}U_{N-M+1} \\
Q_N & = \; 0 & + \; 0 & + \ldots + & 0 & + \; 0 & + \ldots + & 0 & + P_MU_{N-M+1}
\end{array}
$$

$$U_1 = \frac{Q_1}{P_1} = \frac{428}{1.06} = 404 \text{ cfs/in}$$

$$U_2 = \frac{Q_2 - P_2U_1}{P_1} = \frac{1923 - 1.93 \times 404}{1.06} = 1079 \text{ cfs/in}$$

$$U_3 = \frac{Q_3 - P_3U_1 - P_2U_2}{P_1} = \frac{5297 - 1.81 \times 404 - 1.93 \times 1079}{1.06} = 2343 \text{ cfs/in}$$

and similarly for the remaining ordinates

$$U_4 = \frac{9131 - 1.81 \times 1079 - 1.93 \times 2343}{1.06} = 2506 \text{ cfs/in}$$

$$U_5 = \frac{10625 - 1.81 \times 2343 - 1.93 \times 2506}{1.06} = 1460 \text{ cfs/in}$$

$$U_6 = \frac{7834 - 1.81 \times 2506 - 1.93 \times 1460}{1.06} = 453 \text{ cfs/in}$$

TABLE 7.4.2

Excess rainfall hyetograph and direct runoff hydrograph for Example 7.4.1

Time ($\frac{1}{2}$ h)	Excess rainfall (in)	Direct runoff (cfs)
1	1.06	428
2	1.93	1923
3	1.81	5297
4		9131
5		10625
6		7834
7		3921
8		1846
9		1402
10		830
11		313

TABLE 7.4.3
Unit hydrograph derived in Example 7.4.1

n	1	2	3	4	5	6	7	8	9
U_n (cfs/in)	404	1079	2343	2506	1460	453	381	274	173

$$U_7 = \frac{3921 - 1.81 \times 1460 - 1.93 \times 453}{1.06} = 381 \text{ cfs/in}$$

$$U_8 = \frac{1846 - 1.81 \times 453 - 1.93 \times 381}{1.06} = 274 \text{ cfs/in}$$

$$U_9 = \frac{1402 - 1.81 \times 381 - 1.93 \times 274}{1.06} = 173 \text{ cfs/in}$$

The derived unit hydrograph is given in Table 7.4.3. Solutions may be similarly obtained by focusing on other rainfall pulses. The depth of direct runoff in the unit hydrograph can be checked and found to equal 1.00 inch as required. In cases where the derived unit hydrograph does not meet this requirement, the ordinates are adjusted by proportion so that the depth of direct runoff is 1 inch (or 1 cm).

In general the unit hydrographs obtained by solutions of the set of equations in Table 7.4.1 for different rainfall pulses are not identical. To obtain a unique solution a *method of successive approximation* (Collins, 1939) can be used, which involves four steps: (1) assume a unit hydrograph, and apply it to all excess-rainfall blocks of the hyetograph except the largest; (2) subtract the resulting hydrograph from the actual DRH, and reduce the residual to unit hydrograph terms; (3) compute a weighted average of the assumed unit hydrograph and the residual unit hydrograph, and use it as the revised approximation for the next trial; (4) repeat the previous three steps until the residual unit hydrograph does not differ by more than a permissible amount from the assumed hydrograph.

The resulting unit hydrograph may show erratic variations and even have negative values. If this occurs, a smooth curve may be fitted to the ordinates to produce an approximation of the unit hydrograph. Erratic variation in the unit hydrograph may be due to nonlinearity in the effective rainfall–direct runoff relationship in the watershed, and even if this relationship is truly linear, the observed data may not adequately reflect this. Also, actual storms are not always uniform in time and space, as required by theory, even when the excess rainfall hyetograph is broken into pulses of short duration.

7.5 UNIT HYDROGRAPH APPLICATION

Once the unit hydrograph has been determined, it may be applied to find the direct runoff and streamflow hydrographs. A rainfall hyetograph is selected, the abstractions are estimated, and the excess rainfall hyetograph is calculated as described in Sec. 5.4. The time interval used in defining the excess rainfall hyetograph ordinates must be the same as that for which the unit hydrograph was specified. The discrete convolution equation

$$Q_n = \sum_{m=1}^{n \leq M} P_m U_{n-m+1} \tag{7.5.1}$$

may then be used to yield the direct runoff hydrograph. By adding an estimated baseflow to the direct runoff hydrograph, the streamflow hydrograph is obtained.

Example 7.5.1. Calculate the streamflow hydrograph for a storm of 6 in excess rainfall, with 2 in in the first half-hour, 3 in in the second half-hour and 1 in in the third half-hour. Use the half-hour unit hydrograph computed in Example 7.4.1 and assume the baseflow is constant at 500 cfs throughout the flood. Check that the total depth of direct runoff is equal to the total excess precipitation (watershed area = 7.03 mi^2).

Solution. The calculation of the direct runoff hydrograph by convolution is shown in Table 7.5.1. The unit hydrograph ordinates from Table 7.4.3 are laid out along the top of the table and the excess precipitation depths down the left side. The time interval is in $\Delta t = 0.5$ h intervals. For the first time interval, $n = 1$ in Eq. (7.5.1), and

$$\begin{aligned} Q_1 &= P_1 U_1 \\ &= 2.00 \times 404 \\ &= 808 \text{ cfs} \end{aligned}$$

For the second time interval,

$$\begin{aligned} Q_2 &= P_2 U_1 + P_1 U_2 \\ &= 3.00 \times 404 + 2.00 \times 1079 \\ &= 1212 + 2158 \end{aligned}$$

TABLE 7.5.1
Calculation of the direct runoff hydrograph and streamflow hydrograph for Example 7.5.1

Time	Excess	Unit hydrograph ordinates (cfs/in)									Direct	Streamflow*
	Precipitation	1	2	3	4	5	6	7	8	9	runoff	(cfs)
($\frac{1}{2}$-h)	(in)	404	1079	2343	2506	1460	453	381	274	173	(cfs)	
$n =$ 1	2.00	808									808	1308
2	3.00	1212	2158								3370	3870
3	1.00	404	3237	4686							8327	8827
4			1079	7029	5012						13,120	13,620
5				2343	7518	2920					12,781	13,281
6					2506	4380	906				7792	8292
7						1460	1359	762			3581	4081
8							453	1143	548		2144	2644
9								381	822	346	1549	2049
10									274	519	793	1293
11										173	173	673
										Total	54,438	

*Baseflow = 500 cfs.

$$= 3370 \text{ cfs}$$

as shown in the table. For the third time interval,

$$Q_3 = P_3U_1 + P_2U_2 + P_1U_3$$
$$= 1.00 \times 404 + 3.00 \times 1079 + 2.00 \times 2343$$
$$= 404 + 3237 + 4686$$
$$= 8327 \text{ cfs}$$

The calculations for $n = 4, 5, \ldots,$ follow in the same manner as shown in Table 7.5.1 and graphically in Fig. 7.5.1. The total direct runoff volume is

$$V_d = \sum_{n=1}^{N} Q_n \Delta t$$
$$= 54{,}438 \times 0.5 \text{ cfs}\cdot\text{h}$$
$$= 54{,}438 \times 0.5 \frac{\text{ft}^3\cdot\text{h}}{\text{s}} \times \frac{3600 \text{ s}}{1 \text{ h}}$$
$$= 9.80 \times 10^7 \text{ ft}^3$$

and the corresponding depth of direct runoff is found by dividing by the watershed area $A = 7.03 \text{ mi}^2 = 7.03 \times 5280^2 \text{ ft}^2 = 1.96 \times 10^8 \text{ ft}^2$:

$$r_d = \frac{V_d}{A}$$
$$= \frac{9.80 \times 10^7}{1.96 \times 10^8} \text{ft}$$
$$= 0.500 \text{ ft}$$
$$= 6.00 \text{ in}$$

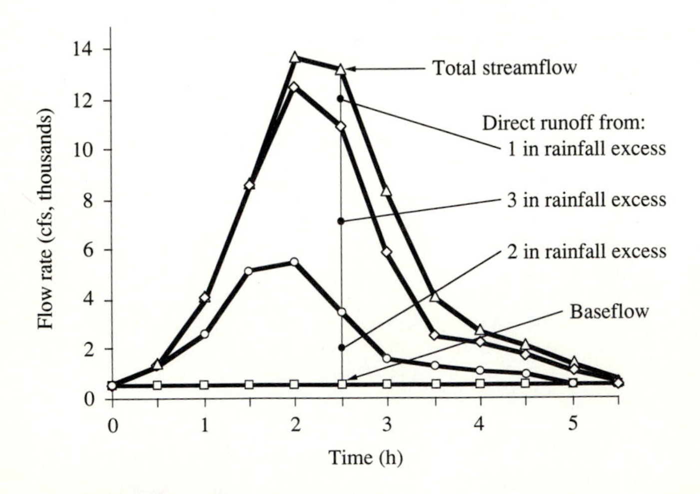

FIGURE 7.5.1
Streamflow hydrograph from a storm with excess rainfall pulses of duration 0.5 h and amount 2 in, 3 in, and 1 in, respectively. Total streamflow = baseflow + direct runoff (Example 7.5.1).

which is equal to the total depth of excess precipitation as required.

The streamflow hydrograph is found by adding the 500 cfs baseflow to the direct runoff hydrograph, as shown on the right-hand side of Table 7.5.1 and graphically in Fig. 7.5.1.

7.6 UNIT HYDROGRAPH BY MATRIX CALCULATION

Deconvolution may be used to derive the unit hydrograph from a complex multipeaked hydrograph, but the possibility of errors or nonlinearity in the data is greater than for a single-peaked hydrograph. Least-squares fitting or an optimization method can be used to minimize the error in the fitted direct runoff hydrograph. The application of these techniques is facilitated by expressing Eq. (7.4.1) in matrix form:

$$\begin{bmatrix} P_1 & 0 & 0 & \dots 0 & 0 & \dots 0 & 0 \\ P_2 & P_1 & 0 & \dots 0 & 0 & \dots 0 & 0 \\ P_3 & P_2 & P_1 & \dots 0 & 0 & \dots 0 & 0 \\ \cdot & & & \cdot & & & \cdot \\ \cdot & & & \cdot & & & \cdot \\ \cdot & & & \cdot & & & \cdot \\ P_M & P_{M-1} & P_{M-2} & \dots P_1 & 0 & \dots 0 & 0 \\ 0 & P_M & P_{M-1} & \dots P_2 & P_1 & \dots 0 & 0 \\ \cdot & & & \cdot & & & \cdot \\ \cdot & & & \cdot & & & \cdot \\ \cdot & & & \cdot & & & \cdot \\ 0 & 0 & 0 & \dots 0 & 0 & \dots P_M & P_{M-1} \\ 0 & 0 & 0 & \dots 0 & 0 & \dots 0 & P_M \end{bmatrix} \cdot \begin{bmatrix} U_1 \\ U_2 \\ U_3 \\ \vdots \\ U_{N-M+1} \end{bmatrix} = \begin{bmatrix} Q_1 \\ Q_2 \\ Q_3 \\ \cdot \\ \cdot \\ \cdot \\ Q_M \\ Q_{M+1} \\ \cdot \\ \cdot \\ \cdot \\ Q_{N-1} \\ Q_N \end{bmatrix} \qquad (7.6.1)$$

or

$$[P][U] = [Q] \qquad (7.6.2)$$

Given $[P]$ and $[Q]$, there is usually no solution for $[U]$ that will satisfy all N equations (7.6.1). Suppose that a solution $[U]$ is given that yields an estimate $[\hat{Q}]$ of the DRH as

$$[P][U] = [\hat{Q}] \qquad (7.6.3a)$$

or

$$\hat{Q}_n = P_n U_1 + P_{n-1} U_2 + \dots + P_{n-M+1} U_M \qquad n = 1, \dots, N \qquad (7.6.3b)$$

with all equations now satisfied. A solution is sought which minimizes the error $[Q] - [\hat{Q}]$ between the observed and estimated DRH's.

Solution by Linear Regression

The solution by linear regression produces the least-squares error between $[Q]$ and $[\hat{Q}]$ (Snyder, 1955). To solve Eq. (7.6.2) for $[U]$, the rectangular matrix $[P]$

is reduced to a square matrix $[Z]$ by multiplying both sides by the transpose of $[P]$, denoted by $[P]^T$, which is formed by interchanging the rows and columns of $[P]$. Then both sides are multiplied by the inverse $[Z]^{-1}$ of matrix $[Z]$, to yield

$$[U] = [Z]^{-1}[P]^T[Q] \tag{7.6.4}$$

where $[Z] = [P]^T[P]$. However, the solution is not easy to determine by this method, because the many repeated and blank entries in $[P]$ create difficulties in the inversion of $[Z]$ (Bree, 1978). Newton and Vinyard (1967) and Singh (1976) give alternative methods of obtaining the least-squares solution, but these methods do not ensure that all the unit hydrograph ordinates will be nonnegative.

Solution by Linear Programming

Linear programming is an alternative method of solving for $[U]$ in Eq. (7.6.2) that minimizes the absolute value of the error between $[Q]$ and $[\hat{Q}]$ and also ensures that all entries of $[U]$ are nonnegative (Eagleson, Mejia, and March, 1966; Deininger, 1969; Singh, 1976; Mays and Coles, 1980).

The general linear programming model is stated in the form of a linear *objective function* to be optimized (maximized or minimized) subject to linear *constraint equations*. Linear programming provides a method of comparing all possible solutions that satisfy the constraints and obtaining the one that optimizes the objective function (Hillier and Lieberman, 1974; Bradley, Hax, and Magnanti, 1977).

Example 7.6.1. Develop a linear program to solve Eq. (7.6.2) for the unit hydrograph given the ERH $P_m, m = 1, 2, \ldots, M$, and the DRH $Q_n, n = 1, 2, \ldots, N$.

Solution. The objective is to minimize $\sum_{n=1}^{N} |\epsilon_n|$ where $\epsilon_n = Q_n - \hat{Q}_n$. Linear programming requires that all the variables be nonnegative; to accomplish this task, ϵ_n is split into two components, a *positive deviation* θ_n and a *negative deviation* β_n. In the case where $\epsilon_n > 0$, that is, when the observed direct runoff Q_n is greater than the calculated value $\hat{Q}_n$, $\theta_n = \epsilon_n$ and $\beta_n = 0$; where $\epsilon_n < 0$, $\beta_n = -\epsilon_n$ and $\theta_n = 0$ (see Fig. 7.6.1). If $\epsilon_n = 0$ then $\theta_n = \beta_n = 0$ also. Hence, the solution must obey

$$Q_n = \hat{Q}_n - \beta_n + \theta_n \qquad n = 1, 2, \ldots, N \tag{7.6.5}$$

and the objective is

$$\text{minimize} \sum_{n=1}^{N} (\theta_n + \beta_n) \tag{7.6.6}$$

The constraints (7.6.5) can be written

$$[\hat{Q}_n] + [\theta_n] - [\beta_n] = [Q_n] \tag{7.6.7}$$

or, expanding as in Eq. (7.6.3*b*),

$$P_n U_1 + P_{n-1} U_2 + \ldots + P_{n-M+1} U_M + \theta_n - \beta_n = Q_n \qquad n = 1, \ldots, N \tag{7.6.8}$$

To ensure that the unit hydrograph represents one unit of direct runoff an additional

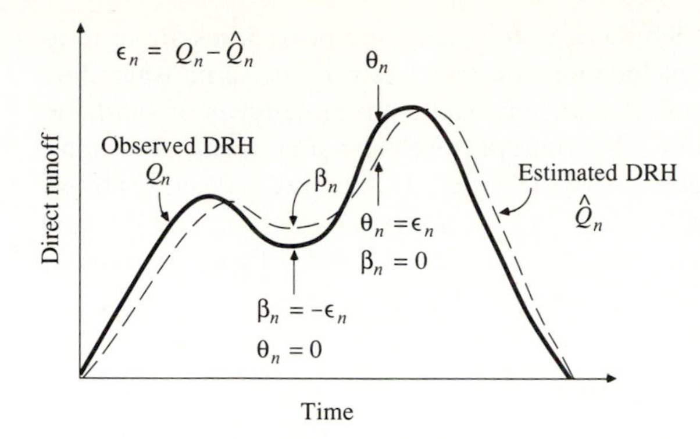

FIGURE 7.6.1
Deviation ϵ_n between observed and estimated direct runoff hydrographs is the sum of a positive deviation θ_n and a negative deviation β_n for solution by linear programming.

constraint equation is added:

$$\sum_{m=1}^{M} U_m = K \tag{7.6.9}$$

where K is a constant which converts the units of the ERH into the units of the DRH. Equations (7.6.6) to (7.6.9) constitute a linear program with *decision variables* (or unknowns) U_m, θ_n and β_n which may be solved using standard linear programming computer programs to produce the unit hydrograph. Linear programming requires all the decision variables to be non-negative, thereby ensuring the unit hydrograph ordinates will be non-negative.

The linear programming method developed in Example 7.6.1 is not limited in application to a single storm. Several ERHs and their resulting DRHs can be linked together as if they comprised one event and used to find a *composite unit hydrograph* best representing the response of the watershed to this set of storms. Multistorm analysis may also be carried out using the least-squares method (Diskin and Boneh, 1975; Mawdsley and Tagg, 1981).

In determination of the unit hydrograph from complex hydrographs, the abstractions are a significant source of error—although often assumed constant, the loss rate is actually a time-varying function whose value is affected by the moisture content of the watershed prior to the storm and by the storm pattern itself. Different unit hydrographs result from different assumptions about the pattern of losses. Newton and Vinyard (1967) account for errors in the loss rate by iteratively adjusting the ordinates of the ERH as well as those of the unit hydrograph so as to minimize the error in the DRH. Mays and Taur (1982) used nonlinear programming to simultaneously determine the loss rate for each storm period and the composite unit hydrograph ordinates for a multistorm event. Unver and Mays (1984) extended this nonlinear programming method to determine the optimal parameters for the loss-rate functions, and the composite unit hydrograph.

7.7 SYNTHETIC UNIT HYDROGRAPH

The unit hydrograph developed from rainfall and streamflow data on a watershed applies only for that watershed and for the point on the stream where the streamflow data were measured. Synthetic unit hydrograph procedures are used to develop unit hydrographs for other locations on the stream in the same watershed or for nearby watersheds of a similar character. There are three types of synthetic unit hydrographs: (1) those relating hydrograph characteristics (peak flow rate, base time, etc.) to watershed characteristics (Snyder, 1938; Gray, 1961), (2) those based on a dimensionless unit hydrograph (Soil Conservation Service, 1972), and (3) those based on models of watershed storage (Clark, 1943). Types (1) and (2) are described here and type (3) in Chap. 8.

Snyder's Synthetic Unit Hydrograph

In a study of watersheds located mainly in the Appalachian highlands of the United States, and varying in size from about 10 to 10,000 mi^2 (30 to 30,000 km^2), Snyder (1938) found synthetic relations for some characteristics of a *standard unit hydrograph* [Fig. 7.7.1*a*]. Additional such relations were found later (U.S. Army Corps of Engineers, 1959). These relations, in modified form are given below. From the relations, five characteristics of a *required unit hydrograph* [Fig. 7.7.1*b*] for a given excess rainfall duration may be calculated: the peak discharge per unit of watershed area, q_{pR}, the basin lag t_{pR} (time difference between the centroid of the excess rainfall hyetograph and the unit hydrograph peak), the base time t_b, and the widths W (in time units) of the unit hydrograph at 50 and 75 percent of the peak discharge. Using these characteristics the required unit hydrograph may be drawn. The variables are illustrated in Fig. 7.7.1.

Snyder defined a standard unit hydrograph as one whose rainfall duration t_r is related to the basin lag t_p by

$$t_p = 5.5t_r \tag{7.7.1}$$

For a standard unit hydrograph he found that:

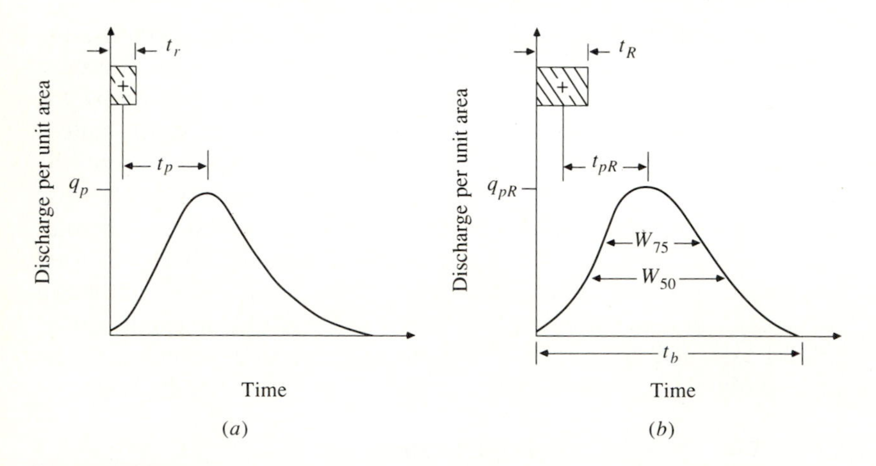

FIGURE 7.7.1
Snyder's synthetic unit hydrograph. (*a*) Standard unit hydrograph (t_p = $5.5t_r$). (*b*) Required unit hydrograph ($t_{pR} \neq 5.5t_R$).

1. The basin lag is

$$t_p = C_1 C_t (L L_c)^{0.3} \tag{7.7.2}$$

where t_p is in hours, L is the length of the main stream in kilometers (or miles) from the outlet to the upstream divide, L_c is the distance in kilometers (miles) from the outlet to a point on the stream nearest the centroid of the watershed area, $C_1 = 0.75$ (1.0 for the English system), and C_t is a coefficient derived from gaged watersheds in the same region.

2. The peak discharge per unit drainage area in $m^3/s \cdot km^2$ (cfs/mi^2) of the standard unit hydrograph is

$$q_p = \frac{C_2 C_p}{t_p} \tag{7.7.3}$$

where $C_2 = 2.75$ (640 for the English system) and C_p is a coefficient derived from gaged watersheds in the same region.

To compute C_t and C_p for a gaged watershed, the values of L and L_c are measured from the basin map. From a *derived unit hydrograph* of the watershed are obtained values of its effective duration t_R in hours, its basin lag t_{pR} in hours, and its peak discharge per unit drainage area, q_{pR}, in $m^3/s \cdot km^2 \cdot cm$ ($cfs/mi^2 \cdot in$ for the English system). If $t_{pR} = 5.5 t_R$, then $t_R = t_r$, $t_{pR} = t_p$, and $q_{pR} = q_p$, and C_t and C_p are computed by Eqs. (7.7.2) and (7.7.3). If t_{pR} is quite different from $5.5 t_R$, the standard basin lag is

$$t_p = t_{pR} + \frac{t_r - t_R}{4} \tag{7.7.4}$$

and Eqs. (7.7.1) and (7.7.4) are solved simultaneously for t_r and t_p. The values of C_t and C_p are then computed from (7.7.2) and (7.7.3) with $q_{pR} = q_p$ and $t_{pR} = t_p$.

When an ungaged watershed appears to be similar to a gaged watershed, the coefficients C_t and C_p for the gaged watershed can be used in the above equations to derive the required synthetic unit hydrograph for the ungaged watershed.

3. The relationship between q_p and the peak discharge per unit drainage area q_{pR} of the required unit hydrograph is

$$q_{pR} = \frac{q_p t_p}{t_{pR}} \tag{7.7.5}$$

4. The base time t_b in hours of the unit hydrograph can be determined using the fact that the area under the unit hydrograph is equivalent to a direct runoff of 1 cm (1 inch in the English system). Assuming a triangular shape for the unit hydrograph, the base time may be estimated by

$$t_b = \frac{C_3}{q_{pR}} \tag{7.7.6}$$

where $C_3 = 5.56$ (1290 for the English system).

5. The width in hours of a unit hydrograph at a discharge equal to a certain percent of the peak discharge q_{pR} is given by

$$W = C_w q_{pR}^{-1.08} \qquad (7.7.7)$$

where $C_w = 1.22$ (440 for English system) for the 75-percent width and 2.14 (770, English system) for the 50-percent width. Usually one-third of this width is distributed before the unit hydrograph peak time and two-thirds after the peak.

Example 7.7.1. From the basin map of a given watershed, the following quantities are measured: $L = 150$ km, $L_c = 75$ km, and drainage area $= 3500$ km^2. From the unit hydrograph derived for the watershed, the following are determined: $t_R = 12$ h, $t_{pR} = 34$ h, and peak discharge $= 157.5$ m^3/s·cm. Determine the coefficients C_t and C_p for the synthetic unit hydrograph of the watershed.

Solution. From the given data, $5.5t_R = 66$ h, which is quite different from t_{pR} (34 h). Equation (7.7.4) yields

$$t_p = t_{pR} + \frac{t_r - t_R}{4}$$

$$= 34 + \frac{t_r - 12}{4} \qquad (7.7.8)$$

Solving (7.7.1) and (7.7.8) simultaneously gives $t_r = 5.9$ h and $t_p = 32.5$ h. To calculate C_t, use (7.7.2):

$$t_p = C_1 C_t (L L_c)^{0.3}$$

$$32.5 = 0.75 C_t (150 \times 75)^{0.3}$$

$$C_t = 2.65$$

The peak discharge per unit area is $q_{pR} = 157.5/3500 = 0.045$ m^3/s·km^2·cm. The coefficient C_p is calculated by Eq. (7.7.3) with $q_p = q_{pR}$, and $t_p = t_{pR}$:

$$q_{pR} = \frac{C_2 C_p}{t_{pR}}$$

$$0.045 = \frac{2.75 C_p}{34.0}$$

$$C_p = 0.56$$

Example 7.7.2. Compute the six-hour synthetic unit hydrograph of a watershed having a drainage area of 2500 km^2 with $L = 100$ km and $L_c = 50$ km. This watershed is a sub–drainage area of the watershed in Example 7.7.1.

Solution. The values $C_t = 2.64$ and $C_p = 0.56$ determined in Example 7.7.1 can also be used for this watershed. Thus, Eq. (7.7.2) gives $t_p = 0.75 \times 2.64 \times (100 \times 50)^{0.3} = 25.5$ h, and (7.7.1) gives $t_r = 25.5/5.5 = 4.64$ h. For a six-hour unit hydrograph, $t_R = 6$ h, and Eq. (7.7.4) gives $t_{pR} = t_p - (t_r - t_R)/4 = 25.5 - (4.64-6)/4 = 25.8$ h. Equation (7.7.3) gives $q_p = 2.75 \times 0.56/25.5 = 0.0604$

m^3/s·km^{2}cm and (7.7.5) gives $q_{pR} = 0.0604 \times 25.5/25.8 = 0.0597$ m^3/s·km^2·cm; the peak discharge is $0.0597 \times 2500 = 149.2$ m^3/s·cm. The widths of the unit hydrograph are given by Eq. (7.7.7). At 75 percent of peak discharge, $W = 1.22 q_{pR}^{-1.08} = 1.22 \times 0.0597^{-1.08} = 25.6$ h. A similar computation gives a $W = 44.9$ h at 50 percent of peak. The base time, given by Eq. (7.7.6), is $t_b = 5.56/q_{pR} = 5.56/0.0597 = 93$h. The hydrograph is drawn, as in Fig. 7.7.2, and checked to ensure that it represents a depth of direct runoff of 1 cm.

A further innovation in the use of Snyder's method has been the regionalization of unit hydrograph parameters. Espey, Altman and Graves (1977) developed a set of generalized equations for the construction of 10-minute unit hydrographs using a study of 41 watersheds ranging in size from 0.014 to 15 mi^2, and in impervious percentage from 2 to 100 percent. Of the 41 watersheds, 16 are located in Texas, 9 in North Carolina, 6 in Kentucky, 4 in Indiana, 2 each in Colorado and Mississippi, and 1 each in Tennessee and Pennsylvania. The equations are:

$$T_p = 3.1 L^{0.23} S^{-0.25} I^{-0.18} \Phi^{1.57} \tag{7.7.9}$$

$$Q_p = 31.62 \times 10^3 A^{0.96} T_p^{-1.07} \tag{7.7.10}$$

$$T_B = 125.89 \times 10^3 A Q_p^{-0.95} \tag{7.7.11}$$

$$W_{50} = 16.22 \times 10^3 A^{0.93} Q_p^{-0.92} \tag{7.7.12}$$

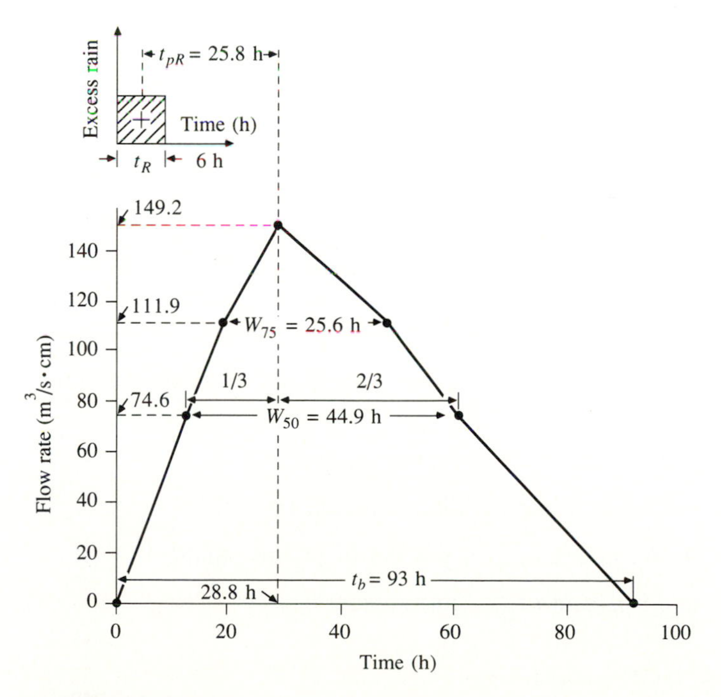

FIGURE 7.7.2
Synthetic unit hydrograph calculated by Snyder's method in Example 7.7.2

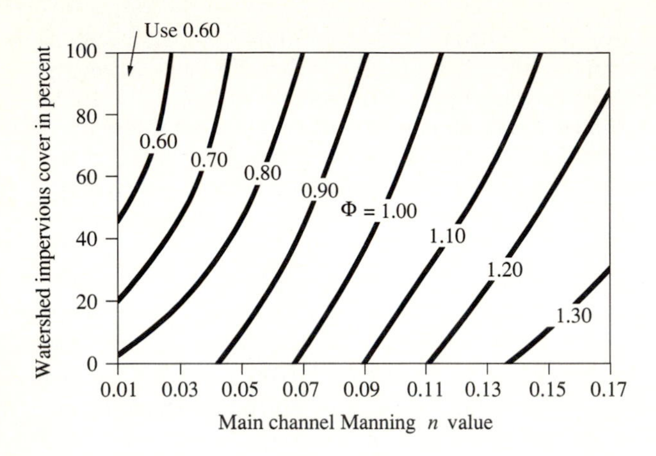

FIGURE 7.7.3
Watershed conveyance factor Φ as a function of channel roughness and watershed imperviousness. (Adapted with permission from Espey, Altman, and Graves, 1977.)

$$W_{75} = 3.24 \times 10^3 A^{0.79} Q_p^{-0.78} \tag{7.7.13}$$

where

L = the total distance (in feet) along the main channel from the point being considered to the upstream watershed boundary

S = the main channel slope (in feet per foot), defined by $H/0.8L$, where H is the difference in elevation between A and B. A is the point on the channel bottom at a distance of $0.2L$ downstream from the upstream watershed boundary; B is a point on the channel bottom at the downstream point being considered

I = the impervious area within the watershed (in percent), assumed equal to 5 percent for an undeveloped watershed

Φ = the dimensionless watershed conveyance factor, which is a function of percent impervious and roughness (Fig. 7.7.3)

A = the watershed drainage area (in square miles)

T_p = the time of rise to the peak of the unit hydrograph from the beginning of runoff (in minutes)

Q_p = the peak flow of the unit hydrograph (in cfs/in)

T_B = the time base of the unit hydrograph (in minutes)

W_{50} = the width of the hydrograph at 50 percent of Q_p (in minutes)

W_{75} = the width of at 75 percent of Q_p (in minutes)

SCS Dimensionless Hydrograph

The SCS dimensionless hydrograph is a synthetic unit hydrograph in which the discharge is expressed by the ratio of discharge q to peak discharge q_p and the time by the ratio of time t to the time of rise of the unit hydrograph, T_p. Given the peak discharge and lag time for the duration of excess rainfall, the unit hydrograph can be estimated from the synthetic dimensionless hydrograph for the given basin. Figure 7.7.4(*a*) shows such a dimensionless hydrograph, prepared from the unit hydrographs of a variety of watersheds. The values of q_p and T_p may be estimated using a simplified model of a triangular unit hydrograph as shown in Figure 7.7.4(*b*), where the time is in hours and the discharge in $m^3/s \cdot cm$ (or cfs/in) (Soil Conservation Service, 1972).

From a review of a large number of unit hydrographs, the Soil Conservation Service suggests the time of recession may be approximated as 1.67 T_p. As the area under the unit hydrograph should be equal to a direct runoff of 1 cm (or 1 in), it can be shown that

$$q_p = \frac{CA}{T_p} \tag{7.7.14}$$

where $C = 2.08$ (483.4 in the English system) and A is the drainage area in square kilometers (square miles).

Further, a study of unit hydrographs of many large and small rural watersheds indicates that the basin lag $t_p \simeq 0.6T_c$, where T_c is the time of concentration of the watershed. As shown in Fig. 7.7.4(*b*), time of rise T_p can be expressed in terms of lag time t_p and the duration of effective rainfall t_r

$$T_p = \frac{t_r}{2} + t_p \tag{7.7.15}$$

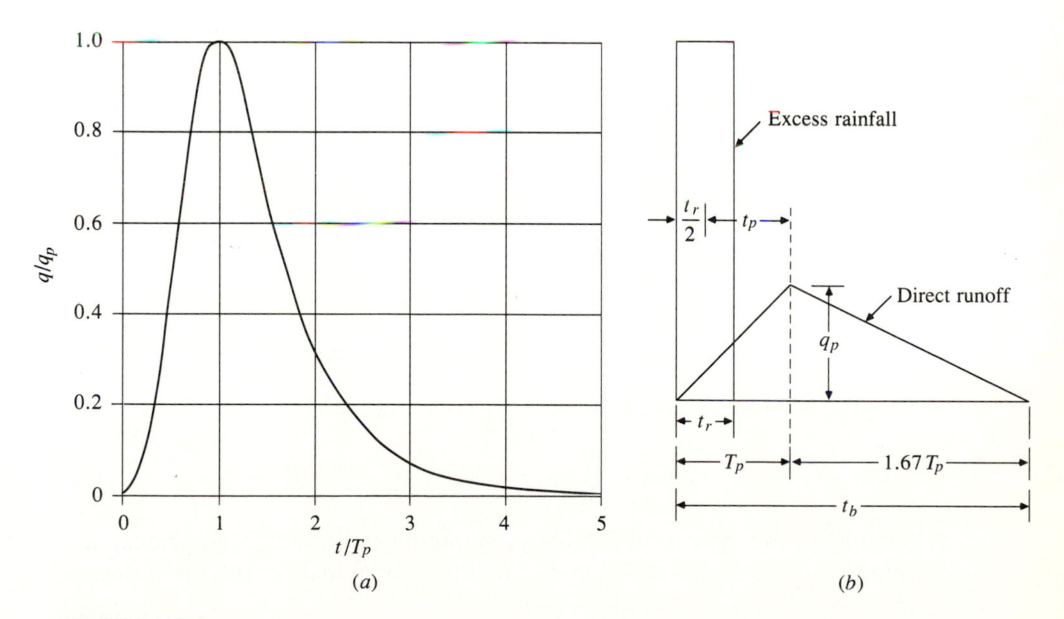

FIGURE 7.7.4
Soil Conservation Service synthetic unit hydrographs (*a*) Dimensionless hydrograph and (*b*) triangular unit hydrograph. (*Source*: Soil Conservation Service, 1972.)

Example 7.7.3. Construct a 10-minute SCS unit hydrograph for a basin of area 3.0 km^2 and time of concentration 1.25 h.

Solution. The duration t_r = 10 min = 0.166 h, lag time $t_p = 0.6T_c = 0.6 \times 1.25 =$ 0.75 h, and rise time $T_p = t_r/2 + t_p = 0.166/2 + 0.75 = 0.833$ h. From Eq. (7.7.14), $q_p = 2.08 \times 3.0/0.833 = 7.49$ m^3/s·cm. The dimensionless hydrograph in Fig. 7.7.4 may be converted to the required dimensions by multiplying the values on the horizontal axis by T_p and those on the vertical axis by q_p. Alternatively, the triangular unit hydrograph can be drawn with $t_b = 2.67T_p = 2.22$ h. The depth of direct runoff is checked to equal 1 cm.

7.8 UNIT HYDROGRAPHS FOR DIFFERENT RAINFALL DURATIONS

When a unit hydrograph of a given excess-rainfall duration is available, the unit hydrographs of other durations can be derived. If other durations are integral multiples of the given duration, the new unit hydrograph can be easily computed by application of the principles of superposition and proportionality. However, a general method of derivation applicable to unit hydrographs of any required duration may be used on the basis of the principle of superposition. This is the *S-hydrograph method*.

The theoretical *S-hydrograph* is that resulting from a continuous excess rainfall at a constant rate of 1 cm/h (or 1 in/h) for an indefinite period. This is the unit step response function of a watershed system. The curve assumes a deformed S shape and its ordinates ultimately approach the rate of excess rainfall at a time of equilibrium. This step response function $g(t)$ can be derived from the unit pulse response function $h(t)$ of the unit hydrograph, as follows.

From Eq. (7.2.4), the response at time t to a unit pulse of duration Δt beginning at time 0 is

$$h(t) = \frac{1}{\Delta t}[g(t) - g(t - \Delta t)] \tag{7.8.1}$$

Similarly, the response at time t to a unit pulse beginning at time Δt is equal to $h(t - \Delta t)$, that is, $h(t)$ lagged by Δt time units:

$$h(t - \Delta t) = \frac{1}{\Delta t}[g(t - \Delta t) - g(t - 2\,\Delta t)] \tag{7.8.2}$$

and the response at time t to a third unit pulse beginning at time $2\Delta t$ is

$$h(t - 2\Delta t) = \frac{1}{\Delta t}[g(t - 2\Delta t) - g(t - 3\,\Delta t)] \tag{7.8.3}$$

Continuing this process indefinitely, summing the resulting equations, and rearranging, yields the unit step response function, or S-hydrograph, as shown in Fig. 7.8.1(a):

$$g(t) = \Delta t\,[h(t) + h(t - \Delta t) + h(t - 2\,\Delta t) + \ldots] \tag{7.8.4}$$

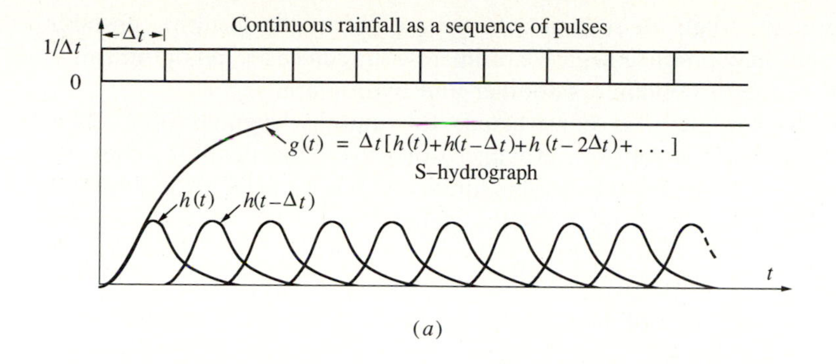

(a)

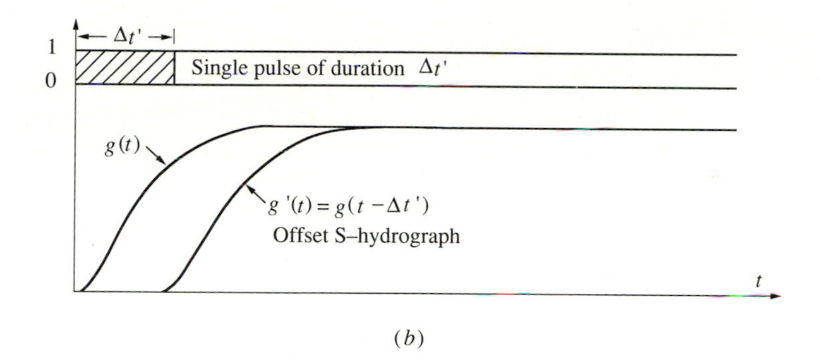

(b)

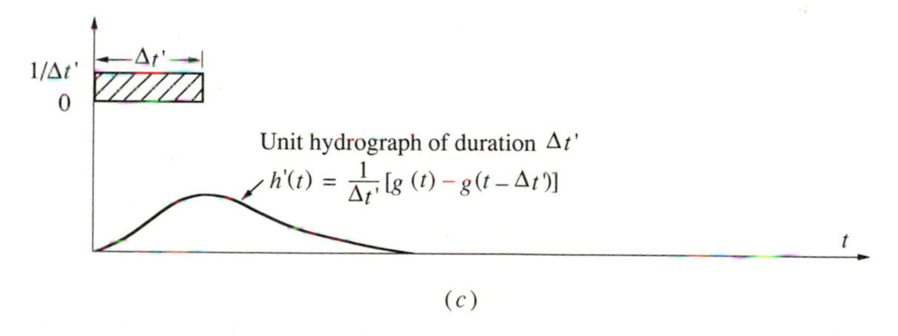

(c)

FIGURE 7.8.1
Using the S-hydrograph to find a unit hydrograph of duration $\Delta t'$ from a unit hydrograph of duration Δt.

where the summation is multiplied by Δt so that $g(t)$ will correspond to an input rate of 1, rather than $1/\Delta t$ as used for each of the unit pulses.

Theoretically, the S-hydrograph so derived should be a smooth curve, because the input excess rainfall is assumed to be at a constant, continuous rate. However, the summation process will result in an undulatory form if there are errors in the rainfall abstractions or baseflow separation, or if the actual duration of excess rainfall is not the derived duration for the unit hydrograph. A duration which produces minimum undulation can be found by trial. Undulation of the curve may be also caused by nonuniform temporal and areal distribution of rainfall; furthermore, when the natural data are not linear, the resulting unstable system oscillations may produce negative ordinates. In such cases, an optimization technique may be used to obtain a smoother unit hydrograph.

After the S-hydrograph is constructed, the unit hydrograph of a given duration can be derived as follows: Advance, or offset, the position of the S-hydrograph by a period equal to the desired duration $\Delta t'$ and call this S-hydrograph an *offset S-hydrograph,* $g'(t)$ [Fig. 7.8.1(*b*)], defined by

$$g'(t) = g(t - \Delta t') \tag{7.8.5}$$

The difference between the ordinates of the original S-hydrograph and the offset S-hydrograph, divided by $\Delta t'$, gives the desired unit hydrograph [Fig. 7.8.1(*c*)]:

$$h'(t) = \frac{1}{\Delta t'}[g(t) - g(t - \Delta t')] \tag{7.8.6}$$

Example 7.8.1. Use the 0.5-hour unit hydrograph in Table 7.4.3 (from Example 7.4.1) to produce the S-hydrograph and the 1.5-h unit hydrograph for this watershed.

Solution. The 0.5-h unit hydrograph is shown in column 2 of Table 7.8.1. The S-hydrograph is found using (7.8.4) with $\Delta t = 0.5$ h. For $t = 0.5$ h, $g(t) = \Delta t\, h(t) = 0.5 \times 404 = 202$ cfs; for $t = 1$ h, $g(t) = \Delta t[h(t) + h(t-0.5)] = 0.5 \times (1079 + 404) = 742$ cfs; for $t = 1.5$ h, $g(t) = \Delta t\,[h(t) + h(t-0.5) + h(t-1.0)] = 0.5 \times (2343 + 1079 + 404) = 1913$ cfs; and so on, as shown in column 3 of Table 7.8.1. The S-hydrograph is offset by $\Delta t' = 1.5$ h (column 4) to give $g(t - \Delta t')$, and the difference divided by $\Delta t'$ to give the 1.5-h unit hydrograph $h'(t)$ (column 5). For example, for $t = 2.0$ h, $h(t) = (3166 - 202)/1.5 = 1976$ cfs/in.

TABLE 7.8.1
Calculation of a 1.5-h unit hydrograph by the S-hydrograph method (Example 7.8.1)

1 Time t (h)	2 0.5-h unit hydrograph $h(t)$ (cfs/in)	3 S-hydrograph $g(t)$ (cfs)	4 Lagged S-hydrograph $g(t-\Delta t')$ (cfs)	5 1.5-h unit hydrograph $h'(t)$ (cfs/in)
0.5	404	202	0	135
1.0	1079	742	0	495
1.5	2343	1913	0	1275
2.0	2506	3166	202	1976
2.5	1460	3896	742	2103
3.0	453	4123	1913	1473
3.5	381	4313	3166	765
4.0	274	4450	3896	369
4.5	173	4537	4123	276
5.0	0	4537	4313	149
5.5	0	4537	4450	58
6.0	0	4537	4537	0

Instantaneous Unit Hydrograph

If the excess rainfall is of unit amount and its duration is infinitesimally small, the resulting hydrograph is an impulse response function (Sec. 7.2) called the instantaneous unit hydrograph (IUH). For an IUH, the excess rainfall is applied to the drainage area in zero time. Of course, this is only a theoretical concept and cannot be realized in actual watersheds, but it is useful because the IUH characterizes the watershed's response to rainfall without reference to the rainfall duration. Therefore, the IUH can be related to watershed geomorphology (Rodriguez-Iturbe and Valdes, 1979; Gupta, Waymire, and Wang, 1980).

The convolution integral (7.2.1) is

$$Q(t) = \int_0^t u(t-\tau)\,I(\tau)\,d\tau \tag{7.8.7}$$

If the quantities $I(\tau)$ and $Q(t)$ have the same dimensions, the ordinate of the IUH must have dimensions $[T^{-1}]$. The properties of the IUH are as follows, with $l = t - \tau$:

$$\begin{aligned} &0 \le u(l) \le \text{some positive peak value} && \text{for } l > 0 \\ &u(l) = 0 && \text{for } l \le 0 \\ &u(l) \to 0 && \text{as } l \to \infty \end{aligned} \tag{7.8.8}$$

$$\int_0^\infty u(l)\,dl = 1 \qquad \text{and} \qquad \int_0^\infty u(l)\,l\,dl = t_L$$

The quantity t_L is the lag time of the IUH. It can be shown that t_L gives the time interval between the centroid of an excess rainfall hyetograph and that of the corresponding direct runoff hydrograph. Note the difference between t_L and the variable t_p used for synthetic unit hydrograph lag time—t_p measures the time from the centroid of the excess rainfall to the peak, not the centroid, of the direct runoff hydrograph. The ideal shape of an IUH as described above resembles that of a single-peaked direct-runoff hydrograph, however, an IUH can have negative and undulating ordinates.

There are several methods to determine an IUH from a given ERH and DRH. For an approximation, the IUH ordinate at time t is simply set equal to the slope at time t of an S-hydrograph constructed for an excess rainfall intensity of unit depth per unit time. This procedure is based on the fact that the S-hydrograph is an integral curve of the IUH; that is, its ordinate at time t is equal to the integral of the area under the IUH from 0 to t. The IUH so obtained is in general only an approximation because the slope of an S-hydrograph is difficult to measure accurately.

The IUH can be determined by various methods of mathematical inversion, using, for example, orthogonal functions such as Fourier series (O'Donnell, 1960) or Laguerre functions (Dooge, 1973); integral transforms such as the Laplace transform (Chow, 1964), the Fourier transform (Blank, Delleur, and Giorgini,

1971), and the Z transform (Bree, 1978); and mathematical modeling related to watershed geomorphology (Sec. 8.5).

REFERENCES

Blank, D., J. W. Delleur, and A. Giorgini, Oscillatory kernel functions in linear hydrologic models, *Water Resour. Res.*, vol. 7, no. 5, pp. 1102–1117, 1971.

Bradley, S. P., A. C. Hax, and T. L. Magnanti, *Applied Mathematical Programming*, Addison-Wesley, Reading, Mass., 1977.

Bree, T., The stability of parameter estimation in the general linear model, *J. Hydrol.*, vol. 37, no. 1/2, pp. 47–66, 1978.

Chow, V. T., Runoff, in *Handbook of Applied Hydrology*, sec. 14, pp. 14-24 to 14-27, McGraw-Hill, New York, 1964.

Chow, V. T., and V. C. Kulandaiswamy, General hydrologic system model, *J. Hyd. Div., Am. Soc. Civ. Eng.*, vol. 97, no. HY6, pp. 791–804, 1971.

Clark, C. O., Storage and the unit hydrograph, *Proc. Am. Soc. Civ. Eng.*, vol. 9, pp. 1333–1360, 1943.

Collins, W. T., Runoff distribution graphs from precipitation occurring in more than one time unit, *Civ. Eng.*, vol. 9, no. 9, pp. 559–561, 1939.

Deininger, R. A., Linear programming for hydrologic analyses, *Water Resour. Res.*, vol. 5, no. 5, pp. 1105–1109, 1969.

Diskin, M. H., and A. Boneh, Determination of an optimal IUH for linear time invariant systems from the multi-storm records, *J. Hydrol.*, vol. 24, pp. 57–76, 1975.

Dooge, J. C. I., Linear theory of hydrologic systems, *Tech. Bull. No. 1468, Agric. Res. Serv.*, pp. 117–124, October, U. S. Department of Agriculture, Washington, D. C., 1973.

Eagleson, P. S., R. Mejia-R., and F. March, Computation of optimum realizable unit hydrographs, *Water Resour. Res.*, vol. 2, no. 4, pp. 755–764, 1966.

Espey, W. H., Jr., D. G. Altman, and C. B. Graves, Nomographs for ten-minute unit hydrographs for small urban watersheds, *Tech. Memo. No. 32, Urban Water Resources Research Prog., Am. Soc. Civ. Eng.*, New York, Dec. 1977

Gray, D. M., Synthetic unit hydrographs for small watersheds, *J. Hyd. Div., Am. Soc. Civ. Eng.*, vol. 87, no. HY4, pp. 33–54, 1961.

Gupta, V. K., E. Waymire, and C. T. Wang, A representation of an instantaneous unit hydrograph from geomorphology, *Water Resour. Res.*, vol. 16, no. 5, pp. 855–862, 1980.

Heerdegen, R. G., The unit hydrograph: a satisfactory model of watershed response? *Water Resour. Bul.* vol. 10, no. 6, pp. 1143–1161, 1974.

Hillier, F. S., and G. J. Lieberman, *Operations Research*, 2nd ed., Holden Day Inc., San Francisco, Calif., 1974.

Kreyszig, E., *Advanced Engineering Mathematics*, Wiley, New York, 1968.

Linsley, R. K., M. A. Kohler, and J. L. H. Paulhus, *Hydrology for engineers*, 3rd ed., McGraw-Hill, New York, 1982.

Mawdsley, J. A., and A. F. Tagg, Identification of unit hydrographs from multi-event analysis, *J. Hydrol.*, vol. 49, no. 3/4, pp. 315–327, 1981.

Mays, L. W., and S. L. Coles, Optimization of unit hydrograph determination, *J. Hyd. Div., Am. Soc. Civ. Eng.*, vol. 106, no. HY1, pp. 85–97, 1980.

Mays, L. W., and C. K. Taur, Unit hydrographs via nonlinear programming, *Water Resour. Res.*, vol. 18, no. 4, pp. 744–752, 1982.

Newton, D. S., and J. W. Vinyard, Computer-determined unit hydrograph from floods, *J. Hyd. Div., Am. Soc. Civ. Eng.*, vol. 93, no. HY5, pp. 219–235, 1967.

O'Donnell, T., Instantaneous unit hydrograph by harmonic analysis, *Proc. Gen. Ass. Helsinki, IASH*, publ. 51, pp. 546–557, 1960.

Rodriguez-Iturbe, I., and J. B. Valdes, The geomorphologic structure of hydrologic response, *Water Resour. Res.*, vol. 15, no. 6, pp. 1409–1420, 1979.

Sherman, L. K., Streamflow from rainfall by the unit-graph method, *Eng. News Rec.*, vol. 108, pp. 501–505, April 7, 1932.

Singh, K. P., Unit hydrographs—a comparative study, *Water Resour. Bul.*, vol. 12, no. 2, pp. 381–392, April, 1976.

Snyder, W. M., Hydrograph analysis by the method of least squares, *Proc. Amer. Soc. Civ. Eng.*, vol. 81, 1–24, 1955.

Snyder, F. F., Synthetic unit-graphs, *Trans. Am. Geophys. Union*, vol. 19, pp. 447–454, 1938.

Soil Conservation Service, Hydrology, sec. 4 of *National Engineering Handbook*, Soil Conservation Service, U. S. Department of Agriculture, Washington, D.C., 1972.

Unver, O., and L. W. Mays, Optimal determination of loss rate functions and unit hydrographs, *Water Resour. Res.*, vol. 20, no. 2, pp. 203–214, 1984.

U. S. Army Corps of Engineers, Flood hydrograph analysis and computations, *Engineering and Design Manual, EM 1110-2-1405*, U. S. Government Printing Office, Washington, D.C., August 31, 1959.

BIBLIOGRAPHY

Amorocho, J., Discussion of predicting storm runoff on small experimental watersheds by N.E. Minshall, *J. Hyd. Div., Am. Soc. Civ. Eng.*, vol. 87, no. HY2, pp. 185–191, 1961.

Amorocho, J., Nonlinear hydrologic analysis, in *Advances in Hydroscience*, ed. by V. T. Chow, Academic Press, New York, vol. 9, pp. 203–251, 1973.

Amorocho, J., and A. Brandstetter, Determination of nonlinear functions in rainfall-runoff processes, *Water Resour. Res.*, vol. 7, no. 1, pp. 1087–1101, 1971.

Bidwell, V. J., Regression analysis of nonlinear catchment systems, *Water Resour. Res.*, vol. 7, no. 5, pp. 1118–1126, 1971.

Broome, P., and R. H. Spigel, A linear model of storm runoff from some urban catchments in New Zealand, *J. Hydrol.* (New Zealand), vol. 21, no. 1, pp. 13–33, 1982.

Croley, T. E., II, Gamma synthetic hydrographs, *J. Hydrol.*, vol. 47, pp. 41–52, 1980.

Diskin, M. H., On the derivation of linkage equations for Laguerre function coefficients, *J. Hydrol.*, vol. 32, pp. 321–327, 1977.

Diskin, M. H., and A. Boneh, The kernel function of linear nonstationary surface runoff systems, *Water Resour. Res.*, vol. 10, no. 4, pp. 753–761, 1974.

Gray, D. M., Synthetic unit hydrographs for small watersheds, *J. Hyd. Div., Am. Soc. Civ. Eng.*, vol. 87, no. HY4, pp. 33–54, 1961.

Hall, M. J., On the smoothing of oscillations in finite-period unit hydrographs derived by the harmonic method, *Hydrol. Sci. Bull.*, vol. 22, no. 2, pp. 313–324, 1977.

Hino, M., and K. Nadaoka, Mathematical derivation of linear and nonlinear runoff kernels, *Water Resour. Res.*, vol. 15, no. 4, pp. 918–928, 1979.

Hossain, A., A. R. Rao, and J. W. Delleur, Estimation of direct runoff from urban watersheds, *J. Hyd. Div., Am. Soc. Civ. Eng.*, vol. 104, no. HY2, pp. 169–188, 1978.

Jacoby, S., A mathematical model for nonlinear hydrologic systems, *J. Geophys. Res.*, vol. 71, no. 20, pp. 4811–4824, 1966.

Levin, A. G., Flash flood forecasts with hydrograph separation by individual time intervals, *Sov. Hydrol.*, vol. 18, no. 1, pp. 40–45, 1979.

Minshall, N. E., Predicting storm runoff on small experimental watersheds, *J. Hyd. Div., Am. Soc. Civ. Eng.*, vol. 86, no. HY8, pp. 17–38, 1960.

Naef, F., Can we model the rainfall-runoff process today? *Hydrol. Sci. Bull.*, vol. 26, no. 3, pp. 281–289, 1981.

Neuman, S. P., and G. de Marsily, Identification of linear systems response by parametric programming, *Water Resour. Res.*, vol. 12, no. 2, pp. 253–262, 1976.

O'Connor, K. M., A discrete linear cascade model for hydrology, *J. Hydrol.*, vol. 29, pp. 203–242, 1976.

Papazafiriou, Z. G., Polynomial approximation of the kernels of closed linear hydrologic systems, *J. Hydrol.*, vol. 27, pp. 319–329, 1975.

Papazafiriou, Z. G., Linear and nonlinear approaches for short-term runoff estimations in time-invariant open hydrologic systems, *J. Hydrol.*, vol. 30, pp. 63–80, 1976.

Raudkivi, A. J., *Hydrology*, Chap. 7, Pergamon Press, Oxford, 1979.

Rusin, S. A., Nonstationary model of the transformation of effective precipitation into runoff, *Sov. Hydrol.*, vol. 4, pp. 289–300, 1973.

Sarma, P. B. S., J. W. Delleur, and A. R. Rao, Comparison of rainfall-runoff models for urban areas, *J. Hydrol.*, vol. 18, pp. 329–347, 1973.

Viessman, W., Jr., J. W. Knapp, G. L. Lewis, and T. E. Harbaugh, *Introduction to Hydrology*, 2nd ed., Chap. 4, Harper and Row, New York, 1977.

Zand, S. M., and J. A. Harder, Application of nonlinear system identification to the lower Mekong River, Southeast Asia, *Water Resour. Res.*, vol. 9, no. 2, pp. 290–297, 1973.

PROBLEMS

7.2.1 A system has a discrete pulse response function with ordinates 0.1, 0.5, 0.3, and 0.1 units. Calculate the output from this system if it has a pulse input of (*a*) 3 units, (*b*) 4 units, (*c*) 3 units in the first time interval followed by 4 units in the second.

7.2.2 A system has the following unit pulse response function: 0.27, 0.36, 0.18, 0.09, 0.05, 0.03, 0.01, 0.01. Calculate the output from this system if it has input (*a*) 2 units, (*b*) 3 units, (*c*) 2 units in the first time interval followed by 3 units in the second time interval.

7.2.3 Calculate and plot the impulse response function $u(t)$, the step response function $g(t)$, the continuous pulse response function $h(t)$, and the discrete pulse response function U_n for a linear reservoir having $k = 1$ h and $\Delta t = 2$ h.

7.2.4 A watershed is modeled as a linear reservoir with $k = 1$ h. Calculate its impulse response function and its pulse response functions for unit pulses of durations 0.5, 1.0, 1.5 and 2.0 h. Plot the response functions for $0 < t < 6$ h.

7.2.5 A watershed modeled as a linear reservoir with $k = 3$ h receives 3 in of excess rainfall in the first two hours of a storm and 2 in of excess rainfall in the second two hours. Calculate the direct runoff hydrograph from this watershed.

7.2.6 Show that the lag time t_L between the centroids of the excess rainfall hyetograph and the direct runoff hydrograph is equal to the storage constant k for a watershed modeled as a linear reservoir.

7.3.1 A watershed has a drainage area of 450 km^2, and its three-hour unit hydrograph has a peak discharge of 150 m^3/s·cm. For English units, what is the peak discharge in cfs/in of the three-hour unit hydrograph?

7.4.1 The excess rainfall and direct runoff recorded for a storm are as follows:

Time (h)	1	2	3	4	5	6	7	8	9
Excess rainfall (in)	1.0	2.0		1.0					
Direct runoff (cfs)	10	120	400	560	500	450	250	100	50

Calculate the one-hour unit hydrograph.

7.4.2 What is the area of the watershed in Prob. 7.4.1?

7.4.3 Derive by deconvolution the six-hour unit hydrograph from the following data for a watershed having a drainage area of 216 km^2, assuming a constant rainfall abstraction rate and a constant baseflow of 20 m^3/s.

Six-hour period	1	2	3	4	5	6	7	8	9	10	11
Rainfall (cm)	1.5	3.5	2.5	1.5							
Streamflow (m^3/s)	26	71	174	226	173	99	49	33	26	22	21

7.4.4 Given below is the flood hydrograph from a storm on a drainage area of 2.5 mi^2.

Hour	1	2	3	4	5	6	7
Discharge (cfs)	52	48	44	203	816	1122	1138
Hour	8	9	10	11	12	13	
Discharge (cfs)	685	327	158	65	47	34	

Excess rainfall of nearly uniform intensity occurred continuously during the fourth, fifth, and sixth hours. Baseflow separation is accomplished by plotting the logarithm of the discharge against time. During the rising flood, the logarithm of baseflow follows a straight line with slope determined from the flow in hours 1–3. From the point of inflection of the falling limb of the flood hydrograph (hour 8), the logarithm of baseflow follows a straight line with slope determined from the flow in hours 11–13. Between the peak of the flood hydrograph and the point of inflection, the logarithm of baseflow is assumed to vary linearly. Derive the one-hour unit hydrograph by deconvolution.

7.4.5 An intense storm with approximately constant intensity lasting six hours over a watershed of area 785 km^2 produced the following discharges Q in m^3/s:

Hour	0	2	4	6	8	10	12	14	16	18	20
Q	18	21	28	44	70	118	228	342	413	393	334
Q_b	18	20	25	32	40	47	54	61	68	75	79
Hour	22	24	26	28	30	32	34	36	38	40	
Q	270	216	171	138	113	97	84	75	66	59	
Q_b	77	73	69	66	63	60	57	55	52	49	
Hour	42	44	46	48	50	52	54	56	58	60	
Q	54	49	46	42	40	38	36	34	33	33	
Q_b	47	44	42	40	38	37	35	34	33	33	

The baseflow Q_b has been estimated from the appearance of the observed hydrograph. Use deconvolution to determine the two-hour unit hydrograph.

7.5.1 Use the unit hydrograph developed in Prob. 7.4.3 to calculate the streamflow hydrograph from a 12-hour-duration storm having 2 cm of rainfall excess in the first six hours and 3 cm in the second six hours. Assume a constant baseflow rate of 30 m^3/s.

7.5.2 Use the one-hour unit hydrograph developed in Prob. 7.4.4 to calculate the streamflow hydrograph for a three-hour storm with a uniform rainfall intensity of in/h. Assume abstractions are constant at 0.5 in/h and baseflow is the same as determined in Prob. 7.4.4.

7.5.3 Use the two-hour unit hydrograph determined in Prob. 7.4.5 to calculate the streamflow hydrograph from a four-hour storm in which 5 cm of excess rainfall fell in the first two hours and 6 cm in the second two hours. Assume the same baseflow rate as given in Prob. 7.4.5.

7.5.4 The six-hour unit hydrograph of a watershed having a drainage area equal to 393 km^2 is as follows:

Time (h)	0	6	12	18	24	30	36	42
Unit hydrograph ($m^3/s{\cdot}cm$)	0	1.8	30.9	85.6	41.8	14.6	5.5	1.8

For a storm over the watershed having excess rainfall of 5 cm for the first six hours and 15 cm for the second six hours, compute the streamflow hydrograph, assuming constant baseflow of 100 m^3/s.

7.5.5 The one-hour unit hydrograph for a watershed is given below. Determine the runoff from this watershed for the storm pattern given. The abstractions have a constant rate of 0.3 in/h. What is the area of this watershed?

Time (h)	1	2	3	4	5	6
Precipitation (in)	0.5	1.0	1.5	0.5		
Unit hydrograph (cfs/in)	10	100	200	150	100	50

7.5.6 Use the same unit hydrograph as in Prob. 7.5.5 and determine the direct runoff hydrograph for a two-hour storm with 1 in of excess rainfall the first hour and 2 in the second hour. What is the area of this watershed?

7.5.7 An agricultural watershed was urbanized over a period of 20 years. A triangular unit hydrograph was developed for this watershed for an excess rainfall duration of one hour. Before urbanization, the average rate of infiltration and other losses was 0.30 in/h, and the unit hydrograph had a peak discharge of 400 cfs/in at 3 h and a base time of 9 h. After urbanization, because of the increase in impervious surfaces, the loss rate dropped to 0.15 in/h, the peak discharge of the unit hydrograph was increased to 600 cfs/in, occurring at 1 h, and the base time was reduced to 6 h. For a two-hour storm in which 1.0 in of rain fell the first hour and 0.50 in the second hour, determine the direct runoff hydrographs before and after urbanization.

7.5.8 The ordinates at one-hour intervals of a one-hour unit hydrograph are (in cfs/in): 269, 538, 807, 645, 484, 323, and 161. Calculate the direct runoff hydrograph from a two-hour storm in which 4 in of excess rainfall occurs at a constant rate. What is the watershed area (mi^2)?

7.5.9 The 10-minute triangular unit hydrograph from a watershed has a peak discharge of 100 cfs/in at 40 min and a total duration of 100 min. Calculate the streamflow hydrograph from this watershed for a storm in which 2 in of rain falls in the first 10 minutes and 1 in in the second 10 minutes, assuming that the loss rate is $\phi = 0.6$ in/h and the baseflow rate is 20 cfs.

7.6.1 The July 19–20, 1979, storm on the Shoal Creek watershed at Northwest Park in Austin, Texas, resulted in the following rainfall-runoff values.

Time(h)	0.5	1.0	1.5	2.0	2.5	3.0	3.5
Rainfall (in)	1.17	0.32	0.305	0.67	0.545	0.10	0.06
Direct runoff (cfs)	11.0	372.0	440.0	506.0	2110.0	1077.0	429.3
Time (h)	4.0	4.5	5.0	5.5	6.0	6.5	7.0
Direct runoff (cfs)	226.6	119.0	64.7	39.7	28.0	21.7	16.7
Time (h)	7.5	8.0	8.5	9.0			
Direct runoff (cfs)	13.3	9.2	9.0	7.3			

Determine the half-hour unit hydrograph using linear programming. Assume that a uniform loss rate is valid. The watershed area is 7.03 mi^2. Compare the unit hydrograph with that determined in Example 7.4.1 for this watershed.

7.6.2 A storm on April 16, 1977, on the Shoal Creek watershed at Northwest Park in Austin, Texas, resulted in the following rainfall-runoff values:

Time (h)	0.5	1.0	1.5	2.0	2.5	3.0	3.5	4.0	4.5
Rainfall (in)	0.28	0.12	0.13	0.14	0.18	0.14	0.07		
Direct runoff (cfs)	32	67	121	189	279	290	237	160	108
Time (h)	5.0	5.5	6.0	6.5	7.0	7.5	8.0	8.5	9.0
Direct runoff (cfs)	72	54	44	33	28	22	20	18	16

Determine the half-hour unit hydrograph by linear programming. Assume that a uniform loss rate is valid. The watershed area is 7.03 mi^2. Compare the unit hydrograph with that developed in Example 7.4.1 for this watershed.

7.6.3 Combine the data from Probs. 7.6.1 and 7.6.2 and calculate a composite unit hydrograph from this watershed by linear programming. Compare the composite unit hydrograph with those determined from the individual storms.

7.6.4 Solve Prob. 7.6.1 by linear regression.

7.6.5 Solve Prob. 7.6.2 by linear regression.

7.6.6 Solve Prob. 7.6.3 by linear regression.

7.7.1 The City of Austin, Texas, uses generalized equations (7.7.9)–(7.7.13) to determine the parameters for 10-minute-duration unit hydrographs for small watersheds. Determine the 10-minute unit hydrographs for levels of imperviousness 10, 40, and 70 percent, on a watershed that has an area of 0.42 mi^2 with a main channel length of 5760 ft. The main channel slope is 0.015 ft/ft as defined in Sec. 7.7. Assume $\Phi = 0.8$. Plot the three unit hydrographs on the same graph.

7.7.2 Using the 10-minute unit hydrograph equations (7.7.9)–(7.7.13), develop the unit hydrograph for a small watershed of 0.3 mi^2 that has a main channel slope of 0.009 ft/ft. The main channel area is 2000 feet long and the percent imperviousness is 25. Next, develop the 10-minute unit hydrograph for the same watershed assuming the main channel length is 6000 feet long. Plot and compare the two unit hydrographs. Assume $n = 0.05$ for the main channel.

7.7.3 Determine direct runoff hydrographs using the two 10-minute unit hydrographs derived in the previous problem for the watersheds with main channel lengths of 2000 ft and 6000 ft. Consider a storm having 1.2 inches rainfall uniformly distributed over the first 30 minutes and 1.5 inches in the second 30 minutes. The infiltration losses are to be determined using the SCS method described in Chap. 5 for curve number CN = 85.

7.7.4 The 10-minute unit hydrograph for a 0.86-mi^2 watershed has 10-minute ordinates in cfs/in of 134, 392, 475, 397, 329, 273, 227, 188, 156, 129, 107, 89, 74, 61, 51, 42, 35, 29, 24, 10, 17, 14, 11, ... Determine the peaking coefficient C_p for Snyder's method. The main channel length is 10,500 ft, and $L_c = 6000$ ft. Determine the coefficient C_t.

7.7.5 Several equations for computing basin lag have been reported in the literature. One such equation that also considers the basin slope was presented by Linsley, Kohler, and Paulhus (1982):

$$t_p = C_t\left(\frac{LL_c}{\sqrt{S}}\right)^n$$

For a basin slope of $S = 0.008$ and $n = 0.4$, determine the coefficient C_t for the unit hydrograph in the previous problem.

7.7.6 The following information for watershed A and its two-hour unit hydrograph has been determined: area = 100 mi^2, $L_c = 10$ mi, $L = 24$ mi, $t_R = 2$ h, $t_{pR} = 6$ h, $Q_p = 9750$ cfs/in, $W_{50} = 4.1$ h, and $W_{75} = 2$ h. Watershed B, which is assumed to be hydrologically similar to watershed A, has the following characteristics: area = 70 mi^2, $L = 15.6$ mi, and $L_c = 9.4$ mi. Determine the one-hour synthetic unit hydrograph for watershed B.

7.7.7 (*a*) Determine the coefficients C_p and C_t for a watershed of area 100 mi^2 with $L = 20$ mi and $L_c = 12$ mi, for $t_R = 2$ h and $t_{pR} = 5$ h. The peak of the unit hydrograph is 9750 cfs/in. Assume Snyder's synthetic unit hydrograph applies.
(*b*) Determine the two-hour unit hydrograph for the upper 70-mi^2 area of the same watershed, which has $L = 12.6$ mi and $L_c = 7.4$ mi. The values of W_{75} and W_{50} for the entire 100-mi^2-area watershed are 2.0 h and 4.2 h, respectively.

7.7.8 The Gimlet Creek watershed at Sparland, Illinois, has a drainage area of 5.42 mi^2; the length of the main stream is 4.45 mi and the main channel length from the watershed outlet to the point opposite the center of gravity of the watershed is 2.0 mi. Using $C_t = 2.0$ and $C_p = 0.625$, determine the standard synthetic unit hydrograph for this basin. What is the standard duration? Use Snyder's method to determine the 30-minute unit hydrograph for this watershed.

7.7.9 The Odebolt Creek watershed near Arthur, Ohio, has a watershed area of 39.3 mi^2; the length of the main channel is 18.10 mi, and the main channel length from the watershed outlet to the point opposite the centroid of the watershed is 6.0 mi. Using $C_t = 2.0$ and $C_p = 0.625$, determine the standard synthetic unit hydrograph and the two-hour unit hydrograph for this watershed.

7.7.10 An 8-mi^2 watershed has a time of concentration of 1.0 h. Calculate a 10-minute unit hydrograph for this watershed by the SCS triangular unit hydrograph method.

Determine the direct runoff hydrograph for a 20-minute storm having 0.6 in of excess rainfall in the first 10 minutes and 0.4 in in the second 10 minutes.

7.7.11 A triangular synthetic unit hydrograph developed by the Soil Conservation Service method has $q_p = 2900$ cfs/in, $T_p = 50$ min, and $t_r = 10$ min. Compute the direct runoff hydrograph for a 20-minute storm, having 0.66 in rainfall in the first 10 minutes and 1.70 in in the second 10 minutes. The rainfall loss rate is $\phi = 0.6$ in/h throughout the storm.

7.8.1 For the data given in Prob. 7.4.4, use the assumption of constant rainfall intensity in hours 4–6 to construct the S-hydrograph. Use the S-hydrograph to calculate the one-, three-, and six-hour unit hydrographs.

7.8.2 For the data given in Prob. 7.4.5, use the assumption of constant rainfall intensity for six hours to construct the S-hydrograph for this watershed. From the S-hydrograph, determine the 2-, 6-, and 12-hour unit hydrographs for this watershed.

7.8.3 The ordinates of a one-hour unit hydrograph specified at one-hour intervals are (in cfs/in): 45, 60, 22, 8, and 1. Calculate the watershed area, the S-hydrograph and the two-hour unit hydrograph for this watershed.

CHAPTER

8

LUMPED FLOW ROUTING

Flow routing is a procedure to determine the time and magnitude of flow (i.e., the flow hydrograph) at a point on a watercourse from known or assumed hydrographs at one or more points upstream. If the flow is a flood, the procedure is specifically known as *flood routing*. In a broad sense, flow routing may be considered as an analysis to trace the flow through a hydrologic system, given the input. The difference between *lumped* and *distributed* system routing is that in a lumped system model, the flow is calculated as a function of time alone at a particular location, while in a distributed system routing the flow is calculated as a function of space and time throughout the system. Routing by lumped system methods is sometimes called *hydrologic routing,* and routing by distributed systems methods is sometimes referred to as *hydraulic routing*. Flow routing by distributed-system methods is described in Chaps. 9 and 10. This chapter deals with lumped system routing.

8.1 LUMPED SYSTEM ROUTING

For a hydrologic system, input $I(t)$, output $Q(t)$, and storage $S(t)$ are related by the continuity equation (2.2.4):

$$\frac{dS}{dt} = I(t) - Q(t) \tag{8.1.1}$$

If the inflow hydrograph, $I(t)$, is known, Eq. (8.1.1) cannot be solved directly to obtain the outflow hydrograph, $Q(t)$, because both Q and S are unknown. A second relationship, or *storage function,* is needed to relate S, I, and Q; coupling the storage function with the continuity equation provides a solvable combination of two equations and two unknowns. In general, the storage function may be written as an arbitrary function of I, Q, and their time derivatives as shown by Eq. (7.1.2):

$$S = f\left(I, \frac{dI}{dt}, \frac{d^2I}{dt^2}, \ldots, Q, \frac{dQ}{dt}, \frac{d^2Q}{dt^2}, \ldots\right) \tag{8.1.2}$$

In Chapter 7, these two equations were solved by differentiating a linearized form of Eq. (8.1.2), substituting the result for dS/dt into Eq. (8.1.1), then integrating the resulting differential equation to obtain $Q(t)$ as a function of $I(t)$. In this chapter, a finite difference solution method is applied to the two equations. The time horizon is divided into finite intervals, and the continuity equation (8.1.1) is solved recursively from one time point to the next using the storage function (8.1.2) to account for the value of storage at each time point.

The specific form of the storage function to be employed in this procedure depends on the nature of the system being analyzed. In this chapter, three particular systems are analyzed. First, reservoir routing by the *level pool method,* in which storage is a nonlinear function of Q only:

$$S = f(Q) \tag{8.1.3}$$

and the function $f(Q)$ is determined by relating reservoir storage and outflow to reservoir water level. Second, storage is linearly related to I and Q in the *Muskingum method* for flow routing in channels. Finally, several *linear reservoir models* are analyzed in which (8.1.2) becomes a linear function of Q and its time derivatives.

The relationship between the outflow and the storage of a hydrologic system has an important influence on flow routing. This relationship may be either *invariable* or *variable,* as shown in Fig. 8.1.1. An invariable storage function has the form of Eq. (8.1.3) and applies to a reservoir with a horizontal water surface. Such reservoirs have a pool that is wide and deep compared with its length in the direction of flow. The velocity of flow in the reservoir is very low. The invariable storage relationship requires that there be a fixed discharge from the reservoir for a given water surface elevation, which means that the reservoir outlet works must be either uncontrolled, or controlled by gates held at a fixed position. If the control gate position is changed, the discharge and water surface elevation change at the dam, and the effect propagates upstream in the reservoir to create a sloping water surface temporarily, until a new equilibrium water surface elevation is established throughout the reservoir.

When a reservoir has a horizontal water surface, its storage is a function of its water surface elevation, or depth in the pool. Likewise, the outflow discharge is a function of the water surface elevation, or head on the outlet works. By combining these two functions, the reservoir storage and discharge can be related to produce an invariable, single-valued storage function, $S = f(Q)$, as shown in Fig. 8.1.1(a). For such reservoirs, the peak outflow occurs when the outflow hydrograph intersects the inflow hydrograph, because the maximum storage occurs

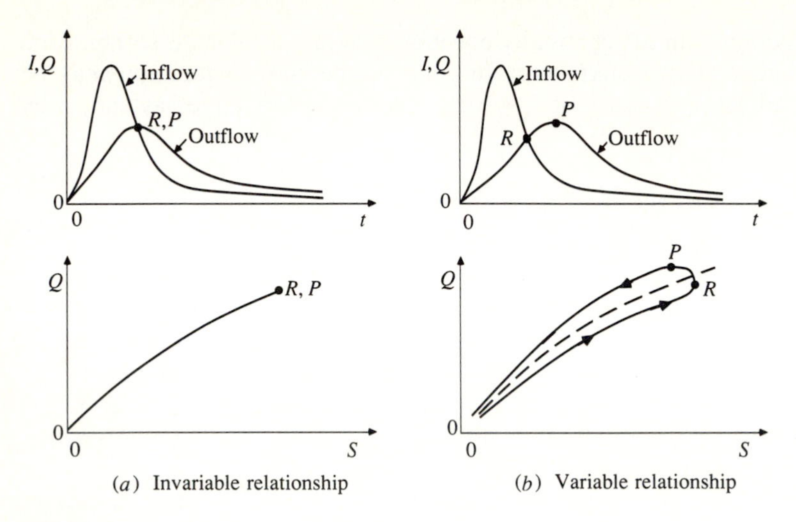

FIGURE 8.1.1
Relationships between discharge and storage.

when $dS/dt = I - Q = 0$, and the storage and outflow are related by $S = f(Q)$. This is indicated in Fig. 8.1.1(*a*) where the points denoting the maximum storage, *R,* and maximum outflow, *P,* coincide.

A variable storage-outflow relationship applies to long, narrow reservoirs, and to open channels or streams, where the water surface profile may be significantly curved due to backwater effects. The amount of storage due to backwater depends on the time rate of change of flow through the system. As shown in Fig. 8.1.1(*b*), the resulting relationship between the discharge and the system storage is no longer a single-valued function but exhibits a curve usually in the form of a single or twisted loop, depending on the storage characteristics of the system. Because of the retarding effect due to backwater, the peak outflow usually occurs later than the time when the inflow and outflow hydrographs intersect, as indicated in Fig. 8.1.1(*b*), where the points *R* and *P* do not coincide. If the backwater effect is not very significant, the loop shown in Fig. 8.1.1(*b*) may be replaced by an average curve shown by the dashed line. Thus, level pool routing methods can also be applied in an approximate way to routing with a variable discharge-storage relationship.

The preceding discussion indicates that the effect of storage is to redistribute the hydrograph by shifting the centroid of the inflow hydrograph to the position of that of the outflow hydrograph in a *time of redistribution*. In very long channels, the entire flood wave also travels a considerable distance and the centroid of its hydrograph may then be shifted by a time period longer than the time of redistribution. This additional time may be considered as *time of translation*. As shown in Fig. 8.1.2, the total *time of flood movement* between the centroids of the outflow and inflow hydrographs is equal to the sum of the time of redistribution and the time of translation. The process of redistribution modifies the shape of the hydrograph, while translation changes its position.

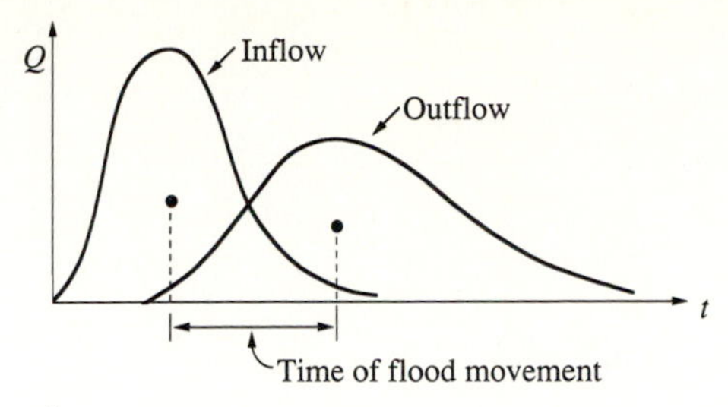

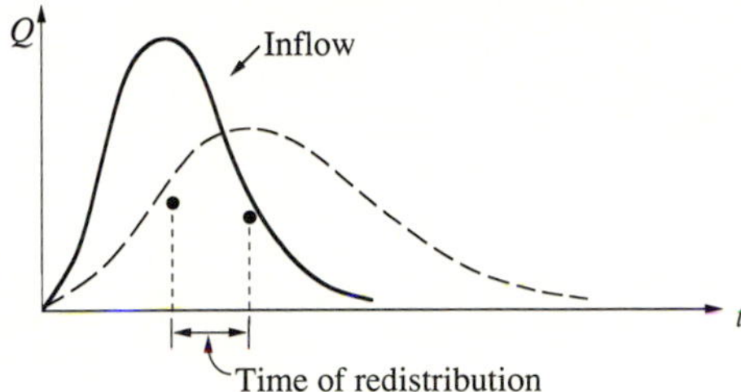

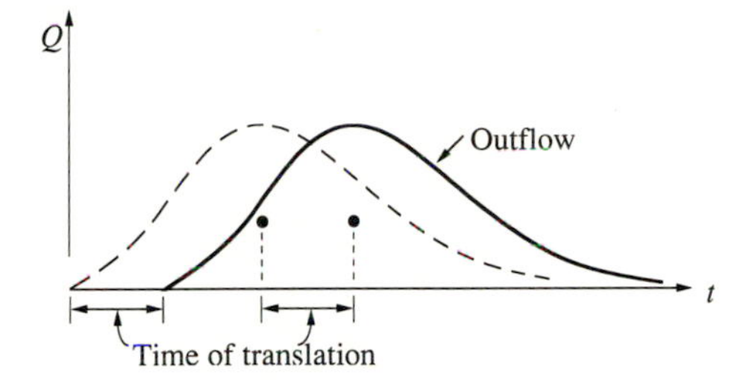

FIGURE 8.1.2
Conceptual interpretation of the time of flood movement.

8.2 LEVEL POOL ROUTING

Level pool routing is a procedure for calculating the outflow hydrograph from a reservoir with a horizontal water surface, given its inflow hydrograph and storage-outflow characteristics. A number of procedures have been proposed for this purpose (e.g., Chow, 1951, 1959), and with the advance of computerization, graphical procedures are being replaced by tabular or functional methods so that the computational procedure can be automated.

The time horizon is broken into intervals of duration Δt, indexed by j, that is, $t = 0, \Delta t, 2\Delta t, \ldots, j\Delta t, (j+1)\Delta t, \ldots$, and the continuity equation (8.1.1) is integrated over each time interval, as shown in Fig. 8.2.1. For the j-th time interval:

$$\int_{S_j}^{S_{j+1}} dS = \int_{j\Delta t}^{(j+1)\Delta t} I(t)dt - \int_{j\Delta t}^{(j+1)\Delta t} Q(t)dt \tag{8.2.1}$$

The inflow values at the beginning and end of the j-th time interval are I_j and I_{j+1}, respectively, and the corresponding values of the outflow are Q_j and Q_{j+1}. Here, both inflow and outflow are flow rates measured as sample data, rather than inflow being pulse data and outflow being sample data as was the case for the unit hydrograph. If the variation of inflow and outflow over the interval is

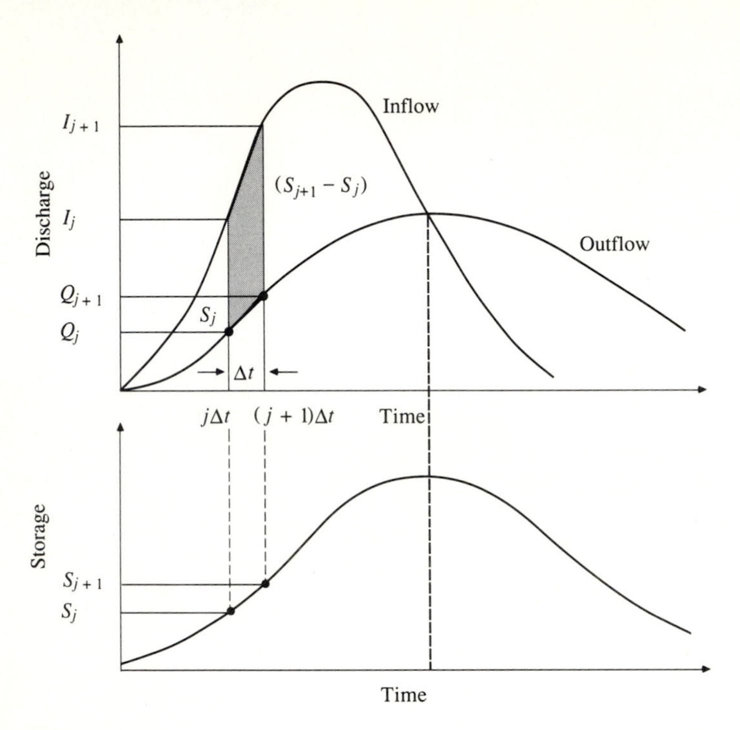

FIGURE 8.2.1
Change of storage during a routing period Δt.

approximately linear, the change in storage over the interval, $S_{j+1} - S_j$, can be found by rewriting (8.2.1) as

$$S_{j+1} - S_j = \frac{I_j + I_{j+1}}{2}\Delta t - \frac{Q_j + Q_{j+1}}{2}\Delta t \tag{8.2.2}$$

The values of I_j and I_{j+1} are known because they are prespecified. The values of Q_j and S_j are known at the jth time interval from calculation during the previous time interval. Hence, Eq. (8.2.2) contains two unknowns, Q_{j+1} and S_{j+1}, which are isolated by multiplying (8.2.2) through by $2/\Delta t$, and rearranging the result to produce:

$$\left(\frac{2S_{j+1}}{\Delta t} + Q_{j+1}\right) = (I_j + I_{j+1}) + \left(\frac{2S_j}{\Delta t} - Q_j\right) \tag{8.2.3}$$

In order to calculate the outflow, Q_{j+1}, from Eq. (8.2.3), a *storage-outflow function* relating $2S/\Delta t + Q$ and Q is needed. The method for developing this function using elevation-storage and elevation-outflow relationships is shown in Fig. 8.2.2. The relationship between water surface elevation and reservoir storage can be derived by planimetering topographic maps or from field surveys. The elevation-discharge relation is derived from hydraulic equations relating head and discharge, such as those shown in Table 8.2.1, for various types of spillways

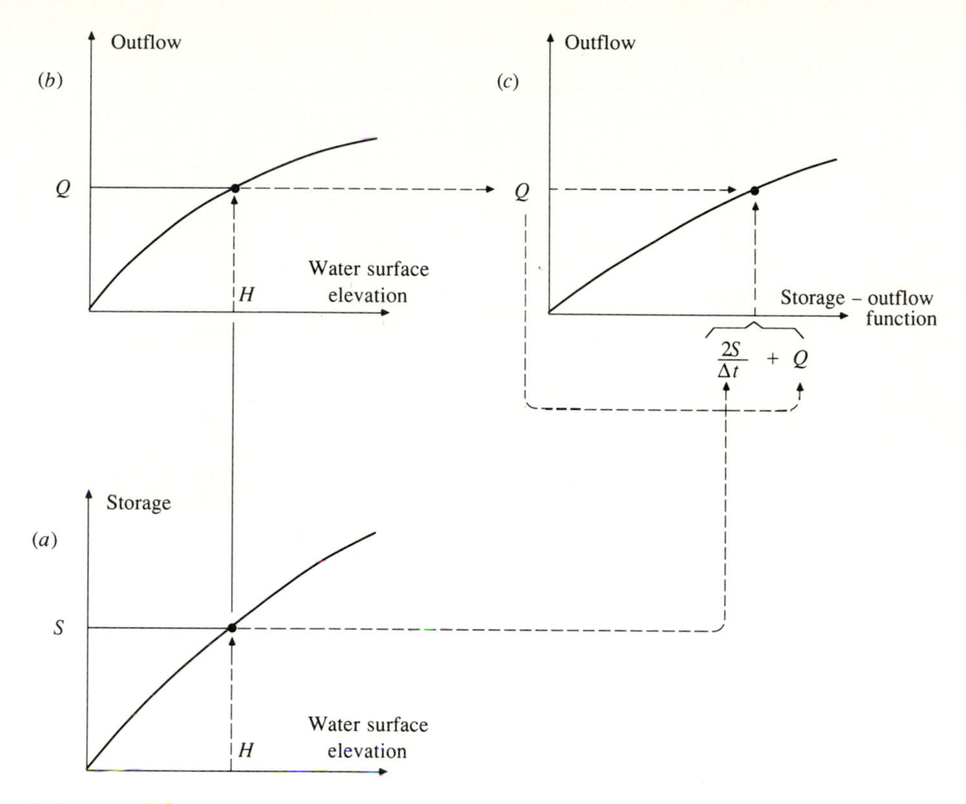

FIGURE 8.2.2
Development of the storage-outflow function for level pool routing on the basis of storage-elevation and elevation-outflow curves.

and outlet works. The value of Δt is taken as the time interval of the inflow hydrograph. For a given value of water surface elevation, the values of storage S and discharge Q are determined [parts (*a*) and (*b*) of Fig. 8.2.2], then the value of $2S/\Delta t + Q$ is calculated and plotted on the horizontal axis of a graph with the value of the outflow Q on the vertical axis [part (*c*) of Fig. 8.2.2].

In routing the flow through time interval j, all terms on the right side of Eq. (8.2.3) are known, and so the value of $2S_{j+1}/\Delta t + Q_{j+1}$ can be computed. The corresponding value of Q_{j+1} can be determined from the storage-outflow function $2S/\Delta t + Q$ versus Q, either graphically or by linear interpolation of tabular values. To set up the data required for the next time interval, the value of $2S_{j+1}/\Delta t - Q_{j+1}$ is calculated by

$$\left(\frac{2S_{j+1}}{\Delta t} - Q_{j+1}\right) = \left(\frac{2S_{j+1}}{\Delta t} + Q_{j+1}\right) - 2Q_{j+1} \tag{8.2.4}$$

The computation is then repeated for subsequent routing periods.

Example 8.2.1. A reservoir for detaining flood flows is one acre in horizontal area, has vertical sides, and has a 5-ft diameter reinforced concrete pipe as the

TABLE 8.2.1
Spillway discharge equations

Spillway type	Equation	Notation
Uncontrolled overflow ogee crest	$Q = CLH^{3/2}$	Q = discharge, cfs C = variable coefficient of discharge L = effective length of crest H = total head on the crest including velocity of approach head.
Gate controlled ogee crest	$Q = \frac{2}{3}\sqrt{2g}CL\left(H_1^{3/2} - H_2^{3/2}\right)$	H_1 = total head to bottom of the opening H_2 = total head to top of the opening C = coefficient which differs with gate and crest arrangement
Morning glory spillway	$Q = C_o(2\pi R_s)H^{3/2}$	C_o = coefficient related to H and R_s R_s = radius of the overflow crest H = total head
Culvert (submerged inlet control)	$Q = C_d WD\sqrt{2gH}$	W = entrance width D = height of opening C_d = discharge coefficient

Source: Design of Small Dams, Bureau of Reclamation, U. S. Department of the Interior, 1973.

outlet structure. The headwater-discharge relation for the outlet pipe is given in columns 1 and 2 of Table 8.2.2. Use the level pool routing method to calculate the reservoir outflow from the inflow hydrograph given in columns 2 and 3 of Table 8.2.3. Assume that the reservoir is initially empty.

Solution. The inflow hydrograph is specified at 10-min time intervals, so $\Delta t = 10$ min $= 600$ s. For all elevations, the horizontal area of the reservoir water surface is 1 acre = 43,560 ft^2, and the storage is calculated as 43,560 × (depth of water). For example, for a depth of 0.5 ft, $S = 0.5 \times 43,560 = 21,780$ ft^3, as shown in column 3 of Table 8.2.2. The corresponding value of $2S/\Delta t + Q$ can then be determined. For a depth 0.5 ft, the discharge is given in column 2 of Table 8.2.2 as 3 cfs, so the storage-outflow function value is

$$\frac{2S}{\Delta t} + Q = \frac{2 \times 21,780}{600} + 3 = 76 \text{ cfs}$$

TABLE 8.2.2
Development of the storage-outflow function for a detention reservoir (Example 8.2.1).

Column:	1 Elevation H (ft)	2 Discharge Q (cfs)	3 Storage S (ft³)	4 $(2S/\Delta t)^* + Q$ (cfs)
	0.0	0	0	0
	0.5	3	21,780	76
	1.0	8	43,560	153
	1.5	17	65,340	235
	2.0	30	87,120	320
	2.5	43	108,900	406
	3.0	60	130,680	496
	3.5	78	152,460	586
	4.0	97	174,240	678
	4.5	117	196,020	770
	5.0	137	217,800	863
	5.5	156	239,580	955
	6.0	173	261,360	1044
	6.5	190	283,140	1134
	7.0	205	304,920	1221
	7.5	218	326,700	1307
	8.0	231	348,480	1393
	8.5	242	370,260	1476
	9.0	253	392,040	1560
	9.5	264	413,820	1643
	10.0	275	435,600	1727

*Time interval $\Delta t = 10$ min.

as shown in column 4 of Table 8.2.2. The storage-outflow function is plotted in Fig. 8.2.3.

The flow routing calculations are carried out using Eq. (8.2.3). For the first time interval, $S_1 = Q_1 = 0$ because the reservoir is initially empty; hence $(2S_1/\Delta t - Q_1) = 0$ also. The inflow values are $I_1 = 0$ and $I_2 = 60$ cfs, so $(I_1 + I_2) = 0 + 60 = 60$ cfs. The value of the storage-outflow function at the end of the time interval is calculated from (8.2.3) with $j = 1$:

$$\left(\frac{2S_2}{\Delta t} + Q_2\right) = (I_1 + I_2) + \left(\frac{2S_1}{\Delta t} - Q_1\right)$$

$$= 60 + 0$$

$$= 60 \text{ cfs}$$

The value of Q_{j+1} is found by linear interpolation given $2S_{j+1}/\Delta t + Q_{j+1}$. If there is a pair of variables (x,y), with known pairs of values (x_1, y_1) and (x_2, y_2), then the interpolated value of y corresponding to a given value of x in the range $x_1 \leq x \leq x_2$ is

$$y = y_1 + \frac{(y_2 - y_1)}{(x_2 - x_1)}(x - x_1)$$

TABLE 8.2.3
Routing of flow through a detention reservoir by the level pool method (Example 8.2.1). The computational sequence is indicated by the arrows in the table.

Column: 1 Time index j	2 Time (min)	3 Inflow (cfs)	4 $I_j + I_{j+1}$ (cfs)	5 $\frac{2S_j}{\Delta t} - Q_j$ (cfs)	6 $\frac{2S_{j+1}}{\Delta t} + Q_{j+1}$ (cfs)	7 Outflow (cfs)
1	0	0		0.0		0.0
2	10	60	60	55.2	60.0	2.4
3	20	120	180	201.1	235.2	17.1
4	30	180	300	378.9	501.1	61.1
5	40	240	420	552.6	798.9	123.2
6	50	300	540	728.2	1092.6	182.2
7	60	360	660	927.5	1388.2	230.3
8	70	320	680	1089.0	1607.5	259.3
9	80	280	600	1149.0	1689.0	270.0
10	90	240	520	1134.3	1669.0	267.4
11	100	200	440	1064.4	1574.3	254.9
12	110	160	360	954.1	1424.4	235.2
13	120	120	280	820.2	1234.1	206.9
14	130	80	200	683.3	1020.2	168.5
15	140	40	120	555.1	803.3	124.1
16	150	0	40	435.4	595.1	79.8
17	160		0	338.2	435.4	48.6
18	170			272.8	338.2	32.7
19	180			227.3	272.8	22.8
20	190			194.9	227.3	16.2
21	200			169.7	194.9	12.6
22	210				169.7	9.8

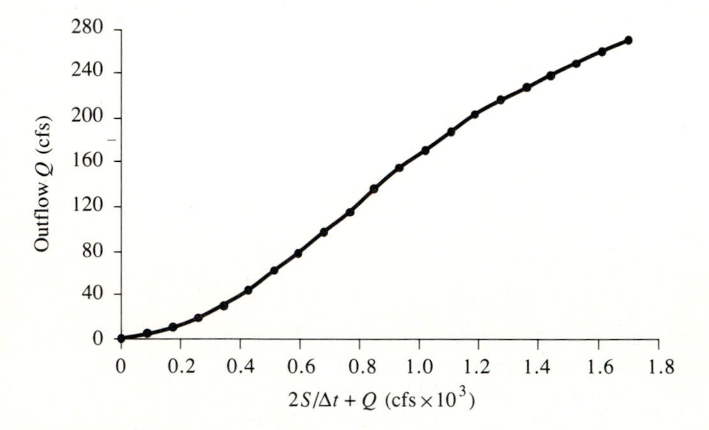

FIGURE 8.2.3
Storage-outflow function for a detention reservoir (Example 8.2.1).

In this case, $x = 2S/\Delta t + Q$ and $y = Q$. Two pairs of values around $2S/\Delta t + Q = 60$ are selected from Table 8.2.2; they are $(x_1, y_1) = (0, 0)$ and $(x_2, y_2) = (76, 3)$. The value of y for $x = 60$ is, by linear interpolation,

$$y = 0 + \frac{(3-0)}{(76-0)}(60-0)$$
$$= 2.4 \text{ cfs}$$

So, $Q_2 = 2.4$ cfs, and the value of $2S_2/\Delta t - Q_2$ needed for the next iteration is found using Eq. (8.2.4) with $j = 2$:

$$\left(\frac{2S_2}{\Delta t} - Q_2\right) = \left(\frac{2S_2}{\Delta t} + Q_2\right) - 2Q_2$$
$$= 60 - 2 \times 2.4$$
$$= 55.2 \text{ cfs}$$

The sequence of computations just described is indicated by the arrows in the first two rows of Table 8.2.3.

Proceeding to the next time interval, $(I_2 + I_3) = 60 + 120 = 180$ cfs, and the routing is performed with $j = 2$ in (8.2.3).

$$\left(\frac{2S_3}{\Delta t} + Q_3\right) = (I_2 + I_3) + \left(\frac{2S_2}{\Delta t} - Q_2\right)$$
$$= 180 + 55.2$$
$$= 235.2 \text{ cfs}$$

By linear interpolation in Table 8.2.2, the value of $Q_3 = 17.1$ cfs and by Eq. (8.2.4), $2S_3/\Delta t - Q_3 = 201.1$ cfs, as shown in the third row of Table 8.2.3. The calculations for subsequent time intervals are performed in the same way, with the results tabulated in Table 8.2.3 and plotted in Fig. 8.2.4. The peak inflow is 360 cfs and occurs at 60 min; the detention reservoir reduces the peak outflow to 270 cfs and delays it until 80 min. As discussed in Sec. 8.1, the outflow is maximized at the point where the inflow and outflow are equal, because storage is also maximized

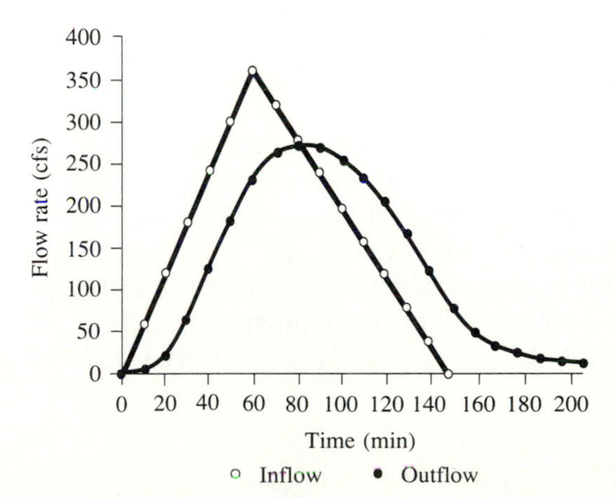

FIGURE 8.2.4
Routing of flow through a detention reservoir (Example 8.2.1).

at that time, and there is a single-valued function relating storage and outflow for a reservoir with a level pool.

The maximum depth in the storage reservoir is calculated by linear interpolation from Table 8.2.2 as 9.77 ft at the peak discharge of 270 cfs. If this depth is too great, or if the discharge of 270 cfs in the 5-ft outlet pipe is too large, either the outlet structure or the surface area of the basin must be enlarged. An equivalent size of elliptical or arch pipe would also tend to lower the headwater elevation.

8.3 RUNGE-KUTTA METHOD

An alternative method for level pool routing can be developed by solving the continuity equation using a numerical method such as the Runge-Kutta method. The Runge-Kutta method is more complicated than the method described in the previous section, but it does not require the computation of the special storage-outflow function, and it is more closely related to the hydraulics of flow through the reservoir. Various orders of Runge-Kutta schemes can be adopted (Carnahan, et al., 1969). A *third order scheme* is described here; it involves breaking each time interval into three increments and calculating successive values of water surface elevation and reservoir discharge for each increment.

The continuity equation is expressed as

$$\frac{dS}{dt} = I(t) - Q(H) \tag{8.3.1}$$

where S is the volume of water in storage in the reservoir; $I(t)$ is the inflow into the reservoir as a function of time; and $Q(H)$ is the outflow from the reservoir, which is determined by the head or elevation (H) in the reservoir. The change in volume, dS, due to a change in elevation, dH, can be expressed as

$$dS = A(H)dH \tag{8.3.2}$$

where $A(H)$ is the water surface area at elevation H. The continuity equation is then rewritten as

$$\frac{dH}{dt} = \frac{I(t) - Q(H)}{A(H)} \tag{8.3.3}$$

The solution is extended forward by small increments of the independent variable, time, using known values of the dependent variable H. For a third order scheme, there are three such increments in each time interval Δt, and three successive approximations are made for the change in head elevation, dH.

Fig. 8.3.1 illustrates how the three approximate values $\Delta H_1, \Delta H_2$, and ΔH_3 are defined for the j-th interval. The slope, dH/dt, approximated by $\Delta H/\Delta t$, is first evaluated at (H_j, t_j), then at $(H_j + \Delta H_1/3, t_j + \Delta t/3)$, and finally at $(H_j + 2\Delta H_2/3, t_j + 2\Delta t/3)$. In equations,

$$\Delta H_1 = \frac{I(t_j) - Q(H_j)}{A(H_j)} \Delta t \tag{8.3.4a}$$

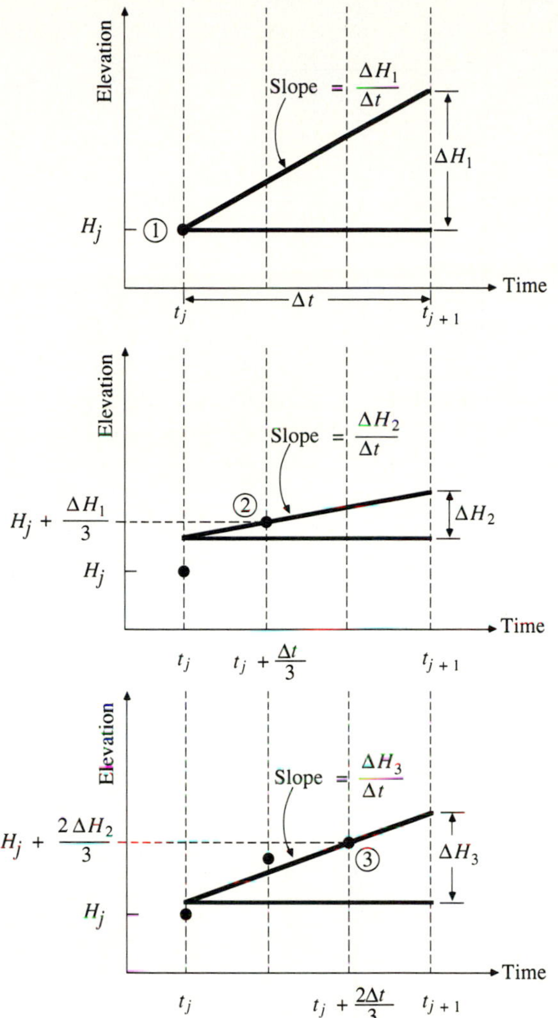

FIGURE 8.3.1
Steps to define elevation increments in the third-order Runge-Kutta method.

$$\Delta H_2 = \frac{I\left(t_j + \frac{\Delta t}{3}\right) - Q\left(H_j + \frac{\Delta H_1}{3}\right)}{A\left(H_j + \frac{\Delta H_1}{3}\right)} \Delta t \tag{8.3.4b}$$

$$\Delta H_3 = \frac{I\left(t_j + \frac{2\Delta t}{3}\right) - Q\left(H_j + \frac{2\Delta H_2}{3}\right)}{A\left(H_j + \frac{2\Delta H_2}{3}\right)} \Delta t \tag{8.3.4c}$$

The value of H_{j+1} is given by

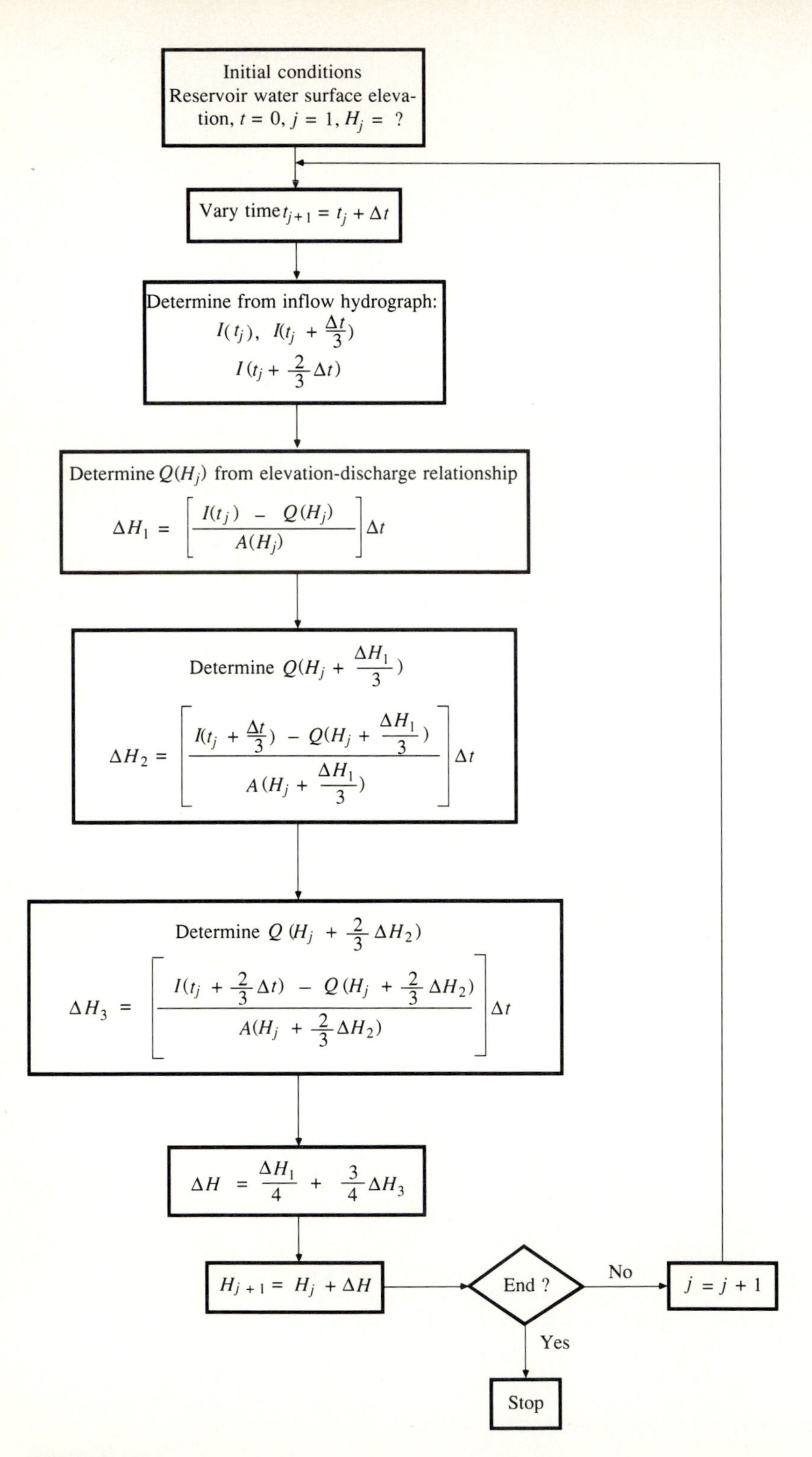

FIGURE 8.3.2
Flowchart of detention basin routing using the third-order Runge-Kutta technique.

$$H_{j+1} = H_j + \Delta H \tag{8.3.5}$$

where

$$\Delta H = \frac{\Delta H_1}{4} + \frac{3\Delta H_3}{4} \tag{8.3.6}$$

A flowchart of the third order Runge-Kutta method is shown in Fig. 8.3.2.

Example 8.3.1. Use the third order Runge-Kutta method to perform the reservoir routing through the one-acre detention reservoir with vertical walls, as described in Example 8.2.1. The elevation-discharge relationship is given in columns 1 and 2 of Table 8.2.2 and the inflow hydrograph in columns 1 and 2 of Table 8.3.1.

Solution. The function $A(H)$ relating the water surface area to the reservoir elevation is simply $A(H) = 43,560$ ft^2 for all values of H because the reservoir has a base area of one acre and vertical sides. A routing interval of Δt = 10 min is used. The procedure begins with the determination of $I(0), I(0 + 10/3)$, and $I(0 + (2/3) \times (10))$, which are found by linear interpolation between the values of 0 and 60 cfs found in column 2 of Table 8.3.1; they are 0, 20, and 40 cfs, respectively. Next, ΔH_1 is computed using Eq. (8.3.4*a*) with Δt = 10 min = 600 s, A = 43,560 ft^2, and $I(0)$ = 0 cfs; since the reservoir is initially empty, $H_j = 0$ and $Q(H_j) = 0$:

$$\begin{aligned}\Delta H_1 &= \frac{I(t_j) - Q(H_j)}{A(H_j)} \Delta t \\ &= \frac{(0 - 0)}{43,560} \times 600 \\ &= 0\end{aligned}$$

TABLE 8.3.1
Routing an inflow hydrograph through a detention reservoir by the Runge-Kutta method (Example 8.3.1).

Column:	1 Time (min)	2 Inflow (cfs)	3 ΔH_1	4 ΔH_2	5 ΔH_3	6 Depth (ft)	7 Outflow (cfs)
	0	0	–	–	–	0	0
	10	60	0	0.28	0.54	0.40	2.4
	20	120	0.79	1.04	1.24	1.53	17.9
	30	180	1.41	1.51	1.59	3.08	62.8
	40	240	1.61	1.62	1.61	4.69	124.5
	50	300	1.59	1.58	1.60	6.28	182.6
	60	360	1.62	1.66	1.72	7.98	230.4
	70	320	1.79	1.42	1.13	9.27	259.0
	80	280	0.84	0.57	0.36	9.75	269.5
	90	240	0.15	−0.05	−0.21	9.63	266.8
	100	200	−0.37	−0.52	−0.63	9.06	254.3
	110	160	−0.75	−0.86	−0.94	8.17	234.7
	120	120	−1.03	−1.10	−1.14	7.05	206.4
	130	80	−1.19	−1.21	−1.21	5.85	167.8
	140	40	−1.21	−1.20	−1.18	4.66	123.5
	150	0	−1.15	−1.12	−1.11	3.54	80.0

For the next time increment, using (8.3.4*b*) with $I(0 + 10/3) = 20$ ft^3/s,

$$\Delta H_2 = \frac{I\left(t_j + \frac{\Delta t}{3}\right) - Q\left(H_j + \frac{\Delta H_1}{3}\right)}{A\left(H_j + \frac{\Delta H_1}{3}\right)} \Delta t$$

$$= \frac{(20 - 0)}{43,560} \times 600$$

$$= 0.28 \text{ ft}$$

For the last increment, $H_j + (2/3)\Delta H_2 = 0 + (2/3)(0.28) = 0.18$ ft. By linear interpolation from Table 8.2.2, $Q(0.18) = 1.10$ cfs. By substitution in (8.3.4*c*)

$$\Delta H_3 = \frac{I\left(t_j + \frac{2\Delta t}{3}\right) - Q\left(H_j + \frac{2\Delta H_2}{3}\right)}{A\left(H_j + \frac{2\Delta H_2}{3}\right)} \Delta t$$

$$= \frac{(40 - 1.10)}{43,560} \times 600$$

$$= 0.54 \text{ ft}$$

The values of ΔH_1, ΔH_2 and ΔH_3 are found in columns 3, 4, and 5 of Table 8.3.1. Then, for the whole ten-minute time interval, ΔH is computed using Eq. (8.3.6):

$$\Delta H = \frac{\Delta H_1}{4} + \frac{3\Delta H_3}{4}$$

$$= \frac{0}{4} + \frac{3}{4}(0.54) = 0.40 \text{ ft}$$

So, H at 10 min is given by $H_2 = H_1 + \Delta H_1 = 0 + 0.40 = 0.40$ ft (column 6), and the corresponding discharge from the pipe is interpolated from Table 8.2.2 as $Q = 2.4$ cfs (column 7).

The routing calculations for subsequent periods follow the same procedure, and the solution, extended far enough to cover the peak outflow, is presented in Table 8.3.1. The result is very similar to that obtained in Example 8.2.1 by the

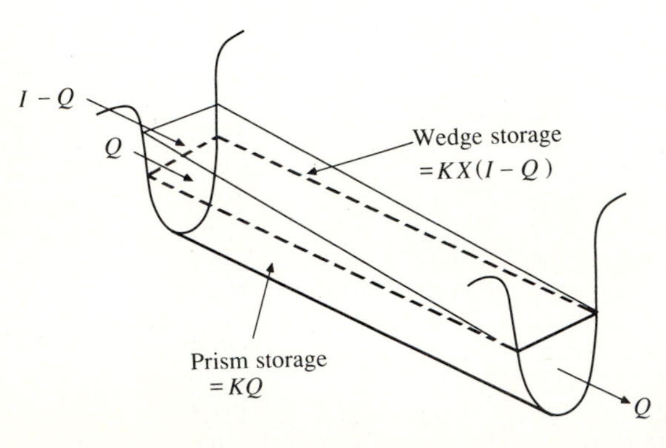

FIGURE 8.4.1
Prism and wedge storages in a channel reach.

level pool routing method. As before, the peak inflow of 360 cfs at 60 min is reduced to 270 cfs occurring at 80 minutes.

8.4 HYDROLOGIC RIVER ROUTING

The *Muskingum method* is a commonly used hydrologic routing method for handling a variable discharge-storage relationship. This method models the storage volume of flooding in a river channel by a combination of wedge and prism storages (Fig. 8.4.1). During the advance of a flood wave, inflow exceeds outflow, producing a *wedge* of storage. During the recession, outflow exceeds inflow, resulting in a negative wedge. In addition, there is a *prism* of storage which is formed by a volume of constant cross section along the length of prismatic channel.

Assuming that the cross-sectional area of the flood flow is directly proportional to the discharge at the section, the volume of prism storage is equal to KQ where K is a proportionality coefficient, and the volume of wedge storage is equal to $KX(I-Q)$, where X is a weighting factor having the range $0 \leq X \leq 0.5$. The total storage is therefore the sum of two components,

$$S = KQ + KX(I - Q) \tag{8.4.1}$$

which can be rearranged to give the storage function for the Muskingum method

$$S = K[XI + (1 - X)Q] \tag{8.4.2}$$

and represents a linear model for routing flow in streams.

The value of X depends on the shape of the modeled wedge storage. The value of X ranges from 0 for reservoir-type storage to 0.5 for a full wedge. When $X = 0$, there is no wedge and hence no backwater; this is the case for a level-pool reservoir. In this case, Eq. (8.4.2) results in a linear-reservoir model, $S = KQ$. In natural streams, X is between 0 and 0.3 with a mean value near 0.2. Great accuracy in determining X may not be necessary because the results of the method are relatively insensitive to the value of this parameter. The parameter K is the time of travel of the flood wave through the channel reach. A procedure called the *Muskingum-Cunge* method is described in Chapter 9 for determining the values of K and X on the basis of channel characteristics and flow rate in the channel. For hydrologic routing, the values of K and X are assumed to be specified and constant throughout the range of flow.

The values of storage at time j and $j + 1$ can be written, respectively, as

$$S_j = K[XI_j + (1 - X)Q_j] \tag{8.4.3}$$

and

$$S_{j+1} = K[XI_{j+1} + (1 - X)Q_{j+1}] \tag{8.4.4}$$

Using Eqs. (8.4.3) and (8.4.4), the change in storage over time interval Δt (Fig. 8.2.1) is

$$S_{j+1} - S_j = K\{[XI_{j+1} + (1 - X)Q_{j+1}] - [XI_j + (1 - X)Q_j]\} \tag{8.4.5}$$

The change in storage can also be expressed, using Eq. (8.2.2), as

$$S_{j+1} - S_j = \frac{(I_j + I_{j+1})}{2}\Delta t - \frac{(Q_j + Q_{j+1})}{2}\Delta t \tag{8.4.6}$$

Combining (8.4.5) and (8.4.6) and simplifying gives

$$Q_{j+1} = C_1 I_{j+1} + C_2 I_j + C_3 Q_j \tag{8.4.7}$$

which is the routing equation for the Muskingum method where

$$C_1 = \frac{\Delta t - 2KX}{2K(1-X) + \Delta t} \tag{8.4.8}$$

$$C_2 = \frac{\Delta t + 2KX}{2K(1-X) + \Delta t} \tag{8.4.9}$$

$$C_3 = \frac{2K(1-X) - \Delta t}{2K(1-X) + \Delta t} \tag{8.4.10}$$

Note that $C_1 + C_2 + C_3 = 1$.

If observed inflow and outflow hydrographs are available for a river reach, the values of K and X can be determined. Assuming various values of X and using known values of the inflow and outflow, successive values of the numerator and denominator of the following expression for K, derived from (8.4.5) and (8.4.6), can be computed.

$$K = \frac{0.5\,\Delta t[(I_{j+1} + I_j) - (Q_{j+1} + Q_j)]}{X(I_{j+1} - I_j) + (1-X)(Q_{j+1} - Q_j)} \tag{8.4.11}$$

The computed values of the numerator and denominator are plotted for each time interval, with the numerator on the vertical axis and the denominator on the horizontal axis. This usually produces a graph in the form of a loop. The value of X that produces a loop closest to a single line is taken to be the correct value for the reach, and K, according to Eq. (8.4.11), is equal to the slope of the line. Since K is the time required for the incremental flood wave to traverse the reach, its value may also be estimated as the observed time of travel of peak flow through the reach.

If observed inflow and outflow hydrographs are not available for determining K and X, their values may be estimated using the Muskingum-Cunge method described in Sec. 9.7.

Example 8.4.1. The inflow hydrograph to a river reach is given in columns 1 and 2 of Table 8.4.1. Determine the outflow hydrograph from this reach if $K = 2.3$ h, $X = 0.15$, and $\Delta t = 1$ h. The initial outflow is 85 ft^3/s.

Solution. Determine the coefficients C_1, C_2, and C_3 using Eqs. (8.4.8) – (8.4.10):

$$C_1 = \frac{1 - 2(2.3)(0.15)}{2(2.3)(1-0.15) + 1} = \frac{0.31}{4.91} = 0.0631$$

$$C_2 = \frac{1 + 2(2.3)(0.15)}{4.91} = \frac{1.69}{4.91} = 0.3442$$

$$C_3 = \frac{2(2.3)(1 - 0.15) - 1}{4.91} = \frac{2.91}{4.91} = 0.5927$$

Check to see that the sum of the coefficients C_1, C_2, and C_3 is equal to 1.

$$C_1 + C_2 + C_3 = 0.0631 + 0.3442 + 0.5927 = 1.0000$$

For the first time interval, the outflow is determined using values for I_1 and I_2 from Table 8.4.1, the initial outflow $Q_1 = 85$ cfs, and Eq. (8.4.7) with $j = 1$.

$$\begin{aligned} Q_2 &= C_1 I_2 + C_2 I_1 + C_3 Q_1 \\ &= 0.0631(137) + 0.3442(93) + 0.5927(85) \\ &= 8.6 + 32.0 + 50.4 \\ &= 91 \text{ cfs} \end{aligned}$$

as shown in columns (3) to (6) of Table 8.4.1. Computations for the following time intervals use the same procedure with $j = 2, 3, \ldots$ to produce the results shown in Table 8.4.1. The inflow and outflow hydrographs are plotted in Fig. 8.4.2. It can be seen that the outflow lags the inflow by approximately 2.3 h, which was the value of K used in the computations and represents the travel time in the reach.

TABLE 8.4.1
Flow routing through a river reach by the Muskingum method (Example 8.4.1).

Column:	(1) Routing period j (h)	(2) Inflow I (cfs)	(3) $C_1 I_{j+1}$ ($C_1 = 0.0631$)	(4) $C_2 I_j$ ($C_2 = 0.3442$)	(5) $C_3 Q_j$ ($C_3 = 0.5927$)	(6) Outflow Q (cfs)
	1	93				85
	2	137	8.6	32.0	50.4	91
	3	208	13.1	47.2	54.0	114
	4	320	20.2	71.6	67.7	159
	5	442	27.9	110.1	94.5	233
	6	546	34.5	152.1	137.8	324
	7	630	39.8	187.9	192.3	420
	8	678	42.8	216.8	248.9	509
	9	691	43.6	233.4	301.4	578
	10	675	42.6	237.8	342.8	623
	11	634	40.0	232.3	369.4	642
	12	571	36.0	218.2	380.4	635
	13	477	30.1	196.5	376.1	603
	14	390	24.6	164.2	357.3	546
	15	329	20.8	134.2	323.6	479
	16	247	15.6	113.2	283.7	413
	17	184	11.6	85.0	244.5	341
	18	134	8.5	63.3	202.2	274
	19	108	6.8	46.1	162.4	215
	20	90	5.7	37.2	127.6	170

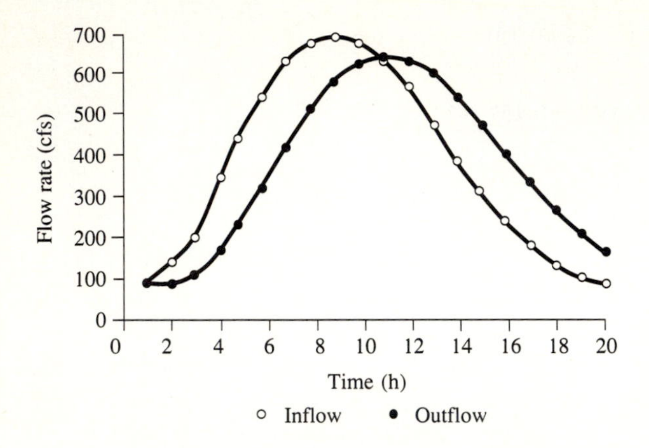

FIGURE 8.4.2
Routing of flow through a river reach by the Muskingum method (Example 8.4.1).

8.5 LINEAR RESERVOIR MODEL

A *linear reservoir* is one whose storage is linearly related to its output by a *storage constant k,* which has the dimension of time because S is a volume while Q is a flow rate.

$$S = kQ \tag{8.5.1}$$

The linear reservoir model can be derived from the general hydrologic system model [Eq. (7.1.6)] by letting $M(D) = 1$ and letting $N(D)$ have a root of $-1/k$ by making $N(D) = 1 + kD$. It can be shown further that if, in Eq. (7.1.6), $M(D) = 1$ and $N(D)$ has n real roots $-1/k_1, -1/k_2, \ldots, -1/k_n$, the system described is a cascade of n linear reservoirs in series, having storage constants $k_1, k_2, \ldots, k_n$,

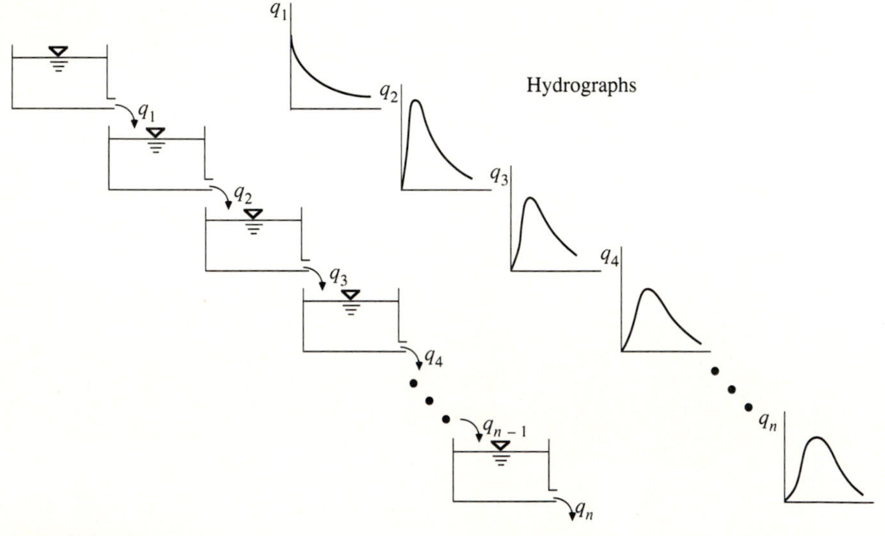

FIGURE 8.5.1
Linear reservoirs in series.

respectively. The concept of a linear reservoir was first introduced by Zoch (1934, 1936, 1937) in an analysis of the rainfall and runoff relationship. A single linear reservoir is a simplified case of the Muskingum model with $X = 0$. The impulse, pulse, and step response functions of a linear reservoir are plotted in Fig. 7.2.4.

Linear Reservoirs in Series

A watershed may be represented by a series of n identical linear reservoirs (Fig. 8.5.1) each having the same storage constant k (Nash, 1957). By routing a unit-volume inflow through the n linear reservoirs, a mathematical model for the instantaneous unit hydrograph (IUH) of the series can be derived. The impulse response function of a linear reservoir was derived in Ex. (7.2.1) as $u(t - \tau) = (1/k) \exp[-(t - \tau)/k]$. This will be the outflow from the first reservoir, and constitutes the inflow to the second reservoir with τ substituted for $t - \tau$, that is for the second reservoir $I(\tau) = (1/k) \exp(-\tau/k)$. The convolution integral (7.2.1) gives the outflow from the second reservoir as

$$
\begin{aligned}
q_2(t) &= \int_0^t I(\tau)u(t - \tau)d\tau \\
&= \int_0^t \left(\frac{1}{k}\right) e^{-\tau/k} \frac{1}{k} e^{-(t-\tau)/k} d\tau \\
&= \frac{t}{k^2} e^{-t/k}
\end{aligned}
\tag{8.5.2}
$$

This outflow is then used as the inflow to the third reservoir. Continuing this procedure will yield the outflow q_n from the n-th reservoir as

$$
u(t) = q_n(t) = \frac{1}{k\Gamma(n)} \left(\frac{t}{k}\right)^{n-1} e^{-t/k} \tag{8.5.3}
$$

where $\Gamma(n) = (n-1)!$ When n is not an integer, $\Gamma(n)$ can be interpolated from tables of the gamma function (Abramowitz and Stegun, 1965). This equation expresses the instantaneous unit hydrograph of the proposed model; mathematically, it is a gamma probability distribution function. The integral of the right side of the equation over t from zero to infinity is equal to 1.

It can be shown that the first and second moments of the IUH about the origin $t = 0$ are respectively

$$
M_1 = nk \tag{8.5.4}
$$

and

$$
M_2 = n(n + 1)k^2 \tag{8.5.5}
$$

The first moment, M_1, represents the lag time of the centroid of the area under the IUH. Applying the IUH in the convolution integral to relate the excess rainfall hyetograph (ERH) to the direct runoff hydrograph (DRH), the principle of linearity requires each infinitesimal element of the ERH to yield its corresponding DRH with the same lag time. In other words, the time difference between the centroids of areas under the ERH and the DRH should be equal to M_1.

By the method of moments, the values of k and n can be computed from a given ERH and DRH, thus providing a simple but approximate calculation of the IUH as expressed by Eq. (8.5.3). If M_{I_1} is the first moment of the ERH about the time origin divided by the total effective rainfall, and M_{Q_1} is the first moment of the DRH about the time origin divided by the total direct runoff, then

$$M_{Q_1} - M_{I_1} = nk \tag{8.5.6}$$

If M_{I_2} is the second moment of the ERH about the time origin divided by the total excess rainfall, and M_{Q_2} is the second moment of the DRH about the time origin divided by the total direct runoff, it can be shown that

$$M_{Q_2} - M_{I_2} = n(n + 1)k^2 + 2nkM_{I_1} \tag{8.5.7}$$

Since the values of M_{I_1}, M_{Q_1}, M_{I_2} and M_{Q_2} can be computed from given hydrologic data, the values of n and k can be found using Eqs. (8.5.6) and (8.5.7), thus determining the IUH. It should be noted that the computed values of n and k may vary somewhat even for small errors in the computed moments; for accuracy, a small time interval and many significant figures must be used in the computation.

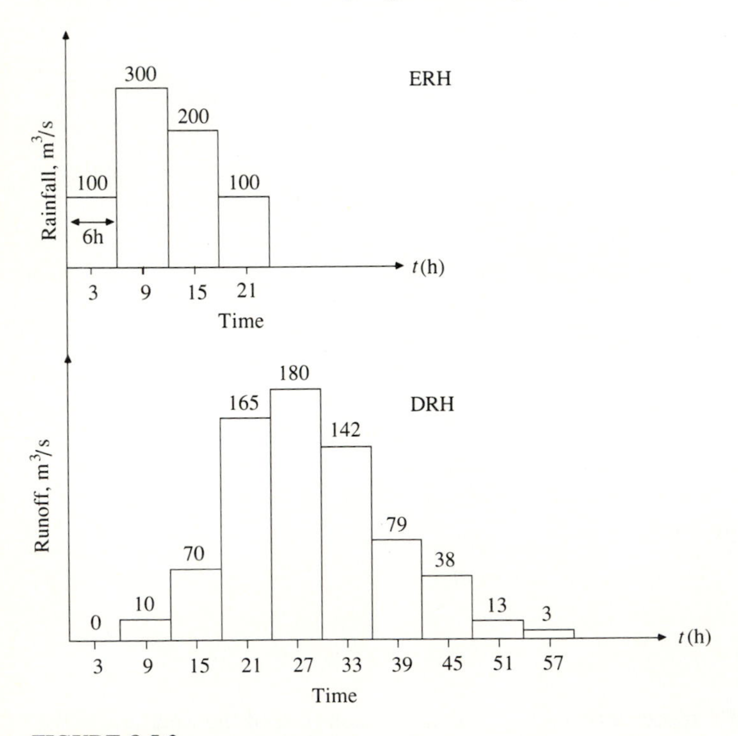

FIGURE 8.5.2
Excess rainfall hyetograph (ERH) and direct runoff hydrograph (DRH) for calculation of n and K in a linear reservoir model (Example 8.5.1).

Example 8.5.1. Given the ERH and the DRH shown in Figure 8.5.2, determine n and k for the IUH.

Solution. Determine the moments of the excess rainfall hyetograph and the direct runoff hydrograph. Each block in the ERH and DRH has duration 6 h $= 6 \times 3600$ s $=$ 21,600 s. The rainfall has been converted to units of m^3/s by multiplying by the watershed area to be dimensionally consistent with the runoff. The sum of the ordinates in the ERH and in the DRH is 700 m^3/s, so the area under each graph $= 700 \times 6 = 4200$ $(m^3/s)\cdot h$.

$$
\begin{aligned}
M_{I_1} &= \sum \left[\frac{\text{incremental area} \times \text{moment arm}}{\text{total area}} \right] \\
&= \frac{6}{4200}[100 \times 3 + 300 \times 9 + 200 \times 15 + 100 \times 21] \\
&= 11.57 \text{ h}
\end{aligned}
$$

The second moment of area is calculated using the parallel axis theorem.

$$
\begin{aligned}
M_{I_2} &= \left\{ \sum [\text{incremental area} \times (\text{moment arm})^2] \right. \\
&\quad \left. + \sum [\text{second moment about centroid of each increment}] \right\} \Big/ \text{total area} \\
&= \frac{1}{4200} \left\{ 6[100 \times 3^2 + 300 \times 9^2 + 200 \times 15^2 + 100 \times 21^2] \right. \\
&\quad \left. + \frac{1}{12} 6^3 [100 + 300 + 200 + 100] \right\} \\
&= 166.3 \text{ h}^2
\end{aligned}
$$

By a similar calculation for the direct runoff hydrograph

$$
\begin{aligned}
M_{Q_1} &= 28.25 \text{ h} \\
M_{Q_2} &= 882.8 \text{ h}^2
\end{aligned}
$$

Solve for nk using (8.5.6):

$$
\begin{aligned}
nk &= M_{Q_1} - M_{I_1} \\
&= 28.25 - 11.57 \\
&= 16.68
\end{aligned}
$$

Solve for n and k using (8.5.7):

$$
\begin{aligned}
M_{Q_2} - M_{I_2} &= n(n+1)k^2 + 2nkM_{I_1} \\
&= n^2k^2 + nk \times k + 2nkM_{I_1}
\end{aligned}
$$

Hence

$$882.8 - 166.3 = (16.68)^2 + 16.68k + 2 \times 16.68 \times 11.57$$

and solving yields

$$k = 3.14\ \text{h}$$

Thus

$$n = \frac{16.68}{k}$$

$$= \frac{16.68}{3.14}$$

$$= 5.31$$

These values of n and k can be substituted into Eq. (8.5.3) to determine the IUH of this watershed. By using the methods described in Sec. 7.2, the corresponding unit hydrograph can be determined for a specified rainfall duration.

Composite Models

In hydrologic modeling, linear reservoirs may also be linked in parallel. Linear reservoirs may be used to model subsurface water in a saturated phase (Kraijenhoff van der Leur, 1958), as well as surface water problems. Diskin et al. (1978) presented a parallel cascade model for urban watersheds. The input to the model is the total rainfall hyetograph, which feeds two parallel cascades of linear reservoirs, for the impervious and pervious areas of the watershed, respectively. Separate excess rainfall hyetographs are determined for the impervious and pervious areas, and used as input to the two cascades of linear reservoirs.

Linear reservoirs in series and parallel may be combined to model a hydrologic system. The use of linear reservoirs in series represents the storage effect of a hydrologic system, resulting in a time shift of nk between the centroid of the inflow and that of the outflow as given by Eq. (8.5.6). A linear channel is an idealized channel in which the time required to translate a discharge through the channel is constant (Chow, 1964). To model the combined effect of storage and translation, the linear reservoir may be used jointly with a linear channel. Other more elaborate composite models have been proposed. Dooge (1959) suggested a series of alternating linear channels and linear reservoirs. For this model, the drainage area of a watershed is divided into a number of subareas, by *isochrones,* which are lines of constant travel time to the watershed outlet. Each subarea is represented by a linear channel in series with a linear reservoir. The outflow from the linear channel is represented by the portion of a time-area diagram corresponding to the subarea. This outflow, together with outflow from the preceding subareas, serves as the inflow to the linear reservoir.

Randomized linear reservoir models have also been developed, in which the storage constant k is related to the Horton stream order (Sec. 5.8) of the subarea being drained. By considering the network of streams draining a watershed as being a random combination of linear reservoirs, with the mechanism of the combination being governed by Horton's stream ordering laws, it is possible to develop a *geomorphic instantaneous unit hydrograph,* the shape of which is related to the stream pattern of the watershed (Boyd, et al., 1979; Rodriguez-

Iturbe, and Valdes, 1979; Gupta, et al., 1980; Gupta, Rodriguez-Iturbe, and Wood, 1986).

REFERENCES

Abramowitz, M., and I. A. Stegun, *Handbook of Mathematical Functions*, Dover, New York, 1965.

Boyd, M. J., D. H. Pilgrim, and I. Cordery, A storage routing model based on catchment geomorphology, *J. Hydrol.*, vol. 42, pp. 42, 209–230, 1979.

Carnahan, B., H. A. Luther, and J. O. Wilkes, *Applied Numerical Methods*, Wiley, New York, 1969.

Chow, V. T., A practical procedure of flood routing, *Civ. Eng. and Public Works Rev.*, vol. 46, no. 542, pp. 586–588, August 1951.

Chow, V. T., *Open-channel Hydraulics*, McGraw-Hill, New York, p. 529, 1959.

Chow, V. T., Runoff, in *Handbook of Applied Hydrology*, sec. 14, pp. 14–30, McGraw-Hill, New York, 1964.

Diskin, M. H., S. Ince, and K. Oben-Nyarko, Parallel cascades model for urban watersheds, *J. Hyd. Div. Am. Soc. Civ. Eng.*, vol. 104, no. HY2, pp. 261–276, February 1978.

Dooge, J. C. I., A general theory of the unit hydrograph, *J. Geophys. Res.*, vol. 64, no. 1, pp. 241–256, 1959.

Gupta, V. K., E. Waymire, and C. T. Wang, A representation of an instantaneous unit hydrograph from geomorphology, *Water Resour. Res.*, vol. 16, no. 5, pp. 855–862, 1980.

Gupta, V. K., I. Rodriguez-Iturbe, and E. F. Wood, *Scale Problems in Hydrology*, D. Reidel, Dordrecht, Holland, 1986.

Kraijenhoff van der Leur, D. A., A study of non-steady groundwater flow with special reference to a reservoir-coefficient, *De Ingenieur*, vol. 70, no. 19, pp. B87–B94, 1958.

Nash, J. E., The form of the instantaneous unit hydrograph, IASH publication no. 45, vol. 3–4, pp. 114–121, 1957.

Rodriguez-Iturbe, I., and J. B. Valdes, The geomorphological structure of hydrologic response, *Water Resour. Res.*, vol. 15, no. 6, pp. 1409–1420, 1979.

Zoch, R. T., On the relation between rainfall and stream flow, *Monthly Weather Rev.*, vols. 62, 64, 65; pp. 315–322, 105–121, 135–147; 1934, 1936, 1937, respectively.

PROBLEMS

8.2.1 Storage vs. outflow characteristics for a proposed reservoir are given below. Calculate the storage-outflow function $2S/\Delta t + Q$ vs. Q for each of the tabulated values if $\Delta t = 2$ h. Plot a graph of the storage-outflow function.

Storage (10^6 m^3)	75	81	87.5	100	110.2
Outflow (m^3/s)	57	227	519	1330	2270

8.2.2 Use the level pool routing method to route the hydrograph given below through the reservoir whose storage-outflow characteristics are given in Prob. 8.2.1. What is the maximum reservoir discharge and storage? Assume that the reservoir initially contains 75×10^6 m^3 of storage.

Time (h)	0	2	4	6	8	10	12	14	16	18
Inflow (m^3/sec)	60	100	232	300	520	1,310	1,930	1,460	930	650

8.2.3 Solve Prob. 8.2.2 assuming the initial reservoir storage is 87.5×10^6 m^3.

8.2.4 Solve Example 8.2.1 in the text if the initial depth in the reservoir is 2 ft. How much higher does this make the maximum water level in the reservoir compared with the level found in Example 8.2.1?

8.2.5 The storage capacity and stage-outflow relationship of a flood-control reservoir are given in the following tables. Route the design flood hydrograph given below through the reservoir up to time 6:00. The initial reservoir level is 3.15 m. Use a routing interval of Δt = 15 min.

Stage (m)	3.15	3.30	3.45	3.60	3.75	3.90	4.05
Storage (m^3)	15	49	110	249	569	1180	2180
Discharge (m^3/s)	0	0.21	0.72	1.25	1.89	2.61	3.40
Stage	4.08	4.15	4.20	4.27	4.35	4.50	
Storage	2440	3140	4050	5380	8610	18600	
Discharge	3.57	3.91	4.25	4.62	5.21	6.20	

Time (h:min)	0:00	0:15	0:30	0:45	1:00	1:15	1:30	1:45	2:00
Inflow (m^3/s)	0	0.04	0.12	0.25	0.53	1.10	3.00	6.12	8.24
Time	2:15	2:30	2:45	3:00	3:15	3:30	3:45	4:00	4:15
Inflow	9.06	9.20	8.75	8.07	7.36	6.66	5.98	5.32	4.67
Time	4:30	4:45	5:00	5:15	5:30	5:45	6:00		
Inflow	4.11	3.65	3.29	3.00	2.73	2.49	2.27		

8.2.6 Consider a 2-acre detention basin with vertical walls. The triangular inflow hydrograph increases linearly from zero to a peak of 540 cfs at 60 min and then decreases linearly to a zero discharge at 180 min. Route the inflow hydrograph through the detention basin using the head-discharge curve for the 5-ft pipe spillway in Table 8.2.2. The pipe is located at the bottom of the basin. Assuming the basin is initially empty, use the level pool routing procedure with a 10-minute time interval to determine the maximum depth in the detention basin.

8.3.1 Solve Prob. 8.2.6 using the third order Runge-Kutta method, with a 10-minute time interval, to determine the maximum depth.

8.3.2 Write a computer program to perform routing using the third order Runge-Kutta method. Then solve Prob. 8.3.1.

8.3.3 Use the third order Runge-Kutta method to route the inflow hydrograph given below through an urban detention basin site with the following characteristics. Use a 3-minute time interval for the routing.

Elevation above MSL (ft)	1000	1010	1020	1030	1040	1050
Surface area (acres)	1	10	15	20	25	30

The detention basin has a conduit spillway 20 ft^2 in area with the inlet elevation at 1002 ft, and an overflow spillway 80 ft in length at elevation 1011 ft. The discharge equations for conduit and overflow spillways are given in Table 8.2.1. Assume the conduit spillway functions as a culvert with submerged inlet control having discharge coefficient $C_d = 0.7$, and the overflow spillway has the coefficients $C(Q = CLH^{3/2})$ tabulated below.

Head H (ft)	0.0–0.2	0.2–0.4	0.4–0.6	0.6–0.8	0.8–1.0
Spillway Coefficient C	2.69	2.72	2.95	2.85	2.98
Head H	1.0–1.2	1.2–1.4	1.4–1.6	1.6–1.8	>1.8
Spillway Coefficient C	3.08	3.20	3.28	3.31	3.35

Inflow hydrograph

Time (min)	0	3	6	9	12	15	18	21	24
Inflow (cfs)	0	60	133	222	321	427	537	650	772
Time	27	30	33	36	39	42	45	48	51
Inflow	902	1036	1174	1312	1451	1536	1571	1580	1568
Time	54	57	60	63	66	69	72	75	78
Inflow	1548	1526	1509	1493	1479	1464	1443	1417	1384
Time	81	84	87	90	93	96	99	102	105
Inflow	1345	1298	1244	1184	1120	1051	979	904	827
Time	108	111	114	117	120	123	126	129	132
Flow	748	669	588	508	427	373	332	302	278
Time	135	138	141	144	147				
Inflow	260	246	235	225	217				

8.3.4 Solve Prob. 8.3.3 for time intervals of 6 and 12 minutes. Compare the results for the 3-, 6-, and 12-minute routing time intervals.

8.3.5 Write a computer program for detention basin routing using the fourth order Runge-Kutta method by Gill (Carnahan et al., 1969). The continuity equation is approximated as:

$$\frac{\Delta H}{\Delta t} = \frac{I(t) - Q(H)}{A(H)} = f(t, H)$$

The unknown depth $H_{t+\Delta t}$ at time $t + \Delta t$, is expressed as

$$H_{t+\Delta t} = H_t + \frac{\Delta t}{6}\left[k_1 + 2\left(1 - \frac{1}{\sqrt{2}}\right)k_2 + 2\left(1 + \frac{1}{\sqrt{2}}\right)k_3 + k_4\right]$$

where

$$k_1 = \left[\frac{I(t) - Q(H_t)}{A(H_t)}\right]$$

$$k_2 = \left[\frac{I\left(t + \frac{\Delta t}{2}\right) - Q\left(H_t + \frac{\Delta t}{2}k_1\right)}{A\left(H_t + \frac{\Delta t}{2}\right)}\right]$$

$$k_3 = \frac{I\left(t + \frac{\Delta t}{2} - Q(H_1)\right)}{A(H_1)}$$

$$k_4 = \left[\frac{I(t + \Delta t) - Q(H_2)}{A(H_2)}\right]$$

where

$$H_1 = H_t + \Delta t\left[\left(-0.5 + \frac{1}{\sqrt{2}}\right)k_1 + \left(1 - \frac{1}{\sqrt{2}}\right)k_2\right]$$

$$H_2 = H_t - \frac{\Delta t}{\sqrt{2}}k_2 + \left(1 + \frac{1}{\sqrt{2}}\right)(\Delta t)k_3$$

$A(H)$ is interpolated from the elevation–water surface area relationship.

8.3.6 Using the computer program written in Prob. 8.3.5, solve Prob. 8.3.1.

8.3.7 Using the computer program written in Prob. 8.3.5, solve Prob. 8.3.3.

8.3.8 In this problem, you are to determine the runoff from a particular watershed and route the runoff hydrograph through a reservoir at the downstream end of the watershed. The reservoir has the following storage-outflow characteristics:

Storage (ac·ft)	0	200	300	400	500	600	700	1100
Outflow (cfs)	0	2	20	200	300	350	450	1200

The rainfall is:

Time (h)	0	0.5	1.0	1.5	2.0
Accumulated rainfall depth (in)	0	1.0	3.0	4.0	4.5

The half-hour unit hydrograph is:

Time (h)	0	0.5	1.0	1.5	2.0	2.5	3.0	3.5
Discharge (cfs/in)	0	200	500	800	700	600	500	400
Time	4.0	4.5	5.0	5.5	6.0			
Discharge	300	200	100	50	0			

The ϕ-index of 0.8 in/h is to be used to account for losses. Determine the peak discharge from the reservoir assuming zero baseflow. What is the area in square miles of the watershed?

8.4.1 Show that the interval between the centroids of the input and the output in the Muskingum method is a constant having the dimension of time.

8.4.2 Assuming K = 24 h and X = 0.2, route a hypothetical flood having a constant flow rate of 1000 units and lasting one day, through a reservoir whose storage is simulated by the Muskingum equation. Plot the inflow and outflow hydrographs. Assume initial outflow is zero.

8.4.3 Using the inflow and outflow hydrograph given below for a channel, determine K and X.

Time (min)	0	3	6	9	12	15	18	21
Channel inflow (cfs)	0	60	120	180	240	300	364	446
Channel outflow (cfs)	0	0	13	42	81	127	178	231
Time	24	27	30	33	36	39	42	45
Channel inflow	530	613	696	776	855	932	948	932
Channel outflow	293	363	437	514	593	672	757	822
Time	48	51	54	57	60	63	66	69
Channel inflow	914	911	921	941	958	975	982	980
Channel outflow	861	879	888	897	910	924	940	954
Time	72	75	78	81	84	87	90	93
Channel inflow	969	951	925	890	852	810	767	717
Channel outflow	964	968	965	956	938	919	884	851
Time	96	99	102	105	108	111	114	117
Channel inflow	668	618	566	514	462	410	359	309
Channel outflow	812	769	725	677	629	579	528	478
Time	120	123	126	129	132	135	138	141
Channel inflow	261	248	238	229	222	216	210	205
Channel outflow	427	373	332	302	278	260	246	235
Time	144	147						
Channel inflow	199	194						
Channel outflow	225	217						

8.4.4 A 4400-foot reach of channel has a Muskingum $K = 0.24$ h and $X = 0.25$. Route the following inflow hydrograph through this reach. Assume the initial outflow $= 739$ cfs.

Time (h)	0	0.5	1.0	1.5	2.0	2.5	3.0
Inflow (cfs)	819	1012	1244	1537	1948	2600	5769
Time	3.5	4.0	4.5	5.0	5.5	6.0	6.5
Inflow	12866	17929	20841	21035	20557	19485	14577
Time	7.0	7.5	8.0				
Inflow	9810	6448	4558				

8.4.5 A watershed is divided into two subareas A and B. The surface runoff from subarea A enters a channel at point 1 and flows to point 2 where the runoff from subarea B is added to the hydrograph and the combined flow routed through a reservoir. Determine the discharge hydrograph from the reservoir, assuming that the reservoir is initially empty. What are the areas of subareas A and B in square miles?

The reservoir has the following storage-outflow characteristics:

Storage (ac·ft)	0	220	300	400	500	600	700	1100
Outflow (cfs)	0	2	20	200	300	350	450	1200

The channel from point 1 to point 2 has Muskingum parameters $K=0.5$ hours and $X = 0.25$. Subarea A is undeveloped and subarea B has residential development. As a result the ϕ-index for subarea A is 0.8 in/h and the ϕ-index for B is 0.2 in/h. The storm is

Time (h)	0	0.5	1.0	1.5	2.0
Accumulated rainfall depth (in)	0	1.0	3.0	4.0	4.5

The half-hour unit hydrographs for subareas A and B are

Time (h)	Subarea A unit hydrograph (cfs/in)	Subarea B unit hydrograph (cfs/in)
0	0	0
0.5	100	200
1.0	200	500
1.5	300	800
2.0	400	700
2.5	350	600
3.0	300	500
3.5	250	400
4.0	200	300
4.5	150	200
5.0	100	100
5.5	50	50
6.0	0	0

8.5.1 For a linear hydrologic system, it is assumed that the system storage $S(t)$ is directly proportional to the output $Q(t)$, or $S(t) = kQ(t)$ where k is a storage constant. At the initial condition, the output is zero. Derive an equation for the output $Q(t)$ in terms of the input $I(t)$ and the storage constant k.

8.5.2 What is the dimension of the storage constant in Prob. 8.5.1? Taking $k = 1$ unit, construct a curve showing the relationship between the ratio Q/I and time. Assume inflow is constant.

8.5.3 Assuming that the input $I(t)$ to a linear reservoir terminates at t_o, derive an equation for the output for $t > t_o$.

8.5.4 Show that the peak discharge of the IUH of a hydrologic system modeled by a series of n linear reservoirs, each having storage constant k, is

$$u(t)_{\max} = \frac{1}{k\Gamma(n)} e^{1-n}(n-1)^{n-1}$$

8.5.5 Show that the first and second moments of the area of the IUH modeled by a series of n linear reservoirs, each having storage constant k, about the time origin are

$$M_1 = nk$$

and

$$M_2 = n(n+1)k^2$$

8.5.6 If C_{I_2}, C_{Q_2} and C_2 are the second moments about the centroids of the areas of the ERH, DRH and IUH, respectively, show that

$$C_{Q_2} = C_{I_2} + C_2$$

8.5.7 If the first and second moments of the areas of the ERH and DRH about the time origin are M_{I_1}, M_{I_2}, M_{Q_1}, and M_{Q_2}, respectively, show that for n linear reservoirs in series

$$M_{Q_2} - M_{I_2} = n(n+1)k^2 + 2nkM_{I_1}$$

8.5.8 Determine the IUH by the n-linear-reservoir method for a watershed having a drainage area of 36 km^2 assuming abstractions of 0.5 cm/h and a constant base flow of 5 m^3/s. Use the following data.

6-h period	1	2	3	4	5	6	7	8	9	10
Rainfall cm/h	1.5	3.5	2.5	1.5						
Sreamflow m^3/s	15	75	170	185	147	84	43	18	8	

8.5.9 Formulate the IUH for a hydrologic system model composed of two linear reservoirs with respective constants k_1 and k_2 (*a*) in series; and (*b*) in parallel, having the system input divided between the reservoirs in the ratio of x to y where $x + y = 1$. Determine their centroids.

8.5.10 Show that the following is a solution of the Muskingum equation:

$$Q(t) = \frac{1}{1-X} I(0) e^{-t/K(1-X)} - \frac{X}{1-X} I(t) + \frac{1}{K(1-X)^2} \int_0^t e^{-\tau/K(1-X)} I(t-\tau) d\tau$$

with τ equal to the duration of $I(t)$ and $I(0) = Q(0)$. Show that the IUH is

$$u(t) = \frac{1}{K(1-X)^2} e^{-t/K(1-X)} - \frac{X}{1-X} \delta(t)$$

where $\delta(t)$ is the unit-impulse input, that is, the limit of $I(\tau)$ as τ approaches zero.

CHAPTER 9

DISTRIBUTED FLOW ROUTING

The flow of water through the soil and stream channels of a watershed is a distributed process because the flow rate, velocity, and depth vary in space throughout the watershed. Estimates of the flow rate or water level at important locations in the channel system can be obtained using a *distributed flow routing* model. This type of model is based on partial differential equations (the Saint-Venant equations for one-dimensional flow) that allow the flow rate and water level to be computed as functions of space and time, rather than of time alone as in the lumped models described in Chaps. 7 and 8.

The computation of flood water level is needed because this level delineates the flood plain and determines the required height of structures such as bridges and levees; the computation of flood flow rate is also important; first, because the flow rate determines the water level, and second, because the design of any flood storage structure such as a detention pond or reservoir requires an estimate of its inflow hydrograph. The alternative to using a distributed flow routing model is to use a lumped model to calculate the flow rate at the desired location, then compute the corresponding water level by assuming steady nonuniform flow along the channel at the site. The advantage of a distributed flow routing model over this alternative is that the distributed model computes the flow rate and water level simultaneously instead of separately, so that the model more closely approximates the actual unsteady nonuniform nature of flow propagation in a channel.

Distributed flow routing models can be used to describe the transformation of storm rainfall into runoff over a watershed to produce a flow hydrograph for the watershed outlet, and then to take this hydrograph as input at the upstream end of a river or pipe system and route it to the downstream end. Distributed models can also be used for routing low flows, such as irrigation water deliveries through a canal or river system. The true flow process in either of these applications varies in all three space dimensions; for example, the velocity in a river varies along the river, across it, and also from the water surface to the river bed. However, for many practical purposes, the spatial variation in velocity across the channel and with respect to the depth can be ignored, so that the flow process can be approximated as varying in only one space dimension—along the flow channel, or in the direction of flow. The *Saint-Venant equations*, first developed by Barre de Saint-Venant in 1871, describe one-dimensional unsteady open channel flow, which is applicable in this case.

9.1 SAINT-VENANT EQUATIONS

The following assumptions are necessary for derivation of the Saint-Venant equations:

1. The flow is one-dimensional; depth and velocity vary only in the longitudinal direction of the channel. This implies that the velocity is constant and the water surface is horizontal across any section perpendicular to the longitudinal axis.
2. Flow is assumed to vary gradually along the channel so that hydrostatic pressure prevails and vertical accelerations can be neglected (Chow, 1959).
3. The longitudinal axis of the channel is approximated as a straight line.
4. The bottom slope of the channel is small and the channel bed is fixed; that is, the effects of scour and deposition are negligible.
5. Resistance coefficients for steady uniform turbulent flow are applicable so that relationships such as Manning's equation can be used to describe resistance effects.
6. The fluid is incompressible and of constant density throughout the flow.

Continuity Equation

The *continuity equation* for an unsteady variable-density flow through a control volume can be written as in Eq. (2.2.1):

$$0 = \frac{d}{dt}\iiint_{\text{c.v.}} \rho \, d\forall + \iint_{\text{c.s.}} \rho \mathbf{V} \cdot \mathbf{dA} \tag{9.1.1}$$

Consider an elemental control volume of length dx in a channel. Fig. 9.1.1 shows three views of the control volume: (*a*) an elevation view from the side, (*b*) a plan view from above, and (*c*) a channel cross section. The inflow to the control volume is the sum of the flow Q entering the control volume at the upstream end of the channel and the *lateral inflow* q entering the control volume as a distributed flow along the side of the channel. The dimensions of q are those of flow per unit length of channel, so the rate of lateral inflow is $q\,dx$ and the mass inflow rate is

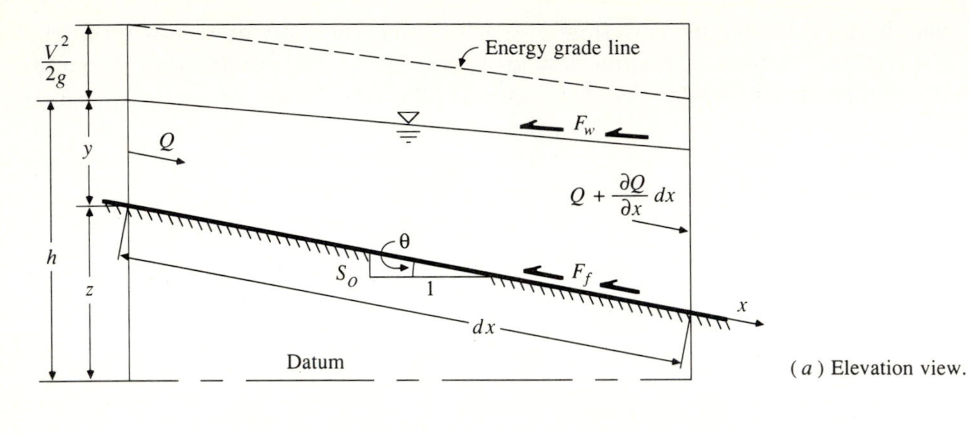

(*a*) Elevation view.

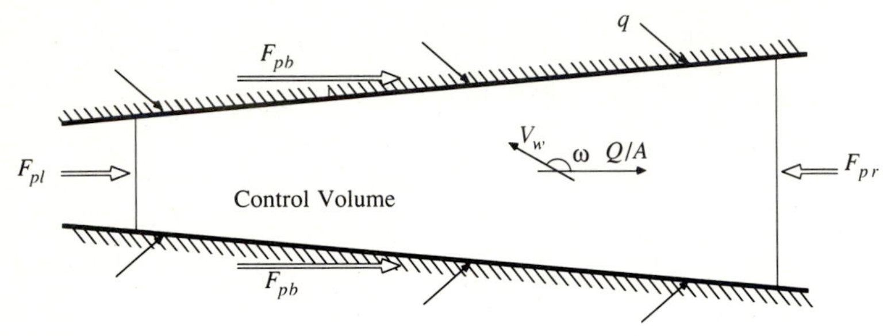

(*b*) Plan view.

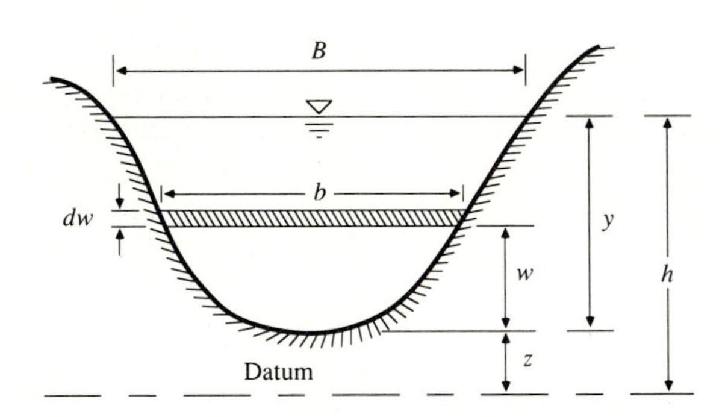

(*c*) Cross section.

FIGURE 9.1.1
An elemental reach of channel for derivation of the Saint-Venant equations.

$$\iint_{\text{inlet}} \rho \mathbf{V} \cdot \mathbf{dA} = -\rho(Q + q\,dx) \tag{9.1.2}$$

This is negative because inflows are considered to be negative in the Reynolds transport theorem. The mass outflow from the control volume is

$$\iint_{\text{outlet}} \rho \mathbf{V} \cdot \mathbf{dA} = \rho\left(Q + \frac{\partial Q}{\partial x} dx\right) \tag{9.1.3}$$

where $\partial Q/\partial x$ is the rate of change of channel flow with distance. The volume of the channel element is $A\,dx$, where A is the average cross-sectional area, so the rate of change of mass stored within the control volume is

$$\frac{d}{dt}\iiint_{c.v.} \rho\, d\forall = \frac{\partial(\rho A dx)}{\partial t} \tag{9.1.4}$$

where the partial derivative is used because the control volume is defined to be fixed in size (though the water level may vary within it). The net outflow of mass from the control volume is found by substituting Eqs. (9.1.2) to (9.1.4) into (9.1.1):

$$\frac{\partial(\rho A dx)}{\partial t} - \rho(Q + q dx) + \rho\left(Q + \frac{\partial Q}{\partial x}dx\right) = 0 \tag{9.1.5}$$

Assuming the fluid density ρ is constant, (9.1.5) is simplified by dividing through by ρdx and rearranging to produce the *conservation form* of the continuity equation,

$$\frac{\partial Q}{\partial x} + \frac{\partial A}{\partial t} - q = 0 \tag{9.1.6}$$

which is applicable at a channel cross section. This equation is valid for a *prismatic* or a *nonprismatic* channel; a prismatic channel is one in which the cross-sectional shape does not vary along the channel and the bed slope is constant.

For some methods of solving the Saint-Venant equations, the *nonconservation form* of the continuity equation is used, in which the average flow velocity V is a dependent variable, instead of Q. This form of the continuity equation can be derived for a unit width of flow within the channel, neglecting lateral inflow, as follows. For a unit width of flow $A = y \times 1 = y$ and $Q = VA = Vy$. Substituting into (9.1.6),

$$\frac{\partial(Vy)}{\partial x} + \frac{\partial y}{\partial t} = 0 \tag{9.1.7}$$

or

$$V\frac{\partial y}{\partial x} + y\frac{\partial V}{\partial x} + \frac{\partial y}{\partial t} = 0 \tag{9.1.8}$$

Momentum Equation

Newton's second law is written in the form of Reynold's transport theorem as in Eq. (2.4.1):

$$\Sigma \mathbf{F} = \frac{d}{dt}\iiint_{c.v.} \mathbf{V}\rho\, d\forall + \iint_{c.s.} \mathbf{V}\rho\mathbf{V}\cdot d\mathbf{A} \tag{9.1.9}$$

This states that the sum of the forces applied is equal to the rate of change of momentum stored within the control volume plus the net outflow of momentum across the control surface. This equation, in the form $\Sigma F = 0$, was applied to steady uniform flow in an open channel in Chap. 2. Here, unsteady nonuniform flow is considered.

FORCES. There are five forces acting on the control volume:

$$\Sigma F = F_g + F_f + F_e + F_w + F_p \tag{9.1.10}$$

where F_g is the *gravity force* along the channel due to the weight of the water in the control volume, F_f is the *friction force* along the bottom and sides of the control volume, F_e is the *contraction/expansion force* produced by abrupt changes in the channel cross section, F_w is the *wind shear force* on the water surface, and F_p is the *unbalanced pressure force* [see Fig. 9.1.1 (*b*)]. Each of these five forces is evaluated in the following paragraphs.

Gravity. The volume of fluid in the control volume is $A\,dx$ and its weight is $\rho g A\,dx$. For a small angle of channel inclination θ, $S_o \approx \sin\theta$ and the gravity force is given by

$$F_g = \rho g A dx \sin\theta \approx \rho g A S_o dx \tag{9.1.11}$$

where the channel bottom slope S_o equals $-\partial z/\partial x$.

Friction. Frictional forces created by the shear stress along the bottom and sides of the control volume are given by $-\tau_0 P dx$, where τ_0 is the bed shear stress and P is the wetted perimeter. From Eq. (2.4.9), $\tau_0 = \gamma R S_f = \rho g (A/P) S_f$, hence the friction force is written as

$$F_f = -\rho g A S_f dx \tag{9.1.12}$$

where the friction slope S_f is derived from resistance equations such as Manning's equation.

Contraction/expansion. Abrupt contraction or expansion of the channel causes energy loss through eddy motion. Such losses are similar to minor losses in a pipe system. The magnitude of eddy losses is related to the change in velocity head $V^2/2g = (Q/A)^2/2g$ through the length of channel causing the losses. The drag forces creating these eddy losses are given by

$$F_e = -\rho g A S_e dx \tag{9.1.13}$$

where S_e is the eddy loss slope

$$S_e = \frac{K_e}{2g}\frac{\partial (Q/A)^2}{\partial x} \tag{9.1.14}$$

in which K_e is the nondimensional expansion or contraction coefficient, negative for channel expansion [where $\partial (Q/A)^2/\partial x$ is negative] and positive for channel contraction.

Wind Shear. The wind shear force is caused by frictional resistance of wind against the free surface of the water and is given by

$$F_w = \tau_w B dx \tag{9.1.15}$$

where τ_w is the wind shear stress. The shear stress of a boundary on a fluid may be written in general as

$$\tau_w = \frac{-\rho C_f |V_r| V_r}{2} \tag{9.1.16}$$

where V_r is the velocity of the fluid relative to the boundary, the notation $|V_r|V_r$ is used so that τ_w will act opposite to the direction of V_r, and C_f is a shear stress coefficient. As shown in Fig. 9.1.1(*b*), the average water velocity is Q/A, and the wind velocity is V_w in a direction at angle ω to the water velocity, so the velocity of the water relative to the air is

$$V_r = \frac{Q}{A} - V_w \cos \omega \tag{9.1.17}$$

The wind force is, from above,

$$\begin{aligned} F_w &= \frac{-\rho C_f |V_r| V_r B dx}{2} \\ &= -W_f B \rho \, dx \end{aligned} \tag{9.1.18}$$

where the *wind shear factor* W_f equals $C_f|V_r|V_r/2$. Note that from this equation the direction of the wind force will be opposite to the direction of the water flow.

Pressure. Referring to Fig. 9.1.1(*b*), the unbalanced pressure force is the resultant of the hydrostatic force on the left side of the control volume, F_{Pl}, the hydrostatic force on the right side of the control volume, F_{pr}, and the pressure force exerted by the banks on the control volume, F_{pb}:

$$F_p = F_{pl} - F_{pr} + F_{pb} \tag{9.1.19}$$

As shown in Fig. 9.1.1(*c*), an element of fluid of thickness dw at elevation w from the bottom of the channel is immersed at depth $y - w$, so the hydrostatic pressure on the element is $\rho g(y - w)$ and the hydrostatic force is $\rho g(y - w)b \, dw$, where b is the width of the element across the channel. Hence, the total hydrostatic force on the left end of the control volume is

$$F_{pl} = \int_0^y \rho g(y - w) b \, dw \tag{9.1.20}$$

The hydrostatic force on the right end of the control volume is

$$F_{pr} = \left(F_{pl} + \frac{\partial F_{pl}}{\partial x} dx \right) \tag{9.1.21}$$

where $\partial F_{pl}/\partial x$ is determined using the Leibnitz rule for differentiation of an integral (Abramowitz and Stegun, 1972):

$$\frac{\partial F_{pl}}{\partial x} = \int_0^y \rho g \frac{\partial y}{\partial x} b\, dw + \int_0^y \rho g (y - w) \frac{\partial b}{\partial x} dw$$

$$= \rho g A \frac{\partial y}{\partial x} + \int_0^y \rho g (y - w) \frac{\partial b}{\partial x} dw \qquad (9.1.22)$$

because $A = \int_0^y b\, dw$. The force due to the banks is related to the rate of change in width of the channel, $\partial b / \partial x$, through the element dx as

$$F_{pb} = \left[\int_0^y \rho g (y - w) \frac{\partial b}{\partial x} dw \right] dx \qquad (9.1.23)$$

Substituting Eq. (9.1.21) into (9.1.19) gives

$$F_p = F_{pl} - \left(F_{pl} + \frac{\partial F_{pl}}{\partial x} dx \right) + F_{pb}$$

$$= -\frac{\partial F_{pl}}{\partial x} dx + F_{pb} \qquad (9.1.24)$$

Now substituting Eqs. (9.1.22) and (9.1.23) into (9.1.24) and simplifying gives

$$F_p = -\rho g A \frac{\partial y}{\partial x} dx \qquad (9.1.25)$$

The sum of the five forces in Eq. (9.1.10) can be expressed, after substituting (9.1.11), (9.1.12), (9.1.13), (9.1.18), and (9.1.25), as

$$\Sigma F = \rho g A S_o\, dx - \rho g A S_f\, dx - \rho g A S_e\, dx - W_f B \rho\, dx - \rho g A \frac{\partial y}{\partial x} dx \qquad (9.1.26)$$

MOMENTUM. The two momentum terms on the right-hand side of Eq. (9.1.9) represent the rate of change of storage of momentum in the control volume, and the net outflow of momentum across the control surface, respectively.

Net momentum outflow. The mass inflow rate to the control volume [Eq. (9.1.2)] is $-\rho(Q + q\, dx)$, representing both stream inflow and lateral inflow. The corresponding momentum is computed by multiplying the two mass inflow rates by their respective velocities and a momentum correction factor β:

$$\iint_{\text{inlet}} \mathbf{V} \rho \mathbf{V} \cdot \mathbf{dA} = -\rho(\beta V Q + \beta v_x q\, dx) \qquad (9.1.27)$$

where $\rho \beta V Q$ is the momentum entering from the upstream end of the channel, and $\rho \beta v_x q\, dx$ is the momentum entering the main channel with the lateral inflow, which has a velocity v_x in the x direction. The term β is known as the *momentum coefficient* or *Boussinesq coefficient*; it accounts for the nonuniform distribution of velocity at a channel cross section in computing the momentum. The value of β is given by

$$\beta = \frac{1}{V^2A}\iint v^2 dA \tag{9.1.28}$$

where v is the velocity through a small element of area dA in the channel cross section. The value of β ranges from 1.01 for straight prismatic channels to 1.33 for river valleys with floodplains (Chow, 1959; Henderson, 1966).

The momentum leaving the control volume is

$$\iint_{\text{outlet}} \mathbf{V}\rho\mathbf{V}\cdot\mathbf{dA} = \rho\left[\beta VQ + \frac{\partial(\beta VQ)}{\partial x}dx\right] \tag{9.1.29}$$

The net outflow of momentum across the control surface is the sum of (9.1.27) and (9.1.29):

$$\begin{aligned}\iint_{\text{c.s.}} \mathbf{V}\rho\mathbf{V}\cdot\mathbf{dA} &= -\rho[\beta VQ + \beta v_x q\,dx] + \rho\left[\beta VQ + \frac{\partial(\beta VQ)}{\partial x}dx\right] \\ &= -\rho\left[\beta v_x q - \frac{\partial(\beta VQ)}{\partial x}\right]dx\end{aligned} \tag{9.1.30}$$

Momentum storage. The time rate of change of momentum stored in the control volume is found by using the fact that the volume of the elemental channel is $A\,dx$, so its momentum is $\rho A\,dx\,V$, or $\rho Q\,dx$, and then

$$\frac{d}{dt}\iiint_{\text{c.v.}} \mathbf{V}\rho\,d\forall = \rho\frac{\partial Q}{\partial t}dx \tag{9.1.31}$$

After substituting the force terms from (9.1.26), and the momentum terms from (9.1.30) and (9.1.31) into the momentum equation (9.1.9), it reads

$$\begin{aligned}&\rho g A S_0\,dx - \rho g A S_f\,dx - \rho g A S_e\,dx - W_f B\rho\,dx - \rho g A\frac{\partial y}{\partial x}dx \\ &\quad = -\rho\left[\beta v_x - \frac{\partial(\beta VQ)}{\partial x}\right]dx + \rho\frac{\partial Q}{\partial t}dx\end{aligned} \tag{9.1.32}$$

Dividing through by $\rho\,dx$, replacing V with Q/A, and rearranging produces the conservation form of the momentum equation:

$$\frac{\partial Q}{\partial t} + \frac{\partial(\beta Q^2/A)}{\partial x} + gA\left(\frac{\partial y}{\partial x} - S_o + S_f + S_e\right) - \beta q v_x + W_f B = 0 \tag{9.1.33}$$

The depth y in Eq. (9.1.33) can be replaced by the water surface elevation h, using [see Fig. 9.1.1(a)]:

$$h = y + z \tag{9.1.34}$$

where z is the elevation of the channel bottom above a datum such as mean sea level. The derivative of Eq. (9.1.34) with respect to the longitudinal distance x along the channel is

$$\frac{\partial h}{\partial x} = \frac{\partial y}{\partial x} + \frac{\partial z}{\partial x} \tag{9.1.35}$$

But $\partial z / \partial x = -S_o$, so

$$\frac{\partial h}{\partial x} = \frac{\partial y}{\partial x} - S_o \tag{9.1.36}$$

The momentum equation can now be expressed in terms of h by using (9.1.36) in (9.1.33):

$$\frac{\partial Q}{\partial t} + \frac{\partial(\beta Q^2/A)}{\partial x} + gA\left(\frac{\partial h}{\partial x} + S_f + S_e\right) - \beta q v_x + W_f B = 0 \tag{9.1.37}$$

The Saint-Venant equations, (9.1.6) for continuity and (9.1.37) for momentum, are the governing equations for one-dimensional, unsteady flow in an open channel. The use of the terms S_f and S_e in (9.1.37), which represent the rate of energy loss as the flow passes through the channel, illustrates the close relationship between energy and momentum considerations in describing the flow. Strelkoff (1969) showed that the momentum equation for the Saint-Venant equations can also be derived from energy principles, rather than by using Newton's second law as presented here.

The nonconservation form of the momentum equation can be derived in a similar manner to the nonconservation form of the continuity equation. Neglecting eddy losses, wind shear effect, and lateral inflow, the nonconservation form of the momentum equation for a unit width in the flow is

$$\frac{\partial V}{\partial t} + V\frac{\partial V}{\partial x} + g\left(\frac{\partial y}{\partial x} - S_o + S_f\right) = 0 \tag{9.1.38}$$

9.2 CLASSIFICATION OF DISTRIBUTED ROUTING MODELS

The Saint-Venant equations have various simplified forms, each defining a one-dimensional distributed routing model. Variations of Eqs. (9.1.6) and (9.1.37) in conservation and nonconservation forms, neglecting lateral inflow, wind shear, and eddy losses, are used to define various one-dimensional distributed routing models as shown in Table 9.2.1.

The momentum equation consists of terms for the physical processes that govern the flow momentum. These terms are: the *local acceleration* term, which describes the change in momentum due to the change in velocity over time, the *convective acceleration* term, which describes the change in momentum due to change in velocity along the channel, the *pressure force* term, proportional to the change in the water depth along the channel, the *gravity force* term, proportional to the bed slope S_o, and the *friction force* term, proportional to the friction slope

TABLE 9.2.1
Summary of the Saint-Venant equations*

Continuity equation

Conservation form

$$\frac{\partial Q}{\partial x} + \frac{\partial A}{\partial t} = 0$$

Nonconservation form

$$V\frac{\partial y}{\partial x} + y\frac{\partial V}{\partial x} + \frac{\partial y}{\partial t} = 0$$

Momentum equation

Conservation form

$$\frac{1}{A}\frac{\partial Q}{\partial t} + \frac{1}{A}\frac{\partial}{\partial x}\left(\frac{Q^2}{A}\right) + g\frac{\partial y}{\partial x} - g(S_o - S_f) = 0$$

Local acceleration term	Convective acceleration term	Pressure force term	Gravity force term	Friction force term

Nonconservation form (unit width element)

$$\frac{\partial V}{\partial t} + V\frac{\partial V}{\partial x} + g\frac{\partial y}{\partial x} - g(S_o - S_f) = 0$$

Kinematic wave: $- g(S_o - S_f) = 0$
Diffusion wave: $g\frac{\partial y}{\partial x} - g(S_o - S_f) = 0$
Dynamic wave: all terms

* Neglecting lateral inflow, wind shear, and eddy losses, and assuming $\beta = 1$.

S_f. The local and convective acceleration terms represent the effect of inertial forces on the flow.

When the water level or flow rate is changed at a particular point in a channel carrying a subcritical flow, the effects of these changes propagate back upstream. These *backwater effects* can be incorporated into distributed routing methods through the local acceleration, convective acceleration, and pressure terms. Lumped routing methods may not perform well in simulating the flow conditions when backwater effects are significant and the river slope is mild, because these methods have no hydraulic mechanisms to describe upstream propagation of changes in flow momentum.

As shown in Table 9.2.1, alternative distributed flow routing models are produced by using the full continuity equation while eliminating some terms of the momentum equation. The simplest distributed model is the *kinematic wave model*, which neglects the local acceleration, convective acceleration, and pressure terms in the momentum equation; that is, it assumes $S_o = S_f$ and the friction and gravity forces balance each other. The *diffusion wave model* neglects the local and convective acceleration terms but incorporates the pressure term. The *dynamic wave model* considers all the acceleration and pressure terms in the momentum equation.

The momentum equation can also be written in forms that take into account whether the flow is steady or unsteady, and uniform or nonuniform, as shown in Eqs. (9.2.1). In the continuity equation, $\partial A/\partial t = 0$ for a steady flow, and the lateral inflow q is zero for a uniform flow.

Conservation form:

$$-\frac{1}{gA}\frac{\partial Q}{\partial t} - \frac{1}{gA}\frac{\partial(Q^2/A)}{\partial x} - \frac{\partial y}{\partial x} + S_o = S_f \qquad (9.2.1a)$$

Nonconservation form:

$$-\frac{1}{g}\frac{\partial V}{\partial t} - \frac{V}{g}\frac{\partial V}{\partial x} - \frac{\partial y}{\partial x} + S_o = S_f \qquad (9.2.1b)$$

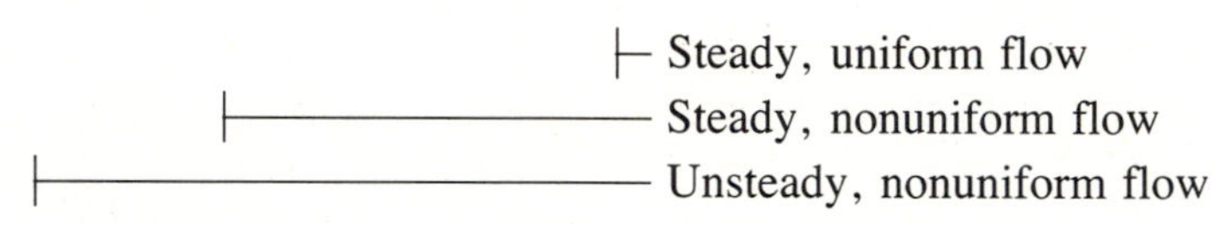

9.3 WAVE MOTION

Kinematic waves govern flow when inertial and pressure forces are not important. Dynamic waves govern flow when these forces are important, such as in the movement of a large flood wave in a wide river. In a kinematic wave, the gravity and friction forces are balanced, so the flow does not accelerate appreciably. Fig. 9.3.1 illustrates the difference between kinematic and dynamic wave motion within a differential element from the viewpoint of a stationary observer on the

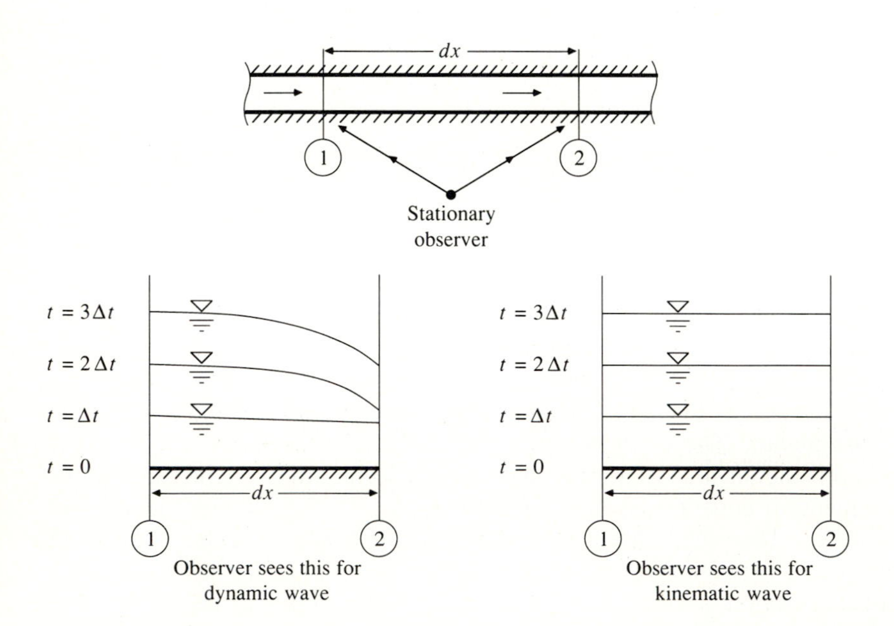

FIGURE 9.3.1
Kinematic and dynamic waves in a short reach of channel as seen by a stationary observer.

river bank. For a kinematic wave, the energy grade line is parallel to the channel bottom and the flow is steady and uniform ($S_o = S_f$) within the differential length, while for a dynamic wave the energy grade line and water surface elevation are not parallel to the bed, even within a differential element.

Kinematic Wave Celerity

A wave is a variation in a flow, such as a change in flow rate or water surface elevation, and the *wave celerity* is the velocity with which this variation travels along the channel. The celerity depends on the type of wave being considered and may be quite different from the water velocity. For a kinematic wave the acceleration and pressure terms in the momentum equation are negligible, so the wave motion is described principally by the equation of continuity. The name kinematic is thus applicable, as *kinematics* refers to the study of motion exclusive of the influence of mass and force; in *dynamics* these quantities are included.

The kinematic wave model is defined by the following equations.

Continuity:

$$\frac{\partial Q}{\partial x} + \frac{\partial A}{\partial t} = q \tag{9.3.1}$$

Momentum:

$$S_o = S_f \tag{9.3.2}$$

The momentum equation can also be expressed in the form

$$A = \alpha Q^{\beta} \tag{9.3.3}$$

For example, Manning's equation written with $S_o = S_f$ and $R = A/P$ is

$$Q = \frac{1.49 S_o^{1/2}}{n P^{2/3}} A^{5/3} \tag{9.3.4}$$

which can be solved for A as

$$A = \left(\frac{n P^{2/3}}{1.49 \sqrt{S_o}} \right)^{3/5} Q^{3/5} \tag{9.3.5}$$

so $\alpha = [nP^{2/3}/(1.49\sqrt{S_o})]^{0.6}$ and $\beta = 0.6$ in this case.

Equation (9.3.1) contains two dependent variables, A and Q, but A can be eliminated by differentiating (9.3.3):

$$\frac{\partial A}{\partial t} = \alpha \beta Q^{\beta - 1} \left(\frac{\partial Q}{\partial t} \right) \tag{9.3.6}$$

and substituting for $\partial A/\partial t$ in (9.3.1) to give

$$\frac{\partial Q}{\partial x} + \alpha \beta Q^{\beta - 1} \left(\frac{\partial Q}{\partial t} \right) = q \tag{9.3.7}$$

Kinematic waves result from changes in Q. An increment in flow, dQ, can be written as

$$dQ = \frac{\partial Q}{\partial x}dx + \frac{\partial Q}{\partial t}dt \tag{9.3.8}$$

Dividing through by dx and rearranging produces:

$$\frac{\partial Q}{\partial x} + \frac{dt}{dx}\frac{\partial Q}{\partial t} = \frac{dQ}{dx} \tag{9.3.9}$$

Equations (9.3.7) and (9.3.9) are identical if

$$\frac{dQ}{dx} = q \tag{9.3.10}$$

and

$$\frac{dx}{dt} = \frac{1}{\alpha\beta Q^{\beta-1}} \tag{9.3.11}$$

Differentiating Eq. (9.3.3) and rearranging gives

$$\frac{dQ}{dA} = \frac{1}{\alpha\beta Q^{\beta-1}} \tag{9.3.12}$$

and by comparing (9.3.11) and (9.3.12), it can be seen that

$$\frac{dx}{dt} = \frac{dQ}{dA} \tag{9.3.13}$$

or

$$c_k = \frac{dQ}{dA} = \frac{dx}{dt} \tag{9.3.14}$$

where c_k is the kinematic wave celerity. This implies that an observer moving at a velocity $dx/dt = c_k$ with the flow would see the flow rate increasing at a rate of $dQ/dx = q$. If $q = 0$, the observer would see a constant discharge. Eqs. (9.3.10) and (9.3.14) are the *characteristic equations* for a kinematic wave, two ordinary differential equations that are mathematically equivalent to the governing continuity and momentum equations.

The kinematic wave celerity can also be expressed in terms of the depth y as

$$c_k = \frac{1}{B}\frac{dQ}{dy} \tag{9.3.15}$$

where $dA = B\,dy$.

Both kinematic and dynamic wave motion are present in natural flood waves. In many cases the channel slope dominates in the momentum equation (9.2.1); therefore, most of a flood wave moves as a kinematic wave. Lighthill and Whitham (1955) proved that the velocity of the main part of a natural flood wave approximates that of a kinematic wave. If the other momentum terms [$\partial V/\partial t$, $V(\partial V/\partial x)$, and $(1/g)\partial y/\partial x$] are not negligible, then a dynamic wave front exists which can propagate both upstream and downstream from the main body of the flood wave, as shown in Fig. 9.3.2. Miller (1984) summarizes several criteria for determining when the kinematic wave approximation is applicable, but there is no single, universal criterion for making this decision.

As previously shown, if a wave is kinematic ($S_f = S_o$) the kinematic wave celerity varies with dQ/dA. For Manning's equation, wave celerity increases as Q increases. As a result, the kinematic wave theoretically should advance downstream with its rising limb getting steeper. However, the wave does not get longer, or attenuate, so it does not subside, and the flood peak stays at the same maximum depth. As the wave becomes steeper the other momentum equation terms become more important and introduce dispersion and attenuation. The celerity of a flood wave departs from the kinematic wave celerity because the discharge is not a function of depth alone, and, at the wave crest, Q and y do not remain constant.

Lighthill and Whitham (1955) illustrated that the profile of a wave front can be determined by combining the Chezy equation (2.5.5)

$$Q = CA\sqrt{RS_f} \tag{9.3.16}$$

with the momentum equation (9.2.1b) to produce

$$Q = CA\sqrt{R\left(S_o - \frac{\partial y}{\partial x} - \frac{V}{g}\frac{\partial V}{\partial x} - \frac{1}{g}\frac{\partial V}{\partial t}\right)} \tag{9.3.17}$$

in which C is the Chezy coefficient and R is the hydraulic radius.

Dynamic Wave Celerity

The dynamic wave celerity can be found by developing the characteristic equations for the Saint-Venant equations. Beginning with the nonconservation form of

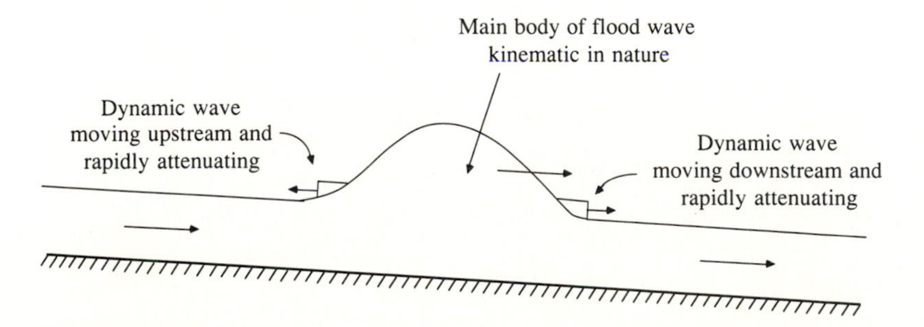

FIGURE 9.3.2
Motion of a flood wave.

the Saint-Venant equations (Table 9.2.1), it may be shown that the corresponding characteristic equations are (Henderson, 1966):

$$\frac{dx}{dt} = V \pm c_d \tag{9.3.18}$$

and

$$\frac{d}{dt}(V \pm 2c_d) = g(S_o - S_f) \tag{9.3.19}$$

in which c_d is the dynamic wave celerity, given for a rectangular channel by

$$c_d = \sqrt{gy} \tag{9.3.20}$$

where y is the depth of flow. For a channel of arbitrary cross section, $c_d = \sqrt{gA/B}$. This celerity c_d measures the velocity of a dynamic wave with respect to still water. As shown in Fig. 9.3.2, in moving water there are two dynamic waves, one proceeding upstream with velocity $V - c_d$ and the other proceeding downstream with velocity $V + c_d$. For the upstream wave to move up the channel requires $V < c_d$, or, equivalently, that the flow be subcritical, since $V = \sqrt{gy}$ is the *critical velocity* of a rectangular, open-channel flow.

Example 9.3.1. A rectangular channel is 200 feet wide, has bed slope 1 percent and Manning roughness 0.035. Calculate the water velocity V, the kinematic and dynamic wave celerities c_k and c_d, and the velocity of propagation of dynamic waves $V \pm c_d$ at a point in the channel where the flow rate is 5000 cfs.

Solution. Manning's equation with $R \approx y$, $S_o = S_f$, and channel width B is written

$$Q = \frac{1.49}{n} S_f^{1/2} A R^{2/3}$$

$$= \frac{1.49}{n} S_o^{1/2} (By) y^{2/3}$$

which is solved for y as

$$y = \left(\frac{nQ}{1.49 S_o^{1/2} B} \right)^{3/5}$$

$$= \left(\frac{0.035 \times 5000}{1.49 \times 0.01^{1/2} \times 200} \right)^{3/5}$$

$$= 2.89 \text{ ft}$$

Hence, the water velocity is

$$V = \frac{Q}{By}$$

$$= \frac{5000}{200 \times 2.89}$$

$$= 8.65 \text{ ft/s}$$

The kinematic wave celerity c_k is given by (9.3.15):

$$
\begin{aligned}
c_k &= \frac{1}{B}\frac{dQ}{dy} \\
&= \frac{1}{B}\frac{d}{dy}\left(\frac{1.49S_o^{1/2}B}{n}y^{5/3}\right) \\
&= \left(\frac{1.49S_o^{1/2}}{n}\right)\left(\frac{5}{3}\right)y^{2/3} \\
&= \frac{1.49 \times 0.01^{1/2} \times 5 \times (2.89)^{2/3}}{0.035 \times 3} \\
&= 14.4 \text{ ft/s}
\end{aligned}
$$

The dynamic wave celerity is

$$
\begin{aligned}
c_d &= \sqrt{gy} \\
&= \sqrt{32.2 \times 2.89} \\
&= 9.65 \text{ ft/s}
\end{aligned}
$$

The velocity of propagation of the upstream dynamic wave is

$$V - c_d = 8.65 - 9.65 = -1.0 \text{ ft/s}$$

and that of the downstream dynamic wave is

$$V + c_d = 8.65 + 9.65 = 18.3 \text{ ft/s}$$

In interpreting these results with Fig. 9.3.2, it can be seen that a flood wave traveling at the kinematic wave celerity (14.4 ft/s) will move down the channel faster than the water velocity (8.65 ft/s), while the dynamic waves move upstream (–1.0 ft/s) and downstream (18.3 ft/s) at the same time.

In the event that the approximation $S_o = S_f$ is not valid, the various velocities and celerities can be determined using the full momentum equation to describe S_f as in Eq. (9.3.17).

9.4 ANALYTICAL SOLUTION OF THE KINEMATIC WAVE

The solution of the kinematic wave equations specifies the distribution of the flow as a function of distance x along the channel and time t. The solution may be obtained numerically by using finite difference approximations to Eq. (9.3.7), or analytically by solving simultaneously the characteristic equations (9.3.10) and (9.3.14). In this section the analytical method is presented for the special case when lateral inflow is negligible; numerical solution is discussed in Sec. 9.6.

The solution for $Q(x,t)$ requires knowledge of the *initial condition* $Q(x,0)$, or the value of the flow along the channel at the beginning of the calculations, and the *boundary condition* $Q(0,t)$, the inflow hydrograph at the upstream end of the channel. The objective is to determine the outflow hydrograph at the downstream end of the channel, $Q(L, t)$, as a function of the inflow hydrograph, any lateral flow occurring along the sides of the channel, and the dynamics of flow in the channel as expressed by the kinematic wave equations.

If the lateral flow is neglected, (9.3.10) reduces to $dQ/dx = 0$, or $Q =$ a constant. Thus, if the flow rate is known at a point in time and space, this flow value can be propagated along the channel at the kinematic wave celerity, as given by

$$c_k = \frac{dQ}{dA} = \frac{dx}{dt} \tag{9.4.1}$$

The solution can be visualized on an x–t plane, as shown in Fig. 9.4.1(b), where distance is plotted on the horizontal axis, and time on the vertical axis. Each point in the x–t plane has a value of Q associated with it, which is the flow rate at that location along the channel, at that point in time. These values of Q may be thought of as being plotted on an axis coming out of the page perpendicular to the x–t plane. In particular, the inflow hydrograph $Q(0, t)$ is shown in Fig. 9.4.1(a) folded down to the left, and the outflow hydrograph $Q(L, t)$ is shown in Fig 9.4.1(c) folded down to the right of the x–t plane. These two

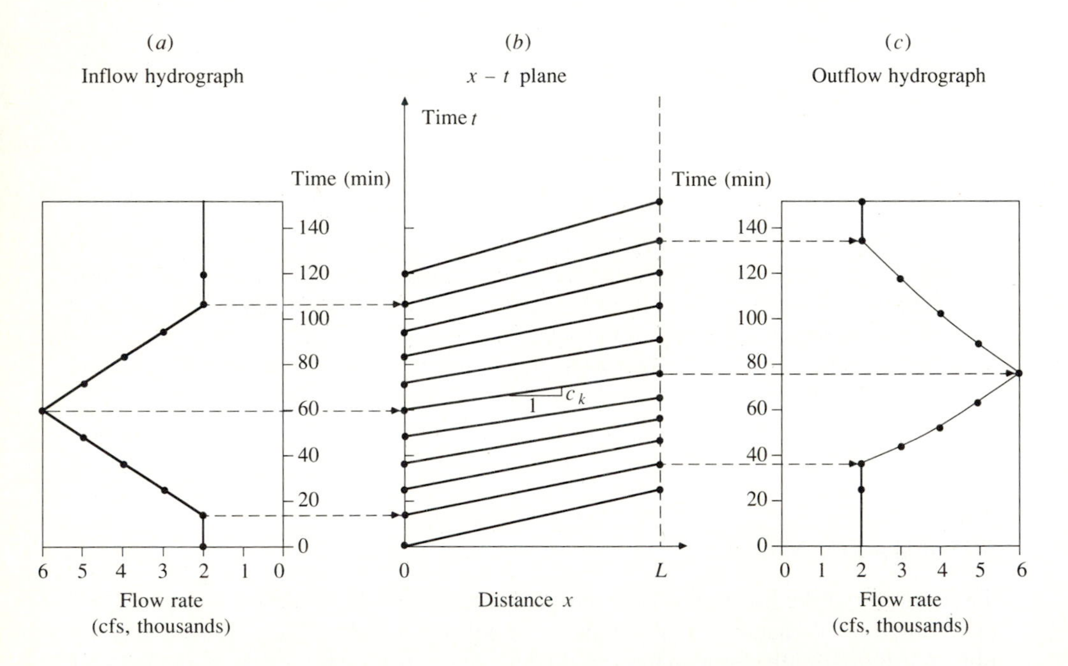

FIGURE 9.4.1
Kinematic wave routing of a flow hydrograph through a channel reach of length L using propagation of the flow along characteristic lines in the x–t plane. If flow rate were plotted on a third axis, perpendicular to the x–t plane (b), then the inflow hydrograph (a) is the variation of flow through time at $x = 0$ folded down to the left of the x–t plane; the outflow hydrograph (c) is the variation of flow rate through time at $x = L$ and is folded down to the right of the x–t plane in the figure. The dashed lines indicate the propogation of specific flow rates along characteristic lines in the x–t plane.

hydrographs are connected by the *characteristic lines* shown in part (*b*) of the figure. The equations for these lines are found by solving (9.4.1):

$$\int_0^x dx = \int_{t_o}^t c_k \, dt$$

or

$$x = c_k(t - t_o) \tag{9.4.2}$$

so the time at which a discharge Q entering a channel of length L at time t_o will appear at the outlet is

$$t = t_o + \frac{L}{c_k} \tag{9.4.3}$$

The slope of the characteristic line is $c_k = dQ/dA$ for the particular value of flow rate being considered. The lines shown in Fig. 9.4.1(*b*) are straight because $q = 0$, and Q is constant along them. If $q \neq 0$, Q and c_k vary along the characteristic lines, which then become curved.

Rainfall-runoff Process

The kinematic wave method has been applied to describe flow over planes, as a model of the rainfall-runoff process. In this application the lateral flow is equal to the difference between the rates of rainfall and infiltration, and the channel flow is taken to be flow per unit width of plane. The characteristic equations can be solved analytically to simulate the outflow hydrograph in response to rainfall of a specified duration. By accumulating the flow from many such planes laid out over a watershed, an approximate model can be developed for the conversion of rainfall into streamflow at the watershed outlet.

The kinematic wave model of the rainfall-runoff process offers the advantage over the unit hydrograph method that it is a solution of the physical equations governing the surface flow, but the solution is only for one-dimensional flow, whereas the actual watershed surface flow is two-dimensional as the water follows the land surface contour. As a consequence, the kinematic wave parameters, such as Manning's roughness coefficient, must be adjusted to produce a realistic outflow hydrograph. Eagleson (1970), Overton and Meadows (1976), and Stephenson and Meadows (1986) present detailed information on kinematic wave models for the rainfall-runoff process.

Example 9.4.1. A 200-foot-wide rectangular channel is 15,000 feet long, has a bed slope of 1 percent, and a Manning's roughness factor of 0.035. The inflow hydrograph to the channel is given in columns 1 and 2 of Table 9.4.1. Calculate the outflow hydrograph by analytical solution of the kinematic wave equations.

Solution. The kinematic wave celerity for a given value of the flow rate is calculated in the same manner as shown in Example 9.3.1, where it was shown that for this channel, $c_k = 14.4$ ft/s for $Q = 5000$ cfs. The corresponding values for the other flow

TABLE 9.4.1
Routing of a flow hydrograph by analytical solution of the kinematic wave (Example 9.4.1).

Column:	1	2	3	4	5
	Inflow Time	Inflow Rate	Kinematic wave celerity	Travel time	Outflow* time
	(min)	(cfs)	(ft/s)	(min)	(min)
	0	2000	10.0	25.1	25.1
	12	2000	10.0	25.1	37.1
	24	3000	11.7	21.3	45.3
	36	4000	13.2	19.0	55.0
	48	5000	14.4	17.4	65.4
	60	6000	15.5	16.1	76.1
	72	5000	14.4	17.4	89.4
	84	4000	13.2	19.0	103.0
	96	3000	11.7	21.3	117.3
	108	2000	10.0	25.1	133.1
	120	2000	10.0	25.1	145.1

*Outflow time = Inflow time + Travel time.

rates on the inflow hydrograph are shown in column 3 of Table 9.4.1. The travel time through a reach of length L is L/c_k so for $L = 15,000$ ft and $c_k = 14.4$ ft/s, the travel time is $15,000/14.4 = 1042$ s $= 17.4$ min, as shown in column 4 of the table. The time when this discharge on the rising limb of the hydrograph will arrive at the outlet of the channel is, by Eq. (9.4.3) $t = t_o + L/c_k = 48 + 17.4 = 65.4$ min, as shown in column 5. The inflow and outflow hydrographs for this example are plotted in Fig. 9.4.1. It can be seen that the kinematic wave is a wave of translation without attenuation; the maximum discharge of 6000 cfs is undiminished by passage through the channel.

9.5 FINITE-DIFFERENCE APPROXIMATIONS

The Saint-Venant equations for distributed routing are not amenable to analytical solution except in a few special simple cases. They are partial differential equations that, in general, must be solved using numerical methods. Methods for solving partial differential equations may be classified as *direct numerical methods* and *characteristic methods*. In direct methods, finite-difference equations are formulated from the original partial differential equations for continuity and momentum. Solutions for the flow rate and water surface elevation are then obtained for incremental times and distances along the stream or river. In characteristic methods, the partial differential equations are first transformed to a characteristic form, and the characteristic equations are solved analytically, as shown previously for the kinematic wave, or by using a finite-difference representation.

In numerical methods for solving partial differential equations, the calculations are performed on a grid placed over the x–t plane. The *x–t grid* is a network of points defined by taking distance increments of length Δx and time increments of duration Δt. As shown in Fig. 9.5.1, the distance points are denoted by index

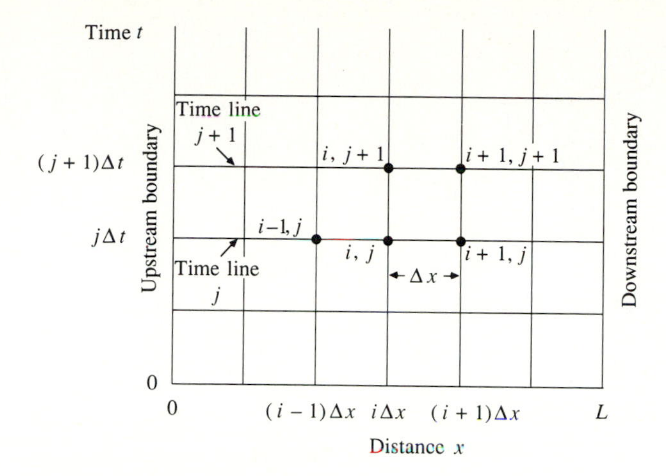

FIGURE 9.5.1
The grid on the x–t plane used for numerical solution of the Saint-Venant equations by finite differences.

i and the time points by index j. A *time line* is a line parallel to the x axis through all the distance points at a given value of time.

Numerical schemes transform the governing partial differential equations into a set of algebraic finite-difference equations, which may be linear or nonlinear. The finite-difference equations represent the spatial and temporal derivatives in terms of the unknown variables on both the current time line, $j + 1$, and the preceding time line, j, where all the values are known from previous computation (see Fig. 9.5.1). The solution of the Saint-Venant equations advances from one time line to the next.

Finite Differences

Finite-difference approximations can be derived for a function $u(x)$ as shown in Fig. 9.5.2. A Taylor series expansion of $u(x)$ at $x + \Delta x$ produces

$$u(x + \Delta x) = u(x) + \Delta x\, u'(x) + \frac{1}{2}\Delta x^2 u''(x) + \frac{1}{6}\Delta x^3 u'''(x) + \dots \qquad (9.5.1)$$

where $u'(x) = \partial u/\partial x$, $u''(x) = \partial^2 u/\partial x^2, \dots$, and so on. The Taylor series expansion at $x - \Delta x$ is

$$u(x - \Delta x) = u(x) - \Delta x\, u'(x) + \frac{1}{2}\Delta x^2 u''(x) - \frac{1}{6}\Delta x^3 u'''(x) + \dots \qquad (9.5.2)$$

A *central-difference* approximation uses the difference defined by subtracting (9.5.2) from (9.5.1)

$$u(x + \Delta x) - u(x - \Delta x) = 2\,\Delta x\, u'(x) + 0(\Delta x^3) \qquad (9.5.3)$$

where $0(\Delta x^3)$ represents a residual containing the third and higher order terms. Solving for $u'(x)$ assuming $0(\Delta x^3) \approx 0$ results in

$$u'(x) \approx \frac{u(x + \Delta x) - u(x - \Delta x)}{2\,\Delta x} \qquad (9.5.4)$$

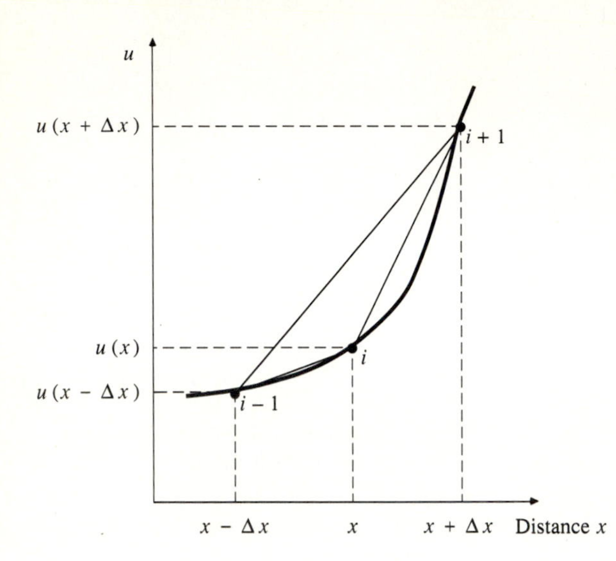

FIGURE 9.5.2
Finite difference approximations for the function $u(x)$.

which has an error of approximation of order Δx^2. This approximation error, due to dropping the higher order terms, is also referred to as a *truncation error*.

A *forward difference* approximation is defined by subtracting $u(x)$ from (9.5.1):

$$u(x + \Delta x) - u(x) = \Delta x\, u'(x) + 0(\Delta x^2) \tag{9.5.5}$$

Assuming second and higher order terms are negligible, solving for $u'(x)$ gives

$$u'(x) \approx \frac{u(x + \Delta x) - u(x)}{\Delta x} \tag{9.5.6}$$

which has an error of approximation of order Δx.

The *backward-difference* approximation uses the difference defined by subtracting (9.5.2) from $u(x)$,

$$u(x) - u(x - \Delta x) = \Delta x\, u'(x) + 0(\Delta x^2) \tag{9.5.7}$$

so that solving for $u'(x)$ gives

$$u'(x) \approx \frac{u(x) - u(x - \Delta x)}{\Delta x} \tag{9.5.8}$$

A finite-difference method may employ either an *explicit scheme* or an *implicit scheme* for solution. The main difference between the two is that in the explicit method, the unknown values are solved *sequentially* along a time line from one distance point to the next, while in the implicit method the unknown values on a given time line are all determined *simultaneously*. The explicit method is simpler but can be unstable, which means that small values of Δx and Δt are required for convergence of the numerical procedure. The explicit method is convenient because results are given at the grid points, and it can treat slightly varying channel geometry from section to section, but it is less efficient than the implicit method and hence not suitable for routing flood flows over a long time period.

The implicit method is mathematically more complicated, but with the use of computers this is not a serious problem once the method is programmed. The method is stable for large computation steps with little loss of accuracy and hence works much faster than the explicit method. The implicit method can also handle channel geometry varying significantly from one channel cross section to the next.

Explicit Scheme

The finite-difference representation is shown by the mesh of points on the time-distance plane shown in Fig. 9.5.1. Assuming that at time t (time line j), the hydraulic quantities u are known, the problem is to determine the unknown quantity at point $(i, j+1)$ at time $t+\Delta t$, that is, u_i^{j+1}.

The simplest scheme determines the partial derivatives at point $(i, j+1)$ in terms of the quantities at adjacent points $(i-1, j), (i, j)$, and $(i+1, j)$ using

$$\frac{\partial u_i^{j+1}}{\partial t} = \frac{u_i^{j+1} - u_i^j}{\Delta t} \tag{9.5.9}$$

and

$$\frac{\partial u_i^j}{\partial x} = \frac{u_{i+1}^j - u_{i-1}^j}{2\,\Delta x} \tag{9.5.10}$$

A forward-difference scheme is used for the time derivative and a central-difference scheme is used for the spatial derivative.

Note that the spatial derivative is written using known terms on time line j. Implicit schemes on the other hand use finite-difference approximations for both the temporal and spatial derivatives in terms of the unknown time line $j+1$.

The discretization of the x–t plane into a grid for the integration of the finite-difference equations introduces numerical errors into the computation. A finite-difference scheme is stable if such errors are not amplified during successive computation from one time line to the next. The numerical stability of the computation depends on the relative grid size. A necessary but insufficient condition for stability of an explicit scheme is the *Courant condition* (Courant and Friedrichs, 1948). For the kinematic wave equations, the Courant condition is

$$\Delta t \leq \frac{\Delta x_i}{c_k} \tag{9.5.11}$$

where c_k is the kinematic wave celerity. For the dynamic wave equations, c_k is replaced by $V + c_d$ in (9.5.11). The Courant condition requires that the time step be less than the time for a wave to travel the distance Δx_i. If Δt is so large that the Courant condition is not satisfied, then there is, in effect, an accumulation or piling up of water. The Courant condition does not apply to the implicit scheme.

For computational purposes in an explicit scheme, Δx is specified and kept fixed throughout the computations, while Δt is determined at each time step. To do this, a Δt_i just meeting the Courant condition is computed at each grid point i on time line j, and the smallest Δt_i is used. Because the explicit method is unstable unless Δt is small, it is sometimes advisable to determine the minimum Δt_i at a time line j then reduce it by some percentage. The Courant condition does not guarantee stability, and therefore is only a guideline.

Implicit Scheme

Implicit schemes use finite-difference approximations for both the temporal and spatial derivative in terms of the dependent variable on the unknown time line. As a simple example the space and time derivatives can be written for the unknown point $(i+1, j+1)$ as

$$\frac{\partial u_{i+1}^{j+1}}{\partial x} = \frac{u_{i+1}^{j+1} - u_i^{j+1}}{\Delta x} \tag{9.5.12}$$

and

$$\frac{\partial u_{i+1}^{j+1}}{\partial t} = \frac{u_{i+1}^{j+1} - u_{i+1}^{j}}{\Delta t} \tag{9.5.13}$$

This scheme is used in Sec. 9.6 for the kinematic wave model. In Chap. 10 a more complex implicit scheme, referred to as a weighted 4-point implicit scheme, is used for the full dynamic wave model.

9.6 NUMERICAL SOLUTION OF THE KINEMATIC WAVE

As shown in Eq. (9.3.7), the continuity and momentum equations for the kinematic wave can be combined to produce an equation with Q as the only dependent variable:

$$\frac{\partial Q}{\partial x} + \alpha\beta Q^{\beta-1}\frac{\partial Q}{\partial t} = q \tag{9.6.1}$$

The objective of the numerical solution is to solve (9.6.1) for $Q(x, t)$ at each point on the x–t grid, given the channel parameters α and β, the lateral inflow $q(t)$, and the initial and boundary conditions. In particular, the purpose of the solution is to determine the outflow hydrograph $Q(L,t)$. The numerical solution of the kinematic wave equation is more flexible than the analytical solution described in Sec. 9.4; it can more easily handle variation in the channel properties and in the initial and boundary conditions, and it serves as an introduction to numerical solution of the dynamic wave equations, presented in Chap. 10.

To solve Eq. (9.6.1) numerically, the time and space derivatives of Q are approximated on the x–t grid as shown in Fig. 9.6.1. The unknown value is Q_{i+1}^{j+1}. The values of Q on the jth time line have been previously determined, and so has Q_i^{j+1}. Two schemes for setting up the finite difference equations are described in this section: a *linear scheme*, in which Q_{i+1}^{j+1} is computed as a linear function of the known values of Q, and a *nonlinear scheme*, in which the finite-difference form of (9.6.1) is a nonlinear equation.

Linear Scheme

The backward-difference method is used to set up the finite-difference equations. The finite-difference form of the space derivative of Q_{i+1}^{j+1} is found by substituting the values of Q on the $(j+1)$th time line into (9.5.12):

$$\frac{\partial Q}{\partial x} \approx \frac{Q_{i+1}^{j+1} - Q_i^{j+1}}{\Delta x} \tag{9.6.2}$$

The finite-difference form of the time derivative is found likewise by substituting

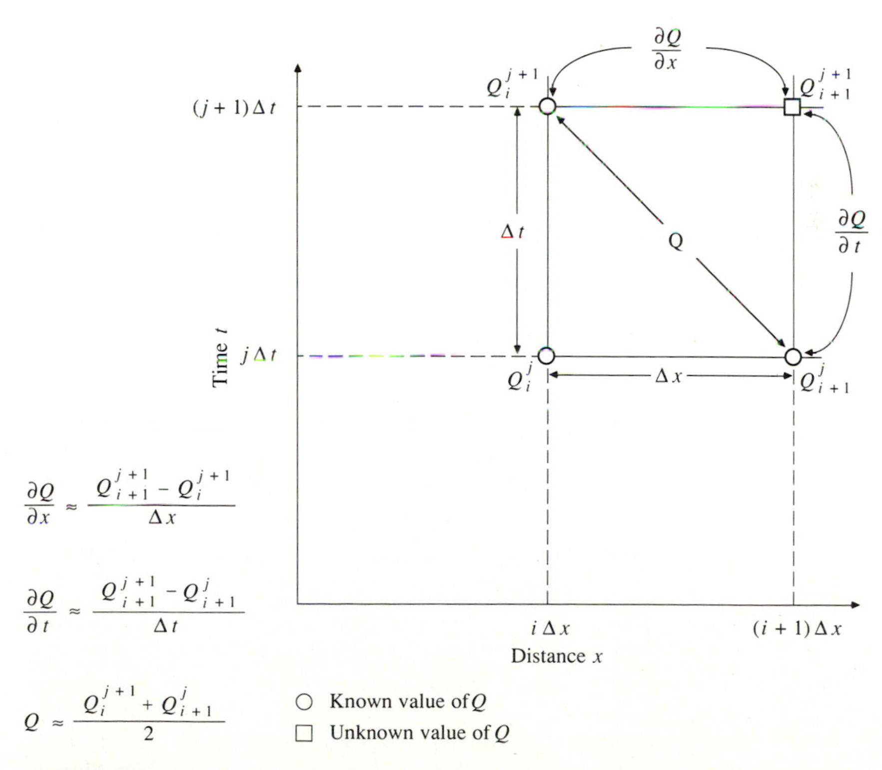

FIGURE 9.6.1
Finite difference box for solution of the linear kinematic wave equation showing the finite difference equations.

the values of Q on the $(i + 1)$th distance line into (9.5.13):

$$\frac{\partial Q}{\partial t} \approx \frac{Q_{i+1}^{j+1} - Q_{i+1}^{j}}{\Delta t} \tag{9.6.3}$$

If the value of Q_{i+1}^{j+1} were used for Q in the term $\alpha\beta Q^{\beta-1}$ in Eq. (9.6.1), the resulting equation would be nonlinear in Q_{i+1}^{j+1}. To create a linear equation, the value of Q used in $\alpha\beta Q^{\beta-1}$ is found by averaging the values across the diagonal in the box shown in Fig. 9.6.1:

$$Q \approx \frac{Q_{i+1}^{j} + Q_{i}^{j+1}}{2} \tag{9.6.4}$$

The value of lateral inflow q is found by averaging the values on the $(i + 1)$th distance line (these are assumed to be given in the problem).

$$q \approx \frac{q_{i+1}^{j+1} + q_{i+1}^{j}}{2} \tag{9.6.5}$$

By substituting Eqs. (9.6.2) to (9.6.5) into (9.6.1), the finite-difference form of the linear kinematic wave is obtained:

$$\frac{Q_{i+1}^{j+1} - Q_{i}^{j+1}}{\Delta x} + \alpha\beta\left(\frac{Q_{i+1}^{j} + Q_{i}^{j+1}}{2}\right)^{\beta-1}\left(\frac{Q_{i+1}^{j+1} - Q_{i+1}^{j}}{\Delta t}\right) = \frac{q_{i+1}^{j+1} + q_{i+1}^{j}}{2} \tag{9.6.6}$$

This equation, solved for the unknown Q_{i+1}^{j+1}, is

$$Q_{i+1}^{j+1} = \frac{\left[\dfrac{\Delta t}{\Delta x}Q_{i}^{j+1} + \alpha\beta Q_{i+1}^{j}\left(\dfrac{Q_{i+1}^{j} + Q_{i}^{j+1}}{2}\right)^{\beta-1} + \Delta t\left(\dfrac{q_{i+1}^{j+1} + q_{i+1}^{j}}{2}\right)\right]}{\left[\dfrac{\Delta t}{\Delta x} + \alpha\beta\left(\dfrac{Q_{i+1}^{j} + Q_{i}^{j+1}}{2}\right)^{\beta-1}\right]} \tag{9.6.7}$$

A flow chart for routing a kinematic wave using this scheme is given in Fig. 9.6.2. Q was chosen as the dependent variable because this results in smaller relative errors than if cross-sectional area A were chosen as the dependent variable (Henderson, 1966). This is shown by taking the logarithm of (9.3.3):

$$\ln A = \ln \alpha + \beta \ln Q \tag{9.6.8}$$

and differentiating:

$$\frac{dQ}{Q} = \frac{1}{\beta}\left(\frac{dA}{A}\right) \tag{9.6.9}$$

to define the relationship between the relative errors dA/A and dQ/Q. Using either Manning's equation or the Darcy-Weisbach equation, β is generally less than 1, and it follows that the discharge estimation error would be magnified by the ratio

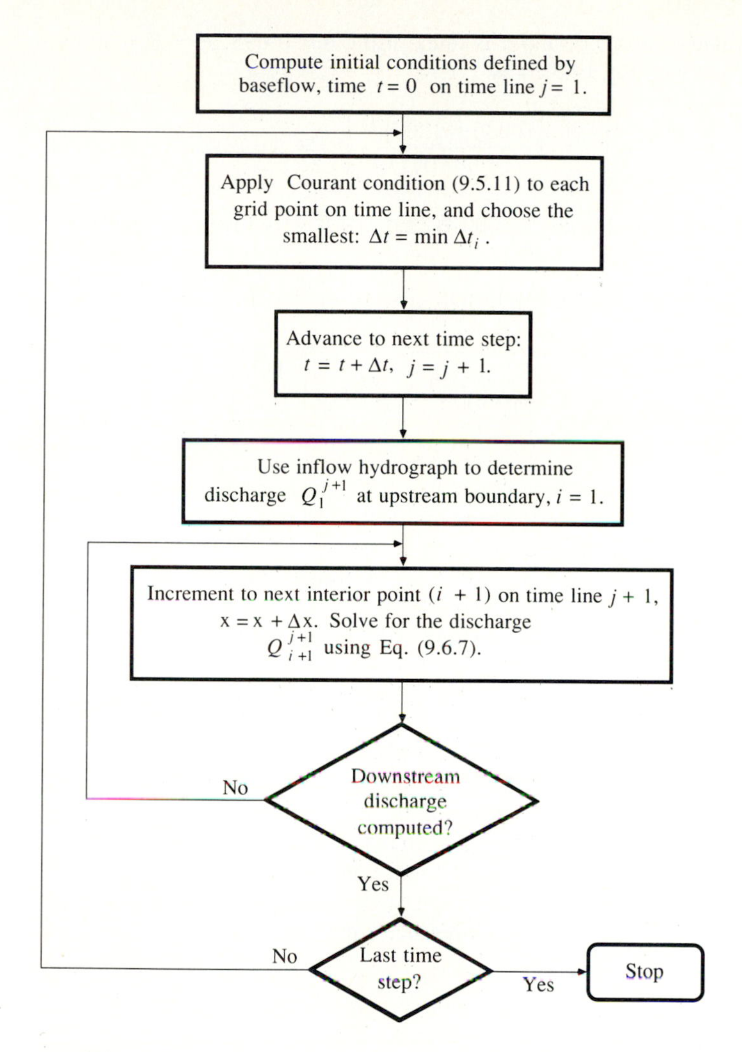

FIGURE 9.6.2
Flowchart for linear kinematic wave computation.

$1/\beta$ if the cross-sectional area were the dependent variable instead of the flow rate.

Example 9.6.1. Using the same data for the rectangular channel in Example 9.4.1 (width $= 200$ ft, length $= 15{,}000$ ft, slope $= 1$ percent, and Manning's $n = 0.035$), develop a linear kinematic wave model and route the inflow hydrograph given in columns 1 and 2 of Table 9.4.1 through the channel using $\Delta x = 3000$ ft and $\Delta t = 3$ min. There is no lateral inflow. The initial condition is a uniform flow of 2000 cfs along the channel.

Solution. The value of β is 0.6, and α is found using $n = 0.035$, $P \approx B = 200$ ft, and $S_o = 0.01$ following Eq. (9.3.5) as

$$\alpha = \left(\frac{nP^{2/3}}{1.49S_o^{1/2}}\right)^{0.6} = \left[\frac{0.035 \times (200)^{2/3}}{1.49(0.01)^{1/2}}\right]^{0.6} = 3.49$$

For $\Delta t = 3$ min $= 180$ s and $\Delta x = 3000$ ft, Eq. (9.6.7) with $q = 0$ gives

$$Q_{i+1}^{j+1} = \frac{\left[\frac{180}{3000}Q_i^{j+1} + (3.49)(0.6)Q_{i+1}^j\left(\frac{Q_{i+1}^j + Q_i^{j+1}}{2}\right)^{(0.6-1)}\right]}{\left[\frac{180}{3000} + (3.49)(0.6)\left(\frac{Q_{i+1}^j + Q_i^{j+1}}{2}\right)^{(0.6-1)}\right]}$$

This problem is solved following the algorithm given in Fig. 9.6.2. Computations proceed from upstream to downstream as shown in Table 9.6.1, in which the distance axis is laid out horizontally, $i = 1, 2, \ldots, 6$, while the time axis runs vertically down the page, $j = 1, 2, \ldots$. The initial condition is $Q_i^1 = 2000$ cfs, covering the first row of discharge values. The upstream boundary condition is the inflow hydrograph Q_1^j in the first column of discharge values. The inflows at $t = 0, 12, 24, \ldots$ min are obtained from Table 9.4.1, and the remainder are filled in by linear interpolation between the tabulated values.

The first time the inflow departs from 2000 cfs is after $t = 12$ min, so the calculations for Q on the 15-min time line are used as an illustration. The computational sequence is indicated in Table 9.6.1 by the sequence of boxes. With $j = 5$ and $i = 1$, the first unknown value is $Q_{i+1}^{j+1} = Q_2^6$, which is the discharge at distance 3000 ft on the 15-min time line. It is found as a function of $Q_{i+1}^j = Q_2^5 =$ 2000 cfs, the discharge at 3000 ft on the 12-min time line, and of $Q_i^{j+1} = Q_1^6 =$ 2250 cfs, the value of the inflow hydrograph at $x = 0$ on the 15-min time line. Substituting these values into the finite-difference equation:

$$Q_2^6 = \frac{\left[\frac{180}{3000}(2250) + (3.49)(0.6)(2000)\left(\frac{2000 + 2250}{2}\right)^{(0.6-1)}\right]}{\left[\frac{180}{3000} + (3.49)(0.6)\left(\frac{2000 + 2250}{2}\right)^{(0.6-1)}\right]} = 2095 \text{ cfs}$$

as shown in the table. Moving along the 15-min time line ($j = 6$), the second unknown is Q_3^6, the value at distance 6000 ft, calculated as a function of $Q_{i+1}^j =$ $Q_3^5 = 2000$ cfs, now at 6000 ft on the 12-min time line, and of $Q_i^{j+1} = Q_2^6 =$ 2095 cfs, the value just computed for 3000 ft at 15 min. The same procedure as above produces $Q_3^6 = 2036$ cfs as shown in Table 9.6.1. All the unknown values are determined in the same manner. The outflow hydrograph is the column of flow rates for $i = 6$ at 15,000 ft.

The values of $\Delta t = 3$ min and $\Delta x = 3000$ ft were chosen so that the Courant condition (9.5.11) would be satisfied everywhere in the x–t plane. As shown in Table 9.4.1, the maximum wave celerity is 15.5 ft/s, for a discharge of 6000 cfs. Here $\Delta x/\Delta t = 3000/180 = 16.7$ ft/s, which is greater than the maximum wave celerity, thus satisfying the Courant condition throughout.

TABLE 9.6.1
Numerical solution of the linear kinematic wave (Example 9.6.1.). Values given in the table are flow rates in cfs. The italicized values show the propagation of the peak discharge. The boxes show the computational sequence for obtaining the flows along the 15-min time line.

Time (min)	Time index	Distance along channel (ft)					
		0	3000	6000	9000	12000	15000
	j	$i=1$	2	3	4	5	6
0	1	2000	2000	2000	2000	2000	2000
3	2	2000	2000	2000	2000	2000	2000
6	3	2000	2000	2000	2000	2000	2000
9	4	2000	2000	2000	2000	2000	2000
12	5	2000	2000	2000	2000	2000	2000
15	6	2250	2095	2036	2013	2005	2002
18	7	2500	2252	2118	2053	2023	2010
21	8	2750	2449	2246	2127	2062	2030
24	9	3000	2672	2414	2238	2129	2067
27	10	3250	2910	2613	2385	2228	2129
30	11	3500	3158	2836	2566	2360	2218
.	.	.	.	.	.	.	.
.	.	.	.	.	.	.	.
.	.	.	.	.	.	.	.
48	17	5000	4695	4374	4037	3694	3358
51	18	5250	4952	4638	4307	3965	3620
54	19	5500	5209	4902	4578	4239	3892
57	20	5750	5465	5165	4848	4516	4171
60	21	***6000***	5720	5427	5118	4793	4452
63	22	5750	***5734***	5573	5332	5043	4723
66	23	5500	5623	***5597***	5457	5237	4961
69	24	5250	5447	5526	***5489***	5356	5145
72	25	5000	5238	5390	5443	***5397***	5263
75	26	4750	5012	5213	5335	5368	***5312***
78	27	4500	4777	5011	5184	5281	5298
.	.	.	.	.	.	.	.
.	.	.	.	.	.	.	.
.	.	.	.	.	.	.	.
144	49	2000	2001	2008	2028	2067	2133
147	50	2000	2001	2005	2019	2049	2101
150	51	2000	2001	2004	2013	2036	2076

Fig. 9.6.3(a) shows plots of the columns of Table 9.6.1, the flow hydrographs at the various points along the channel. It can be seen that the peak discharge diminishes as the wave passes down the channel, also marked by the italicized values in the table. Fig. 9.6.3(b) shows plots of rows from Table 9.6.1, representing the distribution of flow along the channel for various points in time, which shows the rise and fall of the flow as the wave passes down the channel. Fig. 9.6.4 is a comparison of the analytical solution calculated in Example 9.4.1 with two

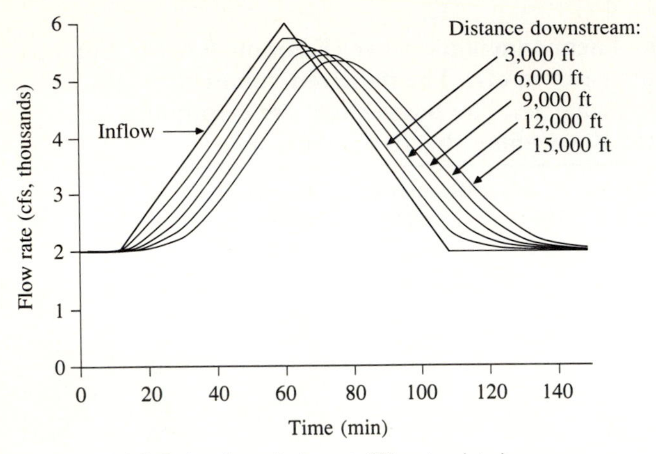

(*a*) Solution through time at different points in space.

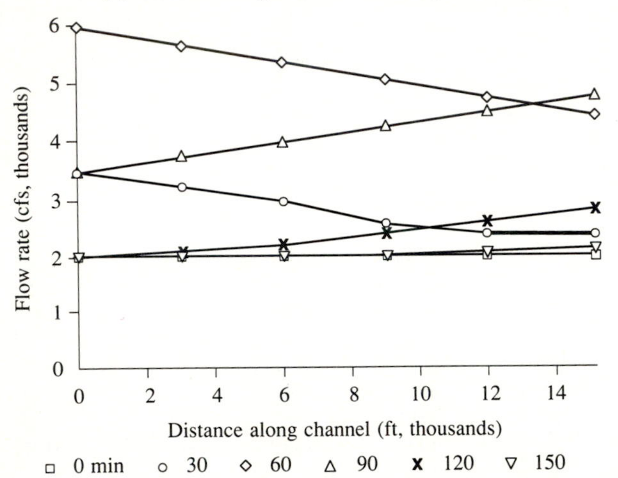

(*b*) Solution through space at different points in time.

FIGURE 9.6.3
Numerical solution of the linear kinematic wave equation in space and time (Example 9.6.1).

numerical solutions, the one calculated here with $\Delta x = 3000$ ft and $\Delta t = 3$ min, and another solution similarly computed with $\Delta x = 1000$ ft and $\Delta t = 1$ min. It can be seen that the numerical scheme introduces dispersion of the flood wave into the solution, with the degree of dispersion increasing with the size of the Δx and Δt increments.

Nonlinear Kinematic Wave Scheme

The finite-difference form of Eq. (9.6.1) can also be expressed as

$$\frac{Q_{i+1}^{j+1} - Q_i^{j+1}}{\Delta x} + \frac{A_{i+1}^{j+1} - A_{i+1}^{j}}{\Delta t} = \frac{q_{i+1}^{j+1} + q_{i+1}^{j}}{2} \tag{9.6.10}$$

As for the linear scheme, Q is taken as the independent variable; using Eq. (9.3.3),

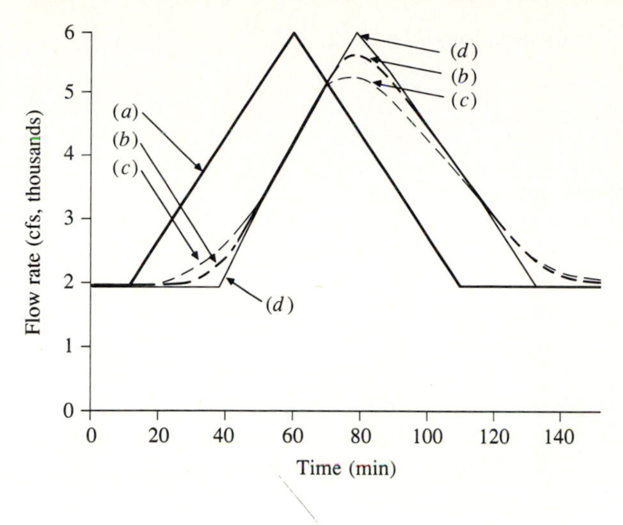

FIGURE 9.6.4
Routing of the kinematic wave by analytical and numerical methods. The analytical solution shows no wave attenuation, while the numerical solutions disperse the wave, the degree of dispersion increasing with the size of the time and distance steps. (*a*) Inflow. (*b*) Numerical solution, using $\Delta t = 1$ min and $\Delta x = 1000$ ft. (*c*) Numerical solution, $\Delta t = 3$ min, $\Delta x = 3000$ ft. (*d*) Analytical solution.

$$A_{i+1}^{j+1} = \alpha\left(Q_{i+1}^{j+1}\right)^{\beta} \tag{9.6.11}$$

and

$$A_{i+1}^{j} = \alpha\left(Q_{i+1}^{j}\right)^{\beta} \tag{9.6.12}$$

Eqs. (9.6.11) and (9.6.12) are substituted into (9.6.10) to obtain, after rearranging,

$$\frac{\Delta t}{\Delta x} Q_{i+1}^{j+1} + \alpha\left(Q_{i+1}^{j+1}\right)^{\beta} = \frac{\Delta t}{\Delta x} Q_{i}^{j+1} + \alpha\left(Q_{i+1}^{j}\right)^{\beta} + \Delta t\left(\frac{q_{i+1}^{j+1} + q_{i+1}^{j}}{2}\right) \tag{9.6.13}$$

This equation has been arranged so that the unknown discharge Q_{i+1}^{j+1} is on the left-hand side, and all the known quantities are on the right-hand side. It is nonlinear in Q_{i+1}^{j+1}; so a numerical solution scheme such as Newton's method will be required (see Sec. 5.6 for an introduction to Newton's method).

The known right-hand side at each finite-difference grid point is

$$C = \frac{\Delta t}{\Delta x} Q_{i}^{j+1} + \alpha\left(Q_{i+1}^{j}\right)^{\beta} + \Delta t\left(\frac{q_{i+1}^{j+1} + q_{i+1}^{j}}{2}\right) \tag{9.6.14}$$

from which a *residual error* $f(Q_{i+1}^{j+1})$ is defined using Eq. (9.6.13) as

$$f\left(Q_{i+1}^{j+1}\right) = \frac{\Delta t}{\Delta x} Q_{i+1}^{j+1} + \alpha\left(Q_{i+1}^{j+1}\right)^{\beta} - C \tag{9.6.15}$$

The first derivative of $f(Q_{i+1}^{j+1})$ is

$$f'\left(Q_{i+1}^{j+1}\right) = \frac{\Delta t}{\Delta x} + \alpha\beta\left(Q_{i+1}^{j+1}\right)^{\beta-1} \tag{9.6.16}$$

The objective is to find Q_{i+1}^{j+1} that forces $f(Q_{i+1}^{j+1})$ to equal 0.

Using Newton's method with iterations $k = 1, 2, \ldots$

$$\left(Q_{i+1}^{j+1}\right)_{k+1} = \left(Q_{i+1}^{j+1}\right)_k - \frac{f\left(Q_{i+1}^{j+1}\right)_k}{f'\left(Q_{i+1}^{j+1}\right)_k} \tag{9.6.17}$$

The convergence criterion for the iterative process is

$$\left| f\left(Q_{i+1}^{j+1}\right)_{k+1} \right| \leq \epsilon \tag{9.6.18}$$

where ϵ is an error criterion. A flowchart for the nonlinear kinematic wave scheme is presented in Fig. 9.6.5.

The initial estimate for Q_{i+1}^{j+1} is important for the convergence of the iterative scheme. One approach is to use the solution from the linear scheme, Eq. (9.6.7), as the first approximation to the nonlinear scheme. Li, Simons, and Stevens (1975) performed a stability analysis indicating that the scheme using Eq. (9.6.13) is unconditionally stable. They also showed that a wide range of values of $\Delta t/\Delta x$ could be used without introducing large errors in the shape of the discharge hydrograph.

9.7 MUSKINGUM-CUNGE METHOD

Several variations of the kinematic wave routing method have been proposed. Cunge (1969) proposed a method based on the Muskingum method, a method traditionally applied to linear hydrologic storage routing. Referring to the time-space computational grid shown in Fig. 9.6.1, the Muskingum routing equation (8.4.7) can be written for the discharge at $x = (i + 1)\Delta x$ and $t = (j + 1)\Delta t$:

$$Q_{i+1}^{j+1} = C_1 Q_i^{j+1} + C_2 Q_i^j + C_3 Q_{i+1}^j \tag{9.7.1}$$

in which C_1, C_2, and C_3 are as defined in Eqs. (8.4.8) through (8.4.10). In those equations, K is a storage constant having dimensions of time, and X is a factor expressing the relative influence of inflow on storage levels. Cunge showed that when K and Δt are taken as constant, Eq. (9.7.1) is an approximate solution of the kinematic wave equations [Eqs. (9.3.1) and (9.3.2)]. He further demonstrated that (9.7.1) can be considered an approximate solution of a modified diffusion equation (Table 9.2.1) if

$$K = \frac{\Delta x}{c_k} = \frac{\Delta x}{dQ/dA} \tag{9.7.2}$$

and

$$X = \frac{1}{2}\left(1 - \frac{Q}{Bc_k S_o \Delta x}\right) \tag{9.7.3}$$

where c_k is the celerity corresponding to Q and B, and B is the width of the water surface. The right-hand side of (9.7.2) represents the time of propagation

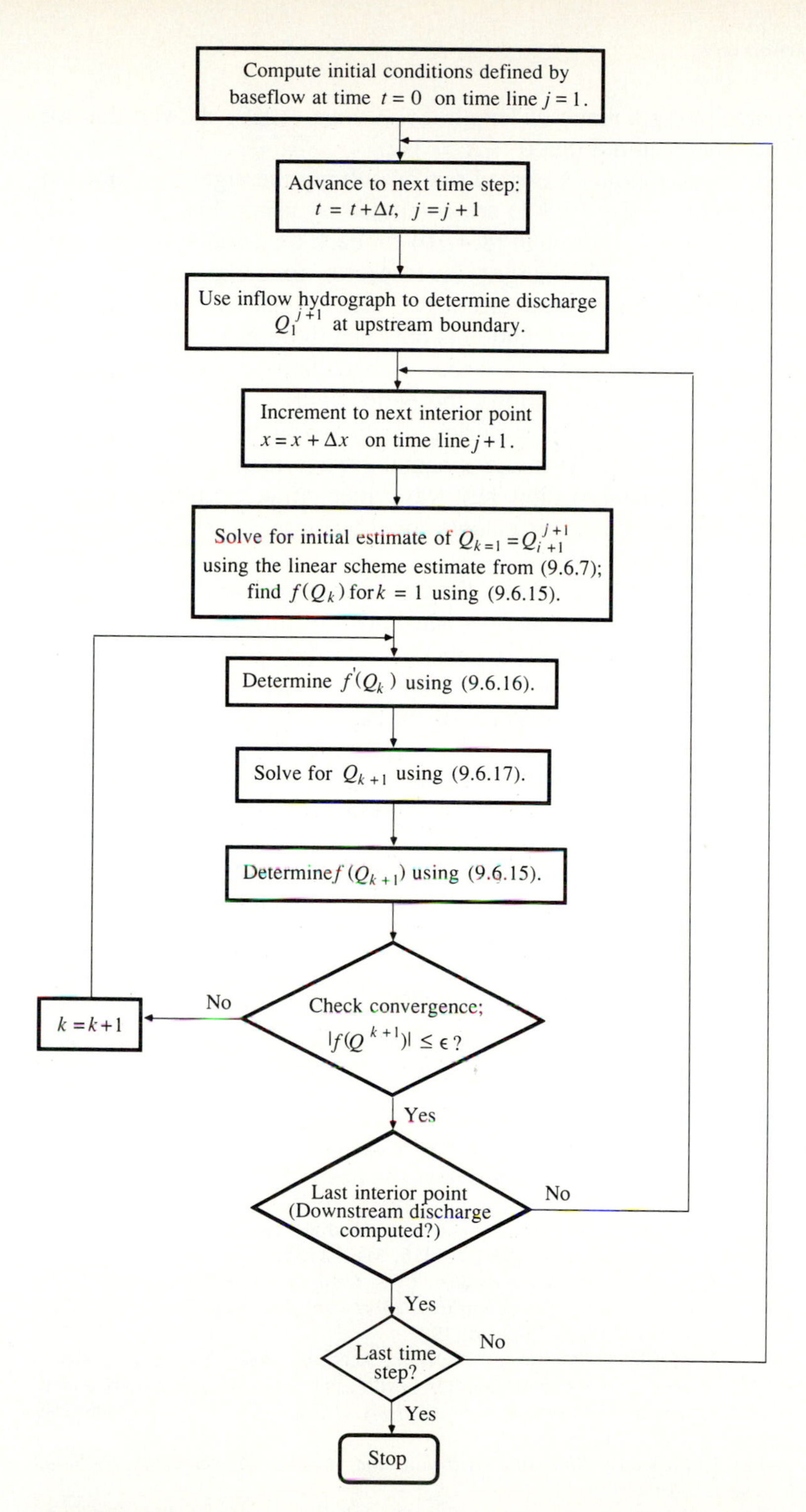

FIGURE 9.6.5
Flowchart for nonlinear kinematic wave computation.

of a given discharge along a reach of length Δx. Cunge (1969) showed that for numerical stability it is required that $0 \leq X \leq 1/2$.

Muskingum-Cunge routing is carried out by solving the algebraic equation (9.7.1). The coefficients in Eq. (9.7.1) are computed by using Eqs. (9.7.2) and (9.7.3) along with Eqs. (8.4.8) through (8.4.10) for each time and space point of computation, since K and X both change with respect to time and space.

The Muskingum-Cunge method offers two advantages over the standard kinematic wave methods. First, the solution is obtained through a linear algebraic equation (9.7.1) instead of a finite difference or characteristic approximation of a partial differential equation; this allows the entire hydrograph to be obtained at required cross sections instead of requiring solution over the entire length of the channel for each time step, as in the kinematic wave method. Second, the solution using (9.7.1) will tend to show less wave attenuation, permitting a more flexible choice of time and space increments for the computations as compared to the kinematic wave method.

The comprehensive British Flood Studies report (Natural Environment Research Council, 1975) concluded that the Muskingum-Cunge method is preferable to methods using a diffusion wave model (see Table 9.2.1) because of its simplicity; its accuracy is similar. Disadvantages of the Muskingum-Cunge method are that it cannot handle downstream disturbances that propagate upstream and that it does not accurately predict the discharge hydrograph at a downstream boundary when there are large variations in the kinematic wave speed such as those which result from the inundation of large flood plains.

REFERENCES

Abramowitz, M., and I. A. Stegun, *Handbook of Mathematical Functions*, Dover, New York, 1972.

Chow, V. T., *Open-channel Hydraulics*, McGraw-Hill, New York, 1959.

Courant, R., and K. O. Friedrichs, *Supersonic Flow and Shock Waves*, Interscience Publishers, New York, 1948.

Cunge, J. A., On the subject of a flood propagation method (Muskingum method), *J. Hydraulics Research*, International Association of Hydraulics Research, vol. 7, no. 2, pp. 205–230, 1969.

Eagleson, P. S., *Dynamic Hydrology*, McGraw-Hill, New York, 1970.

Henderson, F. M., *Open Channel Flow*, Macmillan, New York, 1966.

Li, R.-M., D. B. Simons, and M. A. Stevens, Nonlinear kinematic wave approximation for water routing, *Water Resour. Res.*, vol. 11, no. 2, pp. 245–252, 1975.

Lighthill, M. J., and G. B. Whitham, On kinematic waves, I: flood movement in long rivers, *Proc. R. Soc. London A*, vol. 229, no. 1178, pp. 281–316, May, 1955.

Miller, J. E., Basic concepts of kinematic-wave models, *U. S. Geol. Surv. Prof. Pap. 1302*, 1984.

Natural Environment Research Council, *Flood Studies Report, Vol. III, Flood Routing Studies*, Institute of Hydrology, Wallingford, England, 1975.

Overton, D. E., and M. E. Meadows, *Stormwater Modeling*, Academic Press, New York, 1976.

Saint-Venant, Barre de, Theory of unsteady water flow, with application to river floods and to propagation of tides in river channels, *French Academy of Science*, vol. 73. pp. 148–154, 237–240, 1871.

Stephenson, D., and M. E. Meadows, *Kinematic Hydrology and Modeling*, Developments in Water Science 26, Elsevier, Amsterdam, 1986.

Strelkoff, T., One-dimensional equations of open-channel flow, *J. Hydr. Div., Am. Soc. Civ. Eng.*, vol. 95, no. HY3, pp. 861–876, 1969.

BIBLIOGRAPHY

Explicit solution of the Saint-Venant equations

Amein, M., and C. S. Fang, Streamflow routing (with application to North Carolina rivers), rep. no. 17, Water Resources Res. Inst. of the Univ. of North Carolina, Raleigh, N. C., 1969.

Garrison, J. M., J.-P. P. Granju, and J. T. Price, Unsteady flow simulation in rivers and reservoirs, *J. Hydr. Div., Am. Soc. Civ. Eng.*, vol. 95, no. HY5, pp. 1559–1576, 1969.

Isaacson, E., J. J. Stoker, and A. Troesch, Numerical solution of flood prediction and river regulation problems, reps. IMM-205, IMM-235, Inst. for Math. and Mech., New York Univ., New York, 1954, 1956.

Isaacson, E., J. J. Stoker, and A. Troesch, Numerical solution of flow problems in rivers, *J. Hyd. Div., Am. Soc. Civ. Eng.*, vol. 84, no. HY5, pp. 1-18, 1958.

Johnson, B., Unsteady flow computations on the Ohio-Cumberland-Tennessee-Mississippi river system, tech. rep. H-74-8, Hyd. Lab., U. S. Army Eng. Waterways Exper. Sta., Vicksburg, Miss., 1974.

Liggett, J. A., and D. A. Woolhiser, Difference solutions of the shallow-water equations, *J. Eng. Mech. Div., Am. Soc. Civ. Eng.*, vol. 93, no. EM2, pp. 39–71, 1967.

Martin, C. S., and F. G. De Fazio, Open channel surge simulation by digital computer, *J. Hyd. Div., Am. Soc. Civ. Eng.*, vol. 95, no. HY6, pp. 2049–2070, 1969.

Ragan, R. M., Synthesis of hydrographs and water surface profiles for unsteady open channel flow with lateral inflows, *Ph. D. dissertation*, Cornell University, Ithaca, New York, 1965.

Strelkoff, T., Numerical solution of Saint-Venant equations, *J. Hyd. Div., Am. Soc. Civ. Eng.*, vol. 96, no. HY1, pp. 223–252, 1970.

Stoker, J. J., Numerical solution of flood prediction and river regulation problems. rep. IMM-200, Inst. for Math. and Mech., New York Univ., New York, 1953.

Kinematic wave models of overland flow and river flow

Borah, D. K., S. N. Prasad, and C. V. Alonso, Kinematic wave routing incorporating shock fitting, *Water Resour. Res.*, vol. 16, no. 3, pp. 529–541, 1980.

Brakensiek, D. L., A simulated watershed flow system for hydrograph prediction: a kinematic application, *Proceedings*, International Hydrology Symposium, Fort Collins, Colo., vol. 1, pp. 3.1–3.7, September 1967.

Constantinides, C. A., Two-dimensional kinematic overland flow modeling, *Proceedings*, Second International Conference on Urban Storm Drainage, University of Illinois at Urbana-Champaign, ed. by B. C. Yen, vol. 1, pp. 49–58, June 1981.

Dawdy, D. R., J. C. Schaake, Jr., and W. W. Alley, User's guide for distributed routing rainfall-runoff model, *Water Resources Investigation 78-90*, U. S. Geological Survey, September 1978.

DeVries, J. J., and R. C. MacArthur, Introduction and application of kinematic wave routing techniques using HEC-1, *training document no. 10*, Hydrologic Engineering Center, U. S. Army Corps of Engineers, Davis, Calif., May 1979.

Gburek, W. J., and D. E. Overton, Subcritical kinematic flow in a stable stream, *J. Hyd. Div., Am. Soc. Civ. Eng.*, vol. 99, no. HY9, pp. 1433–1447, 1973.

Harley, B. M., F. E. Perkins, and P. S. Eagleson, A modular distributed model of catchment dynamics, report 133, Ralph M. Parsons Lab., Mass. Inst. Tech., December 1970.

Henderson, F. M., and R. A. Wooding, Overland flow and groundwater from a steady rainfall of finite duration, *J. Geophys. Res.*, vol. 69, no. 8, pp. 1531–1540, April, 1964.

Kibler, D. F., and D. A. Woolhiser, The kinematic cascade as a hydrologic model, hydrology paper no. 39, Colorado State University, Fort Collins, Colo., March 1970.

Leclerc, G., and J. C. Schaake, Methodology for assessing the potential impact of urban development on urban runoff and the efficiency of runoff control alternatives, report no. 167, Ralph M. Parsons Lab., Mass. Inst. Tech., 1973.

Morris, E. M., and D. A. Woolhiser, Unsteady one-dimensional flow over a plane: partial equilibrium and recession hydrographs, *Water Resour. Res.*, vol. 16, no. 2, pp. 355–360, 1980.

Overton, D. E., Route or convolute? *Water Resour. Res.*, vol. 6, no. 1, pp. 43–52, 1970.
Overton, D. E., Estimation of surface water lag time from the kinematic wave equations, *Water Resour. Bull.*, vol. 7, no. 3, pp. 428–440, 1971.
Overton, D. E., Kinematic flow on long impermeable planes, *Water Resour. Bull.*, vol. 8, no. 6, pp. 1198–1204, 1972.
Overton, D. E., and D. L. Brakensiek, A kinematic model of surface runoff response, *Proceedings Symp.*, Results of Research on Representative and Experimental Basins, Wellington, New Zealand, December 1970, UNESCO-IAHS, vol. 1, Paris, 1973.
Ponce, V. M., R.-M. Li, and D. B. Simons, Applicability of kinematic and diffusion models. *J. Hyd. Div., Am. Soc. Civ. Eng.*, vol. 104, no. HY3, pp. 353–360, 1978.
Schaake, J. C. , Jr., Deterministic urban runoff model, *Treatise on Urban Water Systems*, Colorado State University, pp. 357–383, July 1971.
Sherman, B., and V. P. Singh, A kinematic model for surface irrigation, *Water Resour. Res.*, vol. 14, no. 2, pp. 357–364, 1978.
Sherman, B., and V. P. Singh, A kinematic model for surface irrigation: an extension, *Water Resour. Res.*, vol. 18, no. 3, pp. 659–667, 1982.
Wooding, R. A., A hydraulic model for the catchment-stream problem, I. kinematic wave theory, *J. Hydrol.*, vol. 3, pp. 254–267, 1965a.
Wooding, R. A. , A hydraulic model for the catchment-stream problem, II. numerical solutions, *J. Hydrol.*, vol. 3, no. 1, pp. 268–282, 1965b.
Wooding, R. A., A hydraulic model for the catchment-stream problem, III. comparison with runoff observations, *J. Hydrol.*, vol. 4, no. 1, pp. 21–37, 1966.
Woolhiser, D. A., and J. A. Liggett, Unsteady one-dimensional flow over a plane—the rising hydrograph, *Water Resour. Res.*, vol. 3, no. 3, pp. 753–771, 1967.

Kinematic wave models of pipe flow

Book, D. E. , J. W. Labadie, and D. M. Morrow, Dynamic vs. kinematic routing in modeling urban storm drainage, *Proceedings*, Second International Conference on Urban Storm Drainage, University of Illinois at Urbana-Champaign, ed. by B. C. Yen, vol. 1, pp. 154-164, June 1981.
Brandstetter, A., Comparative analysis of urban stormwater models, Battelle Institute, Washington, D.C. 1974.
Brandstetter, A., R. L. Engel, and D. B. Cearlock, A mathematical model for optimum design and control of metropolitan wastewater management systems, *Water Resour. Bull.*, vol. 9, no. 6, pp. 1188–1200, 1973.
Huber, W. C., et al., Storm water management model—User's Manual Version II, Environmental Protection Series, EPA-670/2-75-017, March 1975.
Joliffe, I. B., Unsteady free surface flood wave movement in pipe networks, *Ph.D. thesis*, University of Newcastle, New South Wales, Australia, 1980.
Yen, B. C., and N. Pansic, Surcharge of sewer systems, research report no. 149, Water Resources Center, University of Illinois at Urbana-Champaign, March 1980.
Yen, B. C., and A. S. Sevuk, Design of storm sewer networks, *J. Env. Eng. Div., Am. Soc. Civ. Eng.*, vol. 101, no. EE4, pp. 535–553, 1975.
Yen, B. C., H.G. Wenzel, Jr., L. W. Mays, and W. H. Tang, Advanced methodologies for design of storm sewer systems, research report no. 112, Water Resources Center, University of Illinois at Urbana-Champaign, August 1976.

PROBLEMS

9.1.1 Derive the nonconservation form of the momentum equation (9.2.1*b*) for a unit width of flow in a channel from the conservation form (9.2.1*a*).

9.1.2 (*a*) Describe the advantages and disadvantages of lumped system (hydrologic) routing vs. distributed system (hydraulic) routing.

(*b*) What are the limitations of the kinematic wave approximation?

(*c*) In what type of situation would the kinematic wave model be justified as compared to the dynamic wave model?

(*d*) Describe the difference between a linear and a nonlinear kinematic wave model.

9.1.3 Determine the momentum coefficient β, defined by Eq. (9.1.28), for the streamflow data given in Prob. 6.3.1.

9.1.4 Determine the momentum coefficient β defined by Eq. (9.1.28), for the streamflow data given in Prob. 6.3.5.

9.3.1 Calculate the water velocity V, the kinematic wave celerity c_k, the dynamic wave celerity c_d, and the velocities of propagation of dynamic waves $V \pm c_d$ for the channel described in Example 9.3.1 in the text and flow rates of 10, 50, 100, 500, 1000, 5000, and 10,000 cfs. Plot the results to show the variation of the velocities and celerities as a function of the flow rate.

9.4.1 Compare the analytical and numerical methods for solving the kinematic wave equations and indicate where each may be applicable.

9.4.2 Prove that the kinematic wave celerity is $c_k = 5V/3$, where V is the average velocity, when Manning's equation is used to describe the flow resistance in a wide, rectangular channel.

9.4.3 Prove that the travel time T of a kinematic wave in a wide rectangular channel of width B, length L, slope S_o, and Manning roughness n carrying a flow of Q is given approximately by

$$T = \frac{3}{5}\left(\frac{nB^{2/3}}{1.49S_o^{1/2}}\right)^{3/5} Q^{-2/5}L$$

If $B = 200$ ft, $L = 265$ mi, $S_o = 0.00035$, $n = 0.045$, and $Q = 2000$ cfs, calculate the travel time in days.

9.4.4 You are in charge of releasing water from a reservoir into a river with channel properties given in the previous problem. There are four downstream water users whose daily withdrawals during a one-week period are forecast as shown below. Calculate the amount of water you would release from the reservoir on the first day of this period to supply these users and to have a surplus of 200 cfs flowing past the last user. Assume the release was constant at 2500 cfs for the previous week, the withdrawls were constant during that week at the values shown for day 1 in the table, and that there is no lateral inflow.

User	Distance downstream (mi)	Withdrawal on day (cfs)						
		1	2	3	4	5	6	7
1	183	531	531	531	479	407	383	383
2	187	409	395	378	360	341	285	239
3	228	79	79	154	150	157	80	82
4	265	698	698	702	702	672	674	674

Discuss your answer. What assumptions have you made? How might these assumptions affect the result?

9.4.5 A flood of 100,000 cfs peak discharge has just passed a gaging station on a river. There is a community adjacent to the river 100 miles downstream, for which a flood warning must be issued. How long will it be before the flood peak reaches this community? Assume that the channel is rectangular, with width 500 ft, slope 1 percent, and Manning roughness 0.040.

9.6.1 Develop the finite-difference equations for the linear kinematic wave model for flood-wave routing in a trapezoidal channel. Assume no lateral inflow.

9.6.2 Develop an algorithm to solve the routing scheme for the kinematic wave model in a trapezoidal channel. Describe the step-by-step procedure that you would use to route an inflow hydrograph through a given reach. Divide the reach into n sections, each of length Δx. Use flowcharts or any other guides that you wish to explain the algorithm. This procedure would be the first step in developing a computer program for the routing procedure.

9.6.3 Take the inflow hydrograph given below and use the analytical kinematic wave solution method to route it through a uniform rectangular concrete channel 300 feet wide and 10,000 feet long with a bed slope of 0.015. Assume Manning's $n = 0.020$ and the initial condition is a uniform flow of 500 cfs.

Time (min)	0	20	40	60	80	100	120	140	160
Flow (cfs)	500	1402	9291	11576	10332	5458	2498	825	569

9.6.4 Calculate the solution to Prob. 9.6.3 by the linear numerical kinematic wave solution method, using $\Delta t = 1$ min, and $\Delta x = 2000$ ft. Consider only $0 \leq t \leq 20$ min.

9.6.5 Calculate the complete solution to Prob. 9.6.3 for a time horizon of 160 minutes by the linear kinematic wave method, using $\Delta t = 1$ min and $\Delta x = 2000$ ft.

9.6.6 Write a computer program for the linear kinematic wave model developed for a rectangular channel. The upstream boundary condition is an inflow hydrograph and the initial condition is uniform flow.

9.6.7 Consider a concrete, rectangular, 100-foot-wide drainage channel that is 8000 feet long and has a slope of 0.006 ft/ft and a Manning's roughness factor of $n = 0.015$. Use the computer program developed in Prob. 9.6.6 for the kinematic wave model to route through the reach the hypothetical flood described by

$$Q = Q_b + \frac{Q_p}{2}\left(1 - \cos\frac{2\pi t}{T}\right)$$

in which Q is the discharge, Q_b is the base flow, Q_p is the peak discharge (amplitude), and T is the duration of the flood wave. Use the values $Q_p = 6000$ cfs, $Q_b = 2000$ cfs, and $T = 120$ min. Assume $Q = Q_b$ for $t > T$.

9.6.8 Consider a rectangular channel, 100 feet wide, with bed slope of 0.015 and Manning's $n = 0.035$. In a numerical routing scheme, $\Delta x = 5000$ ft and $\Delta t = 10$ min. Given the following flow rates:

Point	$i, j+1$	i, j	$i+1, j$
Q (cfs)	1040	798	703

determine Q_{i+1}^{j+1} using an implicit finite-difference scheme for a linear kinematic model. Assume $R = y$ in development of the kinematic wave model.

9.6.9 Solve Prob. 9.6.8 using the nonlinear kinematic wave model with Newton's method.

9.7.1 Write a computer program for the Muskingum-Cunge model to route flow through a circular storm sewer pipe. Consider a pipe which is 6 ft in diameter, 1000 ft long, has Manning's n 0.015 and slope 0.001. Route through this pipe an inflow hydrograph described by the equation given in problem (9.6.7) with $Q_b = 20$ cfs, $Q_p = 60$ cfs and $T = 20$ min. Assume inflow Q_b for $t > T$.

9.7.2 Write a computer program for the Muskingum-Cunge model to route flood waves through a rectangular channel. Route the hydrograph in Prob. 9.6.3 through the rectangular channel described in that problem.

9.7.3 Solve Prob. 9.6.7 by the Muskingum-Cunge method.

CHAPTER 10

DYNAMIC WAVE ROUTING

The propagation of flow in space and time through a river or a network of rivers is a complex problem. The desire to build and live along rivers creates the necessity for accurate calculation of water levels and flow rates and provides the impetus to develop complex flow routing models, such as dynamic wave models. Another impetus for developing dynamic wave models is the need for more accurate hydrologic simulation, in particular, simulation of flow in urban watersheds and storm drainage systems. The dynamic wave model can also be used for routing low flows through rivers or irrigation channels to provide better control of water distribution. The propagation of flow along a river channel or an urban drainage system is an unsteady nonuniform flow, unsteady because it varies in time, nonuniform because flow properties such as water surface elevation, velocity, and discharge are not constant along the channel.

One-dimensional distributed routing methods have been classified in Chap. 9 as kinematic wave routing, diffusion wave routing, and dynamic wave routing. Kinematic waves govern the flow when the inertial and pressure forces are not important, that is, when the gravitational force of the flow is balanced by the frictional resistance force. Chapter 9 demonstrated that the kinematic wave approximation is useful for applications where the channel slopes are steep and backwater effects are negligible. When pressure forces become important but inertial forces remain unimportant, a diffusion wave model is applicable. Both the kinematic wave model and the diffusion wave model are helpful in describing downstream wave propagation when the channel slope is greater than about 0.5 ft/mi (0.01 percent) and there are no waves propagating upstream due to disturbances such as tides, tributary inflows, or reservoir operations. When both inertial and pressure forces are important, such as in mild-sloped rivers, and backwater effects from downstream disturbances are not negligible, then both the inertial force and pressure force terms in the momentum equation are needed. Under these circumstances the dynamic wave routing method is required, which involves numerical solution of the full Saint-Venant equations. Dynamic routing was first used by Stoker (1953) and by Isaacson, Stoker, and Troesch (1954, 1956) in their pioneering investigation of flood routing for the Ohio River. This chapter describes the theoretical development of dynamic wave routing models using *implicit finite-difference approximations* to solve the Saint-Venant equations.

10.1 DYNAMIC STAGE-DISCHARGE RELATIONSHIPS

The momentum equation is written in the conservation form [from (9.1.33)] as

$$\frac{\partial Q}{\partial t} + \frac{\partial(\beta Q^2/A)}{\partial x} + gA\left(\frac{\partial y}{\partial x} - S_o + S_f + S_e\right) - \beta q v_x + W_f B = 0 \qquad (10.1.1)$$

Uniform flow occurs when the bed slope S_o is equal to the friction slope S_f and all other terms are negligible, so that the relationship between discharge, or flow rate, and stage height, or water surface elevation, is a single-valued function derived from Manning's equation, as shown by the uniform flow rating curve in Fig. 10.1.1. When other terms in the momentum equation are not negligible, the stage-discharge relationship forms a loop as shown by the outer curve in Fig. 10.1.1, because the depth or stage is not just a function of discharge, but also a function of a variable energy slope. For a given stage, the discharge is usually higher on the rising limb of a flood hydrograph than on the recession limb. As the discharge rises and falls, the rating curve may even exhibit multiple loops as shown in Fig. 10.1.2 for the Red River (Fread, 1973c). The rating curve for uniform flow is typical of lumped or hydrologic routing methods in which

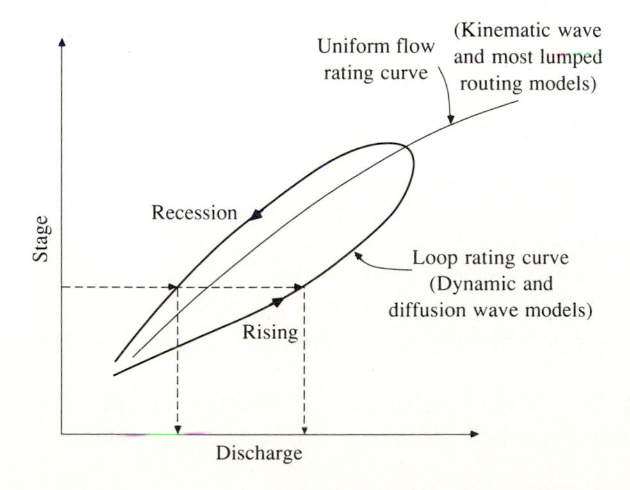

FIGURE 10.1.1
Loop rating curves. The uniform flow rating curve does not reflect backwater effects, whereas the looped curve does.

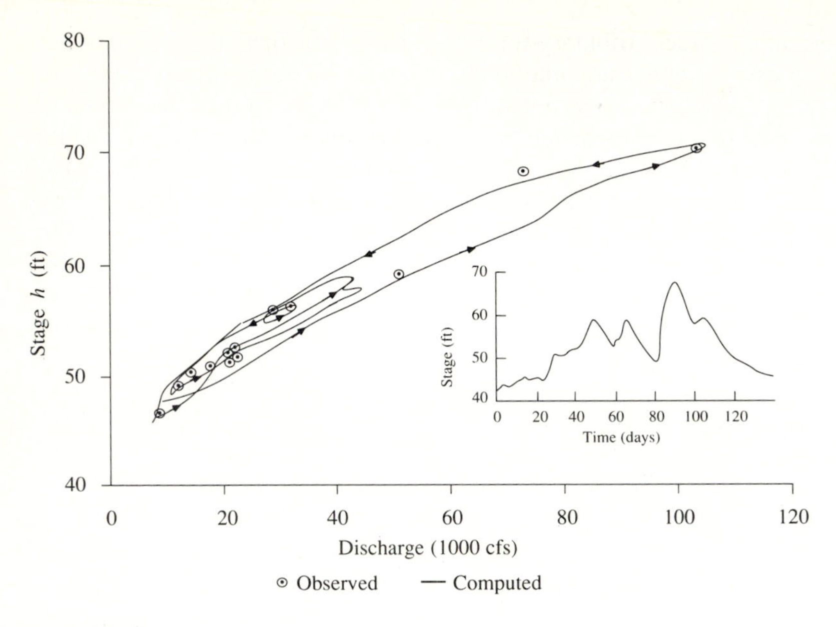

FIGURE 10.1.2
Looped stage-discharge relation for the Red River, Alexandria, Louisiana (May 5–June 17, 1964. *Source:* Fread, 1973c).

$S = f(Q)$, while the loop rating curve is typical of distributed or hydraulic routing methods.

Flow propagation in natural rivers is complicated by several factors: junctions and tributaries, variations in cross section, variations in resistance as a function both of flow depth and of location along the river, inundated areas, and meandering of the river. The interaction between the main channel and the flood plain or inundated valley is one of the most important factors affecting flood propagation. During the rising part of a flood wave, water flows into the flood plain or valley from the main channel, and during the falling flood, water flows from the inundated valley back into the main channel. The effect of the valley storage is to decrease the discharge during the falling flood. Also, some losses occur in the valley due to infiltration and evaporation.

The flood plain has an effect on the wave celerity because the flood wave progresses more slowly in the inundated valley than in the main channel of a river. This difference in wave celerities disperses the flood wave and causes flow from the main channel to the flood plain during the rising flood by creating a transverse water surface slope away from the channel. During the falling flood, the transverse slope is inward from the inundated valley into the main channel, and water then moves from the flood plain back into the main channel [see Fig. 10.1.3(*a*) and (*b*)].

Because the longitudinal axes of the main channel and the flood plain valley are rarely parallel, the situation described above is even more complicated in a

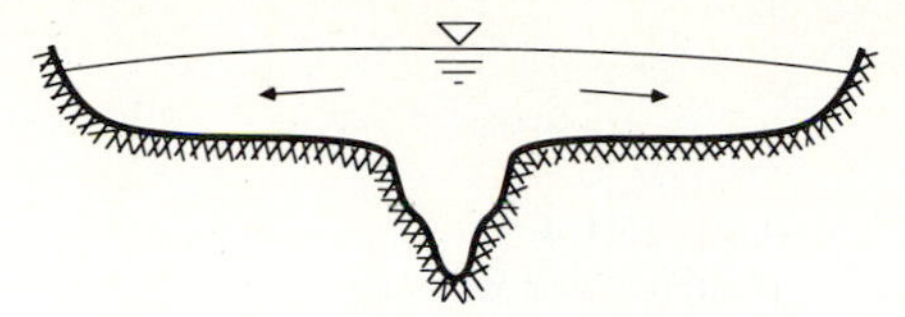
(*a*) Transverse slope during rising flood.

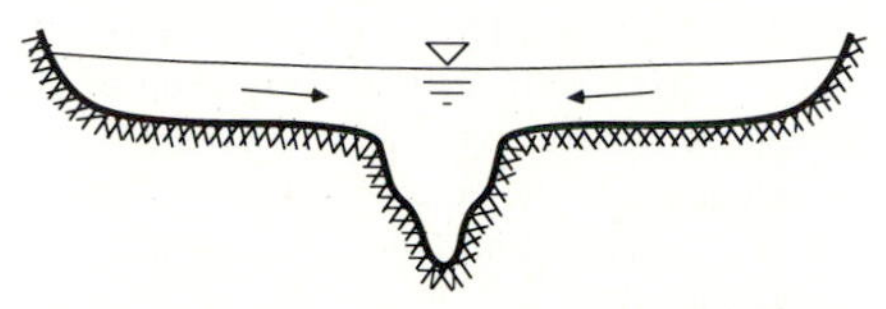
(*b*) Transverse slope during falling flood.

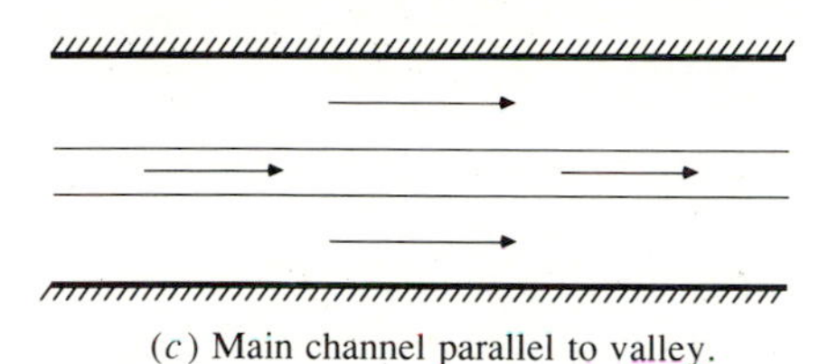
(*c*) Main channel parallel to valley.

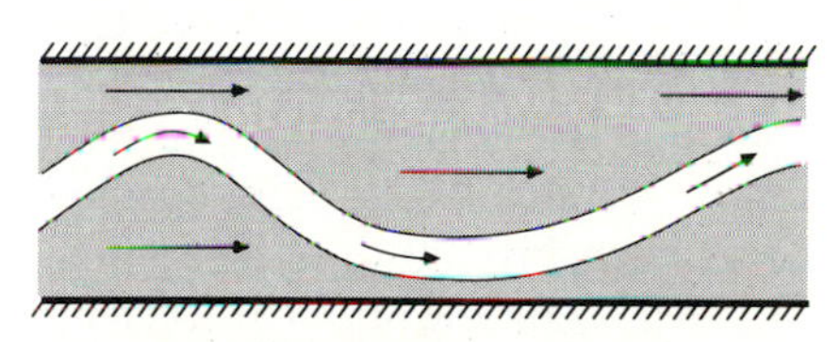
(*d*) Meandering main channel.

FIGURE 10.1.3
Aspects of flow in natural rivers.

meandering river. For a large flood, the axis of the flow becomes parallel to the valley axis [Fig. 10.1.3(*c*) and (*d*)]. The valley water slope and valley water velocity (if depths are sufficient) can be greater than in the main channel, which has a longer flow path than the valley. This situation makes it difficult for flow to go from the main channel to the flood plain valley during the rising flood and vice versa during the falling flood. Flood wave propagation is more complex when the flow is varying rapidly. The description is also more complicated for a branching river system with tributaries and the possibility of flood peaks from different tributaries coinciding. Also, with tributaries, the effects on flood propagation of backwater at the junctions must be considered.

When backwater effects exist, the loop rating curve may consist of a series of loops, each corresponding to a different feature controlling water level in the channel (see Fig. 10.1.4). Backwater effects of reservoirs, channel junctions, narrowing of the natural river channel, and bridges can demonstrate this characteristic.

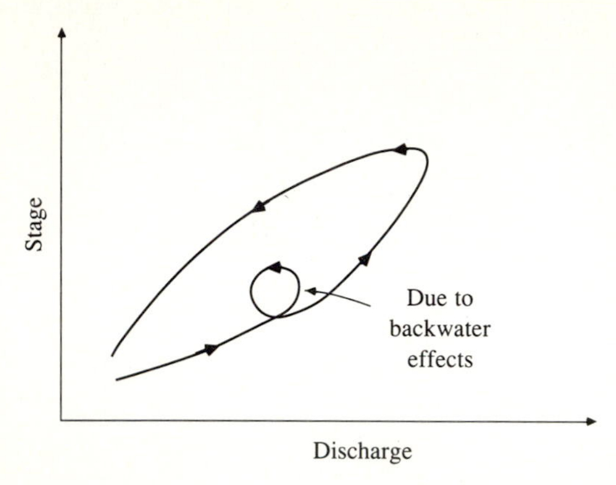

FIGURE 10.1.4
Loop rating curve with significant backwater effects. Backwater effects are due to downstream reservoirs, channel junctions, highway crossings, narrowing of the river section. These produce a series of rating curves with each corresponding to a given backwater level. The backwater effects cause a variable energy slope that can be modeled using the full dynamic wave model.

10.2 IMPLICIT DYNAMIC WAVE MODEL

Implicit finite-difference methods advance the solution of the Saint-Venant equation from one time line to the next simultaneously for all points along the time line. A system of algebraic equations is generated by applying the Saint-Venant equations simultaneously to all the unknown values on a time line. Implicit methods were developed because of the limitation on the time-step size required for numerical stability of explicit methods. For example, an explicit method might require a time step of one minute for stability, while an implicit method applied to the same problem could use a time step of one hour or longer.

The implicit finite-difference scheme uses a weighted four-point method between adjacent time lines at a point M, as shown in Fig. 10.2.1. If a given variable describing the flow, such as flow rate or water surface level, is denoted by u, the time derivative of u is approximated by the average of the finite difference values at distance points i and $i+1$. The value at the ith distance point is $(u_i^{j+1} - u_i^j)/\Delta t$, and that at the $(i+1)$th distance point is $(u_{i+1}^{j+1} - u_{i+1}^j)/\Delta t$, so the approximation is

$$\frac{\partial u}{\partial t} \approx \frac{u_i^{j+1} + u_{i+1}^{j+1} - u_i^j - u_{i+1}^j}{2\,\Delta t} \tag{10.2.1}$$

for the point M located midway between the ith and $(i+1)$th distance points in Fig. 10.2.1.

A slightly different approach is adopted to estimate the spatial derivative $\partial u/\partial x$ and the variable u. For the spatial derivative, the difference terms at the jth and $(j+1)$th time lines are calculated: $(u_{i+1}^j - u_i^j)/\Delta x$, and $(u_{i+1}^{j+1} - u_i^{j+1})/\Delta x$, respectively; then a weighting factor θ is applied to define the spatial derivative as

$$\frac{\partial u}{\partial x} \approx \theta \frac{u_{i+1}^{j+1} - u_i^{j+1}}{\Delta x} + (1-\theta)\frac{u_{i+1}^j - u_i^j}{\Delta x} \tag{10.2.2}$$

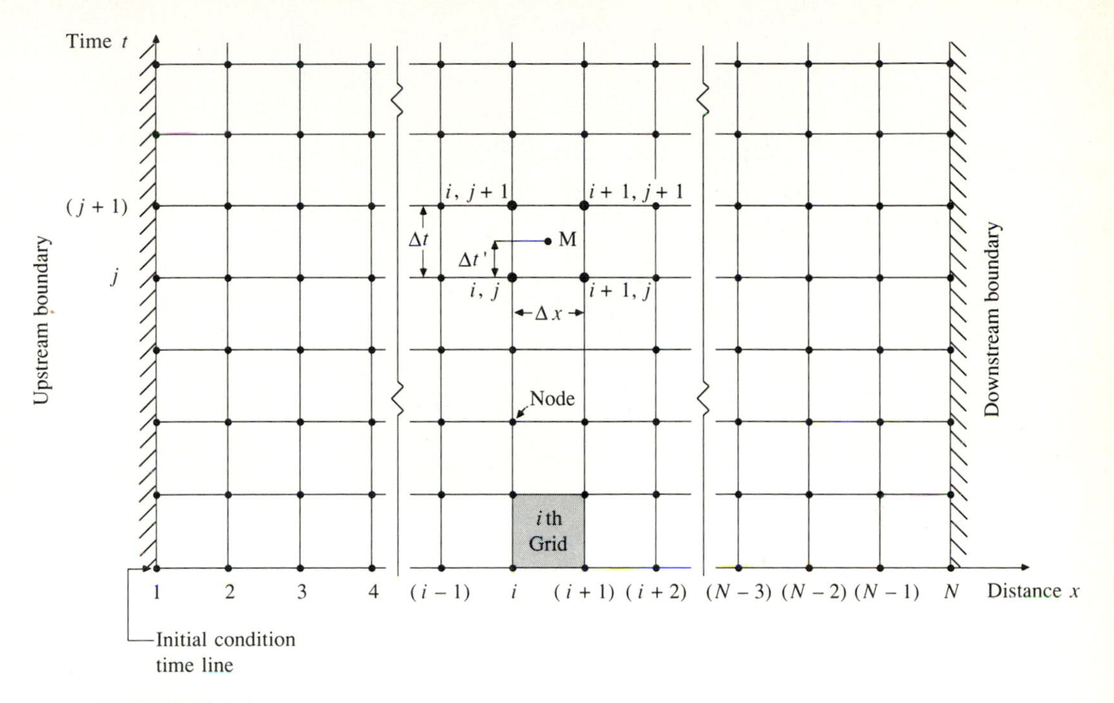

FIGURE 10.2.1
The x–t solution plane. The finite-difference forms of the Saint-Venant equations are solved at a discrete number of points (values of the independent variables x and t) arranged to form the rectangular grid shown. Lines parallel to the time axis represent locations along the channel, and those parallel to the distance axis represent times. (After Fread, 1974a).

and an average value for u is calculated similarly as

$$\bar{u} = \theta \frac{u_i^{j+1} + u_{i+1}^{j+1}}{2} + (1-\theta)\frac{u_i^j + u_{i+1}^j}{2} \tag{10.2.3}$$

The value of $\theta = \Delta t'/\Delta t$ locates point M vertically in the box in Fig. 10.2.1. A scheme using $\theta = 0.5$ is called a *box scheme*. When $\theta = 0$, the point M is located on the jth time line and the scheme is *fully explicit*, while a value of $\theta = 1$ is used in a *fully implicit* scheme with M lying in the $(j + 1)$th time line. Implicit schemes are those with θ in the range 0.5 to 1.0; Fread (1973a, 1974a) recommends a value of 0.55 to 0.6.

A major difference between the explicit and implicit methods is that implicit methods are conditionally stable for all time steps, whereas explicit methods are numerically stable only for time steps less than a critical value determined by the Courant condition. Fread (1973a, 1974a) has shown that the weighted four-point scheme is unconditionally linearly stable for any time step if $0.5 \leq \theta \leq 1.0$. This scheme has a second-order accuracy when $\theta = 0.5$ and a first-order accuracy when $\theta = 1.0$.

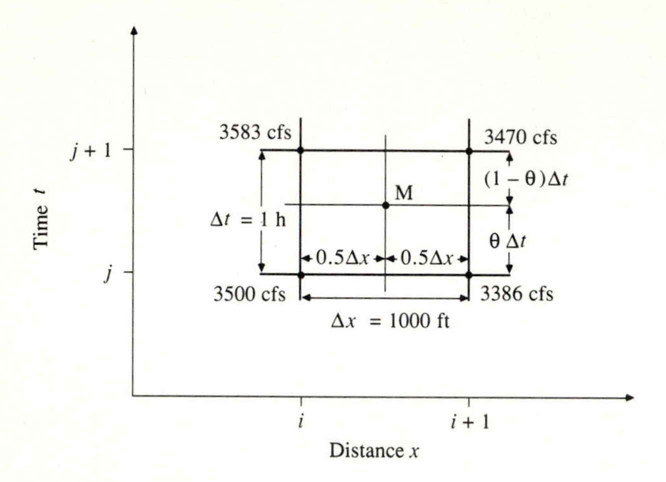

FIGURE 10.2.2
Values of the flow rate at four points in the x–t plane (Example 10.2.1).

Example 10.2.1. The values of flow rate Q at four points in the space-time grid are as shown in Fig. 10.2.2. Using $\Delta t = 1$ h, $\Delta x = 1000$ ft, and $\theta = 0.55$, calculate the values of $\partial Q/\partial t$ and $\partial Q/\partial x$ by the four-point implicit method.

Solution. As shown in Fig. 10.2.2, the values of flow rate at the four points are $Q_i^j = 3500$ cfs, $Q_{i+1}^j = 3386$ cfs, $Q_i^{j+1} = 3583$ cfs, and $Q_{i+1}^{j+1} = 3470$ cfs. The time derivative is calculated using (10.2.1) with $u = Q$ and $\Delta t = 1$ h $= 3600$ s:

$$\frac{\partial Q}{\partial t} = \frac{Q_i^{j+1} + Q_{i+1}^{j+1} - Q_i^j - Q_{i+1}^j}{2\,\Delta t}$$

$$= \frac{3583 + 3470 - 3500 - 3386}{2 \times 3600}$$

$$= 0.023 \text{ cfs/s}$$

The spatial derivative is calculated using (10.2.2):

$$\frac{\partial Q}{\partial x} = \theta \frac{Q_{i+1}^{j+1} - Q_i^{j+1}}{\Delta x} + (1 - \theta)\frac{Q_{i+1}^j - Q_i^j}{\Delta x}$$

$$= (0.55)\frac{(3470 - 3583)}{1000} + (1 - 0.55)\frac{(3386 - 3500)}{1000}$$

$$= -0.113 \text{ cfs/ft}$$

10.3 FINITE DIFFERENCE EQUATIONS

The conservation form of the Saint-Venant equations is used because this form provides the versatility required to simulate a wide range of flows from gradual long-duration flood waves in rivers to abrupt waves similar to those caused by a dam failure. The equations are developed from (9.1.6) and (9.1.37) as follows.

Continuity:

$$\frac{\partial Q}{\partial x} + \frac{\partial(A + A_o)}{\partial t} - q = 0 \tag{10.3.1}$$

Momentum:

$$\frac{\partial Q}{\partial t} + \frac{\partial(\beta Q^2/A)}{\partial x} + gA\left(\frac{\partial h}{\partial x} + S_f + S_e\right) - \beta q v_x + W_f B = 0 \tag{10.3.2}$$

where

x = longitudinal distance along the channel or river

t = time

A = cross-sectional area of flow

A_o = cross-sectional area of off-channel dead storage (contributes to continuity, but not momentum)

q = lateral inflow per unit length along the channel

h = water surface elevation

v_x = velocity of lateral flow in the direction of channel flow

S_f = friction slope

S_e = eddy loss slope

B = width of the channel at the water surface

W_f = wind shear force

β = momentum correction factor

g = acceleration due to gravity.

The weighted four-point finite difference approximations given by Eqs. (10.2.1) – (10.2.3) are used for dynamic routing with the Saint-Venant equations. The spatial derivatives $\partial Q/\partial x$ and $\partial h/\partial x$ are estimated between adjacent time lines according to (10.2.2):

$$\frac{\partial Q}{\partial x} = \theta \frac{Q_{i+1}^{j+1} - Q_i^{j+1}}{\Delta x_i} + (1-\theta)\frac{Q_{i+1}^{j} - Q_i^{j}}{\Delta x_i} \tag{10.3.3}$$

$$\frac{\partial h}{\partial x} = \theta \frac{h_{i+1}^{j+1} - h_i^{j+1}}{\Delta x_i} + (1-\theta)\frac{h_{i+1}^{j} - h_i^{j}}{\Delta x_i} \tag{10.3.4}$$

and the time derivatives are estimated using (10.2.1):

$$\frac{\partial(A + A_o)}{\partial t} = \frac{(A + A_o)_i^{j+1} + (A + A_o)_{i+1}^{j+1} - (A + A_o)_i^{j} - (A + A_o)_{i+1}^{j}}{2\,\Delta t_j} \tag{10.3.5}$$

$$\frac{\partial Q}{\partial t} = \frac{Q_i^{j+1} + Q_{i+1}^{j+1} - Q_i^j - Q_{i+1}^j}{2\,\Delta t_j} \tag{10.3.6}$$

The nonderivative terms, such as q and A, are estimated between adjacent time lines using (10.2.3):

$$\begin{aligned} q &= \theta\frac{q_i^{j+1} + q_{i+1}^{j+1}}{2} + (1-\theta)\frac{q_i^j + q_{i+1}^j}{2} \\ &= \theta\overline{q}_i^{j+1} + (1-\theta)\overline{q}_i^j \end{aligned} \tag{10.3.7}$$

$$\begin{aligned} A &= \theta\frac{A_i^{j+1} + A_{i+1}^{j+1}}{2} + (1-\theta)\frac{A_i^j + A_{i+1}^j}{2} \\ &= \theta\overline{A}_i^{j+1} + (1-\theta)\overline{A}_i^j \end{aligned} \tag{10.3.8}$$

where $\overline{q}_i$ and $\overline{A}_i$ indicate the lateral flow and cross-sectional area averaged over the reach Δx_i.

The finite-difference form of the continuity equation is produced by substituting Eqs. (10.3.3), (10.3.5), and (10.3.7) into (10.3.1):

$$\begin{aligned} &\theta\left(\frac{Q_{i+1}^{j+1} - Q_i^{j+1}}{\Delta x_i} - \overline{q}_i^{j+1}\right) + (1-\theta)\left(\frac{Q_{i+1}^j - Q_i^j}{\Delta x_i} - \overline{q}_i^j\right) \\ &+ \frac{(A+A_o)_i^{j+1} + (A+A_o)_{i+1}^{j+1} - (A+A_o)_i^j - (A+A_o)_{i+1}^j}{2\,\Delta t_j} = 0 \end{aligned} \tag{10.3.9}$$

Similarly, the finite-difference form of the momentum equation is written as:

$$\begin{aligned} &\frac{Q_i^{j+1} + Q_{i+1}^{j+1} - Q_i^j - Q_{i+1}^j}{2\,\Delta t_j} + \theta\left[\frac{(\beta Q^2/A)_{i+1}^{j+1} - (\beta Q^2/A)_i^{j+1}}{\Delta x_i}\right. \\ &\left. + g\overline{A}_i^{j+1}\left(\frac{h_{i+1}^{j+1} - h_i^{j+1}}{\Delta x_i} + \left(\overline{S}_f\right)_i^{j+1} + \left(\overline{S}_e\right)_i^{j+1}\right) - \left(\overline{\beta q v_x}\right)_i^{j+1} + \left(\overline{W_f B}\right)_i^{j+1}\right] \\ &+ (1-\theta)\left[\frac{(\beta Q^2/A)_{i+1}^j - (\beta Q^2/A)_i^j}{\Delta x_i} + g\overline{A}_i^j\left(\frac{h_{i+1}^j - h_i^j}{\Delta x_i} + \left(\overline{S}_f\right)_i^j + \left(\overline{S}_e\right)_i^j\right)\right. \\ &\left. - \left(\overline{\beta q v_x}\right)_i^j + \left(\overline{W_f B}\right)_i^j\right] \\ &= 0 \end{aligned} \tag{10.3.10}$$

The four-point finite-difference form of the continuity equation can be further modified by multiplying Eq. (10.3.9) by Δx_i to obtain

$$\theta(Q_{i+1}^{j+1} - Q_i^{j+1} - \overline{q}_i^{j+1}\Delta x_i) + (1-\theta)(Q_{i+1}^j - Q_i^j - \overline{q}_i^j \Delta x_i)$$

$$+ \frac{\Delta x_i}{2\,\Delta t_j}\left[(A + A_o)_i^{j+1} + (A + A_o)_{i+1}^{j+1} - (A + A_o)_i^j - (A + A_o)_{i+1}^j\right] = 0 \tag{10.3.11}$$

Similarly, the momentum equation can be modified by multiplying by Δx_i to obtain

$$\frac{\Delta x_i}{2\,\Delta t_j}\left(Q_i^{j+1} + Q_{i+1}^{j+1} - Q_i^j - Q_{i+1}^j\right)$$

$$+ \theta\left\{\left(\frac{\beta Q^2}{A}\right)_{i+1}^{j+1} - \left(\frac{\beta Q^2}{A}\right)_i^{j+1} + g\overline{A}_i^{j+1}\left[h_{i+1}^{j+1} - h_i^{j+1} + \left(\overline{S}_f\right)_i^{j+1}\Delta x_i + \left(\overline{S}_e\right)_i^{j+1}\Delta x_i\right]\right.$$

$$\left. - \left(\overline{\beta q v_x}\right)_i^{j+1}\Delta x_i + \left(\overline{W_f B}\right)_i^{j+1}\Delta x_i\right\}$$

$$+ (1-\theta)\left\{\left(\frac{\beta Q^2}{A}\right)_{i+1}^{j} - \left(\frac{\beta Q^2}{A}\right)_i^{j} + g\overline{A}_i^{j}\left[h_{i+1}^{j} - h_i^{j} + \left(\overline{S}_f\right)_i^{j}\Delta x_i + \left(\overline{S}_e\right)_i^{j}\Delta x_i\right]\right.$$

$$\left. - \left(\overline{\beta q v_x}\right)_i^{j}\Delta x_i + \left(\overline{W_f B}\right)_i^{j}\Delta x_i\right\}$$

$$= 0 \tag{10.3.12}$$

where the average values (marked with $^{-}$) over a reach are defined as

$$\overline{\beta}_i = \frac{\beta_i + \beta_{i+1}}{2} \tag{10.3.13}$$

$$\overline{A}_i = \frac{A_i + A_{i+1}}{2} \tag{10.3.14}$$

$$\overline{B}_i = \frac{B_i + B_{i+1}}{2} \tag{10.3.15}$$

$$\overline{Q}_i = \frac{Q_i + Q_{i+1}}{2} \tag{10.3.16}$$

Also,

$$\overline{R}_i = \frac{\overline{A}_i}{\overline{B}_i} \tag{10.3.17}$$

for use in Manning's equation. Manning's equation may be solved for S_f and written in the form shown below, where the term $|Q|Q$ has magnitude Q^2 and sign positive or negative depending on whether the flow is downstream or upstream, respectively:

$$(\overline{S_f})_i = \frac{\overline{n}_i^2|\overline{Q}_i|\overline{Q}_i}{2.208\overline{A}_i^{-2}\overline{R}_i^{-4/3}} \tag{10.3.18}$$

The minor head losses arising from contraction and expansion of the channel are proportional to the difference between the squares of the downstream and upstream velocities, with a contraction/expansion loss coefficient K_e:

$$\left(\overline{S}_e\right)_i = \frac{(K_e)_i}{2g\,\Delta x_i}\left[\left(\frac{Q}{A}\right)_{i+1}^2 - \left(\frac{Q}{A}\right)_i^2\right] \tag{10.3.19}$$

The velocity of the wind relative to the water surface, V_r, is defined by

$$\left(\overline{V}_r\right)_i = \left(\frac{\overline{Q}_i}{\overline{A}_i}\right) - (\overline{V}_w)_i \cos\omega \tag{10.3.20}$$

where ω is the angle between the wind and the water directions. The wind shear factor is then given by

$$\left(\overline{W}_f\right)_i = (C_w)_i\left|\left(\overline{V}_r\right)_i\right|\left(\overline{V}_r\right)_i \tag{10.3.21}$$

where C_w is a friction drag coefficient [$C_w = C_f/2$ given in (9.1.18)].

The terms having superscript j in Eqs. (10.3.11) and (10.3.12) are known either from initial conditions, or from a solution of the Saint-Venant equations for a previous time line. The terms g, Δx_i, β_i, K_e, C_w, and V_w are known and must be specified independently of the solution. The unknown terms are Q_i^{j+1}, Q_{i+1}^{j+1}, h_i^{j+1}, h_{i+1}^{j+1}, A_i^{j+1}, A_{i+1}^{j+1}, B_i^{j+1}, and B_{i+1}^{j+1}. However, all the terms can be expressed as functions of the unknowns, Q_i^{j+1}, Q_{i+1}^{j+1}, h_i^{j+1}, and h_{i+1}^{j+1}, so there are actually four unknowns. The unknowns are raised to powers other than unity, so (10.3.11) and (10.3.12) are nonlinear equations.

The continuity and momentum equations are considered at each of the $N - 1$ rectangular grids shown in Fig. 10.2.1, between the upstream boundary at $i = 1$ and the downstream boundary at $i = N$. This yields $2N - 2$ equations. There are two unknowns at each of the N grid points (Q and h), so there are $2N$ unknowns in all. The two additional equations required to complete the solution are supplied by the upstream and downstream boundary conditions. The upstream boundary condition is usually specified as a known inflow hydrograph, while the downstream boundary condition can be specified as a known stage hydrograph, a known discharge hydrograph, or a known relationship between stage and discharge, such as a rating curve.

10.4 FINITE DIFFERENCE SOLUTION

The following discussion for the solution of a system of finite difference equations follows that of Fread (1976b). The system of nonlinear equations can be expressed in functional form in terms of the unknowns h and Q at time level $j + 1$, as follows:

$$
\begin{array}{lll}
UB(h_1, Q_1) = 0 & \text{upstream boundary condition} & \\
C_1(h_1, Q_1, h_2, Q_2) = 0 & \text{continuity for grid 1} & \\
M_1(h_1, Q_1, h_2, Q_2) = 0 & \text{momentum for grid 1} & \\
\vdots & & \\
C_i(h_i, Q_i, h_{i+1}, Q_{i+1}) = 0 & \text{continuity for grid } i & (10.4.1) \\
M_i(h_i, Q_i, h_{i+1}, Q_{i+1}) = 0 & \text{momentum for grid } i & \\
\vdots & & \\
C_{N-1}(h_{N-1}, Q_{N-1}, h_N, Q_N) = 0 & \text{continuity for grid } N-1 & \\
M_{N-1}(h_{N-1}, Q_{N-1}, h_N, Q_N) = 0 & \text{momentum for grid } N-1 & \\
DB(h_N, Q_N) = 0 & \text{downstream boundary condition} &
\end{array}
$$

This system of $2N$ nonlinear equations in $2N$ unknowns is solved for each time step by the Newton-Raphson method. The computational procedure for each time $j + 1$ starts by assigning trial values to the $2N$ unknowns at that time. These trial values of Q and h can be the values known at time j from the initial condition (if $j = 1$) or from calculations during the previous time step. Using the trial values in the system (10.4.1) results in $2N$ residuals. For the kth iteration these residuals can be expressed as

$$
\begin{array}{lll}
UB(h_1^k, Q_1^k) = RUB^k & \text{residual for upstream boundary condition} & \\
C_1(h_1^k, Q_1^k, h_2^k, Q_2^k) = RC_1^k & \text{residual for continuity at grid 1} & \\
M_1(h_1^k, Q_1^k, h_2^k, Q_2^k) = RM_1^k & \text{residual for momentum at grid 1} & \\
\vdots & & \\
C_i(h_i^k, Q_i^k, h_{i+1}^k, Q_{i+1}^k) = RC_i^k & \text{residual for continuity at grid } i & (10.4.2) \\
M_i(h_i^k, Q_i^k, h_{i+1}^k, Q_{i+1}^k) = RM_i^k & \text{residual for momentum at grid } i & \\
\vdots & & \\
C_{N-1}(h_{N-1}^k, Q_{N-1}^k, h_N^k, Q_N^k) = RC_{N-1}^k & \text{residual for continuity at grid } N-1 & \\
M_{N-1}(h_{N-1}^k, Q_{N-1}^k, h_N^k, Q_N^k) = RM_{N-1}^k & \text{residual for momentum at grid } N-1 & \\
DB(h_N^k, Q_N^k) = RDB^k & \text{residual for downstream boundary condition} &
\end{array}
$$

The solution is approached by finding values of the unknowns Q and h so that the residuals are forced to zero or very close to zero.

The Newton-Raphson method is an iterative technique for solving a system of nonlinear algebraic equations. It uses the same idea as was presented in Chap. 5 for the determination of flow depth in Manning's equation, except that here the solution is for a vector of variables rather than for a single variable. Consider the system of equations (10.4.2) denoted in vector form as

$$f(\boldsymbol{x}) = 0 \tag{10.4.3}$$

where $\boldsymbol{x}=(Q_1, h_1, Q_2, h_2, \ldots, Q_N, h_N)$ is the vector of unknown quantities and for iteration $k, \boldsymbol{x}^k = (Q_1^k, h_1^k, Q_2^k, h_1^k, \ldots, Q_N^k, h_N^k)$. The nonlinear system can be linearized to

$$f(\boldsymbol{x}^{k+1}) \approx f(\boldsymbol{x}^k) + J(\boldsymbol{x}^k)(\boldsymbol{x}^{k+1} - \boldsymbol{x}^k) \tag{10.4.4}$$

where $J(\boldsymbol{x}^k)$ is the *Jacobian*, which is a coefficient matrix made up of the first partial derivatives of $f(\boldsymbol{x})$ evaluated at $\boldsymbol{x}^k$. The right-hand side of Eq. (10.4.4) is the linear vector function of $\bar{\boldsymbol{x}}^k$. Basically, an iterative procedure is used to determine $\boldsymbol{x}^{k+1}$ that forces the residual error $f(\bar{\boldsymbol{x}}^{k+1})$ in Eq. (10.4.4) to zero. This can be accomplished by setting $f(\bar{\boldsymbol{x}}^{k+1}) = 0$ rearranging (10.4.4) to read

$$J(\boldsymbol{x}^k)(\boldsymbol{x}^{k+1} - \boldsymbol{x}^k) = -f(\boldsymbol{x}^k) \tag{10.4.5}$$

This system is solved for $(\boldsymbol{x}^{k+1} - \boldsymbol{x}^k) = \Delta\boldsymbol{x}^k$, and the improved estimate of the solution, $\boldsymbol{x}^{k+1}$, is determined knowing $\Delta\boldsymbol{x}^k$. The process is repeated until $(\boldsymbol{x}^{k+1} - \boldsymbol{x}^k)$ is smaller than some specified tolerance.

The system of linear equations represented by (10.4.5) involves $J(\boldsymbol{x}^k)$, the Jacobian of the set of equations (10.4.1) with respect to h and Q, and $-f(\boldsymbol{x}^k)$, the vector of the negatives of the residuals in (10.4.2). The resulting system of equations is

$$\begin{aligned}
&\frac{\partial UB}{\partial h_1}dh_1 + \frac{\partial UB}{\partial Q_1}dQ_1 = -RUB^k \\
&\frac{\partial C_1}{\partial h_1}dh_1 + \frac{\partial C_1}{\partial Q_1}dQ_1 + \frac{\partial C_1}{\partial h_2}dh_2 + \frac{\partial C_1}{\partial Q_2}dQ_2 = -RC_1^k \\
&\frac{\partial M_1}{\partial h_1}dh_1 + \frac{\partial M_1}{\partial Q_1}dQ_1 + \frac{\partial M_1}{\partial h_2}dh_2 + \frac{\partial M_1}{\partial Q_2}dQ_2 = -RM_1^k \\
&\qquad\vdots \\
&\frac{\partial C_i}{\partial h_i}dh_i + \frac{\partial C_i}{\partial Q_i}dQ_i + \frac{\partial C_i}{\partial h_{i+1}}dh_{i+1} + \frac{\partial C_i}{\partial Q_{i+1}}dQ_{i+1} = -RC_i^k \\
&\frac{\partial M_i}{\partial h_i}dh_i + \frac{\partial M_i}{\partial Q_i}dQ_i + \frac{\partial M_i}{\partial h_{i+1}}dh_{i+1} + \frac{\partial M_i}{\partial Q_{i+1}}dQ_{i+1} = -RM_i^k \\
&\qquad\vdots
\end{aligned} \tag{10.4.6}$$

$$\frac{\partial C_{N-1}}{\partial h_{N-1}}dh_{N-1} + \frac{\partial C_{N-1}}{\partial Q_{N-1}}dQ_{N-1} + \frac{\partial C_{N-1}}{\partial h_N}dh_N + \frac{\partial C_{N-1}}{\partial Q_N}dQ_N = -RC_{N-1}^k$$

$$\frac{\partial M_{N-1}}{\partial h_{N-1}}dh_{N-1} + \frac{\partial M_{N-1}}{\partial Q_{N-1}}dQ_{N-1} + \frac{\partial M_{N-1}}{\partial h_N}dh_N + \frac{\partial M_{N-1}}{\partial Q_N}dQ_N = -RM_{N-1}^k$$

$$\frac{\partial DB}{\partial h_N}dh_N + \frac{\partial DB}{\partial Q_N}dQ_N \quad = -RDB^k$$

In Fig. 10.4.1 these equations are presented in matrix form for a river divided into four reaches (five cross sections). The partial derivative terms are described in detail in App. 10.A.

Gaussian elimination or matrix inversion can be used to solve this set of equations (Conte, 1965). The Jacobian coefficient matrix is a sparse matrix with a band width of at most four elements along the main diagonal. Fread (1971) developed a very efficient solution technique to solve such a system of equations taking advantage of this banded (quad-diagonal) structure. Solving (10.4.6) provides values of dh_i and dQ_i. The values for the unknowns at the

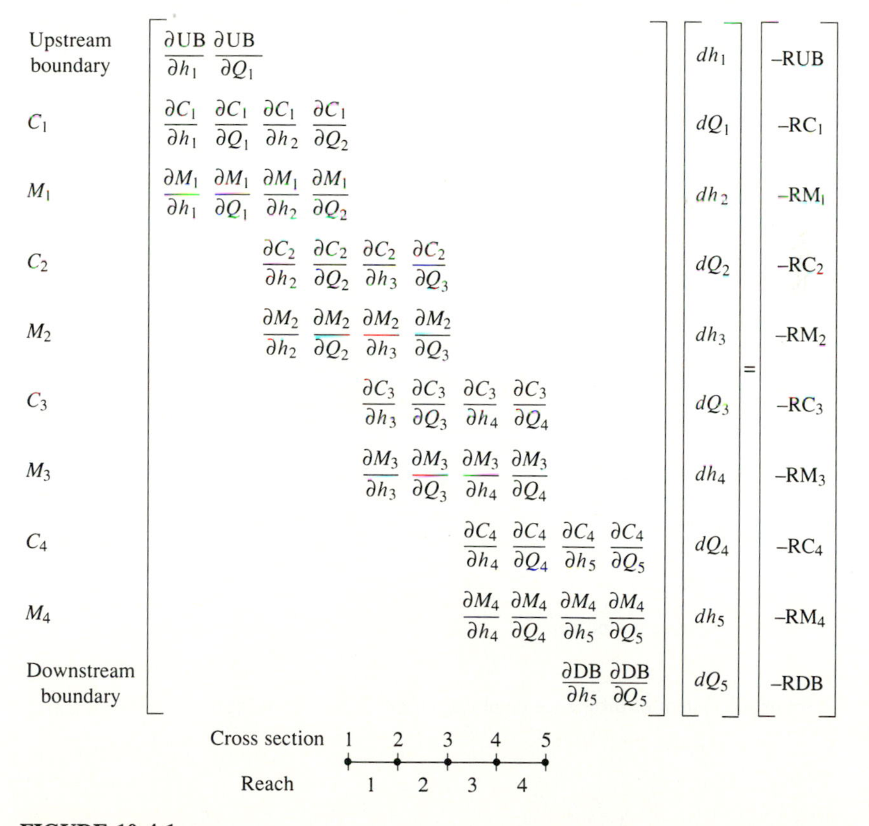

FIGURE 10.4.1.
System of linear equations for an iteration of the Newton-Raphson method for a river with four reaches (five cross sections).

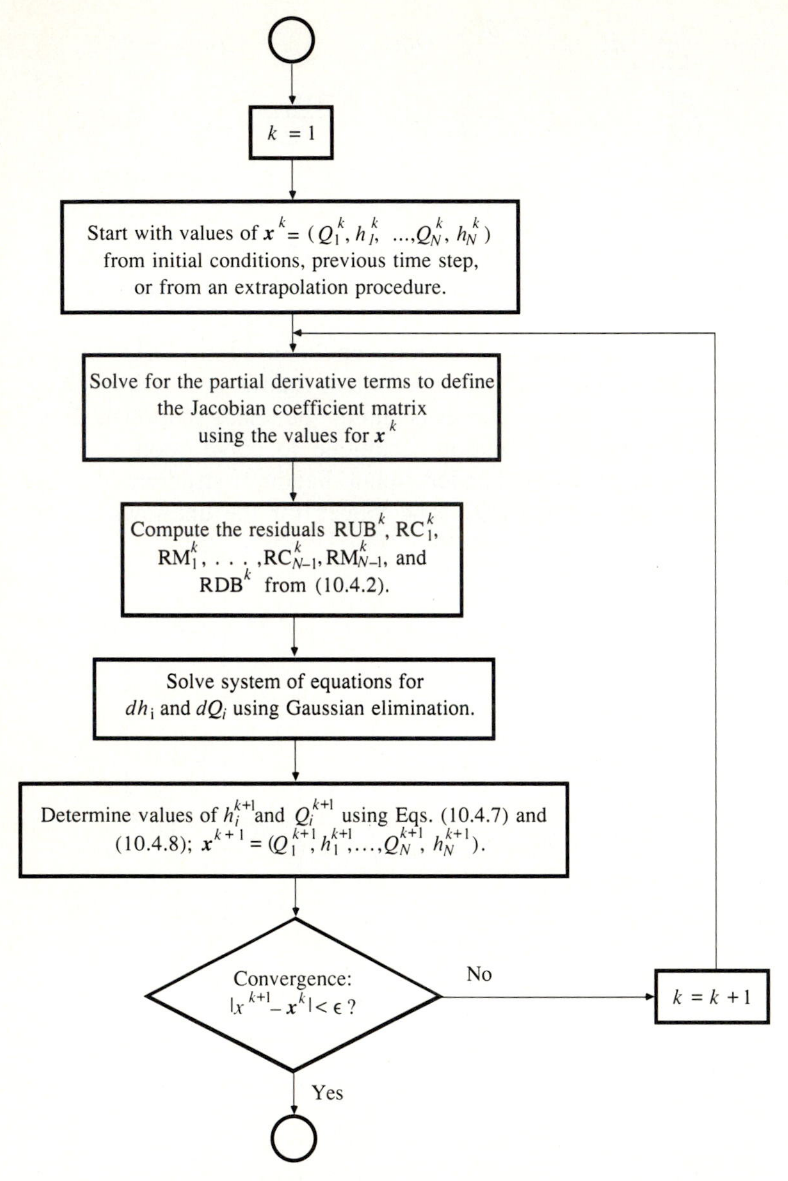

FIGURE 10.4.2
Procedure for solving a system of difference equations at one time step using the Newton-Raphson method.

$(k + 1)$th iteration are then given by

$$h_i^{k+1} = h_i^k + dh_i \tag{10.4.7}$$

$$Q_i^{k+1} = Q_i^k + dQ_i \tag{10.4.8}$$

The flow chart in Fig. 10.4.2 outlines the procedure for solving the system of difference equations for one time step using the Newton-Raphson method.

10.5 DWOPER MODEL

In the early 1970s, the U.S. National Weather Service (NWS) Hydrologic Research Laboratory began to develop a dynamic wave routing model based upon the implicit finite-difference solution of the Saint-Venant equations described in the previous section. This model, known as DWOPER (Dynamic Wave Operational Model) has been implemented on various rivers with backwater effects and mild bottom slopes. It has been applied to the Mississippi, Ohio, Columbia, Missouri, Arkansas, Red, Atchafalaya, Cumberland, Tennessee, Willamette, Platte, Kamar, Verdigris, Ouachita, and Yazoo rivers in the United States (Fread, 1978), and has also been used in many other countries.

One of the DWOPER applications described by Fread (1978) deals with the Mississippi-Ohio-Cumberland-Tennessee system, a branching river system consisting of 393 miles of the Mississippi, Ohio, Cumberland, and Tennessee rivers as shown in Fig. 10.5.1. Eleven gaging stations located at Fords Ferry, Golconda, Paducah, Metropolis, Grand Chain, Cairo, New Madrid, Red Rock, Grand Tower, Cape Girardeau, and Price Landing were used to evaluate the simulation by comparing the observed and calculated water levels and flow rates at those locations. Figure 10.5.2 shows the observed vs. simulated stages at Cape Girardeau, Missouri, and at Cairo, Illinois, for a flood in 1970.

In applying DWOPER to this system, the main stem river is considered to be the Ohio–Lower Mississippi segment, with the Cumberland, Tennessee, and upper Mississippi rivers considered first-order tributaries (Fread, 1973b). The channel bottom slope is mild, varying from about 0.25 to about 0.50 ft/mi (0.005-0.01 percent). Each branch of the river system is influenced by backwater from downstream branches. Total discharge through the system varies from low flows of approximately 120,000 cfs to flood flows of 1,700,000 cfs. A total of 45 cross sections located at unequal intervals ranging from 0.5 to 21 miles were used to describe the system. Three months of simulation time, comparing 20 observed and computed hydrographs using 24-hour time steps, required 15 seconds of CPU time on an IBM 360/195 computer.

Another application by Fread (1974b) on the lower Mississippi illustrates the utility of DWOPER simulating floods resulting from hurricanes. Figure 10.5.3

---

*A computer program for DWOPER can be obtained from the Hydrologic Research Laboratory, Office of Hydrology, NOAA, National Weather Service, Silver Spring, Maryland, 20910.

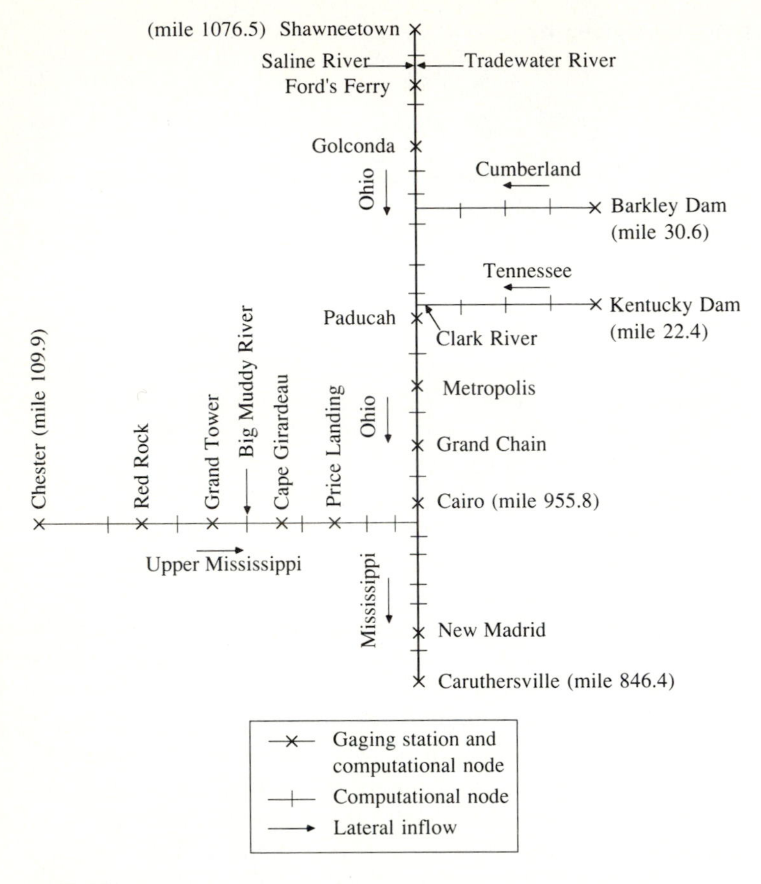

FIGURE 10.5.1
Schematic of the Mississippi-Ohio-Cumberland-Tennessee river system. The numbers shown are river miles from the mouth. (*Source*: Fread, 1978. Used with permission.)

shows the stage and discharge hydrographs on the lower Mississippi River at Carrollton, Louisiana, during hurricane Camille in 1969. The figure shows a brief period of negative discharge resulting from the hurricane-generated flood wave forcing water to flow up the Mississippi River.

10.6 FLOOD ROUTING IN MEANDERING RIVERS

The dynamic wave model developed in the previous section can be expanded to consider flood routing through meandering rivers in wide flood plains (Fig. 10.6.1). The unsteady flow in a river which meanders through a flood plain is complicated by five effects: (1) differences in hydraulic resistances of the main river channel and the flood plain; (2) variation in the cross-sectional geometries of the channel and the plain; (3) short-circuiting effects, in which the flow leaves the meandering main channel and takes a more direct route on the flood plain; (4)

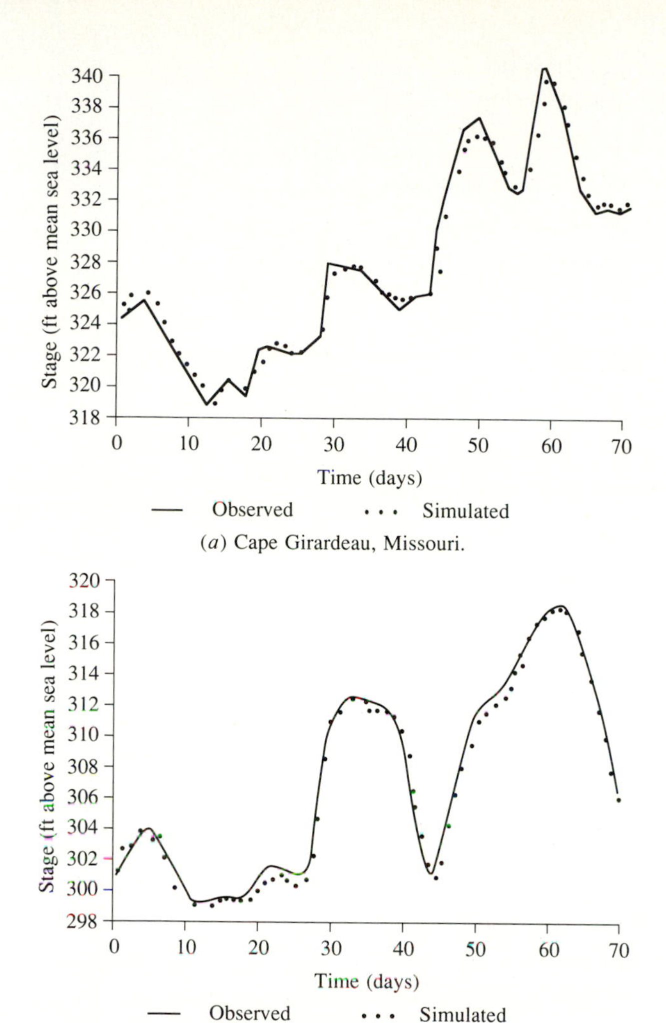

(*a*) Cape Girardeau, Missouri.

(*b*) Cairo, Illinois.

FIGURE 10.5.2
Observed vs. simulated stages at Cape Girardeau, Missouri, and at Cairo, Illinois, for 1970 flood. See Fig. 10.5.1 for location of these stations. Cairo is on the Ohio River and Cape Girardeau on the Mississippi River. (*Source*: Fread, 1978. Used with permission.)

portions of flood plain acting as dead storage areas in which the flow velocity is negligible; and (5) the effect on energy losses of the interaction of flows between the main channel and the flood plain, depending upon the direction of the lateral exchange of flow. Because of these differences, the attenuation and travel time of flow in the channel can differ significantly from that in the flood plain.

Fread (1976a, 1980) developed a model for meandering rivers, distinguishing the left flood plain, the right flood plain, and the channel, denoted by the subscripts l, r, and c respectively. The continuity and momentum equations, neglecting wind shear and lateral flow momentum, are expressed as

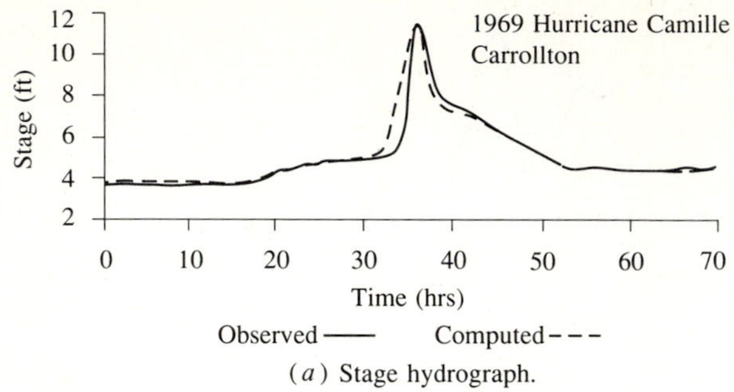

(a) Stage hydrograph.

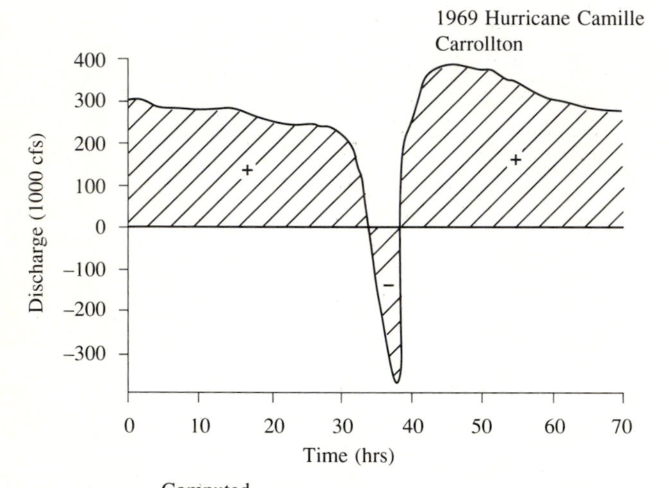

(b) Discharge hydrograph.

FIGURE 10.5.3
Stage and discharge hydrographs for the 1969 hurricane Camille at Carrollton on the lower Mississippi River. Carrollton is at mile 102.4 from the mouth of the Mississippi. The root-mean-square error for the simulation using DWOPER was 0.34 ft. (Fread, 1978. Used with permission.)

$$\frac{\partial(K_c Q)}{\partial x_c} + \frac{\partial(K_l Q)}{\partial x_l} + \frac{\partial(K_r Q)}{\partial x_r} + \frac{\partial(A_c + A_l + A_r + A_o)}{\partial t} - q = 0 \qquad (10.6.1)$$

and

$$\frac{\partial Q}{\partial t} + \frac{\partial(K_c^2 Q^2/A_c)}{\partial x_c} + \frac{\partial(K_l^2 Q^2/A_l)}{\partial x_l} + \frac{\partial(K_r^2 Q^2/A_r)}{\partial x_r} +$$

$$+ gA_c\left(\frac{\partial h}{\partial x_c} + S_{fc} + S_e\right) + gA_l\left(\frac{\partial h}{\partial x_l} + S_{fl}\right) + gA_r\left(\frac{\partial h}{\partial x_r} + S_{fr}\right) = 0 \qquad (10.6.2)$$

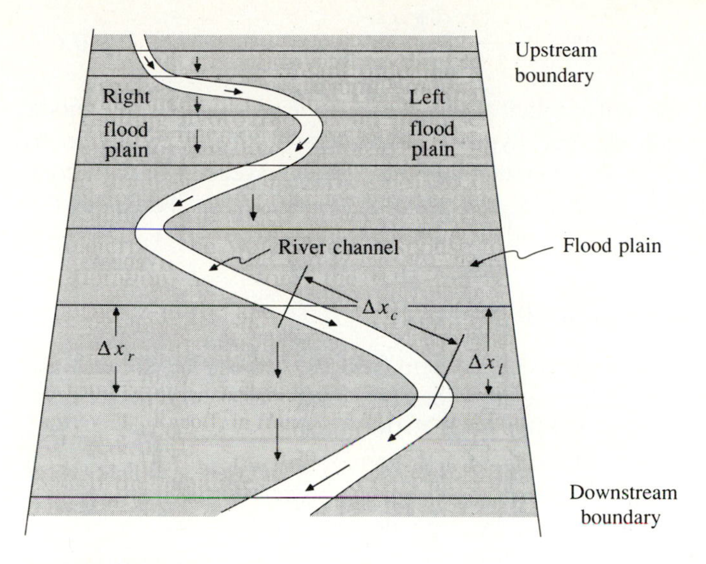

FIGURE 10.6.1
Meandering river in a flood plain. Unsteady flow in natural meandering rivers in wide flood plains is complicated by large differences in resistance and cross-sectional geometries of the river and the flood plain. As shown, further complications are due to the short circuiting of flow along the more direct route of the flood plain. As a result, the wave attenuation and travel time differ between the flood plain and the channel.

The total cross-sectional area of flow is the sum of A_c, A_l, A_r, and A_o. The constants K_c, K_l, and K_r divide the total flow Q into channel flow, left flood plain flow, and right flood plain flow, respectively, and are defined as $K_c =$ $= Q_c/Q$, $K_l = Q_l/Q$, and $K_r = Q_r/Q$.

The flow is assumed one-dimensional, so the water surface is horizontal across the three sections and the head loss h_f incurred in traveling between two river cross sections is the same no matter which flow path is adopted. Hence, $h_f = S_f \Delta x$ in each section of the flow (left, channel, and right), and $S_f = h_f / \Delta x$. By taking the ratio of the flows computed by Manning's equation in this manner, h_f is canceled out; the ratios of the flows in the left and right overbank areas to that in the channel are

$$\frac{Q_l}{Q_c} = \frac{n_c}{n_l} \frac{A_l}{A_c} \left(\frac{R_l}{R_c} \right)^{2/3} \left(\frac{\Delta x_c}{\Delta x_l} \right)^{1/2} \tag{10.6.3}$$

and

$$\frac{Q_r}{Q_c} = \frac{n_c}{n_r} \frac{A_r}{A_c} \left(\frac{R_r}{R_c} \right)^{2/3} \left(\frac{\Delta x_c}{\Delta x_r} \right)^{1/2} \tag{10.6.4}$$

The friction slope is also defined for the left flood plain, (S_{fl}), right flood plain, (S_{fr}), and channel (S_{fc}) using Manning's equation; for example,

$$S_{fl} = \frac{n_l |K_l Q| K_l Q}{2.21 A_l^2 R_l^{4/3}} \tag{10.6.5}$$

The weighted four-point implicit scheme can be used to solve this model for the unknowns h and Q. The dynamic wave model described above by Eqs. (10.6.1) and (10.6.2) is incorporated into the National Weather Service computer program DAMBRK; DAMBRK is a program for analyzing the floods that could result from dam breaks.

10.7 DAM-BREAK FLOOD ROUTING

Forecasting downstream flash floods due to dam failures is an application of flood routing that has received considerable attention. The most widely used dam-breach model is the National Weather Service DAMBRK* model by Fread (1977, 1980, 1981). This model consists of three functional parts: (1) temporal and geometric description of the dam breach; (2) computation of the breach outflow hydrograph; and (3) routing the breach outflow hydrograph downstream.

Breach formation, or the growth of the opening in the dam as it fails, is shown in Fig. 10.7.1. The shape of a breach (triangular, rectangular, or trapezoidal) is specified by the slope z and the terminal width B_w of the bottom of the breach. The DAMBRK model assumes the breach bottom width starts at a point and enlarges at a linear rate until the terminal width is attained at the end of the failure time interval T. The breach begins when the reservoir water surface elevation h exceeds a specified value h_{cr} allowing for overtopping failure or piping failure.

*A computer program for DAMBRK can be obtained from the Hydrologic Research Laboratory, Office of Hydrology, NOAA, National Weather Service, Silver Spring, Maryland, 20910.

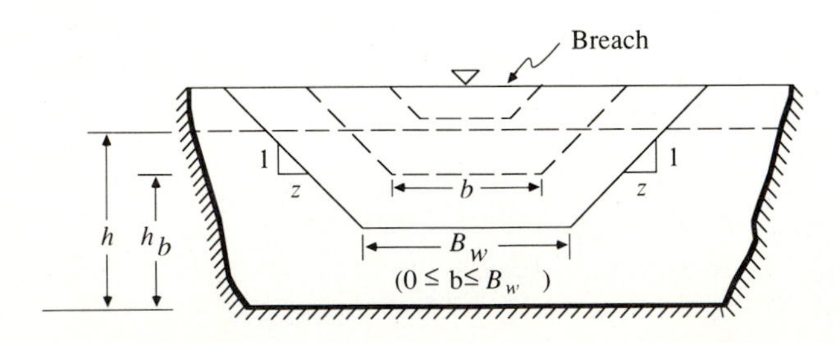

FIGURE 10.7.1
Breach formation. Breach formation in a dam failure is described by the failure time T, the size, and the shape. The shape is specified by z, which defines the side slope of the breach. Typically, $0 \leq z \leq 2$. The bottom width b of the breach is a function of time, with the terminal width being B_w. A triangular breach has $B = 0$ and $z > 0$. A rectangular breach has $B_w > 0$ and $z = 0$. For a trapezoidal breach, $B_w > 0$ and $z > 0$.

Reservoir outflow consists of both the breach outflow Q_b (broad-crested weir flow) and spillway outflow Q_s:

$$Q = Q_b + Q_s \tag{10.7.1}$$

The breach outflow can be computed using a combination of the formulas for a broad-crested rectangular weir, gradually enlarging as the breach widens, and a trapezoidal weir for the breach end slopes (Fread, 1980):

$$Q_b = 3.1 B_w t_b C_v K_s \frac{(h - h_b)^{1.5}}{T} + 2.45 z C_v K_s (h - h_b)^{2.5} \tag{10.7.2}$$

where t_b is the time after the breach starts forming, C_v is the correction factor for velocity of approach, K_s is the submergence correction for tail water effects on weir outflow, and h_b is the elevation of the breach bottom. The spillway outflow can be computed using (Fread, 1980):

$$Q_s = C_s L_s (h - h_s)^{1.5} + \sqrt{2g} C_g A_g (h - h_g)^{0.5} + C_d L_d (h - h_d)^{0.5} + Q_t \tag{10.7.3}$$

where C_s is the uncontrolled spillway discharge coefficient, L_s is the uncontrolled spillway length, h_s is the uncontrolled spillway crest elevation, C_g is the gated spillway discharge coefficient, A_g is the area of gate opening, h_g is the center-line elevation of the gated spillway, C_d is the discharge coefficient for flow over the dam crest, L_d is the length of the crest, h_d is the dam crest elevation, and Q_t is a constant outflow or leakage.

The DAMBRK model uses hydrologic storage routing or the dynamic wave model to compute the reservoir outflow. The reservoir outflow hydrograph is then routed downstream using the full dynamic wave model described in Sec. 10.3; alternatively the dynamic wave model described in Sec. 10.6 for flood routing in meandering rivers with flood plains can be used. The DAMBRK model can simulate several reservoirs located sequentially along a valley with a combination of reservoirs breaching. Highway and railroad bridges with embankments can be treated as *internal boundary conditions*.

Internal boundary conditions are used to describe the flow at locations along a waterway where the Saint-Venant equations are not applicable. In other words, there are locations such as spillways, breaches, waterfalls, bridge openings, highway embankments, and so on, where the flow is rapidly rather than gradually varied. Two equations are required to define an internal boundary condition, because two unknowns (Q and h) are added at the internal boundary. For example, to model a highway stream crossing (Fig. 10.7.2) with flow through a bridge opening Q_{br}, flow over the embankment Q_{em}, and flow through a breach, Q_b, the two internal boundary conditions are:

$$Q_i^{j+1} = Q_{br} + Q_{em} + Q_b \tag{10.7.4}$$

and

$$Q_{i+1}^{j+1} = Q_i^{j+1} \tag{10.7.5}$$

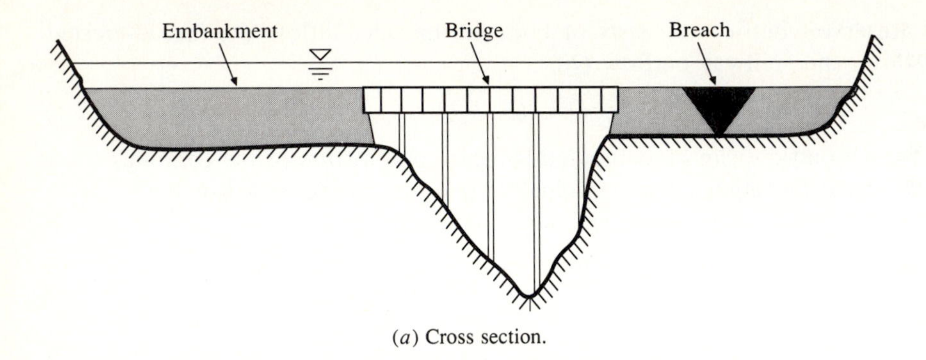

(*a*) Cross section.

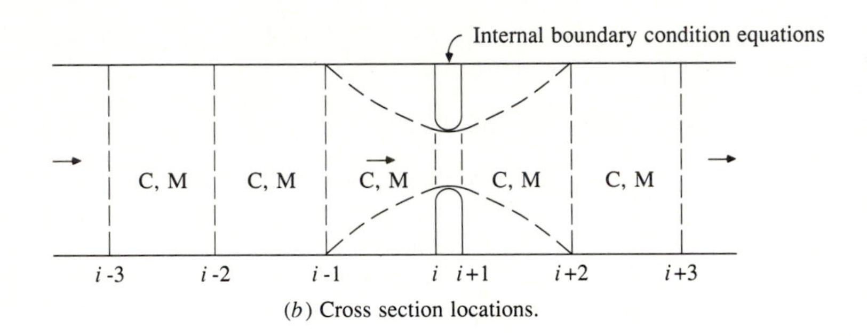

(*b*) Cross section locations.

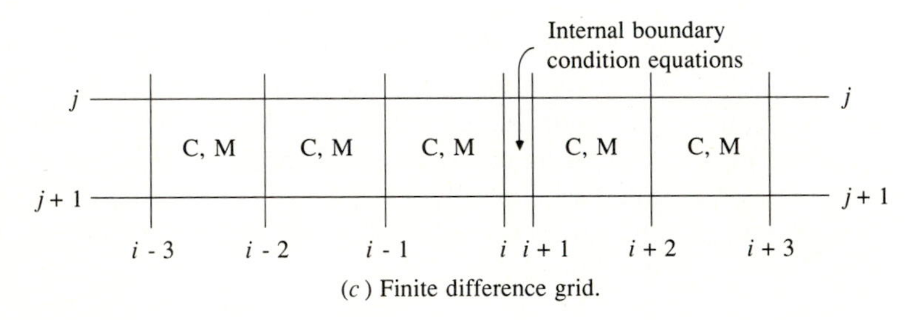

(*c*) Finite difference grid.

FIGURE 10.7.2
Internal boundary condition for highway embankments. C refers to the continuity equation and M to the momentum equation.

The breach flow Q_b is defined by Eq. (10.7.2); Q_{br} is defined by a rating curve or by an orifice equation, such as

$$Q_{\text{br}} = C_b \sqrt{2g}\, A_{i+1}^{j+1} \left(h_i^{j+1} - h_{i+1}^{j+1}\right)^{1/2} \tag{10.7.6}$$

where C_b is a bridge coefficient; the embankment overflow Q_{em} is defined by a broad-crested weir formula

$$Q_{\text{em}} = K_{\text{em}} L_{\text{em}} C_{\text{em}} \left(h_i^{j+1} - h_{\text{em}}\right)^{3/2} \tag{10.7.7}$$

where C_{em}, L_{em}, and K_{em} are the discharge coefficient, length of embankment, and submergence correction factor, and h_{em} is the elevation of the top of the embankment.

Referring to Fig. 10.7.2, the finite-difference grid illustrates that for a given time line, the continuity and momentum equations are written for each grid where the Saint-Venant equations apply, and the internal boundary conditions are written for the grid from i to $i + 1$, where the highway crossing is located.

Teton Dam Failure

The DAMBRK model has been applied (Fread, 1980) to reconstruct the downstream flood wave caused by the 1976 failure of the Teton Dam in Idaho. The Teton Dam was a 300-foot high earthen dam with a 3000-foot long crest. As a result of the breach of this dam, 11 people were killed, 25,000 people were made homeless, and $400 million in damage occurred in the downstream Teton–Snake River valley. The inundated area in the 60-mile reach downstream of the dam is shown with 12 cross sections marked in Fig. 10.7.3. The downstream valley consisted of a narrow canyon approximately 1000 ft wide for the first 5 miles, and thereafter a wide valley that was inundated to a width of about 9 mi. Manning's n values ranged from 0.028 to 0.047 as obtained from field estimates. Interpolated cross sections developed by DAMBRK were used so that cross sections were spaced 0.5 miles apart near the dam and 1.5 miles apart at the downstream end of the reach. A total of 77 cross sections were used.

The computed reservoir outflow hydrograph is shown in Fig. 10.7.3 with a peak value of 1,652,300 cfs. The peak occurred approximately 1.25 hours after the breach opening formed. It is interesting to note that the peak reservoir outflow was about 20 times greater than the largest recorded flood at the site. The computed peak discharges along the downstream valley are shown in Fig. 10.7.4, which illustrates the rapid attenuation of the peak discharge by storage of water in the flood plain as the flood wave progressed downstream through the valley. The computed flood peak travel times are shown in Fig. 10.7.5, and the computed peak elevations are shown in Fig. 10.7.6. The maximum depth of flooding was approximately 60 ft at the dam. The 55-hour simulation of the Teton flood used an initial time step of 0.06 h. Execution time would be less than 10 seconds on most mainframe computers.

FLDWAV Model

The FLDWAV model* (Fread, 1985) is a synthesis of DWOPER and DAMBRK, and adds significant modeling capabilities not available in either of the other models. FLDWAV is a generalized dynamic wave model for one-dimensional

---

*The computer program FLDWAV can be obtained from the Hydrologic Research Laboratory, Office of Hydrology, National Weather Service, NOAA, Silver Spring, MD, 20910.

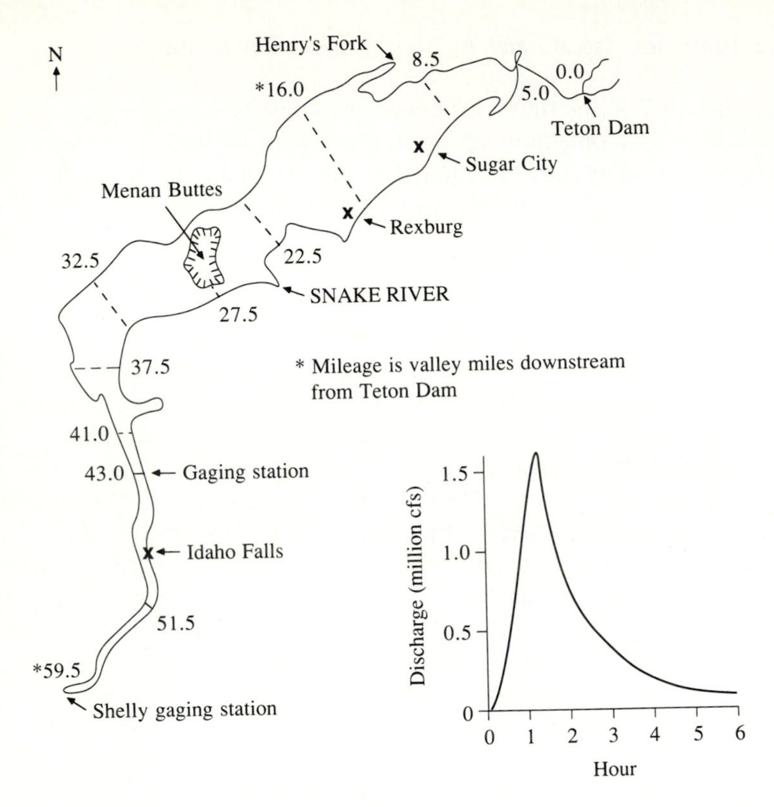

FIGURE 10.7.3
Flooded area downstream of Teton Dam and computed outflow hydrograph at the dam (*Source*: Fread, 1977).

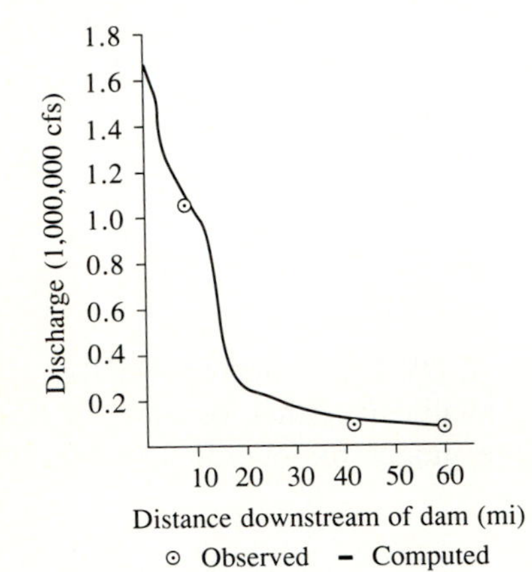

FIGURE 10.7.4
Profile of peak discharge from Teton Dam failure (*Source*: Fread, 1977).

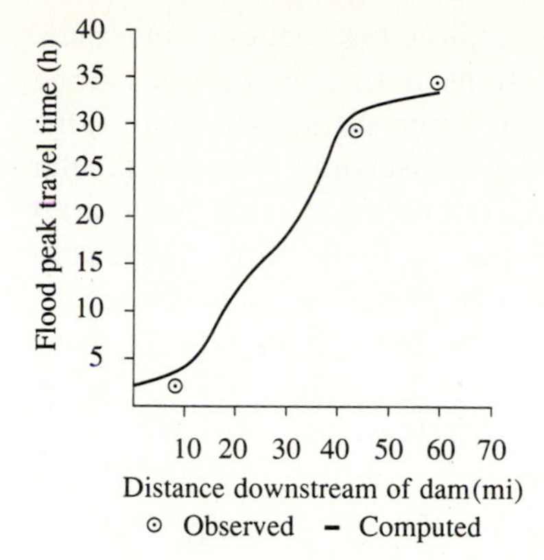

FIGURE 10.7.5
Travel time of flood peak from Teton Dam failure (*Source*: Fread, 1977).

unsteady flows in a single or branched waterway. It is based on an implicit (four-point, nonlinear) finite-difference solution of the Saint-Venant equations. The following special features and capacities are included in FLDWAV: variable Δt and Δx computational intervals; irregular cross-sectional geometry; off-channel storage; roughness coefficients that vary with discharge or water surface elevation, and with distance along the waterway; capability to generate linearly interpolated

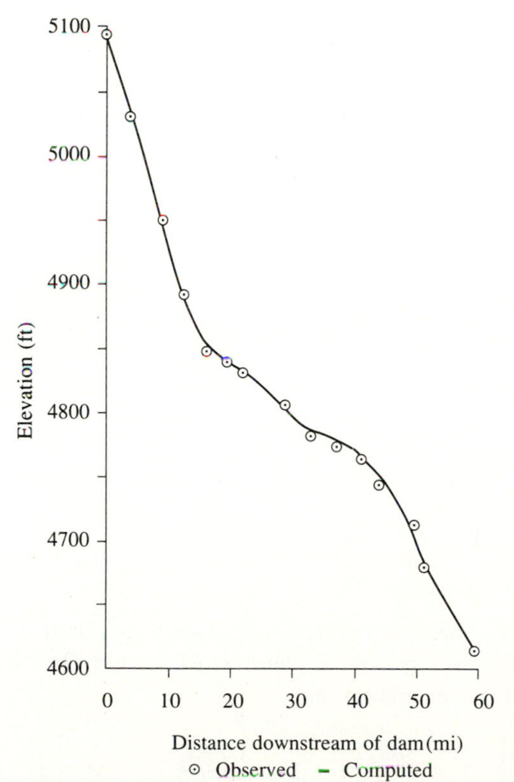

FIGURE 10.7.6
Profile of peak flood elevation from Teton Dam failure (*Source*: Fread, 1977).

cross sections and roughness coefficients between input cross sections; automatic computation of initial steady flow and water elevations at all cross sections along the waterway; external boundaries of discharge or water surface elevation time series (hydrographs), a single-valued or looped depth-discharge relation (tabular or computed); time-dependent lateral inflows (or outflows); internal boundaries enable the treatment of time-dependent dam failures, spillway flows, gate controls, or bridge flows, or bridge-embankment overtopping flow; short-circuiting of flood-plain flow in a valley with a meandering river; levee failure and/or overtopping; a special computational technique to provide numerical stability when treating flows that change from supercritical to subcritical, or conversely, with time and distance along the waterway; and an automatic calibration technique for determining the variable roughness coefficient by using observed hydrographs along the waterway.

FLDWAV is coded in FORTRAN IV, and the computer program is of modular design, with each subroutine requiring less than 64 kilobytes of storage. The overall program storage requirement is approximately 256 kilobytes. Program array sizes are variable, with the size of each array set internally via the input parameters used to describe each particular unsteady flow application. Input data to the program is free- or fixed-format. Program output consists of tabular and/or graphical displays, according to the user's choice.

APPENDIX 10.A

The following equations describe the partial derivative terms in the system of equations (10.4.6) (Fread, 1985). For the continuity equation (C), the terms dependent on h^{j+1} and Q^{j+1} in Eq. (10.3.11) contribute to the derivatives. For $\partial C/\partial h$, the product rule $\partial C/\partial h = \partial C/\partial A \times \partial A/\partial h = B\,\partial C/\partial A$ is used. The derivatives are:

$$\frac{\partial C}{\partial h_i} = \frac{\Delta x_i}{2\,\Delta t_j}(B + B_o)_i^{j+1} \tag{10.A.1}$$

$$\frac{\partial C}{\partial Q_i} = -\theta \tag{10.A.2}$$

$$\frac{\partial C}{\partial h_{i+1}} = \frac{\Delta x_i}{2\,\Delta t_j}(B + B_o)_{i+1}^{j+1} \tag{10.A.3}$$

$$\frac{\partial C}{\partial Q_{i+1}} = \theta \tag{10.A.4}$$

where B_o is the top width of the off-channel dead storage cross-sectional area.

For the momentum equation (M), the terms dependent on h^{j+1} and Q^{j+1} in Eq. (10.3.12) contribute to the derivatives, which are

$$\frac{\partial M}{\partial h_i} = \theta \left\{ \left(\frac{\beta Q^2 B}{A^2} \right)_i^{j+1} + g\bar{A}_i^{j+1} \left[-1 + \left(\frac{\partial S_f}{\partial h} \right)_i^{j+1} \Delta x_i + \left(\frac{\partial S_e}{\partial h} \right)_i^{j+1} \Delta x_i \right] \right.$$

$$+\frac{gB_i^{j+1}}{2}\left[h_{i+1}^{j+1} - h_i^{j+1} + (\overline{S}_f)_i^{j+1}\Delta x_i + (\overline{S}_e)_i^{j+1}\Delta x_i\right] + \frac{1}{2}\left(\overline{W}_f\frac{dB}{dh}\right)_i^{j+1}\Bigg\}$$

(10.A.5)

$$\frac{\partial M}{\partial Q_i} = \frac{\Delta x_i}{2\,\Delta t_j} + \theta\left\{-2\left(\frac{\beta Q}{A}\right)_i^{j+1} + g\overline{A}_i^{j+1}\left[\left(\frac{\partial S_f}{\partial Q}\right)_i^{j+1}\Delta x_i + \left(\frac{\partial S_e}{\partial Q}\right)_i^{j+1}\Delta x_i\right]\right\}$$

(10.A.6)

$$\frac{\partial M}{\partial h_{i+1}} = \theta\Bigg\{-\left(\frac{\beta Q^2 B}{A^2}\right)_{i+1}^{j+1} + g\overline{A}_i^{j+1}\left[1 + \left(\frac{\partial S_f}{\partial h}\right)_{i+1}^{j+1}\Delta x_i + \left(\frac{\partial S_e}{\partial h}\right)_{i+1}^{j+1}\Delta x_i\right]$$

$$+\frac{gB_{i+1}^{j+1}}{2}\left[h_{i+1}^{j+1} - h_i^{j+1} + (\overline{S}_f)_i^{j+1}\Delta x_i + (\overline{S}_e)_i^{j+1}\Delta x_i\right] + \frac{1}{2}\left(\overline{W}_f\frac{dB}{dh}\right)_{i+1}^{j+1}\Delta x_i\Bigg\}$$

(10.A.7)

$$\frac{\partial M}{\partial Q_{i+1}} = \frac{\Delta x_i}{2\,\Delta t_j} + \theta\left\{2\left(\frac{\beta Q}{A}\right)_{i+1}^{j+1} + g\overline{A}_i^{j+1}\left[\left(\frac{\partial S_f}{\partial Q}\right)_{i+1}^{j+1}\Delta x_i + \left(\frac{\partial S_e}{\partial Q}\right)_{i+1}^{j+1}\Delta x_i\right]\right\}$$

(10.A.8)

The derivatives of S_f are found by differentiating Eq. (10.3.18), and are

$$\frac{\partial(\overline{S}_f)_i}{\partial h_i} = 2(\overline{S}_f)_i\left(\frac{1}{\overline{n}_i}\frac{d\overline{n}_i}{dh_i} - \frac{5B_i}{6\overline{A}_i} + \frac{1}{3\overline{B}_i}\frac{dB_i}{dh_i}\right) \tag{10.A.9}$$

$$\frac{\partial(\overline{S}_f)_i}{\partial h_{i+1}} = 2(\overline{S}_f)_i\left(\frac{1}{\overline{n}_i}\frac{d\overline{n}_i}{dh_{i+1}} - \frac{5B_{i+1}}{6\overline{A}_i} + \frac{1}{3\overline{B}_{i+1}}\frac{dB_{i+1}}{dh_{i+1}}\right) \tag{10.A.10}$$

$$\frac{\partial(\overline{S}_f)_i}{\partial Q_i} = (\overline{S}_f)_i\left(\frac{1}{\overline{n}_i}\frac{d\overline{n}_i}{dQ_i} + \frac{1}{\overline{Q}_i}\right) \tag{10.A.11}$$

$$\frac{\partial(\overline{S}_f)_i}{\partial Q_{i+1}} = (\overline{S}_f)_i\left(\frac{1}{\overline{n}_i}\frac{d\overline{n}_i}{dQ_{i+1}} + \frac{1}{\overline{Q}_{i+1}}\right) \tag{10.A.12}$$

The derivatives of S_e, found by differentiating (10.3.19), are

$$\left(\frac{\partial\overline{S}_e}{\partial h}\right)_i = \left(\frac{K_e Q^2 B}{g\,\Delta x A^3}\right)_i \tag{10.A.13}$$

$$\left(\frac{\partial \bar{S}_e}{\partial h}\right)_{i+1} = \left(\frac{-K_e Q^2 B}{g\,\Delta x A^3}\right)_{i+1} \tag{10.A.14}$$

$$\left(\frac{\partial \bar{S}_e}{\partial Q}\right)_{i} = \left(\frac{-K_e Q}{g\,\Delta x A^2}\right)_{i} \tag{10.A.15}$$

$$\left(\frac{\partial \bar{S}_e}{\partial Q}\right)_{i+1} = \left(\frac{K_e Q}{g\,\Delta x A^2}\right)_{i+1} \tag{10.A.16}$$

The partial derivatives for the *UB* and *DB* functions are evaluated as follows:

$$\frac{\partial UB}{\partial h_1} = 0 \tag{10.A.17}$$

$$\frac{\partial UB}{\partial Q_1} = 1 \tag{10.A.18}$$

if the upstream boundary condition is a discharge hydrograph.

$$\frac{\partial DB}{\partial h_N} = 1 \tag{10.A.19}$$

$$\frac{\partial DB}{\partial Q_N} = 0 \tag{10.A.20}$$

if the downstream boundary condition is a stage hydrograph, but

$$\frac{\partial DB}{\partial h_N} = 0 \tag{10.A.21}$$

$$\frac{\partial DB}{\partial Q_N} = 1 \tag{10.A.22}$$

if the downstream boundary condition is a discharge hydrograph, and

$$\frac{\partial DB}{\partial h_N} = -\frac{Q_{k+1} - Q_k}{h_{k+1} - h_k} \tag{10.A.23}$$

$$\frac{\partial DB}{\partial Q_N} = 1 \tag{10.A.24}$$

where k is the iteration number, if the downstream boundary condition is a stage-discharge rating curve.

REFERENCES

Conte, S. D., *Elementary Numerical Analysis*, McGraw-Hill, New York, 1965.

Fread, D. L., Discussion of "Implicit flood routing in natural channels," by M. Amein and C. S. Fang, *J. Hyd. Div., Am. Soc. Civ. Eng.*, vol. 97, no. HY7, pp. 1156–1159, 1971.

Fread, D. L., Effect of time step size in implicit dynamic routing, *Water Resour. Bull.*, vol. 9, no. 2, pp. 338–351, 1973a.

Fread, D. L., Technique for implicit dynamic routing in rivers with major tributaries, *Water Resour. Res.*, vol. 9, no. 4, pp. 918–926, 1973b.

Fread, D. L., A dynamic model of stage-discharge relations affected by changing discharge, Fig. 11, p. 24, *NOAA Technical Memorandum NWS HYDRO-16*, Office of Hydrology, National Weather Service, Washington, D.C., 1973c.

Fread, D. L., Numerical properties of implicit four-point finite difference equations of unsteady flow, *NOAA technical memorandum NWS HYDRO-18*, National Weather Service, NOAA, U. S. Department of Commerce, Silver Spring, Md., 1974a.

Fread, D. L., Implicit dynamic routing of floods and surges in the lower Mississippi, presented at Am. Geophys. Union national meeting, April 1974, Washington, D. C., 1974b.

Fread, D. L., Discussion of "Comparison of four numerical methods for flood routing," by R. K. Price, *J. Hyd. Div., Am. Soc. Civ. Eng.*, vol. 101, no. HY3, pp. 565–567, 1975.

Fread, D. L., Flood routing in meandering rivers with flood plains, *Rivers '76*, Symposium on Inland Waterways for Navigation, Flood Control, and Water Diversions, August 10–12, 1976, Colorado State University, Fort Collins, Colo., vol. I, pp. 16–35, 1976a.

Fread, D. L., Theoretical development of implicit dynamic routing model, Dynamic Routing Seminar at Lower Mississippi River Forecast Center, Slidell, Louisiana, National Weather Service, NOAA, Silver Spring, Md., December 1976b.

Fread, D. L., The development and testing of a dam-break flood forecasting model, *Proceedings*, Dam-break Flood Modeling Workshop, U. S. Water Resources Council, Washington, D.C., pp. 164–197, 1977.

Fread, D. L., NWS operational dynamic wave model, *Proceedings*, 25th Annual Hydraulics Division Specialty Conference, Am. Soc. Civ. Eng., pp. 455–464, August 1978.

Fread, D. L., Capabilities of NWS model to forecast flash floods caused by dam failures, *Preprint Volume, Second Conference on Flash Floods*, March 18–20, Am. Meteorol. Soc., Boston, pp. 171–178, 1980.

Fread, D. L., Some limitations of dam-breach flood routing models, Preprint, Am. Soc. Civ. Eng. Fall Convention, St. Louis, Mo., October 1981.

Fread, D. L., Channel routing, Chap. 14 in *Hydrological Forecasting*, ed. by M. G. Anderson and T. P. Burt, Wiley, New York, pp. 437–503, 1985.

Isaacson, E., J. J. Stoker, and A. Troesch, Numerical solution of flood prediction and river regulation problems, reps. IMM-205, IMM-235, Inst. for Math. and Mech., New York Univ., New York, 1954, 1956.

Stoker, J. J., Numerical solution of flood prediction and river regulation problems, rep. IMM-200, Inst. for Math. and Mech., New York Univ., New York, 1953.

BIBLIOGRAPHY

Implicit solution of the Saint-Venant equations

Abbott, M. B., and F. Ionescu, On the numerical computation of nearly horizontal flows, *J. Hydraul. Res.*, vol. 5, no. 2, pp. 97–117, 1967.

Amein, M., An implicit method for numerical flood routing, Water Resour. Res., vol. 4, no. 4, pp. 719–726, 1968.

Amein, M., and C. S. Fang, Implicit flood routing in natural channels, *J. Hyd. Div., Am. Soc. Civ. Eng.*, vol.96, no. HY12, pp. 2481–2500, 1970.

Amein, M., Computation of flow through Masonboro Inlet, N. C., *J. Waterways, Harbors, and Coastal Eng. Div., Am. Soc. Civ. Eng.*, vol. 101, no. WW1, pp. 93–108, 1975.

Amein, M., and H.-L. Chu, Implicit numerical modeling of unsteady flows, *J. Hyd. Div., Am. Soc. Civ. Eng.*, vol. 101, no. HY6, pp. 717–731, 1975.

Baltzer, R. A., and C. Lai, Computer simulation of unsteady flows in waterways, *J. Hyd. Div., Am. Soc. Civ. Eng.*, vol. 94, no. HY4, pp. 1083–1117, 1968.

Chaudhry, Y. M., and D. N. Contractor, Application of the implicit method to surges in channels, *Water Resour. Res.*, vol. 9, no. 6, pp. 1605–1612, 1973.

Chen, Y. H., Mathematical modeling of water and sediment routing in natural channels, *Ph. D. dissertation*, University of California, Davis, 1973.

Chen, Y. H., and D. B. Simons, Mathematical modeling of alluvial channels, *Modeling 75 Symposium on Modeling Techniques*, Second Annual Symposium of Waterways, Harbors, and Coastal Eng. Div., Am. Soc. Civ. Eng., vol. 1, pp., 466–483, September 1975.

Contractor, D. N., and J. M. Wiggert, Numerical studies of unsteady flow in the James River, VPI-WRRC-Bull. 51, Water Resources Research Center, Virginia Poly. Inst. and State Univ., Blacksburg, Va., 1972.

Dronkers, J. J., Tidal computations for rivers, coastal areas, and seas, *J. Hyd. Div., Am. Soc. Civ. Eng.*, vol. 95, no. HY1. pp. 29–77, 1969.

Greco, F., and L. Panattoni, An implicit method to solve Saint Venant equations, *J. Hydrol.*, vol. 24, no. 1/2, pp. 171–185, 1975.

Gunaratnam, D. J., and F. E. Perkins, Numerical solution of unsteady flows in open channels, *hydrodynamics lab rep. 127*, Dept. of Civ. Eng., Mass. Inst. of Tech., Cambridge, Mass., 1970.

Hoff-Clausen, N. E., K. Havno, and A. Kej, Systems II sewer—a storm sewer model, *Proceedings*, Second International Conference on Urban Storm Drainage, University of Illinois at Urbana-Champaign, ed. by B. C. Yen, vol. 1, pp. 137–145, June 1981.

Isaacson, E., Fluid dynamical calculations, *Numerical Solution of Partial Differential Equations*, Academic Press, New York, pp. 35–49, 1966.

Kamphuis, J. W., Mathematical tidal study of St. Lawrence River, *J. Hyd. Div., Am. Soc. Civ. Eng.,* vol. 96, no. HY3, pp. 643–664, 1970.

Kanda, T., and T. Kitada, An implicit method for unsteady flows with lateral inflows in urban rivers, *Proceedings*, 18th Congress of IAHR, Baden-Baden, Federal Republic of Germany, vol. 2, pp. 213–220, August 1977.

Ponce, V. M., H. Indlekofer, and D. B. Simons, Convergence of four-point implicit water wave models, *J. Hyd. Div., Am. Soc. Civ. Eng.,* vol. 104, pp. 947–958, July 1978c.

Preismann, A., Propagation of translatory waves in channels and rivers, *1st Congress de l'Assoc. Francaise de Calcul*, Grenoble, France, pp. 433–442, 1961.

Preismann, A., and J. A. Cunge, Tidal bore calculation on an electronic computer, *La Houille Blanche*, vol. 5, pp. 588–596, 1961.

Price, R. K., A comparison of four numerical methods for flood routing, *J. Hyd. Div., Am. Soc. Civ. Eng.,* vol. 100, no. HY7, pp. 879–899, 1974.

Price, R. K., FLOUT—a river catchment flood model, report no. IT-168, Hydraulics Research Station, Wallingford, England, 1980.

Quinn, F. H., and Wylie, E. B., Transient analysis of the Detroit River by the implicit method, *Water Resour. Res.*, vol. 8, no. 6, pp. 1461–1469, 1972.

Sjoberg, A., Sewer network models DAGUL-A and DAGUL-DIFF, *Proceedings,* Second International Conference on Urban Storm Drainage, University of Illinois at Urbana-Champaign, ed. by B. C. Yen, vol. 1, pp. 127–136, June 1981.

SOGREAH, Mathematical model of flow simulation in urban sewerage system, CAREDAS Program, Societe Grenobloise d'Etudes et d'Applications Hydrauliques, Grenoble, France, 1973.

Wood, E. F., B. M. Harley, and F. E. Perkins, Transient flow routing in channel networks, *Water Resour. Res.*, vol. 11, no. 3, pp. 423–430, 1975.

Vasiliev, O. F., M. T. Gladyshev, N. A. Pritvits, and V. G. Sudobicher, Numerical methods for the calculation of shock wave propagation in open channels, *Proceedings*, International Association for Hydraulic Research, Eleventh International Congress, Leningrad, U.S.S.R., vol. 44, no. 3, 1965.

Dam-breach flood forecasting

Balloffet, A., One-dimensional analysis of floods and tides in open channels, *J. Hyd. Div., Am. Soc. Civ. Eng.*, vol. 95, no. HY4, pp. 1429–1451, 1969.

Chen, C.-L., and J. T. Armbruster, Dam-break wave model: formulation and verification, *J. Hyd. Div., Am. Soc. Civ. Eng.,* vol. 106, no. HY5, pp. 747–767, 1980.

Cunge, J. A., On the subject of a flood propagation method, *J. Hydraul. Res.*, vol. 7, no. 2, pp. 205–230, 1969.

Keefer, T. M., and R. K. Simons, Qualitative comparison of three dam-break routing models, *Proceedings*, Dam-break Flood Modeling Workshop, U. S. Water Resources Council, Washington, D.C., pp. 292–311, 1977.

Ponce, V. M., and A. J. Tsivoglou, Modeling gradual dam breaches, *J. Hyd. Div., Am. Soc. Civ. Eng.*, vol. 107, no. HY7, pp. 829–838, 1981.

Price, J. T., G. W. Lowe, and J. M. Garrison, Unsteady flow modeling of dam-break waves, *Proceedings*, Dam-break Flood Modeling Workshop, U. S. Water Resources Council, Washington, D.C., pp. 90–130, 1977.

Rajar, R., Mathematical simulation of dam-break flow, *J. Hyd. Div., Am. Soc. Civ. Eng.*, vol. 104, no. HY7, pp. 1011–1026, 1978.

Sakkas, J. G., and T. Strelkoff, Dam-break flood in a prismatic dry channel, *J. Hyd. Div., Am. Soc. Civ. Eng.*, vol. 99, no. HY12, pp. 2195–2216, 1973.

Wetmore, J. N., and D. L. Fread, The NWS simplified dam-break flood forecasting model, *Proceedings*, Fifth Canadian Hydrotechnical Conference, Fredericton, Canada, May 1981.

Xanthopoulous, T., and C. Koutitas, Numerical simulation of a two-dimensional flood wave propagation due to dam failure, *J. Hydraul. Res.*, vol. 14, no. 4, pp. 321–331, 1976.

General

Brutsaert, W., De Saint-Venant equations experimentally verified, *J. Hyd. Div., Am. Soc. Civ. Eng.*, vol. 97, no. HY9, pp. 1387–1401, 1971.

Courant, R., and D. Hilbert, *Methods of Mathematical Physics*, vols. I and II, Interscience Publishers, New York, 1953; original copyright with Julius Springer, Berlin, Germany, 1937.

Cunge, J. A., F. M. Holly, Jr., and A. Verwey, *Practical Aspects of Computational River Hydraulics*, Pitman, London, 1980.

Di Silvio, G., Flood wave modification along channels, *J. Hyd. Div., Am. Soc. Civ. Eng.*, vol. 95, no. HY9, pp. 1589–1614, 1969.

Fread, D. L., and T. E. Harbaugh, Open-channel profiles by Newton's iteration technique, *J. Hydrol.*, vol. 13, no. 1, pp. 70–80 1971.

Harris, G. S., Real time routing of flood hydrographs in storm sewers, *J. Hyd. Div., Am. Soc. Civ. Eng.*, vol. 96, no. HY6, pp. 1247–1260, 1970.

Hayami, S., On the propagation of flood waves, bulletin no. 1, Disaster Prevention Research Institute, Kyoto University, Japan, 1951.

Henderson, F. M., Flood waves in prismatic channels, *J. Hyd. Div., Am. Soc. Civ. Eng.*, vol. 89, no. HY4, pp. 39–67, 1963.

Isaacson, E., and H. B. Keller, *Analysis of Numerical Methods,* Wiley, New York, 1966.

Keefer, T. N., Comparison of linear systems and finite difference flow-routing techniques, *Water Resour. Res.*, vol. 12, no. 5, pp. 997–1006, 1976.

Liggett, J. A., Basic equations of unsteady flow, in *Unsteady Flow in Open Channels,* ed. by K. Mahmood and V. Yevjevich, vol. I, Water Resources Publications, Ft. Collins, Colo., pp. 29–62, 1975.

Liggett, J. A., and J. A. Cunge, Numerical methods of solution of the unsteady flow equations, in *Unsteady Flow in Open Channels,* ed. by K. Mahmood and V. Yevjevich, vol. I, Water Resources Publications, Ft. Collins, Colo., pp. 89–182, 1975.

Mahmood, K. V., V. Yevjevich, and W. A. Miller, Jr. (eds.), *Unsteady Flow in Open Channels,* 3 vols., Water Resources Publications, Ft. Collins, Colo., 1975.

Milne, W. E., *Numerical Solution of Differential Equations,* Dover, New York, 1970.

Natural Environment Research Council, Flood studies report, vol. III, flood routing studies, London, 1975.

Overton, D. E., and M. E. Meadows, *Stormwater Modeling,* Academic Press, New York, 1976.

Ponce, V. M., and D. B. Simons, Shallow wave propagation in open channel flow, *J. Hyd. Div., Am. Soc. Civ. Eng.*, vol. 103, no. HY12, pp. 1461–1476, 1977.

Prandle, D., and N. L. Crookshank, Numerical model of St. Lawrence River estuary, *J. Hyd. Div., Am. Soc. Civ. Eng.*, vol. 100, no. HY4, pp. 517–529, 1974.

Price, R. K., Variable parameter diffusion method for flood routing, report no. INT 115, Hydraulics Research Station, Wallingford, England, 1973.

Price, R. K., Flood routing methods for British rivers, *Proceedings*, Institution of Civil Engineers, vol. 55, pp. 913–930, 1973.

Price, R. K., A simulation model for storm sewer, *Proceedings*, Second International Conference on Urban Storm Drainage, University of Illinois at Urbana-Champaign, vol. 1, pp. 184–192, 1981.

Roache, P. J., *Computational Fluid Dynamics*, Hermosa Publishers, Albuquerque, N. Mex., 1976.

Richtmeyer, R. D., A survey of difference methods for non-steady fluid dynamics, *NCAR technical notes 63-2*, National Center for Atmospheric Research, Boulder, Colo., 1962.

Roesner, L. A., and R. P. Shubinski, Improved dynamic flow routing model for storm drainage systems, *Proceedings*, Second International Conference on Urban Storm Drainage, University of Illinois at Urbana-Champaign, vol. 1, pp. 164–173, 1981.

Sevuk, A. S., and B. C. Yen, A comparative study on flood routing computation, *Proceedings*, International Symposium on River Mechanics of Int. Ass. Hydraul. Res., Bangkok, Thailand, vol. 3, pp. 275–290, January 1973.

Smith, G. D., *Numerical Solution of Partial Differential Equations*, Oxford University Press, London, 1965.

Strelkoff, T., The one-dimensional equations of open-channel flow, *J. Hyd. Div., Am. Soc. Civ. Eng.*, vol. 95, no. HY3, pp. 861–874, 1969.

Thomas, I. E., and P. R. Wormleaton, Finite difference solution of the flood diffusion equation, *J. Hyd. Div., Am. Soc. Civ. Eng.*, vol. 12, pp. 211–220, 1971.

Yen, B. C., Open-channel flow equations revisited, *J. Eng. Mech. Div., Am. Soc. Civ. Engr.*, vol. 99, no. EM5, pp. 979–1009, 1973.

PROBLEMS

10.3.1 Determine the time derivative of the discharge for $Q_i^j = 7000$ cfs, $Q_i^{j+1} = 7166$ cfs, $Q_{i+1}^j = 6772$ cfs, and $Q_{i+1}^{j+1} = 6940$ cfs. Use a time interval of one hour, $\Delta x = 1000$ ft, and $\theta = 0.60$.

10.3.2 Derive Eqs. (10.3.11) and (10.3.12).

10.3.3 Derive the expressions for partial derivatives of the continuity equation given in Appendix 10.A [Eqs. (10.A.1)–(10.A.4)].

10.3.4 Derive the expressions for partial derivatives of the momentum equation given in Appendix 10.A [Eqs. (10.A.5)–(10.A.8)].

10.4.1 Explain the procedure used to solve the system of equations (10.4.1) using the Newton-Rhapson method.

10.4.2 Even though the implicit scheme used for the full dynamic wave model is conditionally stable, explain why the approach is unstable when critical flow conditions are approached.

10.4.3 Explain why instability problems can occur when modeling a river that has cross sections with a main channel and a very wide, flat flood plain. *Hint:* What would the relationships of elevation vs. hydraulic radius, elevation vs. top width, and elevation vs. discharge look like?

10.4.4 Under what conditions would the Manning's n vs. discharge relationship cause instability problems in solving the full dynamic wave model?

10.5.1 The purpose of this problem is to use the U.S. National Weather Service DWOPER or FLDWAV model. Consider a 9000-foot-long trapezoidal irrigation

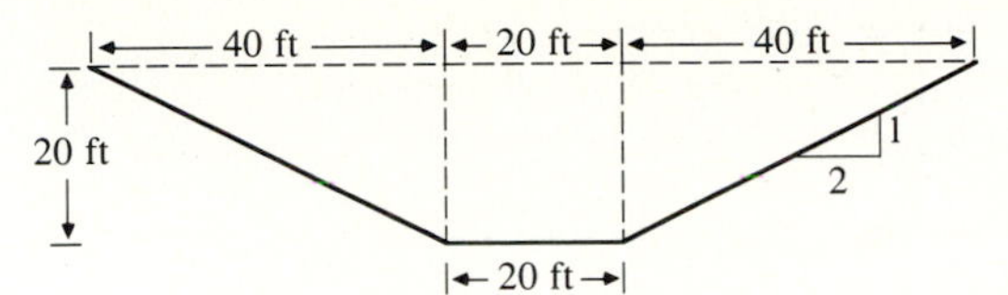

FIGURE 10.P.1
Channel cross section.

channel with cross section as shown in Figure 10.P.1. The upstream inflow hydrograph is shown in Figure 10.P.2. The channel bottom slope is 0.0005 ft/ft. The bottom elevation of the channel at the downstream end is 95.5 ft. The channel has a Manning's roughness factor of $n = 0.025$. In order to model the channel, cross sections (stations) are placed at 1000 ft intervals as shown in Figure 10.P.3. Using a 0.1-h computational time step, simulate the system behavior for the first five hours. Plot the inflow and outflow hydrographs; plot also the distribution of flow along the channel at one-hour intervals. Use an initial condition of 200 cfs along the channel. The downstream boundary condition is the rating curve given below.

Stage (ft)	98.6	100.2	102.6	104.3	106.2	107.7	109.5
Discharge (cfs)	200	550	1000	1700	2200	2600	3200

10.5.2 This problem is an extension of Prob. 10.5.1 to include a tributary channel, as shown in Figure 10.P.4. The shape of the tributary is identical to that of the main stem (Fig. 10.P.1). The tributary is 3000 feet long and has a bottom slope of 0.0005 ft/ft and a bottom elevation of 98.0 ft at the confluence. A Manning's n of 0.025 is assumed. The inflow hydrograph used in Prob. 10.5.1 (Fig. 10.P.2) is the upstream boundary condition for both channels. To model this system, a new station is added to the main stem at distance 4950 ft to account for the tributary channel. The tributary is discretized using six stations, as shown in Fig. 10.P.4. Stations 4 and 5 have no purpose for this problem; Prob. 10.5.4 makes use of them. Plot the new outflow hydrograph at station 11 and compare it with the result obtained in Prob. 10.5.1. Plot the distribution of flow rate along the channel at one-hour time intervals. Use an initial condition of 200 cfs inflow to each of the two channels and 400 cfs downstream of their junction.

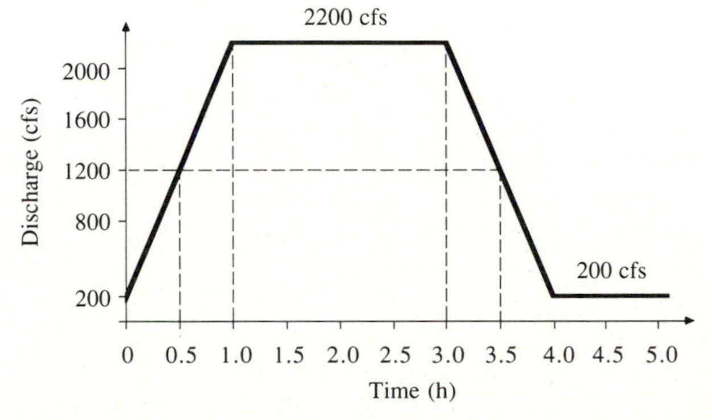

FIGURE 10.P.2
Inflow hydrograph.

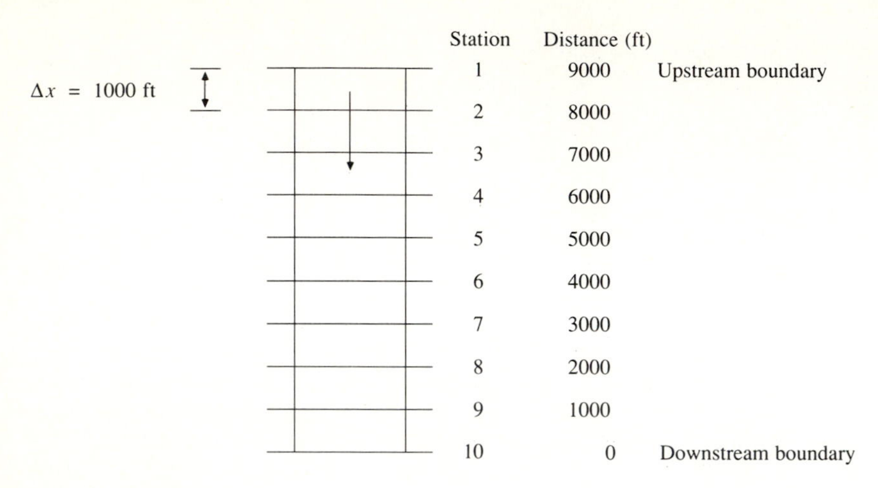

FIGURE 10.P.3
Main irrigation channel.

10.5.3 Solve Prob. 10.5.2 using the tributary inflow hydrograph shown in Figure 10.P.5. This problem has been constructed to illustrate the program's capability of handling backwater effects in channel systems.

10.5.4 Solve Prob. 10.5.2 with a flow-regulating structure placed between stations 5 and 6 on the tributary, just upstream of the confluence. The rating table for the regulating structure is given below.

Stage (ft)	4	6	8	10	13	16	20	25
Discharge (cfs)	100	300	700	1000	1300	1500	1750	2000

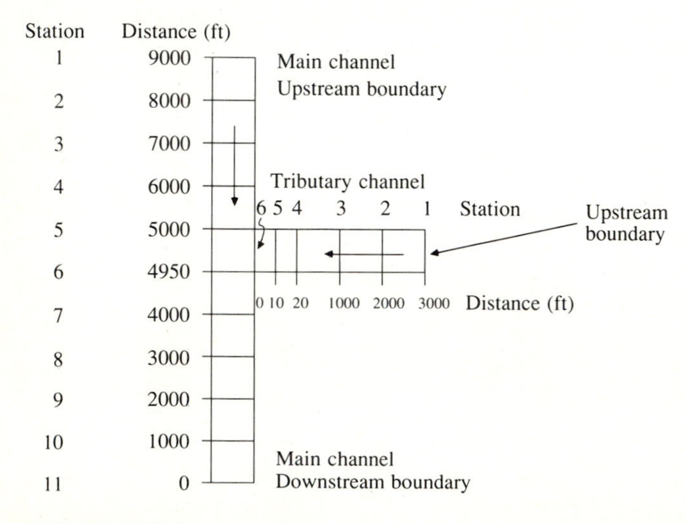

FIGURE 10.P.4
Two-channel irrigation system.

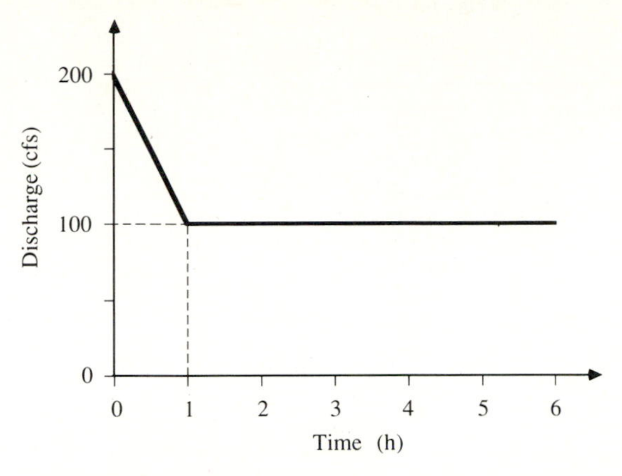

FIGURE 10.P.5
Inflow hydrograph for tributary.

10.6.1 Derive the finite difference equations for continuity and momentum for flood routing in a meandering river. Consider the left and right flood plain, the main channel, and the dead storage in development of these equations.

10.7.1 The purpose of this problem is to illustrate the use of DAMBRK or FLDWAV to perform reservoir routing using the storage routing procedure and dynamic routing in the downstream valley. You are to develop the DAMBRK model input and run it on the computer. The PMF (probable maximum flood) inflow hydrograph to the reservoir is given in Table 10.P.1. Reservoir characteristics are presented in Table 10.P.2, including the elevation-storage relationship and spillway characteristics. Five cross sections labeled *B* through *F* in Table 10.P.3 are used to describe the 12.5-mile-long downstream valley. Manning's roughness factors, as a function of water surface elevation for each cross section, are also given in Table 10.P.3. The slope of the downstream valley is about 10.15 ft/mi. The initial water surface elevation in the reservoir is 2323 ft above mean sea level (MSL). Terminate routing computations after 57 hours of simulation. Use KDMP = 3 and JNK = 1 for the printing instructions. Use a minimum computational distance for interpolated cross sections of 0.5 mi for each reach. Let the initial condition be a flow of 800 cfs.

TABLE 10.P.1
Probable maximum flood for Prob. 10.7.1

Time (h)	0	1	3	5	7	9	11	13
Inflow (cfs)	0.0	12.8	239.2	2000.0	8028.8	20339.5	40906.1	80570.3
Time	15	17	19	21	23	25	27	29
Inflow	156116.8	248330.1	295681.1	279367.0	224737.0	158232.4	107019.0	71600.3
Time	31	33	35	37	39	41	43	45
Inflow	46295.6	29862.1	19363.0	12561.7	8143.1	5275.7	3460.3	2205.5
Time	47	49	51	53	55	57		
Inflow	1218.0	495.8	163.4	63.0	15.7	0.		

TABLE 10.P.2
Characteristics of reservoir

Elevation-storage relationship	
Elevation (ft–MSL)	**Storage (acre–feet)**
2342	160710
2330	85120
2320	58460
2305	30075
2290	13400
2275	4125
2260	335
2245	0
Reservoir	
Top-of-dam elevation (MSL):	2347 ft
Length of reservoir:	5 mi
Elevation (MSL) of bottom of dam:	2245 ft
Normal pool level (MSL):	2323 ft
Spillway	
Elevation of uncontrolled spillway crest:	2333 ft
Discharge of coefficient for uncontrolled spillway:	2.5
Length of uncontrolled spillway:	3900 ft
Turbine	
Discharge:	800 cfs

TABLE 10.P.3
Cross sections of downstream valley

Cross section number	Channel location (mi)						
A	4.95	Elevation (ft)	2250	2260	2300	2330	2350
		Top width (ft)	0	1024	2737	4508	11812
B	5.0	Elevation (ft)	2250	2260	2300	2330	2350
		Top width (ft)	0	1024	2737	4508	11812
C	6.97	Elevation (ft)	2230	2240	2250	2270	2300
		Top width (ft)	0	473	2539	4311	6654
D	10.61	Elevation (ft)	2190	2200	2210	2240	2270
		Top width (ft)	0	1078	2441	3118	3842
E	13.94	Elevation (ft)	2150	2160	2170	2180	2210
		Top width (ft)	0	216	1654	2402	4093
F	17.45	Elevation (ft)	2120	2130	2140	2160	2180
		Top width (ft)	0	945	1772	3091	3858
G	17.5	Elevation (ft)	2120	2130	2140	2160	2180
		Top width (ft)	0	945	1772	3091	3858
H	18.5	Elevation (ft)	2110	2120	2130	2150	2170
		Top width (ft)	0	945	1772	3091	3858
Manning's n values			0.030	0.030	0.035	0.035	0.035

10.7.2 Solve Prob. 10.7.1 if the initial elevation of the reservoir is 2290 feet above mean sea level. By what percentage is the peak downstream flow reduced by the additional flood storage in the reservoir?

10.7.3 The purpose of this problem is to illustrate the use of DAMBRK or FLDWAV to perform a dam-break analysis for a nonflood event. The same reservoir and downstream valley used in Prob. 10.7.1 are used for this application. A nonflood event is used with an inflow of 800 cfs to the reservoir. The initial water surface elevation is at the normal pool level (2323 ft). Failure of the dam occurs at this elevation. The breach is rectangular and the time to maximum breach is 1.25 hours after the beginning of the breach. The breach extends down to the elevation of the bottom of the dam, 2245 feet above MSL, and is 200 feet wide at its base. The computation can be terminated after 10 hours of simulation. Plot the outflow hydrograph from the reservoir.

10.7.4 Solve Prob. 10.7.3 with a time to failure of 2 h and compare the outflow hydrographs for the two failure times.

10.7.5 The purpose of this problem is to illustrate the use of DAMBRK or FLDWAV to perform a dam break analysis for a probable maximum flood (PMF) event. The same reservoir and downstream valley used in Prob. 10.7.1 is used for this application. Use the same breach characteristics as used in Prob. 10.7.3. The PMF inflow into the reservoir (Prob. 10.7.1, Table 10.P.1) is to be used for this problem. Assign an elevation of water when breaching starts (HF) of 2341.48 feet above MSL and a time to maximum breach size (TFM) of 1.25 h. Compare the outflow hydrograph with that obtained in Probs. 10.7.1 and 10.7.3.

10.7.6 This problem illustrates a multiple dam application. Prob. 10.7.5 is extended to include a second dam whose downstream face is at mile 12.5 downstream of the first dam. Three additional cross sections (A, G, and H in Table 10.P.3) are used to define the upstream faces of both dams and the channel downstream of the second dam. Table 10.P.4 is a summary of the characteristics of the second dam. The initial water surface elevation for the second reservoir is at 2150 ft. The breach starts when the water surface elevation reaches the top of the second dam, 2180 ft above MSL, and extends down to elevation 2120 ft above MSL. Assign a width of base of breach of 100 ft. Use the same value for the time to maximum breach size as in the first dam. For the first upstream dam use the same breach characteristics as used in Problem 10.7.3.

10.7.7 Perform a sensitivity analysis for the parameters describing the breach for the situation posed in Prob. 10.7.5. Consider different values of the time to maximum breach size (e.g., 0.5 h, 1.0 h, 1.5 h, 3 h, etc.). Consider different maximum breach widths (e.g., 50 ft, 100 ft, 300 ft, etc.). Plot the flood elevation profile vs. distance downstream for each of your simulations. From the envelope of profiles,

TABLE 10.P.4
Characteristics of second dam

Top-of-dam elevation (MSL):	2180 ft
Elevation (MSL) of bottom of dam:	2120 ft
Elevation of uncontrolled spillway crest:	2165 ft
Discharge of coefficient for uncontrolled spillway:	2.5
Discharge coefficient for uncontrolled weir flow:	4830.8
Length of uncontrolled spillway:	2000 ft

determine the breach parameters that result in the maximum flooding downstream of the dam.

10.7.8 Develop the DAMBRK or FLDWAV model input for the Teton Dam failure. For the Teton Dam failure the water surface at failure was 5288.5 ft MSL, which was below the spillway outlet; therefore, the spillway flows do not need to be modeled. The inflow to the reservoir was considered constant at 13,000 cfs over the entire time of simulation. The reservoir routing is to be performed using the hydrologic routing procedure with the reservoir characteristics defined by the following elevation-storage relationship:

Elevation (ft MSL)	5027	5038.5	5098.5	5228.5	5288.5
Storage (ac·ft)	0	1247.8	25,037.8	137,682.8	230,472.8

A total of 12 cross sections, each with five top widths, are used to model the downstream geometry (Table 10.P.5). Consider a maxium lateral outflow of –0.30 cfs/ft to produce the volume losses experienced by the passage of the dam-break flood wave through the stream reaches between adjacent cross sections (e.g., reach 1 is between cross sections 1 and 2). The Manning's n values, the minimum distance between interpolated cross sections, and the contraction-expansion coefficients for each reach are defined in Table 10.P.6. The following parameters are required for the input: length of reservoir = 17 mi, initial elevation of water surface before failure = 5288.5 ft MSL, side slope of breach $z = 0$,

TABLE 10.P.5
Teton Dam cross section information

Cross section 1 River mile 0 (downstream from dam)					
Water surface elevation (ft MSL)	5027	5037	5051	5107	5125
Top width	0	590	820	1130	1200
Top width, off-channel storage	0	0	0	0	0
Cross section 2 River mile 5.0 mi					
Water surface elevation (ft MSL)	4965	4980	5015	5020	5030
Top width	0	850	1100	1200	1300
Top width, off-channel storage	0	0	3500	4300	5300
Cross section 3 River mile 8.5 mi					
Water surface elevation (ft MSL)	4920	4930	4942	4953	4958
Top width	0	800	4000	11000	15000
Top width, off-channel storage	0	0	0	7000	10000
Cross section 4 River mile 16.0 mi					
Water surface elevation (ft MSL)	4817	4827	4845	4847	4852
Top width	0	884	4000	11000	22000
Top width, off-channel storage	0	0	30000	27000	25000
Cross section 5 River mile 22.5 mi					
Water surface elevation (ft MSL)	4805	4812	4814	4825	4830
Top width	0	1000	1200	11000	16000
Top width, off-channel storage	0	0	0	6000	8000
Cross section 6 River mile 27.5 mi					
Water surface elevation (ft MSL)	4788	4792	4802	4808	4810
Top width	0	286	7000	10000	11000
Top width, off-channel storage	0	0	0	3500	5000

TABLE 10.P.6
Cross sections of downstream valley

Reach	1	2	3	4	5	6	7	8	9	10	11
Manning's n*	0.08	0.05	0.031	0.034	0.038	0.037	0.034	0.034	0.034	0.036	0.036
Minimum Δx (mi)	0.5	0.5	0.5	0.75	1.0	1.0	1.0	1.0	1.0	1.1	1.4
Expansion-contraction coefficients**	0	−0.9	0	0	0.1	−0.5	0	0	0	0	0

*The same Manning's n is used for all depths (water surface elevations).

**Expansion coefficients have a negative value and contraction coefficients have a positive value.

elevation of bottom of breach = 5027 ft MSL, final bottom width of breach = 150 ft, time to maximum breach = 1.25 h, elevation of water when breached = 5288.5 ft MSL, simulation time = 55 h. Because you are simulating a piping failure, let the top of the failure be equal to the initial water surface elevation. The flow for initial conditions can be set at 13,000 cfs. Compare your results with those in Figs. 10.7.3–10.7.6.

10.7.9 Perform a sensitivity analysis of the various input parameters for the Teton Dam application (Prob. 10.7.8). For example, vary the final bottom width and side slope of the breach. Also, vary the maximum time to failure. For each simulation that you make, plot maximum flood elevation vs. distance downstream on the same graph and compare with the results in Fig. 10.7.4.

TABLE 10.P.5 ***(cont.)***
Teton Dam cross section information

Cross section 7 River mile 32.5 mi					
Water surface elevation (ft MSL)	4762	4774	4777	4780	4785
Top width	0	352	5000	10000	18000
Top width, off-channel storage	0	0	9000	10000	24000
Cross section 8 River mile 37.5 mi					
Water surface elevation (ft MSL)	4752	4763	4768	4773	4778
Top width	0	450	3500	6000	9000
Top width, off-channel storage	0	0	4000	8500	12000
Cross section 9 River mile 41.0 mi					
Water surface elevation (ft MSL)	4736	4756	4761	4763	4768
Top width	0	540	2000	4000	6000
Top width, off-channel storage	0	0	3700	3700	5500
Cross section 10 River mile 43.0 mi					
Water surface elevation (ft MSL)	4729	4737	4749	4757	4759
Top width	0	250	587	1750	2000
Top width, off-channel storage	0	0	0	1500	2000
Cross section 11 River mile 51.5 mi					
Water surface elevation (ft MSL)	4654	4659	4668	4678	4683
Top width	0	70	352	400	420
Top width, off-channel storage	0	0	0	0	0
Cross section 12 River mile 59.5 mi					
Water surface elevation (ft MSL)	4601	4604	4606	4615	4620
Top width	0	245	450	500	520
Top width, off-channel storage	0	0	0	0	0

CHAPTER 11

HYDROLOGIC STATISTICS

Hydrologic processes evolve in space and time in a manner that is partly predictable, or deterministic, and partly random. Such a process is called a *stochastic process*. In some cases, the random variability of the process is so large compared to its deterministic variability that the hydrologist is justified in treating the process as purely random. As such, the value of one observation of the process is not *correlated* with the values of adjacent observations, and the statistical properties of all observations are the same.

When there is no correlation between adjacent observations, the output of a hydrologic system is treated as stochastic, space-independent, and time-independent in the classification scheme shown in Fig. 1.4.1. This type of treatment is appropriate for observations of extreme hydrologic events, such as floods or droughts, and for hydrologic data averaged over long time intervals, such as annual precipitation. This chapter describes hydrologic data from pure random processes using statistical parameters and functions. Statistical methods are based on mathematical principles that describe the random variation of a set of observations of a process, and they focus attention on the observations themselves rather than on the physical processes which produced them. Statistics is a science of description, not causality.

11.1 PROBABILISTIC TREATMENT OF HYDROLOGIC DATA

A *random variable* X is a variable described by a *probability distribution*. The distribution specifies the chance that an *observation* x of the variable will fall in a specified range of X. For example, if X is annual precipitation at a specified location, then the probability distribution of X specifies the chance that the observed annual precipitation in a given year will lie in a defined range, such as less than 30 in, or 30 in–40 in, and so on.

A set of observations $x_1, x_2, \ldots, x_n$ of the random variable is called a *sample*. It is assumed that samples are drawn from a hypothetical infinite *population* possessing constant statistical properties, while the properties of a sample may vary from one sample to another. The set of all possible samples that could be drawn from the population is called the *sample space*, and an *event* is a subset of the sample space (Fig. 11.1.1). For example, the sample space for annual precipitation is theoretically the range from zero to positive infinity (though the practical lower and upper limits are closer than this), and an event A might be the occurrence of annual precipitation less than some amount, such as 30 in.

The *probability* of an event, $P(A)$, is the chance that it will occur when an observation of the random variable is made. Probabilities of events can be estimated. If a sample of n observations has n_A values in the range of event A, then the *relative frequency* of A is n_A/n. As the sample size is increased, the relative frequency becomes a progressively better estimate of the probability of the event, that is,

$$P(A) = \lim_{n \to \infty} \frac{n_A}{n} \tag{11.1.1}$$

Such probabilities are called *objective* or *posterior* probabilities because they depend completely on observations of the random variable. People are accustomed to estimating the chance that a future event will occur based on their judgment and experience. Such estimates are called *subjective* or *prior* probabilities.

The probabilities of events obey certain principles:

1. *Total probability*. If the sample space Ω is completely divided into m nonoverlapping areas or events $A_1, A_2, \ldots, A_m$, then

$$P(A_1) + P(A_2) + \ldots + P(A_m) = P(\Omega) = 1 \tag{11.1.2}$$

2. *Complementarity*. It follows that if $\overline{A}$ is the *complement* of A, that is, $\overline{A} = \Omega - A$, then

$$P(\overline{A}) = 1 - P(A) \tag{11.1.3}$$

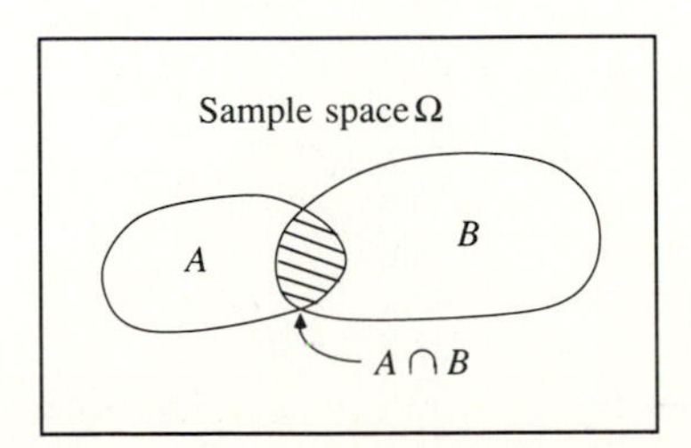

FIGURE 11.1.1
Events A and B are subsets of the sample space Ω.

3. *Conditional probability*. Suppose there are two events A and B as shown in Fig. 11.1.1. Event A might be the event that this year's precipitation is less than 40 in, while B might be the event that next year's precipitation will be less than 40 in. Their overlap is $A \cap B$, the event that A and B both occur, two successive years with annual precipitation less than 40 in/year. If $P(B|A)$ is the *conditional probability* that B will occur given that A has already occurred, then the *joint probability* that A and B will both occur, $P(A \cap B)$, is the product of $P(B|A)$ and the probability that A will occur, that is, $P(A \cap B) = P(B|A)P(A)$, or

$$P(B|A) = \frac{P(A \cap B)}{P(A)} \tag{11.1.4}$$

If the occurrence of B does not depend on the occurrence of A, the events are said to be *independent*, and $P(B|A) = P(B)$. For independent events, from (11.1.4),

$$P(A \cap B) = P(A)P(B) \tag{11.1.5}$$

If, for the example cited earlier, the precipitation events are independent from year to year, then the probability that precipitation is less than 40 in in two successive years is simply the square of the probability that annual precipitation in any one year will be less than 40 in.

The notion of independent events or observations is critical to the correct statistical interpretation of hydrologic data sequences, because if the data are independent they can be analyzed without regard to their order of occurrence. If successive observations are correlated (not independent), the statistical methods required are more complicated because the joint probability $P(A \cap B)$ of successive events is not equal to $P(A)P(B)$.

Example 11.1.1. The values of annual precipitation in College Station, Texas, from 1911 to 1979 are shown in Table 11.1.1 and plotted as a time series in Fig. 11.1.2(a). What is the probability that the annual precipitation R in any year will be less than 35 in? Greater than 45 in? Between 35 and 45 in?

TABLE 11.1.1
Annual Precipitation in College Station, Texas, 1911–1979 (in)

Year	1910	1920	1930	1940	1950	1960	1970
0		48.7	44.8	49.3	31.2	46.0	33.9
1	39.9	44.1	34.0	44.2	27.0	44.3	31.7
2	31.0	42.8	45.6	41.7	37.0	37.8	31.5
3	42.3	48.4	37.3	30.8	46.8	29.6	59.6
4	42.1	34.2	43.7	53.6	26.9	35.1	50.5
5	41.1	32.4	41.8	34.5	25.4	49.7	38.6
6	28.7	46.4	41.1	50.3	23.0	36.6	43.4
7	16.8	38.9	31.2	43.8	56.5	32.5	28.7
8	34.1	37.3	35.2	21.6	43.4	61.7	32.0
9	56.4	50.6	35.1	47.1	41.3	47.4	51.8

Solution. There are $n = 79 - 11 + 1 = 69$ data. Let A be the event $R < 35.0$ in, B the event $R > 45.0$ in. The numbers of values in Table 11.1.1 falling in these ranges are $n_A = 23$ and $n_B = 19$, so $P(A) \approx 23/69 = 0.333$ and $P(B) \approx 19/69 = 0.275$. From Eq. (11.1.3), the probability that the annual precipitation is between 35 and 45 in can now be calculated

$$\begin{aligned} P(35.0 \leq R \leq 45.0 \text{ in}) &= 1 - P(R < 35.0) - P(R > 45.0) \\ &= 1 - 0.333 - 0.275 \\ &= 0.392 \end{aligned}$$

Example 11.1.2. Assuming that annual precipitation in College Station is an independent process, calculate the probability that there will be two successive years of precipitation less than 35.0 in. Compare this estimated probability with the relative frequency of this event in the data set from 1911 to 1979 (Table 11.1.1).

Solution. Let C be the event that $R < 35.0$ in for two successive years. From Example 11.1.1, $P(R < 35.0 \text{ in}) = 0.333$, and assuming independent annual precipitation,

$$\begin{aligned} P(C) &= [P(R < 35.0 \text{ in})]^2 \\ &= (0.333)^2 \\ &= 0.111 \end{aligned}$$

From the data set, there are 9 pairs of successive years of precipitation less than 35.0 in out of 68 possible such pairs, so from a direct count it would be estimated

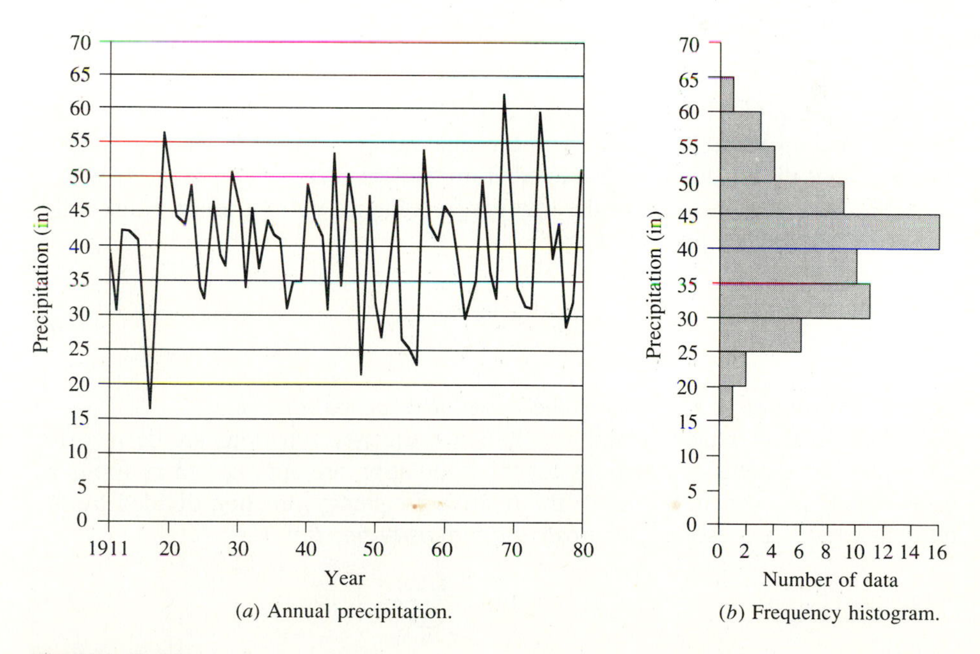

(*a*) Annual precipitation.

(*b*) Frequency histogram.

FIGURE 11.1.2
Annual precipitation in College Station, Texas, 1911–1979. The frequency histogram is formed by adding up the number of observed precipitation values falling in each interval.

that $P(C) \approx n_C/n = 9/68 = 0.132$, approximately the value found above by assuming independence.

Probabilities estimated from sample data, as in Examples 11.1.1 and 11.1.2, are approximate, because they depend on the specific values of the observations in a sample of limited size. An alternative approach is to fit a probability distribution function to the data and then to determine the probabilities of events from this distribution function.

11.2 FREQUENCY AND PROBABILITY FUNCTIONS

If the observations in a sample are identically distributed (each sample value drawn from the same probability distribution), they can be arranged to form a *frequency histogram*. First, the feasible range of the random variable is divided into discrete intervals, then the number of observations falling into each interval is counted, and finally the result is plotted as a bar graph, as shown in Fig. 11.1.2(*b*) for annual precipitation in College Station. The width Δx of the interval used in setting up the frequency histogram is chosen to be as small as possible while still having sufficient observations falling into each interval for the histogram to have a reasonably smooth variation over the range of the data.

If the number of observations n_i in interval i, covering the range $[x_i - \Delta x, x_i]$, is divided by the total number of observations n, the result is called the *relative frequency function* $f_s(x)$:

$$f_s(x_i) = \frac{n_i}{n} \tag{11.2.1}$$

which, as in Eq. (11.1.1), is an estimate of $P(x_i - \Delta x \leq X \leq x_i)$, the probability that the random variable X will lie in the interval $[x_i - \Delta x, x_i]$. The subscript s indicates that the function is calculated from sample data.

The sum of the values of the relative frequencies up to a given point is the *cumulative frequency function* $F_s(x)$:

$$F_s(x_i) = \sum_{j=1}^{i} f_s(x_j) \tag{11.2.2}$$

This is an estimate of $P(X \leq x_i)$, the *cumulative probability* of x_i.

The relative frequency and cumulative frequency functions are defined for a sample; corresponding functions for the population are approached as limits as $n \to \infty$ and $\Delta x \to 0$. In the limit, the relative frequency function divided by the intervai length Δx becomes the *probability density function* $f(x)$:

$$f(x) = \lim_{\substack{n \to \infty \\ \Delta x \to 0}} \frac{f_s(x)}{\Delta x} \tag{11.2.3}$$

The cumulative frequency function becomes the *probability distribution function* $F(x)$,

$$F(x) = \lim_{\substack{n \to \infty \\ \Delta x \to 0}} F_s(x) \tag{11.2.4}$$

whose derivative is the probability density function

$$f(x) = \frac{dF(x)}{dx} \tag{11.2.5}$$

For a given value of x, $F(x)$ is the cumulative probability $P(X \leq x)$, and it can be expressed as the integral of the probability density function over the range $X \leq x$:

$$P(X \leq x) = F(x) = \int_{-\infty}^{x} f(u)\, du \tag{11.2.6}$$

where u is a dummy variable of integration.

From the point of view of fitting sample data to a theoretical distribution, the four functions—relative frequency $f_s(x)$ and cumulative frequency $F_s(x)$ for the sample, and probability distribution $F(x)$ and probability density $f(x)$ for the population—may be arranged in a cycle as shown in Fig. 11.2.1. Beginning in the upper left panel, (*a*), the relative frequency function is computed from the sample data divided into intervals, and accumulated to form the cumulative frequency function shown at the lower left, (*b*). The probability distribution function, at the lower right, (*c*), is the theoretical limit of the cumulative frequency function as the sample size becomes infinitely large and the data interval infinitely small. The probability density function, at the upper right, (*d*), is the value of the slope of the distribution function for a specified value of x. The cycle may be closed by computing a theoretical value of the relative frequency function, called the incremental probability function:

$$\begin{aligned}
p(x_i) &= P(x_i - \Delta x \leq X \leq x_i) \\
&= \int_{x_i - \Delta x}^{x_i} f(x)\, dx \\
&= \int_{-\infty}^{x_i} f(x)\, dx - \int_{-\infty}^{x_i - \Delta x} f(x)\, dx \\
&= F(x_i) - F(x_i - \Delta x) \\
&= F(x_i) - F(x_{i-1})
\end{aligned} \tag{11.2.7}$$

The match between $p(x_i)$ and the observed relative frequency function $f_s(x_i)$ for each x_i can be used as a measure of the degree of fit of the distribution to the data.

The relative frequency, cumulative frequency, and probability distribution functions are all dimensionless functions varying over the range [0,1]. However, since $dF(x)$ is dimensionless and dx has the dimensions of X, the probability density function $f(x) = dF(x)/dx$ has dimensions $[X]^{-1}$ and varies over the range $[0, \infty]$. The relationship $dF(x) = f(x)\, dx$ can be described by saying that $f(x)$

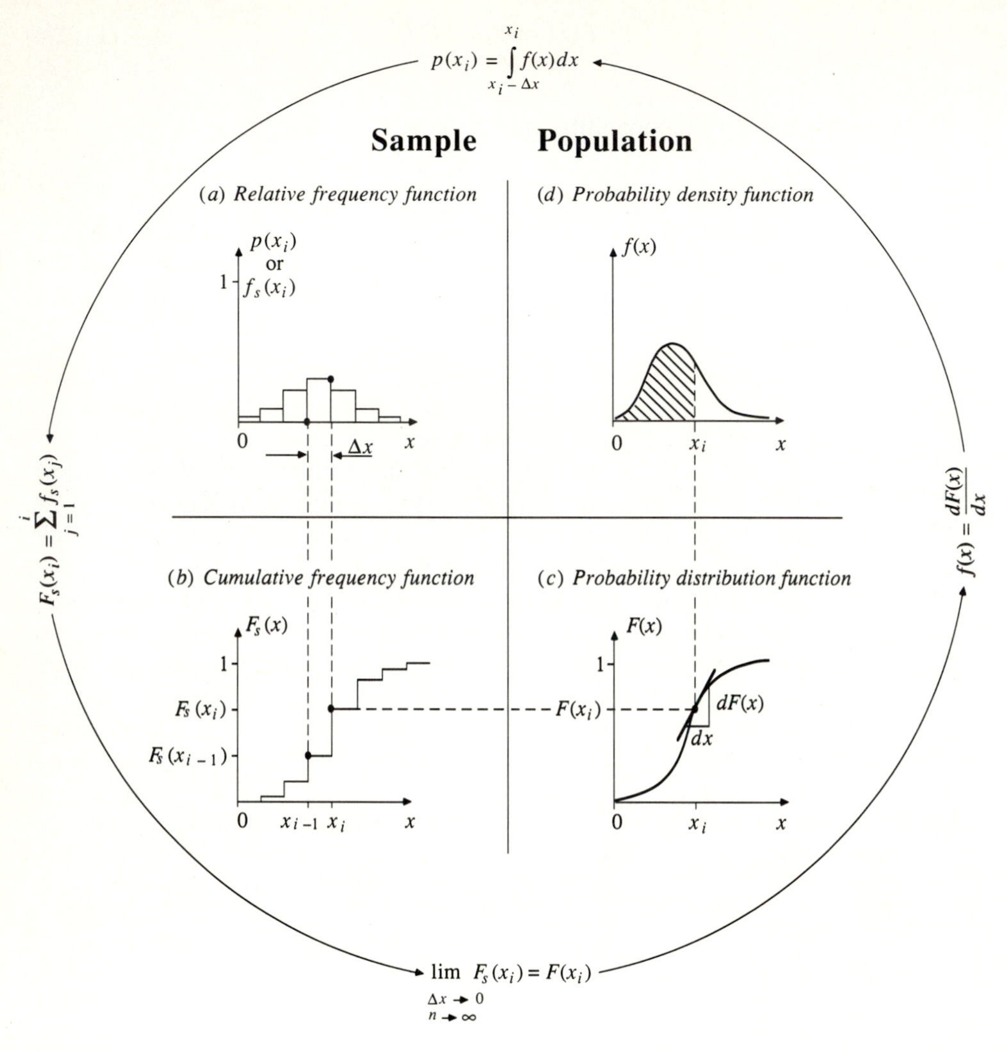

FIGURE 11.2.1
Frequency functions from sample data and probability functions from the population.

represents the "density" or "concentration" of probability in the interval $[x, x + dx]$.

One of the best-known probability density functions is that forming the familiar bell-shaped curve for the normal distribution:

$$f(x) = \frac{1}{\sqrt{2\pi}\sigma} \exp\left[-\frac{(x-\mu)^2}{2\sigma^2}\right] \tag{11.2.8}$$

where μ and σ are parameters. This function can be simplified by defining the *standard normal variable* z as

$$z = \frac{x - \mu}{\sigma} \tag{11.2.9}$$

The corresponding *standard normal distribution* has probability density function

$$f(z) = \frac{1}{\sqrt{2\pi}} e^{-z^2/2} \qquad -\infty \leq z \leq \infty \tag{11.2.10}$$

which depends only on the value of z and is plotted in Fig. 11.2.2. The standard normal probability distribution function

$$F(z) = \int_{-\infty}^{z} \frac{1}{\sqrt{2\pi}} e^{-u^2/2}\, du \tag{11.2.11}$$

where u is a dummy variable of integration, has no analytical form. Its values are tabulated in Table 11.2.1, and these values may be approximated by the following polynomial (Abramowitz and Stegun, 1965):

$$B = \frac{1}{2}[1 + 0.196854|z| + 0.115194|z|^2 + 0.000344|z|^3 + 0.019527|z|^4]^{-4} \tag{11.2.12a}$$

where $|z|$ is the absolute value of z and the standard normal distribution has

$$F(z) = B \qquad \text{for } z < 0 \tag{11.2.12b}$$

$$= 1 - B \quad \text{for } z \geq 0 \tag{11.2.12c}$$

The error in $F(z)$ as evaluated by this formula is less than 0.00025.

Example 11.2.1. What is the probability that the standard normal random variable z will be less than -2? Less than 1? What is $P(-2 < z < 1)$?

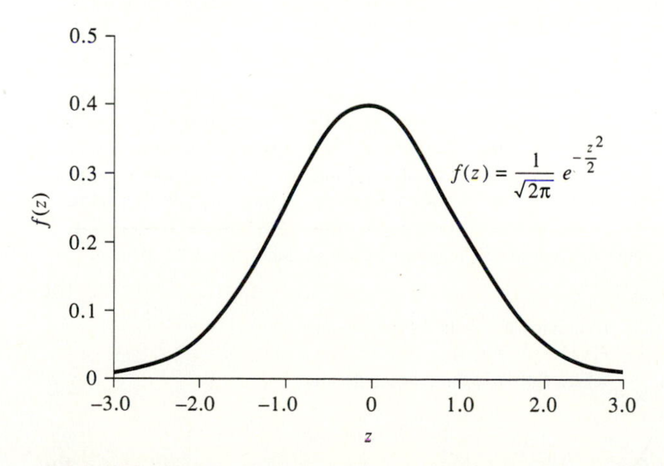

FIGURE 11.2.2
The probability density function for the standard normal distribution ($\mu = 0$, $\sigma = 1$).

TABLE 11.2.1
Cumulative probability of the standard normal distribution

z	.00	.01	.02	.03	.04	.05	.06	.07	.08	.09
0.0	0.5000	0.5040	0.5080	0.5120	0.5160	0.5199	0.5239	0.5279	0.5319	0.5359
0.1	0.5398	0.5438	0.5478	0.5517	0.5557	0.5596	0.5636	0.5675	0.5714	0.5753
0.2	0.5793	0.5832	0.5871	0.5910	0.5948	0.5987	0.6026	0.6064	0.6103	0.6141
0.3	0.6179	0.6217	0.6255	0.6293	0.6331	0.6368	0.6406	0.6443	0.6480	0.6517
0.4	0.6554	0.6591	0.6628	0.6664	0.6700	0.6736	0.6772	0.6808	0.6844	0.6879
0.5	0.6915	0.6950	0.6985	0.7019	0.7054	0.7088	0.7123	0.7157	0.7190	0.7224
0.6	0.7257	0.7291	0.7324	0.7357	0.7389	0.7422	0.7454	0.7486	0.7517	0.7549
0.7	0.7580	0.7611	0.7642	0.7673	0.7704	0.7734	0.7764	0.7794	0.7823	0.7852
0.8	0.7881	0.7910	0.7939	0.7967	0.7995	0.8023	0.8051	0.8078	0.8106	0.8133
0.9	0.8159	0.8186	0.8212	0.8238	0.8264	0.8289	0.8315	0.8340	0.8365	0.8389
1.0	0.8413	0.8438	0.8461	0.8485	0.8508	0.8531	0.8554	0.8577	0.8599	0.8621
1.1	0.8643	0.8665	0.8686	0.8708	0.8729	0.8749	0.8770	0.8790	0.8810	0.8830
1.2	0.8849	0.8869	0.8888	0.8907	0.8925	0.8944	0.8962	0.8980	0.8997	0.9015
1.3	0.9032	0.9049	0.9066	0.9082	0.9099	0.9115	0.9131	0.9147	0.9162	0.9177
1.4	0.9192	0.9207	0.9222	0.9236	0.9251	0.9265	0.9279	0.9292	0.9306	0.9319
1.5	0.9332	0.9345	0.9357	0.9370	0.9382	0.9394	0.9406	0.9418	0.9429	0.9441
1.6	0.9452	0.9463	0.9474	0.9484	0.9495	0.9505	0.9515	0.9525	0.9535	0.9545
1.7	0.9554	0.9564	0.9573	0.9582	0.9591	0.9599	0.9608	0.9616	0.9625	0.9633
1.8	0.9641	0.9649	0.9656	0.9664	0.9671	0.9678	0.9686	0.9693	0.9699	0.9706
1.9	0.9713	0.9719	0.9726	0.9732	0.9738	0.9744	0.9750	0.9756	0.9761	0.9767
2.0	0.9772	0.9778	0.9783	0.9788	0.9793	0.9798	0.9803	0.9808	0.9812	0.9817
2.1	0.9821	0.9826	0.9830	0.9834	0.9838	0.9842	0.9846	0.9850	0.9854	0.9857
2.2	0.9861	0.9864	0.9868	0.9871	0.9875	0.9878	0.9881	0.9884	0.9887	0.9890
2.3	0.9893	0.9896	0.9898	0.9901	0.9904	0.9906	0.9909	0.9911	0.9913	0.9916
2.4	0.9918	0.9920	0.9922	0.9925	0.9927	0.9929	0.9931	0.9932	0.9934	0.9936
2.5	0.9938	0.9940	0.9941	0.9943	0.9945	0.9946	0.9948	0.9949	0.9951	0.9952
2.6	0.9953	0.9955	0.9956	0.9957	0.9959	0.9960	0.9961	0.9962	0.9963	0.9964
2.7	0.9965	0.9966	0.9967	0.9968	0.9969	0.9970	0.9971	0.9972	0.9973	0.9974
2.8	0.9974	0.9975	0.9976	0.9977	0.9977	0.9978	0.9979	0.9979	0.9980	0.9981
2.9	0.9981	0.9982	0.9982	0.9983	0.9984	0.9984	0.9985	0.9985	0.9986	0.9986
3.0	0.9987	0.9987	0.9987	0.9988	0.9988	0.9989	0.9989	0.9989	0.9990	0.9990
3.1	0.9990	0.9991	0.9991	0.9991	0.9992	0.9992	0.9992	0.9992	0.9993	0.9993
3.2	0.9993	0.9993	0.9994	0.9994	0.9994	0.9994	0.9994	0.9995	0.9995	0.9995
3.3	0.9995	0.9995	0.9995	0.9996	0.9996	0.9996	0.9996	0.9996	0.9996	0.9997
3.4	0.9997	0.9997	0.9997	0.9997	0.9997	0.9997	0.9997	0.9997	0.9997	0.9998

Source: Grant, E. L., and R. S. Leavenworth, *Statistical Quality and Control*, Table A, p.643, McGraw-Hill, New York, 1972. Used with permission.

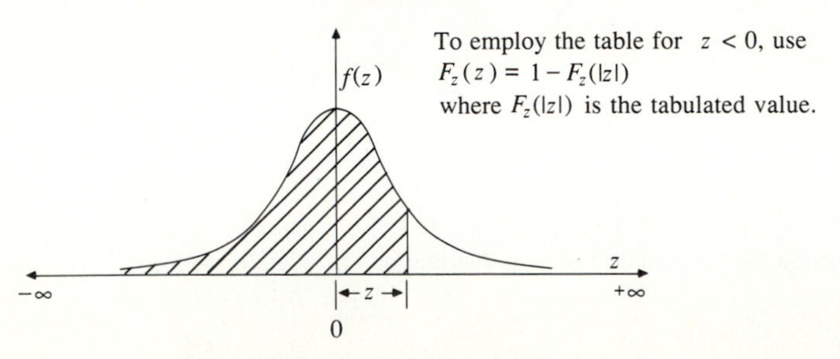

To employ the table for $z < 0$, use
$F_z(z) = 1 - F_z(|z|)$
where $F_z(|z|)$ is the tabulated value.

Solution. $P(Z \leq -2) = F(-2)$, and from Eq. (11.2.12*a*) with $|z| = |-2| = 2$,

$$B = \frac{1}{2}[1 + 0.196854 \times 2 + 0.115194 \times (2)^2 + 0.000344 \times (2)^3 + 0.019527 \times (2)^4]^{-4} = 0.023$$

From (11.2.12*b*), $F(-2) = B = 0.023$.

$P(Z \leq 1) = F(1)$, and from (11.2.12*a*)

$$B = \frac{1}{2}[1 + 0.196854 \times 1 + 0.115194 \times (1)^2 + 0.000344 \times (1)^3 + 0.019527 \times (1)^4]^{-4} = 0.159$$

From (11.2.12*c*), $F(1) = 1 - B = 1 - 0.159 = 0.841$.

Finally,

$$P(-2 < Z < 1) = F(1) - F(-2) = 0.841 - 0.023 = 0.818.$$

11.3 STATISTICAL PARAMETERS

The objective of statistics is to extract the essential information from a set of data, reducing a large set of numbers to a small set of numbers. *Statistics* are numbers calculated from a sample which summarize its important characteristics. Statistical *parameters* are characteristics of a population, such as μ and σ in Eq. (11.2.8).

A statistical parameter is the *expected value E* of some function of a random variable. A simple parameter is the *mean* μ, the expected value of the random variable itself. For a random variable X, the mean is $E(X)$, calculated as the product of x and the corresponding probability density $f(x)$, integrated over the feasible range of the random variable:

$$E(X) = \mu = \int_{-\infty}^{\infty} x f(x)\, dx \tag{11.3.1}$$

$E(X)$ is the first moment about the origin of the random variable, a measure of the midpoint or "central tendency" of the distribution.

The sample estimate of the mean is the average $\bar{x}$ of the sample data:

$$\bar{x} = \frac{1}{n}\sum_{i=1}^{n} x_i \tag{11.3.2}$$

Table 11.3.1 summarizes formulas for some population parameters and their sample statistics.

TABLE 11.3.1
Population parameters and sample statistics

Population parameter	Sample statistic
1. *Midpoint*	
Arithmetic mean	
$\mu = E(X) = \int_{-\infty}^{\infty} x f(x)\, dx$	$\bar{x} = \frac{1}{n}\sum_{i=1}^{n} x_i$
Median	
x such that $F(x) = 0.5$	50th-percentile value of data
Geometric mean	
antilog $[E(\log x)]$	$\left(\prod_{i=1}^{n} x_i\right)^{1/n}$
2. *Variability*	
Variance	
$\sigma^2 = E[(x-\mu)^2]$	$s^2 = \frac{1}{n-1}\sum_{i=1}^{n}(x_i - \bar{x})^2$
Standard deviation	
$\sigma = \{E[(x-\mu)^2]\}^{1/2}$	$s = \left[\frac{1}{n-1}\sum_{i=1}^{n}(x_i - \bar{x})^2\right]^{1/2}$
Coefficient of variation	
$CV = \frac{\sigma}{\mu}$	$CV = \frac{s}{\bar{x}}$
3. *Symmetry*	
Coefficient of skewness	
$\gamma = \frac{E[(x-\mu)^3]}{\sigma^3}$	$C_s = \frac{n\sum_{i=1}^{n}(x_i - \bar{x})^3}{(n-1)(n-2)s^3}$

The *variability* of data is measured by the *variance* σ^2, which is the second moment about the mean:

$$E[(x-\mu)^2] = \sigma^2 = \int_{-\infty}^{\infty} (x-\mu)^2 f(x)\, dx \tag{11.3.3}$$

The sample estimate of the variance is given by

$$s^2 = \frac{1}{n-1} \sum_{i=1}^{n} (x_i - \bar{x})^2 \qquad (11.3.4)$$

in which the divisor is $n - 1$ rather than n to ensure that the sample statistic is *unbiased*, that is, not having a tendency, on average, to be higher or lower than the true value. The variance has dimensions $[X]^2$. The *standard deviation* σ is a measure of variability having the same dimensions as X. The quantity σ is the square root of the variance, and is estimated by s. The significance of the standard deviation is illustrated in Fig. 11.3.1(*a*); the larger the standard deviation, the larger is the spread of the data. The *coefficient of variation* $CV = \sigma/\mu$, estimated by $s/\bar{x}$, is a dimensionless measure of variability.

The *symmetry* of a distribution about the mean is measured by the *skewness* which is the third moment about the mean:

$$E[(x-\mu)^3] = \int_{-\infty}^{\infty} (x-\mu)^3 f(x)\, dx \qquad (11.3.5)$$

The skewness is normally made dimensionless by dividing (11.3.5) by σ^3 to give the *coefficient of skewness* γ:

$$\gamma = \frac{1}{\sigma^3} E[(x-\mu)^3] \qquad (11.3.6)$$

A sample estimate for γ is given by:

$$C_s = \frac{n \sum_{i=1}^{n} (x_i - \bar{x})^3}{(n-1)(n-2)s^3} \qquad (11.3.7)$$

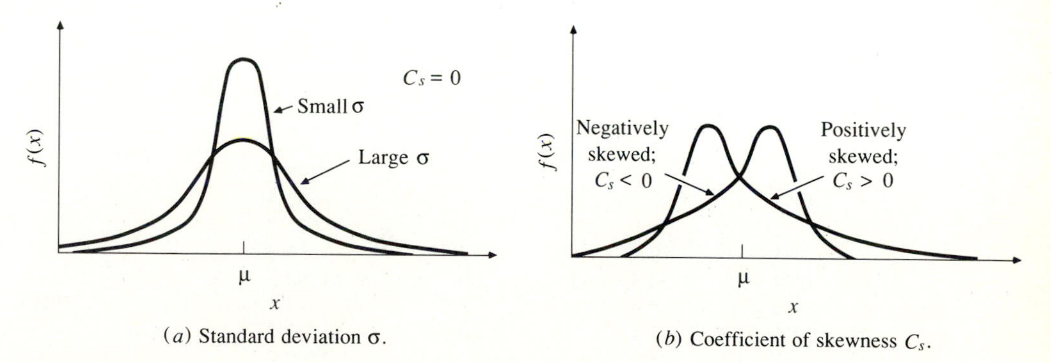

(*a*) Standard deviation σ.

(*b*) Coefficient of skewness C_s.

FIGURE 11.3.1
The effect on the probability density function of changes in the standard deviation and coefficient of skewness.

or

$$C_s = \frac{n^2\left(\sum_{i=1}^{n} x^3\right) - 3n\left(\sum_{i=1}^{n} x\right)\left(\sum_{i=1}^{n} x^2\right) + 2\left(\sum_{i=1}^{n} x^3\right)}{n(n-1)(n-2)s^3} \tag{11.3.8}$$

As shown in Fig. 11.3.1(*b*), for positive skewness ($\gamma > 0$), the data are skewed to the right, with only a small number of very large values; for negative skewness ($\gamma < 0$), the data are skewed to the left. If the data have a pronounced skewness, the small number of extreme values exert a significant effect on the arithmetic mean calculated by Eq. (11.3.2), and alternative measures of central tendency are appropriate, such as the *median* or *geometric mean* as listed in Table 11.3.1.

Example 11.3.1. Calculate the sample mean, sample standard deviation, and sample coefficient of skewness of the data for annual precipitation in College Station, Texas, from 1970 to 1979. The data are given in Table 11.1.1.

Solution. The values of annual precipitation from 1970 to 1979 are copied in column 2 of Table 11.3.2. Using Eq. (11.3.2) the mean is

$$\bar{x} = \frac{1}{n}\sum_{i=1}^{n} x_i$$

$$= \frac{401.7}{10}$$

$$= 40.17 \text{ in}$$

The squares of the deviations from the mean are shown in column 3 of the table,

TABLE 11.3.2
Calculation of sample statistics for College Station annual precipitation, 1970–1979 (in) (Example 11.3.1).

Column:	1 Year	2 Precipitation x	3 $(x-\bar{x})^2$	4 $(x-\bar{x})^3$
	1970	33.9	39.3	–246.5
	1971	31.7	71.7	–607.6
	1972	31.5	75.2	–651.7
	1973	59.6	377.5	7335.3
	1974	50.5	106.7	1102.3
	1975	38.6	2.5	–3.9
	1976	43.4	10.4	33.7
	1977	28.7	131.6	–1509.0
	1978	32.0	66.7	–545.3
	1979	51.8	135.3	1573.0
	Total	401.7	1016.9	6480.3

totaling 1016.9 in^2. From (11.3.4)

$$s^2 = \frac{1}{n-1}\sum_{i=1}^{n}(x_i - \bar{x})^2$$

$$= \frac{1016.9}{9}$$

$$= 113.0 \text{ in}^2$$

The standard deviation is

$$s = (113.0)^{1/2}$$

$$= 10.63 \text{ in}$$

The cubes of the deviation from the mean are shown in column 4 of Table 11.3.2, totaling 6480.3. From (11.3.7)

$$C_s = \frac{n\sum_{i=1}^{n}(x_i - \bar{x})^3}{(n-1)(n-2)s^3}$$

$$= \frac{10 \times 6480.3}{9 \times 8 \times (10.63)^3}$$

$$= 0.749$$

11.4 FITTING A PROBABILITY DISTRIBUTION

A probability distribution is a function representing the probability of occurrence of a random variable. By fitting a distribution to a set of hydrologic data, a great deal of the probabilistic information in the sample can be compactly summarized in the function and its associated parameters. Fitting distributions can be accomplished by the *method of moments* or the *method of maximum likelihood*.

Method of Moments

The method of moments was first developed by Karl Pearson in 1902. He considered that good estimates of the parameters of a probability distribution are those for which moments of the probability density function about the origin are equal to the corresponding moments of the sample data. As shown in Fig. 11.4.1, if the data values are each assigned a hypothetical "mass" equal to their relative frequency of occurrence ($1/n$) and it is imagined that this system of masses is rotated about the origin $x = 0$, then the first moment of each observation x_i about the origin is the product of its moment arm x_i and its mass $1/n$, and the sum of these moments over all the data is

$$\sum_{i=1}^{n}\frac{x_i}{n} = \frac{1}{n}\sum_{i=1}^{n}x_i = \bar{x}$$

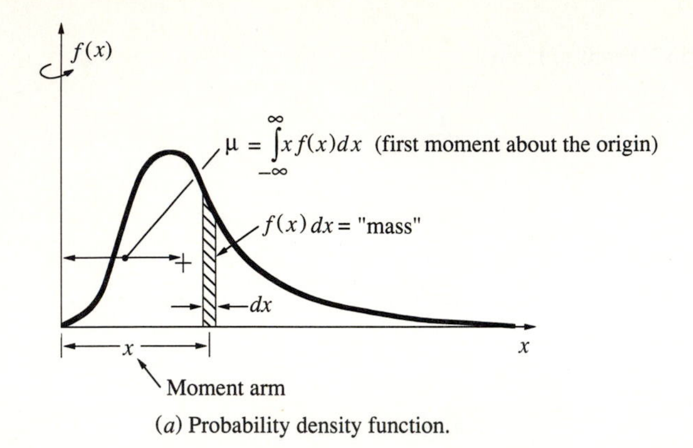

(*a*) Probability density function.

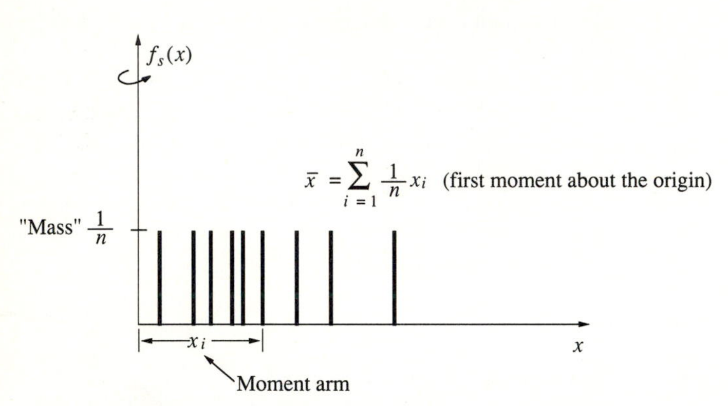

(*b*) Sample data.

FIGURE 11.4.1
The method of moments selects values for the parameters of the probability density function so that its moments are equal to those of the sample data.

the sample mean. This is equivalent to the centroid of a body. The corresponding centroid of the probability density function is

$$\mu = \int_{-\infty}^{\infty} x f(x)\, dx \tag{11.4.1}$$

Likewise, the second and third moments of the probability distribution can be set equal to their sample values to determine the values of parameters of the probability distribution. Pearson originally considered only moments about the origin, but later it became customary to use the variance as the second *central moment*, $\sigma^2 = E[(x-\mu)^2]$, and the coefficient of skewness as the standardized third central moment, $\gamma = E[(x-\mu)^3]/\sigma^3$, to determine second and third parameters of the distribution if required.

Example 11.4.1. The *exponential* distribution can be used to describe various kinds of hydrologic data, such as the interarrival times of rainfall events. Its probability density function is $f(x) = \lambda e^{-\lambda x}$ for $x > 0$. Determine the relationship between the parameter λ and the first moment about the origin, μ.

Solution. Using Eq. (11.4.1),

$$\mu = E(x) = \int_{-\infty}^{\infty} x f(x)\, dx$$

$$= \int_{0}^{\infty} x \lambda e^{-\lambda x}\, dx$$

which may be integrated by parts to yield

$$\mu = \frac{1}{\lambda}$$

In this case $\lambda = 1/\mu$, and the sample estimate for λ is $1/\bar{x}$.

As a matter of interest, it can be seen that the exponential probability density function $f(x) = \lambda e^{-\lambda x}$ and the impulse response function for a linear reservoir (see Ex. 7.2.1) $u(l) = (1/k)e^{-l/k}$ are identical if $x = l$ and $\lambda = 1/k$. In this sense, the exponential distribution can be thought of as describing the probability of the "holding time" of water in a linear reservoir.

Method of Maximum Likelihood

The method of maximum likelihood was developed by R. A. Fisher (1922). He reasoned that the best value of a parameter of a probability distribution should be that value which maximizes the likelihood or joint probability of occurrence of the observed sample. Suppose that the sample space is divided into intervals of length dx and that a sample of independent and identically distributed observations x_1, x_2, . . . , x_n is taken. The value of the probability density for $X = x_i$ is $f(x_i)$, and the probability that the random variable will occur in the interval including x_i is $f(x_i)\, dx$. Since the observations are independent, their joint probability of occurrence is given from Eq. (11.1.5) as the product $f(x_1)\, dx\, f(x_2)\, dx \ldots f(x_n)\, dx = [\prod_{i=1}^{n} f(x_i)]\, dx^n$, and since the interval size dx is fixed, maximizing the joint probability of the observed sample is equivalent to maximizing the *likelihood function*

$$L = \prod_{i=1}^{n} f(x_i) \tag{11.4.2}$$

Because many probability density functions are exponential, it is sometimes more convenient to work with the log-likelihood function

$$\ln L = \sum_{i=1}^{n} \ln\ [f(x_i)] \tag{11.4.3}$$

Example 11.4.2. The following data are the observed times between rainfall events at a given location. Assuming that the interarrival time of rainfall events follows an exponential distribution, determine the parameter λ for this process by the method of maximum likelihood. The times between rainfalls (days) are: 2.40, 4.25, 0.77, 13.32, 3.55, and 1.37.

Solution. For a given value x_i, the exponential probability density is

$$f(x_i) = \lambda e^{-\lambda x_i}$$

so, from Eq. (11.4.3), the log-likelihood function is

$$\begin{aligned} \ln L &= \sum_{i=1}^{n} \ln [f(x_i)] \\ &= \sum_{i=1}^{n} \ln (\lambda e^{-\lambda x_i}) \\ &= \sum_{i=1}^{n} (\ln \lambda - \lambda x_i) \\ &= n \ln \lambda - \lambda \sum_{i=1}^{n} x_i \end{aligned}$$

The maximum value of ln L occurs when $\partial(\ln L)/\partial\lambda = 0$; that is, when

$$\frac{\partial(\ln L)}{\partial\lambda} = \frac{n}{\lambda} - \sum_{i=1}^{n} x_i = 0$$

so

$$\frac{1}{\lambda} = \frac{1}{n} \sum_{i=1}^{n} x_i$$

$$\lambda = \frac{1}{\bar{x}}$$

This is the same sample estimator for λ as was produced by the method of moments. In this case, $\bar{x} = (2.40 + 4.25 + 0.77 + 13.22 + 3.55 + 1.37)/6 = 25.56/6 = 4.28$ days, so $\lambda = 1/4.28 = 0.234$ day^{-1}. Note that $\partial^2(\ln L)/\partial\lambda^2 = -n\lambda^2$, which is negative as required for a maximum.

The value of the log-likelihood function can be calculated for any value of λ. For example, for $\lambda = 0.234$ day^{-1}, the value of the log-likelihood function is

$$\begin{aligned} \ln L &= n \ln \lambda - \lambda \sum_{i=1}^{n} x_i \\ &= 6 \ln (0.234) - 0.234 \times 25.56 \\ &= -14.70 \end{aligned}$$

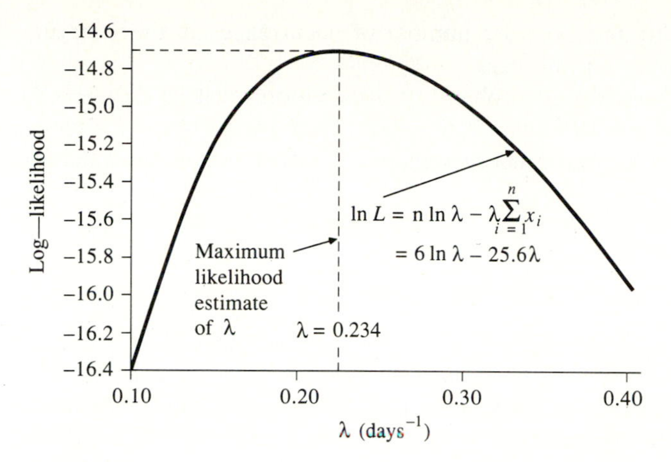

FIGURE 11.4.2
The log-likelihood function for an exponential distribution (Example 11.4.2).

Figure 11.4.2 shows the variation of the log-likelihood function with λ, with the maximum value at $\lambda = 0.234$ day^{-1} as was determined analytically.

The method of maximum likelihood is the most theoretically correct method of fitting probability distributions to data in the sense that it produces the most *efficient* parameter estimates—those which estimate the population parameters with the least average error. But, for some probability distributions, there is no analytical solution for all the parameters in terms of sample statistics, and the log-likelihood function must then be numerically maximized, which may be quite difficult. In general, the method of moments is easier to apply than the method of maximum likelihood and is more suitable for practical hydrologic analysis.

Testing the Goodness of Fit

The goodness of fit of a probability distribution can be tested by comparing the theoretical and sample values of the relative frequency or the cumulative frequency function. In the case of the relative frequency function, the χ^2 *test* is used. The sample value of the relative frequency of interval i is, from Eq. (11.2.1), $f_s(x_i) = n_i/n$; the theoretical value from (11.2.7) is $p(x_i) = F(x_i) - F(x_{i-1})$. The χ^2 *test statistic* χ_c^2 is given by

$$\chi_c^2 = \sum_{i=1}^{m} \frac{n[f_s(x_i) - p(x_i)]^2}{p(x_i)} \tag{11.4.4}$$

where m is the number of intervals. It may be noted that $nf_s(x_i) = n_i$, the observed number of occurrences in interval i, and $np(x_i)$ is the corresponding expected number of occurrences in interval i; so the calculation of Eq. (11.4.4) is a matter of squaring the difference between the observed and expected numbers of occurrences, dividing by the expected number of occurrences in the interval, and summing the result over all intervals.

To describe the χ^2 test, the χ^2 probability distribution must be defined. A χ^2 distribution with ν *degrees of freedom* is the distribution for the sum of squares of ν independent standard normal random variables z_i; this sum is the random variable

$$\chi_\nu^2 = \sum_{i=1}^{\nu} z_i^2 \tag{11.4.5}$$

The χ^2 distribution function is tabulated in many statistics texts (e.g., Haan, 1977). In the χ^2 test, $\nu = m - p - 1$, where m is the number of intervals as before, and p is the number of parameters used in fitting the proposed distribution. A *confidence level* is chosen for the test; it is often expressed as $1 - \alpha$, where α is termed the *significance level*. A typical value for the confidence level is 95 percent. The *null hypothesis* for the test is that the proposed probability distribution fits the data adequately. This hypothesis is rejected (i.e., the fit is deemed inadequate) if the value of χ_c^2 in (11.4.4) is larger than a limiting value, $\chi_{\nu,1-\alpha}^2$, determined from the χ^2 distribution with ν degrees of freedom as the value having cumulative probability $1 - \alpha$.

Example 11.4.3. Using the method of moments, fit the normal distribution to the annual precipitation at College Station, Texas, from 1911 to 1979 (Table 11.1.1). Plot the relative frequency and incremental probability functions, and the cumulative frequency and cumulative probability functions. Use the χ^2 test to determine whether the normal distribution adequately fits the data.

Solution. The range for precipitation R is divided into ten intervals. The first interval is $R \leq 20$ in, the last is $R > 60$ in, and the intermediate intervals each cover a range of 5 in. By scanning Table 11.1.1 the frequency histogram is compiled, as shown in column 2 of Table 11.4.1. The relative frequency function $f_s(x_i)$ (column 3) is calculated by Eq. (11.2.1) with $n = 69$. For example, for $i = 4$ (30–35 in), $n_i = 14$, and

$$\begin{aligned} f_s(x_4) &= \frac{n_4}{n} \\ &= \frac{14}{69} \\ &= 0.203 \end{aligned}$$

The cumulative frequency function (column 4) is found by summing up the relative frequencies as in Eq. (11.2.2). For $i = 4$

$$\begin{aligned} F_s(x_4) &= \sum_{j=1}^{4} f_s(x_j) \\ &= F_s(x_3) + f_s(x_4) \\ &= 0.130 + 0.203 \end{aligned}$$

TABLE 11.4.1
Fitting a normal distribution to annual precipitation at College Station, Texas, 1911–1979 (Example 11.4.3).

Column:	1	2	3	4	5	6	7	8
Interval i	**Range (in)**	n_i	$f_s(x_i)$	$F_s(x_i)$	z_i	$F(x_i)$	$p(x_i)$	χ_c^2
1	< 20	1	0.014	0.014	−2.157	0.015	0.015	0.004
2	20–25	2	0.029	0.043	−1.611	0.053	0.038	0.147
3	25–30	6	0.087	0.130	−1.065	0.144	0.090	0.008
4	30–35	14	0.203	0.333	−0.520	0.301	0.158	0.891
5	35–40	11	0.159	0.493	0.026	0.510	0.209	0.805
6	40–45	16	0.232	0.725	0.571	0.716	0.206	0.222
7	45–50	10	0.145	0.870	1.117	0.868	0.151	0.019
8	50–55	5	0.072	0.942	1.662	0.952	0.084	0.114
9	55–60	3	0.043	0.986	2.208	0.986	0.034	0.163
10	>60	1	0.014	1.000	2.753	1.000	0.014	0.004
Total		69	1.000				1.000	2.377

Mean 39.77
Standard deviation 9.17

$$=0.333$$

It may be noted that this is $P(X \leq 35.0 \text{ in})$ as used in Example 11.1.1.

To fit the normal distribution function, the sample statistics $\bar{x} = 39.77$ in and $s = 9.17$ in are calculated for the data from 1911 to 1979 in the manner shown in Example 11.3.1, and used as estimates for μ and σ. The standard normal variate z corresponding to the upper limit of each of the data intervals is calculated by (11.2.9) and shown in column 5 of the table. For example, for $i = 4$,

$$z = \frac{x - \mu}{\sigma}$$

$$= \frac{35.0 - 39.77}{9.17}$$

$$= -0.520$$

The corresponding value of the cumulative normal probability function is given by (11.2.12) or Table 11.2.1 as 0.301, as listed in column 6 of Table 11.4.1. The incremental probability function is computed by (11.2.7). For $i = 4$,

$$p(x_4) = P(30 \leq X \leq 35 \text{ in})$$

$$= F(35) - F(30)$$

$$= 0.301 - 0.144$$

$$= 0.158$$

and similarly computed values for the other intervals are shown in column 7.

The relative frequency functions $f_s(x_i)$ and $p(x_i)$ from Table 11.4.1 are plotted in Fig. 11.4.3(*a*), and the cumulative frequency and probability distribution functions $F_s(x_i)$ and $F(x)$ in Fig. 11.4.3(*b*). From the similarity of the two functions

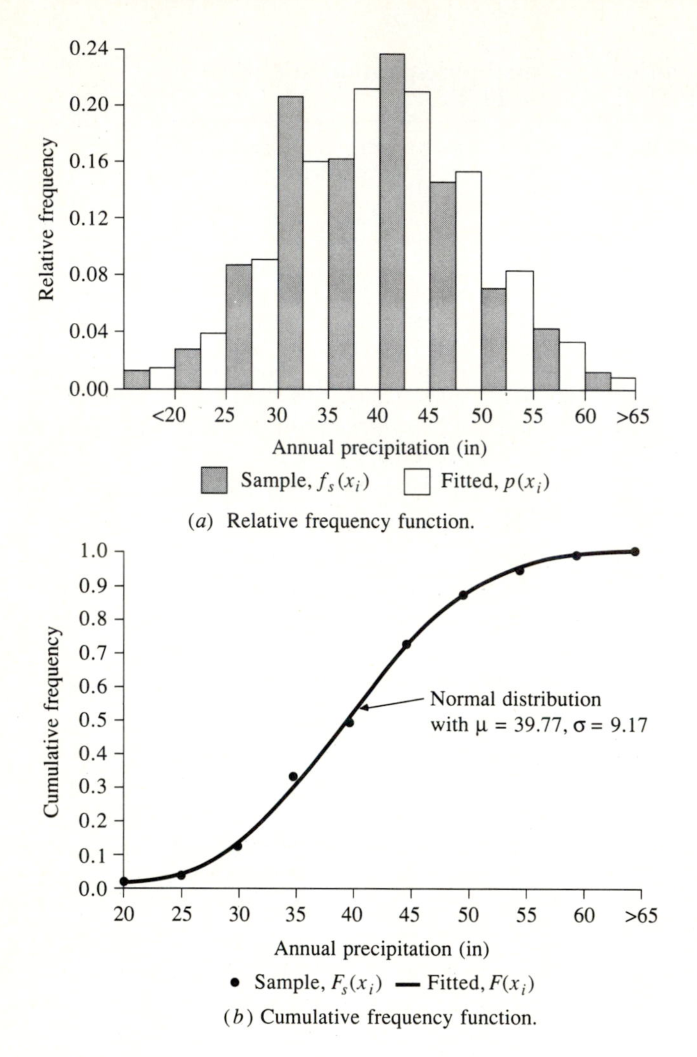

FIGURE 11.4.3
Frequency functions for a normal distribution fitted to annual precipitation in College Station, Texas (Example 11.4.3).

shown in each plot, it is apparent that the normal distribution fits these annual precipitation data very well.

To check the goodness of fit, the χ^2 test statistic is calculated by (11.4.4). For $i = 4$,

$$\frac{n[f_s(x_4) - p(x_4)]^2}{p(x_4)} = \frac{69 \times (0.20290 - 0.15777)^2}{0.15777}$$
$$= 0.891$$

as shown in column 8 of Table 11.4.1. The total of the values in column 8 is $\chi_c^2 =$ 2.377. The value of $\chi^2_{\nu,1-\alpha}$ for a cumulative probability of $1-\alpha = 0.95$ and degrees of freedom $\nu = m - p - 1 = 10 - 2 - 1 = 7$ is $\chi^2_{7,0.95} = 14.1$ (Abramowitz and Stegun, 1965). Since this value is greater than χ^2_c, the null hypothesis (the distribution fits the data) cannot be rejected at the 95 percent confidence level; the fit of the normal distribution to the College Station annual precipitation data is accepted. If the distribution had fitted poorly, the values of $f_s(x_i)$ and $p(x_i)$ would have been quite different from one another, resulting in a value of χ^2_c larger than 14.1, in which case the null hypothesis would have been rejected.

11.5 PROBABILITY DISTRIBUTIONS FOR HYDROLOGIC VARIABLES

In Sec. 11.4, the normal distribution was used to describe annual precipitation at College Station, Texas. Although this distribution fits this set of data particularly well, observations of other hydrologic variables follow different distributions. In this section, a selection of probability distributions commonly used for hydrologic variables is presented, and examples of the types of variables to which these distributions have been applied are given. Table 11.5.1 summarizes, for each distribution, the probability density function and the range of the variable, and gives equations for estimating the distribution's parameters from sample moments.

Normal Distribution

The normal distribution arises from the *central limit theorem*, which states that if a sequence of random variables X_i are independently and identically distributed with mean μ and variance σ^2, then the distribution of the sum of n such random variables, $Y = \sum_{i=1}^{n} X_i$, tends towards the normal distribution with mean $n\mu$ and variance $n\sigma^2$ as n becomes large. The important point is that this is true no matter what the probability distribution function of X is. So, for example, the probability distribution of the sample mean $\bar{x} = 1/n \sum_{i=1}^{n} x_i$ can be approximated as normal with mean μ and variance $(1/n)^2 n\sigma^2 = \sigma^2/n$ no matter what the distribution of x is. Hydrologic variables, such as annual precipitation, calculated as the sum of the effects of many independent events tend to follow the normal distribution. The main limitations of the normal distribution for describing hydrologic variables are that it varies over a continuous range $[-\infty, \infty]$, while most hydrologic variables are nonnegative, and that it is symmetric about the mean, while hydrologic data tend to be skewed.

Lognormal Distribution

If the random variable $Y = \log X$ is normally distributed, then X is said to be lognormally distributed. Chow (1954) reasoned that this distribution is applicable to hydrologic variables formed as the products of other variables since if $X = X_1 X_2 X_3 \ldots X_n$, then $Y = \log X = \sum_{i=1}^{n} \log X_i = \sum_{i=1}^{n} Y_i$, which tends to the normal distribution for large n provided that the X_i are independent and identically distributed. The lognormal distribution has been found to describe the distribution of hydraulic conductivity in a porous medium (Freeze, 1975),

TABLE 11.5.1
Probability distributions for fitting hydrologic data

Distribution	Probability density function	Range	Equations for parameters in terms of the sample moments
Normal	$f(x) = \frac{1}{\sigma\sqrt{2\pi}} \exp\left(-\frac{(x-\mu)^2}{2\sigma^2}\right)$	$-\infty \le x \le \infty$	$\mu = \bar{x},\ \sigma = s_x$
Lognormal	$f(x) = \frac{1}{x\sigma\sqrt{2\pi}} \exp\left(-\frac{(y-\mu_y)^2}{2\sigma_y^2}\right)$	$x > 0$	$\mu_y = \bar{y},\ \sigma_y = s_y$
	where $y = \log x$		
Exponential	$f(x) = \lambda e^{-\lambda x}$	$x \ge 0$	$\lambda = \frac{1}{\bar{x}}$
Gamma	$f(x) = \frac{\lambda^\beta x^{\beta-1} e^{-\lambda x}}{\Gamma(\beta)}$	$x \ge 0$	$\lambda = \frac{\bar{x}}{s_x^2}$
	where Γ = gamma function		$\beta = \frac{\bar{x}^2}{s_x^2} = \frac{1}{CV^2}$

TABLE 11.5.1 (*cont.*)

Probability distributions for fitting hydrologic data

Distribution	Probability density function	Range	Equations for parameters in terms of the sample moments
Pearson Type III (three parameter gamma)	$f(x) = \frac{\lambda^{\beta}(x-\epsilon)^{\beta-1}e^{-\lambda(x-\epsilon)}}{\Gamma(\beta)}$	$x \geq \epsilon$	$\lambda = \frac{s_x}{\sqrt{\beta}}$, $\beta = \left(\frac{2}{C_s}\right)^2$ $\epsilon = \bar{x} - s_x\sqrt{\beta}$
Log Pearson Type III	$f(x) = \frac{\lambda^{\beta}(y-\epsilon)^{\beta-1}e^{-\lambda(y-\epsilon)}}{x\Gamma(\beta)}$ where $y = \log x$	$\log x \geq \epsilon$	$\lambda = \frac{s_y}{\sqrt{\beta}}$, $\beta = \left[\frac{2}{C_s(y)}\right]^2$ $\epsilon = \bar{y} - s_y\sqrt{\beta}$ (assuming $C_s(y)$ is positive)
Extreme Value Type I	$f(x) = \frac{1}{\alpha}\exp\left[-\frac{x-u}{\alpha} - \exp\left(-\frac{x-u}{\alpha}\right)\right]$	$-\infty < x < \infty$	$\alpha = \frac{\sqrt{6}s_x}{\pi}$ $u = \bar{x} - 0.5772\alpha$

the distribution of raindrop sizes in a storm, and other hydrologic variables. The lognormal distribution has the advantages over the normal distribution that it is bounded ($X > 0$) and that the log transformation tends to reduce the positive skewness commonly found in hydrologic data, because taking logarithms reduces large numbers proportionately more than it does small numbers. Some limitations of the lognormal distribution are that it has only two parameters and that it requires the logarithms of the data to be symmetric about their mean.

Exponential Distribution

Some sequences of hydrologic events, such as the occurrence of precipitation, may be considered *Poisson processes*, in which events occur instantaneously and independently on a time horizon, or along a line. The time between such events, or *interarrival time*, is described by the exponential distribution whose parameter λ is the mean rate of occurrence of the events. The exponential distribution is used to describe the interarrival times of random shocks to hydrologic systems, such as slugs of polluted runoff entering streams as rainfall washes the pollutants off the land surface. The advantage of the exponential distribution is that it is easy to estimate λ from observed data and the exponential distribution lends itself well to theoretical studies, such as a probability model for the linear reservoir ($\lambda = 1/k$, where k is the storage constant in the linear reservoir). Its disadvantage is that it requires the occurrence of each event to be completely independent of its neighbors, which may not be a valid assumption for the process under study—for example, the arrival of a front may generate many showers of rain—and this has led investigators to study various forms of *compound Poisson processes*, in which λ is considered a random variable instead of a constant (Kavvas and Delleur, 1981; Waymire and Gupta, 1981).

Gamma Distribution

The time taken for a number β of events to occur in a Poisson process is described by the gamma distribution, which is the distribution of a sum of β independent and identical exponentially distributed random variables. The gamma distribution has a smoothly varying form like the typical probability density function illustrated in Fig. 11.2.1 and is useful for describing skewed hydrologic variables without the need for log transformation. It has been applied to describe the distribution of depth of precipitation in storms, for example. The gamma distribution involves the *gamma function* $\Gamma(\beta)$, which is given by $\Gamma(\beta) = (\beta-1)! = (\beta-1)(\beta-2)\ldots3\cdot2\cdot1$ for positive integer β, and in general by

$$\Gamma(\beta) = \int_0^\infty u^{\beta-1}e^{-u}\,du \tag{11.5.1}$$

(Abramowitz and Stegun, 1965). The two-parameter gamma distribution (parameters β and λ) has a lower bound at zero, which is a disadvantage for application to hydrologic variables that have a lower bound larger than zero.

Pearson Type III Distribution

The Pearson Type III distribution, also called the *three-parameter gamma distribution*, introduces a third parameter, the lower bound ϵ, so that by the method of moments, three sample moments (the mean, the standard deviation, and the coefficient of skewness) can be transformed into the three parameters λ, β, and ϵ of the probability distribution. This is a very flexible distribution, assuming a number of different shapes as λ, β, and ϵ vary (Bobee and Robitaille, 1977).

The Pearson system of distributions includes seven types; they are all solutions for $f(x)$ in an equation of the form

$$\frac{d[f(x)]}{dx} = \frac{f(x)(x-d)}{C_0 + C_1 x + C_2 x^2} \tag{11.5.2}$$

where d is the *mode* of the distribution (the value of x for which $f(x)$ is a maximum) and C_0, C_1, and C_2 are coefficients to be determined. When $C_2 = 0$, the solution of (11.5.2) is a Pearson Type III distribution, having a probability density function of the form shown in Table 11.5.1. For $C_1 = C_2 = 0$, a normal distribution is the solution of (11.5.2). Thus, the normal distribution is a special case of the Pearson Type III distribution, describing a nonskewed variable. The Pearson Type III distribution was first applied in hydrology by Foster (1924) to describe the probability distribution of annual maximum flood peaks. When the data are very positively skewed, a log transformation is used to reduce the skewness.

Log–Pearson Type III Distribution

If log X follows a Pearson Type III distribution, then X is said to follow a log–Pearson Type III distribution. This distribution is the standard distribution for frequency analysis of annual maximum floods in the United States (Benson, 1968), and its use is described in detail in Chap. 12. As a special case, when log X is symmetric about its mean, the log–Pearson Type III distribution reduces to the lognormal distribution.

The location of the bound ϵ in the log–Pearson Type III distribution depends on the skewness of the data. If the data are positively skewed, then $\log X \geq \epsilon$ and

TABLE 11.5.2
Shape and mode location of the log–Pearson Type III distribution as a function of its parameters

Shape parameter β	$\lambda < -\ln 10$	$-\ln 10 < \lambda < 0$	$\lambda > 0$
$0 < \beta < 1$	No mode J-shaped	Minimum mode U-shaped	No mode Reverse J-shaped
$\beta > 1$	Unimodal	No mode Reverse J-shaped	Unimodal

Source: Bobee, 1975.

ϵ is a lower bound, while if the data are negatively skewed, $\log X \leq \epsilon$ and ϵ is an upper bound. The log transformation reduces the skewness of the transformed data and may produce transformed data which are negatively skewed from original data which are positively skewed. In that case, the application of the log–Pearson Type III distribution would impose an artificial upper bound on the data. Depending on the values of the parameters, the log–Pearson Type III distribution can assume many different shapes, as shown in Table 11.5.2 (Bobee, 1975).

As described previously, the log–Pearson Type III distribution was developed as a method of fitting a curve to data. Its use is justified by the fact that it has been found to yield good results in many applications, particularly for flood peak data. The fit of the distribution to data can be checked using the χ^2 test, or by using probability plotting as described in Chap. 12.

Extreme Value Distribution

Extreme values are selected maximum or minimum values of sets of data. For example, the annual maximum discharge at a given location is the largest recorded discharge value during a year, and the annual maximum discharge values for each year of historical record make up a set of extreme values that can be analyzed statistically. Distributions of the extreme values selected from sets of samples of any probability distribution have been shown by Fisher and Tippett (1928) to converge to one of three forms of *extreme value distributions,* called Types I, II, and III, respectively, when the number of selected extreme values is large. The properties of the three limiting forms were further developed by Gumbel (1941) for the Extreme Value Type I (EVI) distribution, Frechet (1927) for the Extreme Value Type II (EVII), and Weibull (1939) for the Extreme Value Type III (EVIII).

The three limiting forms were shown by Jenkinson (1955) to be special cases of a single distribution called the *General Extreme Value* (GEV) distribution. The probability distribution function for the GEV is

$$F(x) = \exp\left[-\left(1 - k\,\frac{x-u}{\alpha}\right)^{1/k}\right] \tag{11.5.3}$$

where k, u, and α are parameters to be determined.

The three limiting cases are (1) for $k = 0$, the Extreme Value Type I distribution, for which the probability density function is given in Table 11.5.1, (2) for $k < 0$, the Extreme Value Type II distribution, for which (11.5.3) applies for $(u + \alpha/k) \leq x \leq \infty$, and (3) for $k > 0$, the Extreme Value Type III distribution, for which (11.5.3) applies for $-\infty \leq x \leq (u + \alpha/k)$. In all three cases, α is assumed to be positive.

For the EVI distribution x is unbounded (Table 11.5.1), while for EVII, x is bounded from below (by $u + \alpha/k$), and for the EVIII distribution, x is similarly bounded from above. The EVI and EVII distributions are also known as the *Gumbel* and *Frechet* distributions, respectively. If a variable x is described by the EVIII distribution, then $-x$ is said to have a *Weibull* distribution.

REFERENCES

Abramowitz, M., and I. A. Stegun, *Handbook of Mathematical Functions,* Dover, New York, p. 932, 1965.

Benson, M. A., Uniform flood-frequency estimating methods for federal agencies, *Water Resour. Res.,* vol. 4, no. 5, pp. 891–908, 1968.

Bobee, B., The log–Pearson Type III distribution and its application in hydrology, *Water Resour. Res.,* vol. 11, no. 5, pp. 681–689, 1975.

Bobee, B. B., and R. Robitaille, The use of the Pearson Type 3 and log Pearson Type 3 distributions revisited, *Water Resour. Res.,* vol. 13, no. 2, pp. 427–443, 1977.

Chow, V. T., The log-probability law and its engineering applications, *Proc. Am. Soc. Civ. Eng.,* vol. 80, pp. 1–25, 1954.

Fisher, R. A., On the mathematical foundations of theoretical statistics, *Trans. R. Soc. London A,* vol. 222, pp. 309–368, 1922.

Fisher, R. A., and L. H. C. Tippett, Limiting forms of the frequency distribution of the largest or smallest member of a sample, *Proc. Cambridge Phil. Soc.,* vol. 24, part II, pp. 180–191, 1928.

Foster, H. A., Theoretical frequency curves and their application to engineering problems, *Trans. Am. Soc. Civ. Eng.,* vol. 87, pp. 142–173, 1924.

Frechet, M., Sur la loi de probabilite de l'ecart maximum ("On the probability law of maximum values"), *Annales de la societe Polonaise de Mathematique,* vol. 6, pp. 93–116, Krakow, Poland, 1927.

Freeze, R. A., A stochastic-conceptual analysis of one-dimensional groundwater flow in nonuniform homogenous media, *Water Resour. Res.,* vol. 11, no. 5, pp. 725–741, 1975.

Gumbel, E. J., The return period of flood flows, *The Annals of Mathematical Statistics,* vol. 12, no. 2, pp. 163–190, June 1941.

Haan, C. T., *Statistical Methods in Hydrology,* Iowa State Univ. Press, Ames, Iowa, 1977.

Jenkinson, A. F., The frequency distribution of the annual maximum (or minimum) values of meteorological elements, *Quart. Jour. Roy. Met. Soc.,* vol. 81, pp. 158–171, 1955.

Kavvas M. L., and J. W. Delleur, A stochastic cluster model of daily rainfall sequences, *Water Resour. Res.,* vol. 17, no. 4, pp. 1151–1160, 1981.

Pearson, K., On the systematic fitting of curves to observations and measurements, *Biometrika,* vol. 1, no. 3, pp. 265–303, 1902.

Waymire, E., and V. K. Gupta, The mathematical structure of rainfall representations I. A review of the stochastic rainfall models, *Water Resour. Res.,* vol. 17, no. 5, pp. 1261–1294, 1981.

Weibull, W., A statistical theory of the strength of materials, *Ingeniors Vetenskaps Akademien* (The Royal Swedish Institute for Engineering Research), proceedings no. 51, pp. 5–45, 1939.

PROBLEMS

11.1.1 The annual precipitation data for College Station, Texas, from 1911 to 1979 are given in Table 11.1.1. Estimate from the data the probability that the annual precipitation will be greater than 50 in in any year. Calculate the probability that annual precipitation will be greater than 50 in in two successive years (*a*) by assuming annual precipitation is an independent process; (*b*) directly from the data. Do the data suggest there is any tendency for years of precipitation > 50 in to follow one another in College Station?

11.1.2 Solve Prob. 11.1.1 for precipitation less than 30 in. Is there a tendency for years of precipitation less than 30 in to follow each other more than independence of events from year to year would suggest?

11.3.1 Calculate the mean, standard deviation, and coefficient of skewness for College Station annual precipitation from 1960 to 1969. The data are given in Table 11.1.1.

11.3.2 Calculate the mean, standard deviation, and coefficient of skewness for College Station annual precipitation for the six 10-year periods beginning in 1920, 1930, 1940, 1950, 1960, 1970 (e.g., 1920-1929). Compare the values of these statistics for the six samples. Calculate the mean and standard deviation of the six sample means and their coefficient of variation. Repeat this exercise for the six sample standard deviations and the six coefficients of skewness. As measured by the coefficient of variation of each sample statistic, which of these three sample statistics (mean, standard deviation, or coefficient of skewness) varies most from sample to sample?

11.4.1 Prove that the mean μ of the exponential distribution $f(x) = \lambda e^{-\lambda x}$ is given by $\mu = 1/\lambda$.

11.4.2 Show that the maximum likelihood estimates of the parameters of the normal distribution are given by

$$\mu = \frac{1}{n}\sum_{i=1}^{n} x_i \qquad \text{and} \qquad \sigma^2 = \frac{1}{n}\sum_{i=1}^{n} (x_i - \bar{x})^2$$

11.4.3 Calculate the value of the maximum likelihood estimates of the parameters of the normal distribution fitted to College Station annual precipitation from 1970 to 1979. Use the formulas given in Prob. 11.4.2 above and the data given in Table 11.1.1. Compare the result with the moment estimates given in Example 11.3.1.

11.4.4 Calculate the value of the log-likelihood function of College Station annual precipitation from 1970 to 1979 with $\mu = 40.17$ in and $\sigma = 10.63$ in. Holding μ constant, recompute and plot the value of the log-likelihood function by varying σ in increments of 0.1 from 9.5 to 11.5. Determine the value of σ that maximizes the log-likelihood function.

11.4.5 Solve Example 11.1.1 in the text using the probabilities for events A and B calculated from a normal distribution with $\mu = 39.77$ in and $\sigma = 9.17$ in (as fitted to the College Station precipitation data in Example 11.4.3). Compare the results you obtain with those in Example 11.1.1. Which method do you think is more reliable?

11.4.6 A reservoir system near College Station, Texas, is experiencing a drought and it is determined that if next year's annual precipitation in the reservoir watershed is less than 35 in, a reduction in the reservoir water supplied for irrigation will be required during the following year. If the annual precipitation is less than 35 in for each of the next two years, a reduction in municipal water supply will also be required. Using the normal distribution fitted to the precipitation data in Example 11.4.3, calculate the probability that these supply reductions will be necessary. Do you think these probabilities are sufficiently high to justify warning the irrigation and municipal water users of possible supply reductions?

11.5.1 The Pearson system of distributions obeys the equation $d[f(x)]/dx = [f(x)(x - d)]/(C_0 + C_1 x + C_2 x^2)$ where d is the mode of the distribution [the value of x where $f(x)$ is maximized] and C_0, C_1, and C_2 are coefficients. By setting $C_2 = 0$, show that the Pearson Type III distribution is obtained.

11.5.2 In Prob. 11.5.1, set $C_1 = C_2 = 0$ and show that the normal distribution is obtained.

11.5.3 The demand on a city's water treatment and distribution system is rising to near system capacity because of a long period of hot, dry weather. Rainfall will avert a situation where demand exceeds system capacity. If the average time between rainfalls in this city at this time of year is 5 days, calculate the chance that there will be no rain (*a*) for the next 5 days, (*b*) 10 days, (*c*) 15 days. Use the exponential distribution.

11.5.4 Data for the annual maximum discharge of the Guadalupe River at Victoria, Texas, are presented in Table 12.1.1. The statistics for the logarithms to base 10 of these data are $\bar{y} = 4.2743$ and $s_y = 0.3981$. Fit the lognormal distribution to these data. Plot the relative frequency and incremental probability functions, and the cumulative frequency and probability distribution functions of the data as shown in Fig. 11.4.3 (use a log scale for the Guadalupe River discharges).

11.5.5 Data for inflow to the site of the proposed Justiceburg reservoir are given in Table 15.P.5. Calculate the mean, standard deviation, and coefficient of skewness of the annual total inflows and fit a probability distribution to the data.

CHAPTER 12

FREQUENCY ANALYSIS

Hydrologic systems are sometimes impacted by extreme events, such as severe storms, floods, and droughts. The magnitude of an extreme event is inversely related to its frequency of occurrence, very severe events occurring less frequently than more moderate events. The objective of frequency analysis of hydrologic data is to relate the magnitude of extreme events to their frequency of occurrence through the use of probability distributions. The hydrologic data analyzed are assumed to be independent and identically distributed, and the hydrologic system producing them (e.g., a storm rainfall system) is considered to be stochastic, space-independent, and time-independent in the classification scheme shown in Fig. 1.4.1. The hydrologic data employed should be carefully selected so that the assumptions of independence and identical distribution are satisfied. In practice, this is often achieved by selecting the annual maximum of the variable being analyzed (e.g., the annual maximum discharge, which is the largest instantaneous peak flow occurring at any time during the year) with the expectation that successive observations of this variable from year to year will be independent.

The results of flood flow frequency analysis can be used for many engineering purposes: for the design of dams, bridges, culverts, and flood control structures; to determine the economic value of flood control projects; and to delineate flood plains and determine the effect of encroachments on the flood plain.

12.1 RETURN PERIOD

Suppose that an extreme event is defined to have occurred if a random variable X is greater than or equal to some level x_T. The *recurrence interval* τ is the time

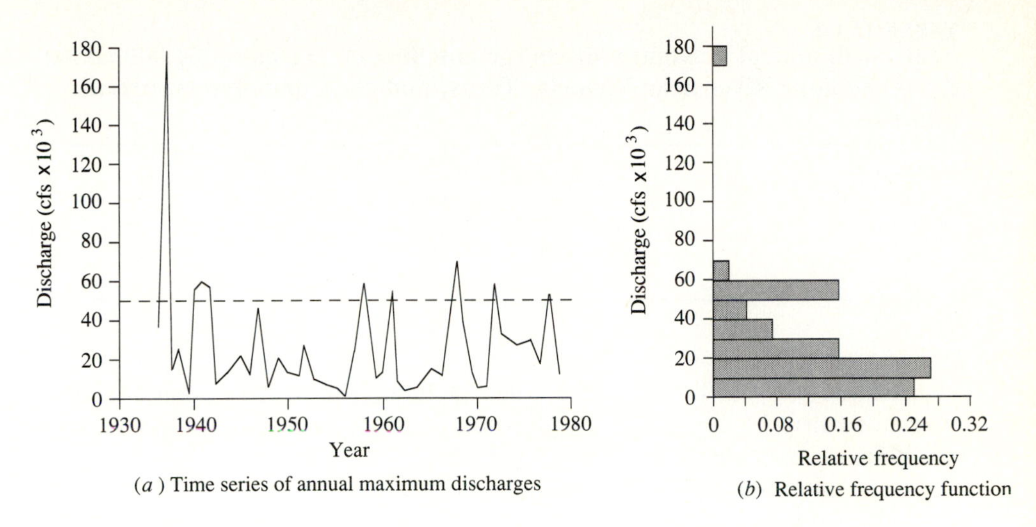

(*a*) Time series of annual maximum discharges

(*b*) Relative frequency function

FIGURE 12.1.1
Annual maximum discharge of the Guadalupe River near Victoria, Texas.

between occurrences of $X \geq x_T$. For example, Fig. 12.1.1 shows the record of annual maximum discharges of the Guadalupe River near Victoria, Texas, from 1935 to 1978, plotted from the data given in Table 12.1.1. If $x_T = 50,000$ cfs, it can be seen that the maximum discharge exceeded this level nine times during the period of record, with recurrence intervals ranging from 1 year to 16 years, as shown in Table 12.1.2.

The *return period* T of the event $X \geq x_T$ is the expected value of τ, $E(\tau)$, its average value measured over a very large number of occurrences. For the Guadalupe River data, there are 8 recurrence intervals covering a total period of 41 years between the first and last exceedences of 50,000 cfs, so the return period of a 50,000 cfs annual maximum discharge on the Guadalupe River is

TABLE 12.1.1
Annual maximum discharges of the Guadalupe River near Victoria, Texas, 1935–1978, in cfs

Year	1930	1940	1950	1960	1970
0		55,900	13,300	23,700	9,190
1		58,000	12,300	55,800	9,740
2		56,000	28,400	10,800	58,500
3		7,710	11,600	4,100	33,100
4		12,300	8,560	5,720	25,200
5	38,500	22,000	4,950	15,000	30,200
6	179,000	17,900	1,730	9,790	14,100
7	17,200	46,000	25,300	70,000	54,500
8	25,400	6,970	58,300	44,300	12,700
9	4,940	20,600	10,100	15,200	

TABLE 12.1.2
Years with annual maximum discharge equaling or exceeding 50,000 cfs on the Guadalupe River near Victoria, Texas, and corresponding recurrence intervals

Exceedence year	1936	1940	1941	1942	1958	1961	1967	1972	1977	Average
Recurrence interval (years)		4	1	1	16	3	6	5	5	5.1

approximately $\bar{\tau} = 41/8 = 5.1$ years. Thus the return period of an event of a given magnitude may be defined as the *average recurrence interval* between events *equalling or exceeding* a specified magnitude.

The probability $p = P(X \geq x_T)$ of occurrence of the event $X \geq x_T$ in any observation may be related to the return period in the following way. For each observation, there are two possible outcomes: either "success" $X \geq x_T$ (probability p) or "failure" $X < x_T$ (probability $1-p$). Since the observations are independent, the probability of a recurrence interval of duration τ is the product of the probabilities of $\tau - 1$ failures followed by one success, that is, $(1-p)^{\tau-1}p$, and the expected value of τ is given by

$$
\begin{aligned}
E(\tau) &= \sum_{\tau=1}^{\infty} \tau(1-p)^{\tau-1}p \\
&= p + 2(1-p)p + 3(1-p)^2p + 4(1-p)^3p + \ldots \\
&= p[1 + 2(1-p) + 3(1-p)^2 + 4(1-p)^3 + \ldots]
\end{aligned}
\tag{12.1.1a}
$$

The expression within the brackets has the form of the power series expansion $(1+x)^n = 1 + nx + [n(n-1)/2]x^2 + [n(n-1)(n-2)/6]x^3 + \ldots$, with $x = -(1-p)$ and $n = -2$, so (12.1.1*a*) may be rewritten

$$
\begin{aligned}
E(\tau) &= \frac{p}{[1-(1-p)]^2} \\
&= \frac{1}{p}
\end{aligned}
\tag{12.1.1b}
$$

Hence $E(\tau) = T = 1/p$; that is, the probability of occurrence of an event in any observation is the inverse of its return period:

$$
P(X \geq x_T) = \frac{1}{T} \tag{12.1.2}
$$

For example, the probability that the maximum discharge in the Guadalupe River will equal or exceed 50,000 cfs in any year is approximately $p = 1/\bar{\tau} = 1/5.1 = 0.195$.

What is the probability that a T-year return period event will occur at least once in N years? To calculate this, first consider the situation where no T-year event occurs in N years. This would require a sequence of N successive "failures," so that

$$P(X < x_T \text{ each year for } N \text{ years}) = (1 - p)^N$$

The complement of this situation is the case required, so by (11.1.3)

$$P(X \geq x_T \text{ at least once in } N \text{ years}) = 1 - (1 - p)^N \tag{12.1.3}$$

Since $p = 1/T$,

$$P(X \geq x_T \text{ at least once in } N \text{ years}) = 1 - \left(1 - \frac{1}{T}\right)^N \tag{12.1.4}$$

Example 12.1.1. Estimate the probability that the annual maximum discharge Q on the Guadalupe River will exceed 50,000 cfs at least once during the next three years.

Solution. From the discussion above, $P(Q \geq 50,000 \text{ cfs in any year}) \approx 0.195$, so from Eq. (12.1.3)

$$\begin{aligned} P(Q \geq 50,000 \text{ cfs at least once during the next 3 years}) &= 1 - (1 - 0.195)^3 \\ &= 0.48 \end{aligned}$$

The problem in Example 12.1.1 could have been phrased, "What is the probability that the discharge on the Guadalupe River will exceed 50,000 cfs at least once during the next three years?" The calculation given used only the annual maximum data, but, alternatively, all exceedences of 50,000 cfs contained in the Guadalupe River record could have been considered. This set of data is called the *partial duration series*. It will contain more than the nine exceedences shown in Table 12.1.2 if there were two or more exceedences of 50,000 cfs within some single year of record.

Hydrologic Data Series

A *complete duration series* consists of all the data available as shown in Fig. 12.1.2(a). A *partial duration series* is a series of data which are selected so that their magnitude is greater than a predefined *base value*. If the base value is selected so that the number of values in the series is equal to the number of years of the record, the series is called an *annual exceedence series;* an example is shown in Fig. 12.1.2(b). An *extreme value series* includes the largest or smallest values occurring in each of the equally-long time intervals of the record. The time interval length is usually taken as one year, and a series so selected is called an *annual series*. Using largest annual values, it is an *annual maximum series* as shown in Fig. 12.1.2(c). Selecting the smallest annual values produces an *annual minimum series*.

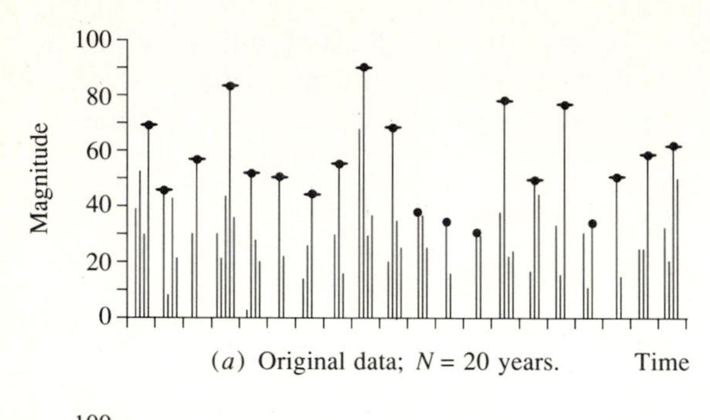

(a) Original data; $N = 20$ years.

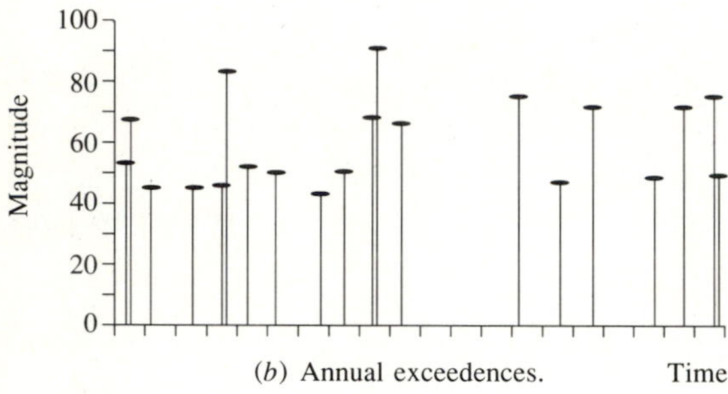

(b) Annual exceedences.

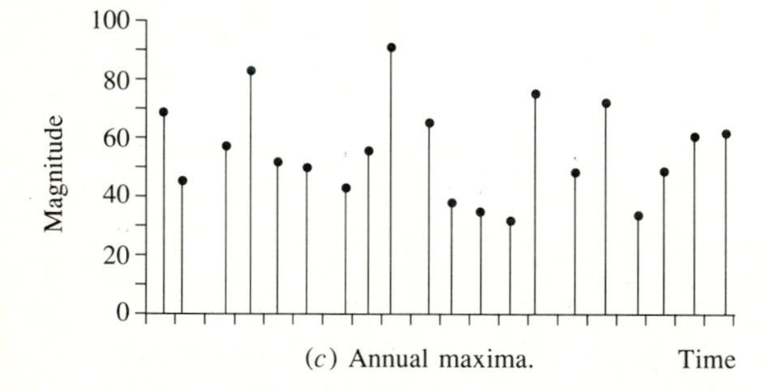

(c) Annual maxima.

FIGURE 12.1.2
Hydrologic data arranged by time of occurrence. (*Source:* Chow, 1964. Used with permission.)

The annual maximum values and the annual exceedence values of the hypothetical data in Fig. 12.1.3(a) are arranged graphically in Fig. 12.1.3(b) in order of magnitude. In this particular example, only 16 of the 20 annual maxima appear in the annual exceedence series; the second largest value in several years outranks some annual maxima in magnitude. However, in the annual maximum series, these second largest values are excluded, resulting in the neglect of their effect in the analysis.

The return period T_E of event magnitudes developed from an annual exceedence series is related to the corresponding return period T for magnitudes derived from an annual maximum series by (Chow, 1964)

$$T_E = \left[\ln\left(\frac{T}{T-1}\right)\right]^{-1} \tag{12.1.5}$$

Although the annual exceedence series is useful for some purposes, it is limited by the fact that it may be difficult to verify that all the observations are

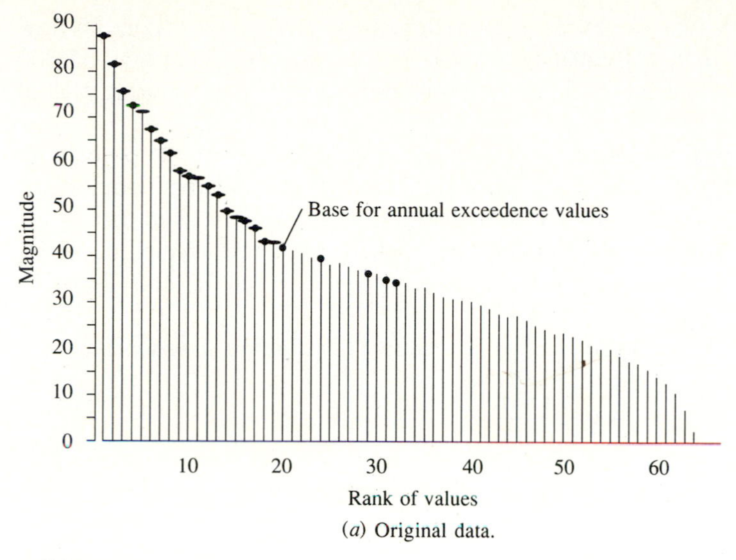

(*a*) Original data.

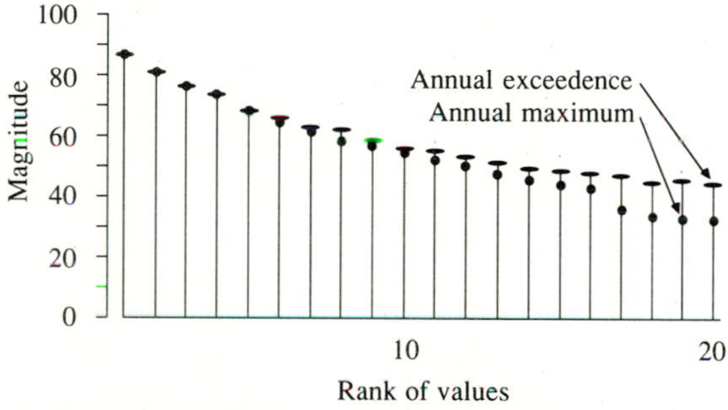

(*b*) Annual exceedence and maximum values.

FIGURE 12.1.3
Hydrologic data arranged in the order of magnitude. (*Source:* Chow, 1964. Used with permission.)

independent—the occurrence of a large flood could well be related to saturated soil conditions produced during another large flood occurring a short time earlier. As a result, it is usually better to use the annual maximum series for analysis. In any case, as the return period of the event being considered becomes large, the results from the two approaches become very similar because the chance that two such events will occur within any year is very small.

12.2 EXTREME VALUE DISTRIBUTIONS

The study of extreme hydrologic events involves the selection of a sequence of the largest or smallest observations from sets of data. For example, the study of peak flows uses just the largest flow recorded each year at a gaging station out of the many thousands of values recorded. In fact, water level is usually recorded every 15 minutes, so there are $4 \times 24 = 96$ values recorded each day, and $365 \times 96 = 35,040$ values recorded each year; so the annual maximum flow event used for flood flow frequency analysis is the largest of more than 35,000 observations during that year. And this exercise is carried out for each year of historical data.

Since these observations are located in the extreme tail of the probability distribution of all observations from which they are drawn (the parent population), it is not surprising that their probability distribution is different from that of the parent population. As described in Sec. 11.5, there are three asymptotic forms of the distributions of extreme values, named Type I, Type II, and Type III, respectively.

The Extreme Value Type I (EVI) probability distribution function is

$$F(x) = \exp\left[-\exp\left(-\frac{x-u}{\alpha}\right)\right] \qquad -\infty \leq x \leq \infty \tag{12.2.1}$$

The parameters are estimated, as given in Table 11.5.1, by

$$\alpha = \frac{\sqrt{6}\, s}{\pi} \tag{12.2.2}$$

$$u = \bar{x} - 0.5772\alpha \tag{12.2.3}$$

The parameter u is the mode of the distribution (point of maximum probability density). A *reduced variate* y can be defined as

$$y = \frac{x-u}{\alpha} \tag{12.2.4}$$

Substituting the reduced variate into (12.2.1) yields

$$F(x) = \exp\left[-\exp(-y)\right] \tag{12.2.5}$$

Solving for y:

$$y = -\ln\left[\ln\left(\frac{1}{F(x)}\right)\right] \tag{12.2.6}$$

Let (12.2.6) be used to define y for the Type II and Type III distributions. The values of x and y can be plotted as shown in Fig. 12.2.1. For the EVI distribution the plot is a straight line while, for large values of y, the corresponding curve for the EVII distribution slopes more steeply than for EVI, and the curve for the EVIII distribution slopes less steeply, being bounded from above. Figure 12.2.1 also shows values of the return period T as an alternate axis to y. As shown by Eq. (12.1.2),

$$\begin{aligned}
\frac{1}{T} &= P(x \geq x_T) \\
&= 1 - P(x < x_T) \\
&= 1 - F(x_T)
\end{aligned}$$

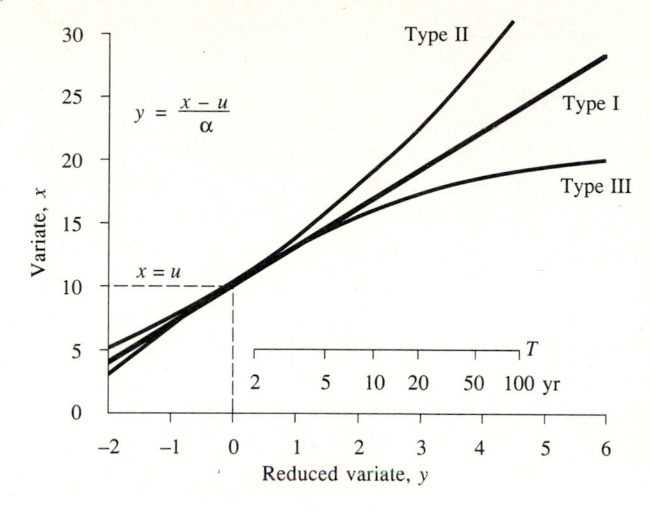

FIGURE 12.2.1
For each of the three types of extreme value distributions the variate x is plotted against a reduced variate y calculated for the Extreme Value Type I distribution. The Type I distribution is unbounded in x, while the Type II distribution has a lower bound and the Type III distribution has an upper bound. (*Source:* Natural Environment Research Council, 1975, Fig. 1.10, p. 41. Used with permission.)

so

$$F(x_T) = \frac{T-1}{T}$$

and, substituting into (12.2.6),

$$y_T = -\ln\left[\ln\left(\frac{T}{T-1}\right)\right] \tag{12.2.7}$$

For the EVI distribution, x_T is related to y_T by Eq. (12.2.4), or

$$x_T = u + \alpha y_T \tag{12.2.8}$$

Extreme value distributions have been widely used in hydrology. They form the basis for the standardized method of flood frequency analysis in Great Britain (Natural Environment Research Council, 1975). Storm rainfalls are most commonly modeled by the Extreme Value Type I distribution (Chow, 1953; Tomlinson, 1980), and drought flows by the Weibull distribution, that is, the EVIII distribution applied to $-x$ (Gumbel, 1954, 1963).

Example 12.2.1. Annual maximum values of 10-minute-duration rainfall at Chicago, Illinois, from 1913 to 1947 are presented in Table 12.2.1. Develop a model for storm rainfall frequency analysis using the Extreme Value Type I distribution and calculate the 5-, 10-, and 50-year return period maximum values of 10-minute rainfall at Chicago.

Solution. The sample moments calculated from the data in Table 12.2.1 are $\bar{x} = 0.649$ in and $s = 0.177$ in. Substituting into Eqs. (12.2.2) and (12.2.3) yields

$$\alpha = \frac{\sqrt{6}\, s}{\pi}$$

$$= \frac{\sqrt{6} \times 0.177}{\pi}$$

$$= 0.138$$

$$u = \bar{x} - 0.5772\alpha$$

$$= 0.649 - 0.5772 \times 0.138$$

$$= 0.569$$

The probability model is

$$F(x) = \exp\left[-\exp\left(-\frac{x - 0.569}{0.138}\right)\right]$$

To determine the values of x_T for various values of return period T, it is convenient to use the reduced variate y_T. For $T = 5$ years, Eq. (12.2.7) gives

$$y_T = -\ln\left[\ln\left(\frac{T}{T-1}\right)\right]$$

$$= -\ln\left[\ln\left(\frac{5}{5-1}\right)\right]$$

$$= 1.500$$

and Eq. (12.2.8) yields

$$x_T = u + \alpha y_T$$

$$= 0.569 + 0.138 \times 1.500$$

$$= 0.78 \text{ in}$$

So the 10-minute, 5 year storm rainfall magnitude at Chicago is 0.78 in. By the same method, the 10- and 50-year values can be shown to be 0.88 in and 1.11 in,

TABLE 12.2.1
Annual maximum 10-minute rainfall in inches at Chicago, Illinois, 1913–1947

Year	1910	1920	1930	1940
0		0.53	0.33	0.34
1		0.76	0.96	0.70
2		0.57	0.94	0.57
3	0.49	0.80	0.80	0.92
4	0.66	0.66	0.62	0.66
5	0.36	0.68	0.71	0.65
6	0.58	0.68	1.11	0.63
7	0.41	0.61	0.64	0.60
8	0.47	0.88	0.52	
9	0.74	0.49	0.64	

Mean = 0.649 in
Standard deviation = 0.177 in

respectively. It may be noted from the data in Table 12.2.1 that the 50-year return period rainfall was equaled once in the 35 years of data (in 1936), and that the 10-year return period rainfall was equaled or exceeded four times during this period, so the frequency of occurrence of observed extreme rainfalls is approximately as predicted by the model.

12.3 FREQUENCY ANALYSIS USING FREQUENCY FACTORS

Calculating the magnitudes of extreme events by the method outlined in Example 12.2.1 requires that the probability distribution function be invertible, that is, given a value for T or $[F(x_T) = T/(T-1)]$, the corresponding value of x_T can be determined. Some probability distribution functions are not readily invertible, including the Normal and Pearson Type III distributions, and an alternative method of calculating the magnitudes of extreme events is required for these distributions.

The magnitude x_T of a hydrologic event may be represented as the mean μ plus the departure Δx_T of the variate from the mean (see Fig. 12.3.1):

$$x_T = \mu + \Delta x_T \tag{12.3.1}$$

The departure may be taken as equal to the product of the standard deviation σ and a *frequency factor* K_T; that is, $\Delta x_T = K_T \sigma$. The departure Δx_T and the frequency factor K_T are functions of the return period and the type of probability distribution to be used in the analysis. Equation (12.3.1) may therefore be expressed as

$$x_T = \mu + K_T \sigma \tag{12.3.2}$$

which may be approximated by

$$x_T = \bar{x} + K_T s \tag{12.3.3}$$

In the event that the variable analyzed is $y = \log x$, then the same method is applied to the statistics for the logarithms of the data, using

$$y_T = \bar{y} + K_T s_y \tag{12.3.4}$$

and the required value of x_T is found by taking the antilog of y_T.

The frequency factor equation (12.3.2) was proposed by Chow (1951), and it is applicable to many probability distributions used in hydrologic frequency analysis. For a given distribution, a K–T relationship can be determined between the frequency factor and the corresponding return period. This relationship can be expressed in mathematical terms or by a table.

Frequency analysis begins with the calculation of the statistical parameters required for a proposed probability distribution by the method of moments from the given data. For a given return period, the frequency factor can be determined from the K–T relationship for the proposed distribution, and the magnitude x_T computed by Eq. (12.3.3), or (12.3.4).

The theoretical K–T relationships for several probability distributions commonly used in hydrologic frequency analysis are now described.

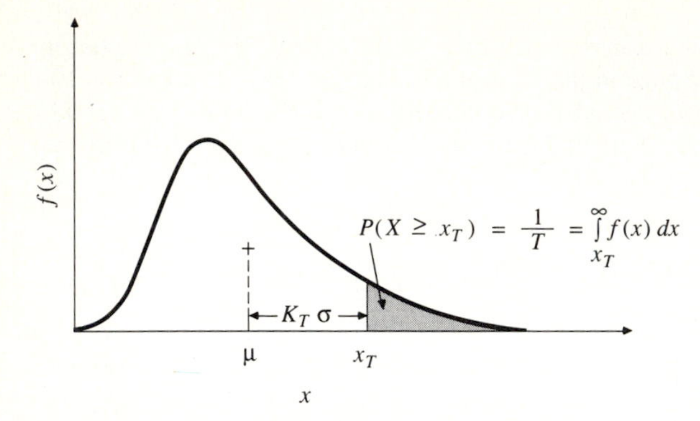

FIGURE 12.3.1
The magnitude of an extreme event x_T expressed as a deviation $K_T\sigma$ from the mean μ, where K_T is the frequency factor.

NORMAL DISTRIBUTION. The frequency factor can be expressed from Eq. (12.3.2) as

$$K_T = \frac{x_T - \mu}{\sigma} \tag{12.3.5}$$

This is the same as the standard normal variable z defined in Eq. (11.2.9).

The value of z corresponding to an exceedence probability of p ($p = 1/T$) can be calculated by finding the value of an intermediate variable w:

$$w = \left[\ln\left(\frac{1}{p^2}\right)\right]^{1/2} \qquad (0 < p \leq 0.5) \tag{12.3.6}$$

then calculating z using the approximation

$$z = w - \frac{2.515517 + 0.802853w + 0.010328w^2}{1 + 1.432788w + 0.189269w^2 + 0.001308w^3} \tag{12.3.7}$$

When $p > 0.5$, $1 - p$ is substituted for p in (12.3.6) and the value of z computed by (12.3.7) is given a negative sign. The error in this formula is less than 0.00045 in z (Abramowitz and Stegun, 1965). The frequency factor K_T for the normal distribution is equal to z, as mentioned above.

For the lognormal distribution, the same procedure applies except that it is applied to the logarithms of the variables, and their mean and standard deviation are used in Eq. (12.3.4).

Example 12.3.1. Calculate the frequency factor for the normal distribution for an event with a return period of 50 years.

Solution. For T = 50 years, p = 1/50 = 0.02. From Eq. (12.3.6)

$$\begin{aligned} w &= \left[\ln\left(\frac{1}{p^2}\right)\right]^{1/2} \\ &= \left[\ln\left(\frac{1}{0.02^2}\right)\right]^{1/2} \\ &= 2.7971 \end{aligned}$$

Then, substituting w into (12.3.7)

$$
\begin{aligned}
K_T &= z \\
&= 2.7971 - \frac{2.51557 + 0.80285 \times 2.7971 + 0.01033 \times (2.7971)^2}{1 + 1.43279 \times 2.7971 + 0.18927 \times (2.7971)^2 + 0.00131 \times (2.7971)^3} \\
&= 2.054
\end{aligned}
$$

EXTREME VALUE DISTRIBUTIONS. For the Extreme Value Type I distribution, Chow (1953) derived the expression

$$K_T = -\frac{\sqrt{6}}{\pi}\left\{0.5772 + \ln\left[\ln\left(\frac{T}{T-1}\right)\right]\right\} \tag{12.3.8}$$

To express T in terms of K_T, the above equation can be written as

$$T = \frac{1}{1 - \exp\left\{-\exp\left[-\left(\gamma + \frac{\pi K_T}{\sqrt{6}}\right)\right]\right\}} \tag{12.3.9}$$

where $\gamma = 0.5772$. When $x_T = \mu$, Eq. (12.3.5) gives $K_T = 0$ and Eq. (12.3.8) gives $T = 2.33$ years. This is the return period of the mean of the Extreme Value Type I distribution. For the Extreme Value Type II distribution, the logarithm of the variate follows the EVI distribution. For this case, (12.3.4) is used to calculate y_T, using the value of K_T from (12.3.8).

Example 12.3.2. Determine the 5-year return period rainfall for Chicago using the frequency factor method and the annual maximum rainfall data given in Table 12.2.1.

Solution. The mean and standard deviation of annual maximum rainfalls at Chicago are $\bar{x} = 0.649$ in and $s = 0.177$ in, respectively. For T=5, Eq. (12.3.8) gives

$$
\begin{aligned}
K_T &= -\frac{\sqrt{6}}{\pi}\left\{0.5772 + \ln\left[\ln\left(\frac{T}{T-1}\right)\right]\right\} \\
&= -\frac{\sqrt{6}}{\pi}\left\{0.5772 + \ln\left[\ln\left(\frac{5}{5-1}\right)\right]\right\} \\
&= 0.719
\end{aligned}
$$

By (12.3.3),

$$
\begin{aligned}
x_T &= \bar{x} + K_T s \\
&= 0.649 + 0.719 \times 0.177 \\
&= 0.78 \text{ in}
\end{aligned}
$$

as determined in Example 12.2.1.

LOG–PEARSON TYPE III DISTRIBUTION. For this distribution, the first step is to take the logarithms of the hydrologic data, $y = \log x$. Usually logarithms to base 10 are used. The mean $\bar{y}$, standard deviation s_y, and coefficient of skewness C_s are calculated for the logarithms of the data. The frequency factor depends on the return period T and the coefficient of skewness C_s. When $C_s = 0$, the frequency factor is equal to the standard normal variable z. When $C_s \neq 0, K_T$ is approximated by Kite (1977) as

$$K_T = z + (z^2 - 1)k + \frac{1}{3}(z^3 - 6z)k^2 - (z^2 - 1)k^3 + zk^4 + \frac{1}{3}k^5 \qquad (12.3.10)$$

where $k = C_s/6$.

TABLE 12.3.1
K_T values for Pearson Type III distribution (positive skew)

Skew coefficient	Return period in years						
	2	5	10	25	50	100	200
	Exceedence probability						
C_s or C_w	0.50	0.20	0.10	0.04	0.02	0.01	0.005
3.0	–0.396	0.420	1.180	2.278	3.152	4.051	4.970
2.9	–0.390	0.440	1.195	2.277	3.134	4.013	4.909
2.8	–0.384	0.460	1.210	2.275	3.114	3.973	4.847
2.7	–0.376	0.479	1.224	2.272	3.093	3.932	4.783
2.6	–0.368	0.499	1.238	2.267	3.071	3.889	4.718
2.5	–0.360	0.518	1.250	2.262	3.048	3.845	4.652
2.4	–0.351	0.537	1.262	2.256	3.023	3.800	4.584
2.3	–0.341	0.555	1.274	2.248	2.997	3.753	4.515
2.2	–0.330	0.574	1.284	2.240	2.970	3.705	4.444
2.1	–0.319	0.592	1.294	2.230	2.942	3.656	4.372
2.0	–0.307	0.609	1.302	2.219	2.912	3.605	4.298
1.9	–0.294	0.627	1.310	2.207	2.881	3.553	4.223
1.8	–0.282	0.643	1.318	2.193	2.848	3.499	4.147
1.7	–0.268	0.660	1.324	2.179	2.815	3.444	4.069
1.6	–0.254	0.675	1.329	2.163	2.780	3.388	3.990
1.5	–0.240	0.690	1.333	2.146	2.743	3.330	3.910
1.4	–0.225	0.705	1.337	2.128	2.706	3.271	3.828
1.3	–0.210	0.719	1.339	2.108	2.666	3.211	3.745
1.2	–0.195	0.732	1.340	2.087	2.626	3.149	3.661
1.1	–0.180	0.745	1.341	2.066	2.585	3.087	3.575
1.0	–0.164	0.758	1.340	2.043	2.542	3.022	3.489
0.9	–0.148	0.769	1.339	2.018	2.498	2.957	3.401
0.8	–0.132	0.780	1.336	1.993	2.453	2.891	3.312
0.7	–0.116	0.790	1.333	1.967	2.407	2.824	3.223
0.6	–0.099	0.800	1.328	1.939	2.359	2.755	3.132
0.5	–0.083	0.808	1.323	1.910	2.311	2.686	3.041
0.4	–0.066	0.816	1.317	1.880	2.261	2.615	2.949
0.3	–0.050	0.824	1.309	1.849	2.211	2.544	2.856
0.2	–0.033	0.830	1.301	1.818	2.159	2.472	2.763
0.1	–0.017	0.836	1.292	1.785	2.107	2.400	2.670
0.0	0	0.842	1.282	1.751	2.054	2.326	2.576

The value of z for a given return period can be calculated by the procedure used in Example 12.3.1. Table 12.3.1 gives values of the frequency factor for the Pearson Type III (and log–Pearson Type III) distribution for various values of the return period and coefficient of skewness.

Example 12.3.3. Calculate the 5- and 50-year return period annual maximum discharges of the Guadalupe River near Victoria, Texas, using the lognormal and log–Pearson Type III distributions. The data from 1935 to 1978 are given in Table 12.1.1.

TABLE 12.3.1 (*cont.*)
K_T values for Pearson Type III distribution (negative skew)

Skew coefficient C_s or C_w	Return period in years						
	2	5	10	25	50	100	200
	Exceedence probability						
	0.50	0.20	0.10	0.04	0.02	0.01	0.005
–0.1	0.017	0.846	1.270	1.716	2.000	2.252	2.482
–0.2	0.033	0.850	1.258	1.680	1.945	2.178	2.388
–0.3	0.050	0.853	1.245	1.643	1.890	2.104	2.294
–0.4	0.066	0.855	1.231	1.606	1.834	2.029	2.201
–0.5	0.083	0.856	1.216	1.567	1.777	1.955	2.108
–0.6	0.099	0.857	1.200	1.528	1.720	1.880	2.016
–0.7	0.116	0.857	1.183	1.488	1.663	1.806	1.926
–0.8	0.132	0.856	1.166	1.448	1.606	1.733	1.837
–0.9	0.148	0.854	1.147	1.407	1.549	1.660	1.749
–1.0	0.164	0.852	1.128	1.366	1.492	1.588	1.664
–1.1	0.180	0.848	1.107	1.324	1.435	1.518	1.581
–1.2	0.195	0.844	1.086	1.282	1.379	1.449	1.501
–1.3	0.210	0.838	1.064	1.240	1.324	1.383	1.424
–1.4	0.225	0.832	1.041	1.198	1.270	1.318	1.351
–1.5	0.240	0.825	1.018	1.157	1.217	1.256	1.282
–1.6	0.254	0.817	0.994	1.116	1.166	1.197	1.216
–1.7	0.268	0.808	0.970	1.075	1.116	1.140	1.155
–1.8	0.282	0.799	0.945	1.035	1.069	1.087	1.097
–1.9	0.294	0.788	0.920	0.996	1.023	1.037	1.044
–2.0	0.307	0.777	0.895	0.959	0.980	0.990	0.995
–2.1	0.319	0.765	0.869	0.923	0.939	0.946	0.949
–2.2	0.330	0.752	0.844	0.888	0.900	0.905	0.907
–2.3	0.341	0.739	0.819	0.855	0.864	0.867	0.869
–2.4	0.351	0.725	0.795	0.823	0.830	0.832	0.833
–2.5	0.360	0.711	0.771	0.793	0.798	0.799	0.800
–2.6	0.368	0.696	0.747	0.764	0.768	0.769	0.769
–2.7	0.376	0.681	0.724	0.738	0.740	0.740	0.741
–2.8	0.384	0.666	0.702	0.712	0.714	0.714	0.714
–2.9	0.390	0.651	0.681	0.683	0.689	0.690	0.690
–3.0	0.396	0.636	0.666	0.666	0.666	0.667	0.667

Source: U. S. Water Resources Council (1981).

Solution. The logarithms of the discharge values are taken and their statistics calculated: $\bar{y} = 4.2743, s_y = 0.4027, C_s = -0.0696$.

Lognormal distribution. The frequency factor can be obtained from Eq. (12.3.7), or from Table 12.3.1 for coefficient of skewness 0. For $T = 50$ years, K_T was computed in Example 12.3.1 as $K_{50} = 2.054$; the same value can be obtained from Table 12.3.1. By (12.3.4)

$$y_T = \bar{y} + K_T s_y$$

$$y_{50} = 4.2743 + 2.054 \times 0.4027$$

$$= 5.101$$

Then

$$x_{50} = (10)^{5.101}$$

$$= 126,300 \text{ cfs}$$

Similarly, $K_5 = 0.842$ from Table 12.3.1, $y_5 = 4.2743 + 0.842 \times 0.4027 = 4.6134$, and $x_5 = (10)^{4.6134} = 41,060$ cfs.

Log–Pearson Type III distribution. For $C_s = -0.0696$, the value of K_{50} is obtained by interpolation from Table 12.3.1 or by Eq. (12.3.10). By interpolation with T = 50 yrs:

$$K_{50} = 2.054 + \frac{(2.00 - 2.054)}{(-0.1 - 0)}(-0.0696 - 0) = 2.016$$

So $y_{50} = \bar{y} + K_{50}s_y = 4.2743 + 2.016 \times 0.4027 = 5.0863$ and $x_{50} = (10)^{5.0863} =$ 121,990 cfs. By a similar calculation, $K_5 = 0.845$, $y_5 = 4.6146$, and $x_5 = 41,170$ cfs.

The results for estimated annual maximum discharges are:

	Return Period	
	5 years	**50 years**
Lognormal ($C_s = 0$)	41,060	126,300
Log-Pearson Type III ($C_s = -0.07$)	41,170	121,990

It can be seen that the effect of including the small negative coefficient of skewness in the calculations is to alter slightly the estimated flow with that effect being more pronounced at $T = 50$ years than at $T = 5$ years. Another feature of the results is that the 50-year return period estimates are about three times as large as the 5-year return period estimates; for this example, the increase in the estimated flood discharges is less than proportional to the increase in return period.

12.4 PROBABILITY PLOTTING

As a check that a probability distribution fits a set of hydrologic data, the data may be plotted on specially designed *probability paper*, or using a plotting scale that linearizes the distribution function. The plotted data are then fitted with a straight line for interpolation and extrapolation purposes.

Probability Paper

The cumulative probability of a theoretical distribution may be represented graphically on probability paper designed for the distribution. On such paper the ordinate usually represents the value of x in a certain scale and the abscissa represents the probability $P(X \geq x)$ or $P(X < x)$, the return period T, or the reduced variate y_T. The ordinate and abscissa scales are so designed that the data to be fitted are expected to appear close to a straight line. The purpose of using the probability paper is to linearize the probability relationship so that the plotted data can be easily used for interpolation, extrapolation, or comparison purposes. In the case of extrapolation, however, the effect of various errors is often magnified; therefore, hydrologists should be warned against such practice if no consideration is given to this effect.

Plotting Positions

Plotting position refers to the probability value assigned to each piece of data to be plotted. Numerous methods have been proposed for the determination of plotting positions, most of which are empirical. If n is the total number of values to be plotted and m is the rank of a value in a list ordered by descending magnitude, the exceedence probability of the mth largest value, x_m, is, for large n,

$$P(X \geq x_m) = \frac{m}{n} \tag{12.4.1}$$

However, this simple formula (known as California's formula) produces a probability of 100 percent for $m = n$, which may not be easily plotted on a probability scale. As an adjustment, the above formula may be modified to

$$P(X \geq x_m) = \frac{m-1}{n} \tag{12.4.2}$$

While this formula does not produce a probability of 100 percent, it yields a zero probability (for $m = 1$), which may not be easily plotted on probability paper either.

The above two formulas represent the limits within which suitable plotting positions should lie. One compromise of the two formulas is

$$P(X \geq x_m) = \frac{m-0.5}{n} \tag{12.4.3}$$

which was first proposed by Hazen (1930). Another compromising formula (known as Chegodayev's) widely used in the U.S.S.R. and Eastern European countries is

$$P(X \geq x_m) = \frac{m-0.3}{n+0.4} \tag{12.4.4}$$

The Weibull formula is a compromise with more statistical justification. If the n values are distributed uniformly between 0 and 100 percent probability, then there must be $n + 1$ intervals, $n - 1$ between the data points and 2 at the ends. This simple plotting system is expressed by the Weibull formula:

$$P(X \geq x_m) = \frac{m}{n + 1} \qquad (12.4.5)$$

indicating a return period one year longer than the period of record for the largest value.

In practice, for a complete duration series (employing all the data, not just selected extreme values), Eq. (12.4.1) is used, with n referring to the number of items in the data rather than to the number of years. For annual maximum series, Eq. (12.4.5), which is equivalent to the following formula for return period, was adopted as the standard plotting position method by the U. S. Water Resources Council (1981):

$$T = \frac{n + 1}{m} \qquad (12.4.6)$$

where n refers to the number of years in the record.

Most plotting position formulas are represented by the following form:

$$P(X \geq x_m) = \frac{m - b}{n + 1 - 2b} \qquad (12.4.7)$$

where b is a parameter. For example, $b = 0.5$ for Hazen's formula, $b = 0.3$ for Chegodayev's, and $b = 0$ for Weibull's. Also, for some other examples $b = 3/8$ for Blom's formula, 1/3 for Tukey's, and 0.44 for Gringorten's (see Chow, 1964).

Cunnane (1978) studied the various available plotting position methods using criteria of *unbiasedness* and *minimum variance*. An unbiased plotting method is one that, if used for plotting a large number of equally sized samples, will result in the average of the plotted points for each value of m falling on the theoretical distribution line. A minimum variance plotting method is one that minimizes the variance of the plotted points about the theoretical line. Cunnane concluded that the Weibull plotting formula is biased and plots the largest values of a sample at too small a return period. For normally distributed data, he found that the Blom (1958) plotting position ($b = 3/8$) is closest to being unbiased, while for data distributed according to the Extreme Value Type I distribution, the Gringorten (1963) formula ($b = 0.44$) is the best. For the log–Pearson Type III distribution, the optimal value of b depends on the value of the coefficient of skewness, being larger than 3/8 when the data are positively skewed and smaller than 3/8 when the data are negatively skewed. The same plotting positions can be applied to the logarithms of the data, when using the lognormal distribution, for example.

Once the data series is identified and ranked, and the plotting positions calculated, a graph of magnitude (x) vs. probability [$(P(X > x), P(X < x)$, or T)] can be plotted to graphically fit a distribution. Alternatively, an analytical fit can be made using the method of moments, and the resulting fitted line compared with the sample data.

Example 12.4.1. Perform a probability plotting analysis of the annual maximum discharges of the Guadalupe River near Victoria, Texas, given in Table 12.1.1. Compare the plotted data with the lognormal distribution fitted to them in Example 12.3.3.

Solution. First the data are ranked from largest ($m = 1$), to smallest ($m = n = 44$), as shown in columns 1 and 2 of Table 12.4.1. Blom's plotting formula is used, since the logarithms of the data are being fitted to a normal distribution. Blom's formula uses $b = 3/8$ in Eq. (12.4.7). For example, for $m = 1$, the exceedence probability $P(Q \geq 179{,}000 \text{ cfs}) \approx (m - 3/8)/(n + 1 - 6/8) = (1 - 3/8)/(44 + 1/4) = 0.014$, as shown in column 3 of Table 12.4.1. The corresponding value of the standard normal variable z is determined using $p = 0.014$ in Eqs. (12.3.6) and (12.3.7) in the manner shown in Example 12.3.1; the result, $z = 2.194$, is listed in column 4 of the table. The event magnitude with the same exceedence probability in the fitted lognormal distribution is found using the frequency factor method with $\bar{y} = 4.2743$, $s_y = 0.4027$, and $K_T = z = 2.194$; the result is $\log Q = 4.2743 + 2.194 \times 0.4027 = 5.158$ (column 5). This value is compared with $\log Q$ from the observed data, that is $\log (179{,}000) = 5.253$, as shown in column 6. The observed data are plotted against the fitted curve in Fig. 12.4.1, in which the value of the standard normal variable is used as the horizontal axis to linearize the plot; this is equivalent to using normal probability plotting paper. The plot shows that the fitted line is consistent with the observed data, even including the largest value of 179,000 cfs, which looks quite different from the rest of the data in Fig. 12.1.1.

TABLE 12.4.1
Probability plotting using the normal distribution and Blom's formula for the annual maximum discharges of the Guadalupe River near Victoria, Texas (Example 12.4.1)

Column:	1 Discharge Q (cfs)	2 Rank m	3 Exceedence probability $\frac{m - 3/8}{n + 1/4}$	4 Standard normal variable z	5 Log Q from lognormal distribution	6 Log Q from data
	179,000	1	0.014	2.194	5.158	5.253
	70,000	2	0.037	1.790	4.995	4.845
	58,500	3	0.059	1.561	4.903	4.767
	58,300	4	0.082	1.393	4.835	4.766
	58,000	5	0.105	1.256	4.780	4.763
	.	.	.	.	.	.
	.	.	.	.	.	.
	.	.	.	.	.	.
	5,720	40	0.895	−1.256	3.768	3.757
	4,950	41	0.918	−1.393	3.714	3.695
	4,940	42	0.941	−1.561	3.646	3.694
	4,100	43	0.963	−1.790	3.553	3.613
	1,730	44	0.986	−2.194	3.391	3.238

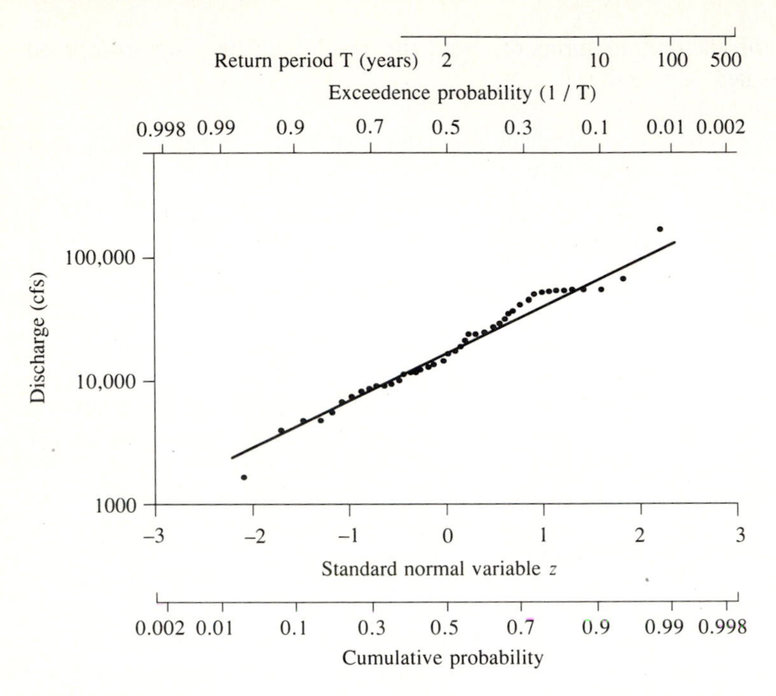

FIGURE 12.4.1
Annual maximum discharge for the Guadalupe River near Victoria, Texas, plotted using Blom's formula on a probability scale for the lognormal distribution.

12.5 WATER RESOURCES COUNCIL METHOD

The U. S. Water Resources Council* recommended that the log–Pearson Type III distribution be used as a base distribution for flood flow frequency studies (U. S. Water Resources Council, 1967, 1976, 1977, and 1981; Benson, 1968). Their decision was an attempt to promote a consistent, uniform approach to flood flow frequency determination for use in all federal planning involving water and related land resources. The choice of the log–Pearson Type III distribution is, however, subjective to some extent, in that there are several criteria that may be employed to select the best distribution, and no single probability distribution is the best under all criteria.

Determination of the Coefficient of Skewness

The coefficient of skewness used in fitting the log–Pearson Type III distribution is very sensitive to the size of the sample and, in particular, is difficult to estimate

---

*The U.S. Water Resources Council was abolished in 1981. The Council's work on guidelines for determining flood flow frequency was taken over by the Interagency Advisory Committee on Water Data, U.S. Geological Survey, Reston, Virginia.

accurately from small samples. Because of this, the Water Resources Council recommended using a generalized estimate of the coefficient of skewness, C_w, based upon the equation

$$C_w = WC_s + (1 - W)C_m \tag{12.5.1}$$

where W is a weighting factor, C_s is the coefficient of skewness computed using the sample data, and C_m is a map skewness, which is read from a map such as Fig. 12.5.1. The weighting factor W is calculated so as to minimize the variance of C_w, as explained next.

The estimates of the sample skew coefficient and the map skew coefficient in Eq. (12.5.1) are assumed to be independent with the same mean and different variances, $V(C_s)$ and $V(C_m)$. The variance of the weighted skew, $V(C_w)$, can be expressed as

$$V(C_w) = W^2V(C_s) + (1 - W)^2V(C_m) \tag{12.5.2}$$

The value of W that minimizes the variance C_w can be determined by differentiating (12.5.2) with respect to W and solving $d[V(C_w)]/dW = 0$ for W to obtain

$$W = \frac{V(C_m)}{V(C_s) + V(C_m)} \tag{12.5.3}$$

The second derivative

$$\frac{d^2V(C_w)}{dW^2} = 2[V(C_s) + V(C_m)] \tag{12.5.4}$$

is greater than zero, confirming that the weight given by (12.5.3) minimizes the variance of the skew, $V(C_w)$.

Determination of W using Eq. (12.5.3) requires knowledge of $V(C_m)$ and $V(C_s)$. $V(C_m)$ is estimated from the map of skew coefficients for the United States as 0.3025. Alternatively, $V(C_m)$ can be derived from a regression study relating the skew to physiographical and meteorological characteristics of the basins (Tung and Mays, 1981).

By substituting Eq. (12.5.3) into Eq. (12.5.1), the weighted skew C_w can be written

$$C_w = \frac{V(C_m)C_s + V(C_s)C_m}{V(C_m) + V(C_s)} \tag{12.5.5}$$

The variance of the station skew C_s for log-Pearson Type III random variables can be obtained from the results of Monte Carlo experiments by Wallis, Matalas, and Slack (1974). They showed that $V(C_s)$ of the logarithmic station skew is a function of record length and population skew. For use in calculating C_w, this function can be approximated with sufficient accuracy as

$$V(C_s) = 10^{A-B\log_{10}(n/10)} \tag{12.5.6}$$

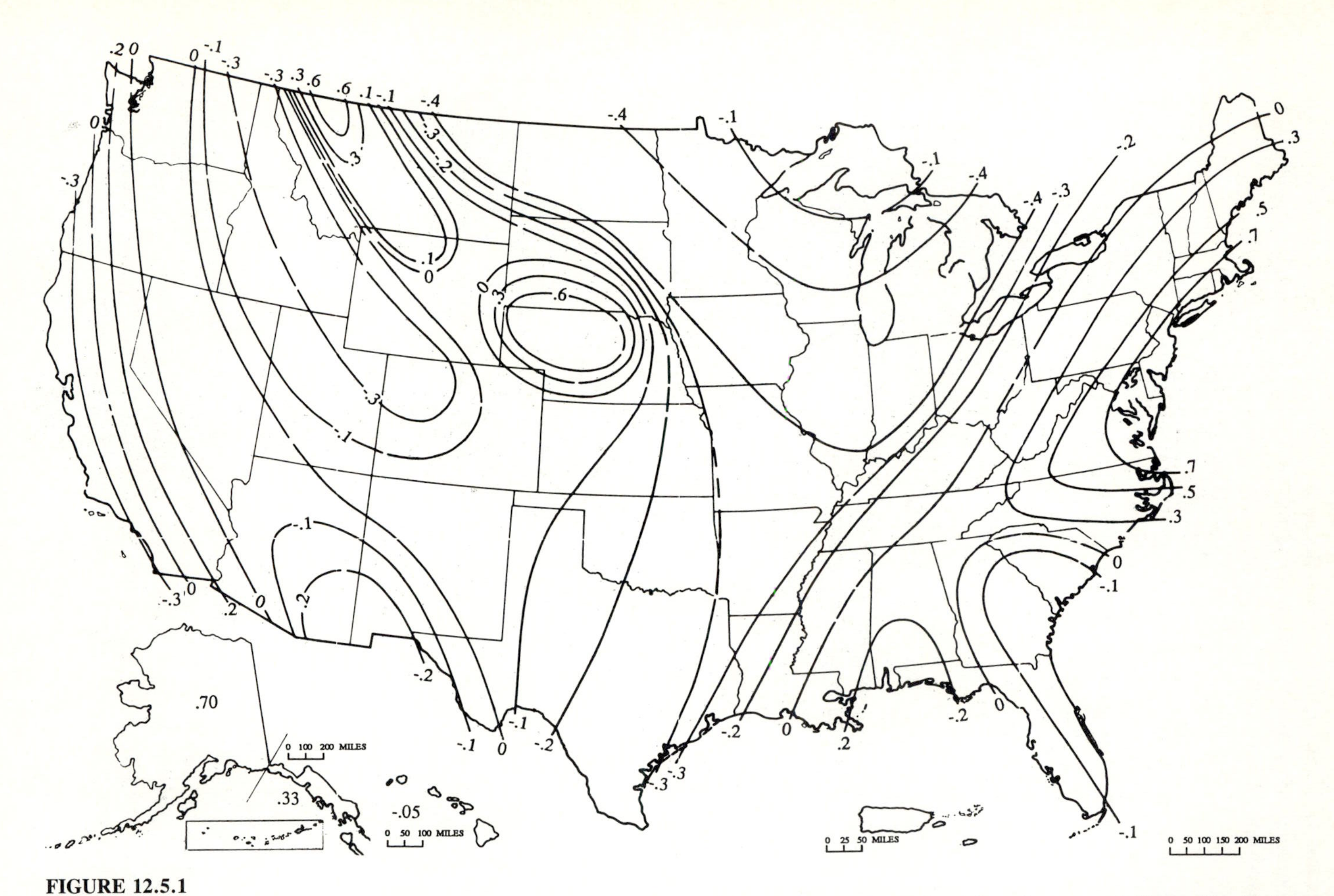

FIGURE 12.5.1
Generalized skew coefficients of annual maximum streamflow. (*Source:* Guidelines for determining flood flow frequency, Bulletin 17B, Hydrology Subcommittee, Interagency Advisory Committee on Water Data, U. S. Geological Survey, Reston, Va. Revised with corrections March 1982.)

where

$$A = -0.33 + 0.08|C_s| \quad \text{if} \quad |C_s| \le 0.90 \qquad (12.5.7a)$$

or

$$A = -0.52 + 0.30|C_s| \quad \text{if} \quad |C_s| > 0.90 \qquad (12.5.7b)$$

$$B = 0.94 - 0.26|C_s| \quad \text{if} \quad |C_s| \le 1.50 \qquad (12.5.7c)$$

or

$$B = 0.55 \quad \text{if} \quad |C_s| > 1.50 \qquad (12.5.7d)$$

in which $|C_s|$ is the absolute value of the station skew (used as an estimate of population skew) and n is the record length in years.

Example 12.5.1. Determine the frequency curve comprising the estimated flood magnitudes for return periods of 2, 5, 10, 25, 50, and 100 years using the Water Resources Council method for data from Walnut Creek at Martin Luther King Blvd. in Austin, Texas, as listed in Table 12.5.1.

Solution. The sample data shown in columns 1 and 2 of Table 12.5.1 cover $n = 16$ years, from 1967 to 1982.

Step 1. Transform the sample data, x_i, to their logarithmic values, y_i; that is, let $y_i = \log x_i$ for $i = 1, \ldots, n$, as shown in column 3 of the table.

TABLE 12.5.1
Calculation of statistics for logarithms of annual maximum discharges for Walnut Creek (Example 12.5.1)

Column:	1	2	3	4	5
	Year	Flow x (cfs)	$y = \log x$	$(y - \bar{y})^2$	$(y - \bar{y})^3$
	1967	303	2.4814	1.3395	−1.5502
	1968	5,640	3.7513	0.0127	0.0014
	1969	1,050	3.0212	0.3814	−0.2356
	1970	6,020	3.7796	0.0198	0.0028
	1971	3,740	3.5729	0.0043	−0.0003
	1972	4,580	3.6609	0.0005	0.0000
	1973	5,140	3.7110	0.0052	0.0004
	1974	10,560	4.0237	0.1481	0.0570
	1975	12,840	4.1086	0.2207	0.1037
	1976	5,140	3.7110	0.0052	0.0004
	1977	2,520	3.4014	0.0564	−0.0134
	1978	1,730	3.2380	0.1606	−0.0644
	1979	12,400	4.0934	0.2067	0.0940
	1980	3,400	3.5315	0.0115	−0.0012
	1981	14,300	4.1553	0.2668	0.1378
	1982	9,540	3.9795	0.1161	0.0396
	Total		58.2206	2.9555	−1.4280
	$n = 16$	$\bar{y} = 3.6388$			

Step 2. Compute the sample statistics. The mean of log–transformed values is

$$\bar{y} = \frac{1}{n}\sum_{i=1}^{n} y_i = \frac{58.22}{16} = 3.639$$

Using column 4 of the table, the standard deviation is

$$\begin{aligned} s_y &= \left(\frac{1}{n-1}\sum_{i=1}^{n}(y_i - \bar{y})^2\right)^{1/2} \\ &= \left(\frac{1}{15}2.9555\right)^{1/2} \\ &= 0.4439 \end{aligned}$$

Using column 5 of the table, the skew coefficient is

$$C_s = \frac{n\sum_{i=1}^{n}(y_i - \bar{y})^3}{(n-1)(n-2)s_y^3} = \frac{16 \times (-1.4280)}{15 \times 14 \times (0.4439)^3} = -1.244$$

Step 3. Compute the weighted skew. The map skew is -0.3 from Fig. 12.5.1 at Austin, Texas. The variance of the station skew can be computed by Eq. (12.5.6) as follows. From (12.5.7*b*) with $|C_s| > 0.90$

$$A = -0.52 + 0.30|-1.244| = -0.147$$

From (12.5.7*c*) with $|C_s| < 1.50$

$$B = 0.94 - 0.26|-1.244| = 0.617$$

Then using (12.5.6)

$$V(C_s) = (10)^{-0.147-0.617\log(16/10)} = 0.533$$

The variance of the generalized skew is $V(C_m) = 0.303$. The weight to be applied to C_s is $W = V(C_m)/[V(C_m) + V(C_s)] = 0.303/(0.303 + 0.533) = 0.362$, and the complementary weight to be applied to C_m is $1 - W = 1 - 0.362 = 0.638$. Then, from (12.5.1)

$$\begin{aligned} C_w &= WC_s + (1 - W)C_m \\ &= 0.362 \times (-1.244) + 0.638 \times (-0.3) \\ &= -0.64 \end{aligned}$$

Step 4. Compute the frequency curve coordinates. The log–Pearson Type III frequency factors K_T for skew coefficient values of -0.6 and -0.7 are found in Table 12.3.1. The values for $C_w = -0.64$ are found by linear interpolation as in Example 12.3.3, with results presented in column 2 of Table 12.5.2. The corresponding value of y_T is found from Eq. (12.3.4), and its antilogarithm is taken to determine the estimated flood magnitude. For example, for $T = 100$ years, $K_T = 1.850$ and

TABLE 12.5.2
Results of frequency analysis using the Water Resources Council method (Examples 12.5.1 and 12.5.2)

Column:	1 Return period T (years)	2 Frequency factor K_T	3 log Q_T	4 Flood Estimates Q_T (cfs)	5 Q'_T (cfs)
	2	0.106	3.686	4,900	5,500
	5	0.857	4.019	10,500	10,000
	10	1.193	4.169	14,700	13,200
	25	1.512	4.310	20,400	17,600
	50	1.697	4.392	24,700	20,900
	100	1.850	4.460	28,900	24,200

The values in column 4 are those computed without adjustment for outliers and those in column 5 after outlier adjustment.

$$
\begin{aligned}
y_T &= \bar{y} + K_T s_y \\
&= 3.639 + 1.850 \times 0.4439 \\
&= 4.460
\end{aligned}
$$

and $Q_T = (10)^{4.460} = 28,900$ cfs, as shown in columns 3 and 4 of the table. Similarly computed flood estimates for the other required return periods are also shown.

As was shown in Example (12.3.3), the increase in flood magnitude is less than directly proportional to the increase in return period. For example, increasing the return period from 10 years to 100 years approximately doubles the estimated flood magnitude in the table. As stated previously, flood magnitudes estimated using the log–Pearson Type III distribution are very sensitive to the value of the skew coefficient. The flood magnitudes for the longer return periods (50 and 100 years) are difficult to estimate reliably from only 16 years of data.

Testing for Outliers

The Water Resources Council method recommends that adjustments be made for outliers. *Outliers* are data points that depart significantly from the trend of the remaining data. The retention or deletion of these outliers can significantly affect the magnitude of statistical parameters computed from the data, especially for small samples. Procedures for treating outliers require judgment involving both mathematical and hydrologic considerations. According to the Water Resources Council (1981), if the station skew is greater than +0.4, tests for high outliers are considered first; if the station skew is less than −0.4, tests for low outliers are considered first. Where the station skew is between ±0.4, tests for both high and low outliers should be applied before eliminating any outliers from the data set.

The following frequency equation can be used to detect high outliers:

$$y_H = \bar{y} + K_n s_y \tag{12.5.8}$$

where y_H is the high outlier threshold in log units and K_n is as given in Table 12.5.3 for sample size n. The K_n values in Table 12.5.3 are used in *one-sided tests* that detect outliers at the 10-percent level of significance in normally distributed data. If the logarithms of the values in a sample are greater than y_H in the above equation, then they are considered high outliers. Flood peaks considered high outliers should be compared with historic flood data and flood information at nearby sites. Historic flood data comprise information on unusually extreme events outside of the systematic record. According to the Water Resources Council (1981), if information is available that indicates a high outlier is the maximum over an extended period of time, the outlier is treated as historic flood data and excluded from analysis. If useful historic information is not available to compare to high outliers, then the outliers should be retained as part of the systematic record.

A similar equation can be used to detect low outliers:

$$y_L = \bar{y} - K_n s_y \tag{12.5.9}$$

where y_L is the low outlier threshold in log units. Flood peaks considered low outliers are deleted from the record and a conditional probability adjustment described by the Water Resources Council (1981) can be applied.

Example 12.5.2. Using the data for the Walnut Creek example (Table 12.5.1), determine if there are any high or low outliers for the sample. If so, omit them from the data set and recalculate the flood frequency curve.

TABLE 12.5.3
Outlier test K_n values

Sample size n	K_n	Sample size n	K_n	Sample size n	K_n	Sample size n	K_n
10	2.036	24	2.467	38	2.661	60	2.837
11	2.088	25	2.486	39	2.671	65	2.866
12	2.134	26	2.502	40	2.682	70	2.893
13	2.175	27	2.519	41	2.692	75	2.917
14	2.213	28	2.534	42	2.700	80	2.940
15	2.247	29	2.549	43	2.710	85	2.961
16	2.279	30	2.563	44	2.719	90	2.981
17	2.309	31	2.577	45	2.727	95	3.000
18	2.335	32	2.591	46	2.736	100	3.017
19	2.361	33	2.604	47	2.744	110	3.049
20	2.385	34	2.616	48	2.753	120	3.078
21	2.408	35	2.628	49	2.760	130	3.104
22	2.429	36	2.639	50	2.768	140	3.129
23	2.448	37	2.650	55	2.804		

Source: U.S. Water Resources Council, 1981. This table contains one-sided 10-percent significance level K_n values for the normal distribution.

Solution.

Step 1. Determine the threshold value for high outliers. From Table 12.5.3, $K_n = 2.279$ for $n = 16$ data. From Eq. (12.5.8) using $\bar{y}$ and s_y from Example 12.5.1

$$y_H = \bar{y} + K_n s_y = 3.639 + 2.279(0.4439) = 4.651$$

Then

$$Q_H = (10)^{4.651} = 44,735 \text{ cfs}$$

The largest recorded value (14,300 cfs in Table 12.5.1) does not exceed the threshold value, so there are no high outliers in this sample.

Step 2. Determine the threshold value for low outliers. The same K_n value is used:

$$y_L = \bar{y} - K_n s_y = 3.639 - 2.279(0.4439) = 2.627$$

$$Q_L = (10)^{2.627} = 424 \text{ cfs}$$

The 1967 peak flow of 303 cfs is less than Q_L and so is considered a low outlier.

Step 3. The low outlier is deleted from the sample and the frequency analysis is repeated using the same procedure as in Example 12.5.1. The statistics for the logarithms of the new data set, now reduced to 15 values, are $\bar{y} = 3.716$, $s_y = 0.3302$, and $C_s = -0.545$. It can be seen that the omission of the 303 cfs value has significantly altered the calculated skewness value (from the -1.24 found in Example 12.5.1). The map skewness remains at -0.3 for Austin, Texas, and the revised weighted skewness is $C_w = -0.41$. Values of K_T are interpolated from Table 12.3.1 at the required return periods, and the corresponding flood flow estimates computed as Q'_T, listed in column 5 of Table 12.5.2. By comparing these values with those given in column 4 for the full data set, it can be seen that the effect of removing the low outlier in this example is to decrease the flood estimates for the longer return periods.

Computer Program HECWRC

The computer program HECWRC (U. S. Army Corps of Engineers, 1982) performs flood flow frequency analysis of annual maximum flood series according to the U. S. Water Resources Council Bulletin 17B (1981). This program is available from the U. S. Army Corps of Engineers Hydrologic Engineering Center in Davis, California, in both a mainframe computer version and a microcomputer version.

12.6 RELIABILITY OF ANALYSIS

The reliability of the results of frequency analysis depends on how well the assumed probabilistic model applies to a given set of hydrologic data.

Confidence Limits

Statistical estimates are often presented with a range, or *confidence interval*, within which the true value can reasonably be expected to lie. The size of the confidence interval depends on the *confidence level* β. The upper and lower boundary values of the confidence interval are called *confidence limits* (Fig. 12.6.1).

Corresponding to the confidence level β is a *significance level* α, given by

$$\alpha = \frac{1 - \beta}{2} \tag{12.6.1}$$

For example, if $\beta = 90$ percent, then $\alpha = (1 - 0.9)/2 = 0.05$, or 5 percent.

For estimating the event magnitude for return period T, the upper limit $U_{T,\alpha}$ and lower limit $L_{T,\alpha}$ may be specified by adjustment of the frequency factor equation:

$$U_{T,\alpha} = \bar{y} + s_y K^U_{T,\alpha} \tag{12.6.2}$$

and

$$L_{T,\alpha} = \bar{y} + s_y K^L_{T,\alpha} \tag{12.6.3}$$

where $K^U_{T,\alpha}$ and $K^L_{T,\alpha}$ are the upper and lower confidence limit factors, which can be determined for normally distributed data using the noncentral t distribution (Kendall and Stuart, 1967). The same factors are used to construct approximate confidence limits for the Pearson Type III distribution. Approximate values for these factors are given by the following formulas (Natrella, 1963; U. S. Water Resources Council, 1981):

$$K^U_{T,\alpha} = \frac{K_T + \sqrt{K_T^2 - ab}}{a} \tag{12.6.4}$$

$$K^L_{T,\alpha} = \frac{K_T - \sqrt{K_T^2 - ab}}{a} \tag{12.6.5}$$

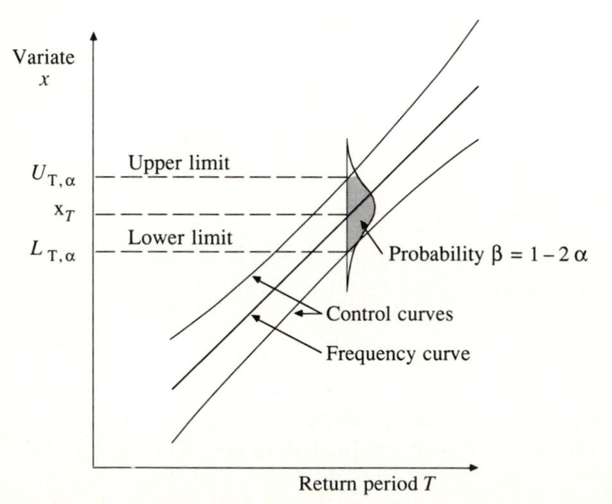

FIGURE 12.6.1
Definition of confidence limits.

in which

$$a = 1 - \frac{z_\alpha^2}{2(n-1)} \tag{12.6.6}$$

and

$$b = K_T^2 - \frac{z_\alpha^2}{n} \tag{12.6.7}$$

The quantity z_α is the standard normal variable with exceedence probability α.

Example 12.6.1. Determine the 90-percent confidence limits for the 100-year discharge for Walnut Creek, using the data presented in Example 12.5.1. The logarithmic mean, standard deviation, and skew coefficient are 3.639, 0.4439, and −0.64, respectively, for 16 years of data.

Solution. For $\beta = 0.9$, $\alpha = 0.05$ and the required standard normal variable z_α has exceedence probability 0.05, or cumulative probability 0.95. From Table 11.2.1, the required value is $z_\alpha = 1.645$. The frequency factor K_T for $T = 100$ years was calculated in Example 12.5.1 as $K_{100} = 1.850$. Hence, by Eqs. (12.6.4) to (12.6.7)

$$a = 1 - \frac{z_\alpha^2}{2(n-1)} = 1 - \frac{(1.645)^2}{2(16-1)} = 0.9098$$

$$b = K_T^2 - \frac{z_\alpha^2}{n} = (1.850)^2 - \frac{(1.645)^2}{16} = 3.253$$

$$K_{100,0.05}^U = \frac{K_T + \sqrt{K_T^2 - ab}}{a} = \frac{1.850 + [(1.850)^2 - 0.9098 \times 3.253]^{1/2}}{0.9098}$$

$$= 2.781$$

$$K_{100,0.05}^L = \frac{K_T - \sqrt{K_T^2 - ab}}{a} = \frac{1.850 - [(1.850)^2 - 0.9098 \times 3.253]^{1/2}}{0.9098}$$

$$= 1.286$$

The confidence limits are computed using Eqs. (12.6.2) and (12.6.3):

$$\begin{aligned} U_{100,0.05} &= \bar{y} + s_y K_{100,0.05}^U \\ &= 3.639 + 0.4439 \times 2.781 \\ &= 4.874 \end{aligned}$$

$$\begin{aligned} L_{100,0.05} &= \bar{y} + s_y K_{100,0.05}^L \\ &= 3.639 + 0.4439 \times 1.286 \\ &= 4.210 \end{aligned}$$

The corresponding discharges for the upper and lower limits are $(10)^{4.874} = 74,820$ cfs, and $(10)^{4.210} = 16,200$ cfs, respectively, as compared to an estimated event magnitude of 28,900 cfs from Table 12.5.2. The confidence interval is quite wide in this case because the sample size is small. As the sample size increases, the width of the confidence interval around the estimated flood magnitude will diminish.

Standard Error

The *standard error of estimate* s_e is a measure of the standard deviation of event magnitudes computed from samples about the true event magnitude. Formulas for the standard error of estimate for the normal and Extreme Value Type I distributions are (Kite, 1977):

Normal

$$s_e = \left(\frac{2 + z^2}{n}\right)^{1/2} s \tag{12.6.8}$$

Extreme Value Type I

$$s_e = \left[\frac{1}{n}(1 + 1.1396K_T + 1.1000K_T^2)\right]^{1/2} s \tag{12.6.9}$$

where s is the standard deviation of the original sample of size n. Standard errors may be used to construct confidence limits in a similar manner to that illustrated in Example 12.6.1, except that in this case the confidence limits for significance level α are defined as $x_T \pm s_e z_\alpha$.

Example 12.6.2. Determine the standard error of estimate and the 90 percent confidence limits of the 5-year-return-period, 10-minute-duration rainfall at Chicago, Illinois. From Example 12.3.2, the estimated 5-year depth is $x_T = 0.78$ in; also, $s = 0.177$ in, $K_T = 0.719$, and $n = 35$.

Solution. The standard error is computed for the Extreme Value Type I distribution using Eq. (12.6.9)

$$\begin{aligned} s_e &= \left[\frac{1}{n}(1 + 1.1396K_T + 1.1000K_T^2)\right]^{1/2} s \\ &= \left\{\frac{1}{35}\left[1 + 1.1396 \times 0.719 + 1.1000 \times (0.719)^2\right]\right\}^{1/2} \times 0.177 \\ &= 0.046 \text{ in} \end{aligned}$$

The 90 percent confidence limits, with $z_\alpha = 1.645$ for $\alpha = 0.05$, are $x_T \pm s_e z_\alpha = 0.78 \pm 0.046 \times 1.645 = 0.70$ and 0.86 in. Thus the 5 year, 10-minute rainfall estimate in Chicago is 0.78 in with 90 percent confidence limits [0.70, 0.86] in.

Expected Probability

Expected probability is defined as the average of the true exceedence probabilities of all magnitude estimates that might be made from successive samples of a specified size for a specified flood frequency (Beard, 1960; U. S. Water Resources

Council, 1981). The flood magnitude estimate computed for a given sample is approximately the median of all possible estimates; that is, there is an approximately equal chance that the true magnitude will be either above or below the estimated magnitude. But the probability distribution of the estimate is positively skewed, so the average of the magnitudes computed from many samples is larger than the median. The skewness arises because flood magnitude has a lower bound at zero but no upper bound.

The consequence of the discrepancy between the median and the mean flood estimate is that, if a very large number of estimates of flood magnitude are made over a region, on average more 100-year floods will occur than expected (Beard, 1978). The expected probability of occurence of flood events in any year can be estimated for events of nominal return period T by the following formulas, which are derived for the normal distribution, and apply approximately to the Pearson Type III distribution (Beard, 1960; Hardison and Jennings, 1972).

The expected probability for the normal distribution is expressed for a sample size of n as

$$E(P_n) = P\left[t_{n-1} > z\left(\frac{n}{n+1}\right)^{1/2}\right] \tag{12.6.10}$$

where z is the standard normal variable for the desired probability of exceedence and t_{n-1} is the student's t-statistic with $n - 1$ degrees of freedom. Calculation can be performed using the appropriate tables for t_{n-1} and z. These computations can also be carried out using the following equations (U. S. Water Resources Council, 1981; U. S. Army Corps of Engineers, 1972).

T (years)	**Exceedence probability**	**Expected probability $E(P_n)$**	
1000	0.001	$0.001\left(1.0+\frac{280}{n^{1.55}}\right)$	(12.6.11*a*)
100	0.01	$0.01\left(1.0+\frac{26}{n^{1.16}}\right)$	(12.6.11*b*)
20	0.05	$0.05\left(1.0+\frac{6}{n^{1.04}}\right)$	(12.6.11*c*)
10	0.10	$0.10\left(1.0+\frac{3}{n^{1.04}}\right)$	(12.6.11*d*)
3.33	0.30	$0.30\left(1.0+\frac{0.46}{n^{0.925}}\right)$	(12.6.11*e*)

Example 12.6.3. Determine the expected probability for the 100-year discharge for the Walnut Creek data given in Example 12.5.1 ($n = 16$).

Solution. For $T = 100$ years, use Eq. (12.6.11b) to obtain

$$E(P_n) = 0.01\left(1 + \frac{26}{n^{1.16}}\right)$$

$$= 0.01\left(1 + \frac{26}{(16)^{1.16}}\right)$$

$$= 0.020$$

The 100-year discharge according to the above adjustment has an expected probability of 0.02 (not 0.01) or a return period of 1/0.02 = 50 years.

REFERENCES

Abramowitz, M., and Stegun, I. A., *Handbook of Mathematical Functions,* Dover, New York, 1965.

Beard, L. R., Probability estimates based on small normal distribution samples, *J. Geophys. Res.,* vol. 65, no. 7, pp. 2143–2148, 1960.

Beard, L. R., Statistical methods in hydrology, U. S. Army Corps of Engineers, January 1962.

Beard, L. R., Impact of hydrologic uncertainties on flood insurance, *J. Hyd. Div., Am. Soc. Civ. Eng.,* vol. 104, no. HY11, pp. 1473–1484, 1978.

Benson, M. A., Uniform flood-frequency estimating methods for federal agencies, *Water Resour. Res.,* vol. 4, no. 5, pp. 891–908, 1968.

Blom, G., *Statistical Estimates and Transformed Beta Variables,* Wiley, New York, pp. 68–75 and 143–146, 1958.

Chow, V. T., A general formula for hydrologic frequency analysis, *Trans. Am. Geophysical Union,* vol. 32, no. 2, pp. 231–237, 1951.

Chow, V. T., Frequency analysis of hydrologic data with special application to rainfall intensities, bulletin no. 414, University of Illinois Eng. Expt. Station, 1953.

Chow, V. T., The log-probability law and its engineering applications, *Proc. Am. Soc. Civ. Eng.,* vol. 80, no. 536, pp. 1–25, September 1954.

Chow, V. T., Statistical and probability analysis of hydrologic data, sec. 8-I in *Handbook of Applied Hydrology,* ed. by V. T. Chow, McGraw-Hill, New York, 1964.

Cunnane, C., Unbiased plotting positions—a review, *J. Hydrol.,* vol. 37, pp. 205–222, 1978.

Gringorten, I. I., A plotting rule for extreme probability paper, *J. Geophys. Res.,* vol. 68, no. 3, pp. 813–814, 1963.

Gumbel, E. J., The statistical theory of droughts, *Proc. Am. Soc. Civ. Eng.,* vol. 80, pp. 439-1 to 439-19, 1954.

Gumbel, E. J., Statistical forecast of droughts, *Bull. Int. Assn. Sci. Hydrol.,* 8th year, no. 1, pp. 5–23, 1963.

Hardison, C. H., and M. E. Jennings, Bias in computed flood risk, *J. Hyd. Div., Am. Soc. Civ. Eng.,* vol. 98, no. HY3, pp. 415–427, March 1972.

Hazen, A., *Flood Flows, a Study of Frequencies and Magnitudes,* Wiley, New York, 1930.

Kendall, M. G., and A. Stuart, *The Advanced Theory of Statistics,* vol. 2, 2nd ed., Hafner, New York, 1967.

Kite, G. W., *Frequency and Risk Analysis in Hydrology,* Water Resources Publications, Fort Collins, Colo., 1977.

Natrella, M. G., *Experimental Statistics,* National Bureau of Standards Handbook 91, 1963.

Natural Environment Research Council, Flood studies report, vol. 1, hydrological studies, Natural Environment Research Council, London (available from Institute of Hydrology, Wallingford, Oxon, England), 1975.

Tomlinson, A. I., The frequency of high intensity rainfalls in New Zealand, *Water and Soil Tech. Publ.*, no. 19, Ministry of Works and Development, Wellington, New Zealand, 1980.

Tung, Y.-K., and L. W. Mays, Reducing hydrologic parameter uncertainty, *J. Water Res. Planning and Management Div., Am. Soc. Civ. Eng.*, vol. 107, no. WR1, pp. 245–262, March 1981.

U. S. Army Corps of Engineers Hydrologic Engineering Center, Regional frequency computation, generalized computer program, Davis, California, July 1972.

U. S. Army Corps of Engineers Hydrologic Engineering Center, Flood flow frequency analysis, computer program 723-X6-L7550 user's manual, Davis, California, February 1982.

U. S. Water Resources Council, A uniform technique for determining flood flow frequencies, bulletin 15, Washington, D.C., 1967.

U. S. Water Resources Council, Guidelines for determining flood flow frequency, bulletin 17, Washington, D.C., 1976.

U. S. Water Resources Council, Guidelines for determining flood flow frequency, bulletin 17A, Washington, D.C., 1977.

U. S. Water Resources Council (now called Interagency Advisory Committee on Water Data), Guidelines for determining flood flow frequency, bulletin 17B, available from Office of Water Data Coordination, U.S. Geological Survey, Reston, VA 22092, 1981.

Wallis, J. R., N. C. Matalas, and J. R. Slack, Just a moment, *Water Resour. Res.*, vol. 10, no. 2, pp. 211–219, April 1974.

PROBLEMS

12.1.1 Estimate the return period of an annual maximum discharge of 40,000 cfs from the data given in Table 12.1.1.

12.1.2 Estimate the return period of annual maximum discharges of 10,000, 20,000, 30,000, 40,000 and 50,000 cfs for the Guadalupe River at Victoria, Texas, from the data given in Table 12.1.1. Plot a graph of flood discharge vs. return period from the results.

12.1.3 Calculate the probability that a 100-year flood will occur at a given site at least once during the next 5, 10, 50, and 100 years. What is the chance that a 100-year flood will not occur at this site during the next 100 years?

12.1.4 What is the probability that a five-year flood will occur (*a*) in the next year, (*b*) at least once during the next five years, and (*c*) at least once during the next 50 years?

12.2.1 Calculate the 20-year and 100-year return period rainfall of 10 minutes duration at Chicago using the data given in Table 12.2.1. Use the Extreme Value Type I distribution.

12.3.1 (*a*) For the annual maximum series given below, determine the 25-, 50-, and 100-year peak discharges using the Extreme Value Type I distribution.

Year	1	2	3	4	5	6	7
Peak discharge (cfs)	4,780	1,520	9,260	17,600	4,300	21,200	12,000
Year	8	9	10	11	12	13	14
Peak discharge	2,840	2,120	3,170	3,490	3,920	3,310	13,200
Year	15	16	17	18	19	20	21
Peak discharge	9,700	3,380	9,540	12,200	20,400	7,960	15,000
Year	22	23	24	25	26	27	
Peak discharge	3,930	3,840	4,470	16,000	6,540	4,130	

(*b*) Determine the risk that a flow equaling or exceeding 25,000 cfs will occur at this site during the next 15 years.

(*c*) Determine the return period for a flow rate of 15,000 cfs.

12.3.2 The maximum discharges as recorded at a river gaging station are as follows:

Date of Occurrence	Discharge (cfs)	Date of Occurrence	Discharge (cfs)
1940 June 23	908	1944 Feb. 26	1610
1941 Feb. 13	1930	1944 March 13	4160
1941 March 20	3010	1945 May 14	770
1941 May 31	2670	1946 Jan. 5	5980
1941 June 3	2720	1946 Jan. 9	2410
1941 June 28	2570	1946 March 5	1650
1941 Sept. 8	1930	1947 March 13	1260
1941 Oct. 23	2270	1948 Feb. 28	4630
1942 June 3	1770	1948 March 15	2690
1942 June 10	1770	1948 March 19	4160
1942 June 11	1970	1949 Jan. 4	1680
1942 Sept. 3	1570	1949 Jan. 15	1640
1942 Dec. 27	3850	1949 Feb. 13	2310
1943 Feb. 20	2650	1949 Feb. 18	3300
1943 March 15	2450	1949 Feb. 24	3460
1943 June 2	1290	1950 Jan. 25	3050
1943 June 20	1200	1950 March 5	2880
1943 Aug. 2	1200	1950 June 2	1450
1944 Feb. 23	1490		

Select the annual maximum series from this data set. By fitting the annual maximum data to an Extreme Value Type I distribution, determine the flood flow for 10-, 50-, and 100-year return periods.

12.3.3 Select the annual exceedence series from the data set given in Prob. 12.3.2 and calculate the 10-, 50-, and 100-year discharge values from these data using the Extreme Value Type I distribution. Compare the computed values with those obtained in Prob. 12.3.2.

12.3.4 Solve Prob. 12.3.2 using the lognormal distribution.

12.3.5 Solve Prob. 12.3.2 using the log–Pearson Type III distribution.

12.3.6 The record of annual peak discharges at a stream gaging station is as follows:

Year	1961	1962	1963	1964	1965	1966	1967	1968	1969
Discharge (m^3/s)	45.3	27.5	16.9	41.1	31.2	19.9	22.7	59.0	35.4

Determine using the lognormal distribution

(*a*) The probability that an annual flood peak of 42.5 m^3/s will not be exceeded.

(*b*) The return period of a discharge of 42.5 m^3/s.

(*c*) The magnitude of a 20-year flood.

12.3.7 Show that the frequency factor for the Extreme Value Type I distribution is given by

$$K_T = -\frac{\sqrt{6}}{\pi}\left[0.5772 + \ln\left(\ln\frac{T}{T-1}\right)\right]$$

12.4.1 Plot the annual maximum discharge data from Walnut Creek given in Table 12.5.1 on a lognormal probability scale using Blom's plotting formula.

12.4.2 Solve Prob. 12.4.1 using the Weibull plotting formula and compare the results of the two plotting formulas.

12.4.3 Plot the data given in Prob. 12.3.1 on an Extreme Value Type I probability scale using the reduced variate y as the horizontal axis and discharge as the vertical axis. Use the Gringorten plotting formula.

12.4.4 Solve Prob. 12.4.3 using the Weibull plotting formula and compare the results of the two plotting formulas.

12.5.1 Perform a frequency analysis for the annual maximum discharge of Walnut Creek using the data given in Table 12.5.1, employing the log–Pearson Type III distribution without the U. S. Water Resources Council corrections for skewness and outliers. Compare your results with those given in Table 12.5.2 for the 2-, 5-, 10-, 25-, 50-, and 100-year events.

12.5.2 Using the log–Pearson Type III distribution and the hydrologic data in the following table, compute the 2-, 5-, 10-, 25-, 50-, and 100-year annual maximum floods at Leaf River, Illinois. Use the U. S. Water Resources Council method for skewness and check for outliers. The map skew for Leaf River is −0.4.

Annual maximum discharges for Leaf River, Illinois

Year	1940	1941	1942	1943	1944	1945	1946	1947	1948	1949	1950
Discharge (cfs)	2160	3210	3070	4000	3830	978	6090	1150	6510	3070	3360

12.5.3 Using the annual maximum flows given below for Mills Creek near Los Molinos, California, determine the 2-, 10-, 25-, 50-, and 100-year flood peaks using the log–Pearson Type III distribution with the U. S. Water Resources Council skewness adjustment. The map skewness at Los Molinas is $C_m = 0$.

Year	1929	1930	1931	1932	1933	1934	1935	1936
Discharge (cfs)	1,520	6,000	1,500	5,440	1,080	2,630	4,010	4,380
Year	1937	1938	1939	1940	1941	1942	1943	1944
Discharge	3,310	23,000	1,260	11,400	12,200	11,000	6,970	3,220
Year	1945	1946	1947	1948	1949	1950	1951	1952
Discharge	3,230	6,180	4,070	7,320	3,870	4,430	3,870	5,280
Year	1953	1954	1955	1956	1957	1958		
Discharge	7,710	4,910	2,480	9,180	6,150	6,880		

The statistics of the logarithms to base 10 of these data are: mean 3.6656, standard deviation 0.3031, coefficient of skewness −0.165.

12.5.4 The station record for Fishkill Creek at Beacon, New York, has a mean of the tranformed flows (log Q) of 3.3684, a standard deviation of transformed flows of 0.2456, and a skew coefficient of the tranformed flows of 0.7300. The station record is in cfs and is based upon 24 values.

(*a*) Determine the flood discharge for 2-, 20-, and 100-year return periods using the lognormal distribution.

(*b*) Determine the flood discharges for the same return periods using the sample skew for the log–Pearson III distribution.

(*c*) Determine the flood discharges using the procedure as recommended by the U. S. Water Resources Council. The map skew is 0.6. Compare the results obtained in parts (*a*), (*b*), and (*c*).

12.5.5 Use the U. S. Water Resources Council method to determine the 2-, 10-, 25-, 50-, and 100-year peak discharges for the station record of the San Gabriel River at Georgetown, Texas. The map skew is −0.3.

Year	1935	1936	1937	1938	1939	1940	1941	1942
Discharge (cfs)	25,100	32,400	16,300	24,800	903	34,500	30,000	18,600
Year	1943	1944	1945	1946	1947	1948	1949	1950
Discharge	7,800	37,500	10,300	8,000	21,000	14,000	6,600	5,080
Year	1951	1952	1953	1954	1955	1956	1957	1958
Discharge	5,350	11,000	14,300	24,200	12,400	5,660	155,000	21,800
Year	1959	1960	1961	1962	1963	1964	1965	1966
Discharge	3,080	71,500	22,800	4,040	858	13,800	26,700	5,480
Year	1967	1968	1969	1970	1971	1972	1973	
Discharge	1,900	21,800	20,700	11,200	9,640	4,790	18,100	

12.5.6 Solve Prob. 12.5.5 using the U. S. Army Corps of Engineers computer program HECWRC for flood flow frequency analysis with the log–Pearson III distribution.

12.5.7 Use the U. S. Water Resources Council method to determine the 2-, 10-, 25-, 50-, and 100-year peak discharges for the station record (Table 12.1.1) for the Guadalupe River at Victoria, Texas. The map coefficient of skewness is −0.3.

12.5.8 Solve Prob. 12.5.7 using the U. S. Army Corps of Engineers computer program HECWRC for flood flow frequency analysis with the log–Pearson III distribution.

12.6.1 Plot the 90-percent confidence limits of the flood flow frequency curve for the Walnut Creek data given in Table 12.5.1. Consider the 2-, 10-, 25-, 50-, and 100-year return periods.

12.6.2 Plot the 90-percent confidence limits of the flood flow frequency curve for the Los Molinos, California station record (Prob. 12.5.3). Consider the 2-, 10-, 25-, 50-, and 100-year return periods.

12.6.3 Plot the 90-percent confidence limits of the flood flow frequency curve for the San Gabriel River at Georgetown, Texas (Prob. 12.5.5). Consider the 2-, 10-, 25-, 50-, and 100-year return periods.

12.6.4 Plot the 90-percent confidence limits of the flood flow frequency curve for the Guadalupe River at Victoria, Texas (Prob. 12.5.7).

12.6.5 Determine the expected probability of a 10-year event for the Walnut Creek data (Table 12.5.1).

12.6.6 Determine the expected probability of a 10-year and a 100-year flood on the Guadalupe River at Victoria, Texas (data given in Table 12.1.1).

12.6.7 Determine the expected probability of a 10-year and a 100-year flood discharge estimated for the San Gabriel River at Georgetown, Texas (Prob. 12.5.5).

CHAPTER
13

HYDROLOGIC DESIGN

Hydrologic design is the process of assessing the impact of hydrologic events on a water resource system and choosing values for the key variables of the system so that it will perform adequately. Hydrologic design may be used to develop plans for a new structure, such as a flood control levee, or to develop management programs for better control of an existing system, for example, by producing a flood plain map for limiting construction near a river. There are many factors besides hydrology that bear on the design of water resource systems; these include public welfare and safety, economics, aesthetics, legal issues, and engineering factors such as geotechnical and structural design. While the central concern of the hydrologist is on the flow of water through a system, he or she must also be aware of these other factors and of how the hydrologic operation of the system might affect them. In this sense hydrologic design is a much broader subject than hydrologic analysis as covered in previous chapters.

13.1 HYDROLOGIC DESIGN SCALE

The purposes of water resources planning and management may be grouped roughly into two categories. One is *water control,* such as drainage, flood control, pollution abatement, insect control, sediment control, and salinity control. The other is *water use* and management, such as domestic and industrial water supply, irrigation, hydropower generation, recreation, fish and wildlife improvement, low-flow augmentation for water quality management, and watershed management. In either case, the task of the hydrologist is the same, namely, to determine a design inflow, to route the flow through the system, and to check whether the output values are satisfactory. The difference between the two cases is that design for water control is usually concerned with extreme events of short duration, such as the instantaneous peak discharge during a flood, or the minimum flow over a period of a few days during a dry period, while design for water use is concerned with the complete flow hydrograph over a period of years.

The *hydrologic design scale* is the range in magnitude of the design variable (such as the design discharge) within which a value must be selected to determine the inflow to the system (see Fig. 13.1.1). The most important factors in selecting the design value are cost and safety. It is too costly to design small structures such as culverts for very large peak discharges; however, if a major hydraulic structure, such as the spillway on a large dam, is designed for too small a flood, the result might be a catastrophe, such as a dam's failure. The optimal magnitude for design is one that balances the conflicting considerations of cost and safety.

Estimated Limiting Value

The practical upper limit of the hydrologic design scale is not infinite, since the global hydrologic cycle is a closed system; that is, the total quantity of water on earth is essentially constant. Some hydrologists recognize no upper limit, but such a view is physically unrealistic. The lower limit of the design scale is zero in most cases, since the value of the design variable cannot be negative. Although the true upper limit is usually unknown, for practical purposes an estimated upper limit may be determined. This *estimated limiting value* (ELV) is defined as *the largest magnitude possible for a hydrologic event at a given location, based on the best available hydrologic information.* The range of uncertainty for the ELV

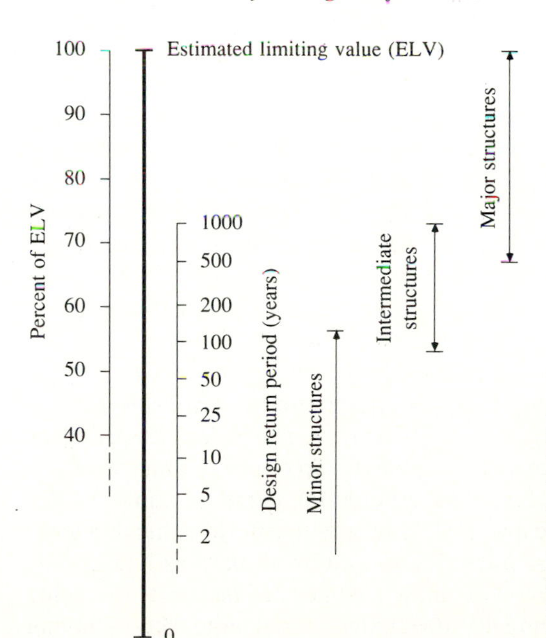

FIGURE 13.1.1
Hydrologic design scale. Approximate ranges of the design level for different types of structures are shown. Design may be based on a percentage of the ELV or on a design return period. The values for the two scales shown in the diagram are illustrative only and do not correspond directly with one another.

depends on the reliability of information, technical knowledge, and accuracy of analysis. As information, knowledge, and analysis improve, the estimate better approximates the true upper limit, and its range of uncertainty decreases. There have been cases in which observed hydrologic events exceeded their previously estimated limiting values.

The concept of an estimated limiting value is implicit in the commonly used *probable maximum precipitation* (PMP) and the corresponding *probable maximum flood* (PMF). The probable maximum precipitation is defined by the World Meteorological Organization (1983) as a "quantity of precipitation that is close to the physical upper limit for a given duration over a particular basin." Based on worldwide records, the PMP can have a return period of as long as 500,000,000 years, corresponding approximately to a frequency factor of 15. However, the return period varies geographically. Some would arbitrarily assign a return period, say 10,000 years, to the PMP or PMF, but this suggestion has no physical basis.

Probability-Based Limits

Because of its unknown probability, the estimated limiting value is used deterministically. Lower down on the design scale, a probability- or frequency-based approach is commonly adopted. The magnitudes of hydrologic events at this level are smaller, usually within or near the range of frequent observations. As a result, their probabilities of occurrence can be estimated adequately when hydrologic records of sufficient length are available for frequency analysis. The probabilistic approach is less subjective and more theoretically manageable than the deterministic approach. Probabilistic methods also lead to logical ways of determining optimum design levels, such as by hydroeconomic and risk analyses, which will be discussed in Sec. 13.2.

For a densely populated area, where the failure of water-control works would result in loss of life and extensive property damage, a design using the ELV might be justified. In a less populous area where failure would result only in minor damage, a design for a much smaller degree of protection is reasonable. Between these extremes on the hydrologic design scale, varying conditions exist and varying design values are required. When the probabilistic behavior of a hydrologic event can be determined, it is usually best to use the event magnitude for a specified return period as a design value.

Based on past experience and judgment, some generalized design criteria for water-control structures have been developed, as summarized in Table 13.1.1. According to the potential consequence of failure, structures are classified as *major*, *intermediate* and *minor*; the corresponding approximate ranges on the design scale are shown in Fig. 13.1.1. The criteria for dams in Table 13.1.1 pertain to the design of spillway capacities, and are taken from the National Academy of Sciences (1983). The Academy defines a small dam as having 50–1000 acre·ft of storage or being 25–40 ft high, an intermediate dam as having 1000–50,000 acre·ft of storage or being 40–100 ft high, and a large dam as having more than 50,000 acre·ft of storage or being more than 100 ft high. In general,

TABLE 13.1.1
Generalized design criteria for water-control structures

Type of structure	Return period (years)	ELV
Highway culverts		
Low traffic	5–10	—
Intermediate traffic	10–25	—
High traffic	50–100	—
Highway bridges		
Secondary system	10–50	—
Primary system	50–100	—
Farm drainage		
Culverts	5–50	—
Ditches	5–50	—
Urban drainage		
Storm sewers in small cities	2–25	—
Storm sewers in large cities	25–50	—
Airfields		
Low traffic	5–10	—
Intermediate traffic	10–25	—
High traffic	50–100	—
Levees		
On farms	2–50	—
Around cities	50–200	—
Dams with no likelihood of loss of life (low hazard)		
Small dams	50–100	—
Intermediate dams	100+	—
Large dams	—	50–100%
Dams with probable loss of life (significant hazard)		
Small dams	100+	50%
Intermediate dams	—	50–100%
Large dams	—	100%
Dams with high likelihood of considerable loss of life (high hazard)		
Small dams	—	50–100%
Intermediate dams	—	100%
Large dams	—	100%

there would be considerable loss of life and extensive damage if a major structure failed. In the case of an intermediate structure, a small loss of life would be possible and the damage would be within the financial capability of the owner. For minor structures, there generally would be no loss of life, and the damage would be of the same magnitude as the cost of replacing or repairing the structure.

Design for Water Use

The above discussion applies to the hydrologic design for the control of excessive waters, such as floods. Design for water use is handled similarly, except that insufficient rather than excessive water is the concern. Because of the long time span of droughts, there are fewer of them in historical hydrologic records than there are extreme floods. It is therefore more difficult to determine drought design levels through frequency analysis, especially if the design event lasts several years, as is sometimes the case in water supply design. A common basis for the design of municipal water supply systems is the *critical drought of record*, that is, the worst recorded drought. The design is considered satisfactory if it will supply water at the required rate throughout an equivalent critical period. The limitation of the critical-period approach is that the risk level associated with basing the design on this single historical event is unknown. To overcome this limitation, methods of synthetic streamflow generation have been developed using computers and random number generation to prepare synthetic streamflow records that are statistically equivalent to the historical record. Together with the historical record, the synthetic records provide a probabilistic basis for design against drought events (Hirsch, 1979; Salas, et al., 1980).

Hydrologic design for water use is closely regulated by the legal framework of water rights, especially in arid regions. The law specifies which users will have their allocations reduced in the event of a shortage. In an effort to protect the fish and wildlife of a stream, methods have been developed in recent years to quantify their need for *instream flow* (Milhous and Grenney, 1980). Unlike flood control and water supply, for which sufficient hydrologic information is provided by flow rate and water level, instream flow needs are influenced also by turbidity, temperature, and other water quality variables in a complex manner varying from one species to another. Water resources systems are subject to the demands of competing users, the need to maintain instream flow, and competing demands related to flood control. Hydrologic design must specify the appropriate design level for each of these factors.

13.2 SELECTION OF THE DESIGN LEVEL

A *hydrologic design level* on the design scale is the magnitude of the hydrologic event to be considered for the design of a structure or project. As it is not always economical to design structures and projects for the estimated limiting value, the ELV is often modified for specific design purposes. The final design value may be further modified according to engineering judgment and the experience of the designer or planner. Three approaches are commonly used to determine a hydrologic design value: an empirical approach, risk analysis, and hydroeconomic analysis.

Empirical Approach

During the early years of hydraulic engineering practice, around the early 1900s, a spillway designed to pass a flood 50 to 100 percent larger than the largest recorded in a period of perhaps 25 years was considered adequate. This design criterion is no more than a rule of thumb involving an arbitrary factor of safety. As an example of the inadequacies of this criterion, the Republican River in Nebraska in 1935 experienced a flood over 10 times as large as any that had occurred on that river during 40 prior years of record. This design practice was found to be entirely inadequate, and hydrologists and hydraulic engineers searched for better methods.

As an empirical approach the most extreme event among past observations is often selected as the design value. The probability that the most extreme event of the past N years will be equaled or exceeded once during the next n years can be estimated as

$$P(N, n) = \frac{n}{N + n} \tag{13.2.1}$$

Thus, for example, the probability that the largest flood observed in N years will be equaled or exceeded in N future years is 0.50.

If a drought lasting m years is the critical event of record over an N-year period, what is the probability $P(N, m, n)$ that a worse drought will occur within the next n years? The number of sequences of length m in N years of record is $N - m + 1$, and in n years of record $n - m + 1$. Thus the chance that the worst event over the past and future spans combined will be contained in the n future years is given approximately by

$$\begin{aligned} P(N, m, n) &= \frac{(n - m + 1)}{(N - m + 1) + (n - m + 1)} \\ &= \frac{n - m + 1}{N + n - 2m + 2} \qquad (n \geq m) \end{aligned} \tag{13.2.2}$$

which reduces to (13.2.1) when $m = 1$.

Example 13.2.1. If the critical drought of record, as determined from 40 years of hydrologic data, lasted 5 years, what is the chance that a more severe drought will occur during the next 20 years?

Solution. Using Eq. (13.2.2),

$$\begin{aligned} P(40, 5, 20) &= \frac{20 - 5 + 1}{40 + 20 - 2 \times 5 + 2} \\ &= 0.308 \end{aligned}$$

Risk Analysis

Water-control design involves consideration of risks. A water-control structure might fail if the magnitude for the design return period T is exceeded within the expected life of the structure. This *natural*, or *inherent*, hydrologic risk of failure can be calculated using Eq. (12.1.4):

$$\overline{R} = 1 - [1 - P(X \geq x_T)]^n \tag{13.2.3}$$

where $P(X \geq x_T) = 1/T$, and n is the expected life of the structure; $\overline{R}$ represents the probability that an event $x \geq x_T$ will occur at least once in n years. This

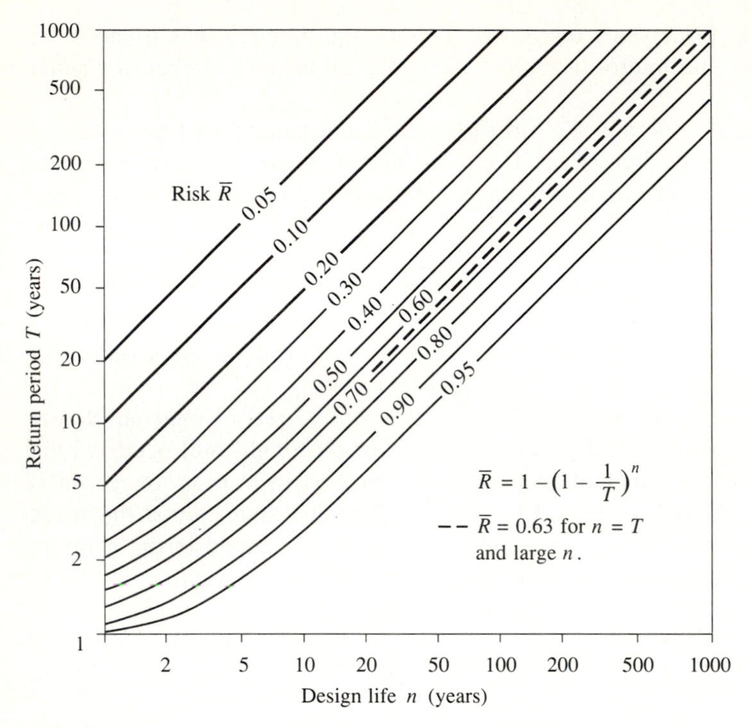

FIGURE 13.2.1
Risk of at least one exceedence of the design event during the design life.

relationship is plotted in Fig. 13.2.1. If, for example, a hydrologist wants to be approximately 90 percent certain that the design capacity of a culvert will not be exceeded during the structure's expected life of 10 years, he or she designs for the 100-year peak discharge of runoff. If a 40-percent risk of failure is acceptable, the design return period can be reduced to 20 years or the expected life extended to 50 years.

Example 13.2.2. A culvert has an expected life of 10 years. If the acceptable risk of at least one event exceeding the culvert capacity during the design life is 10 percent, what design return period should be used? What is the chance that a culvert designed for an event of this return period will not have its capacity exceeded for 50 years?

Solution. By Eq. (13.2.3)

$$\overline{R} = 1 - \left(1 - \frac{1}{T}\right)^n$$

or

$$0.10 = 1 - \left(1 - \frac{1}{T}\right)^{10}$$

and solving yields $T = 95$ years.

If $T = 95$ years, the risk of failure over $n = 50$ years is

$$\bar{R} = 1 - \left(1 - \frac{1}{95}\right)^{50}$$
$$= 0.41$$

So the probability that the capacity will not be exceeded during this 50-year period is $1 - 0.41 = 0.59$, or 59 percent.

It can be seen in Fig. 13.2.1 that, for a given risk of failure, the required design return period T increases linearly with the design life n, as T and n become large. Under these conditions, what is the risk of failure if the design return period is equal to the design life, that is, $T = n$? By expanding Eq. (13.2.3) as a power series, it can be shown that for large values of n, $1 - (1 - 1/T)^n \approx 1 - e^{-n/T}$, so, for $T = n$, the risk is $1 - e^{-1} = 0.632$. For example, there is approximately a 63-percent chance that a 100-year event will be exceeded at least once during the next 100 years.

Although natural hydrologic uncertainty can be accounted for as above, other kinds of uncertainty are difficult to calculate. These are often treated using a *safety factor,* SF, or a *safety margin,* SM. Letting the hydrologic design value be L and the actual capacity adopted for the project be C, the factor of safety is

$$SF = \frac{C}{L} \qquad (13.2.4)$$

and the safety margin is

$$SM = C - L \qquad (13.2.5)$$

The actual capacity is larger than the hydrologic design value because it has to allow for other kinds of uncertainty: technological (hydraulic, structural, construction, operation, etc.), socioeconomic, political, and environmental.

For a specified hydrologic risk $\bar{R}$ and design life n of a structure, Eq. (13.2.3) can be used to compute the relevant return period T. The hydrologic event magnitude L corresponding to this exceedence probability is found by a frequency analysis of hydrologic data. The design value C is then given by L multiplied by an assigned factor of safety, or by L plus an added margin of safety. For example, it is customary to design levees with a safety margin of one to three feet, that is, one to three feet of freeboard above the calculated maximum water surface elevation.

Hydroeconomic Analysis

The optimum design return period can be determined by hydroeconomic analysis if the probabilistic nature of a hydrologic event and the damage that will result if it occurs are both known over the feasible range of hydrologic events. As the design return period increases, the capital cost of a structure increases, but the expected damages decrease because of the better protection afforded. By summing the capital cost and the expected damage cost on an annual basis, a design return period having minimum total cost can be found.

Figure 13.2.2(a) shows the damage that would result if an event, such as a flood, having the specified return period were to occur. If the design event magnitude is x_T, the structure will prevent all damages for events with $x \leq x_T$ but none for $x > x_T$, so the *expected annual damage cost* is found by taking the product of the probability $f(x)\,dx$ that an event of magnitude x will occur in any given year, and the damage $D(x)$ that would result from that event, and

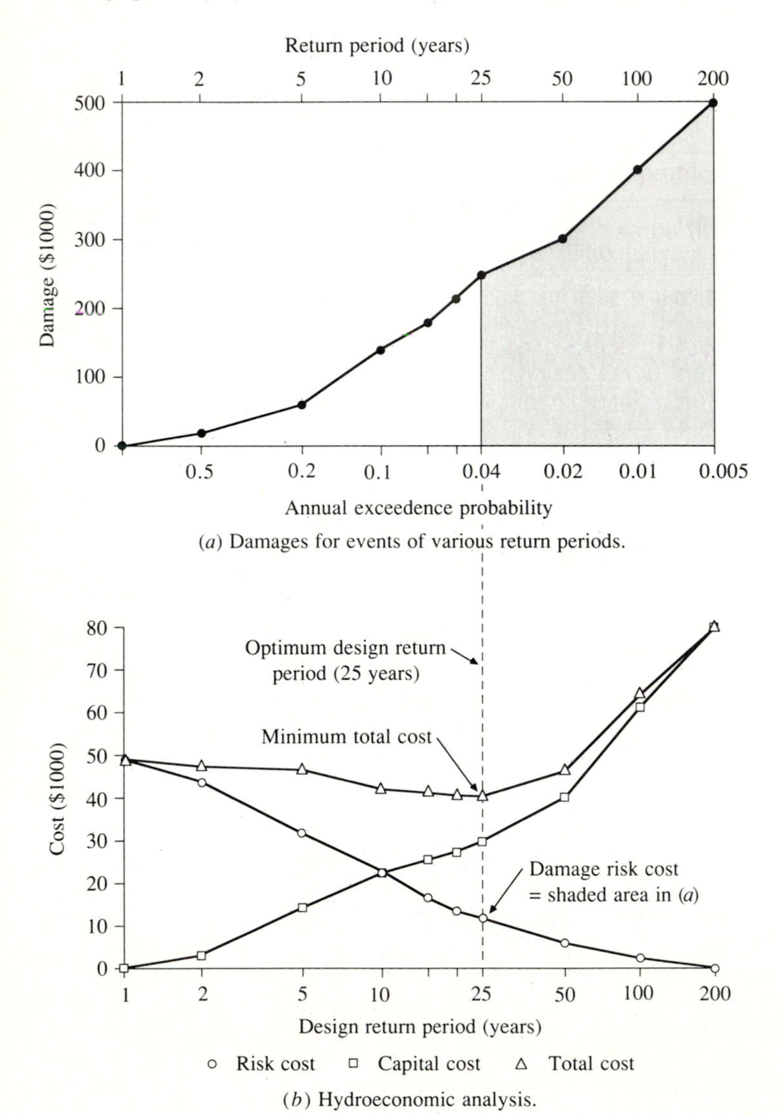

FIGURE 13.2.2
Determination of the optimum design return period by hydroeconomic analysis (Example 13.2.3).

integrating for $x > x_T$ (the design level). That is, the expected annual cost D_T is

$$D_T = \int_{x_T}^{\infty} D(x) f(x)\, dx \tag{13.2.6}$$

which is the shaded area in Fig. 13.2.2(a).

The integral (13.2.6) is evaluated by breaking the range of $x > x_T$ into intervals and computing the expected annual damage cost for events in each interval. For $x_{i-1} \leq x \leq x_i$,

$$\Delta D_i = \int_{x_{i-1}}^{x_i} D(x) f(x)\, dx \tag{13.2.7}$$

which is approximated by

$$\begin{aligned} \Delta D_i &= \left[\frac{D(x_{i-1}) + D(x_i)}{2}\right] \int_{x_{i-1}}^{x_i} f(x)\, dx \\ &= \frac{D(x_{i-1}) + D(x_i)}{2} \left[P(x \leq x_i) - P(x \leq x_{i-1})\right] \end{aligned} \tag{13.2.8}$$

But $P(x \leq x_i) - P(x \leq x_{i-1}) = [1 - P(x \geq x_i)] - [1 - P(x \geq x_{i-1})] = P(x \geq x_{i-1}) - P(x \geq x_i)$, so (13.2.8) can be written

$$\Delta D_i = \frac{D(x_{i-1}) + D(x_i)}{2} \left[P(x \geq x_{i-1}) - P(x \geq x_i)\right] \tag{13.2.9}$$

and the annual expected damage cost for a structure designed for return period T is given by

$$D_T = \sum_{i=1}^{\infty} \left[\frac{D(x_{i-1}) + D(x_i)}{2}\right] \left[P(x \geq x_{i-1}) - P(x \geq x_i)\right] \tag{13.2.10}$$

By adding D_T to the annualized capital cost of the structure, the total cost can be found; the optimum design return period is the one having the minimum total cost.

Example 13.2.3. For events of various return periods at a given location, the damage costs and the annualized capital costs of structures designed to control the events, are shown in columns 4 and 7, respectively, of Table 13.2.1. Determine the expected annual damages if no structure is provided, and calculate the optimal design return period.

Solution. For each return period shown in column 2 of Table 13.2.1, the annual exceedence probability is $P(x \geq x_T) = 1/T$. The corresponding damage cost ΔD is found using Eq. (13.2.9). For example, for the interval $i = 1$ between $T = 1$ year and $T = 2$ years,

$$\Delta D_1 = \left[\frac{D(x_1) + D(x_2)}{2}\right] \left[P(x \geq x_1) - P(x \geq x_2)\right]$$

TABLE 13.2.1
Calculation of the optimum design return period by hydroeconomic analysis (Example 13.2.3)

Column:	1	2	3	4	5	6	7	8
	Increment i	Return period T (years)	Annual exceedence probability	Damage ($)	Incremental expected damage ($/year)	Damage risk cost ($/year)	Capital cost ($/year)	Total cost ($/year)
		1	1.000	0		49,098	0	49,098
	1	2	0.500	20,000	5,000	44,098	3,000	47,098
	2	5	0.200	60,000	12,000	32,098	14,000	46,098
	3	10	0.100	140,000	10,000	22,098	23,000	45,098
	4	15	0.067	177,000	5,283	16,815	25,000	41,815
	5	20	0.050	213,000	3,250	13,565	27,000	40,565
	6	25	0.040	250,000	2,315	11,250	29,000	*40,250*
	7	50	0.020	300,000	5,500	5,750	40,000	45,750
	8	100	0.010	400,000	3,500	2,250	60,000	62,250
	9	200	0.005	500,000	2,250	0	80,000	80,000
			Annual expected damage	=	$49,098			

$$= \left(\frac{0 + 20,000}{2}\right)(1.0 - 0.5)$$

$$= \$5,000/\text{year}$$

as shown in column 5 of the table. Summing these incremental costs yields an annual expected damage cost of $49,098/year if no structure is built. This represents the average annual cost of flood damage over many years, assuming constant economic conditions. This amount is the damage risk cost corresponding to no structure, and is shown in the first line of column 6 of the table.

The damage risk costs diminish as the design return period of the control structure increases. For example, if $T = 2$ years were selected, the damage risk cost would be 49,098 $-\Delta D_1 = 49,098 - 5,000 =$ $44,098/year. The values of damage risk cost and capital cost (column 7) are added to form the total cost (column 8); the three costs are plotted in Fig. 13.2.2(*b*). It can be seen from the table and the figure that the optimum design return period, the one having minimal total cost, is 25 years, for which the total cost is $40,250/year. Of this amount, $29,000/year (72 percent) is capital cost and $11,250/year (28 percent) is damage risk cost.

Hydroeconomic analysis has been applied to the design of flood control reservoirs, levees, channels, and highway stream crossings (Corry, Jones, and Thompson, 1980). For a flood damage study, the duration and extent of flooding must be determined for events of various return periods and economic surveys must be taken to quantify damages for each level of flooding. The social costs of flooding are difficult to quantify. The U. S. Army Corps of Engineers Hydrologic Engineering Center in Davis, California, has available the following computer programs for hydroeconomic analysis (U. S. Army Corps of Engineers, 1986):

DAMCAL (Damage Reach Stage–Damage Calculation), EAD (Expected Annual Flood Damage Computation), SID (Structure Inventory for Damage Analysis), AGDAM (Agricultural Flood Damage Analysis), and SIPP (Interactive Nonstructural Analysis Package).

13.3 FIRST ORDER ANALYSIS OF UNCERTAINTY

Many of the uncertainties associated with hydrologic systems are not quantifiable. For example, the conveyance capacity of a culvert with an unobstructed entrance can be calculated within a small margin of error, but during a flood, debris may become lodged around the entrance to the culvert, reducing its conveyance capacity by an amount that cannot be predetermined. Hydrologic uncertainty may be broken down into three categories: *natural*, or *inherent*, *uncertainty*, which arises from the random variability of hydrologic phenomena; *model uncertainty*, which results from the approximations made when representing phenomena by equations; and *parameter uncertainty*, which stems from the unknown nature of the coefficients in the equations, such as the bed roughness in Manning's equation. Inherent uncertainty in the magnitude of the design event is described by Eq. (13.2.3); in this section, model and parameter uncertainty will be considered.

The *first order analysis of uncertainty* is a procedure for quantifying the expected variability of a dependent variable calculated as a function of one or more independent variables (Ang and Tang, 1975; Kapur and Lamberson, 1977; Ang and Tang, 1984; Yen, 1986). Suppose w is expressed as a function of x:

$$w = f(x) \tag{13.3.1}$$

There are two sources of error in w: first, the function f, or model, may be incorrect; second, the measurement of x may be inaccurate. In the following analysis it is assumed that there is no model error, or *bias*. Kapur and Lamberson (1977) show how to extend the analysis when there is model error. Assuming, then, that $f(\cdot)$ is a correct model, a nominal value of x, denoted $\bar{x}$, is selected as a design input and the corresponding value of w calculated:

$$\bar{w} = f(\bar{x}) \tag{13.3.2}$$

If the true value of x differs from $\bar{x}$, the effect of this discrepancy on w can be estimated by expanding $f(x)$ as a Taylor series around $x = \bar{x}$:

$$w = f(\bar{x}) + \frac{df}{dx}(x - \bar{x}) + \frac{1}{2!}\frac{d^2f}{dx^2}(x - \bar{x})^2 + \dots \tag{13.3.3}$$

where the derivatives df/dx, d^2f/dx^2, . . . , are evaluated at $x = \bar{x}$. If second and higher order terms are neglected, the resulting *first order* expression for the error in w is

$$w - \bar{w} = \frac{df}{dx}(x - \bar{x}) \tag{13.3.4}$$

The variance of this error is $s_w^2 = E[(w - \bar{w})^2]$ where E is the expectation operator [see Eq. (11.3.3)]; that is,

$$s_w^2 = E\left\{\left[\frac{df}{dx}(x-\bar{x})\right]^2\right\}$$

or

$$s_w^2 = \left(\frac{df}{dx}\right)^2 s_x^2 \tag{13.3.5}$$

where s_x^2 is the variance of x.

Equation (13.3.5) gives the variance of a dependent variable w as a function of the variance of an independent variable x, assuming that the functional relationship $w = f(x)$ is correct. The value s_w is the *standard error of estimate* of w.

If w is dependent on several *mutually independent* variables $x_1, x_2, \ldots, x_n$, it can be shown by a procedure similar to the above that

$$s_w^2 = \left(\frac{\partial f}{\partial x_1}\right)^2 s_{x_1}^2 + \left(\frac{\partial f}{\partial x_2}\right)^2 s_{x_2}^2 + \ldots + \left(\frac{\partial f}{\partial x_n}\right)^2 s_{x_n}^2 \tag{13.3.6}$$

Kapur and Lamberson (1977) show how to extend (13.3.6) to account for the effect on s_w^2 of correlation between $x_1, x_2, \ldots, x_n$, if any exists.

First-Order Analysis of Manning's Equation: Depth as the Dependent Variable

Manning's equation is widely applied in hydrology to determine depths of flow for specified flow rates, or to determine discharges for specified depths of flow, taking into account the resistance to flow in channels arising from bed roughness. A common application, such as in channel design or flood plain delineation, is to calculate the depth of flow y in the channel, given the flow rate Q, roughness coefficient n, and the shape and slope of the channel as determined by design or by surveys. Once the depth of flow (or elevation of the water surface) is known, the values of the design variables are determined, such as the channel wall elevation or the flood plain extent. The hydrologist faced with this task is conscious of the uncertainties involved, especially in the selection of the design flow and Manning roughness. Although it is not so obvious, there is also uncertainty in the value of the friction slope S_f, depending on how it is calculated, ranging from the simplest case of uniform flow ($S_o = S_f$) to more complex cases of steady nonuniform flow or unsteady nonuniform flow [see Eq. (9.2.1)]. The first-order analysis of uncertainty can be used to estimate the effect on y of uncertainty in Q, n, and S_f.

Consider, first, the effect on flow depth of variation in the flow rate Q. Manning's equation is written in English units as

$$Q = \frac{1.49}{n} S_f^{1/2} A R^{2/3} \tag{13.3.7}$$

where A is the cross-sectional area and R the hydraulic radius, both dependent on the flow depth y. If variations in y are dependent only on variations in Q, then, by (13.3.5),

$$s_y^2 = \left(\frac{dy}{dQ}\right)^2 s_Q^2 \tag{13.3.8}$$

where dy/dQ is the rate at which the depth changes with changes in Q. Now, in Chap. 5, it was shown [Eq. (5.6.15)] that the inverse of this derivative, namely dQ/dy, is given for Manning's equation by

$$\frac{dQ}{dy} = Q\left[\frac{2}{3R}\frac{dR}{dy} + \frac{1}{A}\frac{dA}{dy}\right] \tag{13.3.9}$$

Table 5.6.1 gives formulas for the channel shape function $(2/3R)(dR/dy) + (1/A)(dA/dy)$ for common channel cross sections. Substituting into (13.3.8),

$$s_y^2 = \frac{s_Q^2}{Q^2\left(\dfrac{2}{3R}\dfrac{dR}{dy} + \dfrac{1}{A}\dfrac{dA}{dy}\right)^2} \tag{13.3.10}$$

But $s_Q/Q = \mathrm{CV}_Q$, the coefficient of variation of the flow rate (see Table 11.3.1), so (13.3.10) can be rewritten

$$s_y^2 = \frac{\mathrm{CV}_Q^2}{\left(\dfrac{2}{3R}\dfrac{dR}{dy} + \dfrac{1}{A}\dfrac{dA}{dy}\right)^2} \tag{13.3.11}$$

which specifies the variance of the flow depth as a function of the coefficient of variation of the flow rate and the value of the channel shape function. To take into account also the uncertainty in Manning's roughness n and the friction slope S_f, it may be similarly shown, using Eq. (13.3.6), that

$$s_y^2 = \frac{\mathrm{CV}_Q^2 + \mathrm{CV}_n^2 + (1/4)\mathrm{CV}_{S_f}^2}{\left(\dfrac{2}{3R}\dfrac{dR}{dy} + \dfrac{1}{A}\dfrac{dA}{dy}\right)^2} \tag{13.3.12}$$

giving the variance of the flow depth y as a function of the coefficients of variation of flow rate, Manning's n and friction slope, and the channel shape function.

Example 13.3.1. A 50-foot wide rectangular channel has a bed slope of one percent. A hydrologist estimates that the design flow rate is 5000 cfs and that the roughness is $n = 0.035$. If the coefficients of variation of the flow estimate and the roughness estimate are 30 percent and 15 percent, respectively, what is the standard error of estimate of the flow depth y? If houses are built next to this channel with floor elevation one foot above the water surface elevation calculated for the design event, estimate the chance that these houses will be flooded during the design event due to uncertainties involved in calculating the water level. Assume uniform flow.

Solution. For a width of 50 feet, $A = 50y$ and $R = 50y/(50 + 2y)$; the flow depth for the base case is calculated from Manning's equation:

$$Q = \frac{1.49}{n} S_f^{1/2} A R^{2/3}$$

$$5000 = \frac{1.49}{0.035}(0.01)^{1/2}(50y)\left(\frac{50y}{50 + 2y}\right)^{2/3}$$

which is solved using Newton's iteration technique (see Sec. 5.6) to yield

$$y = 7.37 \text{ ft}$$

The standard error of the estimate is s_y, calculated by Eq. (13.3.12) with $CV_Q = 0.30$, $CV_n = 0.15$, and $CV_{S_f} = 0$. From Table 5.6.1, for a rectangular channel,

$$\left(\frac{2}{3R}\frac{dR}{dy} + \frac{1}{A}\frac{dA}{dy}\right) = \frac{5B + 6y}{3y(B + 2y)}$$

$$= \frac{5 \times 50 + 6 \times 7.37}{3 \times 7.37(50 + 2 \times 7.37)}$$

$$= 0.206$$

So

$$s_y^2 = \frac{CV_Q^2 + CV_n^2 + (1/4)CV_{S_f}^2}{\left(\frac{2}{3R}\frac{dR}{dy} + \frac{1}{A}\frac{dA}{dy}\right)^2}$$

$$= \frac{(0.30)^2 + (0.15)^2}{(0.206)^2}$$

or $s_y = 1.63$ ft.

If the houses are built with their floors one foot above the calculated water surface elevation, they will be flooded if the actual depth is greater than 7.37 + 1.00 = 8.37 ft. If the water surface elevation y is normally distributed, then the probability that they will be flooded is evaluated by converting y to the standard normal variable z by subtracting the mean value of y (7.37 ft) from both sides of the inequality and dividing by the standard error (1.63 ft):

$$P(y > 8.37) = P\left(\frac{y - 7.37}{1.63} > \frac{8.37 - 7.37}{1.63}\right)$$

$$= P\left(\frac{y - 7.37}{1.63} > 0.613\right)$$

$$= P(z > 0.613)$$

$$= 1 - F_z(0.613)$$

where F_z is the standard normal distribution function. Using Table 11.2.1 or the method employed in Example 11.2.1, the result is $F_z(0.613) = 0.73$, so $P(y > 8.37) = 1 - 0.73 = 0.27$. There is approximately a 27 percent chance that the houses will be flooded during the design event due to uncertainties in calculating the water level for that event.

This example has treated only parameter uncertainty in the calculations. The true probability that the houses will be flooded is greater than that calculated here, because the critical flood may exceed the design magnitude (due to natural uncertainty).

It is clear from Example 13.3.1 that reasonable amounts of uncertainty in the estimation of Q and n can produce significant uncertainty in flow depth. A 15-percent error in estimating $n = 0.035$ is an error of $0.035 \times 0.15 = 0.005$. This would be indicated from a measurement of 0.035 ± 0.005, which is about as accurate as an experienced hydrologist can get from observation of an existing channel. A 30-percent error in estimating Q is $5000 \times 0.30 = 1500$ cfs. An estimate of $Q = 5000 \pm 1500$ cfs may also reflect the correct order of uncertainty, especially if the design return period is large (e.g., $T = 100$ years).

The use of the channel shape function $(2/3R)(dR/dy) + (1/A)(dA/dy)$ in (13.3.12) depends on knowledge of dR/dy and dA/dy, which may be difficult to obtain for irregularly shaped channels. Also, the assumption that y depends on Q alone may not be valid. In such cases, Eq. (13.3.6) can be used to obtain s_y, treating y as a function of Q and n, and a computer program simulating flow in the channel can be used to estimate the required partial derivatives $\partial y/\partial Q$ and $\partial y/\partial n$ by rerunning the program for various values of Q and n and reading off the computed values of flow depth or water surface elevation. Figure 13.3.1 shows the results of such a procedure for the channel and conditions given in Example 13.3.1. The gradients $\partial y/\partial Q$ and $\partial y/\partial n$ are approximately linear for this example; this validates the use of only first-order terms in the analysis of uncertainty (if the lines were significantly curved, analysis would require keeping the second-order terms in the Taylor-series expansion).

Example 13.3.2. For the same conditions as in Example 13.3.1 ($B = 50$ ft, $Q = 5000$ cfs, $S_o = 0.01$, $n = 0.035$), the variation of flow rate with flow depth at the base case level has been found from Fig. (13.3.1) to be $\partial Q/\partial y = 1028$ cfs/ft, and the variation of n with flow depth, $\partial n/\partial y = 0.0072$ ft^{-1}. If $CV_Q = 0.30$ and $CV_n = 0.15$, calculate the standard error of y.

Solution. From Eq. (13.3.6),

$$s_y^2 = \left(\frac{\partial y}{\partial Q}\right)^2 s_Q^2 + \left(\frac{\partial y}{\partial n}\right)^2 s_n^2$$

In this case, $s_Q = 5000 \times 0.30 = 1500$, $s_n = 0.035 \times 0.15 = 0.0053$; also, $\partial y/\partial Q = 1/1028$, $\partial y/\partial n = 1/0.0072$. Thus,

$$s_y^2 = \left(\frac{1}{1028}\right)^2 \times (1500)^2 + \left(\frac{1}{0.0072}\right)^2 \times (0.0053)^2$$

or $s_y = 1.63$ ft as computed in Example 13.3.1.

First-Order Analysis of Manning's Equation: Discharge as the Dependent Variable

Another application of Manning's equation is the calculation of the discharge or capacity C of a stream channel or other conveyance structure for a given depth, roughness coefficient n, bottom slope, and cross-sectional geometry. Manning's equation (13.3.7) can be expressed using $R = A/P$ as

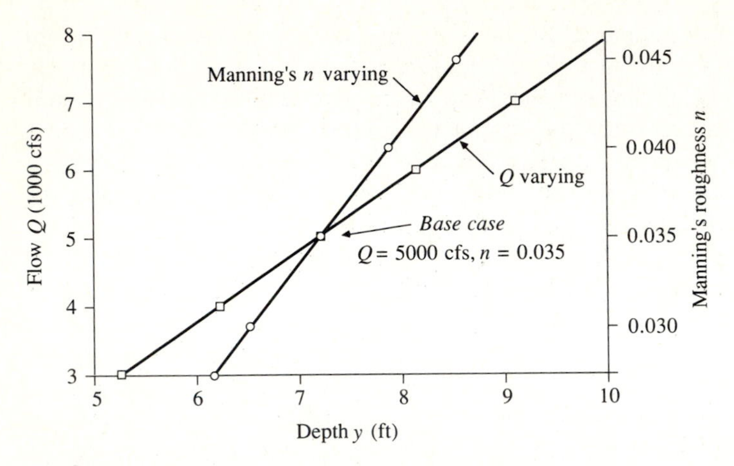

FIGURE 13.3.1
Variation of the flow depth with flow rate and with Manning's n. Rectangular channel with width 50 ft, bed slope 0.01. Uniform flow assumed. (Example 13.3.2).

$$C = Q = \frac{1.49}{n} S_f^{1/2} A^{5/3} P^{-2/3} \tag{13.3.13}$$

in which P is the wetted perimeter. Performing first-order analysis on (13.3.13), the coefficient of variation of the capacity can be expressed as

$$\mathrm{CV}_Q^2 = \mathrm{CV}_n^2 + \frac{1}{4}\mathrm{CV}_{S_f}^2 \tag{13.3.14}$$

assuming $\mathrm{CV}_A \approx 0$ and $\mathrm{CV}_P \approx 0$.

Manning's equation for a channel and flood plain (overbank) can also be expressed as (Chow, 1959)

$$Q = 1.49\left(\frac{1}{n_c} A_c^{5/3} P_c^{-2/3} + \frac{2}{n_b} A_b^{5/3} P_b^{-2/3}\right) S_f^{1/2} \tag{13.3.15}$$

in which n_c and n_b are the roughness coefficients for the channel and the floodplain, respectively and A_c, P_c, A_b, and P_b are the cross-sectional areas and the wetted perimeters of the channel and the overbank flow. Equation (13.3.15) assumes that the cross-sectional shape of the channel and the flood plain are both symmetrical about the channel center line. This equation can be used to evaluate levee capacity (the flow rate the levee can carry without overtopping). The levee capacity can be considered a random variable related to the independent random variables n_c, n_b, and S_f. Applying first-order analysis, the coefficient of variation of the capacity is (Lee and Mays, 1986)

$$\mathrm{CV}_Q^2 = \frac{1}{4}\mathrm{CV}_{S_f}^2 + \frac{1}{\Psi^2}\mathrm{CV}_{n_c}^2 + \left(\frac{\Psi - 1}{\Psi}\right)^2 \mathrm{CV}_{n_b}^2 \tag{13.3.16}$$

where CV_{A_c}, CV_{P_c}, CV_{A_b}, and CV_{P_b} have been assumed negligible, and

$$\Psi = 1 + 2\left(\frac{n_c}{n_b}\right)\left(\frac{A_b}{A_c}\right)^{5/3}\left(\frac{P_c}{P_b}\right)^{2/3} \tag{13.3.17}$$

In studies of flood data on the Ohio River, Lee and Mays (1986) concluded that uncertainties in the roughness coefficients and the friction slope account for 95 percent of the uncertainties in computing the capacity. They presented a method for determining the uncertainty in the friction slope using the observed flood hydrograph of the river.

13.4 COMPOSITE RISK ANALYSIS

The previous sections have introduced the concepts of inherent uncertainty due to the natural variability of hydrologic phenomena, and model and parameter uncertainty arising from the way the phenomena are analyzed. Composite risk analysis is a method of accounting for the risks resulting from the various sources of uncertainty to produce an overall risk assessment for a particular design. The concepts of loading and capacity are central to this analysis.

The *loading*, or *demand*, placed on a system is the measure of the impact of external events. The demand for water supply is determined by the people who use the water. The magnitude of a flash flood depends on the characteristics of the storm producing it and on the condition of the watershed at the time of the storm. The *capacity*, or *resistance*, is the measure of the ability of the system to withstand the loading or meet the demand.

If loading is denoted by L and capacity by C, then the risk of failure $\overline{R}$ is given by the probability that L exceeds C, or

$$\begin{aligned}\overline{R} &= P\left(\frac{C}{L} < 1\right) \\ &= P(C - L < 0)\end{aligned} \tag{13.4.1}$$

The risk depends upon the probability distributions of L and C. Suppose that the probability density function of L is $f(L)$. This function could be, for example, an Extreme Value or log–Pearson Type III probability density function for extreme values, as described earlier. Given $f(L)$, the chance that the loading will exceed a fixed and known capacity C^* is (see Fig. 13.4.1)

$$P(L > C^*) = \int_{C^*}^{\infty} f(L)\, dL \tag{13.4.2}$$

The true capacity is not known exactly, but may be considered to have probability density function $g(C)$, which could be the normal or lognormal distribution arising from the first-order analysis of uncertainty in the system capacity. For example, if Manning's equation has been used to determine the capacity of a hydraulic structure, the uncertainty in C can be evaluated by first-order analysis as described above. The probability that the capacity lies within a small range

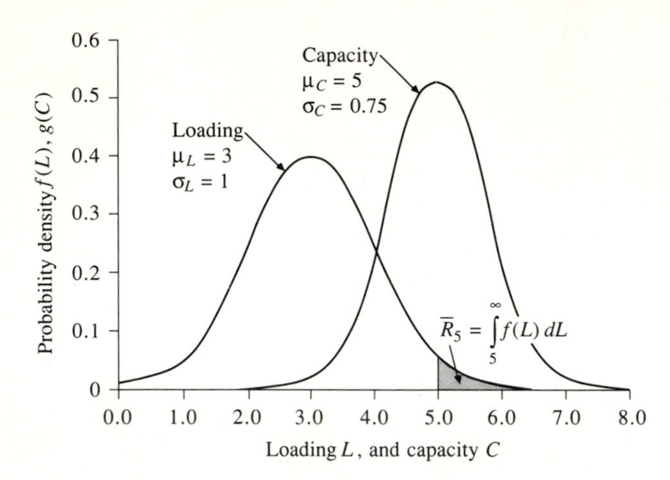

FIGURE 13.4.1
Composite risk analysis. Area shaded is the risk $\overline{R}_5$ of the loading exceeding a fixed capacity of 5 units. The risk that the loading will exceed the capacity when the capacity is random is given by $\overline{R} = \int_{-\infty}^{\infty} [\int_C^\infty f(L)\,dL]\, g(C)\,dC$. The loading and capacity shown are both normally distributed (Example 13.4.1).

dC around a value C is $g(C)\,dC$. Assuming that L and C are independent random variables, the *composite risk* is evaluated by calculating the probability that loading will exceed capacity at each value in the range of feasible capacities, and integrating to obtain

$$\overline{R} = \int_{-\infty}^{\infty} \left[\int_C^{\infty} f(L)\,dL \right] g(C)\,dC \tag{13.4.3}$$

The *reliability* of a system is defined to be *the probability that a system will perform its required function for a specified period of time under stated conditions* (Harr, 1987). Reliability R is the complement of risk, or the probability that the loading will not exceed the capacity:

$$\begin{aligned} R &= P(L \leq C) \\ &= 1 - \overline{R} \end{aligned} \tag{13.4.4}$$

or

$$R = \int_{-\infty}^{\infty} \left[\int_0^{C} f(L)\,dL \right] g(C)\,dC \tag{13.4.5}$$

Example 13.4.1. During the coming year, a city's estimated water demand is three units, with a standard deviation of one unit. Calculate (*a*) the risk of demand exceeding supply if the city's water supply system has an estimated capacity of 5 units; (*b*) the risk of failure if the estimate of the capacity has a standard error of 0.75 units. Assume that loading and capacity are both normally distributed.

Solution. (*a*) The loading is normally distributed with $\mu_L = 3$ and $\sigma_L = 1$. Its probability function, from Eq. (11.2.5), is

$$f(L) = \frac{1}{\sqrt{2\pi}\sigma_L} e^{-(L-\mu_L)^2/2\sigma_L^2}$$

$$= \frac{1}{\sqrt{2\pi}} e^{-(L-3)^2/2}$$

The risk $\overline{R}$ is evaluated using (13.4.2) with $C^* = 5$:

$$\overline{R} = \int_{C^*}^{\infty} f(L)\, dL$$

$$= \int_{5}^{\infty} \frac{1}{\sqrt{2\pi}} e^{-(L-3)^2/2} dL$$

or

$$\overline{R} = 1 - \int_{-\infty}^{5} \frac{1}{\sqrt{2\pi}} e^{-(L-3)^2/2} dL$$

The integral is evalutated by converting the variable of integration to the standard normal variable: $u = (L - \mu_L)/\sigma_L = (L - 3)/1 = L - 3$, so $dL = du$, and $L = 5$ becomes $u = 5 - 3 = 2$; $L = -\infty$ becomes $u = -\infty$, and then

$$\overline{R} = 1 - \int_{-\infty}^{2} \frac{1}{\sqrt{2\pi}} e^{-u^2/2} du$$

$$= 1 - F_z(2)$$

where F_z is the standard normal distribution function. From Table 11.2.1, $F_z(2) = 0.977$, and

$$\overline{R} = 1 - 0.977$$

$$= 0.023$$

The chance that demand will exceed supply for a fixed capacity of 5 is approximately 2 percent.

(*b*) The capacity now has a normal distribution with $\mu_C = 5$ and $\sigma_C = 0.75$. Hence, its probability density is

$$g(C) = \frac{1}{\sqrt{2\pi}\sigma_C} e^{-(C-\mu_C)^2/2\sigma_C^2}$$

$$= \frac{1}{\sqrt{2\pi}(0.75)} e^{-(C-5)^2/2\times(0.75)^2}$$

$$= \frac{1.333}{\sqrt{2\pi}} e^{-(C-5)^2/1.125}$$

and the risk of failure is given by Eq. (13.4.3), with $f(L)$ as before:

$$\overline{R} = \int_{-\infty}^{\infty} \left[\int_{C}^{\infty} f(L)\, dL \right] g(C)\, dC$$

$$= \int_{-\infty}^{\infty} \left[\int_{C}^{\infty} \frac{1}{\sqrt{2\pi}} e^{-(L-3)^2/2} dL \right] \frac{1.333}{\sqrt{2\pi}} e^{-(C-5)^2/1.125} dC$$

The integral is evaluated by computer using numerical integration to yield $\bar{R}=0.052$. Thus, the chance that the city's water demand will exceed its supply during the coming year, assuming the capacity to be normally distributed with mean 5 and standard deviation 0.75 is approximately 5 percent; compare this with the result of 2 percent when the capacity was considered fixed at 5 units.

It is clear from Example 13.4.1 that calculation of the composite risk of failure can be a complicated exercise requiring the use of a computer to perform the necessary integration. This is especially true when more realistic distributions for the loading and capacity are chosen, such as the Extreme Value or log–Pearson Type III distributions for loading, and the lognormal distribution for capacity. Yen and co-workers at the University of Illinois (Yen, 1970; Tang and Yen, 1972; Yen, et al., 1976) and Mays and co-workers at the University of Texas at Austin (Tung and Mays, 1980; Lee and Mays, 1986) have made detailed risk analysis studies for various kinds of open-channel and pipe-flow design problems.

The composite risk analysis described here is a *static* analysis, which means that it estimates the risk of failure under the single worst case loading on the system during its design life. A more complex *dynamic* risk analysis considers the possibility of a number of extreme loadings during the design life, any one of which could cause a failure; the total risk of failure includes the chance of multiple failures during the design life (Tung and Mays, 1980; Lee and Mays 1983).

13.5 RISK ANALYSIS OF SAFETY MARGINS AND SAFETY FACTORS

Safety Margin

The safety margin was defined in Eq. (13.2.5) as the difference between the project capacity and the value calculated for the design loading $SM = C - L$. From (13.4.1), the risk of failure $\bar{R}$ is

$$\begin{aligned}\bar{R} &= P(C - L < 0) \\ &= P(SM < 0)\end{aligned} \tag{13.5.1}$$

If C and L are independent random variables, then the mean value of SM is given by

$$\mu_{SM} = \mu_C - \mu_L \tag{13.5.2}$$

and its variance by

$$\sigma_{SM}^2 = \sigma_C^2 + \sigma_L^2 \tag{13.5.3}$$

so the standard deviation, or standard error of estimate, of the safety margin is

$$\sigma_{SM} = \left(\sigma_C^2 + \sigma_L^2\right)^{1/2} \tag{13.5.4}$$

If the safety margin is normally distributed, then $(SM - \mu_{SM})/\sigma_{SM}$ is a standard normal variate z. By subtracting μ_{SM} from both sides of the inequality in (13.5.1) and dividing both sides by σ_{SM}, it can be seen that

$$\begin{aligned} \overline{R} &= P\left(\frac{SM - \mu_{SM}}{\sigma_{SM}} < \frac{-\mu_{SM}}{\sigma_{SM}}\right) \\ &= P\left(z < -\frac{\mu_{SM}}{\sigma_{SM}}\right) \\ &= F_z\left(-\frac{\mu_{SM}}{\sigma_{SM}}\right) \end{aligned} \tag{13.5.5}$$

where F_z is the standard normal distribution function.

Example 13.5.1. Calculate the risk of failure of the water supply system in Example 13.4.1, assuming that the safety margin is normally distributed, and that $\mu_C = 5$ units, $\sigma_C = 0.75$ units, $\mu_L = 3$ units, and $\sigma_L = 1$ unit.

Solution. From Eq. (13.5.2), $\mu_{SM} = \mu_C - \mu_L = 5 - 3 = 2$. From (13.5.4), $\sigma_{SM} = (\sigma_C^2 + \sigma_L^2)^{1/2} = (1^2 + 0.75^2)^{1/2} = 1.250$. Using (13.5.5),

$$\begin{aligned} \overline{R} &= F_z\left(-\frac{\mu_{SM}}{\sigma_{SM}}\right) \\ &= F_z\left(-\frac{2}{1.250}\right) \\ &= F_z(-1.60) \end{aligned}$$

which is evaluated using Table 11.2.1 to yield $\overline{R} = 0.055$, which is very close to the value obtained in Example 13.4.1 by numerical integration (an inherently approximate procedure). The risk of failure under the stated conditions is $\overline{R} = 0.055$, or 5.5%.

Note that this method of analysis assumes that the safety margin is normally distributed but does not specify what the distributions of loading and capacity must be. Ang (1973) indicates that, provided $\overline{R} > 0.001$, $\overline{R}$ is not greatly influenced by the choice of distributions for L and C, and the assumption of a normal distribution for SM is satisfactory. For lower risk than this (e.g., $\overline{R} = 0.00001$), the shapes of the tails of the distributions for L and C become critical, and in this case, the full composite risk analysis described in Sec. 13.4 should be used to evaluate the risk of failure.

Safety Factor

The safety factor SF is given by the ratio C/L and the risk of failure can be expressed as $P(SF < 1)$. By taking logarithms of both sides of this inequality

$$\begin{aligned}\overline{R} &= P(\text{SF} < 1) \\ &= P\big(\ln(\text{SF}) < 0\big) \\ &= P\left(\ln\frac{C}{L} < 0\right)\end{aligned} \tag{13.5.6}$$

If the capacity and loading are independent and lognormally distributed, then the risk can be expressed (Huang, 1986)

$$\overline{R} = F_z\left(\frac{-\ln\left[\dfrac{\mu_C}{\mu_L}\left(\dfrac{1+\text{CV}_L^2}{1+\text{CV}_C^2}\right)^{1/2}\right]}{\left\{\ln\left[(1+\text{CV}_C^2)(1+\text{CV}_L^2)\right]\right\}^{1/2}}\right) \tag{13.5.7}$$

Example 13.5.2. Solve Example 13.5.1 assuming capacity and loading are both lognormally distributed.

Solution. From Example 13.5.1, $\mu_C = 5$ and $\sigma_C = 0.75$, and hence $\text{CV}_C = 0.75/5 = 0.15$. Likewise, $\mu_L = 3$ and $\sigma_L = 1$, so $\text{CV}_L = 1/3 = 0.333$. Hence, by Eq. (13.5.7), the risk is

$$\overline{R} = F_z\left(\frac{-\ln\left\{\dfrac{5}{3}\left[\dfrac{1+(0.333)^2}{1+(0.15)^2}\right]^{1/2}\right\}}{\left\{\ln\left[(1+(0.15)^2)(1+(0.333)^2)\right]\right\}^{1/2}}\right)$$

$$= F_z(-1.5463) = 0.061$$

The risk of failure under the above assumptions, then, is 6.1 percent. For the same problem (Example 13.5.1) assuming that the safety margin was normally distributed, the risk was found to be 5.5 percent; the risk level has not changed greatly with use of the lognormal instead of the normal distribution.

Risk–Safety Factor–Return Period Relationship

A common design practice is to choose a return period and determine the corresponding loading L as the design capacity of a hydraulic structure. The safety factor is inherently built into the choice of the return period. Alternatively, the loading value can be multiplied by a safety factor SF; then the structure is designed for capacity $C = \text{SF} \times L$. As discussed in this chapter, there are various kinds of uncertainty associated both with L and with the capacity C of the structure as designed. By composite risk analysis, a risk of failure can be calculated for the selected return period and safety factor. The result of such a calculation is shown

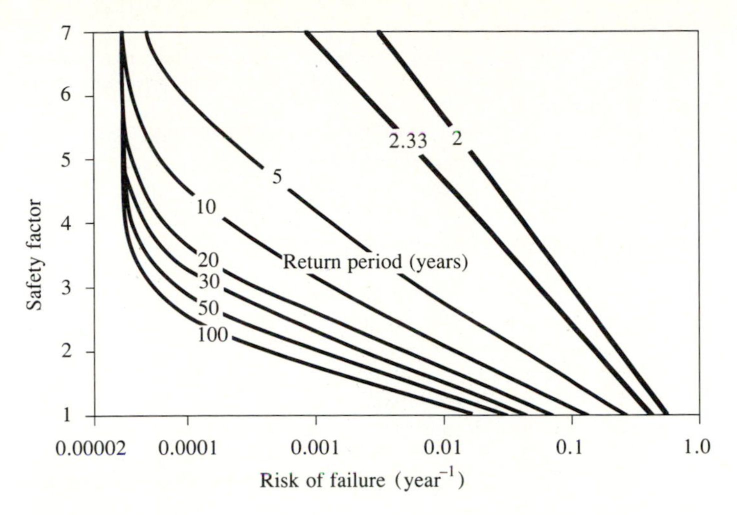

FIGURE 13.5.1
The risk–safety factor–return period relationship for culvert design on the Glade River near Reston, Virginia. The probability distribution for loading used to develop this figure was the Extreme Value Type I distribution of annual maximum floods. A lognormal distribution for the culvert capacity was developed using first-order analysis of uncertainty. The risk level for given return period and safety factor was determined using composite risk analysis. (*Source*: Tung and Mays, 1980.)

in Fig. 13.5.1, which shows a risk chart applying to culvert design on the Glade River near Reston, Virginia. The risk values in the chart represent annual probabilities of failure. For example, if the return period is 100 years and the safety factor 1.0, the risk of failure is 0.015 or 1.5 percent in any given year, while if the safety factor is increased to 2, the risk of failure is reduced to $\overline{R} = 0.006$, or 0.6 percent in any given year.

Current hydrologic design practice copes with the inherent uncertainty of hydrologic phenomena by the selection of the design return period, and with model and parameter uncertainty by the assignment of arbitrary safety factors or safety margins. The risks and uncertainties can be evaluated more systematically using the procedures provided by first-order analysis of uncertainty and composite risk analysis as presented here. However, it must be borne in mind that just as any function of random variables is itself a random variable, the estimates of risk and reliability provided by these methods also have uncertainty associated with them, and their true values can never be determined exactly.

REFERENCES

Ang, A. H.-S., Structural risk analysis and reliability–based design, *J. Structural Div., Am. Soc. Civ. Eng.*, vol. 99, no. ST9, pp. 1891–1910, 1973.

Ang, A. H.-S., and W. H. Tang, *Probability Concepts in Engineering Planning and Design*, vol. I, Basic Principles, and vol. II, Decision, Risk and Reliability, Wiley, New York, 1975 and 1984, respectively.

Chow, V. T., *Open-channel Hydraulics,* McGraw-Hill, New York, 1959.

Corry, M. L., J. S. Jones, and P. L. Thompson, The design of encroachments on flood plains using risk analysis, hydraulic engineering circular no. 17, Federal Highway Administration, U. S. Department of Transportation, 1980.

Harr, M. E., *Reliability-based Design in Civil Engineering,* McGraw-Hill, New York, 1987.

Huang, K.-Z., Reliability analysis on hydraulic design of open channel, in *Stochastic and Risk Analysis in Hydraulic Engineering,* ed. by B. C. Yen, Water Resources Publications, Littleton, Colo., p. 60, 1986.

Hirsch, R. M., Synthetic hydrology and water supply reliability, *Water Resour. Res.,* vol. 15, no. 6, pp. 1603–1615, December 1979.

Kapur, K. C., and L. R. Lamberson, *Reliability in Engineering Design,* Wiley, New York, 1977.

Lee, H.-L., and L. W. Mays, Improved risk and reliability model for hydraulic structures, *Water Resour. Res.,* vol. 19, no. 6, pp. 1415–1422, 1983.

Lee, H.-L., and L. W. Mays, Hydraulic uncertainties in flood levee capacity, *J. Hyd. Div., Am. Soc. Civ. Eng.,* vol. 112, no. 10, pp. 928–934, 1986.

Milhous, R. T., and W. J. Grenney, The quantification and reservation of instream flows, *Prog. Wat. Tech.,* vol. 13, pp. 129–154, 1980.

National Academy of Sciences, *Safety of Existing Dams: Evaluation and Improvement,* National Academy Press, Washington, D. C., 1983.

Salas, J. D., et al., *Applied modelling of hydrologic time series,* Water Resources Publications, Littleton, Colo., 1980.

Tang, W. H., and B. C. Yen, Hydrologic and hydraulic design under uncertainties, *Proceedings,* International Symposium on Uncertainties in Hydrologic and Water Resources Systems, Tucson, Ariz., vol. 2, pp. 868–882, 1972.

Tung, Y.-K., and L. W. Mays, Risk analysis for hydraulic design, *J. Hyd. Div., Am. Soc. Civ. Eng.,* vol. 106, no. HY5, pp. 893–913, 1980.

U. S. Army Corps of Engineers Hydrologic Engineering Center, computer program catalog, Davis, Calif., August 1986.

World Meteorological Organization, *Guide to Hydrological Practices,* vol. II, Analysis, forecasting, and other applications, WMO no. 168, 4th ed., Geneva, Switzerland, 1983.

Yen, B. C., Risks in hydrologic design of engineering projects, *J. Hyd. Div., Am. Soc. Civ. Eng.,* vol. 96, no. HY4, proc. paper 7229, pp. 959–966, April 1970.

Yen, B. C., H. G. Wenzel, L. W. Mays, and W. H. Tang, Advanced methodologies for design of storm sewer systems, research report no. 112, Water Resources Center, University of Illinois at Urbana-Champaign, August 1976.

Yen, B. C., ed., *Stochastic and Risk Analysis in Hydraulic Engineering,* Water Resources Publications, Littleton, Colo., 1986.

PROBLEMS

13.2.1 The critical drought of record as determined from 30 years of hydrologic data is considered to have lasted for 3 years. If a water supply design is based on this drought and the design life is 50 years, what is the chance that a worse drought will occur during the design life?

13.2.2 In Prob. 13.2.1, what is the chance that a worse drought will occur during the first 10 years of the design life? The first 20 years?

13.2.3 What is the chance that the largest flood observed in 50 years of record will be exceeded during the next 10 years? The next 20 years?

13.2.4 If a structure has a design life of 15 years, calculate the required design return period if the acceptable risk of failure is 20 percent (*a*) in any year, (*b*) over the design life.

13.2.5 A flood plain regulation prevents construction within the 25-year flood plain. What is the risk that a structure built just on the edge of this flood plain will be flooded during the next 10 years? By how much would this risk be reduced if construction were limited to the edge of the 100-year flood plain?

13.2.6 A house has a 30-year design life. What is the chance it will be flooded during its design life if it is located on the edge of the 25-year flood plain? The 100-year flood plain?

13.2.7 Determine the optimum scale of development (return period) for the flood-control measure considered in Example 13.2.3 if the annual capital costs given in Table 13.2.1 are doubled. Use the same damage costs as in Table 13.2.1.

13.2.8 Determine the optimum scale of development (return period) for the flood-control measure considered in Example 13.2.3, if the damage costs are doubled. Annual capital costs remain the same as in Table 13.2.1.

13.2.9 Determine the optimum scale of development (return period) for the flood control measure considered in Example 13.2.3, if the damage costs and the annual capital costs are both doubled.

13.3.1 A rectangular channel is 200 feet wide, has bed slope 0.5 percent, an estimated Manning's n of 0.040, and a design discharge of 10,000 cfs. Calculate the design flow depth. If the coefficient of variation of the design discharge is 0.20 and of Manning's n is 0.15, calculate the standard error of estimate of the flow depth. What is the probability that the actual water level will be more than 1 foot deeper than the expected value? Within what range can the water level for the design event be expected in 70 percent of events?

13.3.2 In Prob. 13.3.1, calculate $\partial y/\partial Q$ and $\partial y/\partial n$ for the conditions given ($Q = 10,000$ cfs and $n = 0.040$) and solve the problem using these derivatives.

13.3.3 Solve Prob. 13.3.1 if the channel is trapezoidal with bottom width 150 ft and side slopes 1 vert. = 3 hor.

13.3.4 Flow in a natural stream channel has been modeled by a computer program and found to have a depth of 15 ft for a flow rate of 8000 cfs and Manning's n value of 0.045. Rerunning the program shows that changing the design discharge by 1000 cfs changes the water surface elevation by 0.8 ft, and changing Manning's n by 0.005 changes the water surface elevation by 0.6 ft. If the design discharge is assumed to be accurate to ± 30 percent and Manning's n to ± 10 percent, calculate the corresponding error in the flow depth (or water surface elevation).

13.3.5 Suppose for the conditions given in Example 13.3.1, solved in the text, that the channel wall height adopted is 8.4 ft, that is, the calculated depth of 7.4 ft plus a 1.0 ft freeboard, or safety margin. What safety factor SF is implied by this choice? What would the safety factor be if the true Manning's roughness were 0.045 instead of the 0.035 assumed? Is this a safe design?

13.3.6 Using the first-order analysis of uncertainty for Manning's equation, show that the coefficient of variation of the discharge Q is given by $CV_Q^2 = CV_n^2 + (1/4)CV_{Sf}^2$. What assumptions about the variables in Manning's equation are implied by this equation for CV_Q?

13.3.7 In some instances, flood plain studies are made using channel cross sections determined from topographic maps instead of ground surveys. Extend the first-order analysis of uncertainty for water level in Sec. 13.3 to include uncertainty in the cross-sectional area A and wetted perimeter P. If these variables can be determined with coefficients of variation of 20 percent from topographic maps, calculate the additional risk that the houses in Example 13.3.1 will flood during the design event, resulting from the use of channel cross sections from topographic maps, instead of ground surveys, to delineate the flood plain.

13.4.1 A hydrologic design has a loading with mean value 10 units and standard deviation 2 units. Calculate the risk of failure if the capacity is 12 units. Assume normal distribution for the loading.

13.4.2 Solve Prob. 13.4.1 if the loading is lognormally distributed.

13.4.3 In Prob. 13.4.1, assume that the capacity is normally distributed with mean 12 units and standard deviation 1 unit. Recompute the risk of failure, assuming that the loading is also normally distributed.

13.4.4 About half the total water supply for southern California is provided by long-distance water transfers from northern California and from the Colorado River. The annual demand for these transfers was estimated to be 1.48 MAF (million acre-feet) in 1980, and is projected to rise linearly to 1.77 MAF in 1990. Study of observed annual demands from 1980 to 1985 indicates that the coefficient of variation of observed annual demands around those expected is approximately 0.1 (this variability is due to year-to-year variations in weather and other factors). Estimate the annual demand level that has a 70 percent chance of being equaled or exceeded in 1986 and in 1990. Calculate the chance that observed demands will exceed 2.0 MAF/year in 1986, and in 1990. Assume that the annual demands are normally distributed.

13.4.5 In Prob. 13.4.4, calculate the chance that a limit of 2.0 MAF in water transfers will be exceeded at least once from 1986 to 1990. Assume annual demands are independent from one year to the next.

13.5.1 If capacity and loading are both lognormally distributed, show that risk can be calculated by Eq. (13.5.7):

$$\bar{R} = F_z\left(\frac{-\ln\left[\dfrac{\mu_C}{\mu_L}\left(\dfrac{1+\mathrm{CV}_L^2}{1+\mathrm{CV}_C^2}\right)^{1/2}\right]}{\left\{\ln\left[(1+\mathrm{CV}_C^2)(1+\mathrm{CV}_L^2)\right]\right\}^{1/2}}\right)$$

where F_z denotes the standard normal distribution function.

13.5.2 If capacity and loading are both lognormally distributed, show that risk can be approximated by

$$\bar{R} = F_z\left[\frac{\ln\left(\mu_L/\mu_C\right)}{\left(\mathrm{CV}_L^2+\mathrm{CV}_C^2\right)^{1/2}}\right]$$

where F_z denotes the standard normal distribution function.

13.5.3 Calculate the risk of failure of an open channel, assuming that the safety margin is normally distributed: Manning's equation is used to compute the capacity, and a first-order analysis is used to determine the coefficient of variation of the capacity C. The mean loading is 5000 cfs and the coefficient of variation of loading is

0.2. The slope of the channel is 0.01 with a coefficient of variation $CV_{S_f} = 0.10$. The Manning's roughness factor is 0.035 and has a coefficient of variation of $CV_n = 0.15$. The channel cross-section is rectangular with width 50 ft and wall height 9 ft. Failure is assumed to occur if the walls are overtopped.

13.5.4 Rework Prob. 13.5.3 to compute the risk of failure, assuming the capacity and loading to be lognormally distributed.

13.5.5 Use the risk analysis of safety margins method to determine the probability that the houses will be flooded in Example 13.3.1 in the text.

CHAPTER
14

DESIGN STORMS

A *design storm* is a precipitation pattern defined for use in the design of a hydrologic system. Usually the design storm serves as the system input, and the resulting rates of flow through the system are calculated using rainfall-runoff and flow routing procedures. A design storm can be defined by a value for precipitation depth at a point, by a design hyetograph specifying the time distribution of precipitation during a storm, or by an isohyetal map specifying the spatial pattern of the precipitation.

Design storms can be based upon historical precipitation data at a site or can be constructed using the general characteristics of precipitation in the surrounding region. Their application ranges from the use of point precipitation values in the rational method for determining peak flow rates in storm sewers and highway culverts, to the use of storm hyetographs as inputs for rainfall-runoff analysis of urban detention basins or for spillway design in large reservoir projects. This chapter covers the development of point precipitation data, intensity-duration-frequency relationships, design hyetographs, and estimated limiting storms based on probable maximum precipitation.

14.1 DESIGN PRECIPITATION DEPTH

Point Precipitation

Point precipitation is precipitation occurring at a single point in space as opposed to areal precipitation which is precipitation over a region. For point precipitation frequency analysis, the annual maximum precipitation for a given duration is selected by applying the method outlined in Sec. 3.4. to all storms in a year, for each year of historical record. This process is repeated for each of a series of durations. For each duration, frequency analysis is performed on the data, as described in Sec. 12.2, to derive the design precipitation depths for various return periods; then the design depths are converted to intensities by dividing by the precipitation duration.

By analyzing data in this way, Hershfield (1961) developed isohyetal maps of design rainfall depth for the entire United States; these were published in U. S. Weather Bureau technical paper no. 40, commonly called TP 40. The maps presented in TP 40 are for durations from 30 minutes to 24 hours and return periods from 1 to 100 years. Hershfield also furnished interpolation diagrams for making precipitation estimates for durations and return periods not shown on the maps. Fig. 14.1.1 shows the TP 40 map for 100-year 24-hour rainfall. The U. S. Weather Bureau (1964) later published maps for durations of 2 to 10 days.

In many design situations, such as storm sewer design, durations of 30 minutes or less must be considered. In a publication commonly known as HYDRO 35 (Frederick, Meyers, and Auciello, 1977), the U. S. National Weather Service presented isohyetal maps for events having durations from 5 to 60 minutes, partially superseding TP 40. The maps of precipitation depths for 5-, 15-, and 60-minute durations and return periods of 2 and 100 years for the 37 eastern states

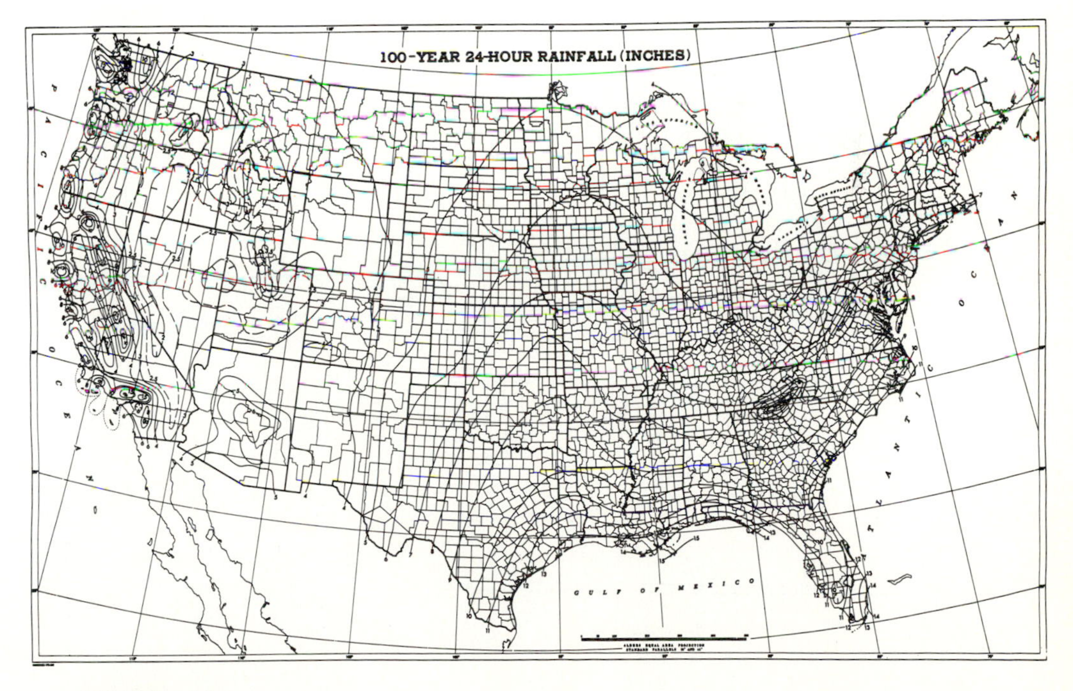

FIGURE 14.1.1
The 100-year 24-hour rainfall (in) in the United States as presented in U. S. Weather Bureau technical paper 40. (*Source*: Hershfield, 1961.)

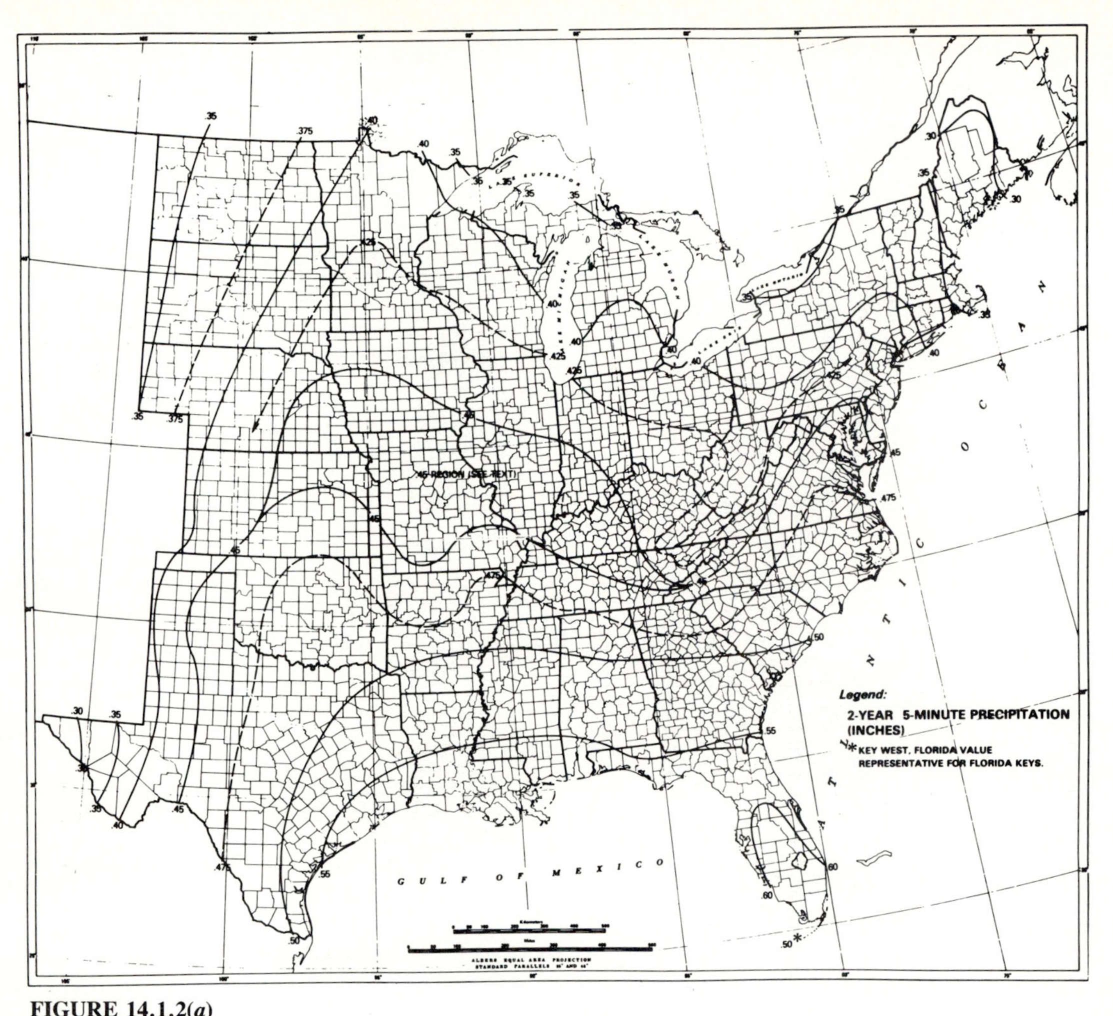

FIGURE 14.1.2(*a*)
2-year 5-minute precipitation (inches). (*Source*: Frederick, Meyers, and Auciello, 1977.)

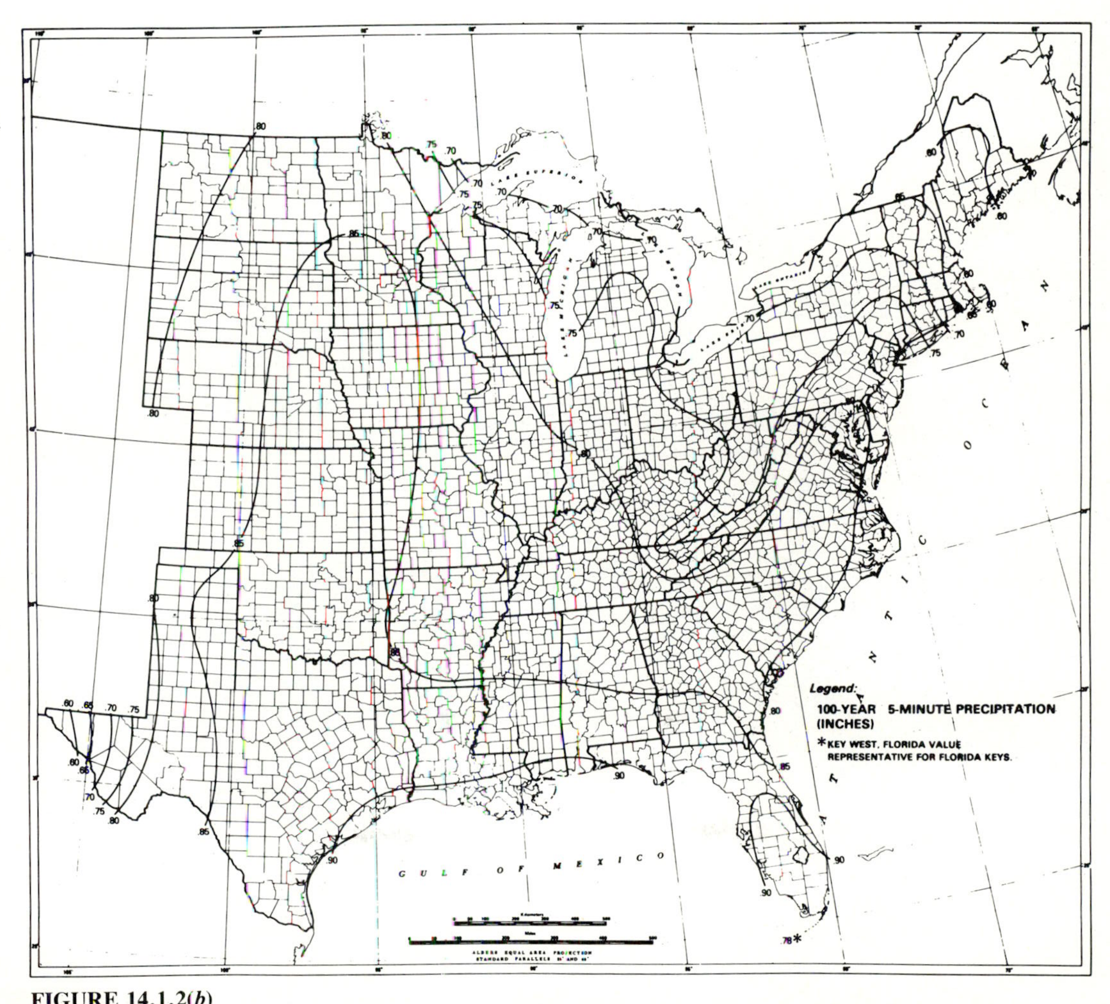

FIGURE 14.1.2(*b*)
100-year 5-minute precipitation (inches). (*Source*: Frederick, Meyers, and Auciello, 1977.)

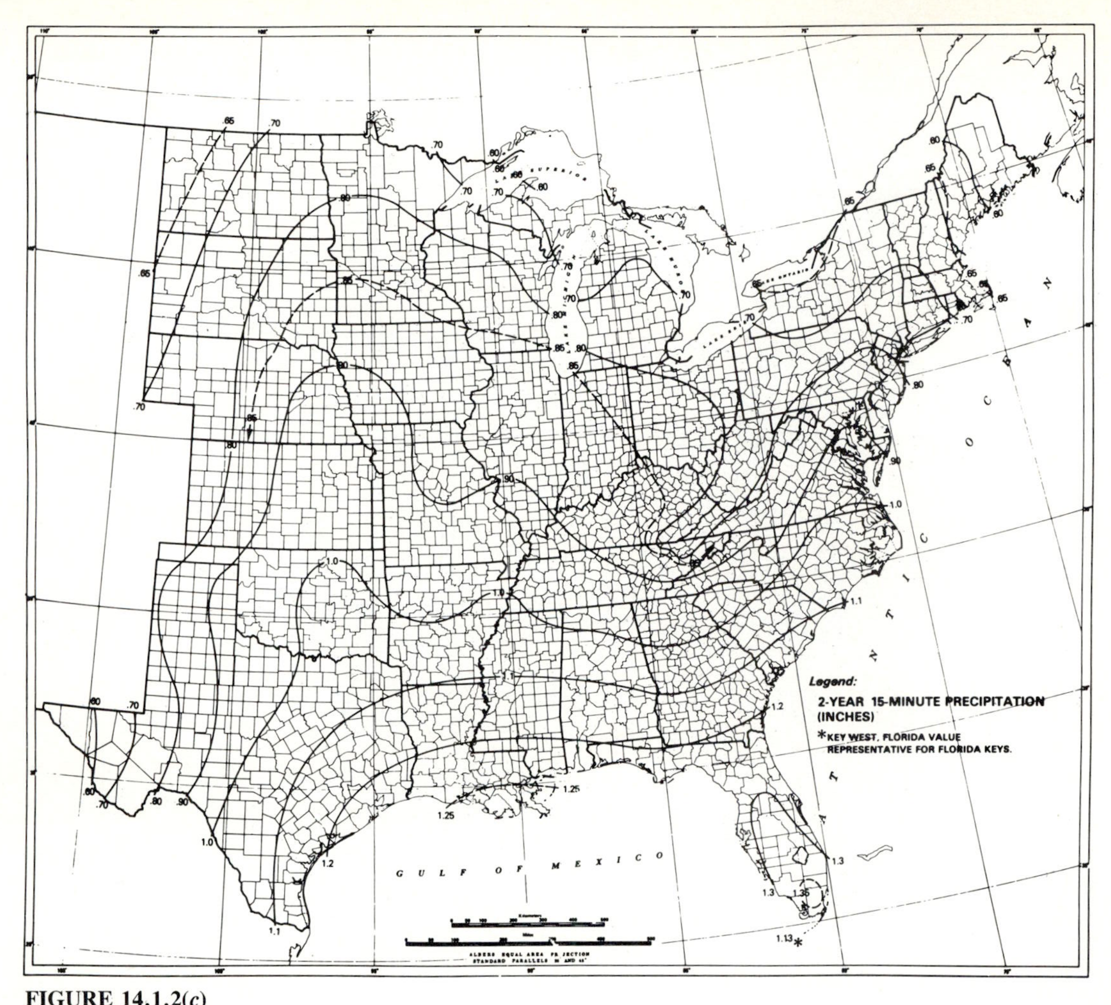

FIGURE 14.1.2(*c*)
2-year 15-minute precipitation (inches). (*Source*: Frederick, Meyers, and Auciello, 1977.)

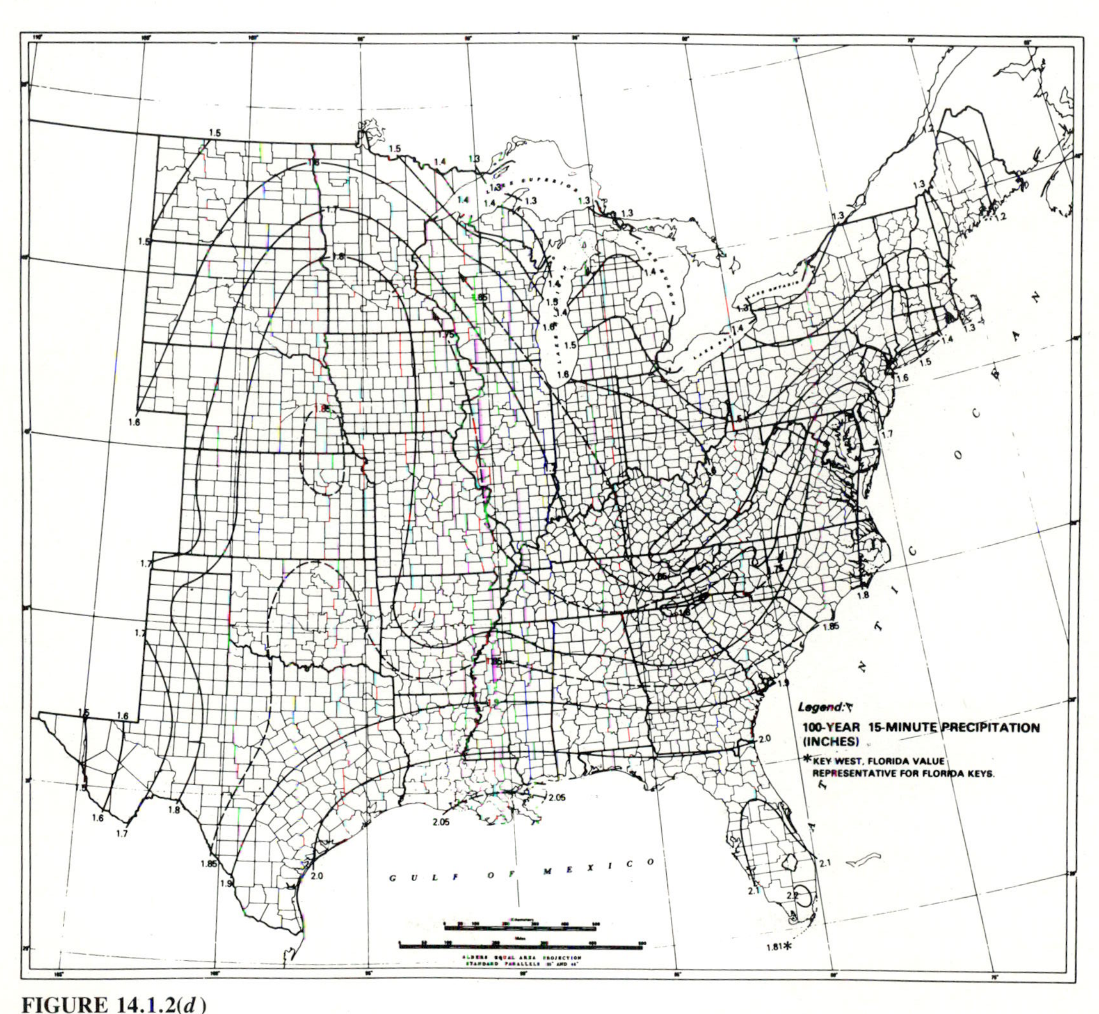

FIGURE 14.1.2(*d*)
100-year 15-minute precipitation (inches). (*Source*: Frederick, Meyers, and Auciello, 1977.)

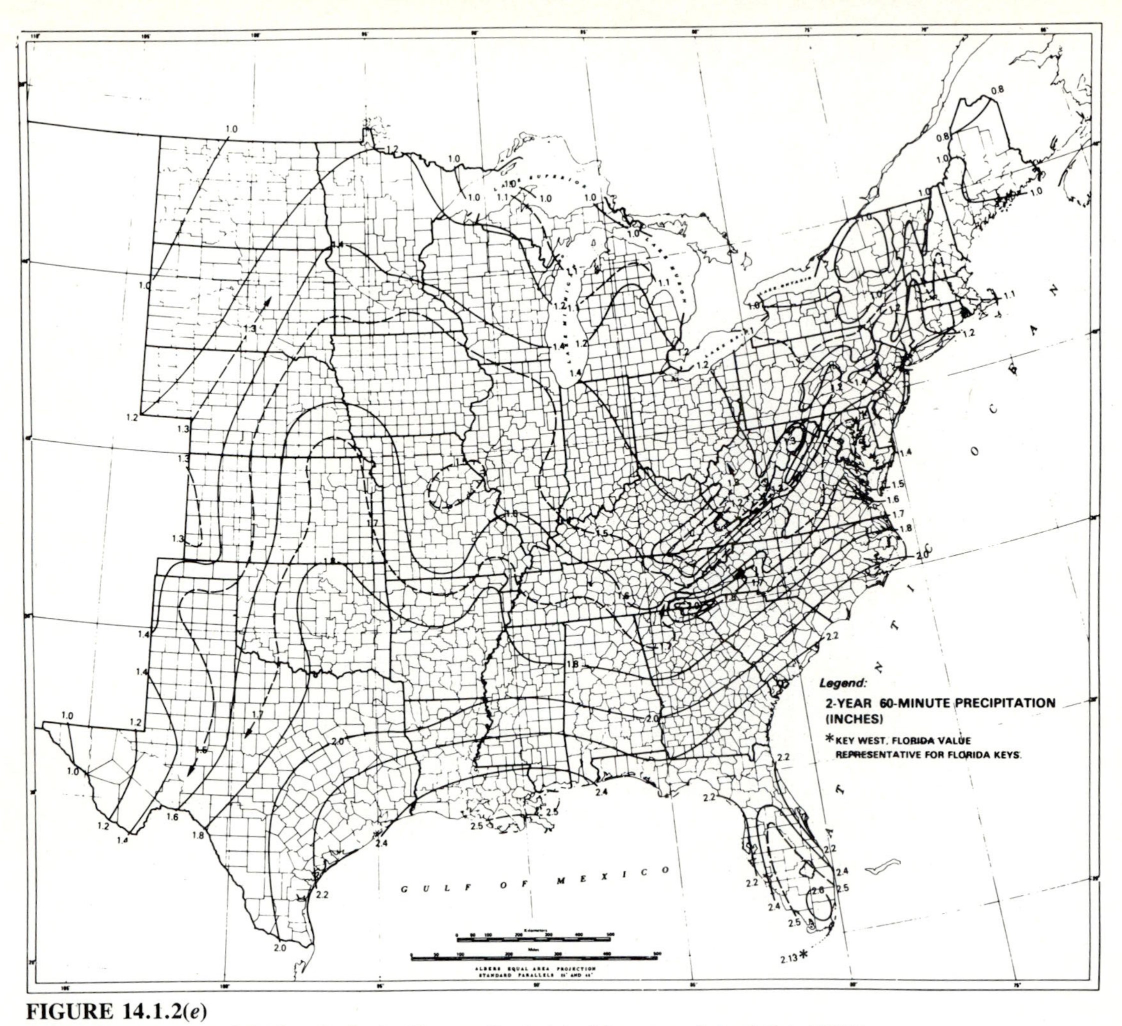

FIGURE 14.1.2(*e*)
2-year 60-minute precipitation (inches). (*Source*: Frederick, Meyers, and Auciello, 1977.)

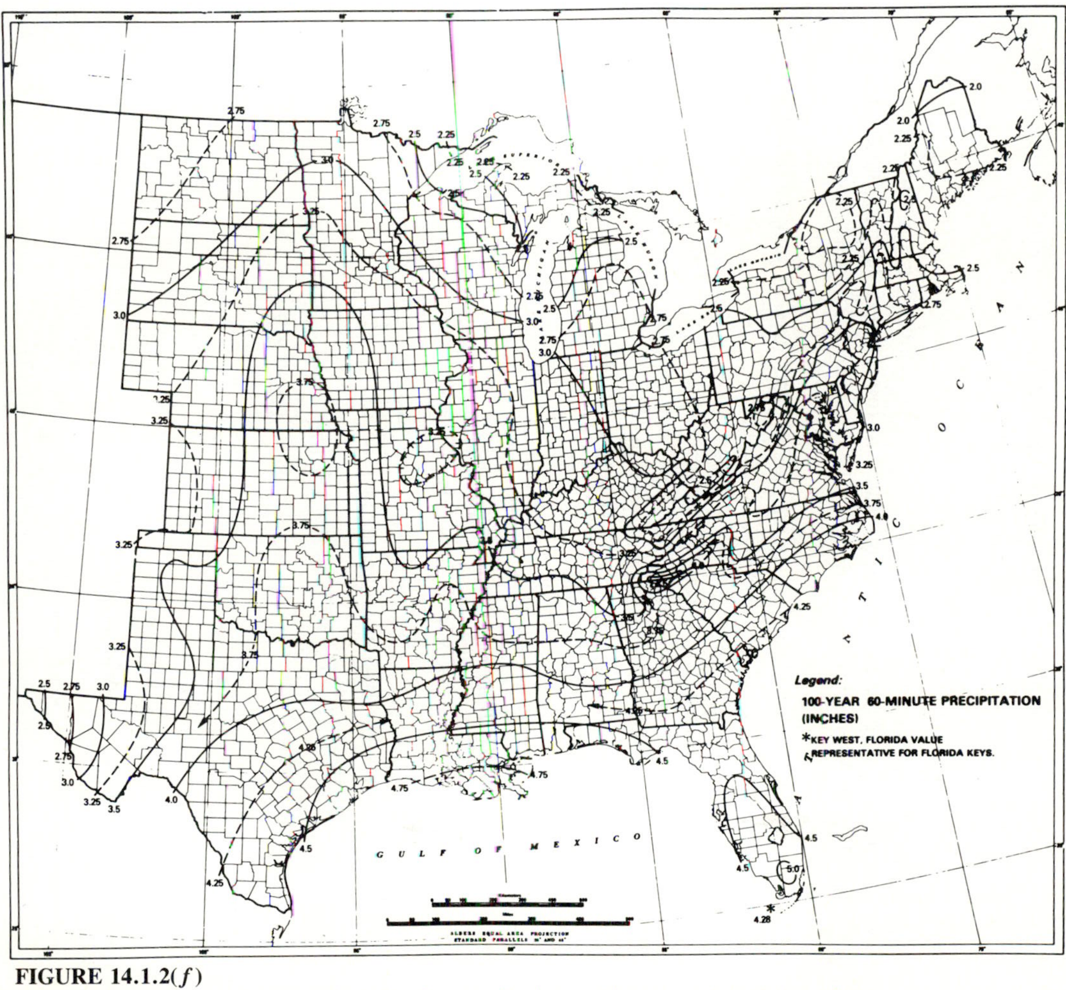

FIGURE 14.1.2(*f*)
100-year 60-minute precipitation (inches). (*Source*: Frederick, Meyers, and Auciello, 1977.)

are shown in Fig. 14.1.2. Depths for 10- and 30-minute durations for a given return period are obtained by interpolation from the 5-, 15-, and 60-minute data for the same return period:

$$P_{10\ \text{min}} = 0.41 P_{5\ \text{min}} + 0.59 P_{15\ \text{min}} \tag{14.1.1a}$$

$$P_{30\ \text{min}} = 0.51 P_{15\ \text{min}} + 0.49 P_{60\ \text{min}} \tag{14.1.1b}$$

For return periods other than 2 or 100 years, the following interpolation equation is used, with the appropriate coefficients a and b from Table 14.1.1.

$$P_{T\ \text{yr}} = aP_{2\ \text{yr}} + bP_{100\ \text{yr}} \tag{14.1.2}$$

Miller, Frederick, and Tracey (1973) present isohyetal maps for 6- and 24-hour durations for the 11 mountainous states in the western United States; these supersede the corresponding maps in TP 40.

Example 14.1.1. Determine the design rainfall depth for a 25-year 30-minute storm in Oklahoma City.

Solution. Oklahoma City is located near the center of the state of Oklahoma and the values of 15- and 60-minute precipitation for 2- and 100-year return periods are read from Fig. 14.1.2 as $P_{2,15} = 1.02$ in, $P_{100,15} = 1.86$ in, $P_{2,60} = 1.85$ in, and $P_{100,60} = 3.80$ in, respectively. Using (14.1.1*b*), the values for 30-minute precipitation depth are calculated

$$P_{30\ \text{min}} = 0.51 P_{15\ \text{min}} + 0.49 P_{60\ \text{min}}$$

For $T = 2$ years, $P_{2,30} = 0.51 \times 1.02 + 0.49 \times 1.85 = 1.43$ in.

For $T = 100$ years, $P_{100,30} = 0.51 \times 1.86 + 0.49 \times 3.80 = 2.81$ in.

Then (14.1.2) is used with coefficients $a = 0.293$ and $b = 0.669$ from Table 14.1.1 to give the 25-year 30-minute precipitation depth:

$$\begin{aligned} P_{25,30} &= aP_{2,30} + bP_{100,30} \\ &= 0.293 \times 1.43 + 0.669 \times 2.81 \\ &= 2.30 \text{ in} \end{aligned}$$

TABLE 14.1.1
Coefficients for interpolating design precipitation depths using Eq. (14.1.2)

Return period T years	a	b
5	0.674	0.278
10	0.496	0.449
25	0.293	0.669
50	0.146	0.835

Source: Frederick, Myers, and Auciello, 1977.

Areal Precipitation Depth

Frequency analysis of precipitation over an area has not been as well developed as has analysis of point precipitation. In the absence of information on the true probability distribution of areal precipitation, point precipitation estimates are usually extended to develop an average precipitation depth over an area. The areal estimate may be either storm-centered or location-fixed. For the location-fixed case, one accounts for the fact that precipitation stations are sometimes near the storm center, sometimes on the outer edges, and sometimes in between the two. An averaging process results in location-fixed depth-area curves relating areal precipitation to point measurements. Fig. 14.1.3 provides curves for calculating areal depths as a percentage of point precipitation values (World Meteorological Organization, 1983).

Depth-area relationships for various durations, such as those shown in Fig. 14.1.3, are derived by a depth-area-duration analysis, in which isohyetal maps are prepared for each duration from the tabulation of maximum n-hour rainfalls recorded in a densely gaged area. The area contained within each isohyet on these maps is determined and a graph of average precipitation depth vs. area is plotted for each duration.

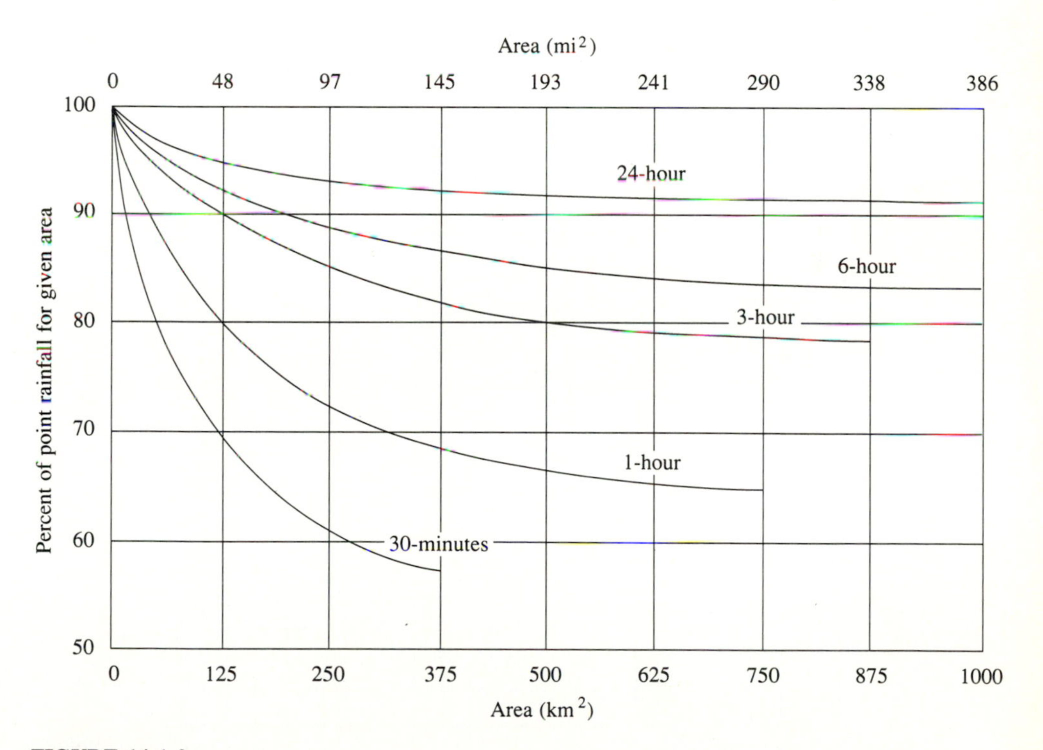

FIGURE 14.1.3
Depth-area curves for reducing point rainfall to obtain areal average values. (*Source*: World Meteorological Organization, 1983; originally published in Technical Paper 29, U. S. Weather Bureau, 1958.)

14.2 INTENSITY-DURATION-FREQUENCY RELATIONSHIPS

One of the first steps in many hydrologic design projects, such as in urban drainage design, is the determination of the rainfall event or events to be used. The most common approach is to use a design storm or event that involves a relationship between rainfall *intensity* (or depth), *duration*, and the *frequency* or return period appropriate for the facility and site location. In many cases, the hydrologist has standard intensity-duration-frequency (IDF) curves available for the site and does not have to perform this analysis. However, it is worthwhile to understand the procedure used to develop the relationships. Usually, the information is presented as a graph, with duration plotted on the horizontal axis, intensity on the vertical axis, and a series of curves, one for each design return period, as illustrated for Chicago in Fig. 14.2.1.

The intensity is the time rate of precipitation, that is, depth per unit time (mm/h or in/h). It can be either the instantaneous intensity or the average intensity over the duration of the rainfall. The average intensity is commonly used and can be expressed as

$$i = \frac{P}{T_d} \tag{14.2.1}$$

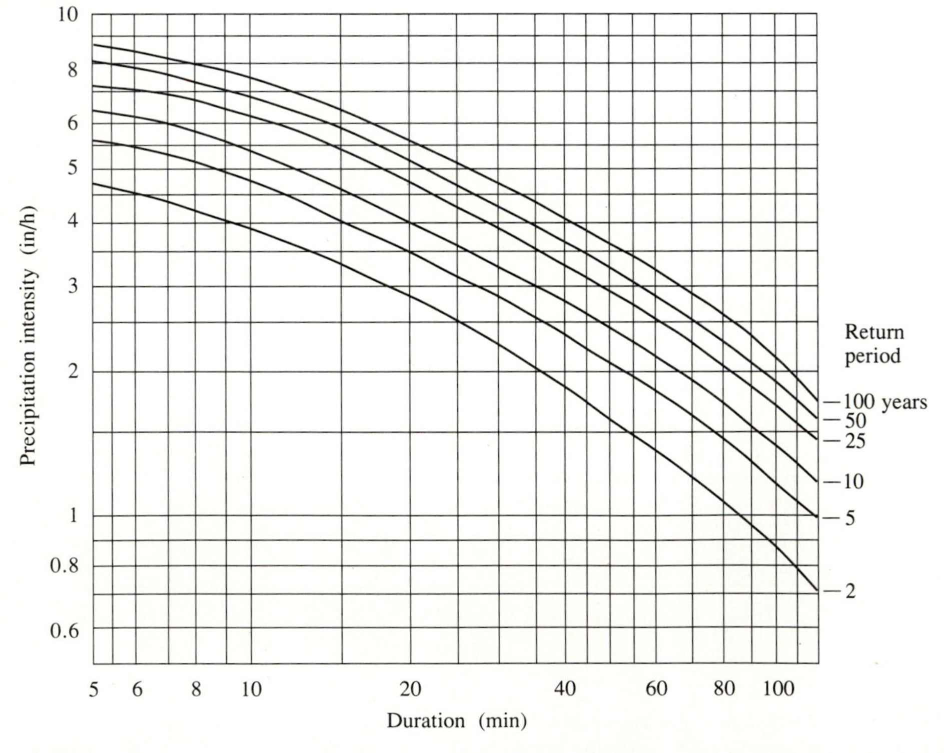

FIGURE 14.2.1
Intensity-duration-frequency curves of maximum rainfall in Chicago, U. S. A.

where P is the rainfall depth (mm or in) and T_d is the duration, usually in hours. The frequency is usually expressed in terms of return period, T, which is the average length of time between precipitation events that equal or exceed the design magnitude.

Example 14.2.1. Determine the design precipitation intensity and depth for a 20-minute duration storm with a 5-year return period in Chicago.

Solution. From the IDF curves for Chicago (Fig. 14.2.1), the design intensity for a 5-year, 20-minute storm is i = 3.50 in/h. The corresponding precipitation depth is given by Eq. (14.2.1) with T_d = 20 min = 0.333 h.

$$\begin{aligned} P &= iT_d \\ &= 3.50 \times 0.333 \\ &= 1.17 \text{ in} \end{aligned}$$

Example 14.2.2. Use the U. S. National Weather Service maps (Fig. 14.1.2) and Eqs. (14.1.1) and (14.1.2) to plot IDF curves for Oklahoma City, Oklahoma, for return periods of 2, 5, 10, 25, 50, and 100 years. Consider rainfall durations ranging from 5 minutes to 1 hour.

Solution. The six maps presented in Fig. 14.1.2 show precipitation for 5-, 15-, and 60-minute durations and 2- and 100-year return periods. The six values for Oklahoma City from these maps are: $P_{2,5}$ = 0.48 in, $P_{100,5}$ = 0.87 in, $P_{2,15}$ = 1.02 in, $P_{100,15}$ = 1.86 in, $P_{2,60}$ = 1.85 in, $P_{100,60}$ = 3.80 in. For T = 2 and 100 yr, the precipitation for 10- and 30-minute durations is obtained by interpolation from the 5-, 15-, and 60-minute values using (14.1.1) as illustrated in Example 14.1.1. For each duration, the values for return period T = 5, 10, 25, and 50 yr are obtained using the values at T = 2 and 100 yr by interpolation using (14.1.2) and Table 14.1.1, as also illustrated in Example 14.1.1. The results are shown in Table 14.2.1 in terms of precipitation depth, and they are converted to intensity by dividing by duration. For example, $P_{25,30}$ = 2.30 in, so the corresponding intensity is $i = P/T_d$ = 2.30 in/0.50 h = 4.60 in/h. The resulting precipitation intensities for each duration and return period are plotted in Fig. 14.2.2.

TABLE 14.2.1
Design precipitation depths (in) at Oklahoma City for various durations and return periods (Example 14.2.1)

Return period T (yr)	Duration T_d (min)				
	5	10	15	30	60
2	*0.48*	0.80	*1.02*	1.43	*1.85*
5	0.57	0.94	1.20	1.74	2.30
10	0.63	1.05	1.34	1.97	2.62
25	0.72	1.21	1.54	2.30	3.08
50	0.80	1.33	1.70	2.56	3.44
100	*0.87*	1.45	*1.86*	2.81	*3.80*

The values in italics are read from Fig. 14.1.2; the remainder are obtained by interpolation using Eqs. (14.1.1) and (14.1.2).

IDF Curves by Frequency Analysis

When local rainfall data are available, IDF curves can be developed using frequency analysis. A commonly used distribution for rainfall frequency analysis is the Extreme Value Type I or Gumbel distribution as discussed in Sec. 12.2. For each duration selected, the annual maximum rainfall depths are extracted from historical rainfall records, then frequency analysis is applied to the annual data. In some situations, particularly when only a few years of data are available (less than 20 to 25 years), an annual exceedence series for each duration may be determined by ranking the depths and choosing the N largest values from a record of N years. Such a series is shown in Table 14.2.2 for a rain gage at Coshocton, Ohio. In the table, the lines connect precipitation data for various durations of the same storm event. The design precipitation depths determined from the annual exceedence series can then be adjusted to match those derived from an annual maximum series by multiplying the depths by 0.88 for the 2-year return period values, 0.96 for the 5-year return period values, and 0.99 for the 10-year return period values (Hershfield, 1961). No adjustment of the estimates is required for longer return periods.

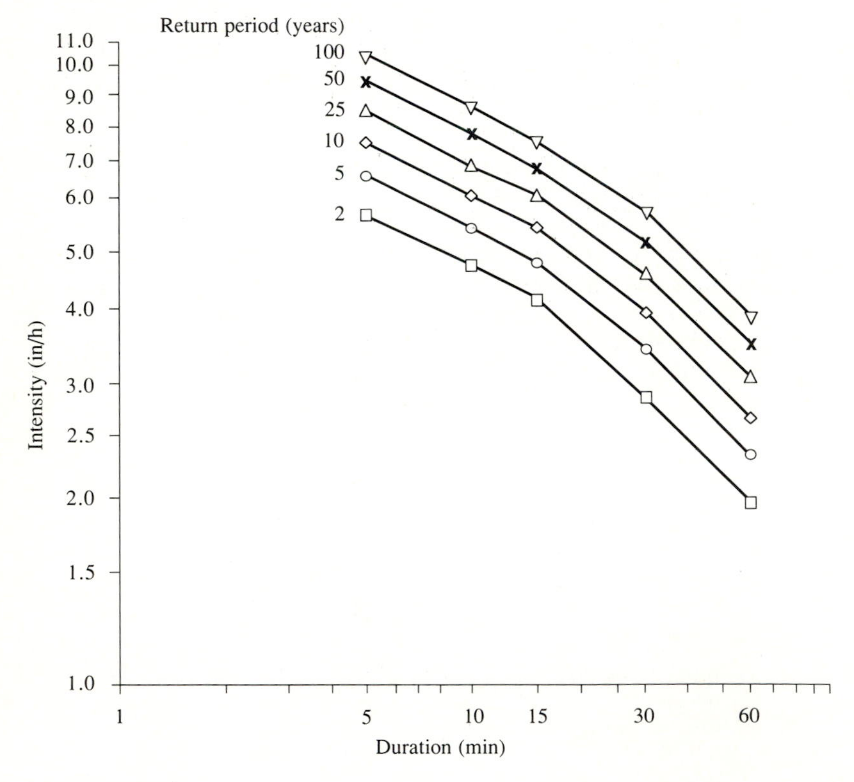

FIGURE 14.2.2
Intensity-duration-frequency curves for Oklahoma City (Example 14.2.1).

TABLE 14.2.2
Annual exceedence series of rainfall data for Coshocton, Ohio

Rank	Return period (yr)	Maximum depth (in) and date for duration shown			
		15 min	30 min	60 min	120 min
1	25.00	1.423 6/12/57	2.625 6/12/57	3.220 6/12/57	3.421 6/12/57
2	12.50	0.940 7/11/51	1.326 7/24/68	1.830 6/27/75	1.900 7/27/69
3	8.33	0.920 6/12/59	1.238 5/13/64	1.756 7/27/69	1.883 8/21/60
4	6.25	0.910 5/13/64	1.177 6/23/52	1.510 8/21/60	1.792 7/4/69
5	5.00	0.890 6/27/75	1.170 7/22/58	1.431 7/24/68	1.733 7/24/68
6	4.17	0.884 6/23/52	1.167 6/27/75	1.375 7/22/58	1.703 8/4/59
7	3.57	0.860 8/14/73	1.149 6/17/70	1.313 6/17/70	1.623 6/12/59
8	3.13	0.810 7/27/69	1.087 6/15/75	1.306 5/13/64	1.609 6/28/57
9	2.78	0.805 6/22/51	1.063 8/22/51	1.290 6/23/52	1.604 6/13/72
10	2.50	0.783 6/24/56	1.060 7/11/51	1.269 4/25/61	1.600 7/28/61
11	2.27	0.770 8/15/75	1.040 6/12/59	1.225 6/12/59	1.570 4/25/61
12	2.08	0.770 7/22/58	1.037 7/19/67	1.213 7/4/69	1.482 7/22/58
13	1.92	0.750 7/10/73	1.027 9/5/75	1.204 6/13/72	1.393 8/11/64
14	1.79	0.750 6/17/70	1.023 7/10/73	1.203 8/11/64	1.353 5/13/64
15	1.67	0.733 7/19/67	1.000 7/10/55	1.200 8/3/63	1.351 9/24/70
16	1.56	0.732 7/30/58	0.975 7/27/69	1.194 8/2/64	1.335 6/23/69
17	1.47	0.710 7/3/52	0.972 7/30/58	1.192 9/12/57	1.310 8/14/57
18	1.39	0.707 8/3/63	0.934 8/27/74	1.174 7/28/61	1.305 6/24/57
19	1.32	0.700 7/24/68	0.919 7/28/61	1.143 6/22/51	1.300 6/11/60
20	1.25	0.700 6/4/63	0.907 9/12/57	1.130 9/24/70	1.300 6/23/52
21	1.19	0.700 6/22/60	0.890 8/14/73	1.130 7/19/67	1.290 8/2/64
22	1.14	0.692 4/3/74	0.880 6/24/56	1.109 9/5/75	1.274 9/12/57
23	1.09	0.688 8/27/74	0.873 6/11/60	1.095 7/6/58	1.230 7/3/52
24	1.04	0.687 9/12/57	0.869 7/4/69	1.094 6/28/57	1.220 7/6/58
25	1.00	0.670 4/13/55	0.850 8/11/64	1.063 8/27/74	1.200 9/5/75

Source: Wenzel, 1982, Copyright by the American Geophysical Union.

Example 14.2.3. Using the data presented in Table 14.2.2, determine the 2-year and 25-year precipitation depth estimates for a 15-minute duration storm in Coshocton, Ohio. Assume the Extreme Value Type I (Gumbel) distribution is applicable.

Solution. The design rainfall depth for a given return period T is determined by Eq. (12.3.3):

$$x_{T,T_d} = \bar{x}_{T_d} + K_T s_{T_d}$$

where $\bar{x}_{T_d}$ and s_{T_d} are the mean and standard deviation of the rainfall depths for a specified duration T_d, and K_T is the frequency factor given by Eq. (12.3.8):

$$K_T = -\frac{\sqrt{6}}{\pi}\left[0.5772 + \ln\left(\ln\frac{T}{T-1}\right)\right]$$

Consider a 15-minute duration for example; the mean and standard deviation of the 15-minute precipitation data in Table 14.2.2 are $\bar{x}_{15} = 0.799$ in and $s_{15} = 0.154$ in, respectively. Using (12.3.8), $K_2 = -(\sqrt{6}/\pi)(0.5772 + \ln\{\ln[T/(T-1)]\}) = -(\sqrt{6}/\pi)\{0.5772 + \ln[\ln[2/1]]\} = -0.164$; the corresponding value for $T = 25$ yr is $K_{25} = 2.044$. Then, using (12.3.3) for a two-year return period,

$$\begin{aligned} x_{2,15} &= \bar{x}_{15} + K_2 s_{15} \\ &= 0.799 - 0.164 \times 0.154 \\ &= 0.774 \text{ in} \end{aligned}$$

Because the data in Table 14.2.2 are an annual exceedence series, this value is multiplied by 0.88 to obtain the design precipitation depth $0.774 \times 0.88 = 0.68$ in for a two-year return period. For a 25-year return period

$$\begin{aligned} x_{25,15} &= \bar{x}_{15} + K_{25} s_{15} \\ &= 0.799 + 2.044 \times 0.154 \\ &= 1.11 \text{ in} \end{aligned}$$

This value is not adjusted because its return period is greater than 10 years.

Equations for IDF Curves

Intensity-duration-frequency curves have also been expressed as equations to avoid having to read the design rainfall intensity from a graph. For example, Wenzel (1982) provided coefficients from a number of cities in the United States for an equation of the form

$$i = \frac{c}{T_d^e + f} \tag{14.2.2}$$

where i is the design rainfall intensity, T_d is the duration, and c, e, and f are coefficients varying with location and return period. Table 14.2.3 shows values of these coefficients for a 10-year return period in 10 U. S. cities.

It is also possible to extend (14.2.2) to include the return period T using the equation

TABLE 14.2.3
Constants for rainfall equation (14.2.2) for 10-year return period storm intensities at various locations

Location	c	e	f
Atlanta	97.5	0.83	6.88
Chicago	94.9	0.88	9.04
Cleveland	73.7	0.86	8.25
Denver	96.6	0.97	13.90
Houston	97.4	0.77	4.80
Los Angeles	20.3	0.63	2.06
Miami	124.2	0.81	6.19
New York	78.1	0.82	6.57
Santa Fe	62.5	0.89	9.10
St. Louis	104.7	0.89	9.44

Constants correspond to i in inches per hour and T_d in minutes. *Source*: Wenzel, 1982, Copyright by the American Geophysical Union.

$$i = \frac{cT^m}{T_d + f} \tag{14.2.3}$$

or

$$i = \frac{cT^m}{T_d^e + f} \tag{14.2.4}$$

Example 14.2.4. Determine and compare the 10-yr, 20-minute design rainfall intensities in Los Angeles and Denver.

Solution. The design rainfall intensity is computed using $T_d = 20$ min, and the values of the coefficients for Los Angeles ($c = 20.3$, $e = 0.63$, and $f = 2.06$) from Table 14.2.3, for which Eq. (14.2.2) gives

$$\begin{aligned} i &= \frac{c}{T_d^e + f} \\ &= \frac{20.3}{20^{0.63} + 2.06} \\ &= 2.34 \text{ in/h} \end{aligned}$$

Similarly, for Denver $i = c/(T_d^e + f) = 96.6/(20^{0.97} + 13.90) = 3.00$ in/h. The design intensity in Denver is greater by $3.00 - 2.34 = 0.66$ in/h, or 28 percent.

14.3 DESIGN HYETOGRAPHS FROM STORM EVENT ANALYSIS

By analysis of observed storm events, the time sequence of precipitation in typical storms can be determined. Huff (1967) developed time distribution relations for heavy storms on areas ranging up to 400 mi^2 in Illinois. Time distribution patterns were developed for four probability groups, from the most severe (first quartile) to the least severe (fourth quartile). Fig. 14.3.1(*a*) shows the probability distribution of first-quartile storms. These curves are smooth, reflecting average rainfall distribution with time; they do not exhibit the burst characteristics of observed storms. Fig. 14.3.1(*b*) shows selected histograms of first-quartile storms for 10-, 50-, and 90-percent cumulative probabilities of occurrence, each illustrating the percentage of total storm rainfall for 10 percent increments of the storm duration. The 50 percent histogram represents a cumulative rainfall pattern that should be exceeded in about half of the storms. The 90 percent histogram can be interpreted as a storm distribution that is equaled or exceeded in 10 percent or less of the storms. The first quartile 50-percent distribution has been used in the ILLUDAS storm drainage simulation model by Terstriep and Stall (1974).

The U. S. Department of Agriculture, Soil Conservation Service (1986)

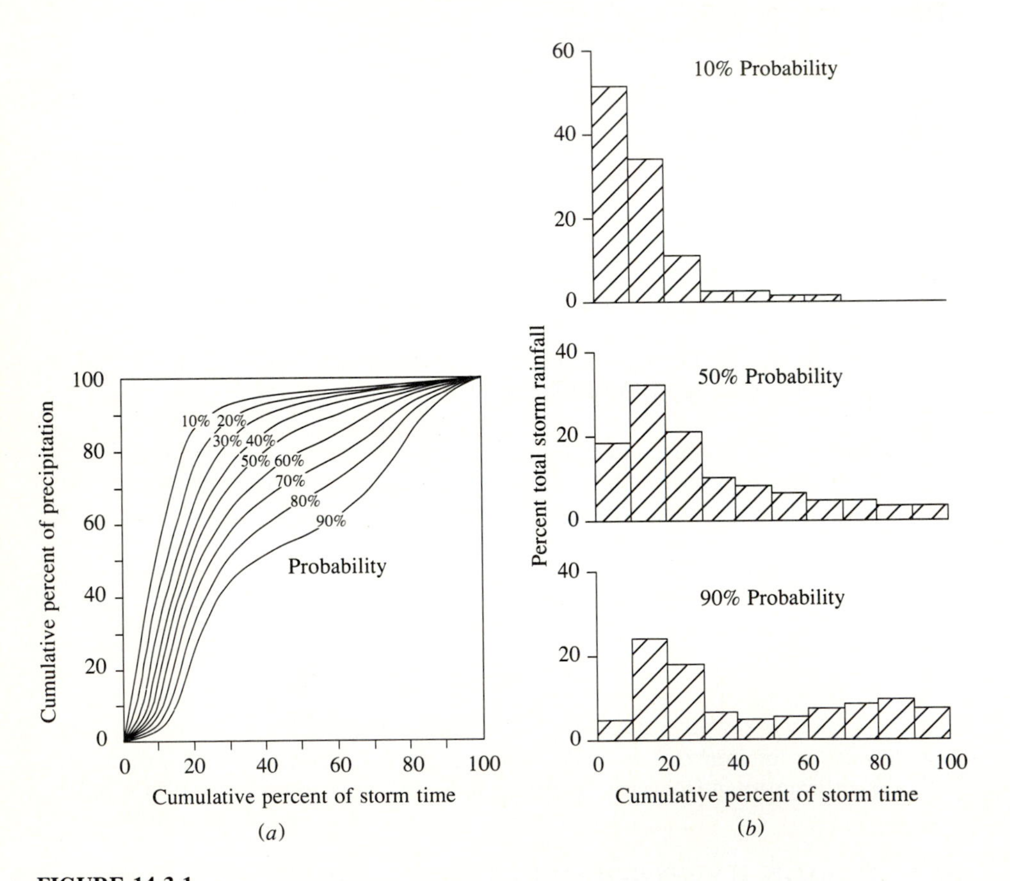

FIGURE 14.3.1
(*a*) Time disribution of first-quartile storms. The probability shown is the chance that the observed storm pattern will lie to the left of the curve. (*b*) Selected histograms for first-quartile storms. (*Source*: Huff, 1967, Copyright by the American Geophysical Union.)

developed synthetic storm hyetographs for use in the United States for storms of 6 and 24 hours duration. These hyetographs were derived from information presented by Hershfield (1961) and Miller, Frederick, and Tracey (1973) and additional storm data. Table 14.3.1 and Fig. 14.3.2 present the cumulative hyetographs. There are four 24-hour duration storms, called Types I, IA, II, and III, respectively; Figure 14.3.3 shows the geographic location within the United States where they should be applied. Types I and IA are for the Pacific maritime climate with wet winters and dry summers. Type III is for the Gulf of Mexico and the Atlantic coastal areas, where tropical storms result in large 24-hour rainfall amounts. Type II is for the remainder of the nation.

Pilgrim and Cordery (1975) developed a hyetograph analysis method that is based on ranking the time intervals in a storm by the depth of precipitation occurring in each, and repeating this exercise for many storms in the region. By summing the ranks for each interval, a typical hyetograph shape can be obtained. This approach is a standard method in Australian hydrologic design (The Institution of Engineers Australia, 1987).

TABLE 14.3.1
SCS rainfall distributions

24-hour storm						6-hour storm		
		P_t/P_{24}						
Hour t	**$t/24$**	**Type I**	**Type IA**	**Type II**	**Type III**	**Hour t**	**$t/6$**	**P_t/P_6**
0	0	0	0	0	0	0	0	0
2.0	0.083	0.035	0.050	0.022	0.020	0.60	0.10	0.04
4.0	0.167	0.076	0.116	0.048	0.043	1.20	0.20	0.10
6.0	0.250	0.125	0.206	0.080	0.072	1.50	0.25	0.14
7.0	0.292	0.156	0.268	0.098	0.089	1.80	0.30	0.19
8.0	0.333	0.194	0.425	0.120	0.115	2.10	0.35	0.31
8.5	0.354	0.219	0.480	0.133	0.130	2.28	0.38	0.44
9.0	0.375	0.254	0.520	0.147	0.148	2.40	0.40	0.53
9.5	0.396	0.303	0.550	0.163	0.167	2.52	0.42	0.60
9.75	0.406	0.362	0.564	0.172	0.178	2.64	0.44	0.63
10.0	0.417	0.515	0.577	0.181	0.189	2.76	0.46	0.66
10.5	0.438	0.583	0.601	0.204	0.216	3.00	0.50	0.70
11.0	0.459	0.624	0.624	0.235	0.250	3.30	0.55	0.75
11.5	0.479	0.654	0.645	0.283	0.298	3.60	0.60	0.79
11.75	0.489	0.669	0.655	0.357	0.339	3.90	0.65	0.83
12.0	0.500	0.682	0.664	0.663	0.500	4.20	0.70	0.86
12.5	0.521	0.706	0.683	0.735	0.702	4.50	0.75	0.89
13.0	0.542	0.727	0.701	0.772	0.751	4.80	0.80	0.91
13.5	0.563	0.748	0.719	0.799	0.785	5.40	0.90	0.96
14.0	0.583	0.767	0.736	0.820	0.811	6.00	1.0	1.00
16.0	0.667	0.830	0.800	0.880	0.886			
20.0	0.833	0.926	0.906	0.952	0.957			
24.0	1.000	1.000	1.000	1.000	1.000			

Source: U. S. Dept. of Agriculture, Soil Conservation Service, 1973, 1986.

Triangular Hyetograph Method

A triangle is a simple shape for a design hyetograph because once the design precipitation depth P and a duration T_d are known, the base length and height of the triangle are determined. Consider a triangular hyetograph as shown in Fig. 14.3.4. The base length is T_d and the height h, so the total depth of precipitation in the hyetograph is given by $P = \frac{1}{2}T_d h$, from which

$$h = \frac{2P}{T_d} \tag{14.3.1}$$

A storm advancement coefficient r is defined as the ratio of the time before the peak t_a to the total duration:

$$r = \frac{t_a}{T_d} \tag{14.3.2}$$

Then the recession time t_b is given by

$$\begin{aligned} t_b &= T_d - t_a \\ &= (1 - r)T_d \end{aligned} \tag{14.3.3}$$

A value for r of 0.5 corresponds to the peak intensity occurring in the middle of the storm, while a value less than 0.5 will have the peak earlier and a value

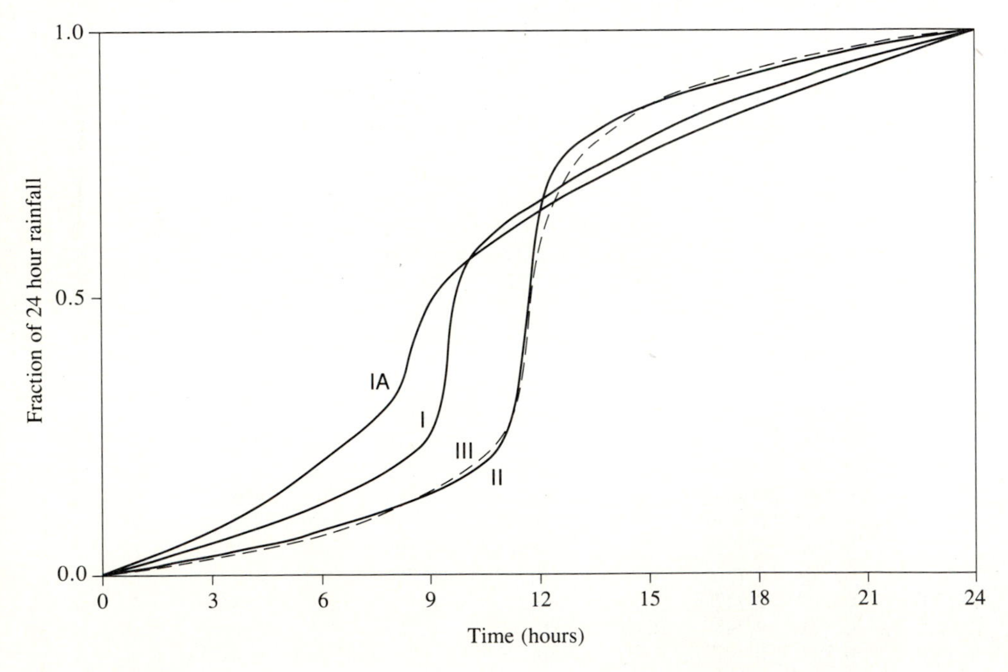

FIGURE 14.3.2
Soil Conservation Service 24-hour rainfall hyetographs. (*Source*: U. S. Dept. of Agriculture, Soil Conservation Service, 1986.)

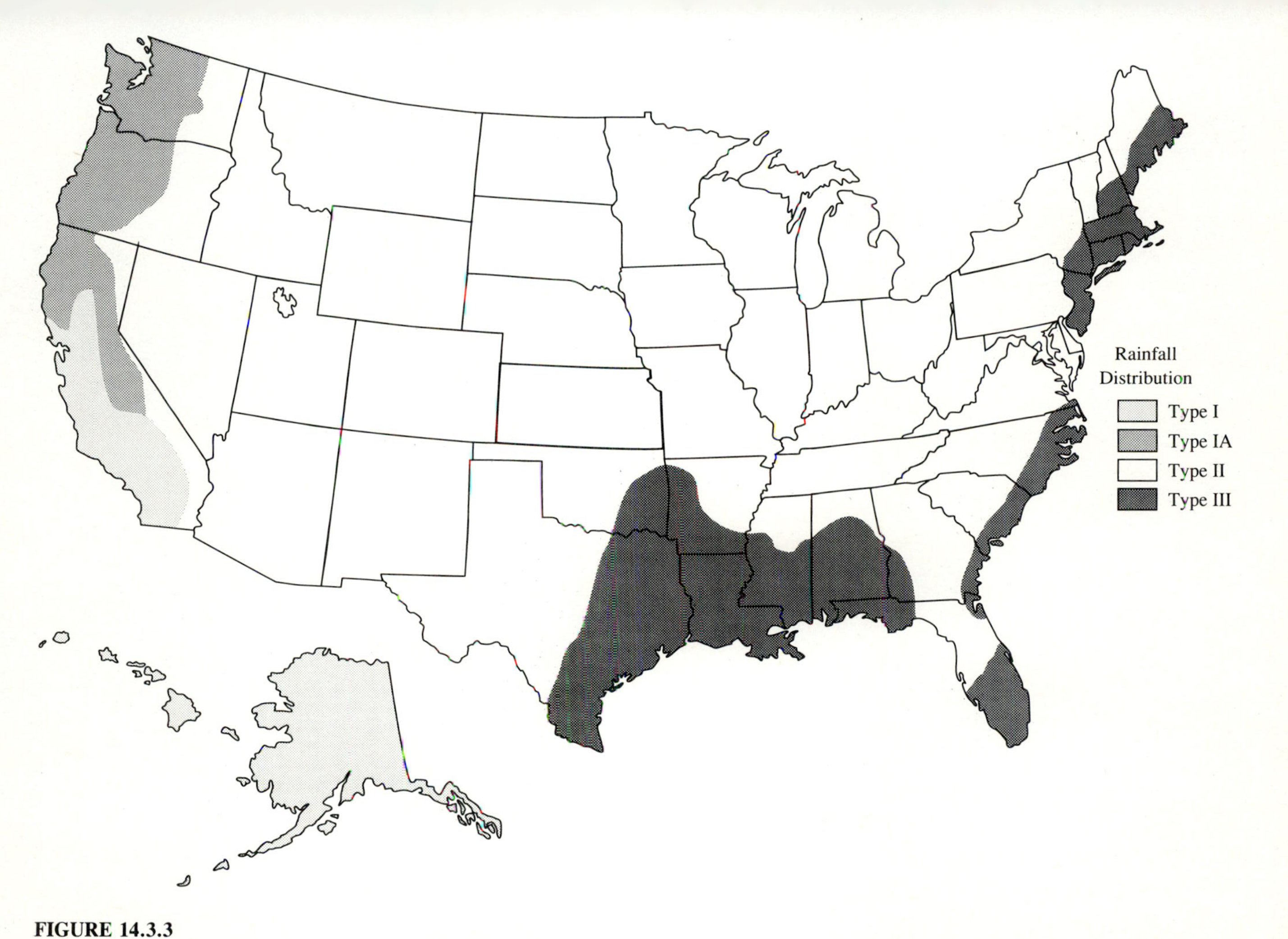

FIGURE 14.3.3
Location within the United States for application of the SCS 24-hour rainfall hyetographs. (*Source*: U. S. Dept. of Agriculture, Soil Conservation Service, 1986).

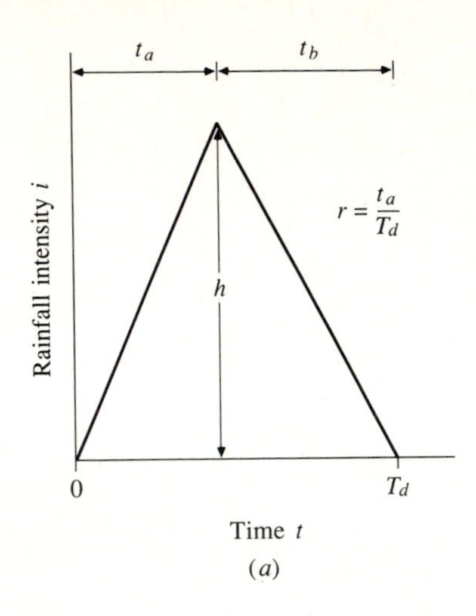

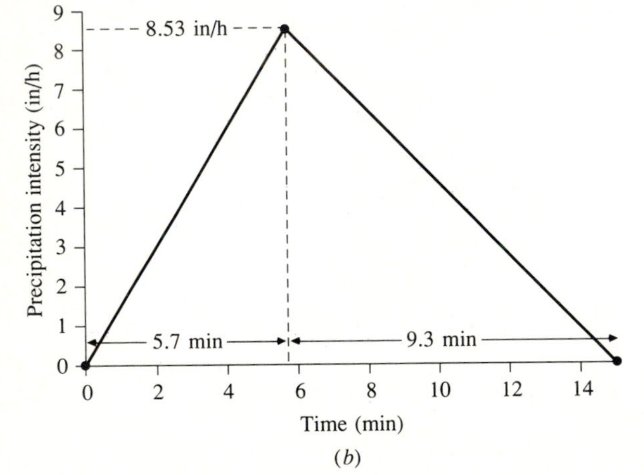

FIGURE 14.3.4
Triangular design hyetographs. (*a*) A general triangular design hyetograph. (*b*) Triangular design hyetograph for a 5-year 15-minute storm in Urbana, Illinois (Example 14.3.1).

greater than 0.5 will have the peak later than the midpoint. A suitable value of r is determined by computing the ratio of the peak intensity time to the storm duration for a series of storms of various durations. The mean of these ratios, weighted according to the duration of each event, is used for r. Values of r reported in the literature are presented in Table 14.3.2, which shows that in many locations storms tend to be of the advanced type, with r less than 0.5.

Yen and Chow (1980) analyzed 9869 storms at four locations: Urbana, Illinois; Boston, Massachusetts; Elizabeth City, New Jersey; and San Luis Obispo, California. Their analysis indicated that the triangular hyetographs for most heavy storms are nearly identical in shape, with only secondary effects from storm duration, measurement inaccuracies, and geographic location.

Example 14.3.1. Determine the triangular hyetograph for the design of an urban storm sewer in Urbana, Illinois. The design return period is 5 years, and the design

TABLE 14.3.2
Values of the storm advancement coefficient *r* for various locations

Location	*r*	Reference
Baltimore	0.399	McPherson (1958)
Chicago	0.375	Keifer and Chu (1957)
Chicago	0.294	McPherson (1958)
Cincinnati	0.325	Preul and Papadakis (1973)
Cleveland	0.375	Havens and Emerson (1968)
Gauhati, India	0.416	Bandyopadhyay (1972)
Ontario	0.480	Marsalek (1978)
Philadelphia	0.414	McPherson (1958)

Source: Wenzel, 1982, Copyright by the American Geophysical Union.

rainstorm duration has been set at 15 minutes. The storm advancement coefficient is $r = 0.38$.

Solution. From Fig. 14.1.2, for precipitation at Urbana (in central Illinois), $P_{2,15} = 0.88$ in, and $P_{100,15} = 1.70$ in. The depth for a 5-year return period is given by Eq. (14.1.2) with $a = 0.674$ and $b = 0.278$ from Table 14.1.1:

$$\begin{aligned} P_{5,15} &= 0.674 P_{2,15} + 0.278 P_{100,15} \\ &= 0.674 \times 0.88 + 0.278 \times 1.70 \\ &= 1.07 \text{ in} \end{aligned}$$

The peak intensity h is calculated using (14.3.1) with $T_d = 15$ min $= 0.25$ h:

$$h = \frac{2P}{T_d} = \frac{2 \times 1.07}{0.25} = 8.56 \text{ in/h}$$

The time t_a to the peak intensity is calculated by (14.3.2):

$$t_a = rT_d = 0.38 \times 0.25 = 0.095 \text{ h} = 5.7 \text{ min}$$

The recession time t_b is

$$t_b = T_d - t_a = 0.25 - 0.095 = 0.155 \text{ h} = 9.3 \text{ min}$$

The resulting design hyetograph is plotted in Fig. 14.3.4(b). Values of precipitation intensity at regular intervals can be calculated and converted to precipitation depth for rainfall-runoff analysis for the storm sewer.

14.4 DESIGN PRECIPITATION HYETOGRAPHS FROM IDF RELATIONSHIPS

In the hydrologic design methods developed many years ago, such as the rational method, only the peak discharge was used. There was no consideration of the time distribution of discharge (the discharge hydrograph) or the time distribution of precipitation (the precipitation hyetograph). However, more recently developed design methods, using unsteady flow analysis, require reliable prediction of the design hyetograph to obtain design hydrographs.

Alternating Block Method

The *alternating block method* is a simple way of developing a design hyetograph from an intensity-duration-frequency curve. The design hyetograph produced by this method specifies the precipitation depth occurring in n successive time intervals of duration Δt over a total duration $T_d = n\Delta t$. After selecting the design return period, the intensity is read from the IDF curve for each of the durations Δt, $2\Delta t$, $3\Delta t$, . . . , and the corresponding precipitation depth found as the product of intensity and duration. By taking differences between successive precipitation depth values, the amount of precipitation to be added for each additional unit of time Δt is found. These increments, or blocks, are reordered into a time sequence with the maximum intensity occurring at the center of the required duration T_d and the remaining blocks arranged in descending order alternately to the right and left of the central block to form the design hyetograph.

Example 14.4.1. Determine, in 10-minute increments, the design precipitation hyetograph for a 2-hour storm in Denver with a 10-year return period.

Solution. As was illustrated in Example 14.2.4, the 10-year precipitation intensity for a given duration can be calculated for Denver by Eq. (14.2.2) with $c = 96.6$, $e = 0.97$, and $f = 13.90$. For a duration of 20 minutes, this computation yields $i = 3.00$ in/h. The values for other durations at intervals of 10 minutes are shown in column 2 of Table 14.4.1. By multiplying intensity and duration, the corresponding precipitation depth is calculated. For a 20-minute duration $P = iT_d = (3.00 \text{ in/h}) \times 0.333 = 1.001$ in; similar calculations for the other durations are shown in column 3 of Table 14.4.1. It can be seen that the cumulative depth increases with duration up to the 2-hour or 120-minute limit.

TABLE 14.4.1
A design precipitation hyetograph developed in 10-min increments for a 10-year 2-hour storm at Denver, using the alternating block method (Example 14.4.1)

Column:	1	2	3	4	5	6
	Duration (min)	Intensity (in/h)	Cumulative depth (in)	Incremental depth (in)	Time (min)	Precipitation (in)
	10	4.158	0.693	0.693	0–10	0.024
	20	3.002	1.001	0.308	10–20	0.033
	30	2.357	1.178	0.178	20–30	0.050
	40	1.943	1.296	0.117	30–40	0.084
	50	1.655	1.379	0.084	40–50	0.178
	60	1.443	1.443	0.063	50–60	0.693
	70	1.279	1.492	0.050	60–70	0.308
	80	1.149	1.533	0.040	70–80	0.117
	90	1.044	1.566	0.033	80–90	0.063
	100	0.956	1.594	0.028	90–100	0.040
	110	0.883	1.618	0.024	100–110	0.028
	120	0.820	1.639	0.021	110–120	0.021

The 10-minute precipitation depth is 0.693 in compared with 1.001 in for 20 minutes, so during the most intense 20 minutes of the design storm, 0.693 in will fall in 10 minutes, while 1.001 − 0.693 = 0.308 in will fall in the remaining 10 minutes. Using this reasoning and taking successive increments of cumulative precipitation depth, as shown in column 4 of Table 14.4.1, the precipitation depth added by each additional increment of storm duration is found. In column 6 the precipitation depths, are ordered so that the maximum block (0.693 in) falls at 50–60 min; the next-largest block (0.308 in) is placed to the right of the maximum block, at 60–70 min (making 1.00 in between 50 and 70 min, as required), the third-largest block (0.178 in) is placed to the left of the maximum block (40–50 min), and so on. The results are plotted in Fig. 14.4.1, and the descending order of the blocks to the left and right of the maximum block can be seen. A design hyetograph built up in this way represents a 10-year event both for the 2-hour total duration and for any period within this duration centered on the maximum block.

Instantaneous Intensity Method

If an equation defining the intensity-duration-frequency curve is known, equations can be developed for the variation of intensity with time in the design hyetograph. The principle is similar to that employed in the alternating block method, namely that the precipitation depth for a period of duration T_d around the storm's peak is equal to the value given by the IDF curve or equation. The difference from the alternating block method is that precipitation intensity is now considered to vary continuously throughout the storm.

Consider a storm hyetograph as shown in Fig. 14.4.2. The dashed horizontal line drawn on the hyetograph for a given precipitation intensity i will intersect the hyetograph before and after the peak. Measured from the time of the peak intensity, the time of intersection before the peak is labeled t_a and that after the peak t_b. The total time between intersections is labeled T_d so that

$$T_d = t_a + t_b \tag{14.4.1}$$

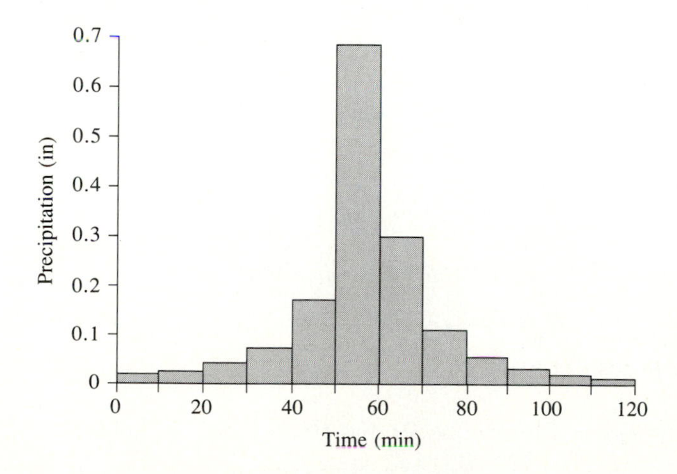

FIGURE 14.4.1
A 10-year 2-hour design hyetograph developed for Denver using the alternating block method (Example 14.4.1).

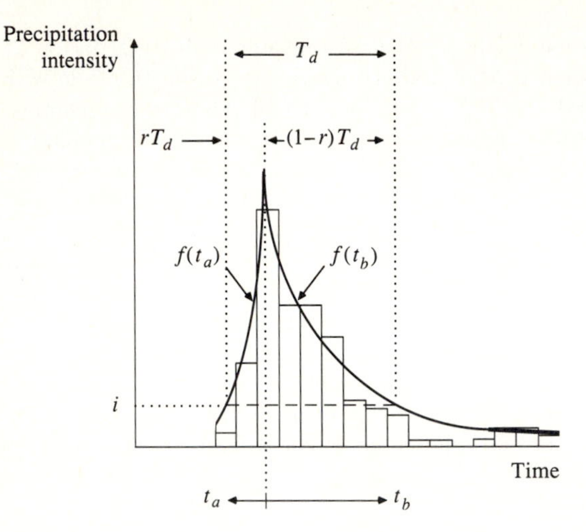

FIGURE 14.4.2
Fitting a hyetograph by curves.

The storm advancement coefficient r is defined as before, as the ratio of the time before the peak to the time between intersections.

$$r = \frac{t_a}{T_d} \tag{14.4.2}$$

It follows from (14.4.1) and (14.4.2) that

$$T_d = \frac{t_a}{r} = \frac{t_b}{1-r} \tag{14.4.3}$$

As shown in Fig. 14.4.2, a pair of curves, $i_a = f(t_a)$ and $i_b = f(t_b)$, are assumed to fit the hyetograph precipitation intensities, where i_a and i_b are precipitation intensities before and after the peak, respectively. Thus, the total amount of rainfall R within time T_d is given by the area under the curves:

$$R = \int_0^{rT_d} f(t_a)\,dt_a + \int_0^{(1-r)T_d} f(t_b)\,dt_b \tag{14.4.4}$$

Noting that $f(t_a) = f(t_b)$ for any given T_d, differentiating Eq. (14.4.4) with respect to T_d gives

$$\frac{dR}{dT_d} = f(t_a) = f(t_b) \tag{14.4.5}$$

If the average rainfall intensity for the duration T_d is i_{ave}, then

$$R = T_d i_{\text{ave}} \tag{14.4.6}$$

Differentiating (14.4.6) with respect to T_d gives

$$\frac{dR}{dT_d} = i_{\text{ave}} + T_d \frac{di_{\text{ave}}}{dT_d} = f(t_a) = f(t_b) \tag{14.4.7}$$

Keifer and Chu (1957) developed a synthetic hyetograph of this type for use in sewer system design for Chicago. They defined the average rainfall intensity i_{ave} as in Eq. (14.2.2):

$$i_{ave} = \frac{c}{T_d^e + f} \quad (14.4.8)$$

By differentiating (14.4.8) and substituting the result into (14.4.7) it can be shown that the intensity i for which the line intersects the hyetograph for a duration T_d is given by

$$i = \frac{c\left[(1-e)T_d^e + f\right]}{\left(T_d^e + f\right)^2} \quad (14.4.9)$$

The equations for the intensities i_a and i_b in terms of t_a and t_b are found by substituting for T_d in (14.4.9) using (14.4.3).

Example 14.4.2. Develop a 2-hour design hyetograph for a 10-year return period storm in Denver using a storm advancement coefficient of $r = 0.5$.

Solution. A value of $r = 0.5$ and a storm duration of 120 minutes will put the peak intensity at time $t = 60$ min measured from the beginning of the storm. The relative times t_a and t_b before and after the peak time are shown in column 2 of Table 14.4.2. A 2-minute time interval is used near the peak for increased accuracy, and a 10-minute time interval for the remainder. The rainfall intensities are calculated by Eq. (14.4.9) using values $c = 96.6$, $e = 0.97$, and $f = 13.9$ from Table 14.2.3; $r = 0.5$; and T_d given by (14.4.3).

$$i = \frac{c\left[(1-e)T_d^e + f\right]}{\left(T_d^e + f\right)^2}$$

Before the peak, $T_d = t_a/r$, so for $t = 50$ min, for example, $t_a = 60 - 50 = 10$ min and $T_d = t_a/r = 10/0.5 = 20$ min; hence,

$$i = i_a = \frac{96.6[(1 - 0.97)20^{0.97} + 13.9]}{(20^{0.97} + 13.9)^2}$$

$$= 1.348 \text{ in/h}$$

Similarly calculated values for all time intervals are shown in column 3 of Table 14.4.2. The values of intensity after the peak are calculated by the same procedure using $T_d = t_b/(1 - r)$. For example, at $t = 70$ min, $t_b = 70 - 60 = 10$ min and $T_d = 10/(1 - 0.5) = 20$ min, so $i_b = 1.348$ in/h for this case also.

The values of precipitation intensity so determined are instantaneous values; the precipitation depth in each time interval may be calculated using the trapezoidal rule. For example, referring to Table 14.4.2, the precipitation depth during the first 10 minutes is $[(0.118 + 0.156)/2] \times (10/60) = 0.023$ in as shown in column 4. The sum of all the precipitation increments is 1.697 in, which is slightly higher than the actual precipitation depth of 1.64 in for a 10-year 2-hour storm in Denver, calculated in Example 14.4.1. The difference arises from the discretization of the continuous time scale in this example and the use of a simple trapezoidal rule to

TABLE 14.4.2
Calculation of the 10-year 2-hour design hyetograph at Denver by the instantaneous intensity method (Example 14.4.2)

Column:	1 Time t (min)	2 Relative time t_a, t_b (min)		3 Intensity (in/h)	4 Incremental precipitation (in)
	0	60		0.118	
	10	50		0.156	0.023
	20	40		0.219	0.031
	30	30		0.334	0.046
	40	20		0.585	0.077
	50	10		1.348	0.161
	52	8		1.691	0.051
	54	6		2.193	0.065
	56	4		2.975	0.086
	58	2	t_a	4.303	0.121
	60	0		6.950	0.188
	62	2	t_b	4.303	0.188
	64	4		2.975	0.121
	66	6		2.193	0.086
	68	8		1.691	0.065
	70	10		1.348	0.051
	80	20		0.585	0.161
	90	30		0.334	0.077
	100	40		0.219	0.046
	110	50		0.156	0.031
	120	60		0.118	0.023
				Total depth (in)	1.697

calculate precipitation depth. The difference would be diminished if two-minute increments were used throughout the time horizon, instead of just around the peak intensity in Table 14.4.2.

Figure 14.4.3 shows the hyetograph developed in this example along with a similarly computed hyetograph for an advancement coefficient of 0.25. It can be seen that the effect of changing the advancement coefficient is simply to change the location of the peak intensity but not change its magnitude.

14.5 ESTIMATED LIMITING STORMS

The estimated limiting values (ELV's) commonly employed for water control design are the *probable maximum precipitation* (PMP), the *probable maximum storm* (PMS), and the *probable maximum flood* (PMF), although many other possible ELV's could be developed as criteria for various types of hydrologic design. The PMP provides only a depth of precipitation, the time distribution of which must be defined to form a PMS. The PMS can be employed as the input

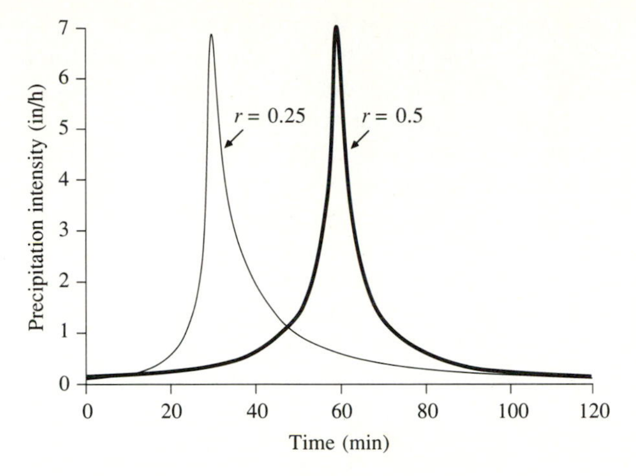

FIGURE 14.4.3
Design hyetographs for a 10-year storm at Denver developed by the instantaneous intensity method with advancement coefficients $r = 0.25$ and 0.5 (Example 14.4.2).

to a rainfall-runoff model of a drainage basin system which can then be used to develop a PMF for the design of runoff control structures.

Probable Maximum Precipitation

The PMP is the estimated limiting value of precipitation. Consequently, the PMP may be defined as the analytically estimated greatest depth of precipitation for a given duration that is physically possible and reasonably characteristic over a particular geographical region at a certain time of year. In practice, no allowance is made in the estimation of the PMP for the effects of long-term climatic change. It must be remembered that the concept of PMP is not completely reliable because it cannot be perfectly estimated and its probability of occurrence is unknown. However, in operational application, the PMP has been found to be useful, and its use will continue because of public concern for safety of projects such as large dams.

There are a variety of methods for determining the PMP. Because of the uncertainties and limitations of data and knowledge, the PMP must be considered an estimate and judgment must be used in setting its value. Methods of estimating PMP are discussed below.

1. *Application of storm models.* Storm models may be employed to estimate PMP where there is insufficient or nonrepresentative storm data, or where there is rugged topography that complicates the storm phenomenon and makes precipitation measurement difficult. The convective cell model for thunderstorms presented in Sec. 3.3 is one example of a storm model, and Wiesner (1970) has presented a similar model for the precipitation arising from orographic lifting of moist air over hills or mountains. When orographic and convective effects occur simultaneously, the two types of models can be superimposed.

The use of storm models is more successful in determining PMP over large areas than over small ones. The principle of continuity is more easily applied using mean velocities of inflow and outflow, and the moisture content of the inflowing air can be defined from the dew-point temperature persisting for a considerable time (usually 12 hours) over a large area. These factors are more difficult to define for extreme precipitation rates that occur locally and over short periods. Therefore, although storm models may indicate the magnitude of the precipitation to be expected, they should be carefully calibrated with data from observed storms in the region of interest before use in design.

2. *Maximization of actual storms*. The chief deficiency of the storm models is their oversimplified representation of the actual storm. If records of actual storms are available, they may be maximized to obtain the PMP values. This process involves increasing the observed storm precipitation by the ratio of the actual moisture inflow to the storm to the maximum moisture inflow theoretically possible at the site.

If there are no adequate records of storms in the project basin, it is possible to transpose storms from other areas to the project basin for the computation of the PMP if these storms could have occurred in the basin. The procedure of storm transposition involves the selection of adequate storms for transposition, determination of the orientation of the storms critical to the basin, and adjustments for differences, if any, in dew-point temperature, elevation, prevailing wind, and orographic effects.

The world's greatest recorded rainfalls, according to the World Meteorological Organization (1983), are approximated by the equation

$$P = 422T_d^{0.475} \tag{14.5.1}$$

where P is the precipitation depth in millimeters and T_d is the duration in hours. The equation was obtained by fitting data from observed extreme rainfalls at many locations for durations ranging from one minute to several months. This equation is an estimate of the precipitation depths that could occur under very extreme circumstances.

3. *Generalized PMP charts*. The estimates of PMP may be made either for individual basins or for large regions encompassing numerous basins of various sizes. In the latter case, the estimates are referred to as *generalized estimates* and are usually displayed as isohyetal maps that depict the regional variation of PMP for some specified duration, basin size, and annual or seasonal variation. These maps are commonly known as *generalized PMP charts*.

The most widely used generalized PMP charts in the United States are those contained in U. S. National Weather Service hydrometeorological report no. 51, commonly known as HMR 51 (Schreiner and Riedel, 1978), for the United States east of the 105th meridian. These maps specify the probable maximum precipitation depth for any time of the year (referred to as an all-season estimate) as a function of storm area ranging from 10 to 20,000 mi^2 and storm duration ranging from 6 to 72 hours. For regions west of the 105th meridian, the National Academy of Sciences (1983) has prepared the diagram shown in Fig. 14.5.1, which specifies the appropriate National Weather Service publication from which probable maximum precipitation data may be obtained.

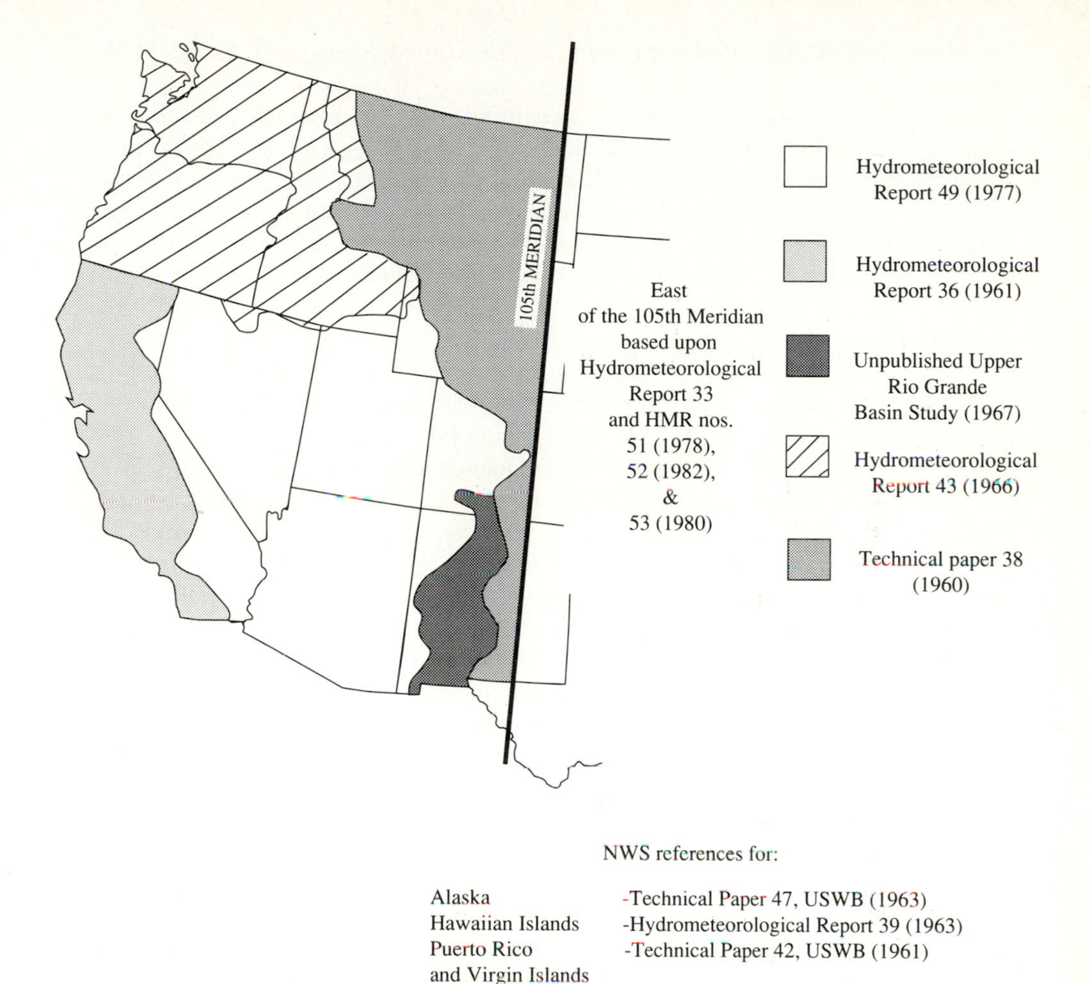

FIGURE 14.5.1
Sources of information for probable maximum precipitation computation in the United States. (*Source*: National Academy of Sciences, 1983. Used with permission.)

The Probable Maximum Storm

The PMS, or probable maximum storm, involves the temporal distribution of rainfall. The PMS values are generally given as maximum accumulated depths for any specified duration. For example, given depths for 6, 12, 18, and 24 hours typically represent the total depth for each duration and not the time sequence in which the precipitation occurs.

In order to develop the hyetograph of a PMS, one needs to know the spatial and temporal distribution of the PMP. One procedure to determine the PMS has been outlined in hydrometeorological report no. 52 (HMR 52) by Hansen,

Schreiner, and Miller (1982) for areas of the United States east of the 105th meridian.

For the purpose of modeling maximum runoff, different critical time sequences of PMP increments should be investigated. In general, the critical time distribution of PMP increments is determined by judgment from experience and available information, such as from weather maps of critical historical storms. A commonly adopted sequence is the most advanced time distribution, that is, beginning with the highest amount and continuing with decreasing increments.

The Probable Maximum Flood

The PMF is the greatest flood to be expected assuming complete coincidence of all factors that would produce the heaviest rainfall and maximum runoff. This is derived from a PMP, and hence its frequency cannot be determined. From the economic viewpoint, it is usually prohibitive to design a structure for PMF, except for large spillways whose failure could lead to excessive damage and loss of life. Hence, a pragmatic approach for many design situations is not to define the design flood as an estimated limiting value, but to scale it downwards by a certain percentage depending on the type of structure and the hazard if it fails. For this purpose, the flood event actually used in design is often the greatest flood that may reasonably be expected, taking into account all pertinent conditions of location, meteorology, hydrology, and topography. This design flood may be determined analytically as the flood caused by a transposed historical largest storm and its magnitude may be a fraction of the ELV.

In practice, the design flood is commonly called the *standard project flood* (SPF). The SPF is estimated using rainfall-runoff modeling by applying the unit-hydrograph method to the *standard project storm* (SPS), which is the greatest storm that may be reasonably expected. The SPS can be derived from a detailed analysis of storm patterns and transposition of storms to a position that would give maximum runoff. For a particular drainage area and season of year in which snow melt is not a major concern, the SPS estimate should represent the most severe flood-producing depth-area-duration relationship and isohyetal pattern of any storm that is considered reasonably characteristic of the region. The watershed runoff characteristics and any water-control structures in the basin should be considered. When melting snow may constitute a substantial volume of runoff to the SPF hydrograph, appropriate allowances for snow melt should be included in the estimate. Where floods are predominantly the result of melting snow, the SPF estimate should be based on the most critical combinations of snow, temperature, and water losses considered reasonably characteristic. Past estimates have indicated that SPS magnitudes and SPF discharges are generally in the range of 40 to 60 percent of the ELV for the same basins.

In some cases the SPF estimate may have a major bearing on the design of a particular project; in other cases the estimate may serve only as an indication of the partial degree of protection proposed for the project. SPF estimates are usually made only for major and intermediate structures since they require considerable effort in preparation.

Computer Programs for PMS and PMF Development

The U. S. Army Corps of Engineers Hydrologic Engineering Center (1984) describes a computer program called HMR 52 which computes the basin-average precipitation for the probable maximum storm in accordance with the criteria in hydrometeorological report No. 52. Input to this program includes the PMP estimates from HMR 51; the program computes the spatially averaged PMP for any of the subbasins or combinations of subbasins in the drainage basin. The program selects the storm area size and orientation so as to produce the maximum basin-average precipitation. Storm centering and time distribution must be provided by the user. This program can be used in conjunction with the U. S. Army Corps of Engineers HEC-1 (Sec. 15.2) rainfall-runoff model to compute the probable maximum flood.

14.6 CALCULATION OF PROBABLE MAXIMUM PRECIPITATION

Estimates of the runoff from the PMP are required in the design of spillways and in dam safety studies. In order to carry out a rainfall-runoff analysis using the PMP both temporal and spatial distribution of the PMP estimates is required. A stepwise procedure to derive the distribution of the PMP for the United States east of the 105th meridian has been developed by Hansen, Schreiner, and Miller (1982) in hydrometeorological report No. 52. This procedure is used for the determination of the probable maximum storm, which is required to determine peak discharge and to develop the probable maximum flood hydrograph through a rainfall-runoff analysis. The procedure is based upon information derived from major storms of record and is applicable to nonorographic regions of the eastern United States.

The procedure for determining the probable maximum precipitation over a watershed has five important elements: (1) *depth-area-duration curves*, which specify the PMP for a specified storm area and duration; (2) a *standard isohyetal pattern* distributing the precipitation spatially in the form of an ellipse; (3) an *orientation adjustment factor*, which reduces the PMP estimates if the standard isohyetal pattern's longitudinal axis is not oriented in the direction of normal atmospheric moisture flow in the region; (4) a *critical storm area*, which generates the largest PMP over the watershed; and (5) an *isohyetal adjustment factor*, which specifies the percentage of the PMP depth that applies on each contour of the standard isohyetal pattern.

Depth-area-duration curves. In probability-based point precipitation estimates, there are three variables to be considered: intensity (or depth), duration, and frequency of occurrence. For the PMP, the frequency of occurrence is replaced by storm area as the third variable. Figure 14.6.1 shows the PMP for the eastern

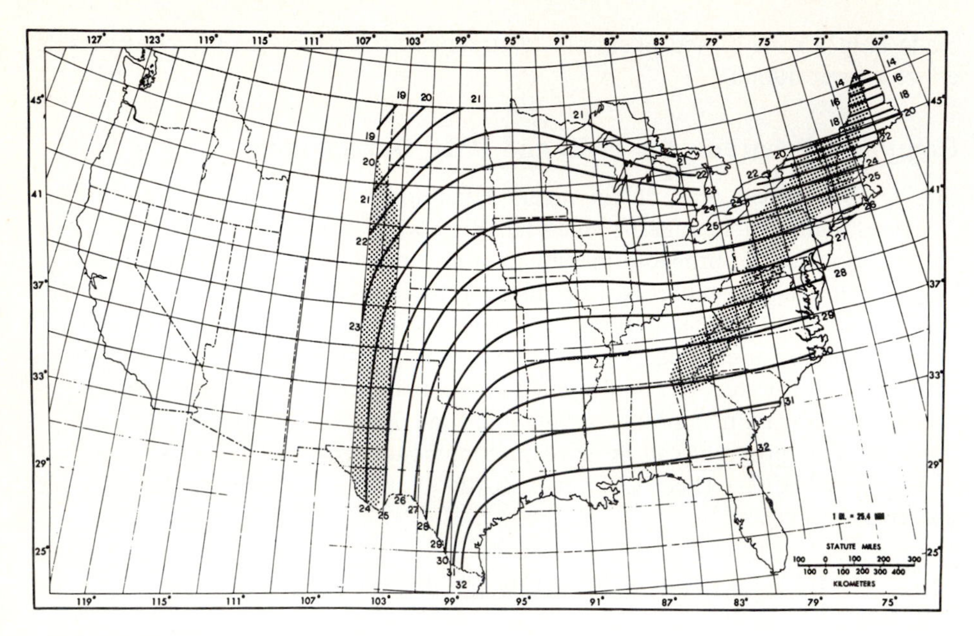

FIGURE 14.6.1(*a*)
All-season PMP (in) for 6 hours, 10 mi^2. (*Source*: Hansen, Schreiner, and Miller, 1982.)

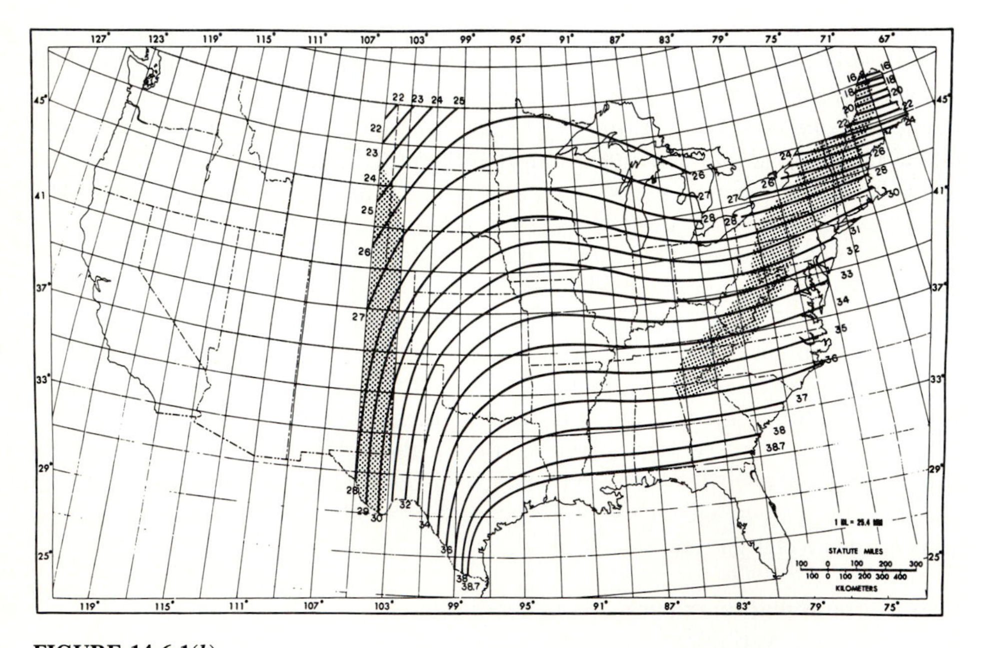

FIGURE 14.6.1(*b*)
All-season PMP (in) for 12 hours, 10 mi^2. (*Source*: Hansen, Schreiner, and Miller, 1982.)

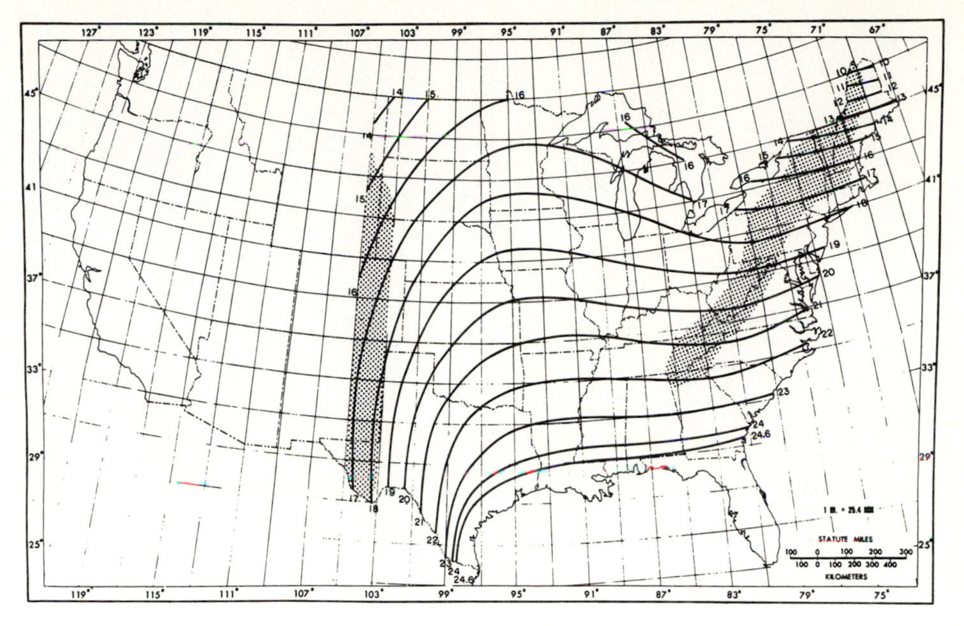

FIGURE 14.6.1(*c*)
All-season PMP (in) for 6 hours, 200 mi². (*Source*: Hansen, Schreiner, and Miller, 1982.)

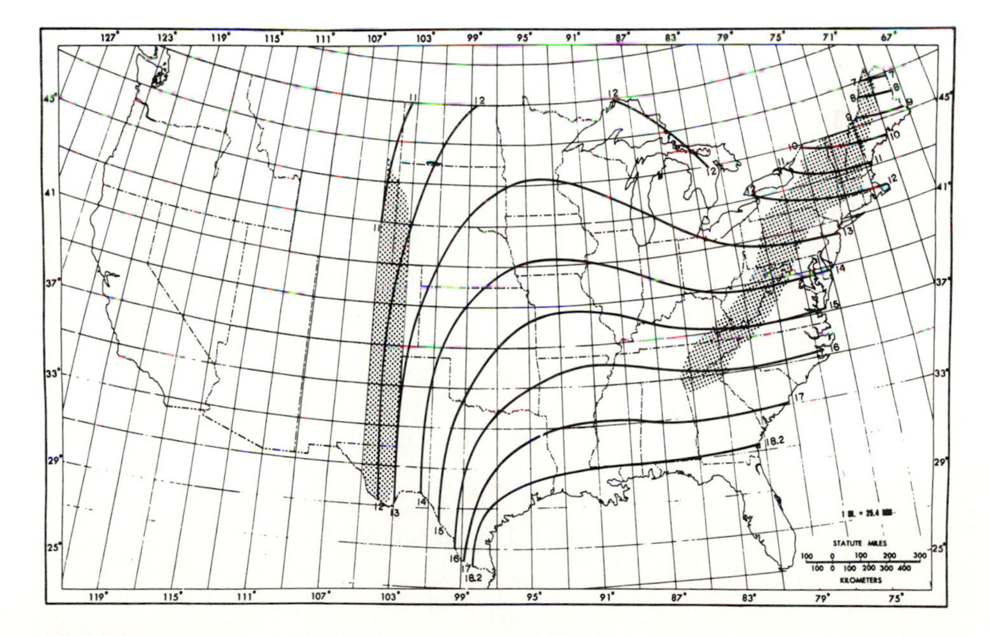

FIGURE 14.6.1(*d*)
All-season PMP (in) for 6 hours, 1000 mi². (*Source*: Hansen, Schreiner, and Miller, 1982.)

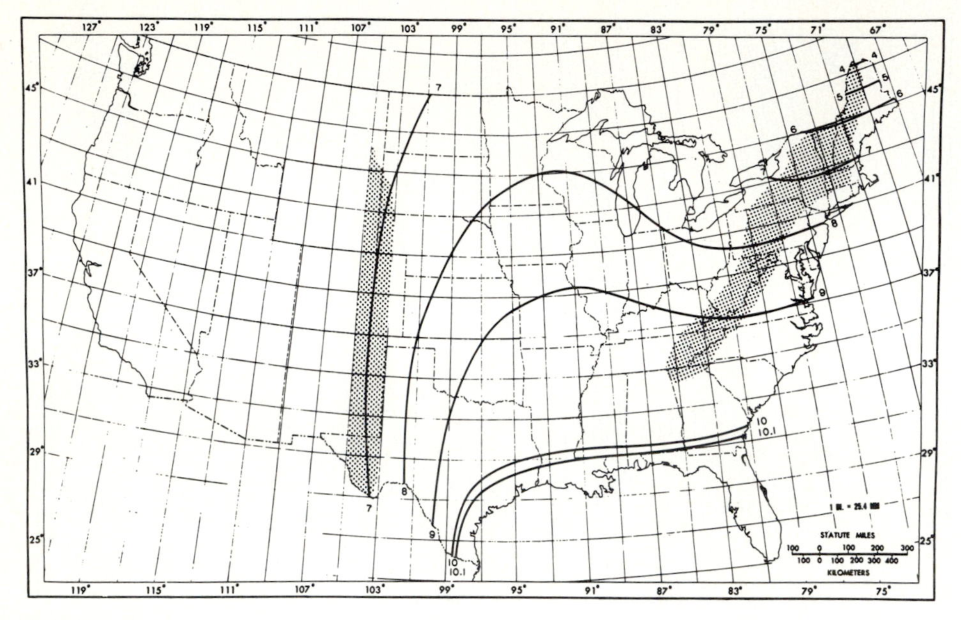

FIGURE 14.6.1(*e*)
All-season PMP (in) for 6 hours, 5000 mi². (*Source*: Hansen, Schreiner, and Miller, 1982.)

United States for a 10-mi² area and durations of 6 and 12 hours. For example, for Chicago, Fig. 14.6.1(*a*) indicates a PMP for 6 hours of approximately 26 in, and Fig. 14.6.1(*b*), gives the PMP for 12 hours as approximately 30 in. In a 12-hour probable maximum storm over a 10-mi² area at Chicago, then, 26 in would occur in the most intense 6 hours and $30-26=4$ in of precipitation would occur in the remaining 6 hours. For a given location, the depth-area-duration graph is a plot of depth vs. storm area with curves for various storm durations.

Standard isohyetal pattern. A standard elliptical storm pattern adopted in HMR 52 is shown in Fig. 14.6.2. It has 14 contours, labeled A through N, each of which bounds a specified area: contour A contains 10 mi², contour B 25 mi², and so on. The ratio of the lengths of the major and minor axes of the ellipses shown is 2.5 to 1. If a and b are the lengths of the semimajor and semiminor axes in Fig. 14.6.2, the area of the ellipse is given by

$$A = \pi a b \tag{14.6.1}$$

Since $a = 2.5b$, substitution into (14.6.1) gives the length of the semiminor axis as:

$$b = \left(\frac{A}{2.5\pi}\right)^{1/2} \tag{14.6.2}$$

For example, for $A = 10$ mi², $b = (10/2.5\pi)^{1/2} = 1.13$ mi, and $a = 2.5b = 2.5 \times 1.13 = 2.82$ mi. The length r for a radius at angle θ with the major axis is given by

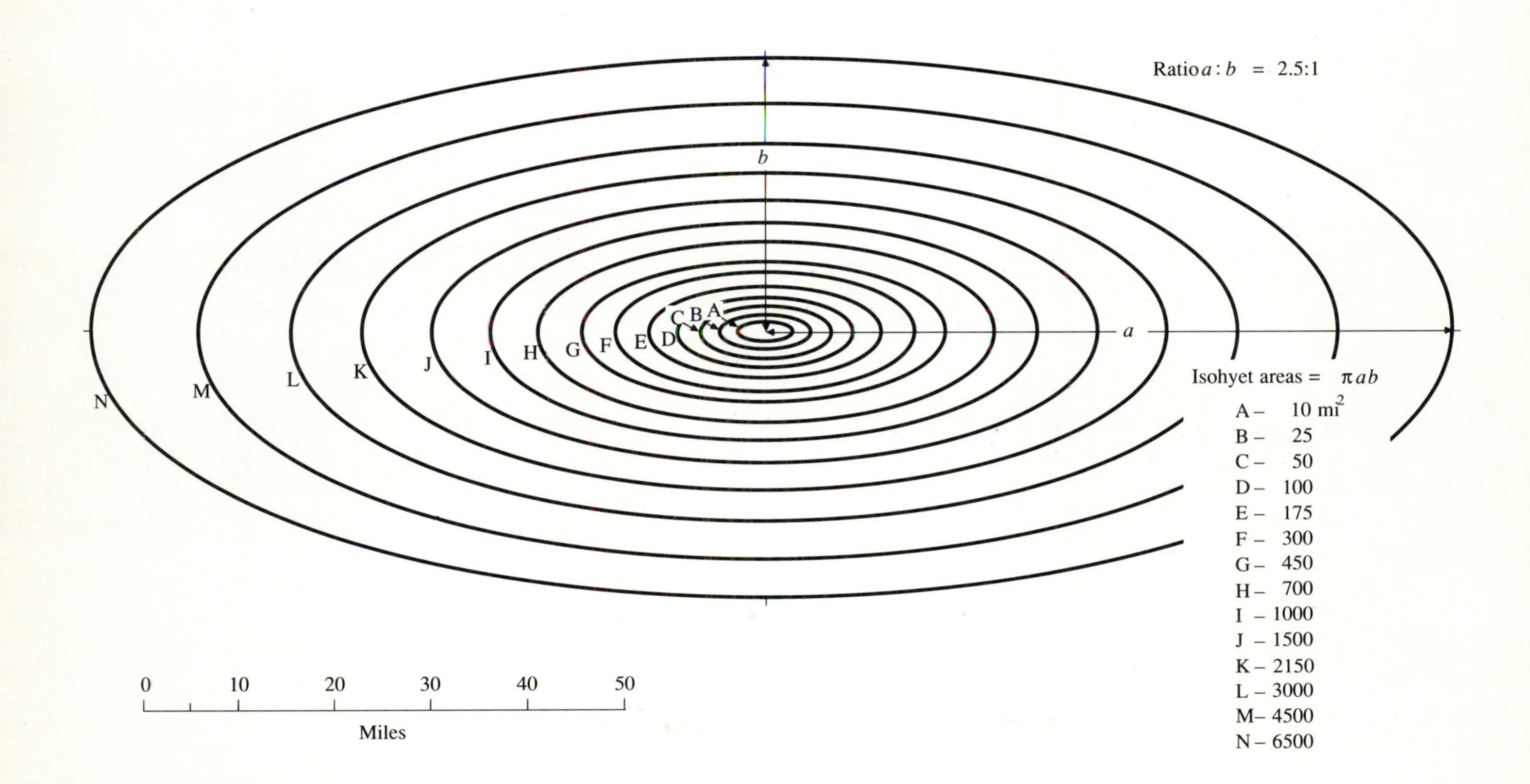

FIGURE 14.6.2
Standard isohyetal pattern recommended for spatial distribution of PMP east of the 105th meridian. (*Source*: Hansen, Schreiner, and Miller, 1982.)

$$r^2 = \frac{a^2 b^2}{a^2 \sin^2\theta + b^2 \cos^2\theta} \tag{14.6.3}$$

For example, for $\theta = 45°$, $a = 2.82$, and $b = 1.13$ mi, $r^2 = 2.82^2 \times 1.13^2/(2.82^2 \sin^2 45° + 1.13^2 \cos^2 45°) = 2.20$ mi.

Orientation adjustment factor. The orientation of the isohyetal pattern most likely to be conducive to a PMP event for any place in the United States east of the 105th meridian is indicated in Fig. 14.6.3. North is considered as 0° for this figure, and angles are measured clockwise from this direction. This figure is based upon averages of isohyetal orientations for major storms in the United States, which predominantly have moisture flowing from the south and west

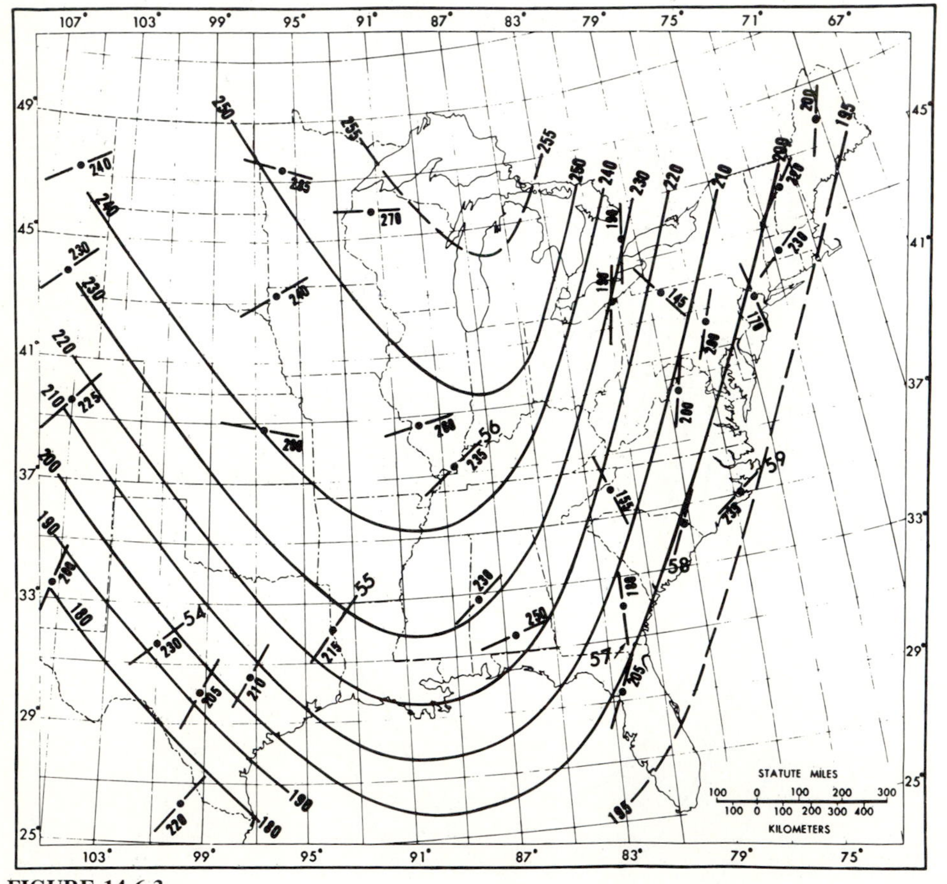

FIGURE 14.6.3
Analysis of isohyetal orientations for selected major storms, adopted as recommended orientation for PMP, within ± 40°. (*Source*: Hansen, Schreiner, and Miller, 1982.)

(180° to 270°). It is important to note that the moisture flow for PMP at any location is not restricted to any one orientation; it is reasonable to expect the moisture flow to occur over a range of orientations centered on the values in Fig. 14.6.3. If the longitudinal axis of the storm pattern is oblique to the recommended orientation, the PMP estimate is reduced by a percentage depending on the angle between the two directions. The adjustment factor for orientation differences is given in Fig. 14.6.4.

Critical storm area. The critical storm area is that for which the precipitation from the depth-area-duration curve yields the greatest PMP over the watershed,

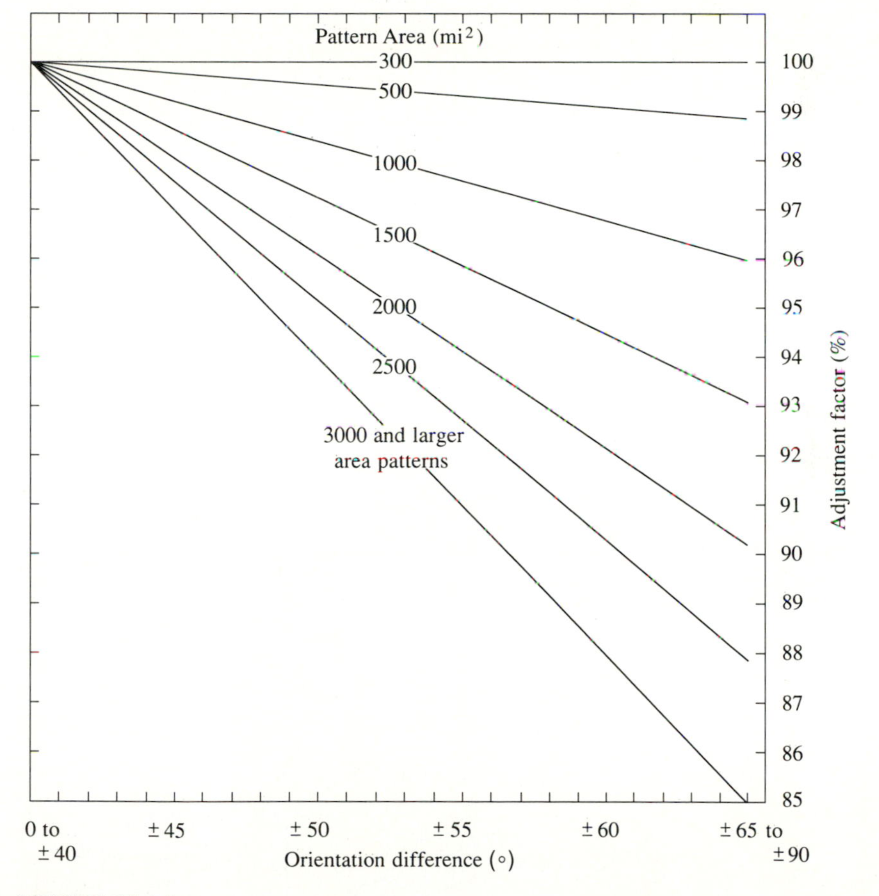

FIGURE 14.6.4
Adjustment factors for isohyetal orientation difference. This figure is used to determine the adjustment factor to apply to isohyet values in Fig. 14.6.2 when the pattern is placed at an orientation more than 40° away from that recommended in Fig. 14.6.3. (*Source*: Hansen, Schreiner, and Miller, 1982.)

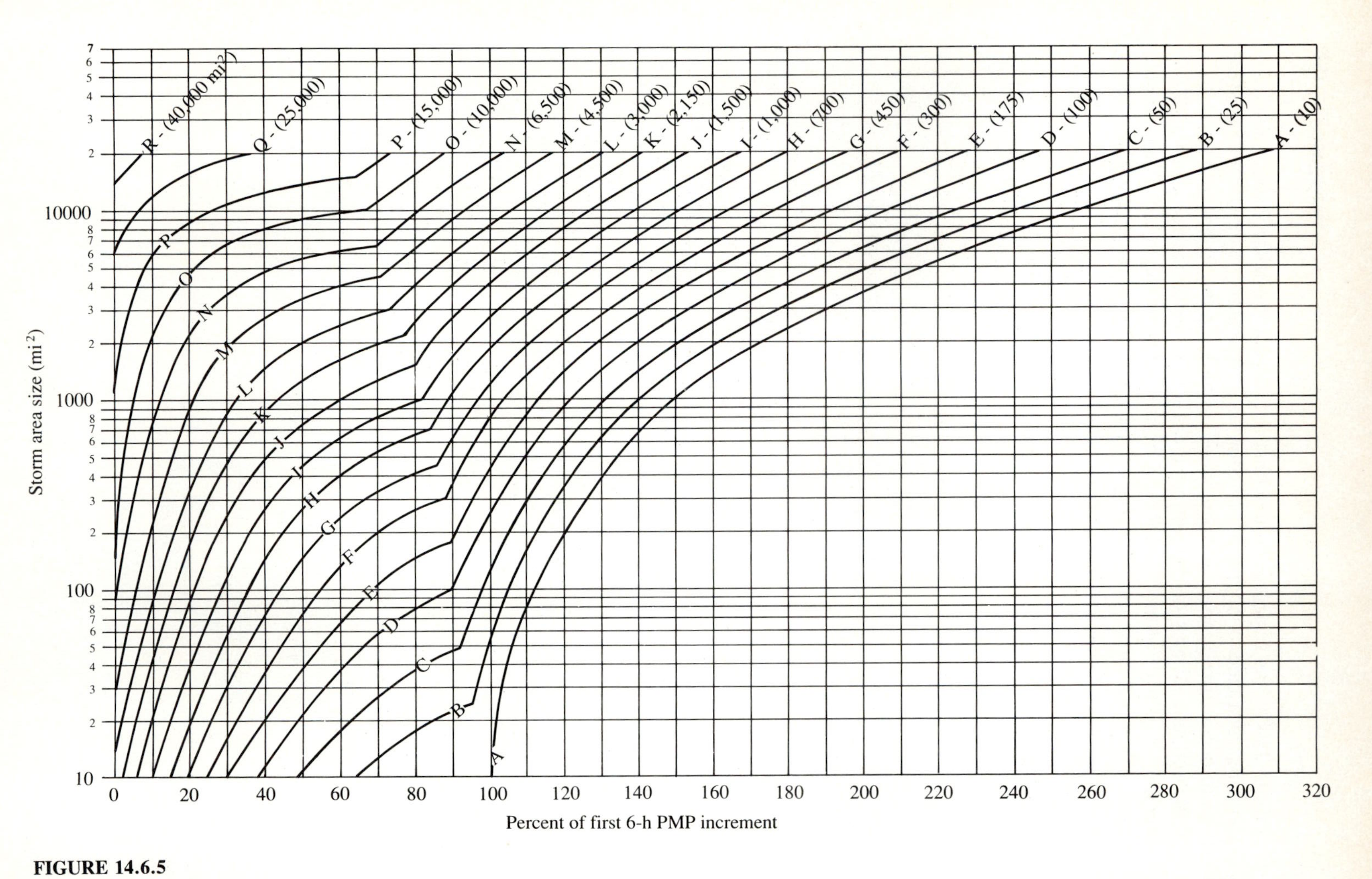

FIGURE 14.6.5
Nomograph for determining isohyet precipitation values from the PMP estimate for a given storm area. (*Source*: Hansen, Schreiner, and Miller, 1982.)

taking into account the fact that the watershed area is not shaped elliptically like the standard isohyetal pattern.

Isohyetal area factor. The value of the PMP represents the average precipitation depth over a specified area for a given duration. When a storm is represented by the standard isohyetal pattern, there will be regions of greater precipitation depth near the center of the pattern and of lesser depth near the edges. Figure 14.6.5 shows, for the most intense 6-hour storm interval, the percent of the specified PMP depth to be applied to each contour of the standard isohyetal pattern to get the correct spatial distribution. For example, using the PMP for an area of 10 mi^2, according to the graph, contour A (area 10 mi^2) gets 100 percent of the PMP depth, and contour B (area 25 mi^2) gets 64 percent. So, since the 6-hour PMP depth for 10 mi^2 at Chicago is 26 in, as found earlier, this storm would have depth $26 \times 0.64 = 16.4$ in over 25 mi^2, which represents a volume of 25 $mi^2 \times 16.4$ in $= 416$ $mi^2 \cdot$in of precipitation, of which 10 $mi^2 \times 26$ in $= 260$ $mi^2 \cdot$in, or 63 percent, occurs in the most intense 10-mi^2 area.

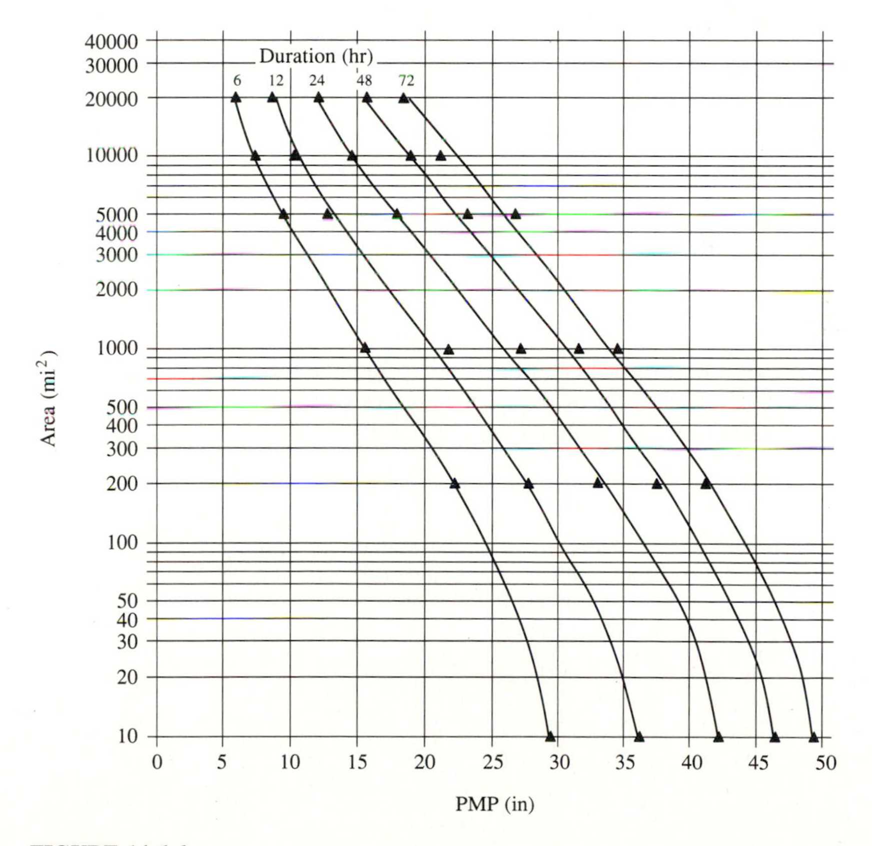

FIGURE 14.6.6
Depth-area-duration curves for 31°45′N, 98°15′W applicable to the Leon River, Texas, drainage. (*Source*: Hansen, Schreiner, and Miller, 1982.)

Example 14.6.1. Determine the spatial distribution and the average depth over the watershed of the 6-hour probable maximum precipitation on the Leon River drainage basin above Belton Reservoir in Texas. The drainage basin area is approximately 3660 mi^2 with the drainage center at 31°45′N, 98°15′W. (This example is a simplified version of a design example presented in HMR 52).

Solution.

1. *Develop the depth-area-duration curve.* The 6-hour PMP depth at Leon River is read for various storm areas from the standard PMP maps, including those shown in Fig. 14.6.1. PMP depth is plotted as a function of storm area and duration as shown in Fig. 14.6.6. In this example, only the 6-hour curve will be considered. The plotted points from the maps are connected with a smooth line to form the depth-area curve for this duration. Then the PMP depths for a number of storm areas are read from this curve as shown in column 2 of Table 14.6.1.

2. *Adjust for watershed orientation.* The standard isohyetal pattern is centered over the Leon River drainage basin as shown in Fig. 14.6.7. The longitudinal axis of the watershed runs southeast (134°) to northwest (314°), oblique to the preferred direction at this location, from the southwest (208°). The map shown in Fig. 14.6.4 applies to an orientation range of 135° to 315°, so the preferred orientation (208°) is subtracted from the watershed axis direction (314°) falling in the range 135° to 315°, i.e. 314° – 208° = 106°. When the orientation difference is greater than 90°, the value of 180° minus the orientation difference is used in reading this figure, that is, in this case 180° − 106° = 74°. The values of the orientation adjustment factor for 74° and the various storm areas are read off Fig. 14.6.4 and entered in column 3 of Table 14.6.1. Multiplication of columns 2 and 3 yields the adjusted depths shown in column 4.

3. *Find the critical storm area.* For a given storm area, the isohyetal adjustment factors for each contour are read from Fig. 14.6.5 by finding the storm

TABLE 14.6.1
Six-hour probable maximum precipitation for Leon River, adjusted for orientation of the watershed relative to the direction of maximum atmospheric moisture flow (Example 14.6.1)

Column:	1 Nominal storm area (mi^2)	2 6-h PMP from depth-area-duration curve (in)	3 Orientation adjustment factor	4 Adjusted 6-h PMP (in)
	1,000	16.1	0.961	15.47
	1,500	14.4	0.933	13.44
	2,150	12.9	0.897	11.57
	3,000	11.5	0.850	9.78
	4,500	9.8	0.850	8.33
	6,500	8.5	0.850	7.23
	10,000	7.1	0.850	6.04
	15,000	5.9	0.850	5.02

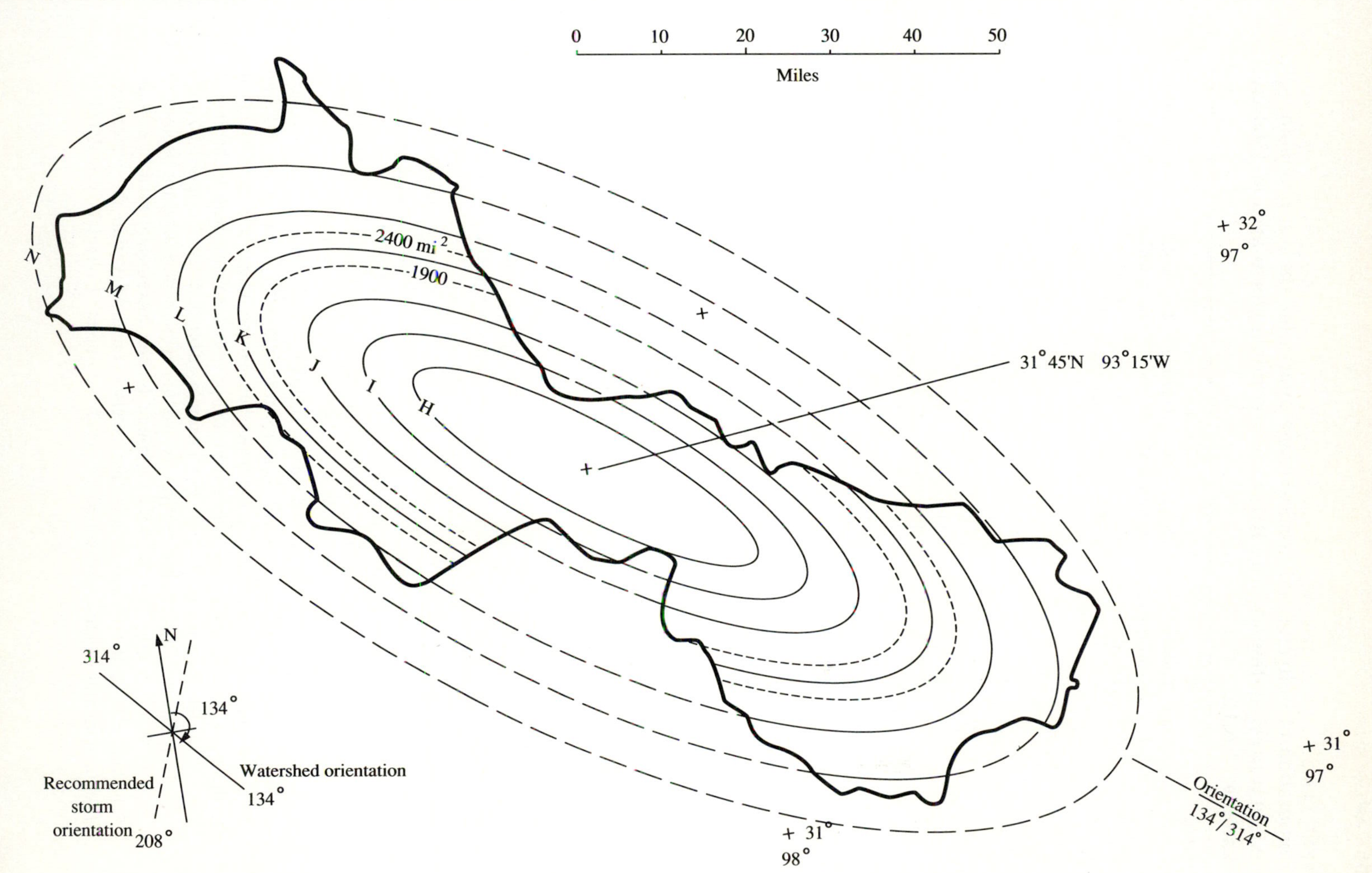

FIGURE 14.6.7
Isohyetal pattern placed on the Leon River, Texas, drainage to give maximum precipitation volume. (*Source*: Hansen, Schreiner, and Miller, 1982.)

TABLE 14.6.2
Computation of isohyetal and areal average depths for the 6-h probable maximum precipitation at Leon River for a nominal storm area of 1500 mi² (Example 14.6.1)

Column:	1 Isohyet label	2 Isohyet adjustment factor	3 Isohyet precipitation depth* (in)	4 Average depth between isohyets P_{av} (in)	5 Incremental area within watershed ΔA (mi²)	6 Incremental volume of precipitation $\Delta V = P_{av}\Delta A$ (in·mi²)
	A	1.62	21.77	21.77	10	217.7
	B	1.52	20.43	21.10	15	316.5
	C	1.42	19.08	19.76	25	493.9
	D	1.32	17.74	18.41	50	920.6
	E	1.22	16.40	17.07	75	1,280.2
	F	1.12	15.05	15.72	125	1,965.6
	G	1.05	14.11	14.58	150	2,187.4
	H	0.96	12.90	13.51	250	3,376.8
	I	0.88	11.83	12.36	271	3,350.9
	J	0.80	10.75	11.29	393	4,436.8
	K	0.56	7.53	9.14	488	4,459.9
	L	0.41	5.51	6.52	582	3,793.7
	M	0.26	3.49	4.50	737	3,318.3
	N	0.16	2.15	2.82	489	1,380.2
				Total	3660	31,498.5

Average PMP depth over watershed (in) = 31,498.5 / 3660 = 8.61 in

*The PMP for 1500 mi² is 13.44 in.

area on the vertical axis and reading horizontally to the curve for that isohyet. For example, for a 1500 mi² area, the graph gives an adjustment factor of 162 percent, or 1.62, for isohyet A. The adjustment factors for all isohyets for a 1500-mi² storm area are shown in column 2 of Table 14.6.2. The adjusted PMP depth for 1500-mi² storm area is 13.44 in from Table 14.6.1, and this value is multiplied by the isohyetal adjustment factor to give the precipitation depth for each isohyet in column 3 of Table 14.6.2; for example, on isohyet A, the depth is $1.62 \times 13.44 = 21.77$ in. The average precipitation depth P_{av} between adjacent isohyets is shown in column 4. For isohyet B, $P_{av} = (21.77 + 20.43)/2 = 21.10$ in, for example.

The incremental areas ΔA added by including each successive isohyet are shown in column 5 of Table 14.6.2. As shown in Fig. 14.6.7, isohyets A through H are completely contained within the watershed, so the incremental area assigned to them is simply the increase in area from the previous isohyet, with the isohyetal areas being shown on Fig. 14.6.2. For example, isohyet B encompasses 25 mi², of which 10 mi² falls within isohyet A, so the incremental area for isohyet B is $25 - 10 = 15$ mi². For isohyets I through N, the incremental area in column 5 is that part of the isohyetal area that is also contained within the watershed. The incremental volume ΔV added for each additional isohyet is calculated as

$\Delta V = P_{av}\Delta A$. For example, the volume within isohyet A is 21.77 in $\times$ 10 mi^2 = 217.7 in·mi^2, while isohyet B includes an additional 21.10 in $\times$ 15 mi^2 = 316.5 in·mi^2 as shown in column 6 of the table. By summing the incremental volumes for all isohyets containing at least part of their area within the watershed, the total volume is found, and it is divided by the watershed area to give the average precipitation depth. In this case, the average depth is 31,489.5 in·mi^2/3660 mi^2 = 8.61 in.

The procedure described above is repeated for all the storm areas in Table 14.6.1 to find the critical storm area: that area which yields the greatest average PMP over the watershed. In this case the 1500-mi^2 area gives the largest average depth and is the critical storm area. So the isohyetal precipitation pattern of the 6-hour PMP at Leon River is given by column 3 of Table 14.6.2, and the watershed average 6-hour PMP is 8.61 in.

This procedure can be repeated for various durations and the corresponding isohyetal and average depths obtained. A storm hyetograph can be developed from these depths by the alternating block method (Sec. 14.4) or other methods. If a percentage of the probable maximum precipitation is required for a design (e.g., 50 percent, as shown in Table 13.1.1 for some dam designs), the values of the isohyet precipitation depths in column 3 of Table 14.6.2 are reduced to the desired percentage of the values shown.

REFERENCES

Bandyopadhyay, M., Synthetic storm pattern and runoff for Gauhati, India, *J. Hyd. Div., Am. Soc. Civ. Eng.*, vol. 98, no. HY5, pp. 845–857, 1972.

Frederick, R. H., V. A. Myers, and E. P. Auciello, Five to 60-minute precipitation frequency for the eastern and central United States, NOAA technical memo NWS HYDRO-35, National Weather Service, Silver Spring, Maryland, June 1977.

Hansen, E. M., L. C. Schreiner, and J. F. Miller, Application of probable maximum precipitation estimates—United States east of the 105th meridian, NOAA hydrometeorological report no. 52, National Weather Service, Washington, D. C., August 1982.

Havens and Emerson, consulting engineers, master plan for pollution abatement, Cleveland, Ohio, July 1968.

Hershfield, D. M., Rainfall frequency atlas of the United States for durations from 30 minutes to 24 hours and return periods from 1 to 100 years, tech. paper 40, U. S. Dept. of Comm., Weather Bureau, Washington, D. C., May 1961.

Huff, F. A., Time distribution of rainfall in heavy storms, *Water Resour. Res.*, vol. 3, no. 4, pp. 1007–1019, 1967.

Institution of Engineers Australia, *Australian Rainfall and Runoff*, vol. 1 ed. by D. H. Pilgrim, vol. 2 ed. by R. P. Canterford, Canberra, Australia, 1987.

Keifer, C. J., and H. H. Chu, Synthetic storm pattern for drainage design, *J. Hyd. Div., Am. Soc. Civ. Eng.*, vol. 83, no. HY4, pp. 1–25, 1957.

Marsalek, J., Research on the design storm concept, tech. memo. 33, Am. Soc. Civ. Eng., Urban Water Resour. Res. Prog., New York, 1978.

McPherson, M. B., Discussion of "Synthetic storm pattern for drainage design," *J. Hyd. Div., Am. Soc. Civ. Eng.*, vol. 84, no. HY1, pp. 49–60, 1958.

Miller, J. F., R. H. Frederick, and R. J. Tracey, Precipitation-frequency atlas of the conterminous western United States (by states), NOAA atlas 2, 11 vols., National Weather Service, Silver Spring, Maryland, 1973.

National Academy of Sciences, *Safety of Existing Dams: Evaluation and Improvement*, National Academy Press, Washington, D. C., 1983.

Pilgrim, D. H., and I. Cordery, Rainfall temporal patterns for design floods, *J. Hyd. Div., Am. Soc. Civ. Eng.*, vol. 101, no. HY1, pp. 81–95, 1975.

Preul, H. D., and C. N. Papadakis, Development of design storm hyetographs for Cincinnati, Ohio, *Water Resour. Bull.*, vol. 9, no. 2, pp. 291–300, 1973.

Schreiner, L. C., and J. T. Riedel, Probable maximum precipitation estimates, United States east of the 105th meridian, NOAA hydrometeorological report no. 51, National Weather Service, Washington, D. C., June 1978.

Terstriep, M. L., and J. B. Stall, The Illinois urban drainage area simulator, ILLUDAS, bulletin 58, Illinois State Water Survey, Urbana, Ill., 1974.

U. S. Army Corps of Engineers Hydrologic Engineering Center, Probable maximum storm (eastern United States), HMR 52, user's manual, CPD-46, March 1984.

U. S. Department of Agriculture Soil Conservation Service, A method for estimating volume and rate of runoff in small watersheds, tech. paper 149, Washington, D. C., April 1973.

U. S. Department of Agriculture Soil Conservation Service, Urban hydrology for small watersheds, tech. release no. 55, June 1986.

U. S. Department of Commerce, Probable maximum precipitation estimates, Colorado River and Great Basin drainages, hydrometeorological report no. 49, NOAA, National Weather Service, Silver Spring, Md., September 1977.

U. S. Department of Commerce, Seasonal variation of 10-square-mile probable maximum precipitation estimates, United States east of the 105th meridian, hydrometeorological report no. 53, NOAA, National Weather Service, Silver Spring, Md., April 1980.

U. S. Weather Bureau, Seasonal variation of the probable maxium precipitation east of the 105th meridian, hydrometeorological report no. 33, Washington, D. C., 1956.

U. S. Weather Bureau, Rainfall-intensity-frequency regime, Part2—Southeastern United States, tech. paper no. 29, March 1958.

U. S. Weather Bureau, Generalized estimates of probable maximum precipitation west of the 105th meridian, tech. paper no. 38, Washington, D. C., 1960.

U. S. Weather Bureau, Generalized estimates of probable maxium precipitation and rainfall-frequency data for Puerto Rico and Virgin Islands, tech. paper no. 42, Washington, D. C., 1961.

U. S. Weather Bureau, Probable maximum precipitation in the Hawaiian Islands, hydrometeorological report no. 39, Washington, D. C., 1963a.

U. S. Weather Bureau, Probable maximum precipitation rainfall-frequency data for Alaska, tech. report no. 47, Washington, D. C., 1963b.

U. S. Weather Bureau, Two- to ten-day precipitation for return periods of 2 to 100 years in the contiguous United States, tech. paper 49, Washington, D. C., 1964.

U. S. Weather Bureau, Meteorological conditions for the probable maximum flood on the Yukon River above Rampart, Alaska, hydrometeorological report no. 42, Environmental Science Services Administration, Washington, D. C., May 1966a.

U. S. Weather Bureau, Probable maximum precipitation, northwest states, hydrometeorological report no. 43, Washington, D. C., 1966b.

U. S. Weather Bureau, Interim report—probable maximum precipitation in California, hydrometeorological report no. 36, Washington, D. C., October 1961, with revisions in October 1969.

Wenzel, H. G., Rainfall for urban stormwater design, in *Urban Storm Water Hydrology*, ed. by David F. Kibler, Water Resources Monograph 7, American Geophysical Union, Washington, D. C., 1982.

Wiesner, C. J., Hydrometeorology, Chapman and Hall, London, 1970.

World Meteorological Organization, *Guide to Hydrological Practices*, vol. II, Analysis, forecasting and other applications, WMO no. 168, 4th ed., Geneva, Switzerland, 1983.

Yen, B. C., Risk-based design of storm sewers, report no. INT 141, Hydraulics Research Station, Wallingford, Oxfordshire, England, July 1975.

Yen, B. C., and V. T. Chow, Design hyetographs for small drainage structures, *J. Hyd. Div., Am. Soc. Civ. Eng.*, vol. 106, no. HY6, pp. 1055–1076, 1980.

PROBLEMS

14.1.1 Determine the 50-year return period precipitation depth for 30 minutes duration in Austin, Texas, from the isohyetal maps given in Fig. 14.1.2.

14.1.2 Determine the 2-, 10-, 25-, and 100-year precipitation depths for a 15 minute duration storm in St. Louis, Missouri.

14.1.3 Determine the 2-, 10-, 25-, and 100-year precipitation depths at Miami, Florida, for 15 minutes duration. How much larger are these values than the corresponding values for St. Louis? Why is the precipitation depth in Miami greater than in St. Louis?

14.2.1 Determine the 10-year, 1-hour design rainfall intensity and depth for Chicago from the IDF curve given in Fig. 14.2.1.

14.2.2 Using the isohyetal maps in Fig. 14.1.2, develop intensity-duration-frequency curves for St. Louis, Missouri, plotting points for durations of 5, 10, 15, 30, and 60 minutes and return periods of 2, 5, 10, 25, 50, and 100 years.

14.2.3 Solve Prob. 14.2.2 for Atlanta, Georgia.

14.2.4 Determine the design rainfall intensities for 10-, 25-, and 100-year storms of 120 minutes duration for a location where the mean rainfall depth is 2.22 in and the standard deviation is 0.823 in. Assume the Gumbel (Extreme Value Type I) distribution applies.

14.2.5 Use the first order analysis of uncertainty and the rainfall intensity-duration equation (14.2.2) to develop an expression for the coefficient of variation of the rainfall intensity i due to the uncertainty of the duration T_d.

14.2.6 Using the first order analysis of uncertainty, determine the coefficient of variation of the rainfall intensity i due to the uncertainty in duration for the Derby catchment in central England (Yen, 1975). The applicable rainfall intensity-duration-frequency relationship is

$$i = \frac{12.1T^{0.25}}{T_d^{0.75} + 0.125} = \frac{C}{T_d^{0.75} + 0.125}$$

with i in millimeters per hour and T_d in hours. Take the coefficient of variation of T_d as 0.20. Determine CV_i for durations of 10, 20, 30 min, 1 hr, and 2 hrs.

14.2.7 Plot the rainfall depth-duration-frequency data for the Coshocton, Ohio, in Table 14.2.2 with return period on the abscissa and rainfall depth (inches) on the ordinate. Using the Extreme Value Type I distribution, fit a line for each of the durations 15, 30, 60, and 120 minutes. Use return periods of 2, 5, 10, and 25 years.

14.2.8 Using the Coshocton, Ohio, data (Table 14.2.2), perform a frequency analysis using the Extreme Value Type I distribution for the 30-minute duration rainfall depths, and identify the 10-, 25-, and 100-year rainfall depths for this duration.

14.2.9 Using the Coshocton, Ohio, data (Table 14.2.2), perform a frequency analysis using the Extreme Value Type I distribution for the 120-minute duration rainfall depths, and identify the 10-, 25-, and 100-year rainfall depths for this duration.

14.2.10 The mean and standard deviation of the annual maximum rainfall depths for various durations in Austin, Texas, are shown below. Determine, for each duration, the design rainfall intensity for return periods of 2, 5, 10, 25, 50, and 100 years. Use the Extreme Value Type I (Gumbel) distribution. Plot the results as a set of intensity-duration-frequency curves.

Duration	Mean depth (in)	Standard deviation (in)
5 min	0.493	0.133
10 min	0.795	0.225
15 min	1.040	0.298
30 min	1.480	0.493
1 h	1.910	0.665
2 h	2.220	0.823
3 h	2.470	0.793
1 day	4.140	2.490

14.3.1 Use the 100-year 24-hour precipitation map for the United States (Fig. 14.1.1) and the SCS storm distribution pattern (Table 14.3.1) to develop a 100-year 24-hour design storm hyetograph for Washington, D. C.

14.3.2 Determine a triangular hyetograph for the design of a culvert in Philadelphia. The design return period is 10 years and the duration is 60 minutes. The value of r is given in Table 14.3.2.

14.3.3 Construct a triangular hyetograph for the design of a culvert in Baltimore, for a design return period of 50 years and a 1-hour duration. The value of r is given in Table 14.3.2.

14.3.4 Show that the time between the beginning of precipitation and centroid of a triangular design hyetograph is given by $(T_d + t_a)/3$, where T_d is the duration of the hyetograph and t_a the time to the peak intensity.

14.4.1 Using the IDF curves for Chicago given in Fig. 14.2.1, develop a 1-hour design hyetograph in 10-minute increments using the alternating block method. Consider a 10-year return period.

14.4.2 Develop and plot the one-hour design rainfall hyetograph for Los Angeles using the instantaneous intensity method with $r = 0.5$.

14.4.3 Solve Prob. 14.4.2 using an advancement coefficient r of 0.375. Plot the design hyetographs for $r = 0.5$ and $r = 0.375$ on the same graph.

14.4.4 Develop the 1-hour design hyetograph for Los Angeles using the alternating block method and 10-minute time increments. Use a 10-year return period.

14.4.5 Develop and plot a 10-year 2-hour design rainfall hyetograph for Miami using the alternating block method and 10-minute time increments.

14.4.6 Develop and plot the 10-year 1-hour design hyetograph for Cleveland using the instantaneous intensity method. The required coefficient values are given in Tables 14.2.3 and 14.3.2.

14.4.7 Derive the intensity equation for a design hyetograph developed by the instantaneous intensity method using the following rainfall intensity-duration-frequency equation:

$$i_{\text{ave}} = \frac{c'}{(T_d + f')^{e'}}$$

where T_d is the rainfall duration. For Los Angeles, $c' = 10.9$, $e' = 0.51$, $f' = 1.15$, and $r = 0.5$. Construct a one-hour design hyetograph for this location.

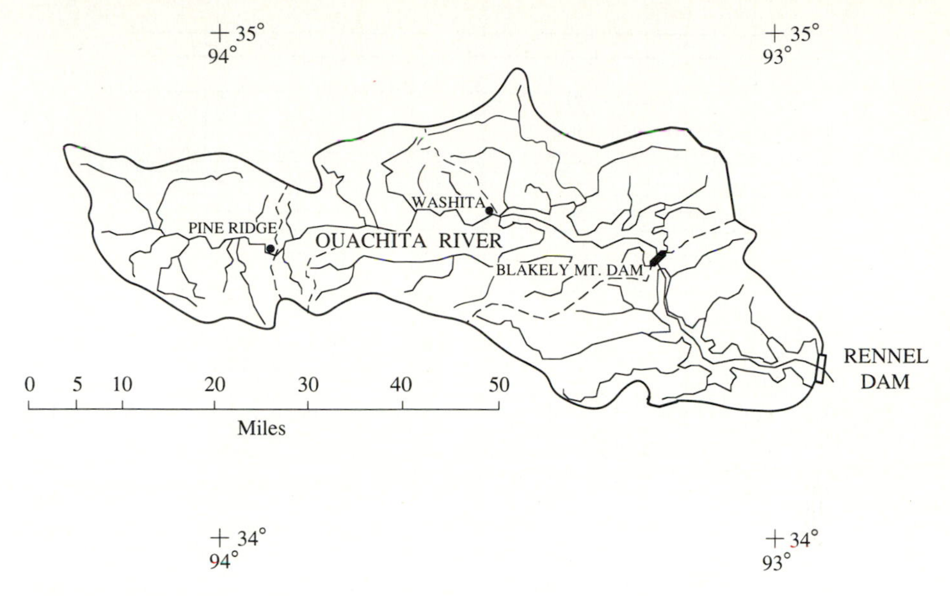

FIGURE 14.P.1
Ouachita River, Arkansas (1600 mi^2) above Rennel Dam, showing drainage. (*Source*: Hansen, Schreiner, and Miller, 1982.)

14.5.1 Use Eq. (14.5.1) for the world's greatest recorded rainfalls to develop and plot a 24-hour design hyetograph in 1-hour time increments by the alternating block method.

14.6.1 Determine the average 6-hour PMP depth over the Leon River watershed for a storm of nominal area 2150 mi^2. See Example 14.6.1 for data.

14.6.2 By trying a number of storm areas different from those shown in Table 14.6.1, determine the critical storm area for the maximum 6-hour PMP depth over the Leon River watershed.

14.6.3 Evaluate the average 6-hour PMP depth over the Leon River watershed assuming no adjustment for orientation. See Example 14.6.1 for data.

14.6.4 Use the Leon River depth-area-duration curves (Fig. 14.6.6) to develop a design PMP hyetograph for average precipitation over a 1000-mi^2 area which has the elliptical shape of the standard isohyetal pattern. Assume that no orientation adjustment is needed.

14.6.5 From the Leon River depth-area-duration curves (Fig. 14.6.6), calculate the additional precipitation depth in a 12-hour PMP beyond that in a 6-hour PMP for storm areas of 10, 100, 1000, and 10,000 mi^2.

14.6.6 Determine the spatial distribution and the average depth of the 6-hour probable maximum precipitation on the Ouachita River, Arkansas, above Rennel Dam (Fig. 14.P.1). The drainage basin area is approximately 1600 mi^2 with the drainage center at approximately 34°36′N, 93°27′W. Consider a storm of area 2150 mi^2. The depth-area-duration curves for this location are given in Fig. 14.P.2. Assume that the major axis of the elliptical storm pattern is in the

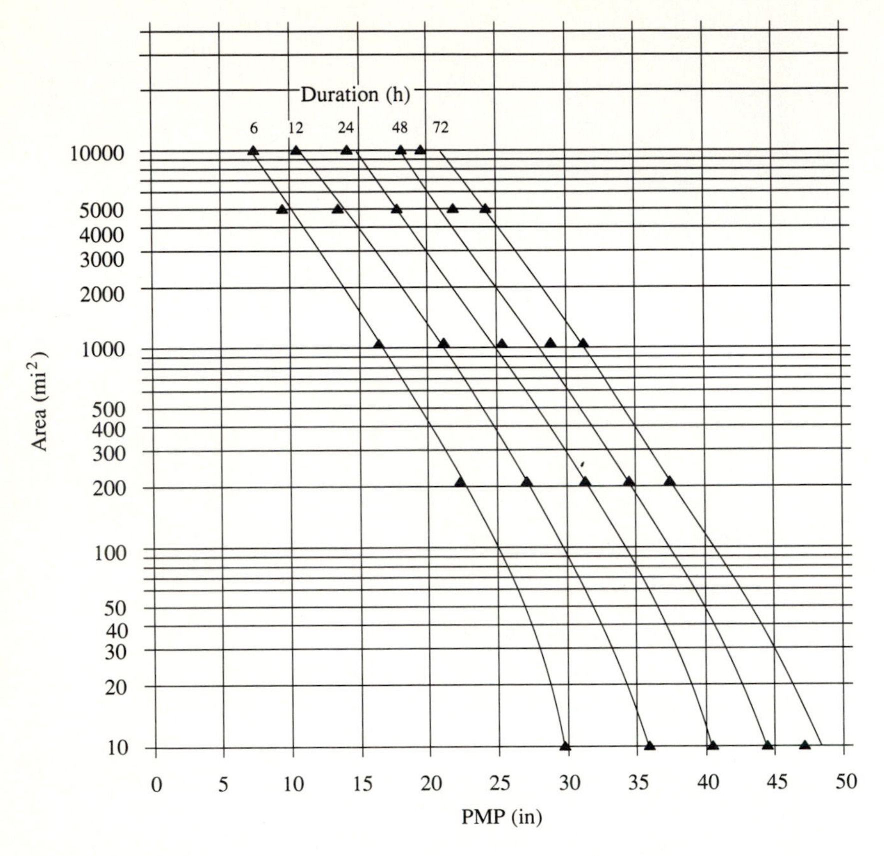

FIGURE 14.P.2
Depth-area-duration curves for 34°36′N, 93°27′W, applicable to the Ouachita River, Arkansas, drainage. (*Source*: Hansen, Schreiner, and Miller, 1982.)

direction 95°/275°. The watershed contains all of the contours A through H of the storm pattern. Contours I through L add the following incremental areas (in mi^2): I, 242; J, 242; K, 224; L, 192.

14.6.7 Solve Prob. 14.6.6 to determine the average depth of the 6-hour probable maximum precipitation of the Ouachita River, Arkansas, above Rennel Dam (Fig. 14.P.1) considering a storm of 1500 mi^2.

14.6.8 Consider a number of storm areas to determine the critical storm area for the 6-hour PMP over the Ouachita River watershed (Prob. 14.6.6).

CHAPTER 15

DESIGN FLOWS

Hydrologic design for water control is concerned with mitigating the adverse effects of high flows or floods. A flood is any high flow that overtops either natural or artificial embankments along a stream. The magnitudes of floods are described by flood discharge, flood elevation, and flood volume. Each of these factors is important in the hydrologic design of different types of flow control structures. A major portion of this chapter deals with development of the design flow or design flood for flow regulation structures (detention basins, flood control reservoirs, etc.) and flow conveyance structures (storm sewers, drainage channels, flood levees, diversion structures, etc.). The purpose of flow regulation structures is to smooth out peak discharges, thereby decreasing downstream flood elevation peaks, and the purpose of flow conveyance structures is to safely convey the flow to downstream points where the adverse effects of flows are controlled or are minimal. This chapter discusses methods and simulation models that can be used in the hydrologic design of flow control structures from urban drainage systems to flood control reservoirs.

Hydrologic design for water use is concerned with the development of water resources to meet human needs and with the conservation of the natural life in water environments. As population and economic activity increase, so do the demands for use of water. But these must be balanced against the finite supply provided by nature and the desire to maintain healthy plant and animal life in rivers, lakes, and estuaries. Hydrologic information plays a vital role in managing the balance between supply and demand for water resources and in planning water resource development projects. In contrast to hydrologic design for water control, which is concerned with mitigating the adverse effects of high flows, hydrologic design for water use is directed at utilizing average flows and with mitigating the effects of extremely low flows.

15.1 STORM SEWER DESIGN

Population growth and urban development can create potentially severe problems in urban water management. One of the most important facilities in preserving and improving the urban water environment is an adequate and properly functioning storm water drainage system. Construction of houses, commercial buildings, parking lots, paved roads, and streets increases the impervious cover in a watershed, and reduces infiltration. Also, with urbanization, the spatial pattern of flow in the watershed is altered and there is an increase in the hydraulic efficiency of flow through artificial channels, curbing, gutters, and storm drainage and collection systems. These factors increase the volume and velocity of runoff and produce larger peak flood discharges from urbanized watersheds than occurred in the preurbanized condition. Many urban drainage systems constructed under one level of urbanization are now operating under a higher level of urbanization and have inadequate capacity.

One view of the typical urban drainage system is shown in Fig. 15.1.1. The system can be considered as consisting of two major types of elements: *location elements* and *transfer elements*. Location elements are the places where the water

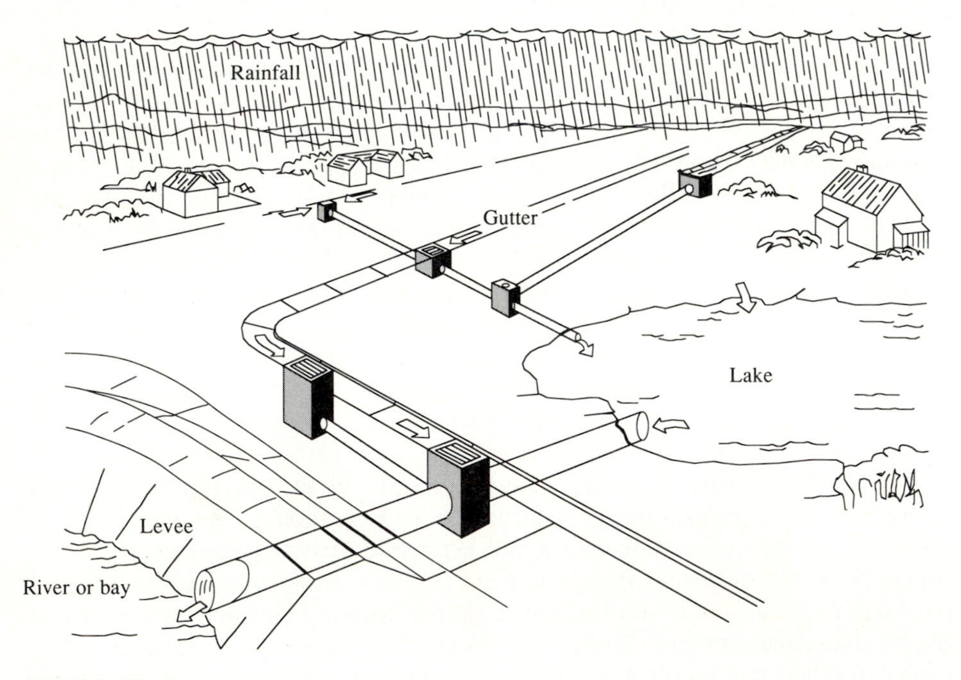

FIGURE 15.1.1
Typical urban drainage system. (*Source*: Roesner, 1982, Copyright by the American Geophysical Union.)

stops and undergoes changes as a result of humanly controlled processes, for example, water storage, water treatment, water use, and wastewater treatment. Transfer elements connect the location elements; these elements include channels, pipelines, storm sewers, sanitary sewers, and streets. The system is fed by rainfall, influent water from various sources, and imported water in the pipes or channels. The receiving water body can be a river, a lake, or an ocean. Figure 15.1.1 shows a *storm sewer system* for collection of storm drainage in a pipe network and discharge to a receiving water body. This section considers the design of a sewer system for storm drainage.

System concepts are increasingly being used as an aid in understanding and developing solutions to complex urban problems. These problems involve distributed systems, and must be analyzed to account for both spatial and temporal variations. Urban watersheds vary in space in that the ground surface slope and cover, and the soil type, change from place to place in the watershed. They vary in time in that hydrologic characteristics change with the process of urbanization. The mathematical formulation of models for urban water systems distributed in both time and space is a complicated task. Consequently, spatial variation is sometimes ignored, and the system is treated as being lumped. Some spatial variation can be introduced by dividing the watershed system into several subsystems that are each considered lumped, and then linking these lumped-system models together to produce a model of the entire system.

Models can be used as tools for planning and management. In particular, a number of computerized watershed simulation models have been proposed. The determination of the runoff volume and peak discharge rate are important issues in urban stormwater management, and methods for calculating these variables range from the well-known rational formula to advanced computer simulation models such as the Storm Water Management Model (SWMM; see Huber, et al., 1975).

Design Philosophy

A *storm sewer system* is a network of pipes used to convey storm runoff in a city. The design of storm sewer systems involves the determination of diameters, slopes, and crown or invert elevations for each pipe in the system. The crown and invert elevations of a pipe are, respectively, the elevations of the top and the bottom of the pipe circumference.

The selection of a *layout*, or network of pipe locations, for a storm sewer system requires a considerable amount of subjective judgment. Hydrologists are usually able to investigate only a few of the possible layouts. Generally, manholes are placed at street intersections and at major changes in grade, or ground surface slope, and the sewers are sloped in the direction of the ground surface, so as to connect with downstream submains and trunk sewers. Once a layout has been selected, the rational method can be used to select pipe diameters. This conventional design approach is based on a set of design standards and criteria, such as those set forth by the American Society of Civil Engineers (1960) and various planning agencies.

Storm drainage design can be divided into two aspects: runoff prediction and system design. In recent years, rainfall-runoff modeling for urban watersheds has been a popular activity and a variety of such rainfall-runoff models are now available, as described by Chow and Yen (1977), Heeps and Mein (1974), Brandstetter (1976), McPherson (1975), Colyer and Pethick (1977), Yen (1978), and Kibler (1982). Computer models are described more fully in Sec. 15.2.

The following constraints and assumptions are commonly used in storm sewer design practice:

1. Free-surface flow exists for the design discharges; that is, the sewer system is designed for "gravity flow"; pumping stations and pressurized sewers are not considered.
2. The sewers are of commercially available circular sizes no smaller than 8 inches in diameter.
3. The design diameter is the smallest commercially available pipe having flow capacity equal to or greater than the design discharge and satisfying all the appropriate constraints.
4. Storm sewers must be placed at a depth such that they will not be susceptible to frost, will be able to drain basements, and will have sufficient cushioning to prevent breakage due to ground surface loading. To these ends, minimum cover depths must be specified.
5. The sewers are joined at junctions such that the crown elevation of the upstream sewer is no lower than that of the downstream sewer.
6. To prevent or reduce excessive deposition of solid material in the sewers, a minimum permissible flow velocity at design discharge or at barely full-pipe gravity flow is specified (e.g., 2.5 ft/s).
7. To prevent scour and other undesirable effects of high-velocity flow, a maximum permissible flow velocity is also specified.
8. At any junction or manhole the downstream sewer cannot be smaller than any of the upstream sewers at that junction.
9. The sewer system is a dendritic, or branching, network converging in the downstream direction without closed loops.

Rational Method

The *rational method*, which can be traced back to the mid-nineteenth century, is still probably the most widely used method for design of storm sewers (Pilgrim, 1986; Linsley, 1986). Although valid criticisms have been raised about the adequacy of this method, it continues to be used for sewer design because of its simplicity. Once the layout is selected and the pipe sizes determined by the rational method, the adequacy of the system can be checked by dynamic routing of flow hydrographs through the system.

The idea behind the rational method is that if a rainfall of intensity i begins instantaneously and continues indefinitely, the rate of runoff will increase until the time of concentration t_c, when all of the watershed is contributing to flow at the outlet. The product of rainfall intensity i and watershed area A is the inflow rate for the system, iA, and the ratio of this rate to the rate of peak discharge Q (which occurs at time t_c) is termed the *runoff coefficient* C ($0 \leq C \leq 1$). This is expressed in the rational formula:

$$Q = CiA \tag{15.1.1}$$

Commonly, Q is in cubic feet per second (cfs), i is in inches per hour, and A is in acres, and the conversion (1 cfs = 1.008 acre·in/hr) is considered to be included in the runoff coefficient. The duration used for the determination of the design precipitation intensity i in (15.1.1) is the time of concentration of the watershed.

In urban areas, the drainage area usually consists of subareas or subcatchments of different surface characteristics. As a result, a composite analysis is required that must account for the various surface characteristics. The areas of the subcatchments are denoted by A_j and the runoff coefficients of each subcatchment are denoted by C_j. The peak runoff is then computed using the following form of the rational formula:

$$Q = i \sum_{j=1}^{m} C_j A_j \tag{15.1.2}$$

where m is the number of subcatchments drained by a sewer.

The assumptions associated with the rational method are:

1. The computed peak rate of runoff at the outlet point is a function of the average rainfall rate during the time of concentration, that is, the peak discharge does not result from a more intense storm of shorter duration, during which only a portion of the watershed is contributing to runoff at the outlet.
2. The time of concentration employed is the time for the runoff to become established and flow from the most remote part of the drainage area to the inflow point of the sewer being designed.
3. Rainfall intensity is constant throughout the storm duration.

Runoff Coefficient

The runoff coefficient C is the least precise variable of the rational method. Its use in the formula implies a fixed ratio of peak runoff rate to rainfall rate for the drainage basin, which in reality is not the case. Proper selection of the runoff coefficient requires judgment and experience on the part of the hydrologist. The proportion of the total rainfall that will reach the storm drains depends on the percent imperviousness, slope, and ponding character of the surface. Impervious surfaces, such as asphalt pavements and roofs of buildings, will produce nearly 100 percent runoff after the surface has become thoroughly wet, regardless of the slope. Field inspection and aerial photographs are useful in estimating the nature of the surface within the drainage area.

The runoff coefficient is also dependent on the character and condition of the soil. The infiltration rate decreases as rainfall continues, and is also influenced by the antecedent moisture condition of the soil. Other factors influencing the runoff coefficient are rainfall intensity, proximity of the water table, degree of soil compaction, porosity of the subsoil, vegetation, ground slope, and depression storage. A reasonable coefficient must be chosen to represent the integrated effects of all these factors. Suggested coefficients for various surface types as used in Austin, Texas are given in Table 15.1.1.

TABLE 15.1.1
Runoff coefficients for use in the rational method

	Return Period (years)						
Character of surface	**2**	**5**	**10**	**25**	**50**	**100**	**500**
Developed							
Asphaltic	0.73	0.77	0.81	0.86	0.90	0.95	1.00
Concrete/roof	0.75	0.80	0.83	0.88	0.92	0.97	1.00
Grass areas (lawns, parks, etc.)							
Poor condition (grass cover less than 50% of the area)							
Flat, 0–2%	0.32	0.34	0.37	0.40	0.44	0.47	0.58
Average, 2–7%	0.37	0.40	0.43	0.46	0.49	0.53	0.61
Steep, over 7%	0.40	0.43	0.45	0.49	0.52	0.55	0.62
Fair condition (grass cover on 50% to 75% of the area)							
Flat, 0–2%	0.25	0.28	0.30	0.34	0.37	0.41	0.53
Average, 2–7%	0.33	0.36	0.38	0.42	0.45	0.49	0.58
Steep, over 7%	0.37	0.40	0.42	0.46	0.49	0.53	0.60
Good condition (grass cover larger than 75% of the area)							
Flat, 0–2%	0.21	0.23	0.25	0.29	0.32	0.36	0.49
Average, 2–7%	0.29	0.32	0.35	0.39	0.42	0.46	0.56
Steep, over 7%	0.34	0.37	0.40	0.44	0.47	0.51	0.58
Undeveloped							
Cultivated Land							
Flat, 0–2%	0.31	0.34	0.36	0.40	0.43	0.47	0.57
Average, 2–7%	0.35	0.38	0.41	0.44	0.48	0.51	0.60
Steep, over 7%	0.39	0.42	0.44	0.48	0.51	0.54	0.61
Pasture/Range							
Flat, 0–2%	0.25	0.28	0.30	0.34	0.37	0.41	0.53
Average, 2–7%	0.33	0.36	0.38	0.42	0.45	0.49	0.58
Steep, over 7%	0.37	0.40	0.42	0.46	0.49	0.53	0.60
Forest/Woodlands							
Flat, 0–2%	0.22	0.25	0.28	0.31	0.35	0.39	0.48
Average, 2–7%	0.31	0.34	0.36	0.40	0.43	0.47	0.56
Steep, over 7%	0.35	0.39	0.41	0.45	0.48	0.52	0.58

Note: The values in the table are the standards used by the City of Austin, Texas. Used with permission.

Rainfall Intensity

The rainfall intensity i is the average rainfall rate in inches per hour for a particular drainage basin or subbasin. The intensity is selected on the basis of the design rainfall duration and return period as described in Sec. 14.2. The design duration is equal to the time of concentration for the drainage area under consideration. The return period is established by design standards or chosen by the hydrologist as a design parameter.

Runoff is assumed to reach a peak at the time of concetration t_c when the entire watershed is contributing to flow at the outlet. The time of concentration is the time for a drop of water to flow from the remotest point in the watershed to the point of interest. A trial and error procedure can be used to determine the critical time of concentration where there are several possible flow paths to consider. The time of concentration to any point in a storm drainage system is the sum of the inlet time t_o (the time it takes for flow from the remotest point to reach the sewer inlet), and the flow time t_f in the upstream sewers connected to the outer point:

$$t_c = t_o + t_f \tag{15.1.3}$$

The flow time is given by Eq. (5.7.3):

$$t_f = \sum_{i=1}^{n} \frac{L_i}{V_i} \tag{15.1.4}$$

where L_i is the length of the ith pipe along the flow path, and V_i is the flow velocity in the pipe.

The inlet time, or time of concentration for the case of no upstream sewers, can be obtained by experimental observations, or it can be estimated by using formulas such as those listed in Table 15.1.2. There may exist several possible flow routes for different catchments drained by a sewer; the longest time of concentration among the times for different routes is assumed to be the critical time of concentration of the area drained.

Because the areas contributing to most storm sewer inlets are relatively small, it is also customary to determine the inlet time on the basis of experience under similar conditions. Inlet time decreases as the slope and imperviousness of the surface increases, and it increases as the distance over which the water has to travel increases and as retention by the contact surfaces increases. All inlet times determined on the basis of experience should be verified by direct overland flow computation.

Drainage Area

The size and shape of the catchment or subcatchment under consideration must be determined. The area may be determined by planimetering topographic maps, or by field surveys where topographic data has changed or where the mapped contour interval is too great to distinguish the direction of flow. The drainage area

TABLE 15.1.2
Summary of time of concentration formulas

Method and Date	Formula for t_c (min)	Remarks
Kirpich (1940)	$t_c = 0.0078L^{0.77}S^{-0.385}$ L = length of channel/ditch from headwater to outlet, ft S = average watershed slope, ft/ft	Developed from SCS data for seven rural basins in Tennessee with well-defined channel and steep slopes (3% to 10%); for overland flow on concrete or asphalt surfaces multiply t_c by 0.4; for concrete channels multiply by 0.2; no adjustments for overland flow on bare soil or flow in roadside ditches.
California Culverts Practice (1942)	$t_c = 60(11.9L^3/H)^{0.385}$ L = length of longest watercourse, mi H = elevation difference between divide and outlet, ft	Essentially the Kirpich formula; developed from small mountainous basins in California (U. S. Bureau of Reclamation, 1973, pp. 67–71).
Izzard (1946)	$t_c = \dfrac{41.025(0.0007i + c)L^{0.33}}{S^{0.333}i^{0.667}}$ i = rainfall intensity, in/h c = retardance coefficient L = length of flow path, ft S = slope of flow path, ft/ft	Developed in laboratory experiments by Bureau of Public Roads for overland flow on roadway and turf surfaces; values of the retardance coefficient range from 0.0070 for very smooth pavement to 0.012 for concrete pavement to 0.06 for dense turf; solution requires iteration; product i times L should be ≤ 500.
Federal Aviation Administration (1970)	$t_c = 1.8(1.1 - C)L^{0.50}/S^{0.333}$ C = rational method runoff coefficient L = length of overland flow, ft S = surface slope, %	Developed from air field drainage data assembled by the Corps of Engineers; method is intended for use on airfield drainage problems, but has been used frequently for overland flow in urban basins.

TABLE 15.1.2 *(cont.)*

Summary of time of concentration formulas

Method and Date	Formula for t_c (min)	Remarks
Kinematic wave formulas Morgali and Linsley (1965) Aron and Erborge (1973)	$t_c = \frac{0.94L^{0.6}n^{0.6}}{(i^{0.4}S^{0.3})}$ L = length of overland flow, ft n = Manning roughness coefficient i = rainfall intensity in/h S = average overland slope ft/ft	Overland flow equation developed from kinematic wave analysis of surface runoff from developed surfaces; method requires iteration since both i (rainfall intensity) and t_c are unknown; superposition of intensity–duration–frequency curve gives direct graphical solution for t_c
SCS lag equation (1973)	$t_c = \frac{100\ L^{0.8}[(1000/\text{CN}) - 9]^{0.7}}{1900\ S^{0.5}}$ L = hydraulic length of watershed (longest flow path), ft CN = SCS runoff curve number S = average watershed slope, %	Equation developed by SCS from agricultural watershed data; it has been adapted to small urban basins under 2000 acres; found generally good where area is completely paved; for mixed areas it tends to overestimate; adjustment factors are applied to correct for channel improvement and impervious area; the equation assumes that $t_c = 1.67 \times$ basin lag.
SCS average velocity charts (1975, 1986)	$t_c = \frac{1}{60}\Sigma \frac{L}{V}$ L = length of flow path, ft V = average velocity in feet per second from Fig. 3-1 of TR 55 for various surfaces	Overland flow charts in Fig. 3-1 of TR 55 show average velocity as function of watercourse slope and surface cover. (See also Table 5.7.1)

Source: Kibler, 1982, Copyright by the American Geophysical Union.

contributing to the system being designed and the drainage subarea contributing to each inlet point must be measured. The outline of the drainage divide must follow the actual watershed boundary, rather than commercial land boundaries, as may be used in the design of sanitary sewers. The drainage divide lines are influenced by pavement slopes, locations of downspouts and paved and unpaved yards, grading of lawns, and many other features introduced by urbanization.

Pipe Capacity

In choosing storm sewer pipe diameters, the mimimum required diameter is computed, and the next larger commercially available size is selected. Commercial pipes are available in diameters of 8, 10, 12, 15, 16, and 18 in, at increments of 3 in between 18 and 36 in, and at increments of 6 in between 3 ft and 10 ft.

Once the design discharge Q entering the sewer pipe has been calculated by the rational formula, the diameter of pipe D required to carry this discharge is determined. It is usually assumed that the pipe is flowing full under gravity but is not pressurized, so the pipe capacity can be calculated by the Manning or Darcy-Weisbach equations for open-channel flow. For Manning's equation, the area is $A = \pi D^2/4$, and the hydraulic radius is $R = A/P = (\pi D^2/4)/\pi D = D/4$. The friction slope S_f is set equal to the bed slope of the pipe, S_0, thus assuming uniform flow, and the discharge is computed for full pipe flow as

$$
\begin{aligned}
Q &= \frac{1.49}{n} S_f^{1/2} A R^{2/3} \\
&= \frac{1.49}{n} S_0^{1/2} \left(\frac{\pi D^2}{4} \right) \left(\frac{D}{4} \right)^{2/3} \\
&= \frac{0.463}{n} S_0^{1/2} D^{8/3}
\end{aligned}
\tag{15.1.5}
$$

This is solved for the required diameter D as

$$
D = \left(\frac{2.16 Q n}{\sqrt{S_0}} \right)^{3/8} \tag{15.1.6}
$$

which is valid for Q in cubic feet per second, and D in feet. When using SI units, with Q in cubic meters per second and D in meters, the coefficient 2.16 is replaced by $2.16 \times 1.49 = 3.21$ in Eq.(15.1.6).

Using the Darcy-Weisbach equation (2.5.4), with A, R, and S_f as for Manning's equation,

$$
\begin{aligned}
Q &= A \left(\frac{8g}{f} R S_f \right)^{1/2} \\
&= \frac{\pi D^2}{4} \left(\frac{8g}{f} \frac{D}{4} S_0 \right)^{1/2}
\end{aligned}
\tag{15.1.7}
$$

Solved for D, this gives

$$D = \left(\frac{0.811 f Q^2}{g S_0} \right)^{1/5} \tag{15.1.8}$$

where f is the Darcy-Weisbach friction factor and g is acceleration due to gravity. Equation (15.1.8) is valid for any dimensionally consistent set of units.

Assessment of the Rational Method

The rational method is criticized by some hydrologists because of its simplified approach to the calculation of design flow rates. Nevertheless, the rational method is still widely used for the design of storm sewer systems in the United States and other countries because of its simplicity and the fact that the required dimensions of the storm sewers are determined as the computation proceeds. More realistic flow simulation procedures involving the routing of flow hydrographs require the dimensions of the flow conveyance structures to be predetermined. The storm sewer system design produced by the rational method can be considered as a preliminary design whose adequacy can be checked by routing flow hydrographs through the system.

The uncertainties involved in the rational method can be examined by the risk analysis procedures described in Chap. 13 (Yen, 1975; Yen, et al., 1976; Yen, 1978). In this case, the loading on the system is described by the rational formula (15.1.2) and the capacity by the pipe conveyance equations (15.1.5) or (15.1.7). Problems 15.1.8 to 15.1.16 address this subject.

Example 15.1.1. A hypothetical drainage basin comprising seven subcatchments is shown in Fig. 15.1.2. Determine the required capacity of the storm sewer EB draining subarea III for a five-year return period storm. This subcatchment has an area of 4 acres, a runoff coefficient of 0.6, and an inlet time of 10 minutes.

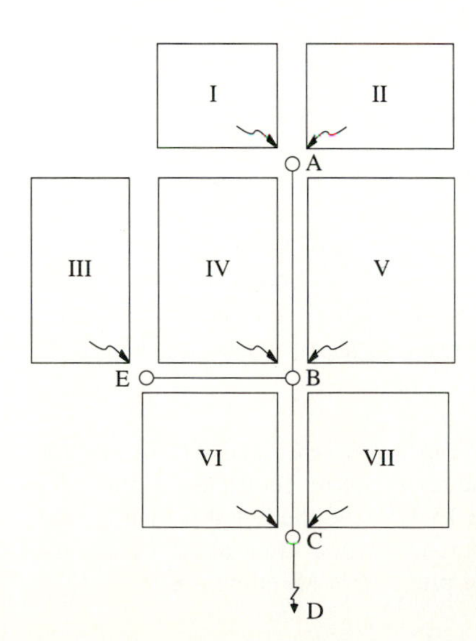

FIGURE 15.1.2
The drainage basin and storm sewer system for Examples 15.1.1 and 15.1.2.

The design precipitation intensity for this location is given by $i = 120T^{0.175}/(T_d + 27)$, where i is the intensity in inches per hour, T is the return period, and T_d is the duration in minutes. The ground elevations at points E and B are 498.43 and 495.55 ft above mean sea level, respectively, and the length of pipe EB is 450 ft. Assume Manning's n is 0.015. Calculate the flow time in the pipe.

Solution. The time of concentration for flow into sewer EB is simply the 10-minute inlet time for flow from subcatchment III to point E. So, $T_d = 10$ min and the design rainfall intensity with $T = 5$ years is

$$i = \frac{120T^{0.175}}{(T_d + 27)}$$

$$= \frac{120(5)^{0.175}}{(10 + 27)}$$

$$= 4.30 \text{ in/h}$$

The design discharge is given by Eq. (15.1.1):

$$Q = CiA$$

$$= 0.6 \times 4.30 \times 4$$

$$= 10.3 \text{ cfs}$$

The slope of the pipe EB is the difference between the ground elevations at points E and B divided by the length of the pipe: $S_0 = (498.43 - 495.55)/450 = 0.0064$. The required pipe diameter is calculated from (15.1.6):

$$D = \left(\frac{2.16Qn}{\sqrt{S_0}}\right)^{3/8}$$

$$= \left(\frac{2.16 \times 10.3 \times 0.015}{\sqrt{0.0064}}\right)^{3/8}$$

$$= 1.71 \text{ ft}$$

The diameter is rounded up to the next commercially available pipe size, 1.75 ft or 21 in.

The flow velocity through pipe EB is found by taking the nominal diameter (1.75 ft), and assuming the pipe is flowing full with $Q = 10.3$ cfs. Hence, $V = Q/A = 10.3/(\pi \times 1.75^2/4) = 4.28$ ft/s. The flow time is $L/V = 450/4.28 = 105$ s $=$ 1.75 min. It should be noted that a slight error in the computed flow time is caused by assuming that the pipe is flowing full. The velocity for partially-full-pipe flow can be determined using Newton's iteration technique presented in Chap. 5, if necessary.

Example 15.1.2. Determine the diameter for pipes AB, BC, and CD in the 27-acre drainage basin shown in Fig. 15.1.2. The area, runoff coefficients, and inlet time for each subcatchment are shown in Table 15.1.3, and the length and slope for each pipe are in columns 2 and 3 of Table 15.1.4. Use the same rainfall intensity equation as in Example 15.1.1 and assume the pipes have Manning's $n = 0.015$.

Solution. The same method as was illustrated in Example 15.1.1 is used for each pipe, except that now the time of concentration must include both inlet time and flow time through upstream sewers. The results obtained in Example 15.1.1 for pipe EB are shown in the first row of Table 15.1.4.

Pipe AB. This pipe drains subcatchments I and II. From Table 15.1.3, $A_I = 2$ acres, $C_I = 0.7$, and the inlet time is $t_I = 5$ min, while $A_{II} = 3$ acres, $C_{II} = 0.7$, and $t_{II} = 7$ min. Hence, the total area drained by pipe AB is 5 acres and $\Sigma CA = C_I A_I + C_{II} A_{II} = 0.7 \times 2 + 0.7 \times 3 = 3.5$. The time of concentration used is 7 min, the larger of the two inlet times. The calculations for the required diameter are carried out in the same way as in Example 15.1.1; the results are shown in the second row of Table 15.1.4. The calculated diameter, 1.94 ft, is rounded up to a commercial size of 2.0 ft (24 in) for pipe AB.

Pipe BC. This pipe drains subcatchments I through V: subcatchments I and II through pipe AB, subcatchment III through pipe EB, and subcatchments IV and V directly. There are thus four possible flow paths for water to reach point B; the time of concentration is the largest of their flow times. The flow time for flow coming from pipe AB is 7 minutes inlet time plus 1.76 minutes travel time, or 8.76 minutes; for the flow from pipe EB it is 10 minutes inlet time plus 1.75 minutes flow time, or 11.75 minutes; and the inlet times for subcatchments IV and V are 10 min and 15 min, respectively. Thus, the time of concentration for pipe BC is taken as 15 min.

For subcatchments I and II, $\Sigma CA = 3.5$, as shown previously. For subcatchments III to V, the values of the runoff coefficient and catchment area are given in Table 15.1.3; at point C, using these values, $\Sigma CA = 3.5 + 0.6 \times 4 + 0.6 \times 4 + 0.5 \times 5 = 10.8$. Proceeding as in Example 15.1.1, the calculated pipe diameter is 2.87 ft, which is rounded up to 3.0 ft (36 in) for pipe BC (third row of Table 15.1.4).

Pipe CD. This pipe drains all seven subcatchments. Using the same method as for the previous pipes, its time of concentration (to point C) is found to be 15 minutes (to point B) plus 1.2 minutes flow time in pipe BC, or 16.2 minutes, and $\Sigma CA = 15.3$. The calculated diameter, 3.22 ft, is rounded up to a commercial size of 3.50 ft (42 in) (fourth row of Table 15.1.4).

The required diameters for pipes AB, BC, and CD are 21, 36, and 42 in, respectively.

TABLE 15.1.3
Characteristics of the drainage basin for Example 15.1.2

Catchment	Area A (acres)	Runoff coefficient C	Inlet time t_i (min)
I	2	0.7	5
II	3	0.7	7
III	4	0.6	10
IV	4	0.6	10
V	5	0.5	15
VI	4.5	0.5	15
VII	4.5	0.5	15

TABLE 15.1.4

Design of sewers by the rational method (Examples 15.1.1 and 15.1.2)

1 Sewer pipe	2 Length L (ft)	3 Slope S_o (ft/ft)	4 Total area drained (acres)	5 $\sum$CA	6 t_c (min)	7 Rainfall intensity i (in/hr)	8 Design discharge Q (cfs)	9 Computed sewer diameter (ft)	10 Pipe size used (ft)	11 Flow velocity Q/A (ft/s)	12 Flow time L/V (min)
EB	450	0.0064	4	2.4	10.0	4.30	10.3	1.71	1.75	4.28	1.75
AB	550	0.0081	5	3.5	7.0	4.68	16.4	1.94	2.00	5.21	1.76
BC	400	0.0064	18	10.8	15.0	3.79	40.9	2.87	3.00	5.78	1.15
CD	450	0.0064	27	15.3	16.2	3.68	56.3	3.22	3.50	5.85	1.28

15.2 SIMULATING DESIGN FLOWS

Since the early 1960s, a host of deterministic hydrologic simulation models have been developed. These models include *event simulation* models for modeling a single rainfall-runoff event and *continuous simulation* models, which have soil moisture accounting procedures to simulate runoff from rainfall in hourly or daily intervals over long time periods. Examples of event simulation models include: the U. S. Army Corps of Engineers (1981) HEC-1 flood hydrograph model; the Soil Conservation Service (1965) TR-20 computer program for project hydrology; the U. S. Environmental Protection Agency (1977) SWMM storm water management model; and the Illinois State Water Survey ILLUDAS model, by Terstriep and Stall (1974). Examples of continuous simulation models include: the U. S. National Weather Service runoff forecast system (Day, 1985); the U. S. Army Corps of Engineers (1976) STORM model; and the U. S. Army Corps of Engineers (1972) SSARR streamflow synthesis and reservoir regulation model. This is by no means a complete list of available models, but it covers most of the models commonly used in hydrologic practice. The HEC-1 model is probably the most widely used hydrologic event simulation model. The acronym HEC stands for Hydrologic Engineering Center, the U. S. Army Corps of Engineers research facility in Davis, California, where this model was developed.

HEC-1 Model

HEC-1 is designed to simulate the surface runoff resulting from precipitation by representing the basin as an interconnected system of components. Each component models an aspect of the rainfall-runoff process within a subbasin or subarea; components include subarea surface runoff, stream channels, and reservoirs. Each component is represented by a set of parameters that specifies the particular characteristics of the component and the mathematical relations describing its physical processes. The end result of the modeling process is the computation of direct runoff hydrographs for various subareas and streamflow hydrographs at desired locations in the watershed.

A subarea *land surface runoff component* is used to represent the movement of water over the land surface and into stream channels. The input to this component is a rainfall hyetograph. Excess rainfall is computed by subtracting infiltration and detention losses, based on an infiltration function that may be chosen from several options, including the SCS curve number loss rate as presented in Sec. 5.5. Rainfall and infiltration are assumed to be uniformly distributed over the subbasin. The resulting rainfall excesses are then applied to the unit hydrograph to derive the subarea outlet runoff hydrograph. Unit hydrograph options include the Snyder's unit hydrograph and the SCS dimensionless unit hydrograph presented in Chap. 7. Alternatively, a kinematic wave model can be used to find subbasin runoff hydrographs.

A *stream routing component* is used to represent flood wave movement in a channel. The input to this component is an upstream hydrograph resulting from individual or combined contributions of subarea runoff, streamflow routings, or diversions. This hydrograph is routed to a downstream point, using the characteristics of the channel. The techniques available to route the runoff hydrograph include the Muskingum method, level-pool routing, and the kinematic wave method.

A suitable combination of subarea runoff and streamflow routing components can be used to represent a rainfall-runoff and stream routing problem. The connectivity of the stream network components is implied by the order in which the input data components are arranged. Simulation must always begin at the uppermost subarea in a branch of the stream network, and proceed downstream until a confluence is reached. Before simulating below the confluence, all flows above it must be routed. The flows are combined at the confluence, and the combined flow is routed downstream.

Use of a *reservoir component* is similar to that of a streamflow routing component. A reservoir component represents the storage-outflow characteristics of a reservoir or flood-retarding structure. The reservoir component functions by receiving upstream inflows and routing them through a reservoir using storage routing methods. The reservoir outflow is solely a function of storage (or water surface elevation) in the reservoir and is not dependent on downstream controls. Spillway characteristics are entered along with top-of-dam characteristics for overtopping. A simplified dam-break option is also available.

Example 15.2.1. (Adapted from Ford, 1986.) A rainfall-runoff model using the HEC-1 computer program is to be developed for the Castro Valley Creek catchment, shown in Fig. 15.2.1 in order to analyze the effects of urbanization. The catchment

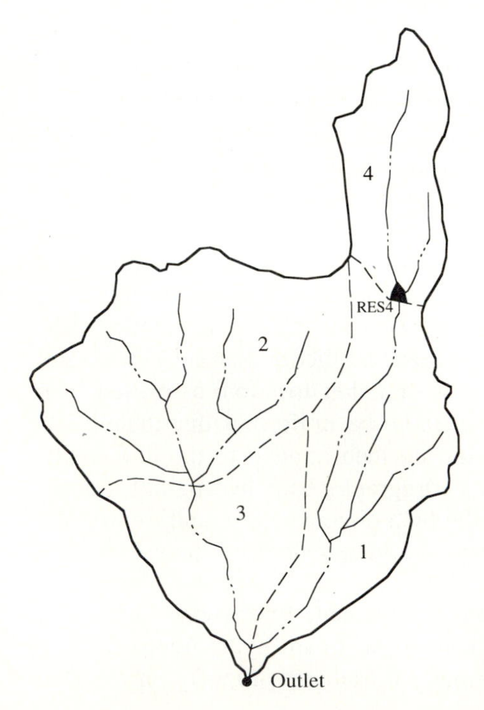

FIGURE 15.2.1
The Castro Valley watershed (Example 15.2.1). (*Source*: Ford, 1986.)

is divided into four subcatchments; a schematic diagram of the watershed is shown in Fig. 15.2.2. Subcatchment 4 is undergoing urbanization through development of a new residential area, and a detention reservoir in subcatchment 4 and downstream channel modifications are being investigated, the purpose of which is to reduce the effects of the additional flow resulting from the development. The objective of the problem is to calculate the runoff hydrograph at the catchment outlet for three different conditons: (1) the existing condition throughout the catchment, (2) the existing condition in subcatchments 1 to 3 with subcatchment 4 urbanized, and (3) the same as (2) but with a modified channel and a reservoir in subcatchment 4. Subarea runoff computations are performed using Snyder's synthetic unit hydrograph with rainfall loss rates determined using the SCS curve number method, channel routing is carried out by the Muskingum method, and routing through the reservoir by the level-pool method.

The following Table presents the existing characteristics of the subcatchments. The total watershed area is 5.51 mi².

Subcatchment	**Area** (mi²)	**Watershed length** L (mi)	**Length to centroid** L_{CA} (mi)	**SCS curve number** CN
1	1.52	2.65	1.40	70
2	2.17	1.85	0.68	84
3	0.96	1.13	0.60	80
4	0.86	1.49	0.79	70

The parameters for Snyder's synthetic unit hydrograph for the existing condition are $C_p = 0.25$ and $C_t = 0.38$. The flood wave travel time (Muskingum coefficient K) for the stream reach passing through subarea 3 is estimated as 0.3 h, and the travel time for subarea 1 is estimated as 0.6 h. The Muskingum X has been approximated as 0.2 for each of the two stream reaches.

The design rainfall is a hypothetical 100-year-return-period storm defined by the following depth-duration data.

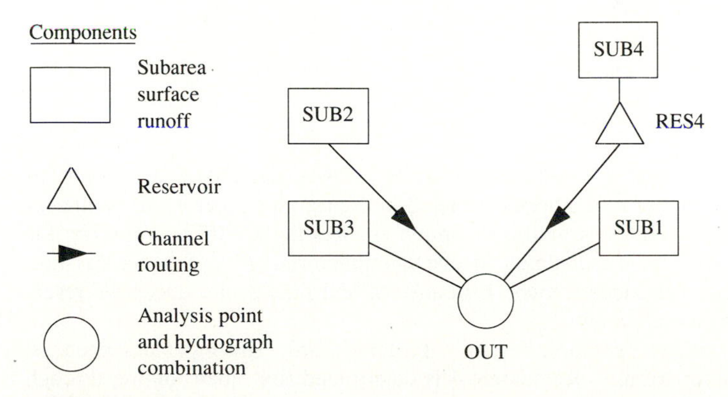

FIGURE 15.2.2
Schematic diagram of Castro Valley watershed showing components of HEC-1 analysis.

Duration	5 min	15 min	1 h	2 h	3 h	6 h	12 h	24 h
Rainfall (in)	0.38	0.74	1.30	1.70	2.10	3.00	5.00	7.00

A residential development in subcatchment 4 will increase the impervious area so that the developed SCS curve number will be 85. The unit hydrograph parameters are expected to change to $C_t = 0.19$ and $C_p = 0.5$. Modification of the channel through subcatchment 1 will change its Muskingum routing parameters to $K = 0.4$ h and $X = 0.3$. The detention reservoir to be constructed at the outlet of subcatchment 4 has the following characteristics:

Low-level outlet		Reservoir capacity (acre·ft)	Elevation (ft above MSL)
Diameter	5 ft	0	388.5
Cross-sectional area	19.63 ft^2	6	394.2
Orifice coefficient	0.71	12	398.2
Centerline elevation	391 ft (above MSL)	18	400.8
Overflow spillway (ogee type)		23	401.8
Length	30 ft	30	405.8
Weir coefficient	2.86		
Crest elevation	401.8 ft (above MSL)		

Solution. The parameters used for Snyder's unit hydrograph in HEC-1 are t_p and C_p; t_p is calculated for the existing condition using Eq. (7.7.2) with $C_1 = 1.0$, and C_t, L, and L_{CA} as given above. For example, for subcatchment 1,

$$t_p = C_t(LL_{CA})^{0.3} = 0.38(2.65 \times 1.40)^{0.3} = 0.56 \text{ h}$$

The results of this calculation for the four subcatchments are:

	Subcatchment				
	1	**2**	**3**	**4**	**4 urbanized**
C_t	0.38	0.38	0.38	0.38	0.19
t_p	0.56	0.41	0.34	0.40	0.20

The HEC-1 input for the Castro Valley Creek catchment is shown in Table 15.2.1. The data file has been annotated in the figure so that it can be understood better. Use of HEC-1's multiplan option enables the runoff hydrographs for all three conditions to be calculated in one computer run. Plan 1 is for existing conditions, plan 2 has subcatchment 4 urbanized, and plan 3 introduces the reservoir and channel modifications.

Each component operation begins with a KK card. The input has been set up so that the runoff from subcatchment 4 is determined first, then routing through the proposed detention reservoir is performed, followed by the Muskingum routing through subcatchment 1. Next, the rainfall-runoff computation is performed for

TABLE 15.2.1

HEC–1 input for the Castro Valley watershed (Example 15.2.1).

Row	Column: 1	8	16	24	32	40	48	56	64	72	80	
1	ID	CASTRO VALLEY CREEK CATCHMENT										[Identification cards]
2	ID	CONSIDER EXISTING CONDITIONS, DEVELOPED CONDITIONS AND										
3	ID	DEVELOPED CONDITIONS WITH IMPROVEMENTS										
4	IT	5	0	0	289							[Time step (min) and duration of computation]
5	JP	3										[Multiplan option with 3 plans]
	* PLAN1 = EXISTING CONDITIONS											
	* PLAN2 = URBANIZED CONDITIONS											
	* PLAN3 = URBANIZED CONDITIONS WITH IMPROVEMENTS											
6	KK	SUB4										[Component identification card]
7	KM	RUNOFF COMPUTATIONS FOR SUBCATCHMENT 4										[Comment card]
8	PH	1	5.51	0.38	0.74	1.30	1.70	2.10	3.00	5.00	7.00	[Precip]
9	BA	0.86										[Basin area in square miles]
10	LS	0	70									[SCS loss rate parameters CN = 70]
11	US	0.40	0.25									[Snyders unit hydrograph parameters, t_p and C_p]
12	KP	2										[Second plan, urbanized conditions]
13	LS	0	85									
14	US	0.20	0.5									
15	KP	3										[Third plan, also urbanized]
16	LS	0	85									
17	US	0.20	0.5									
18	KK	RES4										[Reservoir in subarea 4]
19	KM	ROUTE SUB4 THROUGH RESERVOIR										
20	KP	1										[Runoff not routed through reservoir in plans 1 and 2]
21	RN											
22	KP	2										
23	RN											

TABLE 15.2.1 *(cont.)*

HEC–1 input for the Castro Valley watershed (Example 15.2.1).

Column: 1	8	16	24	32	40	48	56 64 72 80
24 KP	3						[Routing through reservoir in plan 3]
25 RS	1	STOR	0				[Reservoir routing]
26 SV	0	6	12	18	23	30	[Volume in acre·ft]
27 SE	388.5	394.2	398.2	400.8	401.8	405.8	[Elevation in feet above MSL]
28 SL	391	19.63	0.71	0.5			[Pipe outlet characteristics]
29 SS	401.8	30	2.86	1.5			[Spillway characteristics]
30 KK	OUT						[Channel routing component]
31 KM	ROUTE SUBCATCHMENT 4 RUNOFF TO OUTLET						
32 KP	1						[Plans 1 and 2 are the same.]
33 RM	1	0.6	0.2				[Muskingum parameters $K = 0.6$ h, $X = 0.2$]
34 KP	3						
35 RM	1	0.4	0.3				[New Muskingum parameters for plan 3]
36 KK	SUB1						[Runoff from subarea 1]
37 KM	RUNOFF COMPUTATIONS FOR SUBCATCHMENT 1						
38 BA	1.52						
39 LS	0	70					
40 US	0.56	0.25					
41 KK	OUT						[Addition of two hydrographs]
42 KM	COMBINE SUBCATCHMENT 1 RUNOFF WITH SUBCATCHMENT RUNOFF ROUTED TO OUTLET						
43 HC	2						
44 KK	SUB2						[Runoff from subarea 2]
45 KM	RUNOFF COMPUTATIONS FOR SUBCATCHMENT 2						
46 BA	2.17						
47 LS	0	84					
48 US	0.41	0.25					

TABLE 15.2.1 *(cont.)*

HEC–1 input for the Castro Valley watershed (Example 15.2.1).

Column:	1	8	16	24	32	40	48	56	64	72	80
49	KK	OUT						[Channel routing to outlet by the Muskingum method]			
50	KM	ROUTE SUBCATCHMENT 2 RUNOFF TO OUTLET									
51	RM	1	0.3	0.2							
52	KK	OUT						[Addition of two hydrographs]			
53	KM	COMBINE HYDROGRAPHS AT OUTLET									
54	HC	2									
55	KK	SUB3									
56	KM	RUNOFF COMPUTATIONS FOR SUBCATCHMENT 3									
57	BA	0.96									
58	LS	0	80								
59	US	0.34	0.25								
60	KK	OUT						[Addition of two hydrographs]			
61	KM	COMBINE HYDROGRAPHS AT OUTLET									
62	HC	2									
63	ZZ							[Termination card]			

The comments in brackets [ ] are for interpretation of the figure only and are not part of the actual input data.

subcatchment 1 and the resulting runoff hydrograph added to the runoff hydrograph from subcatchment 4. Next, rainfall-runoff computations are performed for subcatchment 2, and this runoff is routed through subcatchment 3 and added to the outlet hydrograph. The final step is to perform the rainfall-runoff computations for subcatchment 3 and to add this result to the outlet hydrograph.

The resulting runoff hydrographs at the outlet of subcatchment 4 and at the outlet of the entire catchment for each of the three plans are shown in Fig. 15.2.3. The peak discharge from subcatchment 4 under existing conditions is 271 cfs, and under urbanized conditions, 909 cfs. The detention reservoir reduces the peak discharge to 482 cfs. The peak water surface elevation in the reservoir is 402.88 ft above mean sea level (MSL) at time 12.67 h. The peak discharges at the outlet are 1906 cfs for existing conditions, 2258 cfs for urbanized conditions, and 2105 cfs for urbanized conditions with the reservoir and channel modifications.

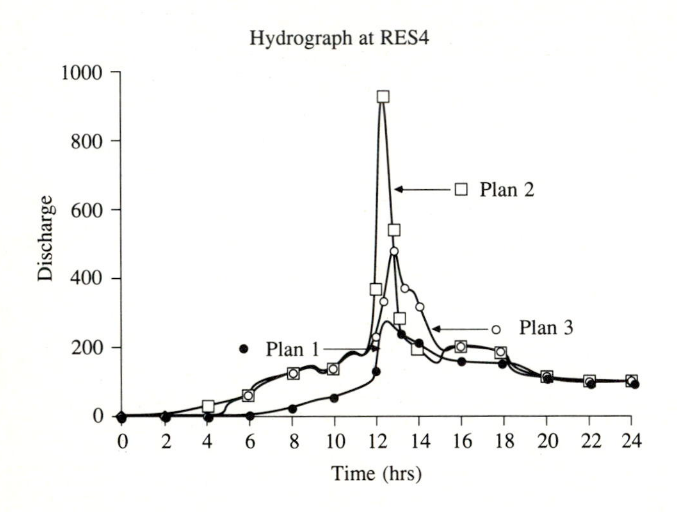

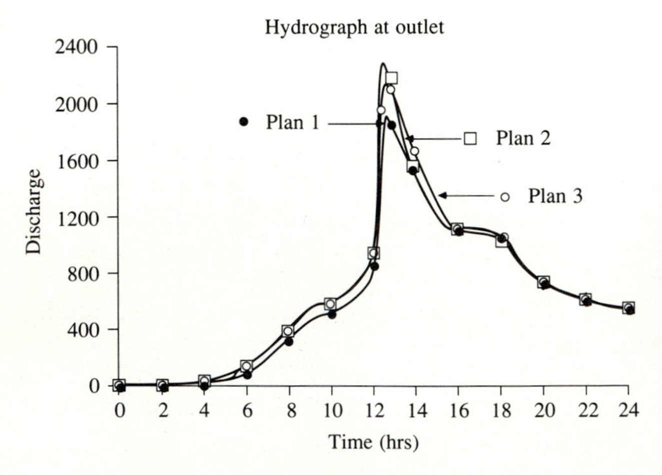

FIGURE 15.2.3
Discharge hydrographs at RES4 and at outlet (Example 15.2.1). Plan 1 is for existing conditions, Plan 2 has subcatchment 4 urbanized, and Plan 3 introduces a reservoir and channel modifications downstream of subcatchment 4.

Urban Storm Drainage Models

The first computerized models of urban storm drainage were developed during the late 1960s, and since that time a multitude of models have been discussed in the literature. The models applicable to design of storm sewer systems can be classified as *design models*, *flow prediction models*, and *planning models*.

Design models. These models determine the sizes and other geometric dimensions of storm sewers (and of other facilities) for a new system or an extension or improvement to an existing system. The design computations are usually carried out for a specified design return period.

Design models may be classified further into *hydraulic design models* and *least-cost optimal design models*. Hydraulic design models range from the simple rational method to much more sophisticated flow simulation models based upon solving the dynamic wave equations. One example of a hydraulic design model is ILLUDAS (Illinois Urban Drainage Area Simulator), developed by Terstriep and Stall (1974), which is popular both in the United States and abroad. This model is an extension of the British TRRL (Transportation and Road Research Laboratory) model (Watkins, 1962) to include both paved-area and grassed-area hydrographs. A flow chart for the ILLUDAS program is given in Fig. 15.2.4.

Least-cost optimal design models are intended for determining the lowest-cost storm sewer layout and pipe diameters that will convey storm drainage adequately. These models are based on optimization techniques such as linear programming, dynamic programming, nonlinear programming, heuristic techniques, or a combination of these. The flow simulation for the sewer network is considered a part of the optimization. One of the more comprehensive models of this type is a dynamic programming model called ILSD (Illinois Sewer Design) developed by Yen, et al. (1976).

Flow prediction models. These models simulate the flow of storm water in existing systems of known geometric sizes or in proposed systems with predetermined geometric sizes. Most flow prediction models simulate the flow for a single rainfall event, but some can simulate the response to a sequence of events. The simulation might be for historical, real-time, or synthetically-generated storm events. At least some simple hydraulics is considered in most models. A model may or may not include water quality simulation. The purpose of a flow simulation may be to check the adequacy and performance of an existing or proposed system for flood mitigation and water pollution control, to provide information for storm water management, or to form part of a real-time operational control system.

An emerging design philosophy is to use either traditional (rational method) or more advanced optimization methods for designing a storm sewer system, then checking the final design by detailed hydraulic simulation and cost analysis. An example of this approach is a British design and analysis procedure called the Wallingford Storm Sewer Package (WASSP; see Price, 1981).

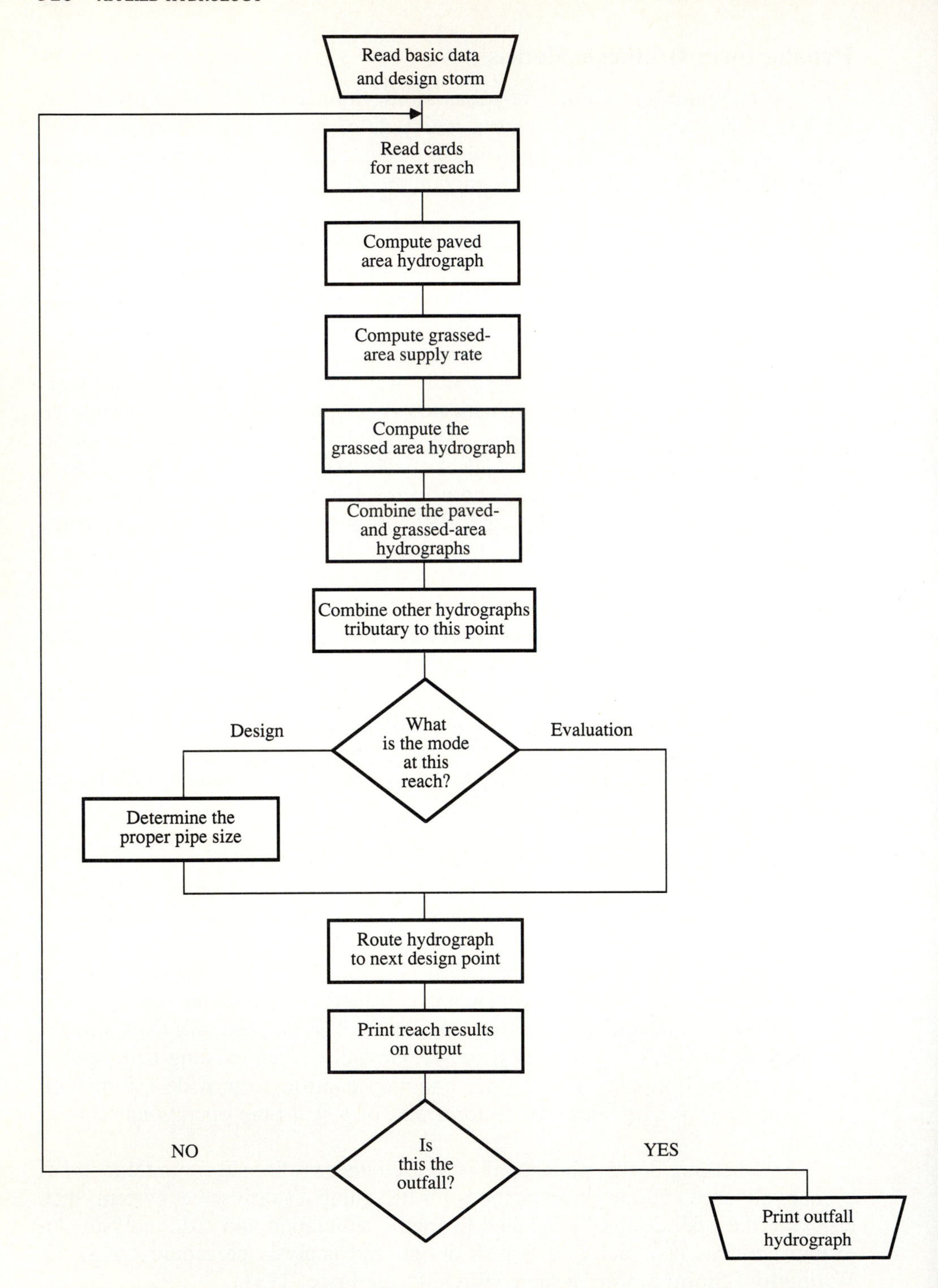

FIGURE 15.2.4
Flowchart for the ILLUDAS urban storm drainage model (*Source*: Terstriep and Stall, 1974).

Planning models. These models are used for broader planning studies of urban stormwater problems, usually for a relatively large space frame and over a relatively long period of time. The quantity and quality of storm water is treated in a gross manner, considering only the mass conservation of water and pollutants without considering the dynamics of their motion through the system. Planning models are employed for such tasks as studies of receiving water quality and treatment facilities. They do not require detailed geometric information on the drainage facilities as do the first two groups of models. Typical examples of planning models are: (1) STORM (Storage, Treatment, Overflow, Runoff Model), created by the U. S. Army Corps of Engineers (1976); (2) SWMM (Storm Water Management Model), developed by Metcalf and Eddy, Inc., the University of Florida, and Water Resources Engineers, Inc., (Metcalf and Eddy, 1971; U. S. Environmental Protection Agency, 1977); (3) RUNQUAL (Runoff Quality), which includes the hydraulic portion of the SWMM RUNOFF model and the stream water quality model QUAL-II (Roesner, Giguere, and Davis, 1977); (4) HSPF (Hydrocomp Simulation Program—Fortran) developed by Johnson, et al. (1980), which is a later version of the Stanford Watershed Model; and (5) MITCAT (MIT catchment model) by Harley, Perkins, and Eagleson (1970).

15.3 FLOOD PLAIN ANALYSIS

A flood plain is the normally dry land area adjoining rivers, streams, lakes, bays, or oceans that is inundated during flood events. The most common causes of flooding are the overflow of streams and rivers and abnormally high tides resulting from severe storms. The flood plain can include the full width of narrow stream valleys, or broad areas along streams in wide, flat valleys. As shown in Fig. 15.3.1, the channel and flood plain are both integral parts of the natural conveyance of a stream. The flood plain carries flow in excess of the channel capacity and the greater the discharge, the further the extent of flow over the flood plain.

The first step in any flood plain analysis is to collect data, including topographic maps, flood flow data if a gaging station is nearby, rainfall data if flood flow data are not available, and surveyed cross sections and channel roughness estimates at a number of points along the stream.

A determination of the flood discharge for the desired return period is required. If gaged flow records are available, a flood flow frequency analysis can be performed. If gaged data are not available, then a rainfall-runoff analysis must be performed to determine the flood discharge. The rainfall hyetograph is determined for the desired return period, a synthetic unit hydrograph is developed for each subarea of the drainage basin, and the direct runoff hydrograph from each subarea is calculated. The subarea direct runoff hydrographs are routed downstream and added to determine the total direct runoff hydrograph at the most downstream part of the drainage basin, as was illustrated in Example 15.2.1 for Castro Valley. The peak discharge of the most downstream hydrograph is used as the design flood discharge.

Once the flood discharge for the desired return period has been determined, the next step is to determine the profile of water surface elevation along the channel. This analysis can be carried out assuming steady, gradually-varied, nonuniform flow using a one-dimensional model such as HEC-2 (U. S. Army Corps of Engineers, 1982), or a two-dimensional model based upon either finite differences or finite elements (Lee and Bennett, 1981; Lee, et al., 1982; Mays and Taur, 1984). One-dimensional models allow the flow properties to vary along the channel only, while two-dimensional models account for changes across the channel as well. Alternatively, an unsteady flow analysis can be performed to identify the maximum water surface elevation at various cross sections during the propagation of the flood wave through a stream or river reach, using DAMBRK, DWOPER, or FLDWAV, as described in Chap. 10. Unsteady flow models are necessary for flood plain delineation in large lakes because the storage in the lake alters the shape and peak discharge of the flood hydrograph as it passes through.

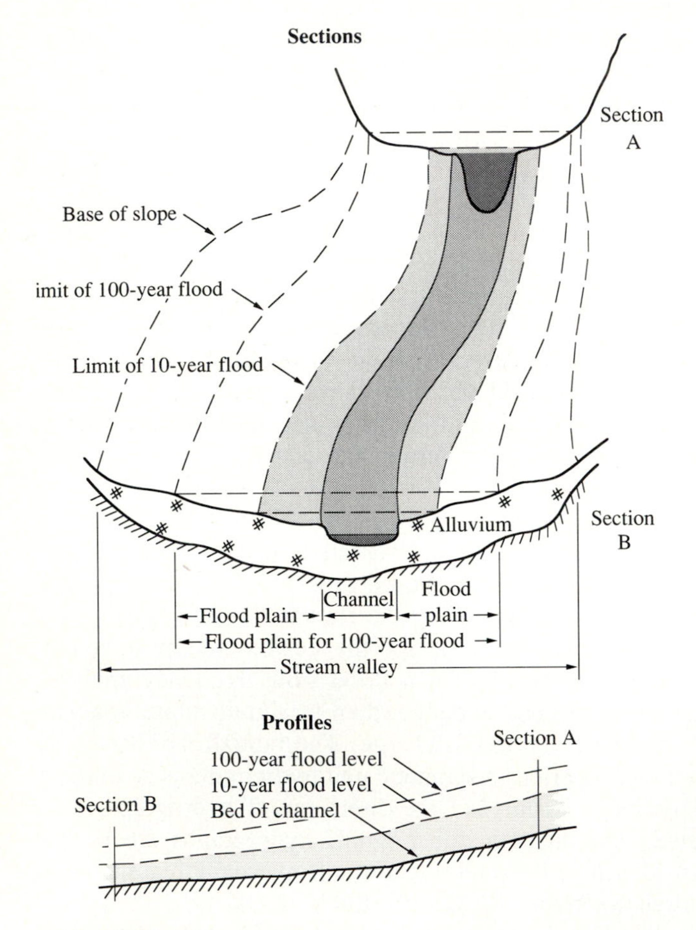

FIGURE 15.3.1
Typical sections and profiles in an unobstructed reach of stream valley. (*Source*: Waananen, et al., 1977. Used by permission.)

After the water surface elevations have been determined, the area covered by the flood plain is delineated. The lateral extent of the flood plain is determined by finding ground points on both sides of the stream that correspond to the flood profile (water surface) elevations. Ground elevations in the flood plain can be determined from topographic maps, street maps, or stereo aerial photos. Topographic maps are the most convenient, with the elevations given by contour lines. The flood plain boundary is determined by following the contour line that corresponds to the flood profile elevation for a particular area. Of course, the flood plain delineation is only as accurate as the topographic maps used. After flood levels have been determined for a particular reach of stream the actual location of the flood plain boundaries should be checked by field surveys.

In order to provide a standard national procedure, the 100-year flood has been adopted by the U. S. Federal Emergency Management Agency (FEMA) as the base flood for purposes of flood plain management measures. The 500-year flood is also employed to indicate additional areas of flood risk in the community. For each stream studied in detail, the boundaries of the 100- and 500-year floods are normally delineated using the flood elevations determined at each cross section. Between cross sections, the boundaries are interpolated using topographic maps at a scale of 1:24,000 with a contour interval of 10 feet or 20 feet. In cases where the 100- and 500-year flood boundaries are close together, only the 100-year boundary is shown.

Encroachment on flood plains, such as by artificial fill material, reduces the flood-carrying capacity, increases the flood heights of streams, and increases flood hazards in areas beyond the encroachment. One aspect of flood plain management involves balancing the economic gain from flood plain development against the resulting increase in flood hazard. For purposes of FEMA studies, the 100-year flood area is divided into a *floodway* and a *floodway fringe*, as shown in Fig. 15.3.2. The floodway is the channel of a stream plus any adjacent flood

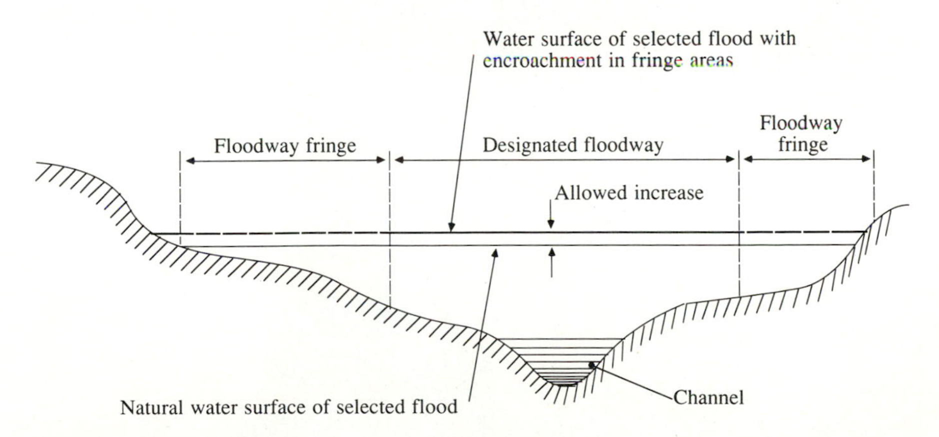

FIGURE 15.3.2
Definition of floodway and floodway fringe. The floodway fringe is the area between the designated floodway limit and the limit of the selected flood. The floodway limit is defined so that encroachment limited to the floodway fringe will not significantly increase flood elevation. The 100-year flood is commonly used and a one-foot allowable increase is standard in the United States.

plain areas that must be kept free of encroachment in order for the 100-year flood to be carried without substantial increases in flood heights. FEMA's minimum standards allow an increase in flood height of 1.0 foot, provided that hazardous velocities are not produced. The floodway fringe is the portion of the flood plain that could be completely obstructed without increasing the water surface elevation of the 100-year flood by more than 1.0 foot at any point.

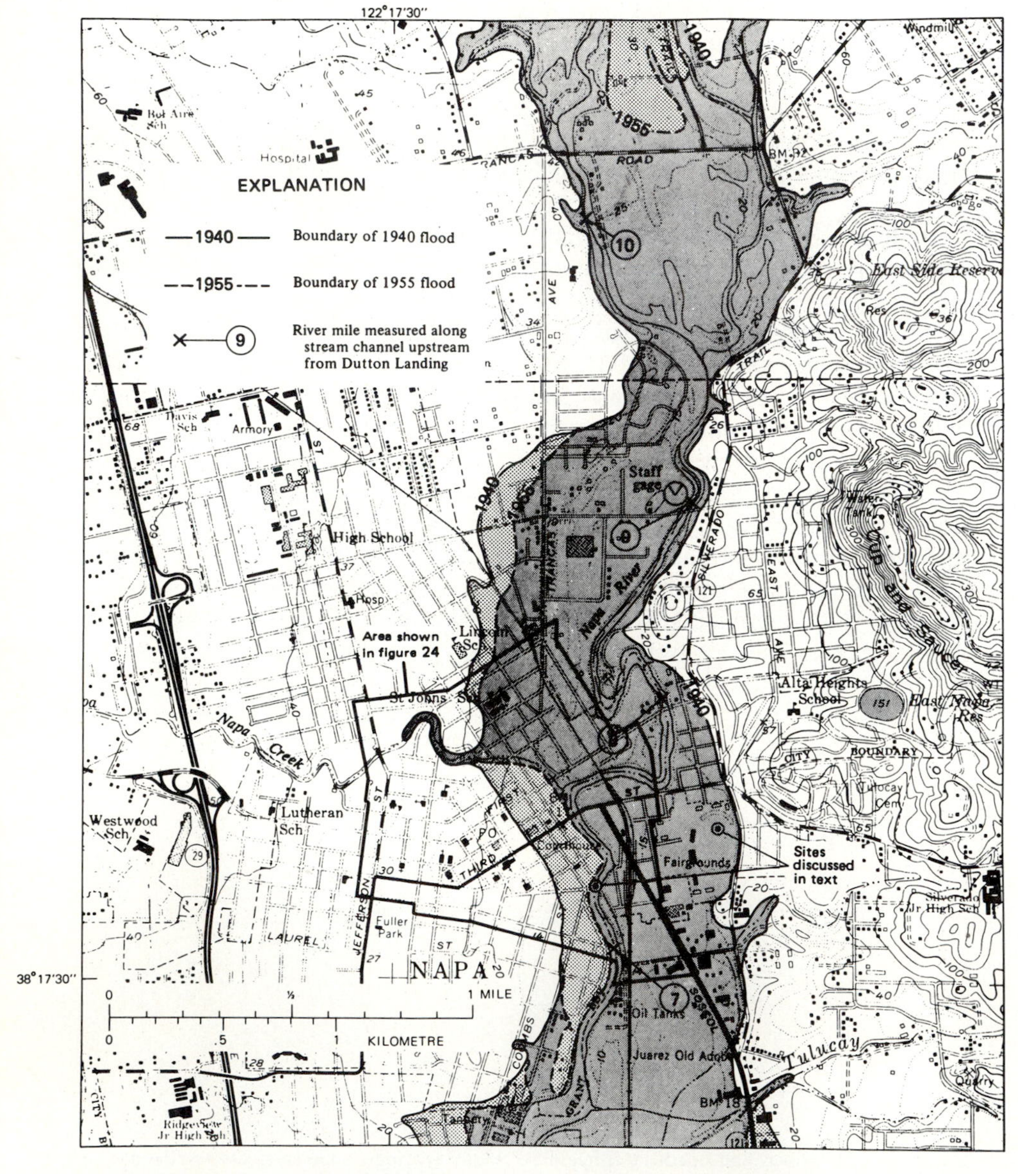

FIGURE 15.3.3
Flood hazard map for Napa, California. (*Source*: Waananen, et al., 1977 Used with permission.)

Two types of flood plain inundation maps, flood-prone area and flood hazard maps, have been used. Flood-prone area maps show areas likely to be flooded by virtue of their proximity to a river, stream, bay, ocean, or other watercourse as determined from readily available information. Flood hazard maps such as Fig. 15.3.3 for Napa, California, show the extent of inundation as determined from a thorough technical study of flooding at a given location. Flood hazard maps are commonly used in flood plain information reports and require updating when changes have occurred in the channels, on the flood plains, and in upstream areas. These changes include structural modifications and channel or flood plain modifications in upstream areas. Development of new buildings on the flood plain, obstructions, or other land use changes can affect the stream discharges, water surface elevation, and flow velocities, thereby changing the elevation profile defining the flood plain.

15.4 FLOOD CONTROL RESERVOIR DESIGN

Urbanization increases both the volume and the velocity of runoff, and efforts have been made in urban areas to offset these effects. Storm water detention basins provide one means of managing storm water. A storm water detention basin can range from as simple a structure as the backwater effect behind a highway or road culvert, up to a large reservoir with sophisticated control devices.

Detention is the holding of runoff for a short period of time before releasing it to the natural water course. The terms "detention" and "retention" are often misused; *retention* is the holding of water in a storage facility for a considerable length of time, for aesthetic, agricultural, consumptive, or other uses. The water might never be discharged to a natural watercourse, but instead be consumed by plants, evaporation, or infiltration into the ground. Detention facilities generally do not significantly reduce the total volume of surface runoff, but simply reduce peak flow rates by redistributing the flow hydrograph. However, there are exceptions: for example, the reduced surface runoff volume from land areas that have been contour-plowed, and the reduced surface runoff from detention basins on granular soils.

On-site detention of storm water is storage of runoff on or near the site where precipitation occurs. In some applications, the runoff may first be conducted short distances by collector sewers located on or adjacent to the site of the detention facility. On-site detention is distinguished from *downstream detention* by its proximity to the upper end of a basin and its use of small detention facilities as opposed to the larger dams normally associated with downstream detention.

The concept of detaining runoff and releasing it at a regulated rate is an important principle in storm water management. In areas having appreciable topographic relief, detention storage attenuates peak flow rates and the high kinetic energy of surface runoff. Such flow attenuation can reduce soil erosion and the amounts of contaminants of various kinds that are assimilated and transported by urban runoff from land, pavements, and other surfaces. Several different methods exist for the detention of storm water, including underground storage, storage in basins and ponds, parking lot storage, and rooftop detention.

There are several considerations involved in the design of storm water detention facilities. These are: (1) the selection of a design rainfall event, (2) the volume of storage needed, (3) the maximum permitted release rate, (4) pollution control requirements and opportunities, and (5) design of the outlet works for releasing the detained water. Flow simulation models such as the HEC-1 model can be used to perform reservoir routing to check the adequacy of detention basin designs.

The hypothetical ponded area in Fig. 15.4.1 serves as an example of a detention pond. Figure 15.4.2 provides a comparison of the outflow hydrographs from this detention pond with the corresponding inflow hydrographs for various flow volumes. In all cases, the detention pond reduces the flood peak discharge, but less so when the volume of runoff is large than when it is small.

Modified Rational Method

The *modified rational method* is an extension of the rational method for rainfalls lasting longer than the time of concentration. This method was developed so that the concepts of the rational method could be used to develop hydrographs for storage design, rather than just flood peak discharges for storm sewer design. The modified rational method can be used for the preliminary design of detention storage for watersheds of up to 20 or 30 acres.

The shape of the hydrograph produced by the modified rational method is a trapezoid, constructed by setting the duration of the rising and recession

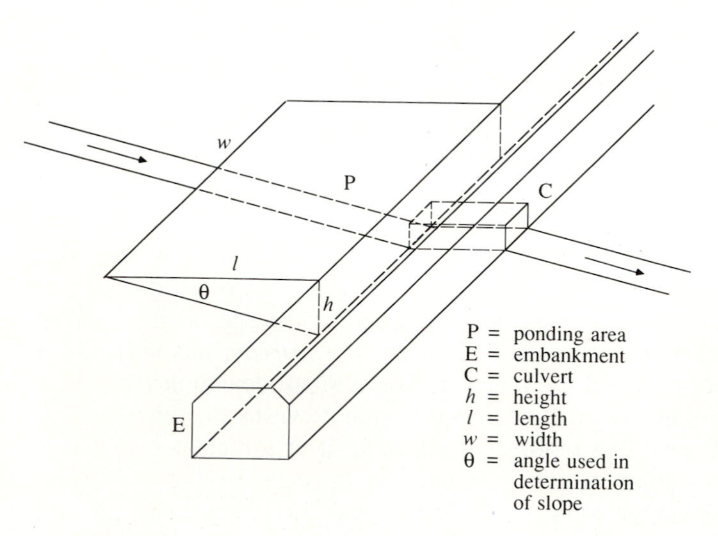

FIGURE 15.4.1
Schematic representation of wedge-shaped ponding area with box culvert outlet. (*Source*: Craig and Rankl, 1978. Used with permission.)

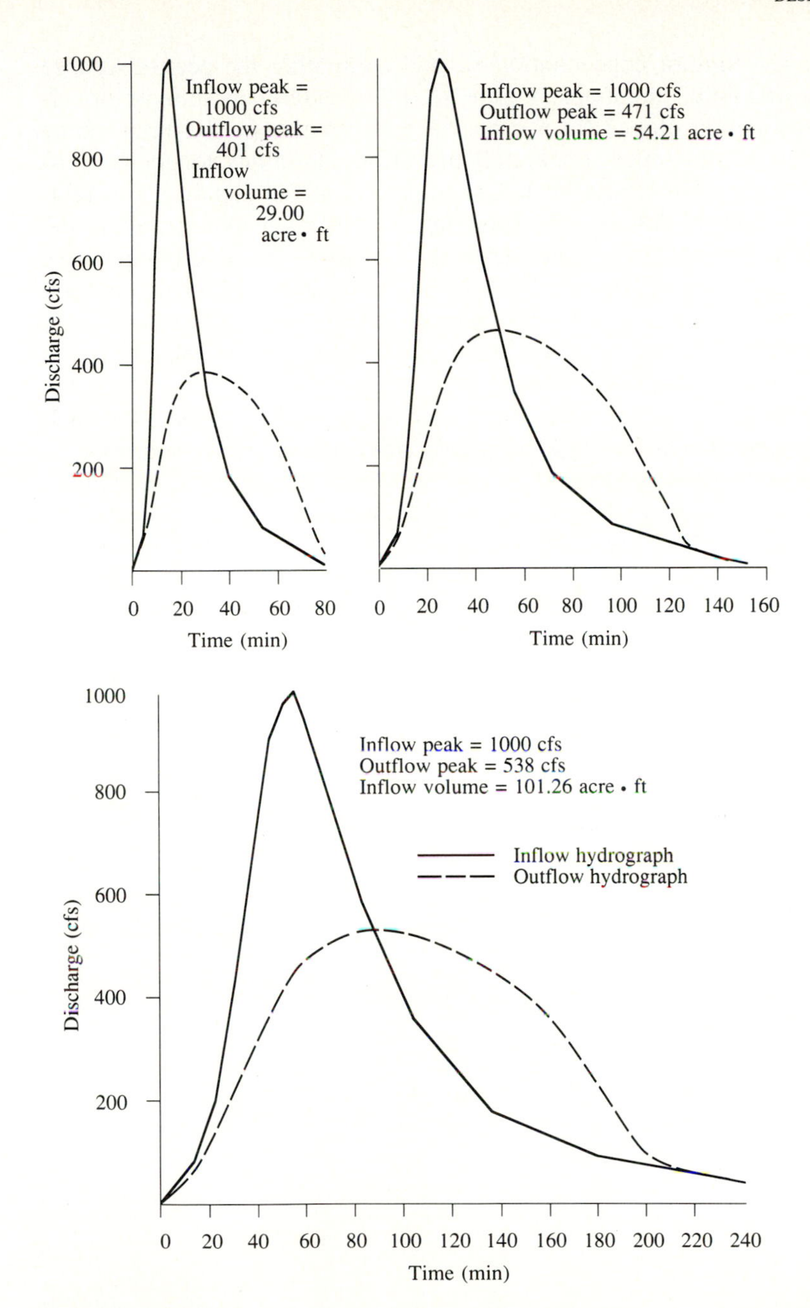

FIGURE 15.4.2
Comparison of inflow and outflow hydrographs for a detention basin. The inflow peaks are all 1000 cfs; however, the inflow volumes vary. The ponding area is a hypothetical wedge-shaped storage area (Fig. 15.4.1), and a 4 ft × 4 ft box culvert serves as the outlet. The pond width is 60 ft with a slope of 0.02 ft/ft. The flow with the largest volume results in the highest outflow rate from the pond. (*Source*: Craig and Rankl, 1978. Used with permission.)

limbs equal to the time of concentration t_c, and computing the peak discharge assuming various rainfall durations. Figure 15.4.3 illustrates modified-rational-method hydrographs developed for a drainage basin that has a 10-minute time of concentration and is subject to rainfall of various durations longer than 10 minutes. For example, consider the tallest trapezoid in the figure. Its rainfall duration is $T_d = 20$ min, and the corresponding rainfall intensity i is used in the rational formula (15.1.1) to compute the peak discharge. The hydrograph rises linearly to this discharge at the time of concentration (10 minutes), is constant until the rainfall ceases (20 minutes), then recedes linearly to zero discharge at 30 minutes. The hydrographs for longer rainfall durations have lower peak discharges because their rainfall intensities are lower.

If an allowable discharge out of a proposed detention basin is known, such as from a requirement that the peak discharge from the detention pond not be greater than the peak discharge from the area under predeveloped conditions, then the required detention storage for each rainfall duration can be approximated by determining the area of the trapezoidal hydrograph above the allowable discharge. By calculating the storage for hydrographs of rainfalls of various durations, the hydrologist can determine the critical duration for the design storm as the one requiring the greatest detention storage. This critical duration can also be determined analytically.

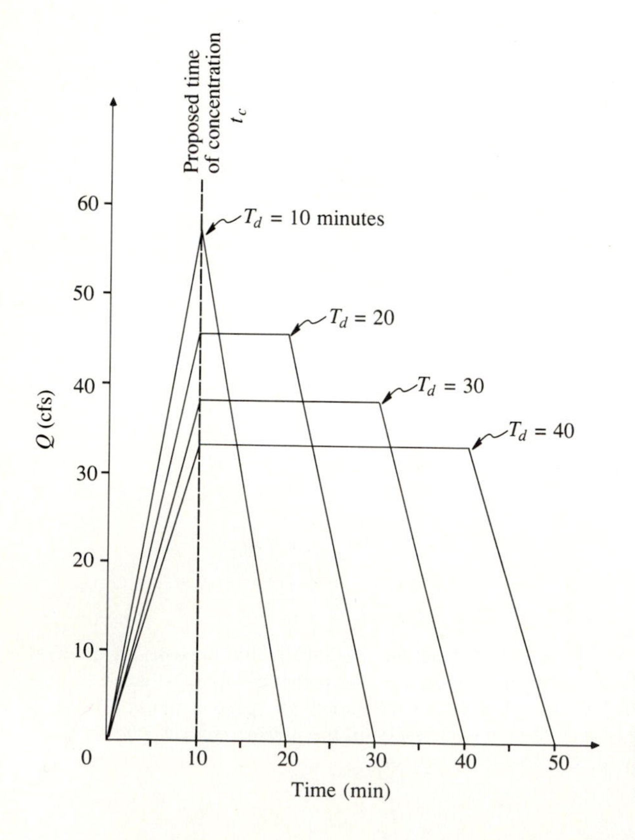

FIGURE 15.4.3
Typical storm water runoff hydrographs for the modified rational method with various rainfall durations.

Figure 15.4.4 is a representation of inflow and outflow hydrographs for a detention basin design. In this figure, α is the ratio of the peak discharge before development, Q_A (or peak discharge from the detention basin), and the peak discharge after development, Q_p:

$$\alpha = \frac{Q_A}{Q_p} \tag{15.4.1}$$

The ratio of the times to peak in the two hydrographs is γ. V_r is the volume of runoff after development. The volume of storage V_s needed in the basin is the accumulated volume of inflow minus outflow during the period when the inflow rate exceeds the outflow rate, shown shaded in the figure.

Using the geometry of the trapezoidal hydrographs, the ratio of the volume of storage to the volume of runoff, V_s/V_r, can be determined (Donahue, McCuen, and Bondelid, 1981):

$$\frac{V_s}{V_r} = 1 - \alpha\left[1 + \frac{T_p}{T_d}\left(1 - \frac{\gamma + \alpha}{2}\right)\right] \tag{15.4.2}$$

where T_d is the duration of the precipitation and T_p is the time to peak of the inflow hydrograph.

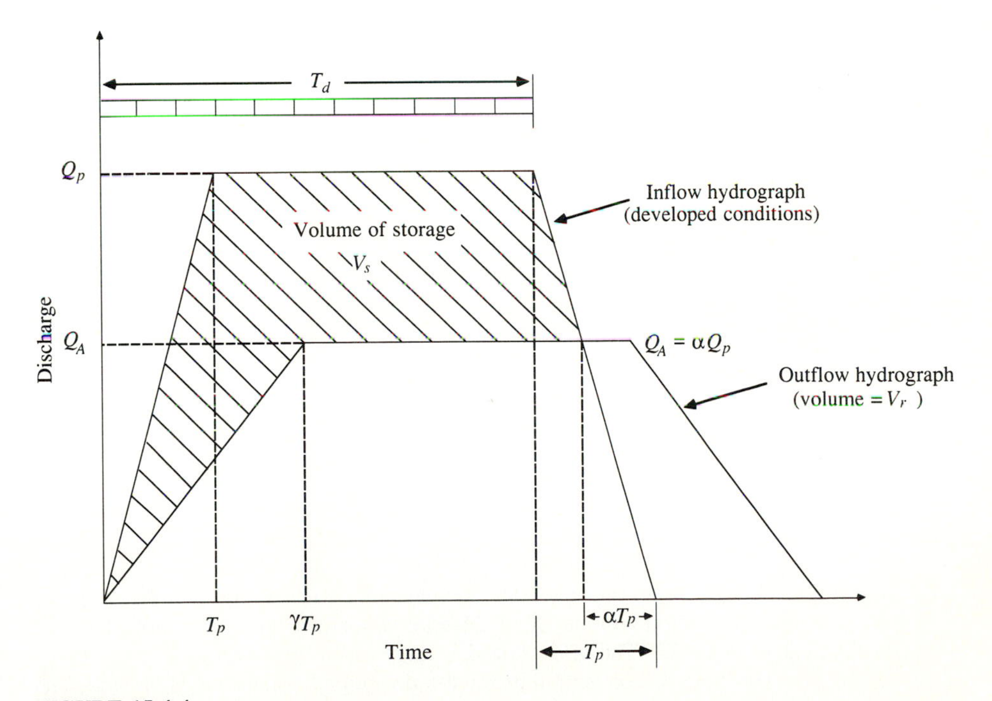

FIGURE 15.4.4
Inflow and outflow hydrographs for detention design. The outflow hydrograph is based on the inflow hydrograph for predeveloped conditions or on other more restrictive outflow criteria. (*Source*: Donahue, McCuen, and Bondelid, 1981. Used with permission.)

Consider a rainfall intensity-duration relationship of the form

$$i = \frac{a}{T_d + b} \tag{15.4.3}$$

where i is rainfall intensity and a and b are coefficients. The volume of runoff after development is equal to the volume under the inflow hydrograph:

$$V_r = Q_p T_d \tag{15.4.4}$$

The volume of storage is determined by substituting (15.4.4) into (15.4.2), and rearranging to get

$$V_s = Q_p T_d \left\{ 1 - \alpha \left[1 + \frac{T_p}{T_d} \left(1 - \frac{\gamma + \alpha}{2} \right) \right] \right\} \tag{15.4.5}$$

$$= T_d Q_p - Q_A T_d - Q_A T_p + \frac{\gamma Q_A T_p}{2} + \frac{Q_A^2 T_p}{2} \frac{1}{Q_p} \tag{15.4.6}$$

where α has been replaced by Q_A/Q_p.

The duration that results in the maximum detention is determined by substituting $Q_p = CiA = CAa/(T_d + b)$, then differentiating (15.4.6) with respect to T_d and setting the derivative equal to zero:

$$\frac{dV_s}{dT_d} = 0 = T_d \frac{dQ_p}{dT_d} + Q_p - Q_A + \frac{Q_A^2 T_p}{2} \left[\frac{d(1/Q_p)}{dT_d} \right]$$

$$= \frac{-T_d CAa}{(T_d + b)^2} + \frac{CAa}{T_d + b} - Q_A + \frac{Q_A^2 T_p}{2CAa}$$

$$= \frac{bCAa}{(T_d + b)^2} - Q_A + \frac{Q_A^2 T_p}{2CAa}$$

where it is assumed that Q_A, T_p, and γ are constants. Solving for T_d,

$$T_d = \left(\frac{bCAa}{Q_A - \dfrac{Q_A^2 T_p}{2CAa}} \right)^{1/2} - b \tag{15.4.7}$$

The time to peak T_p is set equal to the time of concentration.

Example 15.4.1. Determine the critical duration T_d (i.e., the one that requires the maximum detention storage) for a 25-acre watershed with a developed runoff coefficient $C = 0.825$. The allowable discharge is the predevelopment discharge of 18 cfs. The time of concentration for the developed conditions is 20 min, and for undeveloped conditions is 40 min. The applicable rainfall-intensity-duration relationship is

$$i = \frac{96.6}{T_d + 13.9}$$

Solution. The critical duration is found from Eq. (15.4.7):

$$T_d = \left(\frac{(13.9)(0.825)(25.0)(96.6)}{18 - \dfrac{(18)^2(20)}{2(0.825)(25)(96.6)}} \right)^{1/2} - 13.9$$

$$= 27.23 \text{ min}$$

Example 15.4.2. Determine the maximum detention storage for the watershed described in Example 15.4.1 if $\gamma = 40/20 = 2$.

Solution. The peak discharge for the duration of 27.23 min is

$$Q_p = CA\left(\frac{a}{T_d + b}\right)$$

$$= (0.825)(25)\left(\frac{96.6}{27.23 + 13.9}\right)$$

$$= 48.44 \text{ cfs}$$

By Eq. (15.4.6), then,

$$V_s = (27.23)(48.44) - (18)(27.23) - (18)(20) + (18)(20)\left(\frac{2}{2}\right) + \frac{(18)^2(20)}{2}\frac{1}{48.44}$$

$$= 895.77 \text{ cfs} \cdot \text{min} \times 60 \text{ s/min}$$

$$= 53,746. \text{ ft}^3$$

As a comparison, from (15.4.4), $V_r = Q_p T_d = 48.44 \times 27.23 = 1319$ cfs · min = 79, 140 ft^3, so $V_s/V_r = 53,746/79,140 = 0.68$. Hence the detention pond will store 68% of its inflow hydrograph in this example.

15.5 FLOOD FORECASTING

Flood forecasting is an expanding area of application of hydrologic techniques. The goal is to obtain real-time precipitation and stream flow data through a microwave, radio, or satellite communications network, insert the data into rainfall-runoff and stream flow routing programs, and forecast flood flow rates and water levels for periods of from a few hours to a few days ahead, depending on the size of the watershed. Flood forecasts are used to provide warnings for people to evacuate areas threatened by floods, and to help water management personnel operate flood control structures, such as gated spillways on reservoirs. The data collection systems used in flood forecasting are described in Chap. 6.

The components involved in a flood forecasting model for a large reservoir system can be illustrated by considering a model developed at the University of Texas at Austin for the Highland Lakes reservoir system on the lower Colorado River basin in central Texas (Fig. 15.5.1; see Unver, Mays, and Lansey, 1987). This system is characterized by integrated operation of several reservoirs for

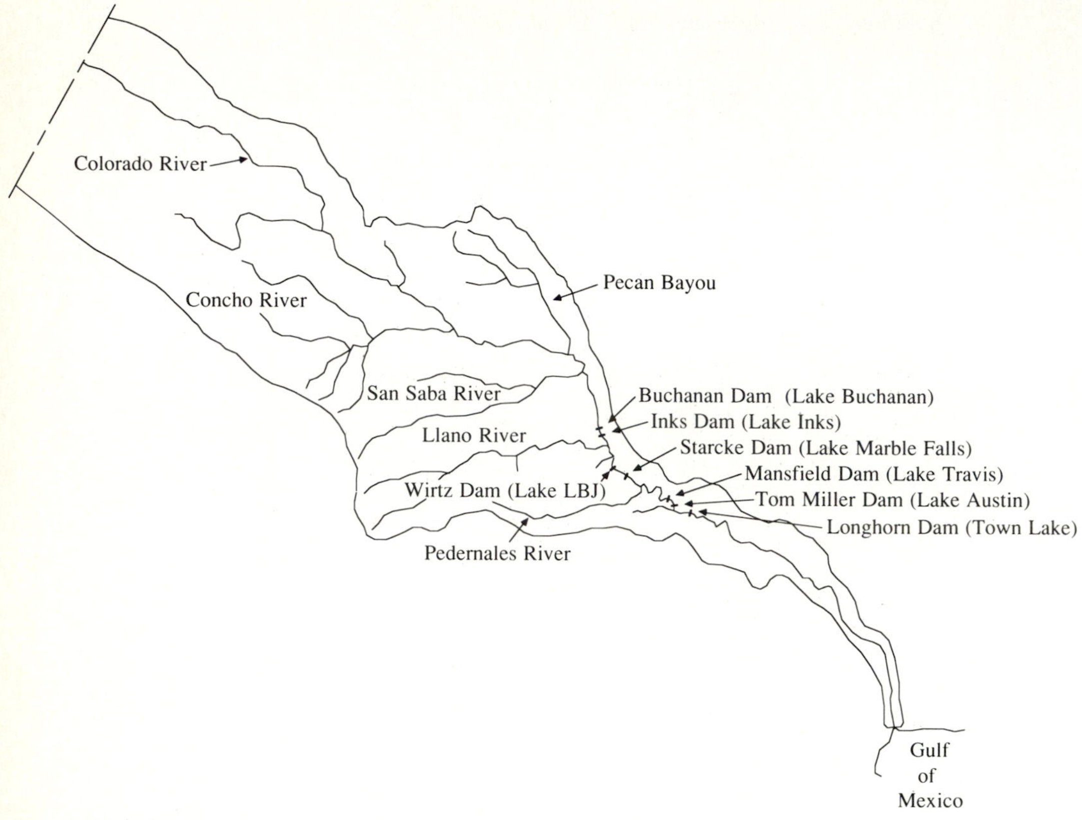

FIGURE 15.5.1
Lower Colorado River basin. (*Source*: Unver, Mays, and Lansey, 1987.)

multiple objectives. The portion of the river basin controlled by the Lower Colorado River Authority (LCRA) extends from the headwaters of Lake Buchanan downstream to the mouth of the Colorado River at the Gulf of Mexico. The Highland Lakes system consists of seven reservoirs that are serially connected.

Urban development in the flood plain of the Highland Lakes has restricted the range within which the reservoirs may be operated during floods. As an example, the original design of Lake Travis in the 1930s called for a release of 90,000 cfs during severe flooding conditions. Subsequent construction in the flood plain downstream of the lake has reduced the safe release (without flooding) to less than 30,000 cfs. Flood control operation of the Highland Lakes is also complicated because only two of the lakes, Buchanan and Travis, can store any significant flood volumes. The other lakes are held at constant level during normal operation.

The flood forecasting model for the Highland Lakes system can be used in a real-time framework to make decisions on reservoir operations during flooding.

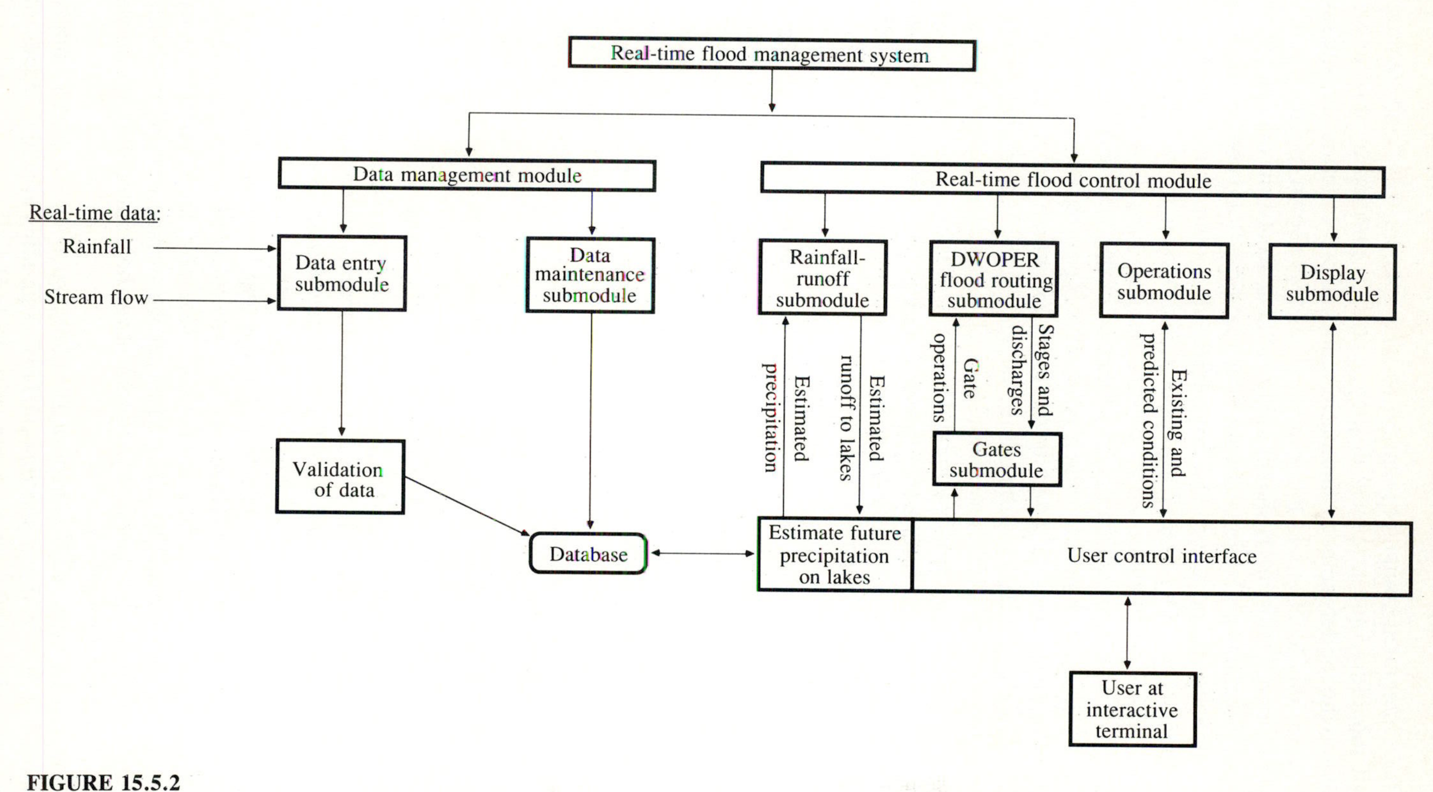

FIGURE 15.5.2
Structure of real-time flood management model. This model is used by the Lower Colorado River Authority to manage the river-lake system shown in Fig. 15.5.1. The model is run and resides on a Micro-vax computer. The real-time data collection system is described in Sec. 6.5. (*Source*: Unver, Mays, and Lansey, 1987.)

This model is an integrated computer program with components for flood routing, rainfall-runoff modeling, and graphical display, and is controlled by interactive software. Input to the model includes automated real-time precipitation and stream flow data from various locations in the watershed, as was shown in Chap. 6.

The overall model structure is shown in Fig. 15.5.2. Real-time data are input to the model from the data collection network. The real-time flood control module includes the following submodules: (1) a DWOPER submodule, that is, the U. S. National Weather Service Dynamic Wave Operational Model for unsteady flow routing; (2) a GATES submodule, which determines gate operation information for DWOPER, such as the gate discharge as a function of the head on the gate; (3) a RAINFALL-RUNOFF submodule which is an SCS-type rainfall-runoff model for the ungaged drainage area surrounding the lakes for which stream flow data is not available; (4) a DISPLAY submodule, which contains graphical display software; and (5) an OPERATIONS submodule which is the user-control software that interactively operates the other submodules and data files.

The input for this flood forecasting model includes both the real-time data and the physical description of system components that remain unchanged during a flood. The physical data include: (1) DWOPER data describing stream cross section information, roughness relationships, and so on; (2) characteristics of the reservoir spillway structures for GATES; and (3) drainage area description and hydrologic parameter estimates for RAINFALL-RUNOFF. The entire river-lake system contains 871 cross sections for DWOPER. It is divided into five subsystems because running the model for the whole system at once is usually not necessary, since floods tend to be localized within one or two of the subsystems. The real-time data include: (1) stream flow data at automated stations and headwater and tailwater elevations at each dam; (2) rainfall data at recording gages; (3) information as to which subsystem of lakes and reservoirs will be considered in the routing; and (4) reservoir operations.

15.6 DESIGN FOR WATER USE

The primary variables to be determined in a water-supply reservoir design are the location and height of the dam, the elevation and capacity of the spillway, and the capacity and mode of operation of the discharge works. Two hydrological variables are paramount: the *storage capacity* in the reservoir and the *firm yield* or mean annual rate of release of water through the dam that can be guaranteed from an analysis of historical data. There can be a great deal of uncertainty in the firm yield, especially for a short historical period. Naturally, the storage capacity and firm yield are interconnected because the larger the storage, the greater is the firm yield, with the limit that the firm yield can never be greater than the mean annual inflow rate to the reservoir.

A reservoir may be a *single-purpose* structure, such as for water supply or flood control, or it may be a *multiple-purpose* reservoir, in which zones of storage are identified for different purposes (Fig. 15.6.1).

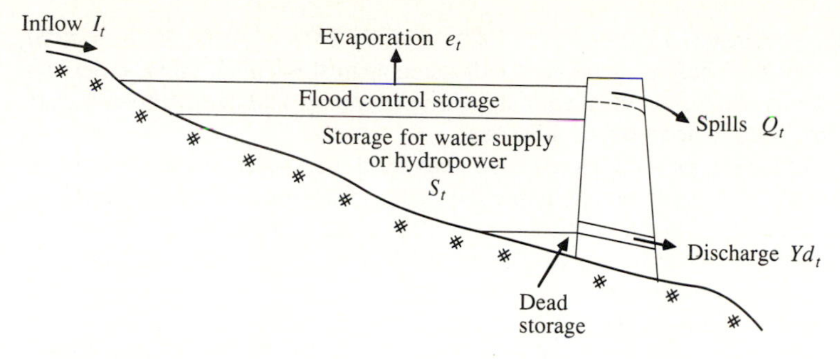

FIGURE 15.6.1
Zones of storage in a multipurpose reservoir.

Overview of Design Process

Hydrologic design of a reservoir for water use involves four steps:

1. Projection into the future of the water demand to be met by the reservoir;
2. Determination of the location and elevation of the dam, and calculation of its surface area–storage capacity curve for present and future conditions;
3. Computation of the firm yield of the reservoir for present and future conditions;
4. Comparison of the water demand and firm yield of the reservoir to determine its *service life,* or period of years during which the reservoir will be adequate to meet the demands.

This process is illustrated by the following examples, which involve the design of a new water-supply reservoir for the city of Winters, Texas (Henningson, Durham, and Richardson, 1979). The reservoir considered here was completed and went into service in 1981.

Projection of Demand.

Example 15.6.1. Project the water supply needed for the city of Winters, Texas, from 1980 to 2030, given the historical population and water use data in the table below.

Year	1910	1920	1930	1940	1950	1960	1970	1980
Population	1347	1509	2423	2335	2676	3266	2907	3061

Solution. A least-squares linear regression equation is fitted to the population data, taking the form

$$\hat{P}(t) = a_0 + a_1 t \qquad (15.6.1)$$

where $\hat{P}(t)$ is the estimated population in year t, and the coefficients are $a_0 = -48,170$ and $a_1 = 26.02\ \text{year}^{-1}$. For example, for 1980, $\hat{P}(t) = -48,170 + 26.02 \times 1980 = 3350$. Using (15.6.1), the population is projected into the future, yielding the following estimates: 1990 — 3610; 2000 — 3870; 2010 — 4130; 2020 — 4390; and 2030 — 4651. These projections are checked against the population projections of local and regional government entities, taking economic and demographic factors into account, and are accepted as adequate.

The population projections are then converted into water use projections. In 1978, a use rate of 175 gal per capita per day (gpcd) was observed; this is projected to increase to 200 gpcd in the year 2000 and to 225 gpcd in 2030. Hence, for a 1980 population of 3350, the water demand is $W = 3350 \times 175 = 0.586$ MGD (million gallons per day); to this demand is added an additional amount to meet the needs of the rural population around the city of Winters. The resulting total demand is 0.66 MGD in 1980, growing to 1.36 MGD in 2030 (Fig. 15.6.2). For reservoir water balance computations, it is more convenient to express the demands in acre·ft/year (1 acre·ft/year $= 8.92 \times 10^{-4}$ MGD); the demands in 1980 and 2030 are projected to be 740 and 1520 acre·ft/year, respectively. The water use forecasting method described here is very simple. More comprehensive methods of accomplishing this task are also available, such as the IWR-Main model (Dziegelewski, Boland, and Baumann, 1981) and statistical time series models (Maidment and Parzen, 1984).

Storage-area curve. Once the location and elevation of the dam are known, its *storage-area curve* can be determined. Associated with each water surface elevation h_j in the reservoir is a surface area A_j and a storage volume S_j. The relationship between S_j and A_j constitutes the storage-area curve.

To determine the storage-area curve, the surface area A_j is determined by measurement on a topographic map of the area enclosed within the contour line at elevation h_j (see columns 2 and 3 of Table 15.6.1). The horizontal slice of storage between elevations h_j and h_{j+1} has an average area of $(A_j + A_{j+1})/2$ and a thickness of $h_{j+1} - h_j$, so the storage at the upper level $j + 1$ is (see Fig. 15.6.3):

$$S_{j+1} = S_j + \frac{(h_{j+1} - h_j)(A_j + A_{j+1})}{2} \tag{15.6.2}$$

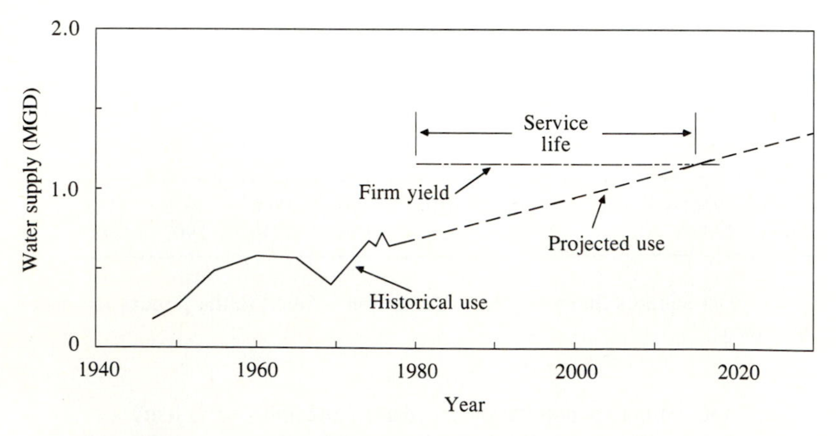

FIGURE 15.6.2
Comparison of projected water use and firm yield of water supply for Winters, Texas.

TABLE 15.6.1
Storage-area computation for reservoir

Column:	1 Level j	2 Elevation h_j (ft above MSL)	3 Surface area A_j (acres)	4 Storage S_j (acre·ft)
	1	1752	0	0
	2	1756	1	2
	3	1760	19	42
	4	1764	77	234
	5	1768	146	680
	6	1772	227	1426
	7	1776	262	2404
	8	1780	341	3610
	9	1784	430	5152
	10	1788	573	7158
	11	1790	643	8374
	12	1792	805	9822

Note: The last two increments use an elevation difference of 2 ft instead of 4 ft because elevation 1790 is the crest of the spillway.

where S_j is the storage at the lower level j of the slice. For example, in Table 15.6.1, $S_3 = 2 + (1760 - 1756)(1 + 19)/2 = 42$ acre·ft. Reservoir storage can also be computed by the *conic method* for which $(A_j + A_{j+1})/2$ in (15.6.2) is replaced by $(A_j + A_{j+1} + \sqrt{A_j A_{j+1}}/3)$ (U. S. Army Corps of Engineers, 1981).

The storage-area curve so computed corresponds to the topographic conditions when the reservoir is first constructed. After the reservoir has been in use for some years, its storage-area curve may be modified by sedimentation in the reservoir, which can reduce both the storage and the area for a given water surface elevation. Specific studies are needed at each site to determine the rate of sedimentation and how the deposited sediment is distributed within the reservoir.

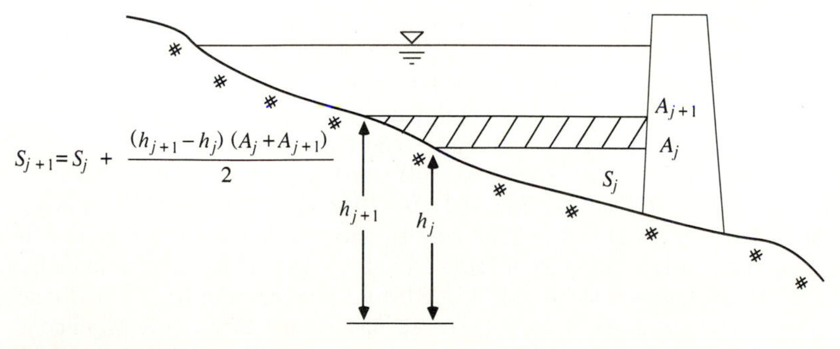

FIGURE 15.6.3
Calculation of storage volume of a reservoir.

Firm yield. The firm yield of a reservoir is the mean annual withdrawal rate that would lower the reservoir to its minimum allowable level just once during the critical drought of record. The critical drought is a period of several years duration that contains sustained low rainfall and stream flows and for which hydrological records of rainfall, stream flow, and evaporation exist at or near the reservoir site. The firm yield is determined by a simulation of the reservoir water balance using monthly time intervals t, $t = 1, 2, \ldots, T$.

The monthly data on reservoir inflow I_t and net evaporation e_t (evaporation minus precipitation on the reservoir water surface) are obtained from measurements at or near the reservoir site for a long period of record, hopefully including the critical drought. Records of water use at the site to be supplied (city, irrigation area, etc.) are analyzed to determine the ratio d_t of mean monthly water use to mean annual water use. The variable d_t, called the *demand factor,* represents the proportion of the annual firm yield needed in month t. Then, starting with a full reservoir, the reservoir water balance is calculated forward in time as

$$S_t = S_{t-1} + I_t - Yd_t - A_t e_t - Q_t \qquad t = 1, 2, \ldots, T \qquad (15.6.3)$$

where S_{t-1} and S_t are the storages at the beginning and end of month t, A_t is the surface area and Q_t the volume of spillway flow in month t, and Y is the withdrawal rate. The surface area A_t is calculated from the storage-area curve given S_{t-1} and S_t. The units of (15.6.3) are acre·ft or m^3.

If the allowable range of operation of storage is from S_{min} to S_{max}, the firm yield is that value of Y giving $S_t = S_{min}$ just once during the period of computations (with $S_t > S_{min}$ for all other months). Normally, spillway flow $Q_t = 0$, but during periods of high inflow it may occur that S_t as computed from Eq. (15.6.3) is greater than S_{max}; in this case, $Q_t = S_t - S_{max}$ and a new value of $S_t = S_{max}$ is used in the next computational step.

Example 15.6.2. Compute the reservoir water balance for the Winters Elm Creek Reservoir for 1940 hydrological conditions as given in Table 15.6.2, if the withdrawal rate Y is 1240 acre·ft/year.

Solution. The given data for monthly inflow and net evaporation for 1940 are given in columns 2 and 5, respectively, of Table 15.6.2. The monthly demand factors d_t are given in column 3. By multiplying these factors by the withdrawal rate $Y = 1240$ acre·ft/year, the monthly withdrawal rates are found, as given in column 4. The net evaporation loss (evaporation less precipitation) from the reservoir surface, in column 7, is the product of data in columns 5 and 6.

At $t = 1$ (January), initial storage is a full reservoir, $S_0 = S_{max} = 8374$ acre·ft; from Eq. (15.6.3), $S_1 = 8374 + 0 - 76 - 102 - 0 = 8196$ acre·ft; for $t = 2$ (February), $S_2 = 8196 + 191 - 68 - 51 - 0 = 8268$ acre·ft, and so on. The surface area A_t is computed by linear interpolation from Table 15.6.1, given average storage in month t; for $t = 1$, the average storage is $(8374 + 8196)/2 = 8285$ acre·ft, and by linear interpolation $A_1 = 638$ acres. It may be noted that A_t and S_t are interdependent, so, in practice, (15.6.3) is calculated iteratively for each month, making small adjustments to A_t until a final storage S_t is found that is consistent with (15.6.3) and Table 15.6.1. Spills occur in May and June to keep the storage from exceeding the maximum level for those months.

TABLE 15.6.2
Simulation of water balance in Winters Elm Creek Reservoir for a withdrawal rate of 1240 acre·ft/yr and hydrologic data from 1940

1	2	3	4	5	6	7	8	9
Month	Inflow	Demand factor	Withdrawal rate	Net evaporation	Surface area	evaporation loss	Spill	Storage
t ($t=1$ is Jan.)	I_t (acre·ft)	d_t	Yd_t (acre·ft)	e_t (ft)	A_t (acres)	$A_t e_t$ (acre·ft)	Q_t (acre·ft)	S_t (acre·ft)
								Initial = 8374
1	0	0.061	76	0.160	638	102	0	8196
2	191	0.055	68	0.080	635	51	0	8268
3	0	0.068	84	0.450	626	282	0	7902
4	706	0.075	93	0.490	625	306	0	8209
5	1334	0.089	111	0.540	643	347	711	8374
6	770	0.114	141	0.520	643	334	295	8374
7	5	0.137	170	0.940	621	584	0	7625
8	933	0.122	151	0.590	612	361	0	8046
9	135	0.086	107	0.790	611	483	0	7591
10	0	0.072	89	0.590	585	345	0	7157
11	143	0.059	73	0.110	573	63	0	7164
12	0	0.062	77	0.200	567	113	0	6974
Total	4217	1.000	1240			3371	1006	

TABLE 15.6.3

Simulation of water balance in Winters Elm Creek Reservoir for a withdrawal rate of 1240 acre·ft/yr and hydrologic data from Jan. to June, 1951

Month	Inflow	Demand factor	Withdrawal rate	Net evaporation	Surface area	Evaporation loss	Spill	Storage
t	I_t (acre·ft)	d_t	Yd_t (acre·ft)	e_t (ft)	A_t (acres)	A_te_t (acre·ft)	Q_t (acre·ft)	S_t (acre·ft)
								$S_{132}=459$
133	0	0.061	76	0.35	103	36	0	347
134	0	0.055	68	0.22	88	19	0	260
135	0	0.068	84	0.29	69	20	0	156
136	0	0.075	93	0.40	37	15	0	48 (minimum)
137	2411	0.089	111	0.36	198	71	0	2277
138	1256	0.114	141	0.42	286	120	0	3272

Note: $t = 1$ is Jan. 1940, $t = 133$ is Jan. 1951; initial storage for Jan. 1951 is $S_{132} = 459$ acre·ft.

Example 15.6.3. Determine the firm yield of the Winters Elm Creek Reservoir give the hydrologic data in Table 15.6.3.

Solution. The water balance simulation illustrated in the previous example for 1940 is carried out sequentially for hydrologic data from the years 1940 to 1969, which contain the critical drought of record for this region, in years 1950 to 1956. There are 30 years × 12 months/year = 360 months of simulation. The minimum storage S_{min} is 48 acre·ft for this reservoir, and is reached in April, 1951 (t = 136). The water balance simulation for the first six months of 1951 is shown in Table 15.6.3. It turns out that the value Y = 1240 acre·ft/year is the correct firm yield, so that the minimum storage is reached just once during the 360-month period of record. In practice, repeated simulations are made using the whole hydrological record with various trial values of Y until the maximum withdrawal rate satisfying the required condition on minimum storage is found.

Balance of supply and demand. The service life of the reservoir is that duration for which the firm yield (or reservoir supply rate) is greater than the expected demand. For the Winters Elm Creek example, the firm yield is Y = 1240 acre·ft/year = 1.106 MGD. As shown by the plot on Fig. 15.6.2, the condition of just-balanced supply and demand is expected to occur in approximately year 2014, or 34 years from the year of reservoir construction.

It may be noted that while the demands are projected forward in time (from 1980 to 2030), the firm yield is calculated on the basis of past hydrologic data (1940 to 1969 in the examples). The underlying assumption of this analysis is that past hydrologic conditions are typical of patterns that could be repeated in any future sequence of years. So, the firm yield can be thought of as a mean annual supply rate that could be withdrawn constantly, year after year, even in the face of future conditions equivalent to the critical drought of record. Of course, the firm yield is not absolutely guaranteed because a future drought may occur which is more severe than the critical drought of record. Studies of the thickness of annual tree rings indicate that in some regions droughts have occurred in previous centuries more severe than those recorded in this century, for which rainfall and flow data are available.

REFERENCES

American Society of Civil Engineers and Water Pollution Control Federation, Design and construction of sanitary and storm sewers, *ASCE Manual and Reports on Engineering Practice*, no. 37, New York, 1960.

Aron, G., and C. E. Egborge, A practical feasibility study of flood peak abatement in urban areas, report, U. S. Army Corps of Engineers, Sacramento District, Sacramento, Calif., March 1973.

Brandstetter, A., Assessment of mathematical models for urban storm and combined sewer management, *Environmental Protection Technology Series*, EPA-600/2-76-175a, Municipal Environmental Research Laboratory, USEPA, August 1976.

Chow, V. T., and B. C. Yen, Urban storm water runoff—determination of volumes and flow rates, *Environmental Protection Technology Series*, EPA-600/2-76-116, Municipal Environmental Research Laboratory, USEPA, May 1976; available from NTIS (PB 253 410), Springfield, Va.

Craig, G. S., and J. G. Rankl, Analysis of runoff from small drainage basins in Wyoming, U. S.

Geological Survey water-supply paper 2056, U. S. Government Printing Office, Washington, D. C., 1978.

Colyer, P. J., and R. W. Pethick, Storm drainage design methods: a literature review, report no. INT 154, Hydraulics Research Station, Wallingford, England, September, 1977.

Day, G. N., Extended streamflow forecasting using NWSRFS, *J. Water Res., Planning and Management Div., Am. Soc. Civ. Eng.*, vol. 111, no. 2, pp. 157–170, 1985.

Donahue, J. R., R. H. McCuen, and T. R. Bondelid, Comparison of detention basin planning and design models, *J. Water Res., Planning and Management Div., Am. Soc. Civ. Eng.*, vol. 107, no. WR2, pp. 385–400, October 1981.

Dziegielewski, B., J. J. Boland, and D. D. Baumann, An annotated bibliography on techniques of forecasting demand of water, report 81-CO3, Engineer Inst. for Wat. Res., U. S. Army Corps of Engineers, Fort Belvoir, Va., 1981.

Federal Aviation Administration, Department of Transportation, circular on airport drainage, report A/C 050-5320-5B, Washington, D. C., 1970.

Ford, D. T., Catchment runoff analysis with computer program HEC-1, in *Flood Plain Hydrology Course Notes*, ed. by L. W. Mays, Continuing Eng. Studies, College of Eng., University of Texas at Austin, 1986.

Harley, B. M., F. E. Perkins, and P. S. Eagleson, A modular distributed model of catchment dynamics, report no. 133, R. M. Parsons Lab for Water Resources and Hydrodynamics, MIT, Cambridge, Massachusetts, December 1970.

Heeps, D. P., and R. G. Mein, Independent comparison of three urban runoff models, *J. Hyd. Div., Am. Soc. Civ. Eng.*, vol. 100, no. HY7, pp. 995–1009, July 1974.

Henningson, Durham, and Richardson, Inc. of Texas, Preliminary Enginering Report, Winters Elm Creek dam and reservoir, Report submitted to the Farmers Home Administration, U.S. Dept. of Agriculture, July 1979.

Huber, W. C., J. P. Heaney, M. A. Medina, W. A. Peltz, H. Sheikhj, and G. F. Smith, Storm water management model user's manual, version II, *Environmental Protection Technology Series,* EPA-670/2-75-017, Municipal Environmental Research Laboratory, USEPA, March 1975.

Izzard, C. F., Hydraulics of runoff from developed surfaces, *Proc. Highway Research Board,* vol. 26, pp. 129–146, 1946.

Johnson, R. C., et al., User's manual for hydrologic simulation program FORTRAN, report, EPA-600/9-80-015, by Hydrocomp, Inc., to USEPA, 1980.

Kibler, D. F., Desk-top methods for urban stormwater calculation, chap. 4 in *Urban Stormwater Hydrology,* ed. by D. F. Kibler, water resources monograph 7, American Geophysical Union, Washington, D. C., 1982.

Kibler, D. F., J. R. Monser, and L. A. Roesner, San Francisco stormwater model users manual and program documentation, prepared for the City and County of San Francisco Department of Public Works, Water Resources Engineers, Walnut Creek, Calif., 1975.

Kirpich, Z. P., Time of concentration of small agricultural watersheds, *Civ. Eng.*, vol. 10, no. 6, p. 362, 1940.

Lee, J. K., D. C. Fruelich, J. J. Gilbert, and G. J. Wiche, Two-dimensional analysis of bridge backwater, proceedings of the SCE Conference, Applying Research to Hydraulic Practice, Jackson, Mississippi, pp. 247–258, August 1982.

Lee, J. K., and C. S. Bennett III, A finite-element model study of the impact of the proposed I-326 crossing on flood stages of the Congaree River near Columbia, South Carolina, U. S. Geological Survey open-file report 81–1194, NSTL Station, Bay St. Louis, Miss., 1981.

Linsley, R. K., Flood estimates: how good are they?, *Water Resour. Res.,* vol. 22, no. 9, supplement, pp. 159S–164S, 1986.

Maidment, D. R., and E. Parzen, Time patterns of water use in six Texas cities, *J. Water Res., Planning and Management Div., Am. Soc. Civ. Eng.*, vol. 110, no. 1, pp. 90–106, 1984.

Mays, L. W., and C. K. Taur, FESWMS-TX two-dimensional analysis of backwater at bridges: user's guide and applications—phase two, research report 314-2F, Center for Transportation Research, University of Texas at Austin, Texas, November 1984.

McPherson, M. B., Urban mathematical modeling and catchment research in the U.S.A., technical memorandum no. IHP-1, ASCE Urban Water Resources Research Program, ASCE, New York, June 1975.

Metcalf & Eddy, Inc., University of Florida, and Water Resource Engineers, Inc., Storm water management model, vol. I, Water Pollution Control Research Series 11024 DOC 10/71, USEPA, October 1971.

Morgali, J. R., and R. K. Linsley, Computer analysis of overland flow, *J. Hyd. Div., Am. Soc. Civ. Eng.,* vol. 91, no. HY3, pp. 81–100, May 1965.

Pilgrim, D. H., Bridging the gap between flood research and design practice, *Water Resources Res.,* vol. 22, no. 9, supplement, pp. 165S–176S, August 1986.

Price, R. K., Wallingford storm sewer design and analysis package, vol. I, *Proceedings,* Second International Conference on Urban Storm Drainage, Urbana, Illinois, pp. 213–220, June 14–19, 1981.

Roesner, L. A., P. R. Giguere, and L. C. Davis, *User's Manual for the Storm Runoff Quality Model, RUNQUAL,* prepared for Southeast Michigan Council of Governments, Detroit, Michigan, July 1977.

Roesner, L. A., Urban runoff process, in *Urban Stormwater Hydrology*, ed. by D. F. Kibler, water resources monograph 7, American Geophysical Union, Washington D.C., p. 138, 1982.

Soil Conservation Service, Computer program for project formulation hydrology, tech. release 20, Washington, D. C., May 1965.

Soil Conservation Service, Urban hydrology for small watersheds, tech. release 55, Washing- ton, D. C., 1975 (updated, 1986).

Terstriep, M. L., and J. B. Stall, The Illinois Urban Drainage Area Simulator, ILLUDAS, bulletin 58, Illinois State Water Survey, Urbana, Illinois, 1974.

U. S. Army Corps of Engineers, Streamflow synthesis and reservoir regulation, user's manual, Engineering Division, North Pacific, Portland, Ore., December 1972.

U. S. Army corps of Engineers, Hydrologic Engineering Center, Storage Treatment, Overflow, Runoff Model, STORM, user's manual, Davis, Calif., 1976.

U. S. Army Corps of Engineers, Hydrologic Engineering Center, HEC-1, flood hydrograph package, user's manual, 1981.

U. S. Army Corps of Engineers, Hydrologic Engineering Center, HEC-2, water surface profiles, user's manual, 1982.

U. S. Bureau of Reclamation, *Design of Small Dams,* 2nd ed., Washington, D. C., 1973.

U. S. Environmental Protection Agency, Stormwater management model, version II, report EPA-600/18-77-014, Washington, D. C., 1977.

Unver, O., L. W. Mays, and K. Lansey, Real-time flood management model for the Highland Lakes system, *J. Water Res., Planning and Management Div., Am. Soc. Civ. Eng.,* Vol. 13, No. 5, pp. 620–638, 1987.

Waananen, A. O., J. T. Limerinos, W. J. Kockelman, W. E. Spangle, and M. L. Blair, Flood-prone areas and land-use planning-selected examples from the San Francisco Bay region, U. S. Geological Survey Professional Paper 942, California 1977.

Watkins, L. H., The design of urban sewer systems, road research technical paper 55, Department of Scientific and Industrial Research, London, Her Majesty's Stationary Office, 1962.

Yen, B. C., Risk based design of storm sewers, report no. INT 141, Hydraulics Research Station, Wallingford, England, July 1975.

Yen, B. C., ed., *Storm Sewer System Design,* Department of Civil Engineering, University of Illinois at Urbana-Champaign, 1978.

Yen, B. C., H. G. Wenzel, Jr., L. W. Mays, and W. H. Tang, Advanced methodologies for design of storm sewer systems, research report 112, Water Resources Center, University of Illinois at Urbana-Champaign, August 1976.

PROBLEMS

15.1.1 Consider the Calder Alley watershed (Kibler, 1982) shown in Fig. 15.P.1. The urbanized watershed has an area of 227 acres of commercial and residential property. The physical characteristics of the various subareas are presented in Table 15.P.1. Determine the overland flow time (to inlets) for each of the subareas A, B_1, C_1, D_1, E, F, and G using the FAA method given in Table 15.1.2.

15.1.2 Determine the time of concentration for subarea F using Kirpich's method and the FAA method.

15.1.3 Determine the overland flow time for subareas A and F for the Calder Alley watershed in Prob. 15.1.1 using the following methods: Kirpich's and Federal Aviation Administration.

15.1.4 Determine the peak discharge from the Calder Alley watershed (Table 15.P.1) using the rational method assuming that there is no storm sewer system. The average slope for the drainage area is 0.0324 ft/ft and the distance from headwater to outlet is 9200 ft. Use Kirpich's method for time of concentration. Use the following 25-year intensity-duration-frequency relationship.

Duration (min)	5	10	20	30	40	50	60
Intensity (in/h)	6.4	5.4	4.0	3.2	2.7	2.35	2.06

15.1.5 Determine the pipe diameters for the storm sewer system for the Calder Alley watershed in Prob. 15.1.1 using the rational method and the intensity-duration-frequency relationship in Prob. 15.1.4. Use the FAA method to determine time of concentration. Manning's n is 0.014. Consider a minimum cover depth of 6.0 ft. The length, slope, and ground elevation are given below:

Pipe	6–5	5–4	4–3	3–2	2–1	
Length (ft)	1732	1400	1480	1440	906	
Inlet	6	5	4	3	2	1
Ground Elevation (ft)	1183	1157	1141	1118.5	1089	1060

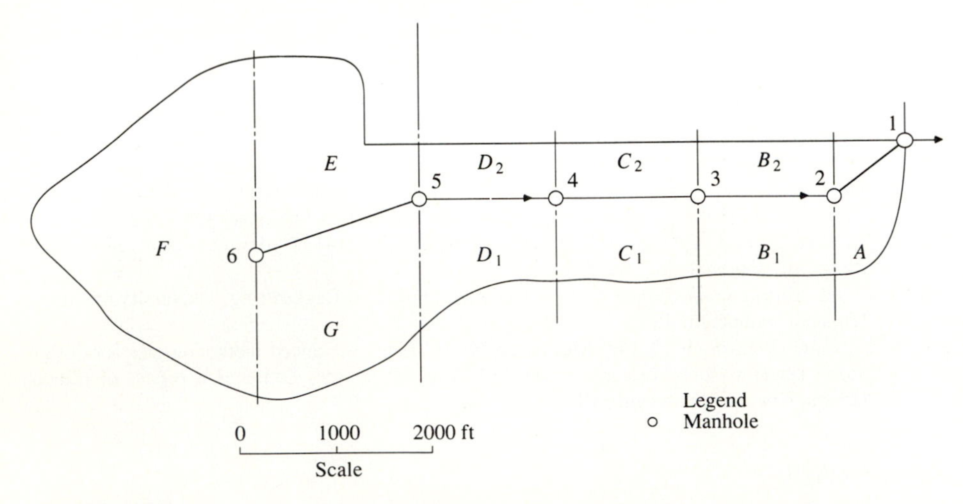

FIGURE 15.P.1
Schematic of Calder Alley storm sewer system. (*Source*: Kibler, 1982. Used with permission.)

TABLE 15.P.1
Data for Calder Alley Basin

Sub-area	Principal land use	Area (acres)	Impervious percent	Impervious acres	Average surface slope (%)	Runoff coefficient C	Inlet number	Distance to inlet (ft)
A	Detached multifamily residential	12.1	4	0.5	3.3	0.33	1	800
B_1	Small business	14.0	45	6.3	4.5	0.66	2	850
B_2	Multifamily residential	8.0	77	6.2	1.2	0.70	2	
C_1	Commercial	17.9	75	13.4	3.4	0.82	3	900
C_2	Commercial	6.1	100	6.1	1.5	0.95	3	
D_1	Commercial	19.0	95	18.1	3.4	0.90	4	1300
D_2	Commercial	4.9	100	4.9	1.5	0.95	4	
E	Residential and small business	47.6	40	19.0	3.5	0.60	5	1600
F	Residential and small business	52.6	20	10.5	3.0	0.42	6	1800
G	Residential and small business	45.0	32	14.4	3.5	0.52	5	1600
Total		227.2		99.4		0.59		

Source: Kibler, 1982.

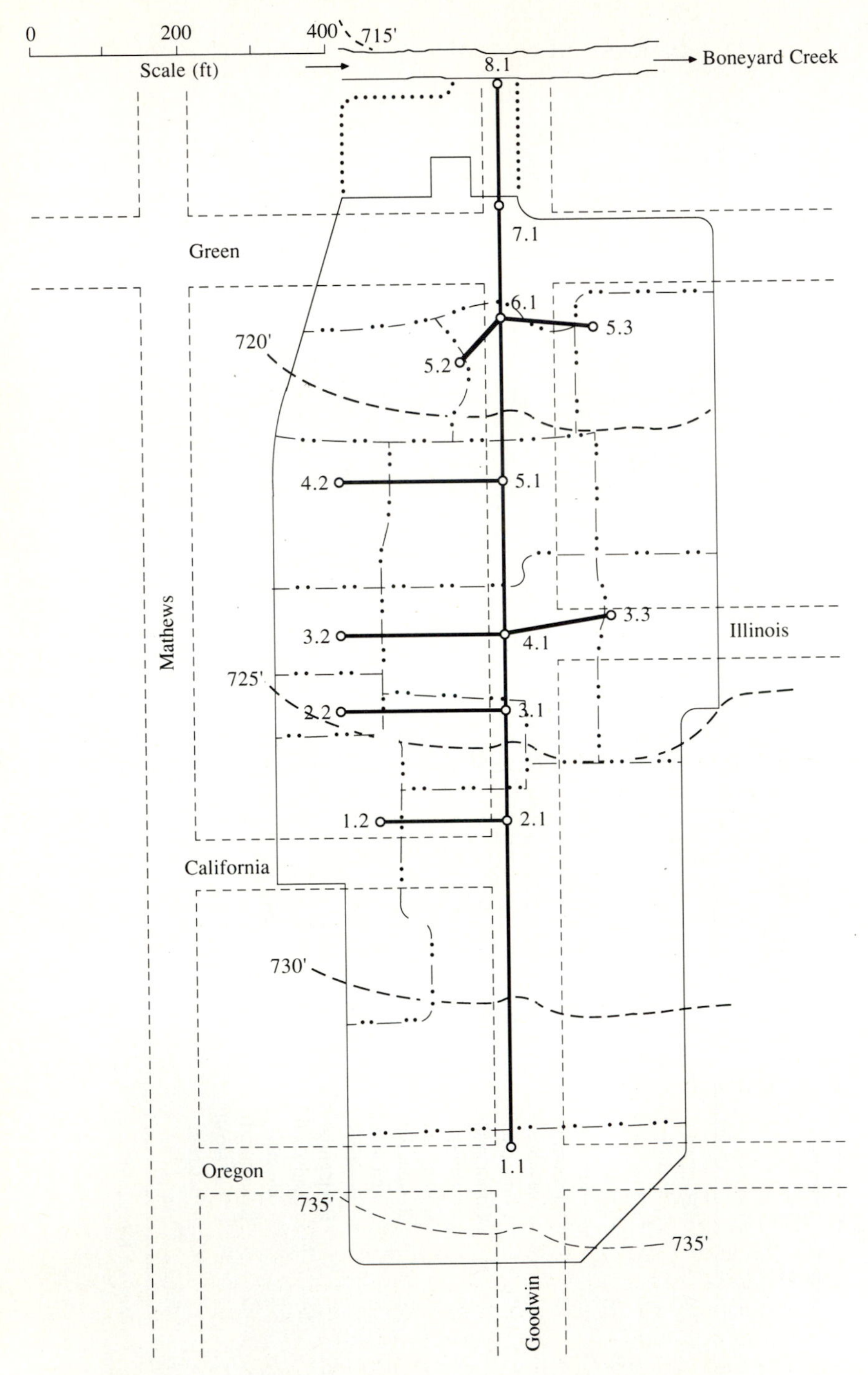

FIGURE 15.P.2
Goodwin Avenue drainage basin and storm sewer system in Urbana, Illinois.

TABLE 15.P.2
Characteristics of catchments of Goodwin Avenue drainage basin

Catchment	Ground elevation at manhole (ft)	Area A (acres)	Runoff coefficient C	Inlet time (min)	Length of outflow pipe from the manhole (ft)
1.1	731.08	2.20	0.65	11.0	390
1.2	725.48	1.20	0.80	9.2	183
2.1	724.27	3.90	0.70	13.7	177
2.2	723.10	0.45	0.80	5.2	200
3.1	722.48	0.70	0.70	8.7	156
3.2	723.45	0.60	0.85	5.9	210
3.3	721.89	1.70	0.65	11.8	130
4.1	720.86	2.00	0.75	9.5	181
4.2	720.64	0.65	0.85	6.2	200
5.1	720.12	1.25	0.70	10.3	230
5.2	721.23	0.70	0.65	11.8	70
5.3	720.26	1.70	0.55	17.6	130
6.1	719.48	0.60	0.75	9.0	160
7.1	715.39	2.30	0.70	12.0	240
8.1	715.10				

15.1.6 Determine the pipe diameters for the storm sewer system for the Goodwin Avenue drainage basin in Urbana, Illinois (Fig. 15.P.2). The catchment characteristics are listed in Table 15.P.2. The rainfall-intensity-duration relationship for a 2-year return period is to be used, as follows:

Duration (min)	5	10	15	20	25
Rainfall intensity (in/h)	5.40	4.18	3.52	3.10	2.76
Duration (min)	30	40	50	60	
Rainfall intensity (in/h)	2.50	2.10	1.76	1.50	

15.1.7 Design the storm sewer system shown in Fig. 15.P.3. The intensity-duration curve is described by

$$i = \frac{120T^{0.175}}{T_d + 27}$$

where i is the intensity in in/h, T is the return period in years, and T_d is the duration in minutes. A return period of 10 years is to be used, and commercial pipe sizes are to be selected. The pipes are concrete, with a Manning's roughness of $n = 0.013$. The constraints on the design are: (1) minimum cover depth of 5 ft; (2) minimum permissible flow velocity of 2 fps; (3) maximum permissible flow velocity of 8 ft/s; and (4) at junctions the downstream sewer cannot be smaller than the upstream sewers. Your design should specify the pipe diameters and their invert elevations at each manhole. Relevant data are as follows:

Pipe	Length (ft)	Ground elevation (ft above MSL) Upstream	Downstream
AC	500	504	499
BC	600	500	499
CD	300	499	485
DE	500	485	481

Catchment	Area (acres)	Runoff coefficient C	Inlet time (min)
I	3.0	0.7	10
II	2.2	0.7	12
III	4.2	0.6	14
IV	2.2	0.5	8
V	4.5	0.6	14

15.1.8 Determine an expression for the coefficient of variation of the discharge computed using the rational formula, $Q = CiA$, as a function of the coefficients of variation of C, i, and A.

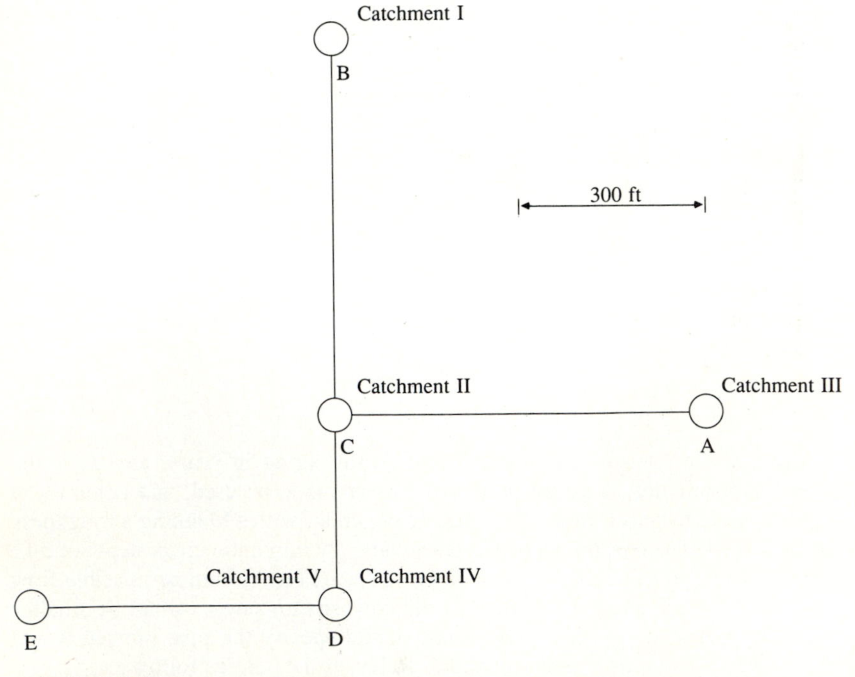

FIGURE 15.P.3
Storm sewer system for Prob. 15.1.7.

15.1.9 Using your results from Prob. 15.1.8, compute the coefficient of variation of the discharge for the rational formula if $CV_C = 0.071$, $CV_i = 0.177$, and $CV_A = 0.05$.

15.1.10 Determine an expression for the coefficient of variation of the discharge computed by Manning's equation (15.1.5) using first-order analysis and assuming full-pipe flow. Consider diameter D, slope S_0, and roughness n to be variables.

15.1.11 Using your results from Prob. 15.1.10, determine the coefficient of variation of the discharge computed using Manning's equation if $CV_n=0.0553$, $CV_D=0.01$, and $CV_{S_0} = 0.068$.

15.1.12 Determine the composite risk of the loading exceeding the capacity of a storm sewer pipe for which the loading is determined using the rational formula and the capacity determined using Manning's equation (15.1.5) for full-pipe flow. Assume that both the loading and capacity are normally distributed. The following data apply:

Parameter	Mean	Coefficient of variation
C	0.825	0.071
i	3.4 in/h	0.177
A	10 acres	0.05
n	0.015	0.0553
d	5 ft	0.010
S_0	0.001	0.068

15.1.13 Perform a first-order analysis of uncertainty for the Darcy-Weisbach equation for full-pipe flow using the following equation:

$$Q = \frac{\pi}{4}\left(\frac{2gS_0}{f}\right)^{1/2} D^{5/2}$$

Consider S_0, f, and D to be variables.

15.1.14 Determine the composite risk of the loading exceeding the capacity of a storm sewer, for a loading condition determined using the rational formula and a capacity determined using the Darcy-Weisbach equation. The following data apply:

Parameter	Mean	Coefficient of variation
C	0.498	0.07
i	48.28 mm/h	0.25
A	87734 m^2	0.05
D	914 mm	0.02
f	0.0297	0.25
S_0	0.005	0.40

15.1.15 A weighted runoff coefficient C for the rational formula can be expressed as $C = \Sigma\ C_j\alpha_j$, in which $\alpha_j = A_j/A$, where A is the total area of the drainage basin and A_j is the subarea having runoff coefficient C_j. Determine the expression for the coefficient of variation of the weighted C in terms of the means and coefficients of variation of the α_j's and C_j's.

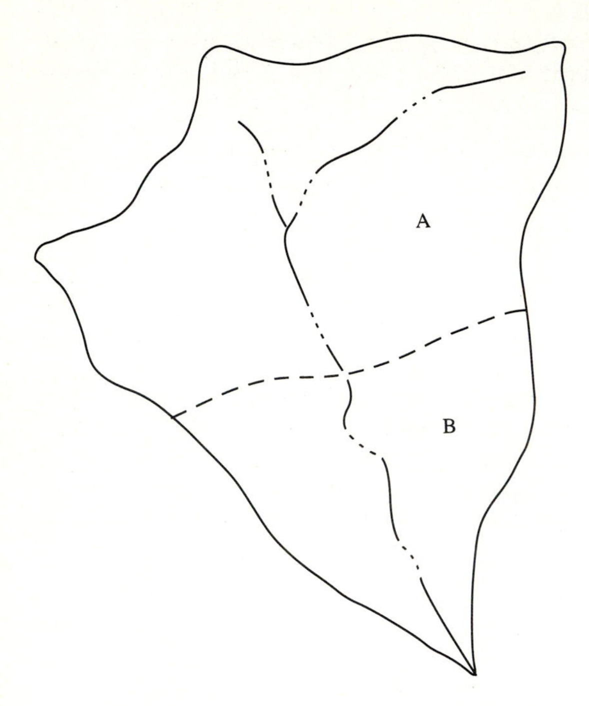

FIGURE 15.P.4
Watershed for Probs. 15.2.1–15.2.3.

15.1.16 Determine the mean weighted runoff coefficient (see Prob. 15.1.15) and its coefficient of variation for the Derby catchment in central England (Yen, 1975). The catchment has a total area of 87,734 m^2 (or 21.68 acres) with the subareas and runoff coefficients given below.

Subarea type	$\alpha_j = \frac{A_j}{A}$	C_j	CV_{α_j}	CV_{C_j}
Roof	0.145	0.85	0.10	0.048
Paths	0.103	0.80	0.15	0.255
Roads (asphalt)	0.179	0.875	0.10	0.0618
Permeable clay	0.573	0.15	0.20	0.0544

15.2.1 Use the U. S. Army Corps of Engineers program HEC-1 to determine the runoff hydrograph for the watershed shown in Fig. 15.P.4. The watershed is divided into two subcatchments, A and B, with the following characteristics:

Subcatchment	Area (mi²)	Watershed length L (mi)	Length to centroid L_{CA} (mi)	SCS curve number CN
A	2.17	1.85	0.68	70
B	0.96	1.13	0.60	75

Use Snyder's synthetic unit hydrograph with $C_p = 0.25$ and $C_t = 0.38$ for both subcatchments. Muskingum routing is to be used for flows through subcatchment B with $K = 0.3$h and $X = 0.25$. Consider a 15-minute unit hydrograph, and use the 12-hour storm defined by the following depth-duration data.

Duration	15 min	1 h	2 h	3 h	6 h	12 h
Rainfall (in)	0.74	1.30	1.70	2.10	3.00	5.00

15.2.2 Solve Prob. 15.2.1 for urbanized conditions, with the area developed so that the curve numbers are 85 for subcatchment A and 90 for subcatchment B. The Muskingum K for the channel through subcatchment B is now 0.2 h. The Snyder parameters C_p and C_t change to 0.35 and 0.30, respectively. Use the U. S. Army Corps of Engineers HEC-1 program to compute the runoff hydrograph for urbanized conditions using the hypothetical storm given in Prob. 15.2.1.

15.2.3 Using the urbanized conditions in Prob. 15.2.2 for the watershed in Prob. 15.2.1, determine the runoff hydrograph if the following detention reservoir is placed at the outlet of subcatchment B. Use the U. S. Army Corps of Engineers HEC-1 computer program to perform the computations.

Low-level outlet	
Diameter	5 ft
Orifice coefficient	0.71
Centerline elevation	391 ft above MSL
Overflow spillway (ogee type)	
Length	30 ft
Weir coefficient	2.86
Crest elevation	400 ft above MSL

Reservoir capacity (acre·ft)	**Elevation (ft above MSL)**
0	388.5
6	394.2
12	398.2
18	400.8
23	401.8
30	405.8

15.2.4 Determine the 100-year discharge from Waller Creek at its confluence with the Colorado River in Austin, Texas (see Fig. 15.P.5), using the U. S. Army Corps of Engineers HEC-1 computer program. Use the 100-year 3-hour design rainfall pattern for Austin given in Prob. 5.5.1. The subcatchment data are presented in Table 15.P.3. The SCS infiltration model is to be used to describe losses with the appropriate curve numbers listed in Table 15.P.3. Use Muskingum routing through the subareas with a weighting factor X of 0.250 and the K values in the table. Use Snyder synthetic 10-minute unit hydrographs for each subarea. Perform computations over a 20-hour time frame. What are the total loss and the total excess volume for each subarea?

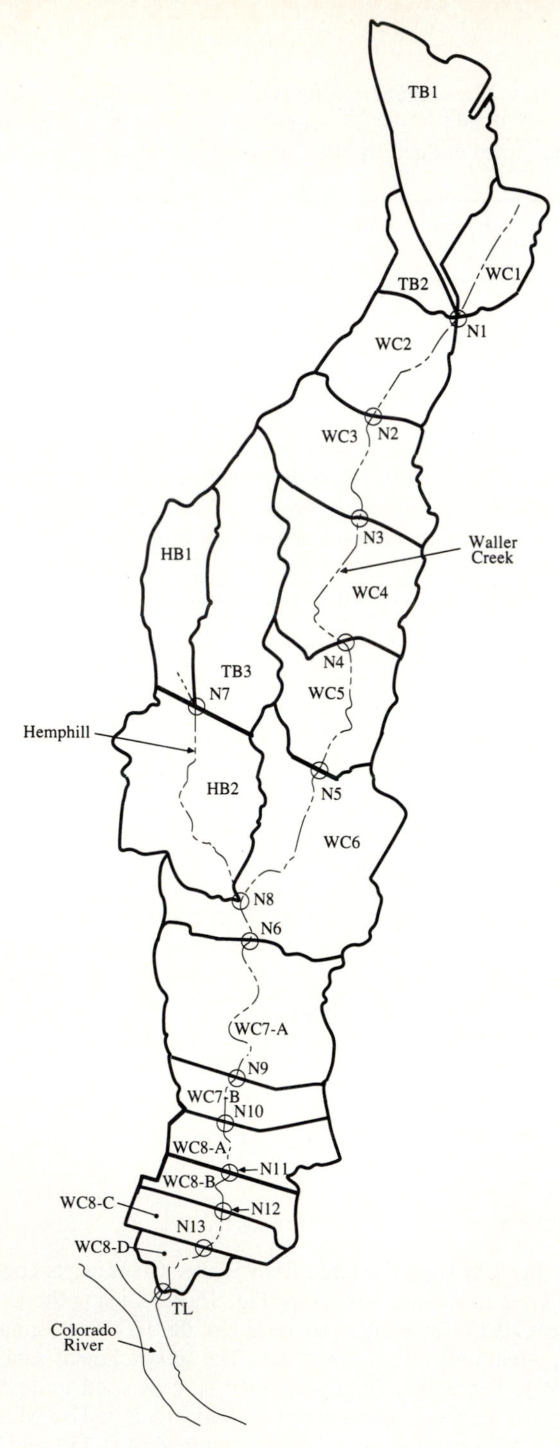

FIGURE 15.P.5
Waller Creek subarea delineation and node locations.

TABLE 15.P.3
Subarea physiographic characteristics for Waller Creek Watershed

Subarea designation	Area mi^2	Length (ft)	Slope (ft/ft)	Curve number	Lag time t_p (h)	Muskingum* K (h)	Peak Factor C_p
WC1	0.19	4800	0.015	87	0.28	–	0.54
WC2	0.32	4250	0.013	86	0.29	0.20	0.54
WC3	0.41	3600	0.015	87	0.26	0.15	0.53
WC4	0.47	5000	0.011	86	0.32	0.24	0.54
WC5	0.32	5700	0.014	85	0.31	0.22	0.54
WC6	0.70	7400	0.012	89	0.32	0.25	0.53
WC7–A	0.53	5400	0.013	89	0.29	0.24	0.54
WC7–B	0.17	2900	0.017	89	0.22	0.07	0.53
WC8–A	0.19	2500	0.025	91	0.18	0.07	0.51
WC8–B	0.16	2750	0.026	91	0.18	0.07	0.50
WC8–C	0.15	3300	0.014	91	0.24	0.06	0.54
WC8–D	0.10	2900	0.013	91	0.23	0.11	0.54
HB1	0.20	6300	0.012	86	0.33	–	0.56
HB2	0.50	7600	0.011	87	0.34	0.34	0.54
TB1	0.49	7700	0.013	86	0.34	–	0.54
TB2	0.11	3700	0.012	90	0.27	–	0.56
TB3	0.45	8000	0.010	86	0.37	–	0.55

*For routing through subarea. Use $K = 0.06$ for routing between nodes N8 and N6 in subarea WC6.

15.2.5 Solve Prob. 15.2.4 using the SCS dimensionless unit hydrograph for each subarea.

15.2.6 Solve Prob. 15.2.4 to determine the runoff hydrograph using the 100-year 24-hour SCS Type II storm given in Table 14.3.1. This storm has a total depth of 10 in. What are the total loss and the total excess volume for each subarea?

15.2.7 Solve Prob. 15.2.4 to determine the runoff hydrograph using the SCS dimensionless unit hydrograph for each subarea and the 100-year 24-hour SCS Type II storm given in Table 14.3.1. What are the total loss and the total excess volume for each subarea?

15.4.1 Determine and draw the modified-rational-method hydrographs for a 27-acre drainage basin in Urbana, Illinois, for which the time of concentration is 20 minutes and the runoff coefficient is 0.65. Use a 25-year return period and the intensity-duration-frequency curve defined by $i = 120T^{0.175}/(T_d + 27)$. Consider rainfall durations of 30, 40, 50, 60, and 120 min.

15.4.2 Solve Prob. 15.4.1 for a 100-year return period.

15.4.3 Determine the maximum storage for a detention pond on a 25-acre watershed for which the developed runoff coefficient is 0.8, and the time of concentration before development is 25 min and after development is 15 min. The allowable discharge is 25 cfs; $a = 96.6$, and $b = 13.9$.

15.4.4 Solve Prob. 15.4.3 considering a developed time of concentration of 10 min.

15.4.5 Develop an equation for estimating the maximum storage of a detention basin, using the modified rational method. Express the storage as a function of duration, then differentiate the storage function with respect to duration and set the derivative equal to zero; then solve this expression for the duration.

Instead of the intensity-duration relation given in Eq. (15.4.3), use

$$i = \frac{a}{(T_d + b)^c}$$

where i is the rainfall intensity in in/h and T_d is the duration in minutes. The allowable discharge is Q_A.

15.4.6 Use the equation developed in Prob. 15.4.5 to estimate the required maximum storage for a detention pond on a 25-acre watershed in Austin, Texas. The time of concentration for predeveloped conditions is 30 minutes and for developed conditions is 10 minutes. The runoff coefficient for predeveloped conditions is 0.44 and for developed conditions is 0.90. Assume that all surface runoff flows into the detention pond. Consider a 25-year storm with $a = 97.86$, $b = 16.4$, and $c = 0.76$ in the rainfall intensity-duration relationship.

15.4.7 Solve Prob. 15.4.6 using the 100-year storm, for which $a = 117.28$, $b = 17.2$, and $c = 0.74$.

15.4.8 Develop an equation for estimating the maximum storage of a detention basin using the modified rational method (see Fig. 15.P.6 for defining the volume of storage). Use the following rainfall equation

$$P = bT_d^a + cT_d + d$$

where P is inches of precipitation and T_d is the duration in hours.

15.4.9 Use the equation developed in Prob. 15.4.8 to estimate the required maximum storage for a detention pond on a 25-acre watershed. The time of concentration for predeveloped conditions is 30 minutes and for developed conditions is 5 minutes. The runoff coefficient for predeveloped conditions is 0.44 and for developed conditions is 0.90. Assume that all surface runoff flows into the detention pond. Consider a 25-year storm where:

$$a = 1.281,\ b = -10.73,\ c = 14.11,\ d = 0.089 \text{ for } T_d < 0.5 \text{ h}$$
$$a = -0.0640,\ b = -21.33,\ c = 0.084,\ d = 24.99 \text{ for } T_d \geq 0.5 \text{ h}$$

15.4.10 Use HEC-1 or another simulation model to route the given inflow hydrograph, which is derived from a 100-year SCS Type II design storm, through Ganzert Lake, located near Round Rock, Texas. The runoff hydrograph (reservoir inflow) from the upstream drainage is:

Time (h:min)	0:00	1:00	1:30	2:00	2:30	3:00	3:30
Reservoir inflow (cfs)	0	201	259	322	394	474	566
Time (h:min)	4:00	4:30	5:00	5:30	6:00	6:30	7:00
Reservoir inflow (cfs)	674	807	976	1221	1615	2332	8969
Time (h:min)	7:30	8:00	8:30	9:00	9:30	10:00	10:30
Reservoir inflow (cfs)	17185	13870	8386	4533	2817	2221	1770
Time (h:min)	11:00	11:30	12:00	12:30	13:00	13:30	14:00
Reservoir inflow (cfs)	1538	1366	1233	1128	1040	970	909
Time (h:min)	14:30	15:00	15:30				
Reservoir inflow (cfs)	853	805	768				

Characteristics of the reservoir are given below:

SCS floodwater retarding structure site no. 8* (Ganzert Lake)

Elevation (ft above MSL)	Reservoir surface area (acres)	Reservoir storage (acre·ft)	Discharge: Principal spillway (cfs)	Discharge: Emergency spillway (cfs)	Discharge: Total (cfs)
824.0	29.0	174.0			
824.8	37.0	200.0	0.0		0.0
826.4	40.0	262.0	32.6		32.6
828.0	48.0	332.0	65.1		65.1
832.0	70.0	568.0	73.3		73.3
836.0	111.0	930.0	80.6		80.6
840.0	168.0	1488.0	87.4		87.4
844.0	218.0	2260.0	93.6		93.6
846.8	264.0	2935.0	97.7	0	97.7
847.8	281.0	3209.0	99.2	280	379.2
848.0	284.0	3264.0	99.5	442	541.5
848.8	299.0	3521.0	100.0	1092	1192.0
849.8	317.0	3842.0	100.0	2142	2242.0
850.8	336.0	4163.0	100.0	3612	3712.0
851.8	354.0	4484.0	100.0	5397	5497.0
852.0	358.0	4548.0	–	–	–

Top-of-dam elevation:	851.8 ft above MSL
Emergency spillway crest elevation:	846.8 ft above MSL
Principal spillway crest elevation:	826.4 ft above MSL
Sediment pool elevation:	824.8 ft above MSL
Emergency spillway crest length:	210.0 ft
Drainage area:	5180.0 acres
Sediment storage:	262.0 acre·ft
Floodwater storage:	2673.0 acre·ft
Maximum emergency spillway capacity:	3850.0 cfs

*Information obtained from the Soil Conservation Service (U. S. Department of Agriculture) in Waco, Texas.

Assume that the initial water surface elevation is at the principal spillway crest elevation. What is the peak storage, the peak stage, and the peak discharge from the reservoir?

15.4.11 Use HEC-1 or another simulation model to route the given inflow hydrograph, which is derived from a 100-year SCS Type II design storm, through Smith Lake, located near Round Rock, Texas. The runoff hydrograph (reservoir inflow) from the upstream drainage is:

Time (h:min)	0:00	1:00	1:30	2:00	2:30	3:00
Reservoir inflow (cfs)	0	215	259	311	373	450
Time (h:min)	3:30	4:00	4:30	5:00	5:30	6:00
Reservoir inflow (cfs)	557	721	1008	3090	6746	7363
Time (h:min)	6:30	7:00	7:30	8:00	8:30	9:00
Reservoir inflow (cfs)	6249	3797	2121	1381	1064	897
Time (h:min)	9:30	10:00	10:30	11:00	11:30	
Reservoir inflow (cfs)	786	702	637	584	541	

Characteristics of the reservoir are given below:

SCS floodwater retarding structure site no. 9* (Smith Lake)

			Discharge			
Elevation (ft above MSL)	Reservoir surface area (acres)	Reservoir storage (acre·ft)	Principal spillway (cfs)	Low-flow outlet (cfs)	Emergency spillway (cfs)	Total (cfs)
770.0	39.7	192.3	0.0			0
774.0	65.0	401.7	40.2	16.0		56.2
778.8	96.9	725.5	45.0	20.0		65.0
782.0	134.7	1188.7	49.3	24.0		73.3
785.4	183.2	1725.0	52.7	27.4	0	80.1
786.4	197.4	1939.0	53.7	28.4	420	502.1
787.4	211.7	2153.0	54.6	29.3	1620	1703.9
788.4	225.9	2367.0	55.5	30.4	3210	3295.9
789.4	240.2	2581.0	–	–	5430	5520.0

Top-of-dam elevation:	791.0 ft above MSL
Emergency spillway crest elevation:	785.4 ft above MSL
Principal spillway crest elevation:	770.0 ft above MSL
Emergency spillway crest length:	300.0 ft
Drainage area:	3616.0 acres
Sediment storage:	182.0 acre·ft
Floodwater storage:	1638.0 acre·ft
Maximum emergency spillway capacity:	5100.0 cfs

*Information obtained from the Soil Conservation Service (U. S. Department of Agriculture) in Waco, Texas.

Assume that the initial water surface elevation is at the principal spillway crest elevation. What is the peak storage, the peak stage, and the peak discharge from the reservoir?

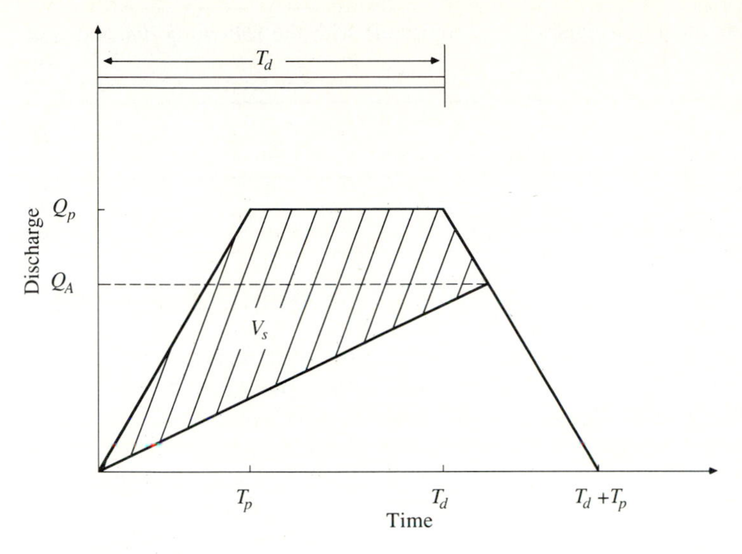

FIGURE 15.P.6
Hydrograph for Probs. 15.4.8–15.4.9.

15.6.1 Write a computer program to perform the water balance computation for a reservoir. The computer program will have the following input: water surface elevation–surface area–capacity characteristics; monthly demand fractions; monthly reservoir inflows; monthly net evaporation data; and beginning storage level. Your output should consist of monthly values for net evaporation loss, demand, inflows, and spills. In addition, print out each end-of-year storage volume.

15.6.2 Compute the monthly water balances for the proposed Justiceburg reservoir site near Lubbock, Texas, for the years 1940–1942. Assume the reservoir is initially at a normal conservation storage level, at elevation 2220 ft above MSL (which is also the elevation of the service spillway). The emergency overflow spillway is at elevation 2240 ft above MSL. The elevation–surface area–capacity characteristics are listed below:

Elevation (ft above MSL)	2,130	2,140	2,150	2,160	2,170	2,180
Area (acres)	108	253	506	765	1,046	1,330
Capacity (acre·ft)	608	2,407	6,187	12,515	21,549	33,417
Elevation (ft above MSL)	2,190	2,200	2,205	2,210	2,215	2,220
Area (acres)	1,682	2,045	2,232	2,437	2,651	2,884
Capacity (acre·ft)	48,485	67,065	77,737	89,414	102,108	115,937
Elevation (ft above MSL)	2,225	2,230	2,235	2,240		
Area (acres)	3,197	3,589	4,094	4,784		
Capacity (acre·ft)	131,153	148,069	167,194	189,268		

Consider an annual demand of 26,100 acre·ft with the following demand fractions.

Month	1	2	3	4	5	6
Fraction	0.05	0.05	0.05	0.06	0.07	0.13
Month	7	8	9	10	11	12
Fraction	0.15	0.17	0.09	0.07	0.06	0.05

The net evaporation data are given in Table 15.P.4 and the runoff data (reservoir inflows) are given in Table 15.P.5.

15.6.3 Use the computer program written for Prob. 15.6.1 to perform the water balance from 1940 to 1978 for the Justiceburg reservoir site described in Prob. 15.6.2. What is the minimum storage in the reservoir for this time period?

15.6.4 Use the computer program written for Prob. 15.6.1 to perform the water balance (1940–1978) for the Justiceburg reservoir site. Use the information presented in Prob. 15.6.2, but assume, in turn, the following reservoir capacities: 50,000 acre·ft, 90,000 acre·ft, and 130,000 acre·ft.

15.6.5 Determine the firm yield for the Justiceburg reservoir site (Prob. 15.6.2). Use your computer program written for Prob. 15.6.1 to perform the water balance. Assume a reservoir capacity of 110,000 acre·ft, and allow complete drawdown of the reservoir.

15.6.6 Solve Prob. 15.6.5 to determine the firm yield for a reservoir capacity of 110,000 acre·ft allowing a minimum storage of 2000 acre·ft.

15.6.7 Solve Prob. 15.6.5 to determine the firm yield for a reservoir capacity of 130,000 acre·ft allowing complete drawdown. Also, determine the firm yield for a minimum storage of 2000 acre·ft.

TABLE 15.P.4
Net evaporation data for the Justiceburg reservoir site in feet

Year	Jan.	Feb.	Mar.	Apr.	May	Jun.	Jul.	Aug.	Sep.	Oct.	Nov.	Dec.	Total
1940	0.09	0.04	0.46	0.37	0.56	0.48	1.03	0.58	0.79	0.39	0.04	0.16	4.99
1941	0.12	0.09	0.02	0.18	–0.48	0.35	0.50	0.63	0.18	–0.28	0.26	0.14	1.71
1942	0.18	0.25	0.38	0.08	0.65	0.50	0.62	0.53	–0.03	0.12	0.34	–0.13	3.49
1943	0.21	0.35	0.35	0.52	0.27	0.56	0.51	1.12	0.59	0.53	0.31	–0.01	5.31
1944	0.03	0.05	0.29	0.48	0.41	0.68	0.45	0.66	0.33	0.32	0.13	0.01	3.84
1945	0.11	0.13	0.34	0.39	0.64	0.78	0.49	0.53	0.61	0.13	0.32	0.19	4.66
1946	0.08	0.28	0.39	0.58	0.52	0.69	0.90	0.73	0.33	0.13	0.26	0.14	5.03
1947	0.05	0.23	0.14	0.36	0.09	0.67	0.79	0.84	1.08	0.60	0.22	0.12	5.19
1948	0.13	0.11	0.41	0.57	0.45	0.58	0.71	0.81	0.69	0.40	0.41	0.33	5.60
1949	–0.12	0.17	0.30	0.24	0.02	0.33	0.55	0.66	0.34	0.22	0.47	0.23	3.41
1950	0.27	0.24	0.44	0.38	0.15	0.43	0.13	0.60	0.18	0.69	0.53	0.33	4.37
1951	0.23	0.18	0.38	0.49	0.38	0.63	0.76	0.50	0.73	0.50	0.33	0.33	5.44
1952	0.20	0.35	0.53	0.26	0.43	0.98	0.62	1.01	0.75	0.80	0.42	0.24	6.59
1953	0.35	0.27	0.36	0.56	0.60	1.02	0.83	0.67	0.84	0.15	0.28	0.31	6.24
1954	0.32	0.42	0.53	0.31	0.03	0.89	1.02	0.78	0.93	0.59	0.47	0.34	6.63
1955	0.18	0.21	0.63	0.60	0.29	0.51	0.56	0.76	0.41	0.33	0.50	0.47	5.45
1956	0.31	0.26	0.60	0.71	0.52	0.64	0.93	1.06	1.02	0.64	0.35	0.40	7.44
1957	0.24	0.12	0.40	0.12	–0.09	0.41	0.78	0.87	0.63	0.10	0.05	0.34	3.97
1958	0.03	0.09	–0.05	0.12	0.15	0.61	0.71	0.81	0.50	0.28	0.27	0.21	3.73
1959	0.17	0.20	0.48	0.30	0.23	–0.06	0.33	0.77	0.78	0.25	0.40	0.28	4.13
1960	0.05	0.09	0.23	0.46	0.46	0.36	0.01	0.87	0.58	0.00	0.37	0.03	3.51
1961	0.08	0.01	0.22	0.51	0.41	0.26	0.14	0.51	0.65	0.52	0.15	0.16	3.62
1962	0.12	0.29	0.33	0.33	0.80	0.54	0.43	0.61	0.14	0.31	0.26	0.15	4.31
1963	0.18	0.15	0.36	0.49	–0.01	0.17	0.73	0.61	0.51	0.57	0.33	0.21	4.30
1964	0.20	0.15	0.45	0.65	0.52	0.49	0.79	0.61	0.41	0.55	0.33	0.20	5.35
1965	0.35	0.30	0.30	0.38	0.30	0.55	0.72	0.46	0.34	0.57	0.38	0.19	4.84
1966	0.09	0.08	0.37	0.15	0.25	0.50	0.48	0.24	0.23	0.51	0.39	0.22	3.51
1967	0.13	0.20	0.25	0.38	0.38	0.60	0.42	0.58	0.16	0.63	0.19	0.14	4.06
1968	0.04	0.05	–0.06	0.25	0.33	0.40	0.42	0.48	0.48	0.33	0.10	0.15	2.97
1969	0.15	0.07	0.07	0.29	0.06	0.54	0.69	0.61	0.21	–0.09	0.18	0.17	2.95
1970	0.17	0.19	0.08	0.48	0.57	0.59	0.83	0.57	0.24	0.35	0.44	0.38	4.89
1971	0.28	0.28	0.50	0.50	0.64	0.58	0.76	0.23	0.30	0.21	0.23	0.15	4.66
1972	0.26	0.25	0.43	0.49	0.31	0.41	0.48	0.29	0.26	0.23	0.16	0.17	3.74
1973	0.07	0.06	0.19	0.26	0.62	0.77	0.49	0.73	0.38	0.46	0.41	0.42	4.86
1974	0.25	0.37	0.38	0.60	0.68	0.76	0.89	0.33	0.17	0.14	0.23	0.19	4.99
1975	0.22	0.09	0.40	0.48	0.46	0.62	0.34	0.56	0.27	0.57	0.27	0.24	4.52
1976	0.40	0.46	0.55	–0.06	0.59	0.83	0.22	0.37	0.19	0.12	0.29	0.30	4.26
1977	0.11	0.23	0.47	0.25	–0.09	0.67	0.94	0.31	0.81	0.54	0.46	0.46	5.16
1978	0.17	0.10	0.46	0.74	0.34	0.56	0.61	0.63	0.47	0.35	0.30	0.22	4.95
Average	0.17	0.19	0.34	0.39	0.34	0.56	0.61	0.63	0.47	0.35	0.30	0.22	4.57

Note: May–December, 1978, data were unavailable and have been replaced by their average values.

TABLE 15.P.5
Runoff data for Justiceburg reservoir site in acre·ft

Year	Jan.	Feb.	Mar.	Apr.	May	Jun.	Jul.	Aug.	Sep.	Oct.	Nov.	Dec.	Total
1940	20	190	0	570	2,650	5,570	100	15,780	6,280	0	2,910	130	34,200
1941	0	870	5,530	30,830	68,820	21,700	11,210	5,670	10,730	54,660	2,500	890	213,410
1942	250	40	10	3,250	760	3,530	650	7,360	9,400	13,950	730	1,430	41,360
1943	700	40	620	970	3,000	4,190	3,050	0	0	0	0	0	12,570
1944	0	20	0	60	3,510	990	6,250	540	220	590	300	920	13,400
1945	60	20	570	170	230	3,910	12,140	0	2,020	10,990	60	0	30,170
1946	0	0	0	10	1,100	2,680	330	4,180	4,840	13,740	320	2,470	29,670
1947	240	0	10	0	47,310	2,700	1,060	100	1,130	540	30	2,350	55,470
1948	10	4,380	290	0	1,560	10,120	15,350	2,300	30	2,040	4,220	10	40,310
1949	0	0	0	1,290	9,420	12,340	120	270	10,450	1,220	140	10	35,260
1950	10	130	0	3,650	19,950	1,790	5,370	1,260	19,770	360	10	10	52,310
1951	10	0	0	20	2,800	9,650	790	5,170	210	10	0	0	18,660
1952	0	0	0	0	4,940	170	2,300	340	170	0	90	20	8,030
1953	0	10	160	540	5,140	670	2,810	4,540	60	22,670	1,020	70	37,690
1954	20	10	0	22,990	25,360	2,030	0	0	0	0	260	0	50,670
1955	10	1,620	6,970	160	46,440	13,480	24,380	910	70,500	31,510	940	460	197,380
1956	280	100	20	10	5,010	690	510	620	0	320	40	20	7,620
1957	0	5,670	110	17,620	43,810	32,360	3,920	1,770	2,810	6,760	5,490	150	120,470
1958	60	80	210	2,560	15,300	3,160	270	1,030	4,100	430	530	30	27,760
1959	10	0	0	170	2,240	28,180	31,380	4,310	0	6,820	260	3,760	77,130
1960	380	240	70	0	1,310	1,590	32,380	460	20	60,090	1,830	1,020	99,390

TABLE 15.P.5 *(cont.)*

Runoff data for Justiceburg reservoir site in acre·ft

Year	Jan.	Feb.	Mar.	Apr.	May	Jun.	Jul.	Aug.	Sep.	Oct.	Nov.	Dec.	Total
1961	1,140	1,200	810	180	50	26,460	31,050	1,780	600	280	2,110	30	65,690
1962	20	10	0	10	0	10,680	3,490	2,210	35,350	480	20	340	52,610
1963	40	20	80	60	12,740	39,630	100	560	900	1,820	1,700	10	57,660
1964	480	20	0	0	2,370	2,920	0	2,120	1,570	0	210	490	10,180
1965	0	0	100	640	26,260	810	30	12,110	1,570	210	10	940	42,680
1966	30	10	20	11,560	2,100	600	30	4,800	720	10	10	10	19,900
1967	0	10	4,440	1,150	920	49,000	23,340	60	2,680	1,560	10	10	83,180
1968	700	1,300	2,750	370	1,070	1,170	1,350	1,740	0	330	3,710	60	14,550
1969	0	10	2,700	2,550	35,450	1,160	10	3,290	17,510	9,480	2,450	50	74,660
1970	20	10	8,110	240	4,670	690	0	320	2,260	400	0	0	16,720
1971	0	0	0	10	3,360	770	1,130	9,530	21,370	1,060	40	80	37,350
1972	20	10	10	0	1,650	4,860	3,990	40,550	3,410	710	100	50	55,360
1973	740	1,730	1,950	110	120	240	2,680	270	2,150	10	0	0	10,000
1974	0	0	0	210	140	40	140	990	5,220	2,740	140	10	9,630
1975	10	40	0	30	360	1,950	6,040	5,390	8,160	10	570	10	22,570
1976	20	0	0	3,070	260	180	9,170	700	1,390	830	30	0	15,650
1977	0	0	0	1,660	7,360	2,870	0	3,800	20	30	0	0	15,740
1978	0	0	0	0	12,130	4,100	970	270	5,640	2,000	860	420	26,390
Average	140	460	910	2,740	10,810	7,940	6,100	3,770	6,490	6,380	860	420	47,020

Note: November and December, 1978, data were unavailable and have been replaced by their average values.

INDEXES

AUTHOR INDEX

SUBJECT INDEX

R0160847307
627
C552